Transform Linear Algebra

Transform Linear Algebra

Frank Uhlig
Auburn University

Prentice Hall

Prentice Hall
Upper Saddle River, New Jersey 07458

Library of Congress Cataloging-in-Publication Data

Uhlig, Frank
Transform linear algebra / Frank Uhlig.
p. cm.
Includes bibliographical references and index.
ISBN 0-13-041535-9
1. Algebras, Linear. 2. Transformations (Mathematics) I. Title.
QA184.2.U48 2002
512′.5—dc21 2001045935

Acquisitions Editor: George Lobell
Editor-in-Chief: Sally Yagan
Vice President/Director of Production and Manufacturing: David W. Riccardi
Executive Managing Editor: Kathleen Schiaparelli
Senior Managing Editor: Linda Mihatov Behrens
Production Editor: Bob Walters
Manufacturing Buyer: Alan Fischer
Manufacturing Manager: Trudy Pisciotti
Marketing Manager: Angela Battle
Marketing Assistant: Richel Beckman
Assistant Editor of Media: Vince Jansen
Assistant Managing Editor, Math Media Production: John Matthews
Editorial Assistant/Supplements Editor: Melanie Van Benthuysen
Art Director: Jonathan Boylan
Interior Designer: John Christiana, Wanda Espana
Cover Designer: John Christiana
Creative Director: Carole Anson
Managing Editor Audio/Video Assets: Grace Hazeldine
Director of Creative Services: Paul Belfanti
Cover Photo: Kaz Chiba/Photodisc, Joe Cotitta/Epic Places Incorporated
Art Studio: Laserwords

Prentice Hall

Printed in the United States of America

10 9 8 7 6 5 4 3 2 1

ISBN 0-13-041535-9

Pearson Education LTD., *London*
Pearson Education Australia PTY. Limited, *Sydney*
Pearson Education Singapore, Pte. Ltd.
Pearson Education North Asia Ltd., *Hong Kong*
Pearson Education Canada, Ltd., *Toronto*
Pearson Educaciûn de Mexico, S.A. de C.V.
Pearson Education – Japan, *Tokyo*
Pearson Education Malaysia, Pte. Ltd.

Frank Uhlig

Born April 2, 1945, Mägdesprung/Harz; grew up in Mülheim/Ruhr, Germany; married, two sons.

Mathematics student at University of Cologne, California Institute of Technology.

Ph.D., CalTech, 1972; Assistant, University of Würzburg, RWTH Aachen, Germany, 1972–1982.

Two Habilitations (Mathematics), University of Würzburg 1977, RWTH Aachen 1978.

Visiting Professor, Oregon State University 1979/1980; Professor of Mathematics, Auburn University 1982.

Two Fulbright Grants; (Co-)organizer of eight research conferences.

Research Areas: linear algebra, matrix theory, numerical analysis, numerical algebra, geometry, Krein spaces, graph theory, mechanics, inverse problems.

40+ papers, 2+ books.

(Photo by Daniel Uhlig)

Menu, please

Contents

*On the Web at http://www.auburn.edu/~uhligfd/TLA/download/C14.pdf

*On the Web at http://www.auburn.edu/~uhligfd/TLA/download/AIPS.pdf

Preface

This book has evolved over many years, through taking and teaching courses on linear algebra to being involved in the linear algebra educational community, and by doing research in several related areas of mathematics.

I started teaching linear algebra in 1965 during my second year at the University of Cologne, Germany, when I was asked to lead a recitation section for Prof. Dr. W. Jehne's freshman class. He, his staff, and Emil Artin's 1960/61 Hamburg lecture notes introduced me to Riesz's lemma, and as you will see, I have finally returned, with a view of my own.

This book emphasizes the concepts of linear algebra and matrix theory. Teaching in this fashion is driven by the desire to convey the structure and nature of linear spaces and of linear transformations. Chapters 1 through 7 deal with linear transformations with respect to the standard unit vector basis. Starting with Chapter 7, we learn how to view and perceive each linear transformation more clearly with respect to its own specific basis.

Almost every chapter has three sections. The first section contains a one- (or two-) hour lecture, followed by problems. The second section deals with theoretical and mathematical enrichment, such as further concepts or alternative developments and proofs. The final section of each chapter details applications to and from linear algebra.

My main aim is to help students develop an intuitive understanding of the subject and to present them with teachable and learnable concepts that lay out the foundations of linear algebra. This book is aimed at beginning or intermediate-level undergraduate students in engineering, science, and mathematics curricula. By stressing the importance of very few deliberate fundamental concepts throughout the lectures, I have tried to bring my students closer to understanding what linear algebra is all about and why there is such a need for learning and teaching it well.

I am deeply indebted to my students for their patience and feedback, to my colleagues and my family, to Darrell for his expertise and help with tricky LaTeX questions, to Rosie for her quick and expert typing of the first draft, to Achim for his critical reading, to Maria and Michael for checking the exercises and examples, and to the reviewers and editors.

Frank Uhlig
Auburn, Alabama, 2002

Lintels and keystones

Notes to Instructors

This book is like no other textbook of linear algebra.

Built upon the concept and characterization of *linear transformations* as fundament for the subject, *Transform Linear Algebra* has a unique and innovative structure. By design, its students develop and master their basic linear algebraic thinking skills early on, before encountering the notion of linear (in)dependence, span, and basis, for example. As each of these concepts is approached via linear transformations, the students benefit from the unifying idea of a *linear transform*, which sustains all other concepts. Our design makes this book a well-rounded and quite comprehensive introduction to linear algebra that is easily taught and readily learned.

It is both a challenge and a boon to study linear transformations and functions in the first week of class—a challenge because most sophomore students do not arrive with a firm notion of functions, and a boon once this challenge has been overcome by learning to work in $\mathbb{R}^n$ and by understanding linear transformations of $\mathbb{R}^n$ as constant matrix times vector products.

The first lecture practices linear algebra thinking by formally characterizing linear functions of $\mathbb{R}^n$. In the middle chapters of the book, formal proofs take a lesser role. Instead, we deduce the theorems intuitively by using linear transformations combined with row reduction. This enables students to grasp the idea of *proof* first, before we rely again on formal proofs in Chapters 10 to 14.

For a more detailed account of my thoughts and findings while writing this book, see

`http://www.auburn.edu/~uhligfd/TLA/download/tlateach.pdf.`

Linear algebra has become an integral part of nearly all undergraduate science and engineering curricula over the past 50 years. From the 1950s on, two approaches to teaching the subject have competed with each other, and the emphasis has swung back and forth from abstract vector spaces based on groups, rings of endomorphisms, etc., to concrete matrix-based computational presentations. Over time, matrix theory has emerged as the unifying and prevalent subject of research and teaching for this branch of mathematics. And with time, the lessons of matrix theory have put their indelible stamp on the way we understand, talk, and think about linear algebra.

The book synthesizes the two differing historic approaches. We follow the inner needs and insights of matrix analysis and develop the conceptual framework of finite-dimensional linear spaces, linear transformations, and all of elementary linear algebra. These concepts are presented here much more concretely than they are in an abstract vector space approach.

The modern concepts of matrix theory go just as deeply into the subject and yet are more powerful and intuitive for students than the abstract algebraic approach of old. They are powerful precisely because they allow for concrete discovery through computations with vectors and matrices. This discovery is nowadays best done with the help of a computer and appropriate software. The two aspects of linear algebra—concepts and computations—are linked together throughout the book in a synergetic way. Concepts set up concrete computations as computations reinforce concepts in each chapter-starting lecture of the book.

Thus, the book deals with linear algebra on the concrete real (or complex) number level. In this realm, linear mappings and linear spaces can be completely and simply described by n-vectors and $m \times n$ matrices, which are customarily defined as rectangular arrays of mathematical entries. See specifically the Linear Algebra Curriculum Study Group's (LACSG's) Recommendations[1] regarding concreteness for a first course on linear algebra, which we wholeheartedly embrace.

Following the dictum that

form follows function,

the book is organized as a "matrix" of text. Fourteen chapters comprise up to eight (sub)sections each. The first section of each chapter draws out an **essential and fundamental lecture** on one subject of linear algebra, followed by general exercises. The second section (from Chapter 6 on) covers theoretical enrichments, again with problems attached; and the third section describes applications for the chapter, with applied exercises and descriptions of the relevant MATLAB software. Each chapter closes with a number of review exercises, a list of standard questions and tasks for the chapter, and lists of subheadings and standard equations for the lecture. Thus in "matrix lingo," the book is organized as a

14×8 *array*, or *matrix, of mathematical information*,

some of whose entries are left blank. There is enough material for a yearlong course on linear algebra and certainly too much for complete coverage in any one-semester or one-quarter course. Thus, each instructor needs to decide which sections should be taught and what is of importance, given the school's syllabus, the levels and interests of the students, and the instructor's own desires.

Typically, an introductory one-semester linear algebra course consists of the fundamental lectures one through nine, augmented by some theoretical or software enhancements. The course terminates with "general matrix eigenvalues." Such a course should allow a day (or two) of discussion and homework following each individual lecture. For best delivery, some of the lectures will take two or more class hours. All lectures are designed to be intuitive introductions to their subject, starting with the chapter-specific concepts and then giving students hands-on

[1]D. Carlson et al., "The Linear Algebra Curriculum Study Group Recommendations for a First Course in Linear Algebra," *Coll. Math. J.*, vol. 24 (1993), pp. 41–46.

methods for understanding and solving associated problems. Our approach places a high emphasis on grouping certain mathematical concepts of linear algebra together in a conscious sequence, while embedding them in and enhancing them with concrete and repeatable methods for problem solving.

Among the basic first nine lectures is Lecture Eight on determinants. This lecture is optional, and the flow of the book separates after Lecture Seven. Instructors have the choice of three paths:

$$\begin{array}{c} \vdots \\ \downarrow \\ \textit{Lecture Seven} \\ \downarrow \\ \textit{Lecture } 9.1 \\ \downarrow \\ \textit{Lecture Ten} \\ \downarrow \\ \vdots \end{array} \quad , \quad \begin{array}{c} \vdots \\ \downarrow \\ \textit{Lecture Seven} \\ \downarrow \\ \textit{Lecture Eight} \\ \downarrow \\ \textit{Lecture } 9.1.\text{D} \\ \downarrow \\ \textit{Lecture Ten} \\ \downarrow \\ \vdots \end{array} \quad , \ \underline{\text{or}} \quad \begin{array}{c} \vdots \\ \downarrow \\ \textit{Lecture Seven} \\ \downarrow \\ \textit{Lecture } 9.1 \\ \downarrow \\ \textit{Lecture Eight} \\ \downarrow \\ \textit{Lecture } 9.1.\text{D} \\ \downarrow \\ \textit{Lecture Ten} \\ \downarrow \\ \vdots \end{array}$$

The first path does not use determinants for matrix eigenvalue problems; thus, it usually leads faster to the later chapters. Determinants are multilinear functions and are somewhat extraneous to the study of linearity. For those who want to cover them, we have included Chapter 8 and Section 9.1.D. Lecture 9.1.D has the same contents as the vector iteration–based eigenvalue Lecture 9.1, but it relies on determinants. The third path covers eigenvalues both without and with determinants. It is well suited for a yearlong course. From Lecture Nine on, the three paths are confluent again; that is, we treat all subsequent eigenvalue problems both with vector iteration as in Section 9.1 and by determinantal means as in Chapter 8 and Section 9.1.D.

It is also possible to use the book in the currently more popular sequence of treating linear equations first. In fact, we have taught beginning with Lectures 2 and 3 on row reduction and linear equations, followed by Lecture One on linear transformations and then Chapters 4, 5 etc. In this approach, one starts the course with Lecture Two and the system of linear equations (2.2) and then introduces its augmented matrix (2.1), etc. After completing Lecture Three, one continues with part (b) of Lecture One and the now well-studied matrix–vector multiplication as an example of a linear transformation. The rest of the Chapters 1, 4, 5, etc then follow naturally. However, we prefer to start with Lecture 1 and linear transformations for the more fundamental and conceptual understanding that this approach gives students. Starting with linear transformations requires no extra sophistication on the part of the students. Rather, it *gives* sophistication.

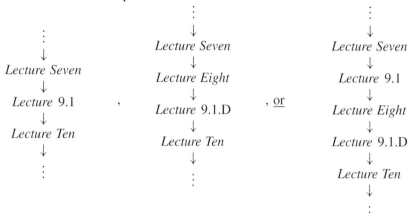

The run of 14 lectures incorporates several natural stopping points: After Lecture Nine in a bare-minimum course, or after Lecture Eleven, Twelve, or Fourteen, any elementary linear algebra course might stop. Stopping the development of an elementary linear algebra course at either of these four natural points involves a **choice** between covering more material or giving more enhancements to less material, a decision that is best left to the instructor.

In the parlance of linear algebra (and of Chapter 5), Lectures One through Nine constitute a *minimal spanning set* for the subject (i.e., they form a coherent, compact, and basic introduction to linear algebra). Depending on how many of the lectures in the book are covered in a course, there is more or less time for enrichment of the material on one or both of the offered levels (i.e., in theory or applications). This enrichment can easily take the form of further lectures, either interspersed with the fundamental ones, or it can be given on a second go-around after the fundamental lectures have been covered. Alternatively, it can involve additional assigned reading, problem solving, or computer lab sessions for mastering the MATLAB software and the subject.

The book can serve for a full-year course in the subject. Clearly, not all sections can be covered in a one-semester course, nor should they be. Regarding the contemporary breadth of the field of linear algebra and matrices, we have not attempted to include many potentially useful topics and elementary chapters of linear algebra here. We omit abstract vector spaces, except for an occasional mention and a short appendix. Likewise, we do not cover inner product spaces, except in an appendix, finitely generated modules over fields, coding theory, linear controls, etc. Following the LACSG recommendations for a first course in linear algebra, the book deliberately omits the more abstract and algebraic aspects of the field. Instead, we include a fair amount of applications in the physical sciences and numerical analysis.

One essential further chapter for a yearlong course, Chapter 14 on nondiagonalizable matrices and the Jordan normal form, as well as an Appendix D on inner product spaces, are posted on the Web at

`http://www.auburn.edu/~uhligfd/TLA/download/c14.pdf`

and

`http://www.auburn.edu/~uhligfd/TLA/download/AIPS.pdf`.

When teaching a year long course from this book, it would be best to proceed from Chapter 12 directly to Chapter 14 on the Web, followed by Lecture 13 in the book.

While following the LACSG recommendations on teaching linear algebra concretely in $\mathbb{R}^n$ at first, the book is deliberately at odds with the LACSG recommendation of proceeding from the concrete (examples) to the abstract (concepts). I have had little success teaching in that way. A bit of soul-searching on what constitutes linear algebra—on what the "coinage" of linear algebra really is—led me to realize that linear transformations and vectors are at the basis of linear algebra. By introducing these fundamental concepts early and clearly, and by building on them throughout the course, both students and instructors have a strong and familiar base to come back to in every instance and every application.

This is, in essence, what I attempt and hope to do with this book.

The book offers a large number of applications of *linear transformations*. Since linear transformations underlie all of linear algebra, every topic of linear algebra can be derived from them. For example, *Transform Linear Algebra* introduces linear dependence and independence of vectors from the image space of an associated matrix transformation. The solvability of systems of linear equations, the composition of linear maps, and the matrix inverse, as well as coordinate vectors, change of basis, and matrix eigenvalues, are each developed from looking at specific aspects of linear transformations. Consequently, a *synergy* develops from chapter to chapter that makes student understanding easier as the book and course progress. Previously taught subjects are repeatedly revisited in solving more advanced problems. Studying linear algebra from differing angles while being guided by its fundamental concepts reinforces overall learning and retention. Students who may have struggled with some aspects of an earlier chapter will generally be led to overcome that difficulty in a later chapter. "Things will become clear" from a new viewpoint and by repetition of the fundamentals.

The one "workhorse" that gets us through most of our subject at the introductory level is the row echelon form reduction of a matrix. This reduction is developed immediately after the first chapter on linear transformations and is used to solve linear equations mechanically before the more theoretical and conceptual studies begin.

The lectures of this book are derived from the author's actual written class notes before and after each lecture. They were written down and corrected while teaching several elementary linear algebra courses, much in the vein of Pete Stewart.[2] The notes have been refined over time and by repeated use. Conceived as lectures, they are frequently more wordy than a textbook chapter would be on the chosen subject; they often lead the reader with questions to guide him or her in a certain way to certain understandings. Different teachers will surely lecture differently. The printed lectures are designed to give students an independent and complete guide to linear algebra, possibly different from the actual in-class presentation of the material in their particular classes. Moreover the lectures serve as a guide to instructors, and the course should follow their approach in introducing and developing the topics. Each lecture gives a balanced account of the underlying concepts <u>and</u> elementary methods to solve related problems concretely by hand. Thus, the lectures provide food for expanding minds and satisfy the practically curious at the same time.

The presentation of the material culminates in nearly 1,000 exercises, half of which are solved in full detail in an appendix. Every instructor, however, would truly like to have an unlimited number of easily computable homework and test problems at his or her disposal. For this purpose, we include "Teacher's Problem Making Exercises" in 10 of the chapters. That is, in 10 problem sections from Section 2.1.P to Section 14.1.P, we explain how to construct an infinity of model integer matrix problems for the preceding lecture. These instructions will generate model problems with integer solutions that can be obtained in integer arithmetic, using the methods of the respective chapter. Consequently, the students can practice the recommended hand computational schemes on such problems nearly without effort.

---

[2]G. W. Stewart, *Afternotes on Numerical Analysis*, SIAM, Philadelphia, 1996, 200 pp.

Finally, there are four appendices: one on complex numbers and complex vectors for students who need reinforcement or have not encountered them before, another on finding integer roots of integer polynomials, a third on abstract vector spaces, and a fourth on inner product spaces, this one on the Web at

`http://www.auburn.edu/~uhligfd/TLA/download/AIPS.pdf.`

A note on the pace of the lectures

Each lecture covers about the same amount of material, yet each requires a different amount of time to teach and to grasp its information.

Chapter One is best taught slowly, because the concept of a function or a transformation is generally not firmly anchored in sophomores.

Lectures Two and Three also benefit from a slow and meticulous approach. Students generally are not prepared to follow through with disciplined computations and mathematical reasoning. They have to be taught.

After these *largo* to *adagio* opening lectures, Lectures Four through Eight can be taught at a quicker pace—say, *allegro*—and they will take roughly the same amount of time as the starting three. With their constant referral back to the nature of linear transformations and to row reduction, students understand the subjects and techniques of these lectures quickly.

Yet Lecture Seven plants a seed with basis changes and matrix similarity that slows the course down again in Chapters Nine through Fourteen. The final matrix eigenvalue lectures require harder computations with more complicated mathematical reasoning and have a slower pace—say, *adagio* again.

Frank Uhlig

Transform Linear Algebra

Join the flow

Introduction

We shall not cease from exploration
And the end of all our exploring
Will be to arrive where we started
and know the place for the first time.

T. S. Eliot,
Excerpt from "Little Gidding"
in FOUR QUARTETS[1]

Linear algebra is a circular subject. Studying Linear algebra feels like exploring a city or a country for the first time. An overwhelming number of concepts, all intertwined and connected, are present in any first encounter with linear algebra. As with a new city, one has to start discovering slowly and deliberately. Of great help is that linear algebra is akin to geometry, and like geometry, many of its insights have been permanently there within us. We must only explore, look around, and awake our intuition with the reality of this magic mathematical place.

The Introduction starts with a survey of mathematical precursors, or "stepping-stones" that a beginning student of linear algebra may or may not have encountered before. A description of real vectors follows that familiarizes the reader with elementary vector notation and geometry. Finally, we describe basic concepts of sets and functions.

Mathematical Stepping-Stones

First for the background:

High school geometry is one of the most helpful prerequisites for our studies. In geometry, students learn to describe two- and three-dimensional space, position, direction, and so forth systematically. Much of linear algebra is directly related to our intuitive notion of space.

High school algebra deals mostly with arithmetic, equations, and functions. Every linear algebra course builds on the elementary algebraic concept of a

---

function, like that in the defining equation $f(x) := 3x^2 - 7x + 14$. In high school, such functions are evaluated and plotted; one finds zeros (i.e., values x_0 for which $f(x_0) = 0$, etc.). Simple functional relations are studied, such as $x = r \cdot \cos\alpha$. This is a first and somewhat simple study of functions, because high school algebra courses are interested mainly in specific values $x, y, x_0, \alpha_1, \alpha_0$ and in their functional relations.

Thereafter, calculus studies the behavior of functions and their slopes via differentiation and areas between and under them using integration. Function extrema, intervals of increase and decrease, and integrals are studied and applied to various models of the physical and economic world. Mostly functions of one variable are examined; later courses in calculus deal with multivariate (two-or three-dimensional) functions as well.

[For some of the applications in Sections 1.3, 7.3, and 11.3, it is advisable to have studied a year of calculus before taking linear algebra.]

A differential equations course then studies the space of all functions with certain properties. In such a course, one searches for a function f defined on a multitude of points (usually an interval $[a, b] \subset \mathbb{R}$ of the reals), that satisfies a functional equation involving one or several derivatives. An example of a problem in differential equations is:

Find a differentiable function $f : [a, b] \longrightarrow \mathbb{R}$ that satisfies
$f' + 2f = \cos x$ for all $x \in [a, b]$.

For this example, it is not enough to have $f'(x_0) + 2f(x_0) = \cos(x_0)$ holding for some specific value $x_0 \in [a, b]$, but f must satisfy the differential equation for all $x \in [a, b]$. (We solve this particular equation and develop a general method for solving such differential equations numerically by using linear algebra in Sections 7.3. and 11.3.)

An elementary course in differential equations thus studies the space of one-dimensional or multidimensional functions with respect to differential equations for functions.

[Knowledge of differential equations is not necessary for a study of linear algebra. In fact, the reverse may be the case, especially with regard to the study of systems of linear differential equations. (See Sections 9.3, 11.3, and 14.3.) The two courses may well be taken concurrently.]

Finally, linear algebra studies functions f between vector spaces that satisfy the **linearity condition**

$$f(\alpha\mathbf{x} + \beta\mathbf{y}) = \alpha f(\mathbf{x}) + \beta f(\mathbf{y}).$$

Here, f is a function, also called a mapping or a **transformation**, $\mathbf{x}$ and $\mathbf{y}$ are **vectors**, and α and β are **scalars** or numbers.

Linear Algebra can be taught and learned by using vectors and the concept of a linear function. This book hinges in no way on having completed a calculus or differential equations course. It contains a few elementary and self-contained applications to these two areas that may be skipped at will.

Real Vectors, Sets, and Symbols

The vectors in this book will most often come from **real n-space** $\mathbb{R}^n$. Complex vectors and numbers come up necessarily with matrix eigenvalues in Chapter 9 and beyond. Thus, we will have to use the complex spaces $\mathbb{C}^n$ as well. (For a short treatment of complex numbers and complex vectors, see Appendix A at the end of the book.)

Everyone is familiar with the low-dimensional vector spaces $\mathbb{R}^1$, $\mathbb{R}^2$, and $\mathbb{R}^3$: a line, a plane, and three-dimensional space, respectively. In general, we work in n-dimensional spaces, where n is an arbitrary positive integer, such as $n = 2, 4, 14$, or 200.

By using the set notation braces $\{.. \mid ..\}$, we can describe n–space.

DEFINITION 2 For each integer $n \geq 1$, the symbol

$$\mathbb{R}^n = \left\{ x \;\middle|\; x = \begin{pmatrix} x_1 \\ \vdots \\ x_n \end{pmatrix}, x_i \in \mathbb{R}, i = 1, \ldots, n \right\}$$

denotes **real n-space**. The **origin** (i.e., the point of $\mathbb{R}^n$, all of whose n **components** are zero) is denoted by the vector $\mathbf{0} = \begin{pmatrix} 0 \\ \vdots \\ 0 \end{pmatrix} \in \mathbb{R}^n$.

For example, $\mathbb{R}^1 = \mathbb{R}$ is the real-number line and $\mathbb{R}^2 = \left\{ \mathbf{x} \;\middle|\; \mathbf{x} = \begin{pmatrix} x_1 \\ x_2 \end{pmatrix}, x_i \in \mathbb{R} \right\}$ is the standard real plane that is spanned or generated by two copies of $\mathbb{R}$. Its origin is the point $\mathbf{0} = \begin{pmatrix} 0 \\ 0 \end{pmatrix} \in \mathbb{R}^2$. (See Figure I-1.)

Note that the reference to vectors $\mathbf{x} \in \mathbb{R}^n$ as **columns** $\mathbf{x} = \begin{pmatrix} x_1 \\ \vdots \\ x_n \end{pmatrix}$ or as **rows** $\mathbf{x} = \begin{pmatrix} x_1, \ldots, x_n \end{pmatrix}$ is somewhat interchangeable. Row vectors are easier to handle in print as they fit better onto a printed line of text. But in most of our applications we consider vectors as columns. If $\mathbf{x} = \begin{pmatrix} x_1 \\ \vdots \\ x_n \end{pmatrix} \in \mathbb{R}^n$ is a

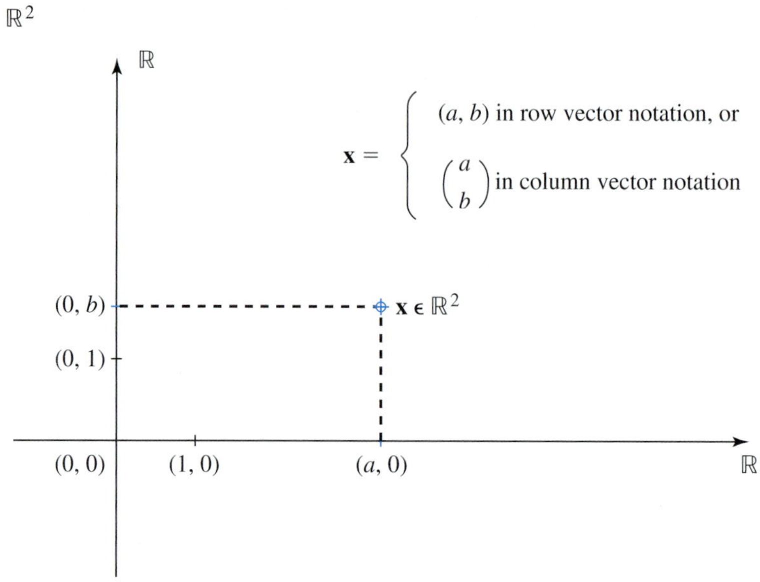

Figure I-1

column vector, then we denote the corresponding row vector in $\mathbb{R}^n$ by the symbol $\mathbf{x}^T := (\,x_1, \ \ldots, \ x_n\,)$ and vice versa. That is, if $\mathbf{y} = (\,y_1, \ \ldots, \ y_m\,) \in \mathbb{R}^m$ is a row vector, then $\mathbf{y}^T = \begin{pmatrix} y_1 \\ \vdots \\ y_m \end{pmatrix} \in \mathbb{R}^m$ denotes the corresponding column vector. This convenient vector operation is called **transposing** a vector, and $\mathbf{x}^T$ is called the **transpose** of $\mathbf{x}$.

$\mathbb{R}^n$ has one privileged point: the **origin**, $\mathbf{0} = \begin{pmatrix} 0 \\ \vdots \\ 0 \end{pmatrix} \in \mathbb{R}^n$. A vector $\mathbf{x} \in \mathbb{R}^n$ can be viewed as an arrow from $\mathbb{R}^n$'s origin $\mathbf{0}$ to the point $\mathbf{x}$ in $\mathbb{R}^n$ with components $\begin{pmatrix} x_1 \\ \vdots \\ x_n \end{pmatrix}$. Alternatively, the vector from one point $\mathbf{x} = \begin{pmatrix} x_1 \\ \vdots \\ x_n \end{pmatrix} \in \mathbb{R}^n$ to another point $\mathbf{y} = \begin{pmatrix} y_1 \\ \vdots \\ y_n \end{pmatrix} \in \mathbb{R}^n$ is described algebraically by the difference vector

$\mathbf{y} - \mathbf{x} = \begin{pmatrix} y_1 - x_1 \\ \vdots \\ y_n - x_n \end{pmatrix} \in \mathbb{R}^n$, see Figure I-2.

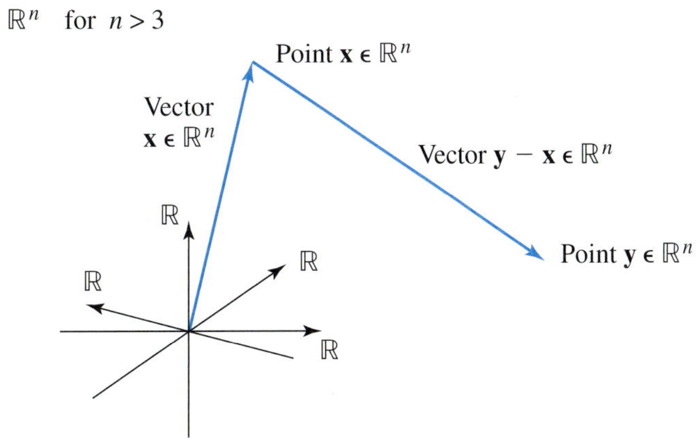

Figure I-2

For any $n > 3$, the geometry of n-space is hard to comprehend in an intuitive fashion. Just look at Figure I-2, with four copies of $\mathbb{R}$ as the coordinate axes in $\mathbb{R}^4$. On the other hand, it is easy to depict $\mathbb{R}^2$ and to describe and draw its geometry pictorially on paper, a blackboard, or a computer screen. Objects in $\mathbb{R}^3$ are much harder to draw. By using perspective and related techniques, we have learned over time to create accurate images of complicated objects in 3-space. Nowadays we use photography and its offshoots, as well as computer-aided design (CAD), for this purpose. However, our geometric understanding and drawing ability are almost completely lacking for higher dimensions, such as in $\mathbb{R}^4$, $\mathbb{R}^5$, $\mathbb{R}^{47}$ etc.

Whereas our intuition fails us for $\mathbb{R}^n$ and $n > 3$, linear algebra helps us to interpret these spaces and their geometry via algebra. Vectors in space are geometric objects that become algebraic ones when we write out their components, such as $\mathbf{x} = \begin{pmatrix} x_1 \\ \vdots \\ x_n \end{pmatrix} \in \mathbb{R}^n$. Linear algebra studies $\mathbb{R}^n$ (and $\mathbb{C}^n$) for any positive integer n in an algebraic setting. In this sense, the study of linear algebra is a higher dimensional continuation of the study of geometry in the space that surrounds us. While we have to learn the language and techniques of linear algebra, of vectors, and of matrices to master $\mathbb{R}^n$, much of our innate geometric knowledge and intuition of $\mathbb{R}^2$ and $\mathbb{R}^3$ will help us deal with $\mathbb{R}^n$ algebraically.

Throughout the book, we use the language of **sets** for brevity. Specifically, we use the following symbols:

Symbol	Meaning
$X = \{\dots\}$	the defining **braces** $\{\dots\}$ of a **set** X. For example, $X = \{1, 2, 3\}$. Every element described by ... inside the braces is a member of X.
$X = \{\dots \mid \dots\}$	the set X consists of **elements**, denoted by the first set of dots, that satisfy a certain **condition**, described by the second set of dots, that follows the vertical line inside the braces. $\mathbb{R}^+ = \{x \in \mathbb{R} \mid x > 0\}$ is the set of positive real numbers, for instance.
$\dots := \dots$	**defined as**. The left-hand term ... is defined by the right-hand term ..., such as changing variables and setting $x := t - 1$ for example, or by defining a set X as $X := \mathbb{R}$.
$x \in X$	x is an **element** of X; x belongs to X; x lies in X; X contains x. An example is $-2 \in \mathbb{R}$, meaning -2 is a real number.
$y \notin X$	y is not an element of X; y does not belong to X; y does not lie in X; X does not contain y. An example is $-2 \notin \mathbb{R}^+$.
$U \subset X$	U is a **subset** of X; every element u that lies in U also lies in X. An example is $\mathbb{R}^+ \subset \mathbb{R}$ and $\mathbb{R} \subset \mathbb{R}$. Note that any set X is a subset of itself (i.e., $X \subset X$ for all sets X).
$U = \emptyset$	U is the **empty set** $\emptyset$, which contains no elements. The empty set $\emptyset$ is a subset of every set X.
$V = X \cap Y$	V is the **intersection** of the sets X and Y; that is, v contains all elements that belong to both X and Y.
$W = X \cup Y$	W is the **union** of the sets X and Y; that is, W contains all elements that belong to X or to Y.
$f : X \longrightarrow Y$	The **function** f **sends**, **transforms**, **maps to**, or **assigns** each element x of its **domain** X an element $f(x)$ in its **range** Y. For example, if $X = \{\textit{all calendar dates}\}$, $Y = \{\textit{all calendar years}\}$, and f maps dates in X to their years, then $f(\textit{October } 14, 1951) = 1951$.
$A, B, X, U, \dots$	Capital Roman letters usually denote **matrices**, **sets**, and **subspaces**.
$\mathbf{x}, \mathbf{y}, \mathbf{u}, \mathbf{v}, \dots$	Lowercase bold Roman letters usually stand for **vectors** in $\mathbb{R}^n$ or $\mathbb{C}^n$.

$\alpha, \beta, \gamma, \lambda, \mu, \ldots$	Lowercase Greek letters denote **scalars** (i.e., real or complex numbers). Pronounce α as "alpha," β "beta," λ "lambda," and μ "mu."
$0, \mathbf{0}$	The real number **zero**, or the zero vector, i.e., the **origin** of any space.
$\Longleftrightarrow$	**If and only if**; shorthand notation for logical equivalence. An example is $x - 2 = 0 \Longleftrightarrow x = 2$.
$\Rightarrow$	**Implies**; logical shorthand notation. An example is $x = 2 \Rightarrow x > 0$.
$\Leftarrow$	**Follows from** or is implied by. An example is $x^2 = 4 \Leftarrow x = -2$.
■	**End-of-proof** symbol. Indicates that the proof of a theorem or a lemma has been completed.
◀	End mark for Examples, extended Theorems, Lemmas, and Propositions

EXAMPLE 1 The sets $\mathbb{R} = \{\text{real numbers}\}$ and $\mathbb{R}^+ = \{x \in \mathbb{R} \mid x > 0\}$ describe two commonly used sets. The number $2 \in \mathbb{R}$ and $2 \in \mathbb{R}^+$; hence, $2 \in \mathbb{R} \cap \mathbb{R}^+$. But $\sqrt{-4} \notin \mathbb{R}$ and $-3 \notin \mathbb{R}^+$. $\mathbb{R}^+$ describes the positive real numbers. $\mathbb{R}^+$ is a subset of all real numbers $\mathbb{R}$, as is the set $\{1, 2, \pi\} \subset \mathbb{R}$.

Continuing in a more abstract vein, the words *fat* and *rat* are a subset U of the set X of words *fat*, *hat*, and *rat*, or $U := \{fat, rat\} \subset \{fat, hat, rat\} =: X$. Clearly $U, X \subset \{xay \mid x, y \text{ any letters of the alphabet }\} =: V$. Here, V contains all three-letter words with the middle letter a (along with all other three-letter English strings of characters with the middle letter a). Note that $U \cap X = U$ and $U \cup X = X$.

The word *map* does not belong to the set of words consisting of *cat*, *nap*, and *fan*, or $map \notin \{cat, nap, fan\} =: W$. However, $map \in V$, $W \cap U = \emptyset$; $W \cap X = \emptyset$, $W \cup X \subset V$, and $W \subset V$. ◀

Functions f describe assignments or mappings of the elements $x \in X$ of one set—f's **domain** X— to the elements $y = f(x) \in Y$ in f's **range** Y. A map f from X to Y is described by the sequence of symbols $f : X \to Y$, which we read as "f maps X to Y", or "f is a function from set X to set Y." Each function $f : X \to Y$ specifies the assignment of a unique element $y = f(x) \in Y$ for each $x \in X$.

EXAMPLE 2 The function $f : \mathbb{R} \to \mathbb{R}$ defined by the assignment $f(x) := 2x - 3x^2$ maps each real number x to the value of $2x - 3x^2 \in \mathbb{R}$. For example $f(2) = 4 - 12 = -8$ and $f(0) = 0$.

The function $g : \mathbb{R}^2 \to \mathbb{R}^2$ defined by $g\left(\begin{pmatrix} x_1 \\ x_2 \end{pmatrix}\right) = \begin{pmatrix} 2x_1 - 3 \\ x_1 + x_2^2 \end{pmatrix} \in \mathbb{R}^2$ maps an arbitrary vector $\mathbf{x} = \begin{pmatrix} x_1 \\ x_2 \end{pmatrix} \in \mathbb{R}^2$ to its image $\begin{pmatrix} 2x_1 - 3 \\ x_1 + x_2^2 \end{pmatrix} \in \mathbb{R}^2$. For example, if $\mathbf{x} = \begin{pmatrix} 1 \\ 2 \end{pmatrix}$, then $g(\mathbf{x}) = g\left(\begin{pmatrix} 1 \\ 2 \end{pmatrix}\right) = \begin{pmatrix} -1 \\ 5 \end{pmatrix} \in \mathbb{R}^2$ and $g(\mathbf{0}) = \left(\begin{pmatrix} 0 \\ 0 \end{pmatrix}\right) = \begin{pmatrix} -3 \\ 0 \end{pmatrix}$ for the origin $\mathbf{0} \in \mathbb{R}^2$. ◀

Linear algebra is an old subject: Linear equations have been studied since antiquity. Yet the main development of the subject has taken place in the last two or three hundred years.

For a short history of the subject, look at the website

```
http://www-history.mcs.st-and.ac.uk/history/HistTopics/
        Matrices_and_determinants.html#s32
```

An Eye towards Applications

Linear algebra and matrix theory are two of the most useful branches of modern mathematics. They serve extensively as tools in almost all numerical mathematical computations. Modern science and engineering relies on them. In this book, we outline several applications.

We apply linear algebra to calculus, interpreting differentiation and integration as linear transformations of the space of real functions in Section 1.3, and transforming integration by parts to the problem of finding the inverse differential operator on an invariant subspace in Section 7.3.

We study how to solve linear differential equations theoretically and numerically in Sections 7.3, 9.3, 11.3, and 14.3.

With the use of linear equations, electrical circuit and network flow problems, as well as the balancing of chemical reactions, are treated in Section 3.1.

Section 9.3 studies the outcome of consumer behavior and cell transport problems via the eigenvalues and eigenvectors of stochastic matrices.

The singular value decomposition of a matrix is used in Section 12.3 for data compression and the economical transmission of data.

Throughout the book are applications to the geometry of linear transformations and linear spaces.

> *Linear Algebra is one of the basic subjects of mathematics. It is fundamental in pure mathematics, computer science, and the applications of mathematics to other disciplines.*
>
> *Although centuries old, the subject is now in a stage of important growth. Part of this growth occurs because linear algebra plays a central role in high-speed computation. But a more fundamental reason is that, in the expanding world of mathematics and its applications, the simplest mathematical descriptions of the interplay between complex phenomena are given by linear relations.*

Linear algebra is thus both a basic subject for scientific education and a frontier subject for mathematical research. And both these roles seem permanently cast by nature. It is almost impossible to imagine a future of science and mathematics in which the importance of linear algebra will be diminished.

A. J. Hoffman,
preface to
Linear Algebra and its Applications
Volume 1, Number 1, 1968[2]

Reflections are linear

CHAPTER

1

Linear Transformations

This chapter introduces n–dimensional vector spaces and linear transformations thereof.

1.1 Lecture One, a Double Lecture

Vectors, Linear Functions, and Matrices

The first lecture defines the three main subjects of linear algebra:

- *the n–dimensional linear space $\mathbb{R}^n$, where n is an arbitrary positive integer,*
- *linear transformations $f : \mathbb{R}^n \longrightarrow \mathbb{R}^m$ between any two such spaces, and*
- *m-by-n matrices.*

The first chapter enters deeply into linear algebra, a broad and multifaceted branch of mathematics.

Not every notion of this field can be explained completely in any first chapter. Here we attempt to explain vectors in $\mathbb{R}^n$ and linear transformations thereof in full detail, while merely touching on other related and important subjects.

Like a first-time traveler in an unfamiliar culture or continent, a student should step lightly, curiously, and confidently through Chapter 1. Valuable and concrete knowledge is to be learned here. At the same time, there is built-in room for healthy beginner's confusion: Many questions will be left open, and many concepts cannot be explained satisfactorily. The natural curiosity of a beginning student of linear algebra is a valuable attribute. It needs to be encouraged, cherished, and fostered. A coherent picture of linear algebra will emerge toward the end of the book with Chapter 9.

(a) Vector Algebra and Geometry

For arbitrary positive integers $n = 1, 2, 3, 4, \ldots$, **real** n**–space**

$$\mathbb{R}^n = \left\{ \mathbf{x} \,\middle|\, \mathbf{x} = \begin{pmatrix} x_1 \\ \vdots \\ x_n \end{pmatrix}, x_i \in \mathbb{R}, i = 1, \ldots, n \right\}$$

behaves similarly to our familiar two-dimensional and three-dimensional spaces, where vectors behave like physical forces.

The **addition** $\mathbf{x}+\mathbf{y}$ of two vectors $\mathbf{x}$ and $\mathbf{y}$ in $\mathbb{R}^2$ or $\mathbb{R}^3$ creates a third vector, the resultant vector, by the parallelogram law, as shown in Figure 1-1.

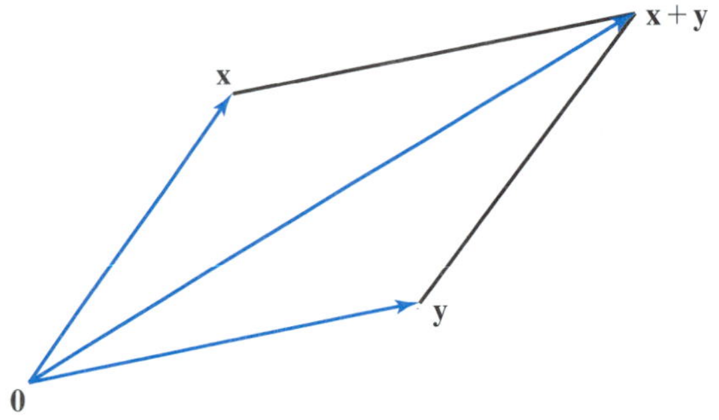

Figure 1-1

Scalar multiplication $\alpha\mathbf{x}$ in $\mathbb{R}^2$ or $\mathbb{R}^3$ stretches or shrinks a given vector $\mathbf{x}$, depending on whether $|\alpha| > 1$ or $0 < |\alpha| < 1$. It changes its direction if $\alpha < 0$. If $\alpha = 0$, then the vector $\alpha\mathbf{x}$ collapses to the origin $\mathbf{0} \in \mathbb{R}^n$, as Figure 1-2 shows.

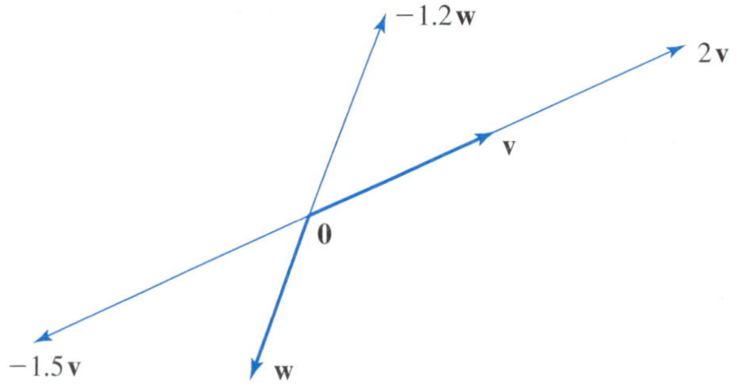

Figure 1-2

The intuitive geometric behaviour of vectors under vector addition and scalar multiplication in $\mathbb{R}^2$ and $\mathbb{R}^3$ can be equivalently described in algebraic terms by using the vector components of $\mathbf{x}$ and $\mathbf{y}$ in $\mathbb{R}^n$.

DEFINITION 1 $\mathbb{R}^n$ is endowed with two vector operations:

(1) **Vector addition**

For any two vectors $\mathbf{x} = \begin{pmatrix} x_1 \\ \vdots \\ x_n \end{pmatrix}$ and $\mathbf{y} = \begin{pmatrix} y_1 \\ \vdots \\ y_n \end{pmatrix} \in \mathbb{R}^n$, the sum is defined as

$$\mathbf{x} + \mathbf{y} = \begin{pmatrix} x_1 \\ \vdots \\ x_n \end{pmatrix} + \begin{pmatrix} y_1 \\ \vdots \\ y_n \end{pmatrix} := \begin{pmatrix} x_1 + y_1 \\ \vdots \\ x_n + y_n \end{pmatrix} \in \mathbb{R}^n.$$

(2) **Scalar multiplication**

For any vector $\mathbf{x} = \begin{pmatrix} x_1 \\ \vdots \\ x_n \end{pmatrix} \in \mathbb{R}^n$ and any scalar $\alpha \in \mathbb{R}$, the **scalar multiple** is defined as

$$\alpha\mathbf{x} = \alpha \begin{pmatrix} x_1 \\ \vdots \\ x_n \end{pmatrix} := \begin{pmatrix} \alpha x_1 \\ \vdots \\ \alpha x_n \end{pmatrix} \in \mathbb{R}^n.$$

These two **vector operations** are performed by adding and scalar multiplying vectors, one component at a time. They give $\mathbb{R}^n$ the "parallelogram law" of physical forces with which we are familiar from $\mathbb{R}^2$ and $\mathbb{R}^3$.

If as in Figure 1-1, two mules pull an object situated at zero, one in the direction of $\mathbf{x}$ with a force indicated by the length of $\mathbf{x}$, the other in direction of $\mathbf{y}$ with a force as indicated by the length of $\mathbf{y}$, then the object at zero experiences being pulled by a force in the direction of $\mathbf{x} + \mathbf{y}$ as drawn in Figure 1-1, with a magnitude equal to the length of the vector $\mathbf{x} + \mathbf{y}$.

Here the resultant vector $\mathbf{x} + \mathbf{y}$ is determined geometrically from $\mathbf{x}$ and $\mathbf{y}$ as the diagonal in the parallelogram with sides $\overline{\mathbf{0x}}$ and $\overline{\mathbf{0y}}$ and corner $\mathbf{0}$. Equivalently $\mathbf{x} + \mathbf{y}$ is determined algebraically by adding the individual components of $\mathbf{x}$ and $\mathbf{y} \in \mathbb{R}^2$ as in Definition 1; see Problem 0.

EXAMPLE 1

(a) If $\mathbf{x} = \begin{pmatrix} 1 \\ 2 \\ 3 \\ 4 \end{pmatrix}$ and $\mathbf{y} = \begin{pmatrix} 4 \\ 3 \\ 2 \\ 1 \end{pmatrix} \in \mathbb{R}^4$, then $2\mathbf{x} = \begin{pmatrix} 2 \\ 4 \\ 6 \\ 8 \end{pmatrix}$ and $-3\mathbf{y} = \begin{pmatrix} -12 \\ -9 \\ -6 \\ -3 \end{pmatrix}$, while $\mathbf{x} + \mathbf{y} = \begin{pmatrix} 1 \\ 2 \\ 3 \\ 4 \end{pmatrix} + \begin{pmatrix} 4 \\ 3 \\ 2 \\ 1 \end{pmatrix} = \begin{pmatrix} 5 \\ 5 \\ 5 \\ 5 \end{pmatrix} = 5 \begin{pmatrix} 1 \\ 1 \\ 1 \\ 1 \end{pmatrix} \in \mathbb{R}^4$.

(b) If $\mathbf{x} = \begin{pmatrix} 1 \\ \vdots \\ 1 \end{pmatrix} \in \mathbb{R}^n$, then $-\mathbf{x} = \begin{pmatrix} -1 \\ \vdots \\ -1 \end{pmatrix}$ and $\mathbf{x} + (-\mathbf{x}) = \begin{pmatrix} 0 \\ \vdots \\ 0 \end{pmatrix}$, the zero vector in $\mathbb{R}^n$. That is, the two forces, or vectors, $\mathbf{x}$ and $-\mathbf{x}$ cancel each other when added.

(c) If $\mathbf{x} \in \mathbb{R}^4$ and $\mathbf{y} \in \mathbb{R}^5$, then their addition is undefined, while scalar multiples such as $3\mathbf{x} \in \mathbb{R}^4$ and $-7\mathbf{y} \in \mathbb{R}^5$ are well defined in their respective spaces.

◀

(b) Linear Transformations

First Examples Our main objects of study in this course are **linear functions** between finite-dimensional real vector spaces.

DEFINITION 2 A function f that maps vectors in $\mathbb{R}^n$ to vectors in $\mathbb{R}^m$, expressed symbolically as $f : \mathbb{R}^n \to \mathbb{R}^m$, is a **linear transformation** if it satisfies the **linearity condition**

$$f(\alpha\mathbf{x} + \beta\mathbf{y}) = \alpha f(\mathbf{x}) + \beta f(\mathbf{y}) \tag{1.1}$$

for all vectors $\mathbf{x}, \mathbf{y} \in \mathbb{R}^n$ and all scalars $\alpha, \beta \in \mathbb{R}$.

The linearity condition (1.1) can be split into two separate conditions, one for vector addition and another for scalar multiplication:

$$\begin{aligned} f(\mathbf{x}+\mathbf{y}) &= f(\mathbf{x}) + f(\mathbf{y}), && \text{for all } \mathbf{x}, \ \mathbf{y} \in \mathbb{R}^n, \quad \underline{\text{and}} \\ f(\alpha\mathbf{x}) &= \alpha f(\mathbf{x}), && \text{for all } \alpha \in \mathbb{R} \text{ and all } \mathbf{x} \in \mathbb{R}^n. \end{aligned} \tag{1.2}$$

EXAMPLE 2 Linear and nonlinear functions occur in many everyday examples.

(a) The exchange rate α between any two currencies, say, the US\$ (\$) and the Euro (€), sets up two linear functions that translate € amounts to \$ and vice versa.

$$\begin{aligned} \$(€) &= \alpha \cdot €, \\ €(\$) &= 1/\alpha \cdot \$. \end{aligned}$$

For example, at the Euro's inception in 1999, 1 € $= 1.1 \cdot \$$; that is, $\alpha = 1.1$ and 22 € were equivalent to $1.1 \cdot 22 = 24.20$ \$.

The reciprocal rate $\dfrac{1}{1.1} = 0.909$ acts to convert \$ amounts to €. We find that $0.909 \cdot 20 = 18.18$ €, for example, was the value of \$20 in 1999.

Find the current exchange rate between the Euro and the dollar and recompute the values of 22 € and \$20 today.

(b) The volume function V of a rectangular three-dimensional box with sides of length $a, b,$ and c is a linear function of any one of its edge lengths where $V(a, b, c) := a \cdot b \cdot c$.

Clearly, the two identities

$$V(a, b, c+d) = a \cdot b \cdot (c+d) = a \cdot b \cdot c + a \cdot b \cdot d = V(a, b, c) + V(a, b, d)$$

and

$$V(a, b, \alpha c) = a \cdot b \cdot \alpha c = \alpha V(a, b, c)$$

verify (1.2), if, for example, the first two edges a and b of V do not vary.

(c) The volume V of a ball in $\mathbb{R}^3$ depends nonlinearly on its radius r. Here $V(r) := \frac{4}{3}\pi r^3$ and $V(2r) = \frac{4}{3}\pi(2r)^3 = 8 \cdot V(r) \neq 2 \cdot V(r)$, violating the second rule of Eq. (1.2).

(d) The function that converts Fahrenheit to Celsius degrees is nonlinear. The conversion formulas are

$$F(C) = \tfrac{9}{5}C + 32,$$
$$C(F) = \tfrac{5}{9}(F - 32).$$

Thus, it is obvious that $F(2C) \neq 2F(C)$ in general or that the Fahrenheit reading of 20°C (68°F) is not double that of 10°C (50°F), violating the second identity in Eq. (1.2). The following is a plot of the Fahrenheit to Celsius conversion function.

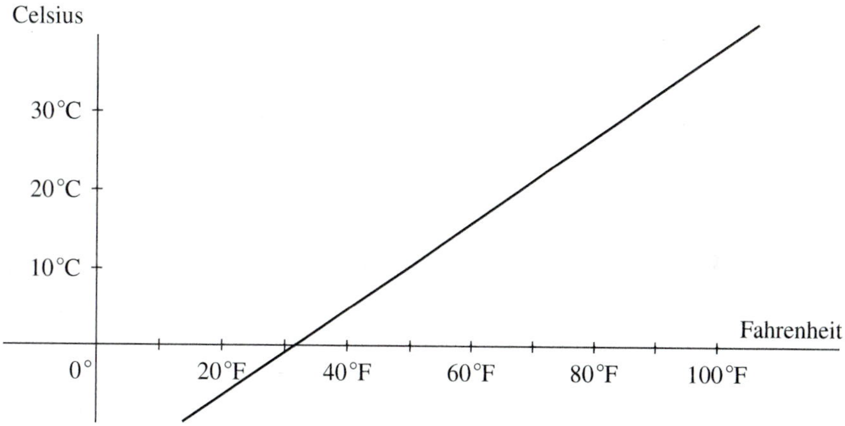

Figure 1-3

The graph is a line. In everyday English, we say that there is a *linear relation* between Fahrenheit and Celsius degrees. Linear algebra, however, uses linearity in a stricter sense. According to Eq. (1.1) or Eq. (1.2), we call functions $f : \mathbb{R} \to \mathbb{R}$ **linear** precisely when

(1) their graph is a line in $\mathbb{R}^2$, <u>and</u>

(2) their graph passes through the origin $\begin{pmatrix} 0 \\ 0 \end{pmatrix} \in \mathbb{R}^2$.

With this understanding of linearity, it is clear that exchange rate functions, as in part (a), are always linear, since different currency descriptions of equal value amounts of money are proportional according to the exchange rate and zero assets in any one currency have zero value in any other currency.

(e) The average A of n real numbers $x_1, \ldots, x_n \in \mathbb{R}$ is defined as

$$A(x_1, \ldots, x_n) := \frac{x_1 + \ldots + x_n}{n}.$$

This is a function that maps $\mathbb{R}^n$ to $\mathbb{R}$ linearly, since $A(\mathbf{x}+\mathbf{y}) = A(\mathbf{x}) + A(\mathbf{y})$ and $A(\alpha\mathbf{x}) = \alpha A(\mathbf{x})$ for all $\mathbf{x}, \mathbf{y} \in \mathbb{R}^n$ and all $\alpha \in \mathbb{R}$.

(f) Selecting one specific component x_j of a vector $\mathbf{x} \in \mathbb{R}^n$ is a linear transformation $T : \mathbb{R}^n \to \mathbb{R}$. Define T formally by setting $T\left(\begin{pmatrix} x_1 \\ \vdots \\ x_n \end{pmatrix}\right) := x_j$ for one specific index $1 \leq j \leq n$. Then

$$T\left(\begin{pmatrix} x_1 \\ \vdots \\ x_n \end{pmatrix} + \begin{pmatrix} y_1 \\ \vdots \\ y_n \end{pmatrix}\right) = T\left(\begin{pmatrix} x_1 + y_1 \\ \vdots \\ x_n + y_n \end{pmatrix}\right) = x_j + y_j$$
$$= T\left(\begin{pmatrix} x_1 \\ \vdots \\ x_n \end{pmatrix}\right) + T\left(\begin{pmatrix} y_1 \\ \vdots \\ y_n \end{pmatrix}\right)$$

and $T(\alpha\mathbf{x}) = \alpha x_j$ for all constants $\alpha \in \mathbb{R}$ and all vectors $\mathbf{x}$ and $\mathbf{y} \in \mathbb{R}^n$.

(g) Rotating all points $\begin{pmatrix} x \\ y \end{pmatrix}$ of the plane $\mathbb{R}^2$ by 45° clockwise around the **origin** $\begin{pmatrix} 0 \\ 0 \end{pmatrix}$ of $\mathbb{R}^2$ is described by the transformation $R\left(\begin{pmatrix} x \\ y \end{pmatrix}\right) := \frac{\sqrt{2}}{2}\begin{pmatrix} x + y \\ -x + y \end{pmatrix}$. To see this, we consider the specific rotations of the vectors $\begin{pmatrix} x \\ 0 \end{pmatrix}$ and $\begin{pmatrix} 0 \\ y \end{pmatrix}$ on the two coordinate axes:

From geometry, $R\left(\begin{pmatrix} x \\ 0 \end{pmatrix}\right) = \frac{\sqrt{2}}{2}\begin{pmatrix} x \\ -x \end{pmatrix}$ and $R\left(\begin{pmatrix} 0 \\ y \end{pmatrix}\right) = \frac{\sqrt{2}}{2}\begin{pmatrix} y \\ y \end{pmatrix}$ as in Figure 1-4. Linearity now helps us to map all vectors $\begin{pmatrix} x \\ y \end{pmatrix} \in \mathbb{R}^2$ by R, namely,

$$R\left(\begin{pmatrix} x \\ y \end{pmatrix}\right) = R\left(\begin{pmatrix} x \\ 0 \end{pmatrix} + \begin{pmatrix} 0 \\ y \end{pmatrix}\right) = \frac{\sqrt{2}}{2}\left(\begin{pmatrix} x \\ -x \end{pmatrix} + \begin{pmatrix} y \\ y \end{pmatrix}\right)$$
$$= \frac{\sqrt{2}}{2}\begin{pmatrix} x + y \\ -x + y \end{pmatrix}.$$

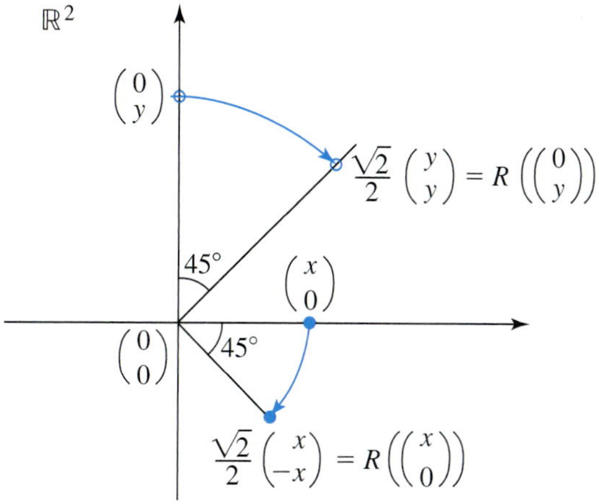

Figure 1-4

This describes the clockwise 45° rotation R of the plane as a linear transformation of $\mathbb{R}^2$ into itself. ◀

EXAMPLE 3 If we apply the linearity condition (1.1) to the mapping of vector sums $R(\mathbf{u}+\mathbf{v}) = R(\mathbf{u}) + R(\mathbf{v})$, then the 45° planar rotation R in Example 2(g) gives rise to the diagram shown in Figure 1-5.

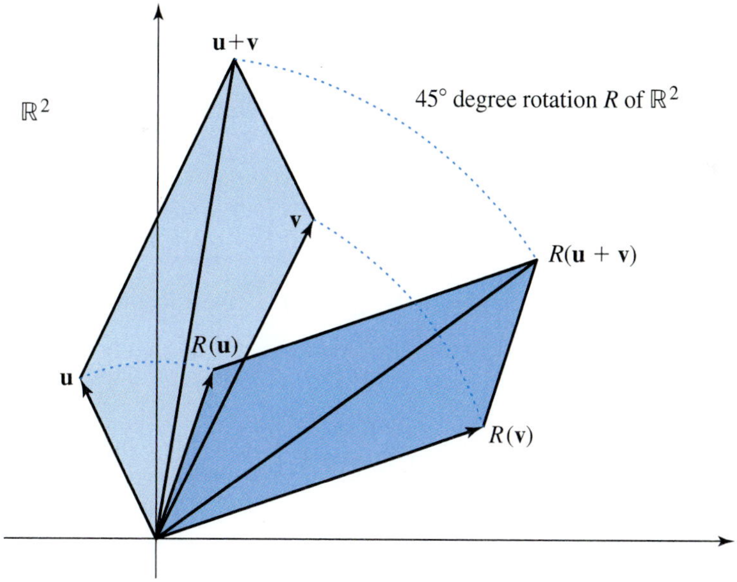

Figure 1-5

Here the left parallelogram describes the original forces $\mathbf{u}$, $\mathbf{v}$, and the resultant $\mathbf{u}+\mathbf{v}$, while the right one depicts $R(\mathbf{u})$, $R(\mathbf{v})$, and $R(\mathbf{u}+\mathbf{v})$ after the 45° rotation R. ◀

A linear transformation f maps the resultant $\alpha\mathbf{x}+\beta\mathbf{y}$ of any two vectors or forces $\alpha\mathbf{x}$ and $\beta\mathbf{y}$ to the resultant of the individual images $f(\alpha\mathbf{x})$ and $f(\beta\mathbf{y})$ of $\alpha\mathbf{x}$ and $\beta\mathbf{y}$ in its range $\mathbb{R}^m$ as exemplified in Example 3.

Thus,

> A linear transformation preserves the structure of the parallelogram law of forces.

What we have just called the resultant vector $\alpha\mathbf{x}+\beta\mathbf{y}$ of any two vectors $\mathbf{x}, \mathbf{y} \in \mathbb{R}^n$ for its analogy with physical forces is called a **linear combination** of the two vectors $\mathbf{x}$ and $\mathbf{y}$ in linear algebra.

In linear algebra, we operate (i.e., change vectors) by scalar multiplication and by vector addition. Any vector $\mathbf{y} = \sum_{i=1}^{k} \alpha_i \mathbf{x}_i \in \mathbb{R}^n$ that can be thus obtained from a set of vectors $\mathbf{x}_1, \ldots, \mathbf{x}_k \in \mathbb{R}^n$ and scalars $\alpha_i \in \mathbb{R}$ is a linear combination of the $\mathbf{x}_i$. For example, $\begin{pmatrix} 1 \\ -2 \end{pmatrix} \in \mathbb{R}^2$ is a linear combination of the vectors $\begin{pmatrix} 1 \\ 0 \end{pmatrix}$ and $\begin{pmatrix} 0 \\ -1 \end{pmatrix}$, since

$$\begin{pmatrix} 1 \\ -2 \end{pmatrix} = \begin{pmatrix} 1 \\ 0 \end{pmatrix} + 2\begin{pmatrix} 0 \\ -1 \end{pmatrix}.$$

Likewise, it is a linear combination of the vectors $\begin{pmatrix} 1 \\ 1 \end{pmatrix}$, $\begin{pmatrix} 0 \\ 0 \end{pmatrix}$, and $\begin{pmatrix} 2 \\ 1 \end{pmatrix}$, since

$$\begin{pmatrix} 1 \\ -2 \end{pmatrix} = -5\begin{pmatrix} 1 \\ 1 \end{pmatrix} + 17\begin{pmatrix} 0 \\ 0 \end{pmatrix} + 3\begin{pmatrix} 2 \\ 1 \end{pmatrix} = \begin{pmatrix} -5+6 \\ -5+3 \end{pmatrix}.$$

But $\mathbf{x} = \begin{pmatrix} 3 \\ 0 \end{pmatrix}$ is not a linear combination of the vectors $\mathbf{y} = \begin{pmatrix} 1 \\ -3 \end{pmatrix}$ and $\mathbf{z} = \begin{pmatrix} -2 \\ 6 \end{pmatrix}$, because any linear combination of these two has the form

$$\alpha\mathbf{y} + \beta\mathbf{z} = \begin{pmatrix} \alpha \\ -3\alpha \end{pmatrix} + \begin{pmatrix} -2\beta \\ 6\beta \end{pmatrix} = (\alpha - 2\beta)\begin{pmatrix} 1 \\ -3 \end{pmatrix},$$

which does not express the given vector $\mathbf{x} = \begin{pmatrix} 3 \\ 0 \end{pmatrix}$.

If we look at the two processes of

(1) mapping vectors of $\mathbb{R}^n$ to vectors in $\mathbb{R}^m$ under a linear map f, and
(2) forming linear combinations of vectors in $\mathbb{R}^n$ or $\mathbb{R}^m$,

then the linearity condition (1.1) $f(\alpha\mathbf{x}+\beta\mathbf{y}) = \alpha f(\mathbf{x}) + \beta f(\mathbf{y})$ for a function $f : \mathbb{R}^n \to \mathbb{R}^m$ assures us that these two processes can be interchanged. That is, the order in which the two operations are performed is irrelevant to the result, or—as we say—these two operations *commute*, since $f(\alpha\mathbf{x}+\beta\mathbf{y}) = \alpha f(\mathbf{x}) + \beta f(\mathbf{y})$ or since the mapping of a linear combination achieves the same result as combining the images linearly does.

Most actions in real life, as well as most operations in mathematics, do not commute; that is, the order in which they are performed is extremely important for the outcome. Small children are cute when they put on their shoes before their socks. Math students get into trouble when they equate $\sin(2x)$ with $2\sin(x)$. It is very rare for two actions or operations to achieve the same result, no matter in which order they are applied. Linear algebra is founded on one of the rare instances of the commutativity of two actions.

> The linear mapping $f(\alpha\mathbf{x}+\beta\mathbf{y})$ of a linear combination $\alpha\mathbf{x}+\beta\mathbf{y}$ of two vectors is identical to $\alpha f(\mathbf{x}) + \beta f(\mathbf{y})$. That is, it is identical to the same coefficient linear combination of the individual vector images $f(\mathbf{x})$ and $f(\mathbf{y})$.

Linear transformations and **vectors** are the fundamental objects of linear algebra.

When studying linear algebra from its fundamental objects, we are naturally led to try and understand those functions $f : \mathbb{R}^n \to \mathbb{R}^m$ that are linear, i.e., those that satisfy the functional equation (1.1) $f(\alpha\mathbf{x}+\beta\mathbf{y}) = \alpha f(\mathbf{x}) + \beta f(\mathbf{y})$ for all vectors $\mathbf{x}, \mathbf{y}$ in the domain of f and for all scalars α and $\beta \in \mathbb{R}$. Our quest is to look for a more practical description of linear mappings than the defining formula (1.1).

Not all functions are linear.

EXAMPLE 4 Is the function $f(x) = 3x^2 - 7x + 14 : \mathbb{R}^1 \to \mathbb{R}^1$ a linear transformation?

f is a real-valued quadratic polynomial, so our hunch is that f may not be linear.

Let us thus try to find specific values x, y, α, and $\beta \in \mathbb{R}$ with

$$f(\alpha x + \beta y) \neq \alpha f(x) + \beta f(y).$$

Note that the linearity condition (1.1) that claims something to be true for all values of the variables can be disproved for a specific function f by pointing out that this "for all . . ." statement does not hold "for some . . ." values of the variables.

For a quick check on the linearity of our given f, we pick

$$\alpha = 1, \beta = -1, \text{ and } x = 1 = y.$$

Then $\alpha x + \beta y = 0$, and the left-hand side of the linearity condition (1.1) evaluates to $f(\alpha x+\beta y) = f(0) = 14$, while the right-hand side is equal to $\alpha f(x)+\beta f(y) = f(1) - f(1) = 0 \neq 14$. Hence this function f is not linear. For a different proof, see Problem 15 in Section 1.1.P. ◀

Splitting a Linear Function into Component Functions Having familiarized ourselves with the linearity condition (1.1), we pose the main question of this chapter:

> What is the **shape** or **nature of a linear function** $f : \mathbb{R}^n \to \mathbb{R}^m$? How can we assert that a given mapping f is linear? How can a linear function be described in a useful and applicable way?

Any function $f : \mathbb{R}^n \to \mathbb{R}^m$ between n space and m space maps vectors $\mathbf{x} = \begin{pmatrix} x_1 \\ \vdots \\ x_n \end{pmatrix} \in \mathbb{R}^n$ with n **components** $x_i \in \mathbb{R}$ to their images $f(\mathbf{x}) = \begin{pmatrix} y_1 \\ \vdots \\ y_m \end{pmatrix} = \begin{pmatrix} f_1(\mathbf{x}) \\ \vdots \\ f_m(\mathbf{x}) \end{pmatrix} \in \mathbb{R}^m$ with m components $f_j(\mathbf{x}) \in \mathbb{R}$. The m individual entries $y_j := f_j(\mathbf{x}) : \mathbb{R}^n \to \mathbb{R}$ of $f(\mathbf{x})$ are the m **component functions** of f.

EXAMPLE 5 The function A in Example 2(e) has only one component function, as does T in Example 2(f), since both functions map $\mathbb{R}^n$ to $\mathbb{R}$.

The rotation $R : \mathbb{R}^2 \to \mathbb{R}^2$ of Example 2(g) has two component functions $f_j : \mathbb{R}^2 \to \mathbb{R}$

$$R\left(\begin{pmatrix} x \\ y \end{pmatrix}\right) = \begin{pmatrix} \frac{\sqrt{2}}{2}(x+y) \\ \frac{\sqrt{2}}{2}(-x+y) \end{pmatrix} = \begin{pmatrix} f_1\left(\begin{pmatrix} x \\ y \end{pmatrix}\right) \\ f_2\left(\begin{pmatrix} x \\ y \end{pmatrix}\right) \end{pmatrix}$$

with $f_1\left(\begin{pmatrix} x \\ y \end{pmatrix}\right) = \frac{\sqrt{2}}{2}(x+y)$ and $f_2\left(\begin{pmatrix} x \\ y \end{pmatrix}\right) = \frac{\sqrt{2}}{2}(-x+y)$.

The function f of Example 4 maps $\mathbb{R}^1$ to $\mathbb{R}^1$ and thus it has one component function $f_1 = f$. ◀

The splitting of a function $f : \mathbb{R}^n \to \mathbb{R}^m$ into component functions f_j gives us the following result for linear functions:

Theorem 1.1 **(Splitting Theorem)** A function $f : \mathbb{R}^n \to \mathbb{R}^m$ is linear if and only if each component function $f_j : \mathbb{R}^n \to \mathbb{R}$ of f is linear for $j = 1, \ldots, m$. ◀

Proof The 'if' part, also called the 'necessary condition' of an equivalence 'if and only if' statement, is proved as follows:

Let us assume that each component function $f_j(\mathbf{x})$ of f is linear, or

$$f_j(\alpha\mathbf{x} + \beta\mathbf{y}) = \alpha f_j(\mathbf{x}) + \beta f_j(\mathbf{y}) \quad \text{for all } \mathbf{x}, \mathbf{y} \in \mathbb{R}^n, \text{ all } \alpha, \beta \in \mathbb{R},$$

and for all $j = 1, \ldots, m$. Then

$$\begin{aligned}
f(\alpha\mathbf{x}+\beta\mathbf{y}) &= \begin{pmatrix} f_1(\alpha\mathbf{x}+\beta\mathbf{y}) \\ \vdots \\ f_m(\alpha\mathbf{x}+\beta\mathbf{y}) \end{pmatrix} \quad \text{(the definition)} \\
&= \begin{pmatrix} \alpha f_1(\mathbf{x}) + \beta f_1(\mathbf{y}) \\ \vdots \\ \alpha f_m(\mathbf{x}) + \beta f_m(\mathbf{y}) \end{pmatrix} \quad \text{(each } f_j \text{ is linear)} \\
&= \begin{pmatrix} \alpha f_1(\mathbf{x}) \\ \vdots \\ \alpha f_m(\mathbf{x}) \end{pmatrix} + \begin{pmatrix} \beta f_1(\mathbf{y}) \\ \vdots \\ \beta f_m(\mathbf{y}) \end{pmatrix} \quad \text{(vector addition)} \\
&= \alpha \begin{pmatrix} f_1(\mathbf{x}) \\ \vdots \\ f_m(\mathbf{x}) \end{pmatrix} + \beta \begin{pmatrix} f_1(\mathbf{y}) \\ \vdots \\ f_m(\mathbf{y}) \end{pmatrix} \quad \text{(scalar multiplication)} \\
&= \alpha f(\mathbf{x}) + \beta f(\mathbf{y}) \text{ (the definition),}
\end{aligned}$$

so that $f : \mathbb{R}^n \to \mathbb{R}^m$ is indeed linear according to Definition 2.

(Proof continues below)

The converse "⇒" part can be proved easily by using the linearity condition (1.1) for f on each of its component functions f_j. However, we shall use a more general result on the composition of linear maps that will be useful in the future.

Lemma 1 If $f : \mathbb{R}^n \to \mathbb{R}^m$ is a linear transformation and $g : \mathbb{R}^m \to \mathbb{R}^k$ is also linear, then the composite function $g \circ f : \mathbb{R}^n \to \mathbb{R}^k$ defined as $g \circ f(\mathbf{x}) := g(f(\mathbf{x})) : \mathbb{R}^n \to \mathbb{R}^k$ is linear as well. ◀

Proof (Lemma 1) From the definition of the composition $g \circ f$ of functions and from the linearity of both f and g, we obtain

$$\begin{aligned}
g \circ f(\alpha\mathbf{x}+\beta\mathbf{y}) &= g(f(\alpha\mathbf{x}+\beta\mathbf{y})) \\
&= g(\alpha f(\mathbf{x}) + \beta f(\mathbf{y})) \\
&= \alpha g(f(\mathbf{x})) + \beta g(f(\mathbf{y})) \\
&= \alpha(g \circ f(\mathbf{x})) + \beta(g \circ f(\mathbf{y})).
\end{aligned}$$

(The students should be able to indicate which property validates each equal sign.) ■

Proof (Theorem 1.1; continued) The 'only if' part, also called the 'sufficient condition' of an equivalence such as the statement of Theorem 1.1 is proved as follows:

Let g be the function that projects vectors $\mathbf{z} \in \mathbb{R}^m$ onto their j^{th} component for any fixed $1 \leq j \leq m$. Namely, let $g : \mathbb{R}^m \to \mathbb{R}$ be defined by $g\left(\begin{pmatrix} z_1 \\ \vdots \\ z_m \end{pmatrix}\right) :=$ $z_j \in \mathbb{R}$ for $\mathbf{z} = \begin{pmatrix} z_1 \\ \vdots \\ z_m \end{pmatrix} \in \mathbb{R}^m$. As seen in Example 2(f), the coordinate projection function g is linear.

For the proof of necessity, we assume that f is linear. Hence by Lemma 1 we know that $g \circ f$ is linear for the component projection function g. Thus,

$$g \circ f(\mathbf{x}) = g(f(\mathbf{x})) = g\left(\begin{pmatrix} f_1(\mathbf{x}) \\ \vdots \\ f_j(\mathbf{x}) \\ \vdots \\ f_m(\mathbf{x}) \end{pmatrix}\right) = f_j(\mathbf{x})$$

must be linear for each j as well. ■

Thus,

> Linear functions $f : \mathbb{R}^n \to \mathbb{R}^m$ are precisely those made up of m linear component functions $f_j : \mathbb{R}^n \to \mathbb{R}$ that map $\mathbb{R}^n$ to $\mathbb{R}$ with $f = \begin{pmatrix} f_1 \\ \vdots \\ f_m \end{pmatrix}$.

We have obtained this result by using the linearity condition for f and by splitting each vector in the **range space** $\mathbb{R}^m$ of f into its m natural components.

Real-Valued Linear Functions of $\mathbb{R}^n$ Now one question remains: Which real valued functions $f : \mathbb{R}^n \to \mathbb{R}$ are linear?

This is solved by looking more closely at the **domain** $\mathbb{R}^n$ of f.

In linear algebra, as well as in many areas of life, problems can often be solved by **staring** at, by *observing*, or by *looking at and visualizing* the problem for a while. For linear algebra this "look and see" approach is particularly appropriate, since linear algebra is very closely related to geometry.

Let us thus look at $f(\mathbf{x}) = f\left(\begin{pmatrix} x_1 \\ \vdots \\ x_n \end{pmatrix}\right) = \begin{pmatrix} y_1 \\ \vdots \\ y_m \end{pmatrix} \in \mathbb{R}^m$ for a vector $\mathbf{x} \in \mathbb{R}^n$ and a given linear map $f : \mathbb{R}^n \to \mathbb{R}^m$.

Note:

From now on, we shall abuse notation a little bit. If $\mathbf{x} = \begin{pmatrix} x_1 \\ \vdots \\ x_n \end{pmatrix}$ is a vector, then formally $f(\mathbf{x})$ must be written as $f\left(\begin{pmatrix} x_1 \\ \vdots \\ x_n \end{pmatrix}\right)$. This we have done so far. To avoid the correct, but awkward, double parenthesis notation, we simply drop one pair of parentheses in expressing $f(\mathbf{x})$ and denote any vector function $f(\mathbf{x})$ from now on simply by $f\begin{pmatrix} x_1 \\ \vdots \\ x_n \end{pmatrix}$ if $\mathbf{x} = \begin{pmatrix} x_1 \\ \vdots \\ x_n \end{pmatrix}$.

We have already seen that each component y_j of f is a linear function of $\mathbf{x}$ if f is linear. That is, $y_j = f_j(\mathbf{x})$, or $f(\mathbf{x}) = \begin{pmatrix} f_1(\mathbf{x}) \\ \vdots \\ f_m(\mathbf{x}) \end{pmatrix}$ with linear component functions $f_j(\mathbf{x}) \in \mathbb{R}$ for each j.

Due to the Splitting Theorem, we need only study each linear component function

$$f_j(\mathbf{x}) : \mathbb{R}^n \to \mathbb{R} \quad \text{of } f.$$

That is, we need to look at the problem for one f_j.

The **argument** of f_j (i.e., the variable on which f_j operates) is $\mathbf{x} \in \mathbb{R}^n$. We can write any vector $\mathbf{x} \in \mathbb{R}^n$ as

$$\mathbf{x} = \begin{pmatrix} x_1 \\ \vdots \\ \vdots \\ \vdots \\ x_n \end{pmatrix} = x_1 \begin{pmatrix} 1 \\ 0 \\ \vdots \\ \vdots \\ 0 \end{pmatrix} + x_2 \begin{pmatrix} 0 \\ 1 \\ 0 \\ \vdots \\ 0 \end{pmatrix} + \cdots + x_n \begin{pmatrix} 0 \\ \vdots \\ \vdots \\ 0 \\ 1 \end{pmatrix} = \sum_{i=1}^{n} x_i \mathbf{e}_i \in \mathbb{R}^n, \tag{1.3}$$

where each $x_i \in \mathbb{R}$ is a scalar and the vectors $\mathbf{e}_1 = \begin{pmatrix} 1 \\ 0 \\ \vdots \\ 0 \end{pmatrix}$, $\mathbf{e}_2 = \begin{pmatrix} 0 \\ 1 \\ 0 \\ \vdots \\ 0 \end{pmatrix}$, and so forth until $\mathbf{e}_n = \begin{pmatrix} 0 \\ \vdots \\ 0 \\ 1 \end{pmatrix} \in \mathbb{R}^n$ are called the **standard unit vectors** $\mathcal{E} = \{\mathbf{e}_1, \ldots, \mathbf{e}_n\}$ of $\mathbb{R}^n$.

For each index $1 \le i \le n$ the standard vector $\mathbf{e}_i \in \mathbb{R}^n$ contains zero entries in every component, except for the ith component, which is one. These vectors $\mathbf{e}_i$ shall help us throughout the book. They form the **standard basis** $\mathcal{E} = \{\mathbf{e}_1, \dots, \mathbf{e}_n\}$ of $\mathbb{R}^n$.

Bases of $\mathbb{R}^n$ will be studied in great detail in Chapter 5 and beyond. For the moment, it suffices to be aware that the standard basis $\mathcal{E} = \{\mathbf{e}_1, \dots, \mathbf{e}_n\}$ allows us to express every vector $\mathbf{x} \in \mathbb{R}^n$ as the unique linear combination $\mathbf{x} = \sum_i x_i \mathbf{e}_i$ of the basic vectors $\mathbf{e}_i$, as in Eq. (1.3).

EXAMPLE 6 The n standard unit vectors $\mathbf{e}_1, \mathbf{e}_2, \dots, \mathbf{e}_n$ can be utilized to express all vectors in $\mathbb{R}^n$:

If $\mathbf{x} = \begin{pmatrix} 2 \\ -1 \\ 0 \\ 3 \end{pmatrix} \in \mathbb{R}^4$, for example, then $\mathbf{x} = 2\mathbf{e}_1 - \mathbf{e}_2 + 3\mathbf{e}_4$ by looking at the ordered coefficients 2, −1, 0, and 3 of $\mathbf{x}$. Likewise, $3\mathbf{e}_2 - \mathbf{e}_1 + 6\mathbf{e}_5 \in \mathbb{R}^5$ is the point $\begin{pmatrix} -1 \\ 3 \\ 0 \\ 0 \\ 6 \end{pmatrix} \in \mathbb{R}^5$ in column vector notation. ◀

Now we scrutinize each component function f_j of the linear mapping $f : \mathbb{R}^n \to \mathbb{R}^m$ in terms of $\mathbf{x} = \sum_i x_i \mathbf{e}_i$. We have

$$\begin{aligned} f_j(\mathbf{x}) &= f_j\left(\sum_{i=1}^n x_i \mathbf{e}_i\right) && \text{(by Eq. (1.3))} \\ &= \sum_{i=1}^n x_i \left(f_j(\mathbf{e}_i)\right) && \text{(by the linearity of } f_j) \\ &= \begin{pmatrix} f_j(\mathbf{e}_1) \\ \vdots \\ f_j(\mathbf{e}_n) \end{pmatrix} \cdot \mathbf{x} && (1.4) \end{aligned}$$

in **dot product notation**. Here the symbol $\mathbf{a} \cdot \mathbf{x} \in \mathbb{R}$ denotes the **dot product** of any two column vectors $\mathbf{a}$ and $\mathbf{x} \in \mathbb{R}^n$.

DEFINITION 3 For two vectors $\mathbf{x}, \mathbf{y} \in \mathbb{R}^n$, the **dot product** is defined as the real number

$$\mathbf{x} \cdot \mathbf{y} = x_1 y_1 + \cdots + x_n y_n (= \mathbf{y} \cdot \mathbf{x}) \in \mathbb{R}.$$

We are ready to state the second fundamental result of Lecture One.

Lemma 2 **(Riesz Lemma)** A mapping $f(\mathbf{x}) : \mathbb{R}^n \to \mathbb{R}$ is linear if and only if $f(\mathbf{x})$ can be expressed as a dot product $\mathbf{a} \cdot \mathbf{x}$ for a fixed vector $\mathbf{a} \in \mathbb{R}^n$ depending on f and every $\mathbf{x} \in \mathbb{R}^n$.

The vector $\mathbf{a}$ is uniquely determined by f, namely, $\mathbf{a} = \begin{pmatrix} f(\mathbf{e}_1) \\ \vdots \\ f(\mathbf{e}_n) \end{pmatrix}$ according to (1.4). This vector is called the **Riesz column vector** of f with respect to the standard basis $\mathcal{E} = \{\mathbf{e}_1, \ldots, \mathbf{e}_n\}$ of $\mathbb{R}^n$. ◀

Proof The proof of necessity was completed in Eq. (1.4).

For sufficiency, we need to verify that the dot product function $f(\mathbf{x}) := \mathbf{a} \cdot \mathbf{x}$ indeed satisfies the linearity condition $f(\alpha\mathbf{x}+\beta\mathbf{y}) = \alpha f(\mathbf{x}) + \beta f(\mathbf{y})$ for any fixed vector $\mathbf{a} \in \mathbb{R}^n$, all $\mathbf{x} \in \mathbb{R}^n$, and all $\alpha, \beta \in \mathbb{R}$. This should be verified by the students. See Problem 28 in Section 1.1.P. ■

EXAMPLE 7 The functions $A\mathbf{x}$ and $T\mathbf{x} : \mathbb{R}^n \to \mathbb{R}$ of Examples 2(e) and 2(f) are both linear. Their Riesz column vectors are $\mathbf{a} = \begin{pmatrix} 1/n \\ \vdots \\ 1/n \end{pmatrix}$ and $\mathbf{t} = \begin{pmatrix} 0 \\ \vdots \\ 0 \\ 1 \\ 0 \\ \vdots \\ 0 \end{pmatrix} \leftarrow j^{\text{th}}$ position,

respectively, since $A\begin{pmatrix} x_1 \\ \vdots \\ x_n \end{pmatrix} = \dfrac{x_1 + \ldots + x_n}{n} = \mathbf{a}\cdot\mathbf{x}$ and $T\begin{pmatrix} x_1 \\ \vdots \\ x_n \end{pmatrix} = x_j = \mathbf{t}\cdot\mathbf{x}$.

For f in Example 4, there clearly is no real constant a with $f(x) = 3x^2 - 7x + 14 = a \cdot x$ for all $x \in \mathbb{R}$. ◀

Dot Products and Matrix Representations In our study of linear transformations, we have just been led to study dot products of pairs of n–vectors.

For two-column vectors $\mathbf{a}$ and $\mathbf{x} \in \mathbb{R}^n$, we rewrite the dot product

$$\mathbf{a} \cdot \mathbf{x} = a_1x_1 + \cdots + a_nx_n$$

as

$$(a_1, \ldots, a_n)\begin{pmatrix} x_1 \\ \vdots \\ x_n \end{pmatrix},$$

i.e., as the <u>more convenient</u> product of a **row vector** representation of $\mathbf{a}$ and a **column vector** representation of $\mathbf{x}$, both in $\mathbb{R}^n$. This is done solely for convenience at this moment. This row-times-column convention for dot products enables us to use a magnificent **shorthand notation** for linear transformations. Namely the row $\times$ column vector representation of the ordinary dot product most naturally gives rise to rectangular arrays of numbers, called **matrices**:

If $f : \mathbb{R}^n \to \mathbb{R}^m$ is linear and $f(\mathbf{x}) = \begin{pmatrix} f_1(\mathbf{x}) \\ \vdots \\ f_m(\mathbf{x}) \end{pmatrix}$ is expressed componentwise, then each component function

$$f_j(\mathbf{x}) = \mathbf{r}_j \cdot \mathbf{x} = \begin{pmatrix} f_j(\mathbf{e}_1) \\ \vdots \\ f_j(\mathbf{e}_n) \end{pmatrix} \cdot \mathbf{x} = \begin{pmatrix} f_j(\mathbf{e}_1) & \dots & f_j(\mathbf{e}_n) \end{pmatrix} \begin{pmatrix} x_1 \\ \vdots \\ x_n \end{pmatrix}$$

has a specific dot product representation $\mathbf{r}_j \cdot \mathbf{x}$ from the Riesz Lemma. Using the latter row $\times$ column dot product representation of the Riesz vector product, we can express

$$f(\mathbf{x}) = \begin{pmatrix} f_1(\mathbf{x}) \\ \vdots \\ \vdots \\ f_m(\mathbf{x}) \end{pmatrix} = \begin{pmatrix} (f_1(\mathbf{e}_1), \dots, f_1(\mathbf{e}_n)) \begin{pmatrix} x_1 \\ \vdots \\ x_n \end{pmatrix} \\ (f_2(\mathbf{e}_1), \dots, f_2(\mathbf{e}_n)) \begin{pmatrix} x_1 \\ \vdots \\ x_n \end{pmatrix} \\ \vdots \\ (f_m(\mathbf{e}_1), \dots, f_m(\mathbf{e}_n)) \begin{pmatrix} x_1 \\ \vdots \\ x_n \end{pmatrix} \end{pmatrix}$$

$$= \begin{pmatrix} f_1(\mathbf{e}_1) & \dots & f_1(\mathbf{e}_n) \\ f_2(\mathbf{e}_1) & \dots & f_2(\mathbf{e}_n) \\ \vdots & & \vdots \\ f_m(\mathbf{e}_1) & \dots & f_m(\mathbf{e}_n) \end{pmatrix} \begin{pmatrix} x_1 \\ \vdots \\ x_n \end{pmatrix} =: \mathbf{A}\mathbf{x}, \tag{1.5}$$

by combining all the respective Riesz row vectors $\mathbf{r}_j$ of f in the left factor $\mathbf{A}$ and writing the common vector $\mathbf{x} = \begin{pmatrix} x_1 \\ \vdots \\ x_n \end{pmatrix}$ on the right. Here

$$\mathbf{A} := \begin{pmatrix} f_1(\mathbf{e}_1) & \dots & f_1(\mathbf{e}_n) \\ \vdots & & \vdots \\ f_m(\mathbf{e}_1) & \dots & f_m(\mathbf{e}_n) \end{pmatrix} = \begin{pmatrix} - & \mathbf{r}_1 & - \\ & \vdots & \\ - & \mathbf{r}_m & - \end{pmatrix} \in \mathbb{R}^{m,n}$$

is a rectangular array of $m \cdot n$ real numbers $f_j(\mathbf{e}_i)$.

That is, $\mathbf{A}$ is the **matrix** that represents the linear transformation f via its **Riesz row vectors**

$$\mathbf{r}_j = \begin{pmatrix} f_j(\mathbf{e}_1) & \dots & f_j(\mathbf{e}_n) \end{pmatrix}.$$

Note that Riesz row vectors and Riesz column vectors contain the same entries. One is written as a row, the other as a column. There is no real difference, except that writing Riesz vectors row–wise allows for the neat **shorthand notation** of matrix × vector multiplication in Eq. (1.5).

Defining a linear mapping $f(\mathbf{x})$ as a **matrix × vector multiplication Ax** via dot products of all Riesz rows with one column vector is to be interpreted as producing the vector $\mathbf{y} = \mathbf{Ax}$ as output of the mapping $\mathbf{x} \mapsto \mathbf{y} = f(\mathbf{x})$. The individual components y_j of $\mathbf{Ax}$ are the ordinary dot products of $\mathbf{x}$ with the Riesz vectors $\mathbf{r}_j$ of f that are stored in the rows of $\mathbf{A}$.

Clearly any $\mathbf{x} \in \mathbb{R}^n$ in the domain of f has n entries, thus $\mathbf{A} = \begin{pmatrix} - & \mathbf{r}_1 & - \\ & \vdots & \\ - & \mathbf{r}_m & - \end{pmatrix}$ must have n columns so that the row × column vector product $\mathbf{r}_j\mathbf{x}$ of its row $\mathbf{r}_j$ with $\mathbf{x} \in \mathbb{R}^n$ is defined for each j.

As the image $\mathbf{y} = f(\mathbf{x}) = \mathbf{Ax}$ lies in $\mathbb{R}^m$, $\mathbf{A}$ must have m rows, each one giving rise to one entry in $\mathbf{y}$ obtained by evaluating one dot product.

DEFINITION 4

(a) A **matrix A** is a rectangular array of **entries**. If $\mathbf{A}$ has m **rows** and n **columns**, we write $\mathbf{A} = \mathbf{A}_{mn}$ and say that $\mathbf{A}$ is an $m \times n$ (or m-by-n) matrix.

An m-by-n matrix $\mathbf{A}$ can be written elementwise as $\mathbf{A} = (a_{ij})$, where an entry of $\mathbf{A}$, denoted by a_{ij}, appears in row i and column j. (**Note**: Rows before columns!)

If both m and n are explicit integers, such as $m = 3$ and $n = 12$, we write $\mathbf{A}_{mn} = \mathbf{A}_{3\times 12}$ for clarity, or, even more simply $\mathbf{A}_{3,12}$. This avoids the possible confusion with the matrix $\mathbf{A}_{31\times 2}(= \mathbf{A}_{31,2})$ that has 31 rows and 2 columns.

(b) The matrix entries a_{ii} for $i = 1, \dots, \min\{m, n\}$ form the **diagonal** of the matrix $\mathbf{A}_{mn}$; the entries a_{ij} with $i > j$ form the **lower triangle** and those a_{ij} with $i < j$ form the **upper triangle** of $\mathbf{A}$. If $m = n$, the matrix $\mathbf{A}_{mn}$ is called **square**.

Note that a column vector $\mathbf{x} \in \mathbb{R}^n$ can be interpreted as an $n \times 1$ matrix in matrix notation. It can thus be denoted by $\mathbf{x} = \mathbf{x}_{n\times 1}$. On the other hand, a row vector $\mathbf{a} \in \mathbb{R}^n$ can be described in matrix notation as $\mathbf{a}_{1\times n}$, having but one row in $\mathbb{R}^n$ and n columns, each in $\mathbb{R}^1 = \mathbb{R}$. If $m = n = 1$ for a real matrix $\mathbf{A}_{mn}$, then everything collapses, since a real one-by-one matrix is a real one-vector, both row and column, and it is a real scalar, depending on one's viewpoint.

Matrix × vector products Ax can only be formed if $\mathbf{A}$ and $\mathbf{x}$ are of **compatible size** (i.e., if $\mathbf{A}_{mn}$ is m by n and $\mathbf{x}_{n\times 1}$ is n by 1, or if and only if the two "inner sizes" n match):

$$\mathbf{A}_{m\underline{\underline{n}}}\mathbf{x}_{\underline{\underline{n}}\times 1} = \mathbf{y}_{m\times 1}.$$

In this case, the resulting vector $\mathbf{Ax} = \mathbf{y}$ inherits the "outer" size $m \times 1$ of the two compatible factors $\mathbf{A}_{mn}$ and $\mathbf{x}_{n\times 1}$ in the expression $\mathbf{y} = \mathbf{Ax}$.

EXAMPLE 8 Matrices and vectors can be multiplied in several different ways if their sizes are compatible:

(a) A row vector $\mathbf{y}_{1\times m}$ can be multiplied on the right by any matrix $\mathbf{A}_{mn}$ with m rows:

For example, if $\mathbf{y} = \begin{pmatrix} 1 & 3 & -2 & 4 \end{pmatrix}_{1\times 4}$ and $\mathbf{A} = \begin{pmatrix} 1 & -1 \\ 2 & 0 \\ 3 & -1 \\ 0 & 0 \end{pmatrix}_{4\times 2}$, then

$$\mathbf{y}_{1\times 4}\mathbf{A}_{4\times 2} = \begin{pmatrix} 1 & 1 \end{pmatrix}_{1\times 2}.$$

(b) A row vector $\mathbf{y}_{1,m}$ and a column vector $\mathbf{x}_{m,1}$ can be multiplied as $\mathbf{y}_{1,m}\mathbf{x}_{m,1} = \mathbf{z}_{1,1} \in \mathbb{R}$ to become a scalar, or, in reverse order, as $\mathbf{x}_{m,1}\mathbf{y}_{1,m} = \mathbf{C}_{mm} \in \mathbb{R}^{m,m}$ to become a square m-by-m matrix:

For example, $\mathbf{y} = \begin{pmatrix} 1 & -1 & 0 & 3 \end{pmatrix}$ and $\mathbf{x} = \begin{pmatrix} 0 \\ -1 \\ 5 \\ 2 \end{pmatrix}$ both lie in $\mathbb{R}^4$, and we have

$$\mathbf{yx} = 1+6 = 7 \text{ and } \mathbf{xy} = \begin{pmatrix} 0 \\ -1 \\ 5 \\ 2 \end{pmatrix} \begin{pmatrix} 1 & -1 & 0 & 3 \end{pmatrix} = \begin{pmatrix} 0 & 0 & 0 & 0 \\ -1 & 1 & 0 & -3 \\ 5 & -5 & 0 & 15 \\ 2 & -2 & 0 & 6 \end{pmatrix}.$$

The first vector product $\mathbf{y}_{1,4}\mathbf{x}_{4,1}$ is the real-valued **dot product** of two compatible vectors, as in Definition 3, while the result of the latter product $\mathbf{x}_{4,1}\mathbf{y}_{1,4}$ of the column vector $\mathbf{x}$ times the row $\mathbf{y}$ is a four-by-four square matrix, called a **dyad**, (See Sections 10.2 and 12.2).

(c) The matrix $\mathbf{A} = \begin{pmatrix} 1 & 2 & 0 \\ -1 & 0 & 4 \end{pmatrix}$ has diagonal entries 1 and 0. Its lower triangle consists of the single entry -1 while its upper triangle is $\begin{matrix} 2 & 0 \\ & 4 \end{matrix}$. $\mathbf{A}$ is not square. ◀

Note that we have been rather imprecise in Definition 4 as far as what the $m \cdot n$ **entries** a_{ij} of a matrix $\mathbf{A}_{mn}$ might be. In most applications, they are scalar real numbers, but complex entries may also occur. Yet there are many applications where some or all of the entries a_{ij} of $\mathbf{A}$ are functions, indeterminants, symbols, $\pm$ signs, matrices or operators, and the like.

The main concept holding the notion of a **matrix** together is that of a **rectangular array** with m rows and n columns, much like a printer's box of old, with $m \cdot n$ small cubicles, each holding whatever math symbol or memento was placed in it.

Using but two theorems and two fundamental lemmas, we have been able to

- **classify all linear transformations** $f : \mathbb{R}^n \to \mathbb{R}^m$
- as **matrix × vector multiplications** $f(\mathbf{x}) = \mathbf{Ax}$

for a fixed matrix $\mathbf{A}_{mn}$ depending on f, an arbitrary input vector $\mathbf{x}_{n\times 1} \in \mathbb{R}^n$ and the image $f(\mathbf{x})_{m\times 1} \in \mathbb{R}^m$ of $\mathbf{x}$ under the linear transformation f.

More specifically, we have learned how to describe the entries in $\mathbf{A}$ precisely in two equivalent ways using the standard unit vectors $\mathbf{e}_i$ of $\mathbb{R}^n$.

The **column representation** from Eq. (1.5) says that the kth column of $\mathbf{A}_{mn}$ is given by $\begin{pmatrix} f_1(\mathbf{e}_k) \\ \vdots \\ f_m(\mathbf{e}_k) \end{pmatrix} = f(\mathbf{e}_k) \in \mathbb{R}^m$ for each $k = 1, \ldots, n$.

The **row representation** from Lemma 2 says that the jth row $\mathbf{r}_j$ of $\mathbf{A}_{mn}$ consists of the Riesz row vector $\mathbf{r}_j = \begin{pmatrix} f_j(\mathbf{e}_1) & \ldots & f_j(\mathbf{e}_n) \end{pmatrix} \in \mathbb{R}^n$ for the jth component function f_j of f and each $j = 1, \ldots, m$.

Thus, we see that $f : \mathbb{R}^n \to \mathbb{R}^m$ is a linear transformation:

$$\iff f(\mathbf{x}) = \mathbf{Ax} \quad \text{for} \quad \mathbf{A} = \begin{pmatrix} | & | & & | \\ f(\mathbf{e}_1) & f(\mathbf{e}_2) & \ldots & f(\mathbf{e}_n) \\ | & | & & | \end{pmatrix}_{mn},$$

$$\iff f(\mathbf{x}) = \mathbf{Ax} \quad \text{for} \quad \mathbf{A} = \begin{pmatrix} f_1(\mathbf{e}_1) & \ldots & f_1(\mathbf{e}_n) \\ & \vdots & \\ f_m(\mathbf{e}_1) & \ldots & f_m(\mathbf{e}_n) \end{pmatrix}_{mn},$$

with $\mathbf{x} \in \mathbb{R}^n$, $f = \begin{pmatrix} f_1 \\ \vdots \\ f_m \end{pmatrix}$, and the standard unit vectors $\mathbf{e}_i$ of $\mathbb{R}^n$.

Theorem 1.2 **(Linear Transformation Theorem)** Every linear transformation $f : \mathbb{R}^n \to \mathbb{R}^m$ can be expressed as

a **<u>constant matrix x vector</u> product**
<u>A x</u>

for the constant m-by-n matrix

$$\mathbf{A}_{mn} = \begin{pmatrix} | & & | \\ f(\mathbf{e}_1) & \ldots & f(\mathbf{e}_n) \\ | & & | \end{pmatrix} = \begin{pmatrix} f_1(\mathbf{e}_1) & \ldots & f_1(\mathbf{e}_n) \\ \vdots & & \vdots \\ f_m(\mathbf{e}_1) & \ldots & f_m(\mathbf{e}_n) \end{pmatrix}$$

$$= \begin{pmatrix} - & \mathbf{r}_1 & - \\ & \vdots & \\ - & \mathbf{r}_m & - \end{pmatrix} \in \mathbb{R}^{m,n}$$

with its m Riesz vectors $\mathbf{r}_j = \begin{pmatrix} f_j(\mathbf{e}_1) & \dots & f_j(\mathbf{e}_n) \end{pmatrix} \in \mathbb{R}^n$ as rows, or, equivalently, with the n unit vector images $f(\mathbf{e}_i) \in \mathbb{R}^m$ as columns.

Here $\mathbf{x} = \begin{pmatrix} x_1 \\ \vdots \\ x_n \end{pmatrix} \in \mathbb{R}^n$ and the $\mathbf{e}_i$ are the standard unit vectors of $\mathbb{R}^n$.

Conversely, for every constant matrix $\mathbf{A} \in \mathbb{R}^{m,n}$ the mapping $\mathbf{x} \in \mathbb{R}^n \mapsto \mathbf{A}\mathbf{x} \in \mathbb{R}^m$ is linear. ◀

DEFINITION 5 The matrix $\mathbf{A}_{mn}$ of Theorem 1.2 is the **standard matrix representation** of the linear transformation $f : \mathbb{R}^n \to \mathbb{R}^m$ with respect to the standard basis $\mathcal{E} = \{\mathbf{e}_1, \dots, \mathbf{e}_n\}$ of $\mathbb{R}^n$. It is customarily denoted by $\mathbf{A}_\mathcal{E}$.

Rephrasing the result, linear transformations between finite dimensional vector spaces correspond precisely to (constant matrix) × vector products in the row × column dot product sense.

EXAMPLE 9 To illustrate, we start with the two simplest linear transformations:

(a) The **identity transformation** id $: \mathbb{R}^n \to \mathbb{R}^n$ maps every vectors $\mathbf{x} \in \mathbb{R}^n$ to itself (i.e., $\text{id}(\mathbf{x}) := \mathbf{x}$). This map id is a linear transformation, since $\text{id}(\alpha\mathbf{x} + \beta\mathbf{y}) = \alpha\mathbf{x} + \beta\mathbf{y}$ and $\alpha\ \text{id}(\mathbf{x}) + \beta\ \text{id}(\mathbf{y}) = \alpha\mathbf{x} + \beta\mathbf{y}$ as well.

What is its standard matrix representation $\mathbf{A}_\mathcal{E}$?

Thinking column-wise, we see that $\mathbf{A}_\mathcal{E} = \begin{pmatrix} | & & | \\ \text{id}(\mathbf{e}_1) & \dots & \text{id}(\mathbf{e}_n) \\ | & & | \end{pmatrix}$

$= \begin{pmatrix} | & & | \\ \mathbf{e}_1 & \dots & \mathbf{e}_n \\ | & & | \end{pmatrix} = \begin{pmatrix} 1 & & 0 \\ & \ddots & \\ 0 & & 1 \end{pmatrix}_{nn}$. This square matrix is called the **identity matrix** $\mathbf{I}_n$, or $\mathbf{I}$ for short if its size is of no relevance.

Alternatively, the rows of $\mathbf{A}_\mathcal{E}$ are the Riesz vectors of the component functions $\text{id}_i(\mathbf{x}) = x_i$, $i = 1, \dots, n$, of id. Since $x_i = \mathbf{e}_i \cdot \mathbf{x}$ for each i, the rows of $\mathbf{A}_\mathcal{E}$ are the standard unit vector rows $\mathbf{e}_i$ and thus we see again that $\mathbf{A}_\mathcal{E} = \mathbf{I}_n$.

(b) Consider the **zero transformation** $O : \mathbb{R}^n \to \mathbb{R}^m$ defined by $O(\mathbf{x}) = \mathbf{0} \in \mathbb{R}^m$ for every $\mathbf{x} \in \mathbb{R}^n$. The zero map is linear since $O(\alpha\mathbf{x} + \beta\mathbf{y}) = \mathbf{0} = \alpha O(\mathbf{x}) + \beta O(\mathbf{y})$ for all $\alpha, \beta \in \mathbb{R}$ and all $\mathbf{x}, \mathbf{y} \in \mathbb{R}^n$. O has the standard matrix representation $\mathbf{B}_\mathcal{E} = \begin{pmatrix} | & & | \\ O(\mathbf{e}_1) & \dots & O(\mathbf{e}_n) \\ | & & | \end{pmatrix} = \begin{pmatrix} | & & | \\ \mathbf{0} & \dots & \mathbf{0} \\ | & & | \end{pmatrix}_{mn} = \mathbf{O}_{mn}$. This is the m-by-n matrix consisting entirely of zero entries, called the **zero matrix**. For $\mathbf{O}$ note that every Riesz vector (i.e., every row of $\mathbf{B}_\mathcal{E}$) is likewise zero, since $O_i(\mathbf{x}) = \mathbf{0} = \mathbf{0} \cdot \mathbf{x}$ for each i and all $\mathbf{x} \in \mathbb{R}^n$. ◀

EXAMPLE 10 The standard matrix representation $\mathbf{A} = \mathbf{A}_{\mathcal{E}}$ of the 45° planar rotation R in Example 2(g) is given by

$$\mathbf{A} = \begin{pmatrix} | & | \\ R(\mathbf{e}_1) & R(\mathbf{e}_2) \\ | & | \end{pmatrix} = \begin{pmatrix} \frac{\sqrt{2}}{2} & \frac{\sqrt{2}}{2} \\ -\frac{\sqrt{2}}{2} & \frac{\sqrt{2}}{2} \end{pmatrix},$$

where the images $R(\mathbf{e}_i)$ of the standard unit vectors $\mathbf{e}_1, \mathbf{e}_2 \in \mathbb{R}^2$ are easily read off Figure 1-4.

Alternatively, the same matrix representation $\mathbf{A}$ of R arises by using the Riesz row vectors $\mathbf{r}_j$ for R's two component functions f_1, f_2 from Example 5. Namely,

$$f_1\begin{pmatrix} x \\ y \end{pmatrix} = \frac{\sqrt{2}}{2}(x+y) = \begin{pmatrix} \frac{\sqrt{2}}{2} & \frac{\sqrt{2}}{2} \end{pmatrix}\begin{pmatrix} x \\ y \end{pmatrix} = \mathbf{r}_1\begin{pmatrix} x \\ y \end{pmatrix},$$

$$f_2\begin{pmatrix} x \\ y \end{pmatrix} = \frac{\sqrt{2}}{2}(-x+y) = \begin{pmatrix} -\frac{\sqrt{2}}{2} & \frac{\sqrt{2}}{2} \end{pmatrix}\begin{pmatrix} x \\ y \end{pmatrix} = \mathbf{r}_2\begin{pmatrix} x \\ y \end{pmatrix}.$$

Viewed from R's two component functions f_1 and f_2, its standard matrix representation is $\mathbf{A} = \begin{pmatrix} — & \mathbf{r}_1 & — \\ — & \mathbf{r}_2 & — \end{pmatrix} = \begin{pmatrix} \frac{\sqrt{2}}{2} & \frac{\sqrt{2}}{2} \\ -\frac{\sqrt{2}}{2} & \frac{\sqrt{2}}{2} \end{pmatrix}$. This matrix representation of R is identical to the earlier one, just as it should be according to Theorem 1.2. ◀

EXAMPLE 11 Assume that the transformation $T : \mathbb{R}^3 \to \mathbb{R}^4$ is defined as $T(\mathbf{x}) = \mathbf{y}$ for $\mathbf{x} = \begin{pmatrix} x_1 \\ x_2 \\ x_3 \end{pmatrix} \in \mathbb{R}^3$ and $\mathbf{y} = \begin{pmatrix} y_1 \\ y_2 \\ y_3 \\ y_4 \end{pmatrix} \in \mathbb{R}^4$ by the explicit rule that

$$\begin{aligned} y_1 &= 3x_1 - 2x_2 + x_3; \\ y_2 &= -x_1 + x_2 - 5x_3; \\ y_3 &= x_1 + x_2 - x_3; \\ y_4 &= 2x_2 + 4x_3. \end{aligned} \tag{1.6}$$

Then the matrix $\mathbf{A} = \begin{pmatrix} | & & | \\ T(\mathbf{e}_1) & \ldots & T(\mathbf{e}_3) \\ | & & | \end{pmatrix} = \begin{pmatrix} 3 & -2 & 1 \\ -1 & 1 & -5 \\ 1 & 1 & -1 \\ 0 & 2 & 4 \end{pmatrix} = \mathbf{A}_{\mathcal{E}}$ represents the linear transformation T. Observe that $\mathbf{A} = \mathbf{A}_{\mathcal{E}}$ contains only the ordered coefficients in Eq. (1.6) that express the image $\mathbf{y} = T(\mathbf{x})$ in terms of $\mathbf{x} \in \mathbb{R}^3$.

The matrix × vector representation $\mathbf{Ax}$ of $T(\mathbf{x})$ is an **abbreviation**, or a shorthand notation, for the more baroque formal definition (1.6) of T. Matrix notation thus helps to describe complicated equations and phenomena such as Eq. (1.6) more simply. ◀

We close the first lecture with a remark:

Remark:

As vectors behave linearly under addition and scalar multiplication, so do linear transformations between the same spaces. Clearly, $\alpha T + \beta S$ is a linear map for all $\alpha, \beta \in \mathbb{R}$ if both $T, S : \mathbb{R}^n \to \mathbb{R}^m$ are linear: Just look at the Riesz vector representation of each component function T_j and S_j of T and S. If $T_j(\mathbf{x}) = \mathbf{r}_j\mathbf{x}$ and $S_j(\mathbf{x}) = \mathbf{s}_j\mathbf{x}$, then obviously $(\alpha T_j + \beta S_j)(\mathbf{x}) = (\alpha\mathbf{r}_j + \beta\mathbf{s}_j)\mathbf{x}$, that is, matrices of the same size add and scalar multiply componentwise just as vectors do. See the solution to Problem 5 of Section 1.1.P.

The composition of linear maps as studied in Lemma 1 will be taken up again in Chapter 6.

The first lecture has touched upon many concepts of linear algebra. Some have been clearly defined and broadly explained, such as vectors, linear transformations, and matrices. Others have been used without much explanation, such as the standard unit vectors, the standard basis, linear combinations, etc. As the Introduction says, "linear algebra is a circular subject. Studying linear algebra feels like exploring a city or a country for the first time. An overwhelming number of concepts are clearly present at any first encounter with linear algebra. These concepts are all intertwined and connected. As with a new city, one has to start discovering slowly and deliberately." Any first encounter with linear algebra involves an immersion into a new landscape, a new language and culture that is full of mathematical vitality. Lecture 1 has established the fundamentals that are key to the entire subject. No one would or should try to understand all of linear algebra in the first week. Some aspects must remain for later studies. Thus linear combinations govern Chapter 4, while bases are discussed in Chapters 5, 7, and 10. We urge our students to move on, a little confused maybe as would be natural, but healthily curious.

We have succeeded in classifying all linear maps $f : \mathbb{R}^n \to \mathbb{R}^m$, i.e., those f with $f(\alpha\mathbf{x} + \beta\mathbf{y}) = \alpha f(\mathbf{x}) + \beta f(\mathbf{y})$ for all vectors $\mathbf{x}, \mathbf{y} \in \mathbb{R}^n$ and all scalars α and β.

1.1.P Problems

The problem sets will usually start out with routine problems and lead to more conceptual ones. There will always be some that are patently unsolvable such as finding a real value of x with $2x = 6$ **and** *$4x = -4$. It is the task of the students to keep their wits, to recognize when this happens, and to learn what can still be done; for example, see Problem 13 in Section 3.2.P for a different opinion on this unsolvable problem. Good luck in learning.*

Each of the problems can and should be solved using only the methods of linearity from this lecture. (Instructors may be wrongly tempted to use or introduce different methods. To do so is unnecessary and will counteract the didactic goals of this book.)

0. (Intended for students with a good knowledge of planar geometry)
Prove that the geometric parallelogram law definition of adding two vectors $\mathbf{a}$ and $\mathbf{b} \in \mathbb{R}^2$ resulting in the diagonal vector of the parallelogram with sides $\overline{\mathbf{0}\,\mathbf{a}}$ and $\overline{\mathbf{0}\,\mathbf{b}}$ is equivalent to the algebraic one of adding the two components of $\mathbf{a} = \begin{pmatrix} a_1 \\ a_2 \end{pmatrix}$ and $\mathbf{b} = \begin{pmatrix} b_1 \\ b_2 \end{pmatrix}$ individually to obtain $\mathbf{a} + \mathbf{b} = \begin{pmatrix} a_1 + b_1 \\ a_2 + b_2 \end{pmatrix}$.

1. To reinforce the students' concept and understanding of functions and the plug–in mechanism tell whether each of the following is TRUE or FALSE:

(a) If $f(x) = 2x$ for all x, then $f(\sqrt{x}) = 2\sqrt{x}$ for all x.

(b) If $f(x) = 2x$ for all x, then $f(x^2) = 4x^2$ for all x.

(c) If $f(x) = 2x$ for all x, then $(f(2x))^2 = (4x)^2$ for all x.

(d) If $f(x) = a$ for all x, then $f(2x) = 2a$ for all x.

(e) If $f(x) = a + x$ for all x, then $f(14x) = 14(a + x)$ for all x.

(f) If $f(x) = x - a$ for all x, then $f(14x) = 14x - a$ for all x.

(g) If $f(a) = a$ for all a, then $f(x) = x$ for all x.

(h) If $f\begin{pmatrix} x \\ y \\ z \end{pmatrix} = \begin{pmatrix} xy \\ yx \\ z \end{pmatrix}$ for all x, y, z, then $f\begin{pmatrix} \text{hat} \\ \text{rat} \\ \text{head} \end{pmatrix} = \begin{pmatrix} \text{hatrat} \\ \text{rathat} \\ \text{head} \end{pmatrix}$.

(i) If $f(a) = x$ for all a, then $f(\sqrt{1 - 300x})$ is not real.

(j) If $f\begin{pmatrix} a \\ B \\ c \end{pmatrix} = \begin{pmatrix} 2a - B \\ Bac \\ c \end{pmatrix}$ for all values of the variables a, B, and c, then $f\begin{pmatrix} \text{mouse} \\ \text{mouse} \\ \text{house} \end{pmatrix} = \begin{pmatrix} \text{mouse} \\ \text{mouse square house} \\ \text{house} \end{pmatrix}$.

2. (a) Express the vector $\mathbf{x} = \begin{pmatrix} 17 \\ 4 \\ -6 \\ 0 \\ 100 \\ \pi \\ \mathbf{e} \end{pmatrix} \in \mathbb{R}^7$ as a sum of the standard unit vectors $\mathbf{e}_i \in \mathbb{R}^7$.

(b) Do the same for $\mathbf{y} = \begin{pmatrix} 1 \\ \vdots \\ 1 \end{pmatrix} \in \mathbb{R}^4$.

(c) Write out the column vector notation for the point $\mathbf{z} = \mathbf{e}_3 - 2\mathbf{e}_1 + 5\mathbf{e}_4 \in \mathbb{R}^7$.

(d) Do this for $\mathbf{w} = \mathbf{e}_6 + 12\mathbf{e}_1 - \mathbf{e}_4 - \mathbf{e}_2 \in \mathbb{R}^6$.

3. Examine the following transformations carefully for linearity or nonlinearity by writing out each individual component function and checking its linearity:

(a) $T\begin{pmatrix} x_1 \\ x_2 \\ x_3 \end{pmatrix} = \begin{pmatrix} 3x_3 - x_1 \\ \cos(1) \cdot |x_1 - x_1| \\ 2x_2 + x_1 - 1 \\ x_1^3 \end{pmatrix}$.

(b) $S(\mathbf{x}) = \begin{pmatrix} x_1 - 3x_5 + 20x_2 \\ (x_5 - 2x_5)^4 \\ x_2x_3 - (x_4 + x_3)x_2 + x_4x_2 \\ (x_2 - x_1)(x_5 + x_5) \\ \sin(0) \end{pmatrix}$.

(c) List all component functions of S and T that are linear.

(*Hint*: Try to rewrite each messy component function as simply as you can.)

4. Use the linearity definition to conclude that the function $f(\mathbf{x}) = -3\mathbf{x}$ is linear for all $\mathbf{x} \in \mathbb{R}^n$.

5. (a) Is the function $f : \mathbb{R}^3 \to \mathbb{R}^3$ defined by $f(\mathbf{x}) :=$

$$\begin{pmatrix} 1 & 2 & -3 \\ 0 & 0 & 7 \\ 0 & 0 & -3 \end{pmatrix}\mathbf{x} + \begin{pmatrix} -1 & 0 & 0 \\ 1 & 3 & 0 \\ 4 & -1 & 3 \end{pmatrix}\mathbf{x}$$

linear? If so, what is its standard matrix representation $\mathbf{A}_\mathcal{E}$? (*Hint*: Look at the three component functions f_i of f.)

(b) What does part (a) mean for representing the sum of two linear transformations as matrices?

6. Consider the linear transformation defined by the rule

$$T\begin{pmatrix} x_1 \\ x_2 \\ x_3 \end{pmatrix} := \begin{pmatrix} x_1 - 2x_2 + x_3 \\ x_2 - x_1 \\ 2x_1 + 3x_3 \end{pmatrix}.$$

Find the standard matrix representation $\mathbf{A}_\mathcal{E}$ for T.

7. Let $\mathbf{A} = \begin{pmatrix} 3 & 4 & -1 & 0 \\ 2 & -1 & 1 & 1 \\ 0 & 2 & 1 & -2 \end{pmatrix} \in \mathbb{R}^{3,4}$, $\mathbf{x} = \begin{pmatrix} 1 \\ 2 \\ 0 \\ -1 \end{pmatrix} \in \mathbb{R}^4$, and $\mathbf{y} = (2\ 1\ 1) \in \mathbb{R}^3$.

Let $\mathbf{z}^T$ denote the **transposed vector** of $\mathbf{z}$ (i.e., a row vector with entries z_i if $\mathbf{z}$ is a column vector and a column vector if $\mathbf{z}$ is a row vector).

Decide which of the following matrix × vector, vector × matrix, or vector × vector products can be formed, and evaluate those that can be formed:

(a) $\mathbf{Ax}$, (b) $\mathbf{Ay}$, (c) $\mathbf{xy}$, (d) $\mathbf{x}^T\mathbf{A}$, (e) $\mathbf{yx}$, (f) $\mathbf{yA}$, (g) $\mathbf{Ay}^T$.

8. For $\mathbf{A} = \begin{pmatrix} 2 & -1 \\ 1 & 3 \end{pmatrix}$ and $\mathbf{x} = \begin{pmatrix} x_1 \\ x_2 \end{pmatrix}$, evaluate
(a) $\mathbf{Ax}$, (b) $\mathbf{x}^T\mathbf{A}$, (c) $(\mathbf{x}^T\mathbf{A})\mathbf{x}$, and (d) $\mathbf{x}^T(\mathbf{Ax})$, if possible.

9. Let $\mathbf{A} = \begin{pmatrix} 1 & -2 & 6 \\ 2 & -1 & 3 \\ 1 & 2 & -1 \end{pmatrix}$. Show that the transformation $\mathbf{x} \in \mathbb{R}^3 \to \mathbf{Ax} \in \mathbb{R}^3$ is linear by verifying that $\mathbf{A}(\alpha\mathbf{x} + \beta\mathbf{y}) = \alpha\mathbf{Ax} + \beta\mathbf{Ay}$ holds for all $\mathbf{x}, \mathbf{y} \in \mathbb{R}^3$ and all $\alpha, \beta \in \mathbb{R}$.

10. Show that with $\mathbf{A} = \begin{pmatrix} 1 & 2 \\ 0 & 1 \\ -1 & 1 \end{pmatrix}$, the mapping $\mathbf{x} \in \mathbb{R}^2 \to \mathbf{Ax} \in \mathbb{R}^3$ is linear by verifying the two linearity conditions of Eq. (1.2).

11. For $\mathbf{A} = \begin{pmatrix} 1 & a & 0 \\ 0 & -1 & a \\ a & 0 & 1 \end{pmatrix}$ and fixed $\mathbf{a} \in \mathbb{R}$, evaluate
$\mathbf{A}\begin{pmatrix} 1 \\ 1 \\ a \end{pmatrix}$ and $\mathbf{A}\begin{pmatrix} a \\ 0 \\ -a \end{pmatrix}$.

12. Let $\mathbf{A}_{mn} = \begin{pmatrix} a_{11} & \dots & a_{1n} \\ \vdots & & \vdots \\ a_{m1} & \dots & a_{mn} \end{pmatrix}$ be a real matrix. For the standard unit column vectors $\mathbf{e}_i$, evaluate

(a) $\mathbf{Ae}_1$, (b) $\mathbf{Ae}_2$, and (c) $\mathbf{Ae}_n$, as well as
(d) $\mathbf{e}_1^T\mathbf{A}$, (e) $\mathbf{e}_2^T\mathbf{A}$, (f) $\mathbf{e}_m^T\mathbf{A}$, and
(g) $\mathbf{e}_i^T\mathbf{Ae}_j$ for any $1 \le i \le m$ and $1 \le j \le n$.
(h) Finally, describe your results verbally in terms of the rows, columns, and entries of $\mathbf{A}$.

13. Let $T : \mathbb{R}^6 \longrightarrow \mathbb{R}^4$ be defined by

$$T(\mathbf{x}) = \begin{pmatrix} x_1 + 2x_2 - x_3 \\ 3x_2 - x_4 \\ 0 \\ -x_6 \end{pmatrix}.$$

(a) Why is T a linear map?
(b) Find $T(\mathbf{e}_1), T(\mathbf{e}_2), T(\mathbf{e}_3), T(\mathbf{e}_4), T(\mathbf{e}_5)$ and $T(\mathbf{e}_6)$.
(c) What is the size of T's standard matrix representation?
(d) Find the standard matrix representation $\mathbf{A}_\mathcal{E}$ of T.

14. Let $T : \mathbb{R}^6 \longrightarrow \mathbb{R}^4$ be defined by

$$T(\mathbf{x}) = \begin{pmatrix} -4x_1 + 2x_2 - x_3 \\ 3^2x_2 - x_4 \\ 1 \\ -5^3x_6 \end{pmatrix}.$$

(a) Find $T(\mathbf{e}_1), T(\mathbf{e}_2), T(\mathbf{e}_3), T(\mathbf{e}_4), T(\mathbf{e}_5)$, and $T(\mathbf{e}_6)$.
(b) Find the standard matrix representation $\mathbf{A}_\mathcal{E}$ of T and its size (i.e., the number of rows and columns), if possible.
(c) Is T a linear map? Explain.

15. Consider the linear transformation $S : \mathbb{R}^4 \to \mathbb{R}^3$ defined by the rule

$$S\begin{pmatrix} x_1 \\ x_2 \\ x_3 \\ x_4 \end{pmatrix} = \begin{pmatrix} -20x_2 + 6x_3 - 5x_4 \\ 3x_2 + x_1 + x_4 \\ 2(x_1 + 3x_3) \end{pmatrix}.$$

Find the standard matrix representation $\mathbf{A}_\mathcal{E}$ for S.

16. Assume that $T : \mathbb{R}^3 \to \mathbb{R}^\ell$ is a function.

(a) If $\ell = 2$ and $T\begin{pmatrix} 1 \\ 0 \\ 0 \end{pmatrix} = \begin{pmatrix} 1 \\ 1 \end{pmatrix}$, $T\begin{pmatrix} 0 \\ 1 \\ 0 \end{pmatrix} = \begin{pmatrix} 1 \\ 1 \end{pmatrix}$, and $T\begin{pmatrix} 0 \\ 0 \\ 1 \end{pmatrix} = \begin{pmatrix} 1 \\ 1 \end{pmatrix}$ can T be a linear transformation? Explain.

(b) If $\ell = 4$ and $T\begin{pmatrix} 1 \\ 1 \\ 1 \end{pmatrix} = \begin{pmatrix} -1 \\ 1 \\ 1 \\ 1 \end{pmatrix}$, $T\begin{pmatrix} 1 \\ 0 \\ 1 \end{pmatrix} = \begin{pmatrix} -1 \\ 1 \\ 0 \\ 1 \end{pmatrix}$ and $T\begin{pmatrix} 0 \\ 1 \\ 0 \end{pmatrix} = \begin{pmatrix} -1 \\ 0 \\ 0 \\ 0 \end{pmatrix}$, can T be a linear transformation?
(*Hint*: Check the linearity condition.)

17. Let $T : \mathbb{R}^3 \longrightarrow \mathbb{R}^3$ be a linear transformation. Find the standard matrix representation $\mathbf{A} = \mathbf{A}_\mathcal{E}$ of T and evaluate $\mathbf{A}\begin{pmatrix} 1 \\ -1 \\ 1 \end{pmatrix}$.

(a) T is determined by $T(\mathbf{e}_3) = \begin{pmatrix} 3 \\ -4 \\ 17 \end{pmatrix}$, $T(\mathbf{e}_2) = \begin{pmatrix} 11 \\ 0 \\ 2 \end{pmatrix}$, and $T(\mathbf{e}_1) = \begin{pmatrix} -1 \\ -1 \\ -1 \end{pmatrix}$.

(b) Do the same for $T(3\mathbf{e}_1) = \mathbf{e}_1$, $T(2\mathbf{e}_3) = \begin{pmatrix} 52 \\ 4 \\ -8 \end{pmatrix}$, and $T(-\mathbf{e}_2) = \begin{pmatrix} -1 \\ 1 \\ 1 \end{pmatrix}$.

(c) Do the same for T defined by $T\begin{pmatrix}1\\0\\0\end{pmatrix}=\begin{pmatrix}4\\0\\1\end{pmatrix}$,

$T\begin{pmatrix}1\\-3\\0\end{pmatrix}=\begin{pmatrix}1\\1\\-1\end{pmatrix}$, and $T\begin{pmatrix}2\\1\\-1\end{pmatrix}=\begin{pmatrix}-1\\6\\3\end{pmatrix}$.

(*Hint*: Use the linearity of T and your knowledge of $T(\mathbf{e}_1)$ to find $T(\mathbf{e}_2)$, and then $T(\mathbf{e}_3)$ from the given data.)

18. Assume that the six vectors $\mathbf{u}_1=\mathbf{e}_1$, $\mathbf{u}_2=\mathbf{e}_1+\mathbf{e}_2$, $\mathbf{u}_3=\mathbf{e}_2-\mathbf{e}_3$ and $\mathbf{v}_1=\mathbf{e}_1+\mathbf{e}_2$, $\mathbf{v}_2=\mathbf{e}_1-\mathbf{e}_2$, $\mathbf{v}_3=\mathbf{e}_1$ all lie in $\mathbb{R}^3$.

(a) Find the standard matrix $\mathbf{A}_{\mathcal{E}}$ that describes the linear transformation $S:\mathbb{R}^3\to\mathbb{R}^3$, defined by $S\mathbf{u}_1=\mathbf{v}_1$, $S\mathbf{u}_2=\mathbf{v}_2$, and $S\mathbf{u}_3=\mathbf{v}_3$.

(b) Find the standard matrix $\mathbf{B}_{\mathcal{E}}$ that describes the linear transformation $T:\mathbb{R}^3\to\mathbb{R}^3$, defined by $T\mathbf{u}_1=\mathbf{v}_1$, $T\mathbf{u}_2=\mathbf{v}_3$, and $T\mathbf{u}_3=\mathbf{v}_2$.

19. Let $T:\mathbb{R}^2\longrightarrow\mathbb{R}^2$ be a linear map, mapping (4, 1) to (−3, 5) and (5, −1) to (−3, 1).

(a) Find $T(\mathbf{e}_1)$ and $T(\mathbf{e}_2)$. (*Hint*: What is $T\left(\begin{pmatrix}4\\1\end{pmatrix}+\begin{pmatrix}5\\-1\end{pmatrix}\right)$?)

(b) Find the standard matrix representation $\mathbf{A}_{\mathcal{E}}$ of T.

(c) Find $T((1,-1))$ and $T((3,2))$

(d) Find a vector $\mathbf{x}$ with $T(\mathbf{x})=\mathbf{0}$.

20. Let $T:\mathbb{R}^n\longrightarrow\mathbb{R}^k$ be a linear transformation. Use the definition of linearity to show that

(a) If $T(\mathbf{x})=\mathbf{b}$ and $T(\mathbf{y})=\mathbf{b}$, then $T(\mathbf{y}-\mathbf{x})=\mathbf{0}$.

(b) If $T(\mathbf{x})=\mathbf{b}$ and $T(\mathbf{y})=\mathbf{c}$, then $T(\mathbf{y}+\mathbf{x})=\mathbf{b}+\mathbf{c}$.

(c) If $T(\mathbf{x})=\mathbf{0}$ and $T(\mathbf{y})=\mathbf{c}$, then $T(\mathbf{y}+\alpha\mathbf{x})=\mathbf{c}$ for all scalars α.

21. Show that

(a) A function $f:\mathbb{R}\to\mathbb{R}$ is linear if and only if f is differentiable on $\mathbb{R}$ with $f'(x)=f(1)$ for all $\mathbf{x}$ and $f(0)=0$.

(b) Use part (a) to show that the function $f:\mathbb{R}\to\mathbb{R}$ of Example 4 is not linear.

(c) Describe all linear functions $g:\mathbb{R}\to\mathbb{R}$ explicitly through their functional equation.

22. If $T:\mathbb{R}^3\longrightarrow\mathbb{R}^3$ maps every vector $\mathbf{x}\in\mathbb{R}^3$ to the difference of $\mathbf{x}=\begin{pmatrix}x_1\\x_2\\x_3\end{pmatrix}$ and the reverse-order vector

$$\widetilde{\mathbf{x}}:=\begin{pmatrix}x_3\\x_2\\x_1\end{pmatrix}.$$

That is, if $T(\mathbf{x})=\mathbf{x}-\widetilde{\mathbf{x}}$, decide whether T is linear. If it is, what is its standard matrix representation $\mathbf{A}_{\mathcal{E}}$?

23. If $T:\mathbb{R}^3\longrightarrow\mathbb{R}^3$ maps every vector $\mathbf{x}\in\mathbb{R}^3$ to the sum of $\mathbf{x}=\begin{pmatrix}x_1\\x_2\\x_3\end{pmatrix}$ and the reverse-order vector $\widetilde{\mathbf{x}}:=\begin{pmatrix}x_3\\x_2\\x_1\end{pmatrix}$.

That is, if $T(\mathbf{x})=\mathbf{x}+\widetilde{\mathbf{x}}$, determine whether T is linear. If it is, what is its standard matrix representation $\mathbf{A}_{\mathcal{E}}$?

24. Let T be the map from second degree polynomials to third degree polynomials defined by $T(a_2x^2+a_1x+a_0):=a_2(x-3)^3+a_1(x-3)^2+a_0$.

(a) Calculate $T(3x^2-4x+7)$.

(b) Is T a linear transformation or not?

(c) Prove that $T(\alpha p+\beta q)=\alpha T(p)+\beta T(q)$ for all second-degree polynomials p, q and all scalars α, β.

25. Consider the projection T of three space onto the second coordinate axis, namely, $T:\mathbb{R}^3\to\mathbb{R}^3$ with $T(x_1,x_2,x_3):=(0,x_2,0)$.

(a) Is T a linear map?

(b) How can you most easily answer this question? (*Hint*: Think about Theorem 1.2.)

(c) Find $T(3,19,-4)$, $T(19,0,-14)$, $T(\mathbf{e}_3)$, $T(3\mathbf{e}_2-4\mathbf{e}_1)$, and $T(\mathbf{0})$.

(d) Write out T in matrix form $\mathbf{A}_{\mathcal{E}}$.

26. Repeat the preceding problem for the reflection R of three space across the x_2-axis (i.e., $R:\mathbb{R}^3\to\mathbb{R}^3$ with $R(x_1,x_2,x_3):=(x_1,-x_2,x_3)$).

27. (a) If $f,g:\mathbb{R}^n\to\mathbb{R}^m$ are two linear maps, show that $(\alpha f+\beta g)(\mathbf{x}):=\alpha f(\mathbf{x})+\beta g(\mathbf{x}):\mathbb{R}^n\to\mathbb{R}^m$ is also a linear map.

(b) If $f:\mathbb{R}^?\to\mathbb{R}^?$ has the standard matrix representation $\mathbf{A}_{\mathcal{E}}=\begin{pmatrix}2&1&-1\\0&-1&2\end{pmatrix}$ and $g:\mathbb{R}^?\to\mathbb{R}^?$ has the matrix representation $\mathbf{B}_{\mathcal{E}}=\begin{pmatrix}1&0&2\\-1&1&-2\end{pmatrix}$, find $f(\mathbf{e}_2)$, $g(\mathbf{e}_1)$, and $f(\mathbf{e}_1+\mathbf{e}_2)$. What is $g(2\mathbf{e}_2-3\mathbf{e}_3)$? What are the dimensions of the range and domain of f?

(c) Find $-2f(\mathbf{e}_3)+3g(\mathbf{e}_3)$ for f and g from part (b).

(d) What is the standard matrix representation $\mathbf{C}_{\mathcal{E}}$ of $3g-2f:\mathbb{R}^2\to\mathbb{R}^3$ for f, g from part (b).

(e) If $f(\mathbf{x})=2\mathbf{e}_1$ and $(3g-2f)(\mathbf{x})=\mathbf{e}_1$, find $g(\mathbf{x})$, if both f and g are linear.

(*Hint*: Look at Problem 5.)

28. Show that a function $f : \mathbb{R} \to \mathbb{R}$ is linear if and only if the **graph** $\{(x, f(x)) \in \mathbb{R}^2 \mid x \in \mathbb{R}\}$ of f is a straight line containing the origin (0,0) of $\mathbb{R}^2$.

29. Which of the following functions $\mathbb{R} \to \mathbb{R}$ are linear?

(a) $f(x) = x - |x|$;

(b) $g(x) = x + |x|$;

(c) $h(x) = 3(f + g)(x)$; f and g from above;

(d) $k(x) = f(x) - g(x)$;

(e) $\ell(x) = \max\{-f(x), g(x)\}$;

(f) $m(x) = \min\{-f(x), g(x)\}$.

30. For any n real numbers x_i, show that

(a) The minimum function $\mathrm{Min}(x_1, \ldots, x_n) := \min\{x_i\} : \mathbb{R}^n \to \mathbb{R}$ is not linear.

(b) The maximum function $\mathrm{Max}(x_1, \ldots, x_n) := \max\{x_i\} : \mathbb{R}^n \to \mathbb{R}$ is not linear.

31. Let $T : \mathbb{R}^2 \to \mathbb{R}^2$ be the clockwise rotation of two space by 90°.

(a) Show that $T(\mathbf{e}_1) = T\begin{pmatrix} 1 \\ 0 \end{pmatrix} = \begin{pmatrix} 0 \\ -1 \end{pmatrix} = -\mathbf{e}_2$.

(b) What is $T(\mathbf{e}_2)$? Is T a linear transformation?

(c) Write out the standard matrix representation $\mathbf{A}_{\mathcal{E}}$ of T.

32. Repeat the foregoing problem for the counter–clockwise rotation R of the plane by 60° degrees.

33. To convert measurements given in yards into feet we use the formula number of feet = 3· number of yards.

(a) How are square yards and cubic yards converted to square feet and cubic feet, respectively?

(b) Which of the three conversion formulas in this problem expresses a linear function? Why?

34. Complete the sufficiency part of the proof of the Riesz Lemma in detail.

35. Assume that f maps $\mathbb{R}^3$ to $\mathbb{R}$. Explain why there is or there is not a linear function f with the respectively given properties:

(a) $f(1, 0, 3) = 0$.

(b) $f(1, 0, 3) = 0$ and $f(2, 1, 1) = 0$.

(c) $f(1, 0, 3) = 0$, $f(2, 1, 1) = 0$, and $f(1, 1, 1) = 0$.

(d) $f(1, 0, 3) = 0$, $f(2, 1, 1) = 0$, and $f(-1, -1, 2) = 1$.

(e) $f(1, 0, 3) = 1$, $f(2, 1, 1) = 2$, and $f(-1, -1, 2) = -1$.

(f) $f(0, 0, 0) = 10$.

36. (a) Show: If $f : \mathbb{R}^n \to \mathbb{R}^m$ is a linear function, then $f(-\mathbf{x}) = -f(\mathbf{x})$ for all $\mathbf{x} \in \mathbb{R}^n$.

(b) Show that $g : \mathbb{R} \to \mathbb{R}$ defined as $g(x) := x^5$ satisfies the condition $g(-x) = -g(x)$ for all real x.

(c) Let $h : \mathbb{R} \to \mathbb{R}$ be given as $h(x) = -2x$. Is h a linear function?

(d) Is g in part (b) a linear function?

37. Let $T : \mathbb{R}^2 \longrightarrow \mathbb{R}^2$ be a linear map, mapping (4, −1) to (3, 2) and (5, −1) to (−5, 8).

(a) Find $T(\mathbf{e}_1)$ and $T(\mathbf{e}_2)$. (*Hint*: What is $T\left(\begin{pmatrix} 4 \\ -1 \end{pmatrix} - \begin{pmatrix} 5 \\ -1 \end{pmatrix}\right)$?)

(b) Find the standard matrix representation $\mathbf{A}_{\mathcal{E}}$ of T. Check your answer by comparing the action of $\mathbf{A}_{\mathcal{E}}$ with that of T in part (a).

(c) Find $T((1, -1))$ and $T((3, 2))$

(d) Find a vector $\mathbf{x}$ such that $T(\mathbf{x}) = \mathbf{0}$.

38. Assume that $T : \mathbb{R}^n \to \mathbb{R}^n$ is a linear transformation. Which of $T^2 : \mathbb{R}^n \to \mathbb{R}^n$, defined as $T^2(\mathbf{x}) := T(T(\mathbf{x}))$ and $T^3 : \mathbb{R}^n \to \mathbb{R}^n$, defined as $T^3(\mathbf{x}) := T(T(T(\mathbf{x})))$ is a linear transformation? Give a proof.

39. Let $T : \mathbb{R}^2 \to \mathbb{R}^2$ be the transformation defined by $T\begin{pmatrix} x_1 \\ x_2 \end{pmatrix} := \begin{pmatrix} 4x_1 - 2x_2 \\ x_2 - x_1 \end{pmatrix}$. Likewise, let S be defined by $S\begin{pmatrix} x_1 \\ x_2 \end{pmatrix} := \begin{pmatrix} 3x_1 + 9x_2 \\ -3x_2 - x_1 \end{pmatrix}$.

(a) Compute the standard matrix representations $\mathbf{A}_{\mathcal{E}}$ and $\mathbf{B}_{\mathcal{E}}$ of T and S, respectively, after establishing that the mappings are linear.

(b) Find the matrix representation of the map $T^2 : \mathbf{x} \mapsto T(T(\mathbf{x}))$ by evaluating $T(T(\mathbf{e}_i))$ for $i = 1$ and 2. Settle first that T^2 is linear.

(c) Repeat part (b) for $S^2 : \mathbf{x} \mapsto S(S(\mathbf{x}))$.

1.2 Tasks and Methods of Linear Algebra

We study linear algebra so that we can understand linear transformations:

$$f : \mathbb{R}^n \to \mathbb{R}^m.$$

Studying linear transformations is the same as trying to understand matrices $\mathbf{A}_{mn}$ and matrix $\times$ vector products $\mathbf{Ax}$ as shown in Section 1.1.

What does this task consist of in an entry-level course? This book deals with three essential and elementary tasks of linear algebra:

(A) **Geometry:**

To study $\mathbb{R}^n$ as a geometric and as a linear space of vectors. What makes it up? What objects lie in it, and what generates it? What linear subspaces does it contain? How does a linear function $f : \mathbb{R}^n \to \mathbb{R}^m$ transform its domain?

(B) **Linear Equations:**

To find *images* and *pre-images* of and for linear transformations or matrix $\times$ vector products.
What is the range (i.e., the set of images $\{\mathbf{Ax}\} \subset \mathbb{R}^m$) of all $\mathbf{x} \in \mathbb{R}^n$ under $\mathbf{A}$?
Where do the images $\mathbf{y} = \mathbf{Ax}$ come from? Which $\mathbf{x}$ map to a given $\mathbf{y}$ under $\mathbf{A}$?
What are the pre-images $\{\mathbf{x} \mid \mathbf{y} = \mathbf{Ax}\} \subset \mathbb{R}^n$ for a given matrix $\mathbf{A}$ and a given $\mathbf{y} \in \mathbb{R}^m$? Are they unique for each $\mathbf{y} \in \mathbb{R}^m$?
Which vectors $\mathbf{x}$ solve a given set of linear equations $\mathbf{Ax} = \mathbf{b}$?
How many $\mathbf{x}$ does $\mathbf{A}$ map to zero, i.e., what vectors $\mathbf{x}$ solve $\mathbf{Ax} = \mathbf{0}$?

(C) **Bases and Matrix Representations:**

In Section 1.1, we have derived the standard matrix representation $\mathbf{A}_{\mathcal{E}}$ for linear mappings $f : \mathbb{R}^n \to \mathbb{R}^m$ with respect to the standard unit vector basis $\mathcal{E} = \{\mathbf{e}_i\}$ of $\mathbb{R}^n$.
What is a basis? Are there other bases of $\mathbb{R}^n$?
What happens to f's matrix representation for different bases of $\mathbb{R}^n$?
Can some other basis of $\mathbb{R}^n$ help us view and understand a given linear transformation f better?

In subsequent chapters, we shall see that the problems mentioned in tasks (A) and (B) involve linear processes themselves, while part (C) is a nonlinear problem involving linear transformations.

The main tool for understanding and solving the problems of (A) and (B) and in parts of (C) is row reduction, or Gaussian elimination, and finding a row echelon form for certain matrices. Part (C) can only be solved by additionally using nonlinear means.

The first six lectures deal with the tasks in parts (A) and (B), and with general preparations for those in part (C). Lecture 7 will be the turning-point lecture, or the half–way point between linearity and nonlinearity in the study of linear algebra. Lectures 8 through 12 and Lecture 14 deal with the nonlinear matrix eigenvalue problem almost exclusively and will give us an understanding and answer to question (C). Finally Lecture 13 sheds light onto the usefulness of the computational methods that we introduce for basic comprehension and hand computations. Our elementary computational methods will generally only work well for low dimensions and with integer matrices. For constructing integer model problems, we include "Teacher's Problem Making Exercises" whenever appropriate. These problems are helpful to students and teachers alike and explain how to generate infinitely many integer exercises that are easily solved in integer arithmetic.

1.3 Applications

Geometry, Calculus, and MATLAB

We view linear transformations geometrically and study the linearity of differentiation and integration in calculus.

All linear transformations $f : \mathbb{R}^n \to \mathbb{R}^m$ are described by matrix times vector multiplication $\mathbf{x} \mapsto \mathbf{Ax}$ for matrices $\mathbf{A}_{mn} = \mathbf{A}_{\mathcal{E}}$ according to Section 1.1. The standard matrix $\mathbf{A}_{mn} = \mathbf{A}_{\mathcal{E}}$ contains the images $\mathbf{Ae}_1, \ldots, \mathbf{Ae}_n$ of the standard unit vectors $\mathbf{e}_1, \ldots, \mathbf{e}_n$ of $\mathbb{R}^n$ in its columns. This allows us to see the effect of a linear transformation visually. More sophisticated graphic descriptions of linear mappings $\mathbf{x} \mapsto \mathbf{Ax}$ will emerge with matrix eigenvalues and eigenvectors in Chapter 9 and with matrix singular values and singular vectors in Chapter 12. For $m = n = 2$, we use the two columns $\mathbf{Ae}_1, \mathbf{Ae}_2$ in the standard matrix representation $\mathbf{A}_{\mathcal{E}}$ to visualize the effect of the linear mapping that sends a point $\mathbf{x} \in \mathbb{R}^2$ to $\mathbf{Ax} \in \mathbb{R}^2$.

EXAMPLE 12 For $m = n = 2$, the matrix of a linear transformation $\mathbb{R}^2 \to \mathbb{R}^2$ has the form $\mathbf{A} = \mathbf{A}_{\mathcal{E}} = \begin{pmatrix} a & c \\ b & d \end{pmatrix}$. Thus, $\mathbf{Ae}_1 = \begin{pmatrix} a \\ b \end{pmatrix}$ and $\mathbf{Ae}_2 = \begin{pmatrix} c \\ d \end{pmatrix}$.

(a) Let $\mathbf{A} = \begin{pmatrix} 2 & -1 \\ -2 & -2 \end{pmatrix}$. Then the images $\mathbf{Ae}_1 = \begin{pmatrix} 2 \\ -2 \end{pmatrix}$ and $\mathbf{Ae}_2 = \begin{pmatrix} -1 \\ -2 \end{pmatrix}$ of the two standard unit vectors $\mathbf{e}_1$ and $\mathbf{e}_2$ indicate how the unit square in the first quadrant with its four corners $\mathbf{0}, \mathbf{e}_1, \mathbf{e}_1 + \mathbf{e}_2$, and $\mathbf{e}_2$ is mapped under $\mathbf{x} \mapsto \mathbf{Ax}$. (See Figure 1-6.)

Note that the unit square is transformed to a larger parallelogram under the linear mapping and that the orientation is reversed: It takes a counterclockwise turn to go from $\mathbf{e}_1$ to $\mathbf{e}_2$ most quickly, but the shortest move from $\mathbf{Ae}_1$ to $\mathbf{Ae}_2$ is clockwise.

(b) For $\mathbf{B} = \begin{pmatrix} 1 & 1 \\ 0 & 1.5 \end{pmatrix}$, we view the linear transformation $\mathbf{x} \mapsto \mathbf{Bx}$ of $\mathbb{R}^2$ as shown in Figure 1-7.

Since $\mathbf{B}\begin{pmatrix} x \\ y \end{pmatrix} = \begin{pmatrix} x + y \\ 1.5y \end{pmatrix}$, the second component $\mathbf{y}$ of $\begin{pmatrix} x \\ y \end{pmatrix}$ is enlarged, or stretched, by a factor of 1.5 under $\mathbf{B}$. The first component of the image contains the sum of the coordinates of the mapped point, resulting in a **shear** of the unit square under $\mathbf{B}$.

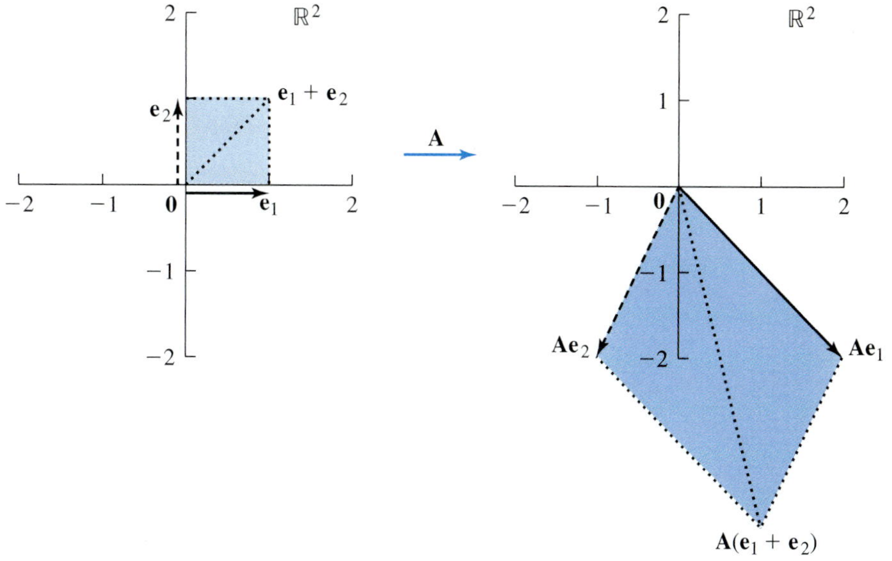

Figure 1-6

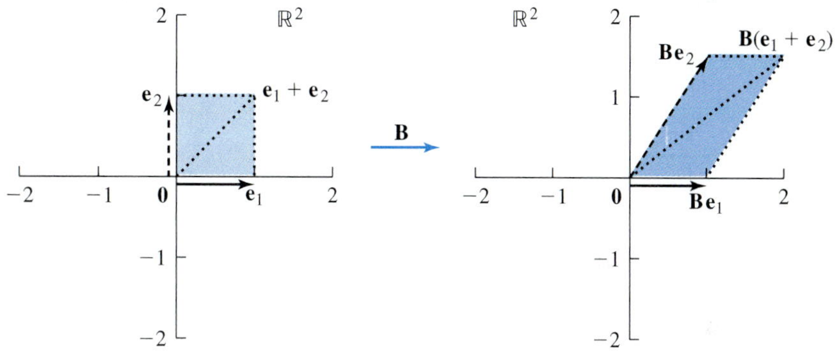

Figure 1-7

(c) For $\mathbf{C} = \begin{pmatrix} a & -b \\ b & a \end{pmatrix}$, we draw the picture for the linear transformation $\mathbf{x} \mapsto \mathbf{Cx}$, which is shown in Figure 1-8.
Here the two images $\mathbf{Ce}_1$ and $\mathbf{Ce}_2$ lie on perpendicular lines through the origin, since the slope of the line segment $\overline{\mathbf{0Ce}_1}$ equals $\frac{b}{a}$, and the slope of $\overline{\mathbf{0Ce}_2}$ is $\left(-\frac{a}{b}\right)$, making their product $\left(\frac{b}{a}\right) \cdot \left(-\frac{a}{b}\right)$ equal to -1 (provided that $a, b \neq 0$). Thus, the linear transformation $\mathbf{x} \mapsto \mathbf{Cx}$ preserves the geometric shape of the unit square. The four corners $\mathbf{C0} = \mathbf{0}$, $\mathbf{Ce}_1$, $\mathbf{C}(\mathbf{e}_1 + \mathbf{e}_2)$, and $\mathbf{Ce}_2$ of its image again form a square with side length $\sqrt{a^2 + b^2}$. However, under $\mathbf{C}$, the unit square is stretched and rotated around the origin by the counterclockwise angle α for which $\tan(\alpha) = \frac{b}{a}$. Thus $\mathbf{x} \mapsto \mathbf{Cx}$

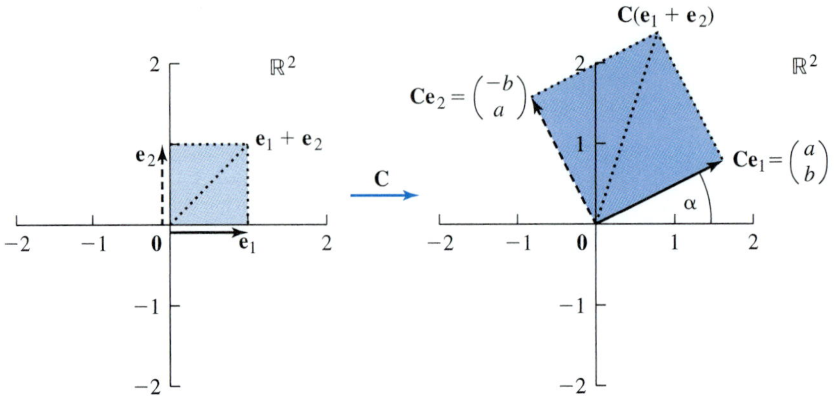

Figure 1-8

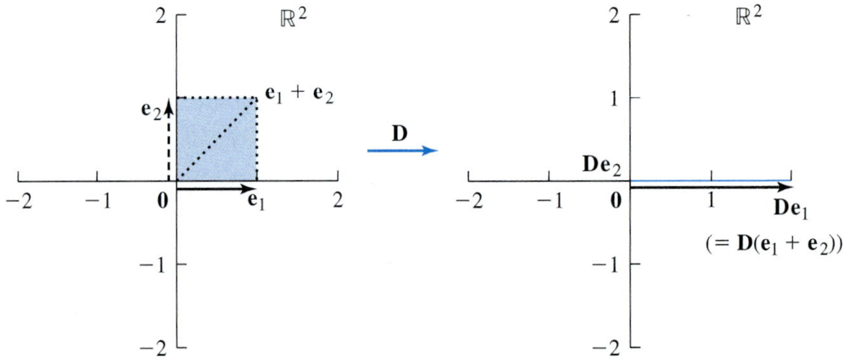

Figure 1-9

is a **rotation** by α, combined with a **dilation** for the stretch factor $d := \sqrt{a^2 + b^2}$. If $d > 1$, then the mapping of $\mathbb{R}^2$ under **C** enlarges lengths and distances by the factor d. If $d = 1$, objects maintain their size; and if $d < 1$, then objects shrink under **C**.

(d) The matrix $\mathbf{D} = \begin{pmatrix} 2 & 0 \\ 0 & 0 \end{pmatrix}$ is diagonal and transforms $\mathbb{R}^2$ as shown in Figure 1-9.
Thus, the set of all images $\{\mathbf{Dx} \mid \mathbf{x} \in \mathbb{R}^2\}$ under **D** is equal to the set of points on the first coordinate axis $\left\{ \begin{pmatrix} \alpha \\ 0 \end{pmatrix} \middle| \alpha \in \mathbb{R} \right\}$ of $\mathbb{R}^2$. Specifically the two–dimensional unit square collapses to the line segment from the origin to $2\mathbf{e}_1$ under **D**.

(e) The matrix $\mathbf{E} = \begin{pmatrix} 0 & 2 \\ 0 & 0 \end{pmatrix}$ has the same image as **D** in part (d). Namely, each image **Ex** is a point on the first coordinate axis, as shown in Figure 1-10.

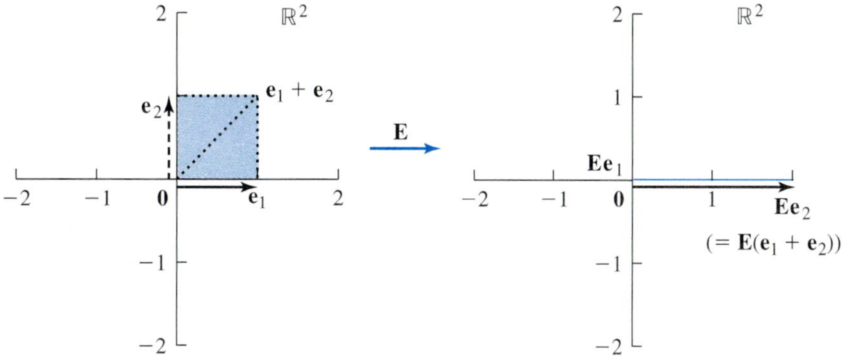

Figure 1-10

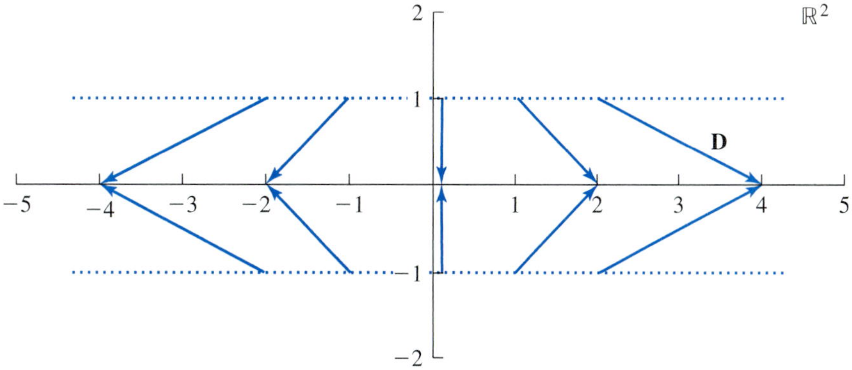

Figure 1-11

But the two matrices $\mathbf{D}$ and $\mathbf{E}$ do not map $\mathbb{R}^2$ in the same way. Note that $\mathbf{D}\begin{pmatrix} -1 \\ \pm 1 \end{pmatrix} = \begin{pmatrix} -2 \\ 0 \end{pmatrix}$, $\mathbf{D}\begin{pmatrix} -2 \\ \pm 1 \end{pmatrix} = \begin{pmatrix} -4 \\ 0 \end{pmatrix}$, $\mathbf{D}\begin{pmatrix} 1 \\ \pm 1 \end{pmatrix} = \begin{pmatrix} 2 \\ 0 \end{pmatrix}$, and $\mathbf{D}\begin{pmatrix} 2 \\ \pm 1 \end{pmatrix} = \begin{pmatrix} 4 \\ 0 \end{pmatrix}$. Thus, the linear transformation $\mathbf{x} \mapsto \mathbf{D}\mathbf{x}$ creates a mapping between two copies of the plane that we superimpose on the plot in Figure 1-11. In it the start of an arrow denotes the original position of $\mathbf{x}$ and its endpoint denotes its image $\mathbf{D}\mathbf{x}$.

On the other hand, under $\mathbf{E}$ all points $\begin{pmatrix} x_1 \\ x_2 \end{pmatrix}$ with $x_2 = 1$ map as $\mathbf{E}\begin{pmatrix} * \\ 1 \end{pmatrix} = \begin{pmatrix} 2 \\ 0 \end{pmatrix}$, regardless of the value of $*$, while those with $x_2 = -1$ map to $\mathbf{E}\begin{pmatrix} * \\ -1 \end{pmatrix} = \begin{pmatrix} -2 \\ 0 \end{pmatrix}$. This gives rise to the overlay plot in Figure 1-12 of the action of $\mathbf{E}$ on $\mathbb{R}^2$.

The two assignment pictures in Figures 1-11 and 1-12 look quite different for the two matrices $\mathbf{D}$ and $\mathbf{E}$ that are different. The differing geometry of

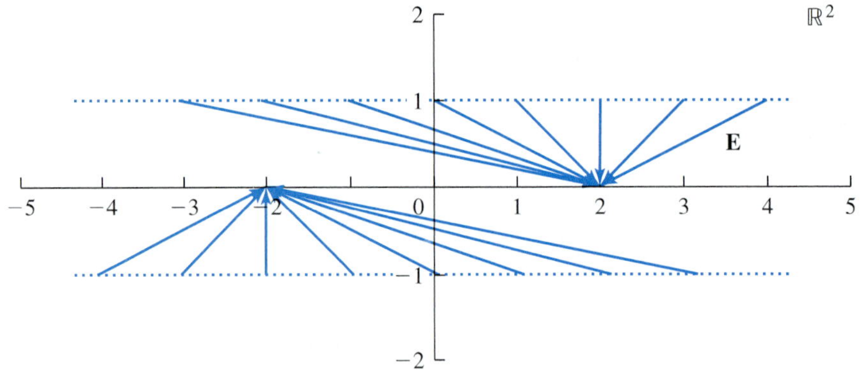

Figure 1-12

the two mappings $\mathbf{x} \mapsto \mathbf{Dx}$ and $\mathbf{x} \mapsto \mathbf{Ex}$ indicates that they are indeed two different linear transformations. ◀

This example will be repeated using the geometry of matrix eigenvalues and eigenvectors in Section 9.2 and that of matrix singular values and singular vectors in Section 12.3.

Now we turn to calculus applications. Several of the processes of calculus such as differentiation and integration can be viewed as linear transformations of the space of functions. For simplicity, we restrict ourselves to real-valued functions defined on an **interval** $[a, b] = \{x \mid a \leq x \leq b\} \subset \mathbb{R}$. Then the set of all functions

$$\mathcal{F} = \{f \mid f : [a, b] \to \mathbb{R}\}$$

becomes a **linear space**, or **vector space**, with the standard definition of **function addition**

$$(f + g)(x) := f(x) + g(x)$$

for which $f + g \in \mathcal{F}$ if $f, g \in \mathcal{F}$, and of **scalar multiplication**

$$(\alpha f)(x) := \alpha f(x)$$

with $\alpha f \in \mathcal{F}$ if $f \in \mathcal{F}$ and $\alpha \in \mathbb{R}$. For an axiomatic introduction to vector spaces, see Appendix C.

We first consider the set of **polynomials** $\mathcal{P}_n \subset \mathcal{F}$ that contains all real polynomials of degree less than or equal to n. Each $p \in \mathcal{P}_n$ has the form

$$p(x) = a_n x^n + a_{n-1} x^{n-1} + \cdots + a_1 x + a_0$$

for $n+1$ real constant coefficients a_j and the real variable x. One can readily see that the powers of x (i.e., the functions x^j) act as generators for this space, since each polynomial $p \in \mathcal{P}_n$ is a certain unique scalar sum, called a **linear combination**, of these basic functions $e_j(x) := x^j$ for $j = 0, 1, \ldots, n$. If we identify $p(x) = a_n x^n + \cdots + a_1 x + a_0$ with its ordered coefficient vector $\begin{pmatrix} a_0 \\ \vdots \\ a_n \end{pmatrix} \in \mathbb{R}^{n+1}$, then

$$\mathcal{P}_n = \left\{ \begin{pmatrix} a_0 \\ \vdots \\ a_n \end{pmatrix} \middle| \, a_i \in \mathbb{R} \right\}$$

looks very much like $\mathbb{R}^{n+1}$.

EXAMPLE 13 The polynomial $p(x) = x^5 - x^3 + 2x^2 + x - 4$ has degree 5 and is represented by its coefficient vector $\begin{pmatrix} -4 \\ 1 \\ 2 \\ -1 \\ 0 \\ 1 \end{pmatrix} \in \mathbb{R}^6$. Conversely, the vector $\begin{pmatrix} 0 \\ 3 \\ -5 \\ 2 \end{pmatrix} \in \mathbb{R}^4$ represents the polynomial $q(x) = 2x^3 - 5x^2 + 3x$ of degree 3, where we read off the polynomial coefficients a_3 of x^3, a_2 of x^2, etc., in decreasing order from the last vector position on up. ◀

Polynomials can be differentiated and integrated. From calculus, recall that the derivative of a polynomial p of degree n is given by the formula

$$p'(x) = (a_n x^n + \cdots + a_2 x^2 + a_1 x + a_0)' = n a_n x^{n-1} + \cdots + 2a_2 x + a_1.$$

Moreover, linear combinations of polynomials differentiate in linear fashion

$$(\alpha p + \beta q)'(x) \;=\; \alpha p'(x) + \beta q'(x) \quad \text{for all } p,\ q \in \mathcal{P}_n \text{ and all } \alpha, \beta \in \mathbb{R}. \quad (1.7)$$

Thus, "differentiation" is a linear function, or a **linear operator** on $\mathcal{P}_n$.

We want to study differentiation and integration in the space of real-valued functions $\mathcal{F}$. If we denote taking derivatives in $\mathcal{F}$ by the operator D, i.e., if $Df := f'$ for all $f \in \mathcal{F}$, then formula (1.7) for polynomials becomes

$$D(\alpha f + \beta g) = \alpha Df + \beta Dg$$

for arbitrary differentiable functions $f, g \in \mathcal{F}$. This describes the linearity of D on $\mathcal{F}$ in the sense of condition (1.1), where the vectors are now functions.

In view of Section 1.1, we may ask the natural question, "What is the matrix representation of the differential operator D with respect to the basic functions $e_j(x) = x^j$ of $\mathcal{P}_n$ for $j = 0, 1, \ldots, n$?"

Using Theorem 1.2 from Section 1.1, we find that this question can be answered by finding De_j for each basis function $e_j(x) = x^j$ and $j = 0, \ldots, n$. Recall that by definition

$$e_0 = x^0 = 1, e_1 = x^1 = x, e_2 = x^2, \ldots, e_n = x^n \quad \text{in } \mathcal{P}_n.$$

From calculus, we know that

$$De_0 = 0,\ De_1 = 1 = e_0,\ De_2 = 2x = 2e_1, \dots, De_j = jx^{j-1} = je_{j-1}, \dots,$$
$$De_n = nx^{n-1} = ne_{n-1} \in \mathcal{P}_n.$$

Thus, according to Theorem 1.2, the matrix representation for the differentiation operator $D : \mathcal{P}_n \to \mathcal{P}_n$ with respect to our chosen polynomial basis $\mathbf{e}_j(x) = x^j$ of $\mathcal{P}_n$ is given by

$$\mathbf{D} = \begin{pmatrix} | & | & & | \\ D\mathbf{e}_0 & D\mathbf{e}_1 & \cdots & D\mathbf{e}_n \\ | & | & & | \end{pmatrix} = \begin{pmatrix} | & | & | & & | \\ \mathbf{0} & \mathbf{e}_0 & 2\mathbf{e}_1 & \cdots & n\mathbf{e}_{n-1} \\ | & | & | & & | \end{pmatrix} =$$

$$= \begin{pmatrix} 0 & 1 & 0 & \dots & 0 \\ \vdots & 0 & 2 & \ddots & \vdots \\ & & 0 & \ddots & 0 \\ & & & & n \\ 0 & \dots & & & 0 \end{pmatrix}_{(n+1)\times(n+1)},$$

where we represent $\mathbf{e}_0 = x^0 = 1$ by the vector $\begin{pmatrix} 1 \\ 0 \\ \vdots \\ 0 \end{pmatrix} \in \mathbb{R}^{n+1}$, $\mathbf{e}_1 = x$ by $\begin{pmatrix} 0 \\ 1 \\ \vdots \\ 0 \end{pmatrix}$, and so forth, until representing $\mathbf{e}_n = x^n$ by $\begin{pmatrix} 0 \\ \vdots \\ 0 \\ 1 \end{pmatrix} \in \mathbb{R}^{n+1}$.

Note that we use the yet undefined term "basis" here for the generators x^j $(j = 0, 1, \dots, n)$ of $\mathcal{P}_n$. The concept of a basis for a linear space will be formally studied in Chapter 5.

EXAMPLE 14 Find $p'(x)$ for the polynomial $p(x) = 3x^5 - 2x^3 + x^2 - x + 4 \in \mathcal{P}_5$ of degree 5 via the differential operator matrix $\mathbf{D}_{6\times 6}$.

In $\mathcal{P}_5$, the given polynomial $p(x)$ has the standard representation $\mathbf{p} = \begin{pmatrix} 4 \\ -1 \\ 1 \\ -2 \\ 0 \\ 3 \end{pmatrix}$ $\in \mathbb{R}^6$, which is found by placing the coefficients of p from the highest power of x on down to the constant term 4 upside down into a vector in $\mathbb{R}^6$. Now evaluate

$\mathbf{p}' = \mathbf{D}_{6\times 6}\mathbf{p}$ as the matrix $\times$ vector product

$$\mathbf{D}_{6\times 6}\mathbf{p} = \begin{pmatrix} 0 & 1 & & & & 0 \\ 0 & 0 & 2 & & & \\ & & 0 & 3 & & \\ & & & 0 & 4 & \\ & & & & 0 & 5 \\ 0 & & & & & 0 \end{pmatrix} \begin{pmatrix} 4 \\ -1 \\ 1 \\ -2 \\ 0 \\ 3 \end{pmatrix} = \begin{pmatrix} -1 \\ 2 \\ -6 \\ 0 \\ 15 \\ 0 \end{pmatrix} = \mathbf{p}' \in \mathbb{R}^6.$$

The highest power coefficient a_5 of $\mathbf{p}'$ occurs at the bottom of this vector and it is zero. Thus $p'(x) = 15x^4 - 6x^2 + 2x - 1$, which agrees with a more standard, calculus style differentiation of $p(x)$. ◀

EXAMPLE 15 The **definite integral** $\text{Int}_a^b(f) := \int_a^b f(x)dx : \mathcal{F}_I \to \mathbb{R}$ on the space of integrable functions $\mathcal{F}_I$ for the interval $I = [a, b]$ is a linear process, since

$$\begin{aligned} \text{Int}_a^b(\alpha f + \beta g) &= \int_a^b (\alpha f(x) + \beta g(x))dx \\ &= \alpha \int_a^b f(x)dx + \beta \int_a^b g(x)dx \\ &= \alpha\, \text{Int}_a^b(f) + \beta\, \text{Int}_a^b(g); \end{aligned}$$

that is, $\text{Int}_a^b(f) : \mathcal{F}_I \to \mathbb{R}$ satisfies the linearity condition (1.1). If we restrict ourselves again to the polynomials $\mathcal{P}_n \subset \mathcal{F}_I$ for a given interval $I = [a, b] \subset \mathbb{R}$ and take the standard basis $\mathbf{e}_j(x) := x^j$ for $j = 0, \ldots, n$ of $\mathcal{P}_n$, then the linear operator $\text{Int}_a^b : \mathcal{P}_n \to \mathbb{R}$ has the matrix form:

$$\begin{aligned} \mathbf{Int}_a^b(p) &= \begin{pmatrix} \text{Int}_a^b(1), & \ldots, & \text{Int}_a^b(x^n) \end{pmatrix} \begin{pmatrix} a_0 \\ \vdots \\ a_n \end{pmatrix} \\ &= \begin{pmatrix} b - a, & \ldots, & \dfrac{b^{n+1} - a^{n+1}}{n+1} \end{pmatrix} \begin{pmatrix} a_0 \\ \vdots \\ a_n \end{pmatrix} \end{aligned}$$

for any $p(x) = a_n x^n + \cdots + a_1 x + a_0$. This follows from the Riesz Lemma, since for the basis functions $\mathbf{e}_j(x) = x^j$ of $\mathcal{P}_n$ elementary calculus shows that $\text{Int}_a^b(x^j) = \int_a^b x^j dx = \dfrac{b^{j+1} - a^{j+1}}{j+1}$ making Int_a^b have the standard matrix representation $\begin{pmatrix} \text{Int}_a^b(1), \ldots, \text{Int}_a^b(x^n) \end{pmatrix}$ as given above. Thus we can evaluate the following definite integral easily via linear algebra as a simple row times column vector product:

$$\int_1^3 x^3 - 3x^2 + 4\,dx = \mathbf{Int}_1^3 \begin{pmatrix} 4 \\ 0 \\ -3 \\ 1 \end{pmatrix}$$

$$= \left(3-1, \frac{3^2-1}{2}, \frac{3^3-1}{3}, \frac{3^4-1}{4}\right) \begin{pmatrix} 4 \\ 0 \\ -3 \\ 1 \end{pmatrix}$$

$$= \left(2, 4, \frac{26}{3}, 20\right) \begin{pmatrix} 4 \\ 0 \\ -3 \\ 1 \end{pmatrix} = 8 - 26 + 20 = 2,$$

which handily agrees with the direct evaluation of the definite integral from calculus rules. ◀

The benefits of representing functions as members of a linear space and of using operator matrix representations to simulate differentiation and integration via matrix × vector products are not immediately obvious, but will become so in Section 7.3, where we shall recast both of these concepts of calculus in matrix notation. This will enable us to solve differential equations and to perform integration by parts easily via linear algebra.

While linear algebra can express problems of very diverse fields and can help solve them, other areas such as numerical analysis and computer science conversely have numerous applications to linear algebra.

There exist a growing number of software packages that can solve applications problems once they have been rephrased as linear algebra ones. We shall introduce the basic principles of one such software, MATLAB™[1], which is highly suited to work on linear algebra problems. MATLAB was developed by Cleve Moler in the 1970s and is widely used by scientists and engineers as the computational matrix software of choice.

MATLAB is short for "Matrix Laboratory" and as such it is extremely well suited to augment our teaching of linear algebra.

Here are its basics:

Whenever MATLAB is installed on a computer, it can be accessed with the system command MATLAB or by clicking on the appropriate icon. It can be exited by entering `quit` or by clicking appropriately.

Matrices and vectors are the lifeblood of MATLAB. Once MATLAB is running, matrices are entered following its command prompt MATLAB >>. For example,

```
MATLAB >> A = [1 2 3; 2 4 0]
```

[1]MATLAB is a registered trade mark of The MathWorks, Inc., 3 Apple Hill Drive, Natick, MA 01760, http://www.mathworks.com.

Square brackets delimit the matrix. The semicolon `;` denotes the end of a row. A is the assigned name of this specific matrix and the computer will keep it in its memory. Entries in one row must be separated by either a blank space or a comma. Thus, the command

```
A = [1,2,3; 2,4 0]
```

produces the same matrix $\mathbf{A} = \begin{pmatrix} 1 & 2 & 3 \\ 2 & 4 & 0 \end{pmatrix}$ as the earlier command. Vectors can be entered as rows or as columns by entering

```
x = [1 2 3]        or        y = [1; -12; 0]
```

for example to obtain the row vector $\mathbf{x} = \begin{pmatrix} 1 & 2 & 3 \end{pmatrix}$ and the column vector $\mathbf{y} = \begin{pmatrix} 1 \\ -12 \\ 0 \end{pmatrix} \in \mathbb{R}^3$, respectively. Vectors given in row notation, such as $\mathbf{x}$ in the preceding example, can be transformed to column notation by the apostrophe command `z = x'` that transposes real vectors and real matrices. With $\mathbf{x}$ as specified, the command `A*x'` produces the answer `ans = 14 10`, while the command `A*y` computes the vector `-23 -46`.

The standard matrix or vector operations '+, −' or 'scalar ·' and 'matrix × vector' are executed by the MATLAB commands

```
A + B, 2 * A, 3*A-2*B, A*x, A*(x+4*y)
```

Note that blank spaces are usually ignored in this language. However, the multiplication operator `*` must be included in all instances. In MATLAB, we must write `A*x` instead of our customary $\mathbf{Ax}$ to express the matrix × vector product of $\mathbf{A}$ and $\mathbf{x}$ correctly. Likewise, a command such as `2pi` will cause consternation in MATLAB, while `2*pi` will generate the real number $2\pi = 6.28\ldots$.

Note also that MATLAB is case sensitive. If x is defined as a vector by the lowercase symbol $\mathbf{x}$, then calling on `X` will generate the error message `??? Undefined function or variable 'X'`.

MATLAB has many built–in matrices and vectors. The commands `eye(n)`, `zeros(n)`, `ones(n)`, and `rand(n)`, for example, create the $n \times n$ **identity matrix** $\mathbf{I}_n := \begin{pmatrix} 1 & & 0 \\ & \ddots & \\ 0 & & 1 \end{pmatrix}$, the $n \times n$ **zero matrix** $\mathbf{O}_n := \begin{pmatrix} 0 & \cdots & 0 \\ \vdots & \ddots & \vdots \\ 0 & \cdots & 0 \end{pmatrix}$, the $n \times n$ matrix $\begin{pmatrix} 1 & \cdots & 1 \\ \vdots & \ddots & \vdots \\ 1 & \cdots & 1 \end{pmatrix}$ of all ones, and a randomly generated $n \times n$ matrix, respectively.

Help is available on screen for all MATLAB functions and commands. For example, by typing

```
help eye
```

one can learn about the identity matrix. The first unit vector $\mathbf{e}_1 \in \mathbb{R}^4$ can be generated by the command

```
e1 = eye(4,1),
```

where the parameters 4 and 1 indicate the number of rows and columns, respectively, that are to be used in `e1` from the 4×4 identity matrix `eye(4)`. Similarly, $\mathbf{e}_2 \in \mathbb{R}^4$ can be generated by the two successive commands

```
e2 = zeros(4,1); e2(2) = 1
```

Here the first command sets up the zero column vector in $\mathbb{R}^4$, while the second command respecifies its second entry as 1.

We note that commands followed by a semicolon generate no screen output in MATLAB, while those followed by a comma or with no termination symbol create screen output.

Alternatively, $\mathbf{e}_2 \in \mathbb{R}^4$ could have been generated from $\mathbf{I}_4$ by the sequence of commands

```
I = eye(4); e2 = I(:,2),
```

where the colon in the first, the row index position of $\mathbf{I}_4$, selects all row entries of $\mathbf{I}_4$, while the 2 used in the column position simultaneously specifies the second column of $\mathbf{I}_4$ for $\mathbf{e}_2$.

In MATLAB, for loops are expressed by string of commands of the form

```
for i= ... ,          , end
```

where i acts as the iteration index. Moreover 'if' and 'while' statements work just as in other languages. The following four sets of MATLAB commands each produce the same second unit vector $\mathbf{e}_2$ of $\mathbb{R}^4$:

```
for i = 1:4, e2(i) = 0, end, e2(2) = 1

for i = 1:4
    e2(i) = 0;
end
e2(2) = 1,

for i = 1:4
    if i = 2
       e2(i) = 1;
    else
       e2(i) = 0;
    end
end,
```

and

```
i = 0
    while i < 4
         i = i + 1
         e2(i) = 0;
         if i = 2
            e2(i) = 1;
         end
    end
```

These lines of code achieve their goal with different degrees of clumsiness, of course.

Subsequent sections on applications will include relevant commands and examples. This will enable students to perform involved computations for various subject areas efficiently on a computer. To do mathematics with the help of a computer is one of the modern developments in teaching and learning mathematics, but students will also need to practice "paper and pencil" math skills throughout the book.

Resources for MATLAB: To find a broad cross-section of currently active tutorials and college Web sites on MATLAB, we recommend to start a search at

`http://www.google.com`

for "MATLAB". To find currently maintained sites with Kermit Sigmon's widely used **MATLAB Primer**, search for "MATLAB PRIMER". The site

`http://www.math.mtu.edu/~msgocken/intro/intro.html`

offers another useful MATLAB tutorial. Moreover, the *ATLAST* download site,

`http://www2.cs.gasou.edu/~atlast/mfiles.html`

an acronym for "Augment the Teaching of Linear Algebra through the use of Software Tools," contains many useful MATLAB functions and problems for teaching enrichment.

1.3.P Problems

1. Find applications of linear algebra in your specific field of study by

(a) Checking the index of standard textbooks in your field.

(b) Asking upperclassmen.

(c) Asking your faculty advisor.

2. Compute the image of the planar unit square under the transformation $\mathbf{x} \mapsto \mathbf{A}\mathbf{x} : \mathbb{R}^2 \to \mathbb{R}^2$ induced by $\mathbf{A} = \begin{pmatrix} 7 & 10 \\ 5 & 7 \end{pmatrix}$ by mapping each unit square corner under $\mathbf{A}$. Draw a picture.

3. Plot the image of the unit square in $\mathbb{R}^2$ under the mapping $\mathbf{x} \mapsto \mathbf{A}\mathbf{x}$ by hand for

(a) $\mathbf{A} = \begin{pmatrix} 2 & -1 \\ -1 & 2 \end{pmatrix}$; (b) $\mathbf{A} = \begin{pmatrix} 2 & 0 \\ 1 & -1 \end{pmatrix}$;

(c) $\mathbf{A} = \begin{pmatrix} 1 & 1 \\ 1 & 1 \end{pmatrix}$; (d) $\mathbf{A} = \begin{pmatrix} -1 & -1 \\ -1 & -1 \end{pmatrix}$;

(e) $\mathbf{A} = \begin{pmatrix} 0.6 & -0.8 \\ 0.8 & 0.6 \end{pmatrix}$; (f) $\mathbf{A} = \begin{pmatrix} 0 & 1 \\ 1 & 0 \end{pmatrix}$.

4. Show that the matrices $\mathbf{A} = \begin{pmatrix} 1 & -1 \\ 3 & -3 \end{pmatrix}$ and $\mathbf{B} = \begin{pmatrix} 0 & 2 \\ 0 & 6 \end{pmatrix}$ have the same set of images $\{\mathbf{Ax} \mid \mathbf{x} \in \mathbb{R}^2\} = \{\mathbf{Bx} \mid \mathbf{x} \in \mathbb{R}^2\}$. How do the two linear transformations $\mathbf{x} \mapsto \mathbf{Ax}$ and $\mathbf{x} \mapsto \mathbf{Bx}$ differ?

5. Write out the standard matrix $\mathbf{A}_{\mathcal{E}}$ for the rotation of the plane by 45 degrees counterclockwise around the origin.

6. Repeat Problem 5 for a clockwise rotation by 30 degrees combined with a dilation for the factor $d = \sqrt{3}$.

7. Let T map $\mathcal{P}_2$ to $\mathcal{P}_3$ and be defined by $T(a_2x^2 + a_1x + a_0) := a_2x^3 + a_1x^2 + a_0$. Is T a linear transformation?

8. Let $T : \mathcal{P}_n \to \mathcal{P}_n$ be defined by $T(a_nx^n + \cdots + a_1x + a_0) := a_0x^n + a_1x^{n-1} + \cdots + a_{n-1}x + a_n$ (i.e., coefficient reversal). Is T a linear transformation? What is its matrix representation with respect to the standard basis $\{x^j\}$ of $\mathcal{P}_n$?

9. Let U be the transformation of $\mathcal{P}_n$ to $\mathbb{R}^3$ defined by $U(p) = (p(1), p(0), p(-2))$.

(a) Find $U(3x^3 - 5x^2 + 22)$, $U(0)$, and $U(x^5 + x^4 + x^3 + x^2)$.

(b) Is U a linear function?

(c) What does a standard matrix representation $\mathbf{A}$ for U look like with respect to the standard basis $\{x_j\}$, and what is its size?

(d) Check your results in part (a) via the matrix representation $\mathbf{A}$ of U in part (c).

10. Let $T : \mathcal{P}_3 \to \mathcal{P}_3$ be defined as $T(p) := 2p'' - p' + 2p$.

(a) Is T a linear transformation?

(b) Find a matrix representation of T with respect to the standard basis $\{x^j\}$ of $\mathcal{P}_3$ by evaluating $T(x^j)$ for each $j = 0, 1, 2, 3$.

(c) Compute $T(-4x^3 + 2x^2 - x + 17)$ from its defining formula using differentiation and via the matrix representation for T in (b). Compare your results.

11. Compute $\int_0^5 3x^4 - 2x^2 + 1dx$ via calculus and via linear algebra.

12. (a) Find a matrix representation $\mathbf{A}_{\mathcal{E}}$ of the second derivative operator D^2 on $\mathcal{P}_n$, defined as $D^2(p) := p''$.
(*Hint*: Look at the second derivatives of the standard basis functions $\{x^j\}$.)

(b) Find a matrix representation $\mathbf{B}_{\mathcal{E}}$ of the third derivative operator D^3 on $\mathcal{P}_2$, $\mathcal{P}_3$, and $\mathcal{P}_5$.

(c) Find a matrix representation $\mathbf{C}_{\mathcal{E}}$ of the differential operator $D^2 - 2D$ on $\mathcal{P}_2$ and $\mathcal{P}_5$.

13. Start up MATLAB on your computer, and when the MATLAB prompt has appeared, enter `A=randn(4)`, `x=randn(1,4)`. (Check out the difference between the MATLAB commands `rand` and `randn` by invoking `help`.)

(a) Look at the screen output. What is the size of $\mathbf{A}$? Is $\mathbf{x}$ a row or a column vector? Transpose the given random row vector $\mathbf{x}$ to the column vector $\mathbf{x}^T$ by invoking `x'` in MATLAB

(b) Which of the matrix products $\mathbf{Ax}$, $\mathbf{Ax}^T$, $\mathbf{xA}$, or $\mathbf{x}^T\mathbf{A}$ can be formed successfully in MATLAB? (Try them all and read the error messages carefully.) Can you form $\mathbf{xx}$, $\mathbf{xx}^T$, or $\mathbf{x}^T\mathbf{x}$?

(c) Can you add $\mathbf{x} + \mathbf{x}$, $2\mathbf{x} + \mathbf{A}$, $\mathbf{A} + 2\mathbf{A}$, $\mathbf{x} + \mathbf{A}$, or $\mathbf{xx}^T - 7\mathbf{A}$? (Have you forgotten to enter any implicit multiplication signs $*$ in your MATLAB commands? Be careful!)

(d) Study the length and size commands in MATLAB on $\mathbf{x}$ and $\mathbf{A}$. (Enter `help length` and `help size` here. Check which size of a matrix is given first by the command `size(A)`, `size(x)`; recall: rows before columns.)

(e) Create $\mathbf{e}_1$ and $\mathbf{e}_3 \in \mathbb{R}^4$ efficiently in MATLAB.

(f) Find $\mathbf{Ae}_1$ and $\mathbf{Ae}_3$ from part (e) and by using the MATLAB colon notation; see `help :`.

14. Find the elementary MATLAB commands that create the following output from the inputs $\mathbf{A} = \begin{pmatrix} 1 & -1 & 2 \\ 0 & 4 & 0 \\ 2 & -1 & 1 \end{pmatrix}$,

$\mathbf{B} = \begin{pmatrix} 1 & 2 \\ -1 & 2 \\ 0 & 3 \end{pmatrix}$, $\mathbf{C} = \begin{pmatrix} 1 & 0 & 1 \\ 1 & 2 & 0 \\ 1 & -1 & 3 \end{pmatrix}$, $\mathbf{x} = \begin{pmatrix} 2 \\ -1 \\ 3 \end{pmatrix}$,

and $\mathbf{y} = \begin{pmatrix} 2 \\ -1 \end{pmatrix}$:

(a)
```
ans =
     9
    -4
     8
```

(b)
```
ans =
    14
    -4
    20
```

(c)
```
ans =
     5
     0
    12
```

(d)
```
??? Error using ==> *
Inner matrix
dimensions must agree.
```

(e) ans =

```
 0  -1   1
-1   2   0
 1   0  -2
```

(f) ans =

```
 9
 0
11
```

15. Given that $\mathbf{A} = \begin{pmatrix} 2 & 0 & -1 \\ 3 & 5 & 2 \\ 0 & 1 & -1 \end{pmatrix}$; $\mathbf{B} = \begin{pmatrix} 3 & 0 & 1 \\ 0 & 3 & -1 \\ 1 & 0 & 3 \end{pmatrix}$, and $\mathbf{x} = \begin{pmatrix} 1 \\ -1 \\ 1 \end{pmatrix}$, which of the following MATLAB commands

(1) `2*A`, (2) `A-B`, (3) `2*(A+B)`,
(4) `2*A-3*B`, (5) `B*x - A*x`,
(6) `B*x`, (7) `2*B*x+3*A*(-x)`

give the following output?

(a) ans =

```
-1   0  -2
 3   2   3
-1   1  -4
```

(b) ans =

```
 5
-8
14
```

(c) ans =

```
 4   0  -2
 6  10   4
 0   2  -2
```

(d) ans =

```
-5   0   -5
 6   1    7
-3   2  -11
```

(e) ans =

```
 3
-4
 6
```

(f) ans =

```
10   0   0
 6  16   2
 2   2   4
```

(g) ans =

```
 4
-4
 4
```

16. (a) Let T be the mapping of the two coefficients a and b of a polynomial $p(x) = ax + b$ of degree one to its zero $x_0 = -\frac{b}{a}$ (if $a \neq 0$).

Is $T : \mathcal{P}_1 \to \mathbb{R}$ a linear map? I.e., is the zero of the sum $p + q$ of two first degree polynomials p and q equal to the sum of the zeros of each individual first-degree polynomial?

(b) Settle the above question for two quadratic polynomials p and q and their roots.

17. Let $T : \mathcal{P}_4 \to \mathcal{P}_5$ be defined as
(a) $T(p) := x^2 p'(x)$,
(b) $T(p) := (2x - 3)p'(x)$,
(c) $T(p) := p'(x) - p(x)$, or
(d) $T(p) := p(1)p'(x) + p(1)$.
Which of the four functions are linear, and which are not?

1.R Review Problems

1. If $T : \mathbb{R}^n \longrightarrow \mathbb{R}^k$ is a linear transformation, how are the columns of the matrix representation $\mathbf{A}_\mathcal{E}$ of T determined with respect to the standard bases $\mathcal{E} = \{\mathbf{e}_i\}$ of the domain and range spaces?

What size does $\mathbf{A}_\mathcal{E}$ have? How many rows? How many columns? How many entries?

2. Which of the following functions are linear transformations (please give complete reasons)?

(a) $f : \mathbb{R}^2 \longrightarrow \mathbb{R}^2$ defined as

$$f(x_1, x_2) := \begin{pmatrix} \sin(x_1) \\ \cos^2(x_1) + \sin^2(x_2) - 1 \end{pmatrix}.$$

(b) $g : \mathbb{R} \longrightarrow \mathbb{R}$ defined as

$$g(x) := \sin^2(x) - (1 - \cos^2(x)).$$

(c) $h : \mathbb{R}^2 \longrightarrow \mathbb{R}^2$ defined as

$$h(x_1, x_2) := \begin{pmatrix} x_1 + 2x_2 \\ 2x_2 + x_1 \end{pmatrix}.$$

(d) $k : \mathbb{R}^3 \longrightarrow \mathbb{R}^3$ defined as $k(x_1, x_2, x_3)$

$$:= \begin{pmatrix} 0 \\ 2x_1 - x_3 + x_2 \\ (x_1 - x_2)(x_1 + 1) + x_1x_2 + x_3 - x_1^2 \end{pmatrix}.$$

(e) $\ell : \mathbb{R}^3 \longrightarrow \mathbb{R}^3$ defined as $\ell(x_1, x_2, x_3)$

$$:= \begin{pmatrix} x_2 - 1 \\ 2x_1 - x_3 + x_2 \\ (x_1 - x_2)(x_1 + 1) + x_1x_2 + x_3 - x_1^2 \end{pmatrix}.$$

(f) If in any of the parts (a) ... (e) the mapping is linear, find its standard matrix representation $\mathbf{A}_\mathcal{E}$.

3. Show that each of the following mappings is linear on $\mathbb{R}^n$:

(a) $M(\mathbf{x}) = \pi\mathbf{x}$;

(b) $K(\mathbf{x}) = -\mathbf{x}$;

(c) $J(\mathbf{x}) = x_n\mathbf{e}_1 - x_1\mathbf{e}_n$.

(d) Find the standard $n \times n$ matrix representations for M, K, and J.

4. Let $\mathbf{b} \in \mathbb{R}^n$ be an arbitrary vector, and define a mapping $T : \mathbb{R}^n \to \mathbb{R}^n$ by $T(\mathbf{x}) := 2\mathbf{x} - 3\mathbf{b}$.

(a) Show that T is a linear mapping if $\mathbf{b} = \mathbf{0}$. What is its matrix representation in terms of the identity matrix $\mathbf{I}_n$?

(b) Is T a linear mapping if $\mathbf{b} \neq \mathbf{0}$ as well? Give a proof.

5. Let F map $\mathbb{R}^2$ to $\mathbb{R}^2$ according to the rule $F(x_1, x_2) := (x_1^2, x_1x_2)$.

(a) Show that $F(\mathbf{0}) = \mathbf{0}$.

(b) Determine why F is not linear.

6. Let $T(\mathbf{x}) = \begin{pmatrix} 1 & 5 & -3 & 7 & 2 \\ 2 & 12 & -3 & 6 & 9 \\ 0 & 0 & 8 & -5 & 0 \end{pmatrix}\begin{pmatrix} x_1 \\ \vdots \\ x_5 \end{pmatrix}$ be a matrix × vector product.

(a) Find $T(\mathbf{e}_1)$, $T(\mathbf{e}_3)$, $T(\mathbf{e}_5)$, and $T(\mathbf{0})$.

(b) How many component functions make up T?

(c) Write out each component function of T as an algebraic expression of the five variables $x_1, \ldots, x_5$.

(d) Express each component function of T in part (c) as a dot product.

(e) Explain why T is a linear function.

7. Let $T : \mathcal{P}_n \to \mathbb{R}$ be defined as $T(p) = m$ if the degree of $p \in \mathcal{P}_n$ is m. Prove or disprove that T is a linear map for all n.
Is there an integer n for which T is linear?

8. Show that

(a) If $f(0) \neq 0$ for a given function f, then f is not linear.

(b) If $f(-x) \neq -f(x)$ for a given function f and some x, then f is not linear.

9. Let $T : U \to V$ be a linear mapping between two linear spaces (that may or may not be in any way like $\mathbb{R}^n$). Prove that $T(\mathbf{0}) = \mathbf{0} \in \mathbf{V}$ simply from the linearity condition for T. (Do not use any matrix representation for T; it may not exist over U.)

10. Assume that f maps $\mathbb{R}^n$ to $\mathbb{R}^m$.

(a) If you are given the n images $f(\mathbf{e}_i)$ for each standard basis vector $\mathbf{e}_i$ in $\mathbb{R}^n$, can you find $f(2\mathbf{e}_1 + 3\mathbf{e}_2 + 4\mathbf{e}_3 + \ldots + (n+1)\mathbf{e}_n)$ from the given information? How?

(b) If $m = n = 3$, f is linear, and if $f(\mathbf{e}_1) = -\mathbf{e}_1 + 2\mathbf{e}_3$, $f(\mathbf{e}_2) = 4\mathbf{e}_2 + 2\mathbf{e}_3$, while $f(\mathbf{e}_3) = \mathbf{0}$, what is $f(-\mathbf{e}_3 + 2\mathbf{e}_1 - 3\mathbf{e}_2)$.

(c) True or false: If $f : \mathbb{R}^n \to \mathbb{R}^m$ is linear, then we can deduce how f maps every point of n–space from knowing how f maps the n standard basis vectors $\mathcal{E} = \{\mathbf{e}_1, \ldots, \mathbf{e}_n\}$ of $\mathbb{R}^n$.

11. Find the standard matrix representation of the linear transformation $T : \mathbb{R}^3 \to \mathbb{R}^2$ that maps $\begin{pmatrix} 2 \\ 1 \\ 0 \end{pmatrix}$ to $\begin{pmatrix} 1 \\ 1 \end{pmatrix}$, $\begin{pmatrix} 1 \\ 0 \\ -1 \end{pmatrix}$ to $\begin{pmatrix} 2 \\ 0 \end{pmatrix}$, and $\begin{pmatrix} 0 \\ 0 \\ 4 \end{pmatrix}$ to $\begin{pmatrix} -1 \\ 8 \end{pmatrix}$.
Check your results.

12. How does the linear transformation $\mathbf{x} \mapsto \mathbf{A}\mathbf{x}$ transform the plane for the matrix $\mathbf{A} = \begin{pmatrix} 0 & -1 \\ -1 & 0 \end{pmatrix}$?

13. Assume that $f : \mathbb{R}^2 \to \mathbb{R}^3$ maps $(2\ ,\ -1)$ to $(1\ ,\ 1\ ,\ 3)$, $(1\ ,\ -1)$ to $(0\ ,\ 1\ ,\ 4)$, and $(1\ ,\ 0)$ to $(1\ ,\ -1\ ,\ 1)$.
Can f be a linear function?

14. Let $\mathbf{B} = \begin{pmatrix} 1 & 0 & -1 \\ 2 & -1 & 3 \end{pmatrix}$.

(a) Show that the function $f(\mathbf{x}) := \mathbf{B}\mathbf{x}$ that maps three-dimensional vectors to two-dimensional vectors is a linear function.

(b) Find $f(\mathbf{0})$, $f(\mathbf{e}_1)$, $f(\mathbf{e}_3)$, $f(\mathbf{e}_1 + \mathbf{e}_3)$, and $f\begin{pmatrix} 2 \\ 3 \\ -2 \end{pmatrix}$.

Standard Questions

1. Is $f : \mathbb{R}^n \to \mathbb{R}^m$ a linear map? (Check each component function.)

2. Which component functions of $f : \mathbb{R}^n \to \mathbb{R}^m$ are linear, which are nonlinear?

3. If f maps a certain set of vectors in a given way, can f be a linear mapping? (Check the given data on f for linearity.)

4. If f maps a certain set of points in a prescribed way and f is linear, what is the standard matrix representation $\mathbf{A}_{\mathcal{E}}$ of f? (That is, how does one find $f(\mathbf{e}_1), \ldots, f(\mathbf{e}_n)$?)

Subheadings of Lecture One

Basic Equations

p. 14 $f(\alpha \mathbf{x} + \beta \mathbf{y}) = \alpha f(\mathbf{x}) + \beta f(\mathbf{y})$ (linear function)

p. 29 $$\mathbf{A}_{\mathcal{E}} = \begin{pmatrix} | & & | \\ f(\mathbf{e}_1) & \dots & f(\mathbf{e}_n) \\ | & & | \end{pmatrix}$$ (standard matrix representation)

Only one way

CHAPTER

2

Row-Echelon Form Reduction

We introduce the simple matrix algorithm of row reduction.

2.1 Lecture Two

Gaussian Elimination and the Echelon Forms

Gaussian elimination and the row-echelon form reduction of a matrix are studied. They are fundamental and powerful techniques for all of linear algebra.

(a) Motivation and Rules of Gaussian Elimination

A given m-by-$n+1$ matrix

$$(\mathbf{A} \mid \mathbf{b}) = \left(\begin{array}{ccc|c} a_{11} & \cdots & a_{1n} & b_1 \\ \vdots & & \vdots & \vdots \\ a_{m1} & \cdots & a_{mn} & b_m \end{array} \right) \tag{2.1}$$

with real or complex entries a_{ij}, $i = 1, \ldots, n$, $j = 1, \ldots, m$, and b_k, $k = 1, \ldots, m$, can be interpreted as a **system of m linear equations $\mathbf{Ax} = \mathbf{b}$**, or as

$$\begin{array}{ccc} a_{11}x_1 + \cdots + a_{1n}x_n & = & b_1 \\ \vdots \qquad\qquad \vdots & & \vdots \\ a_{m1}x_1 + \cdots + a_{mn}x_n & = & b_m \end{array} \tag{2.2}$$

for n unknown entries $x_1, \ldots, , x_n \in \mathbb{R}$. The **augmented matrix** $(\mathbf{A} \mid \mathbf{b})$ in (2.1) of the linear system of equations $\mathbf{Ax} = \mathbf{b}$ consists of the **system matrix** $\mathbf{A}_{mn}$, augmented by the **right-hand-side** column $\mathbf{b}$ on the right. The matrix $(\mathbf{A} \mid \mathbf{b})$ has one more column than $\mathbf{A}$. That is, it has size m by $n+1$ if $\mathbf{A}$ is m by n. The vertical line that separates $\mathbf{A}$ and $\mathbf{b}$ in Eq. (2.1) is for visual guidance only.

The augmented matrix notation $(\mathbf{A} \mid \mathbf{b})$ provides us with a convenient **shorthand notation**. It suppresses the n variables x_i and all plus, minus, and equal signs of the m explicit equations in Eq. (2.2). This shorthand notation transforms finding a solution of Eq. (2.2) into a problem of matrices and vectors (i.e., into one of linear algebra).

To solve the system of linear equations (2.2), we are allowed to modify the individual equations (or augmented matrix rows in Eq. (2.1)) into a simpler and more revealing form as long as the **solution x** of the modified system stays the same. The modifications may consist of any sequence of the following **three legitimate operations**.

1. **Adding Equations:**
   If two sets of quantities are equal, such as the two sides of two linear equations (E_j) and (E_i) in Eq. (2.2), written out as

$$\begin{aligned} (\mathrm{E}_j): \quad & a_{j1}x_1 + \cdots + a_{jn}x_n = b_j, \\ (\mathrm{E}_i): \quad & a_{i1}x_1 + \cdots + a_{in}x_n = b_i, \end{aligned}$$

   then the two left-hand sides and the two right-hand sides of the equations can be added to give the equally valid equation

$$(\mathrm{E}_i + \mathrm{E}_j): \qquad (a_{i1} + a_{j1})x_1 + \cdots + (a_{in} + a_{jn})x_n = b_i + b_j.$$

   In Gaussian elimination, an equation (E_i) is repeatedly replaced by the summed equation $(\mathrm{E}_i + \mathrm{E}_j)$ for $j \neq i$ or by a scaled equations sum such as $(\mathrm{E}_i + c\mathrm{E}_j)$ without affecting the solution set. (Note that if $c = -1$, then we are subtracting (E_j) from (E_i) in the expression $(\mathrm{E}_i + c\mathrm{E}_j)$.)
2. **Scaling Equations:**
   If $(\mathrm{E}_i): a_{i1}x_1 + \cdots + a_{in}x_n = b_i$ holds, then $(c\mathrm{E}_i): ca_{i1}x_1 + \cdots + ca_{in}x_n = cb_i$ is also true for any $c \neq 0$. (If we allow $c = 0$ here, then $(0\mathrm{E}_i)$ is the trivial equation $0 = 0$ and is always true. Replacing (E_i) by $(0\mathrm{E}_i)$ $(0 = 0)$ is not advisable. We may lose some constraints for the solution vector $\mathbf{x}$, and this will possibly give us false information.)
3. **Swapping Equations:**
   The order of the equations in Eq. (2.2) can be changed without affecting the actual solution vector $\mathbf{x}$.

Note that these three legitimate operations also allow for the subtraction of equations and, more generally, the adding or subtracting of arbitrary nonzero multiples of any one equation to or from any other. This is because these more complicated operations can be realized as a sequence of scalings and additions of the equations. Thus, a pair of equations (E_j) and (E_i) is equivalent to (i.e., has the same set of solutions $\mathbf{x}$ as) the pair of equations (E_j) and $(\mathrm{E}_i + c\mathrm{E}_j)$ for all $c \in \mathbb{R}$.

To solve $\mathbf{Ax} = \mathbf{b}$, the three operations are applied systematically to the rows of the augmented matrix $(\mathbf{A} \mid \mathbf{b})$, since each row of $(\mathbf{A} \mid \mathbf{b})$ stands for one equation. This row process will be studied next. It is called **Gaussian elimination** and is designed to create a linear system $\mathbf{Rx} = \tilde{\mathbf{b}}$ with the same solution $\mathbf{x}$ as $\mathbf{Ax} = \mathbf{b}$, but with $\mathbf{R}$'s lower triangle containing only zeros.

The desired outcome of this computation is a matrix $\mathbf{R} = (r_{ij})$ that is **upper triangular** with $r_{ij} = 0$ for all $i > j$, that is, the entries of R below and to the left of its diagonal entries $\{r_{ii}\}$ are all zero. We call a matrix **upper triangular** if its lower triangle consists entirely of zeros. Similarly, a matrix is **lower triangular** if its upper triangle is zero. For example the matrices $\mathbf{R}_1 = \begin{pmatrix} 1 & -2 \\ 0 & -10 \\ 0 & 0 \end{pmatrix}$, $\mathbf{R}_2 = \begin{pmatrix} -4 & 0 & -2 \\ 0 & 0 & 1 \\ 0 & 0 & 0 \end{pmatrix}$, and $\mathbf{R}_3 = \begin{pmatrix} -4 & 1 & 0 & 0 & 1 \\ 0 & 0 & 2 & 1 & -1 \end{pmatrix}$ are all upper triangular; but $\mathbf{C} = \begin{pmatrix} 1 & 0 \\ 2 & -1 \\ 0 & -3 \end{pmatrix}$ is not, because of its nonzero (2,1) and (3,2) entries of 2 and -3, respectively.

After a complete Gaussian elimination has been performed on the augmented matrix $(\mathbf{A} \mid \mathbf{b})$ to obtain $(\mathbf{R} \mid \tilde{\mathbf{b}})$, the equivalent upper triangular system $\mathbf{Rx} = \tilde{\mathbf{b}}$ will have an even more special system matrix $\mathbf{R}$, namely, a matrix in **row-echelon form**. A row-echelon form is a special upper triangular matrix that looks like an **upper staircase matrix**. This shall be introduced and explained formally much later in formula (2.4).

A reduced system $\mathbf{Rx} = \tilde{\mathbf{b}}$ with $\mathbf{R}$ in row-echelon form can be solved easily for the last or trailing unknown entry x_n from the last nonzero row of $(\mathbf{R} \mid \tilde{\mathbf{b}})$, then for x_{n-1} from the last but one nonzero row, and so forth back to x_2 and x_1, the first components of $\mathbf{x}$. This backwards unraveling of $\mathbf{x}$ from the system's row-echelon form $(\mathbf{R} \mid \tilde{\mathbf{b}})$ is called **backsubstitution**.

EXAMPLE 1 (Concretely)

Solve $\mathbf{Ax} = \mathbf{b}$ for $\mathbf{A} = \begin{pmatrix} 1 & -1 \\ 2 & -3 \end{pmatrix}$ and $\mathbf{b} = \begin{pmatrix} 3 \\ -4 \end{pmatrix}$ using Gaussian elimination:

The two linear equations row reduce as follows. Here we keep track of the row operations on the right of the equations:

$$\begin{array}{ll|l}
(E_1) & x_1 - x_2 = 3 & \\
(E_2) & 2x_1 - 3x_2 = -4 & -2\,(E_1) \\
\hline
(E_1) & x_1 - x_2 = 3 & \\
(E_2) - 2(E_1) & -x_2 = -10 &
\end{array}$$

Thus $\mathbf{Ax} = \mathbf{b}$ is equivalent to the upper triangular linear system $\mathbf{Rx} = \tilde{\mathbf{b}}$ with $\mathbf{R} = \begin{pmatrix} 1 & -1 \\ 0 & -1 \end{pmatrix}$ and $\tilde{\mathbf{b}} = \begin{pmatrix} 3 \\ -10 \end{pmatrix}$. To find its solution $\mathbf{x}$, we use backsubstitution from the bottom equation on up.

The last equation of the row-reduced system $\mathbf{Rx} = \tilde{\mathbf{b}}$ makes $x_2 = 10$. Plugging this into the first equation gives $x_1 - x_2 = x_1 - 10 = 3$, or $x_1 = 13$.

As a **check** of our arithmetic, we note by plugging in that $\mathbf{x} = \begin{pmatrix} 13 \\ 10 \end{pmatrix}$ satisfies the originally given second equation (E_2) $2x_1 - 3x_2 = 26 - 30 = -4$, as well as the first original equation (E_1) $x_1 - x_2 = 13 - 10 = 3$. ◀

EXAMPLE 2 (More abstractly, as preparation for the algorithm)
Solve a system of linear equations $\mathbf{A}_{mn}\mathbf{x}_n = \mathbf{b}_m$.
For $m = n = 1$, the single equation $a_{11}x_1 = b_1$ implies $x_1 = b_1/a_{11}$ if $\boxed{a_{11}} \neq 0$.
For $m = n = 2$, we modify the two given equations in Eq. (2.2) as follows:

	System of equations	Operations	
(E_1)	$\boxed{a_{11}}\, x_1 + a_{12}x_2 = b_1$		
(E_2)	$\boxed{a_{21}}\, x_1 + a_{22}x_2 = b_2$	$-\dfrac{\boxed{a_{21}}}{\boxed{a_{11}}} \cdot (E_1)$ (if $\boxed{a_{11}} \neq 0$)	*(create zero below* $\boxed{a_{11}}$*)*
(E_1)	$a_{11}x_1 + \boxed{a_{12}}\, x_2 = b_1$	$-\dfrac{\boxed{a_{12}}}{\boxed{\tilde{a}_{22}}} \cdot (E_2')$ (if $\boxed{\tilde{a}_{22}} \neq 0$)	*(create zero above* $\boxed{\tilde{a}_{22}}$*)*
(E_2')	$0 \cdot x_1 + \boxed{\tilde{a}_{22}}\, x_2 = \tilde{b}_2$		*row echelon form*
(E_1')	$\boxed{a_{11}}\, x_1 + 0 \cdot x_2 = \tilde{b}_1$	$\div a_{11}$ ($\boxed{a_{11}} \neq 0$)	
(E_2')	$0 \cdot x_1 + \boxed{\tilde{a}_{22}}\, x_2 = \tilde{b}_2$	$\div \tilde{a}_{22}$ ($\boxed{\tilde{a}_{22}} \neq 0$)	
(E_1'')	$\boxed{1}\, x_1 + 0 \cdot x_2 = \tilde{\tilde{b}}_1$		*reduced-row echelon*
(E_2'')	$0 \cdot x_1 + \boxed{1}\, x_2 = \tilde{\tilde{b}}_2$		*form for* **A**

Clearly, the original linear system $\mathbf{Ax} = \mathbf{b}$ given by the two equations (E_1) and (E_2) is **equivalent** to the system expressed in the equations (E_1) and (E_2'), and furthermore to the pair of equations (E_1') and (E_2'), and to (E_1'') and (E_2''). That is, the original linear system $\mathbf{Ax} = \mathbf{b}$ has the same solution as its equivalent row echelon form system $\begin{pmatrix} a_{11} & a_{12} \\ 0 & \tilde{a}_{22} \end{pmatrix} \mathbf{x} = \begin{pmatrix} b_1 \\ \tilde{b}_2 \end{pmatrix}$ and as its reduced row echelon form system $\begin{pmatrix} 1 & 0 \\ 0 & 1 \end{pmatrix} \mathbf{x} = \begin{pmatrix} \tilde{\tilde{b}}_1 \\ \tilde{\tilde{b}}_2 \end{pmatrix}$.

If $a_{11} \neq 0$ and $\tilde{a}_{22} = a_{22} - \dfrac{a_{21}}{a_{11}} \cdot a_{12} \neq 0$ in (E_1) and (E_2'), then $x_2 = \tilde{b}_2/\tilde{a}_{22}$ and $x_1 = \tilde{b}_1/a_{11}$ are the two components of the solution $\mathbf{x}$. Thus, the solution $\mathbf{x}$ can be found from the row echelon form of $\mathbf{A}$, i.e., from the two equations (E_1) and (E_2').

Finally, the reduced row echelon form (E_1'') and (E_2'') of $(\mathbf{A} \mid \mathbf{b})$ lets us read off the solution $\mathbf{x}$ most quickly as $x_2 = \tilde{\tilde{b}}_2$ and $x_1 = \tilde{\tilde{b}}_1$.

In Example 1, we have reduced the given augmented system matrix $\left(\begin{array}{cc|c} 1 & -1 & 3 \\ 2 & 3 & -4 \end{array}\right)$ to the equivalent row-echelon form $\left(\begin{array}{cc|c} 1 & -1 & 3 \\ 0 & -1 & -10 \end{array}\right)$. If we subtract the second row of the row-echelon form system from the first, we obtain what was called the intermediate (E_1'), (E_2') pair of equations, namely $\left(\begin{array}{cc|c} 1 & 0 & 13 \\ 0 & -1 & -10 \end{array}\right)$. Multiplying the last equation by -1 now achieves the reduced row-echelon form $\left(\begin{array}{cc|c} 1 & 0 & 13 \\ 0 & 1 & 10 \end{array}\right)$ for the system. For this reduced system, we see that $x_1 = 13$ and $x_2 = 10$ by a most trivial use of backsubstitution.

Note that we always want to solve row reduced systems of equations methodically from the bottom up by backsubstitution. ◀

The Gaussian process is very formal, mechanical, and computer-like. Historically, solutions of simple low-order systems of linear equations can be traced back to the ancients, to Egyptians and Chinese more than 2,000 years ago. A short introduction to the history is available on the Mathematics History Web site mentioned in the Introduction. The process is full of worries whether any boxed $\boxed{}$ entry is zero or not, but rightly so, since we must not divide by zero.

Note that we have used the mnemonic device of a square or rectangular **box** $\boxed{}$ for designating a nonzero **pivot** in the elimination process of Example 2, and an **oval** or **round box** ▢ for framing the element below the pivot that is to be zeroed out, reminding us of the zero that will appear in its position when the elimination is complete. With this notation, we eliminate an entry (a_{ij}) of **A** by using the nonzero pivot $\boxed{a_{kj}}$ in row k and column j and the elimination coefficient

$$-\frac{(a_{ij})}{\boxed{a_{kj}}}.$$

In other words, provided that $\boxed{a_{kj}} \neq 0$, we replace equation (E_i) by

$$\boxed{(E_i) \quad - \quad \frac{(a_{ij})}{\boxed{a_{kj}}} \times (E_k)}$$

to create a zero in place of (a_{ij}).

We will adhere to the pivot $\boxed{}$ and to the to be zeroed out ▢ notation for the rest of the book. However, we will apply the elimination procedure directly to augmented matrices $(\mathbf{A} \mid \mathbf{b})$ instead of to systems of linear equations $\mathbf{Ax} = \mathbf{b}$.

Solving a linear system of equations $\mathbf{Ax} = \mathbf{b}$ becomes much more convenient if we drop the plus, minus, and equal signs as well as the unknown components x_i, and use the three Gaussian row operations directly on the rows of the augmented matrix $(\mathbf{A} \mid \mathbf{b})$.

EXAMPLE 3 To solve

$$\begin{aligned} x_1 + x_2 + 3x_3 &= 4, \\ 2x_1 - x_2 \quad\quad &= 0, \\ x_1 - x_2 + 2x_3 &= -7, \end{aligned} \tag{2.3}$$

we form the equivalent augmented matrix $(\mathbf{A} \mid \mathbf{b})$ and row reduce as follows:

A			**b**	row operations	
$\boxed{1}$	1	3	4		
$\boxed{2}$	-1	0	0	$-\,2 \cdot row_1$	*augmented matrix* $(\mathbf{A} \mid \mathbf{b})$
$\boxed{1}$	-1	2	-7	$-\,row_1$	
$\boxed{1}$	1	3	4		
0	$\boxed{-3}$	-6	-8		
0	$\boxed{-2}$	-1	-11	$-\,\boxed{2}/\boxed{3} \cdot row_2$	
$\boxed{1}$	1	3	4		
0	$\boxed{-3}$	-6	-8		*row-echelon form* $(\mathbf{R} \mid \tilde{\mathbf{b}})$
0	0	$\boxed{3}$	$-17/3$		
x_1	x_2	x_3	$\tilde{\mathbf{b}}$		

Now solve for $\mathbf{x}$ by backsubstitution. The last row of $(\mathbf{R} \mid \tilde{\mathbf{b}})$ makes $x_3 = -17/9$, while the previous row gives $-3x_2 = -8 + 6x_3 = -8 - 6 \cdot \frac{17}{9} = -8 - \frac{34}{3} = -\frac{58}{3}$, or $x_2 = 58/9$. And $x_1 = 29/9$ is found from the first equation $x_1 = 4 - x_2 - 3x_3$ by plugging in the computed values for x_3 and x_2.

It is important to **check** that $\mathbf{Ax} = \begin{pmatrix} 1 & 1 & 3 \\ 2 & -1 & 0 \\ 1 & -1 & 2 \end{pmatrix} \begin{pmatrix} 29/9 \\ 58/9 \\ -17/9 \end{pmatrix}$ is actually equal to $\mathbf{b} = \begin{pmatrix} 4 \\ 0 \\ -7 \end{pmatrix}$ by evaluating the three dot products of the rows of $\mathbf{A}$ with $\mathbf{x}$ that are involved in forming $\mathbf{Ax}$.

Once we have computed a solution vector $\mathbf{x}$ *of a linear system* $\mathbf{Ax} = \mathbf{b}$, *it is highly advisable to check that the matrix* $\times$ *vector product* $\mathbf{Ax}$ *is indeed equal to the original right-hand side* $\mathbf{b}$. ◀

Often, but not always it is possible to continue the Gaussian elimination process until the standard unit vectors $\mathbf{e}_i$ appear in order on the left side of the reduced augmented matrix that is row equivalent to $(\mathbf{A} \mid \mathbf{b})$. We will see in Chapter 6 that this can be achieved precisely for those system matrices $\mathbf{A}$ which are invertible.

EXAMPLE 4 We continue where Example 3 has left off and attempt to find the reduced row echelon form for $\mathbf{R}$ while simultaneously updating the modified right-hand side $\tilde{\mathbf{b}}$.

$\mathbf{R}$			$\tilde{\mathbf{b}}$	row operations	
$\boxed{1}$	1	$\boxed{3}$	4	$-\ row_3$	
0	$\boxed{-3}$	$\boxed{-6}$	−8	$+\ 2 \cdot row_3$	*row echelon form* $(\mathbf{R} \mid \tilde{\mathbf{b}})$
0	0	$\boxed{3}$	−17/3		
$\boxed{1}$	$\boxed{1}$	0	29/3	$+\ \boxed{1}/\boxed{3} \cdot row_2$	
0	$\boxed{-3}$	0	−58/3		
0	0	$\boxed{3}$	−17/3		
$\boxed{1}$	0	0	29/9		
0	$\boxed{-3}$	0	−58/3	$\div(-3)$	
0	0	$\boxed{3}$	−17/3	$\div 3$	
$\boxed{1}$	0	0	29/9		
0	$\boxed{1}$	0	58/9		*reduced row echelon form* $(\mathbf{I}_3 \mid \mathbf{x})$
0	0	$\boxed{1}$	−17/9		
x_1	x_2	x_3	$\tilde{\tilde{\mathbf{b}}} = \mathbf{x}$		

In the reduced row echelon form $(\mathbf{I}_3 \mid \mathbf{x})$ of $(\mathbf{A} \mid \mathbf{b})$, each column of the updated right-hand side corresponds directly to one component x_i of the solution $\mathbf{x}$. Thus, $\mathbf{x} = \tilde{\tilde{\mathbf{b}}} = \frac{1}{9}\begin{pmatrix} 29 \\ 58 \\ -17 \end{pmatrix}$ is the solution for the given system (2.3). This agrees with the solution found in Example 3. Here the matrix $\mathbf{I}_3 = \begin{pmatrix} | & | & | \\ \mathbf{e}_1 & \mathbf{e}_2 & \mathbf{e}_3 \\ | & | & | \end{pmatrix}$ appears on the left side of the reduced row echelon form and contains the ordered unit vectors $\mathbf{e}_i$ of $\mathbb{R}^3$ in its columns. $\mathbf{I}_3$ is diagonal with diagonal entries equal to one. This matrix is called the **identity matrix** of size 3, since $\mathbf{I}\mathbf{x} = \mathbf{x}$ for all $\mathbf{x} \in \mathbb{R}^3$ (i.e., since the linear transformation $\mathbf{x} \in \mathbb{R}^n \mapsto \mathbf{I}_n\mathbf{x} = \mathbf{x} \in \mathbb{R}^n$ does not alter $\mathbf{x}$ for the n-by-n identity matrix $\mathbf{I}_n$). ◀

(b) The Row Reduction Algorithm

We shall try to note the flow of our row reductions consciously:

(a) We designate a nonzero entry in the $(1, 1)$ position of $\mathbf{A}$ as a **pivot** by putting a box $\square$ around it, if such an entry exists for $\mathbf{A}$. If the (1.1) entry is zero, we go to step (f).

(b) This pivot element $\square$ is then used to zero out all nonzero entries ▢ below the pivot entry by row reduction.

(c) Next we disregard the row-reduced first column and the first row and look at the smaller sized rest of the matrix. We designate the leading nonzero entry, if one exists in its shortened first column (i.e., in the second column of **A**), as a pivot by boxing $\boxed{}$ it. If there is no nonzero entry, we go to step (f).

(d) The second pivot is used to zero out all entries below it, again using Gaussian elimination.

(e) We continue analogously with all further rows and columns of A.

(f) If in part (a) or part (c) the leading entry a_{11} or $\tilde{a}_{ij}$ is zero, we try to swap rows in **A** or in the updated matrix $\tilde{\mathbf{A}}$, so that $\tilde{a}_{11} \neq 0$ or $\tilde{\tilde{a}}_{ij} \neq 0$. If this is not possible since all entries of a subcolumn are zero, then we move one column to the right and continue analogously with part (a) or part (c).

In this process, we reach a **row echelon form R** of **A** when we have zeroed out the lower triangle in **A** in step (e).

To shorten the notation, we often abbreviate the term row-echelon form by its acronym **REF**, which is pronounced as the ref of a basketball game. If desired, we can continue from the row echelon form and create zeros above each pivot $\boxed{}$ as well, this time from the last pivot on upwards and back, as in Example 4. If we normalize all pivots to become one by row scaling, then we obtain the **reduced row echelon form** (**RREF**) of **A**. Note that the pivot columns of an RREF $\mathbf{R}_{mn}$ contain different standard unit column vectors $\mathbf{e}_i \in \mathbb{R}^m$. For example, the matrix $\begin{pmatrix} 1 & 0 & 1 & 0 & 0 & 2 \\ 0 & 1 & 2 & 3 & 0 & 0 \\ 0 & 0 & 0 & 0 & 1 & -7 \\ 0 & 0 & 0 & 0 & 0 & 0 \end{pmatrix}$ is in reduced row-echelon form.

Here is a flow chart for row reducing an m-by-n matrix **A**:

$k = 0$ (row index)

For $j = 1, 2, \ldots, n$ do: (column index)

1. Check whether there is a nonzero entry in column j amongst the entries $a_{k+1,j}, a_{k+2,j}, \ldots, a_{m,j}$ below row k.

   If not: Work on the next column (i.e., up j to $j+1$ inside the column loop and keep the same row index k).

   Else: Use row swapping (if needed) to get a nonzero entry into position $(k+1, j)$, so that $\boxed{\tilde{a}_{k+1,j}} \neq 0$ can serve as a pivot.

   (Remember, the rows in $(\mathbf{A} \mid \mathbf{b})$ represent linear equations. Their ordering is irrelevant to the solution **x**.)

   Go to step 2.

2. Eliminate $\tilde{a}_{k+2,j}, \ldots, \tilde{a}_{m,j}$ (i.e., create zero entries in their stead in column j by subtracting $\dfrac{\boxed{\tilde{a}_{\ell,j}}}{\boxed{\tilde{a}_{k+1,j}}} \times row_{(k+1)}$ from row_ℓ for $\ell = k+2, \ldots, m$). Upon elimination,

$$
\mathbf{A} = \begin{pmatrix}
\boxed{*} & & & & * & \\
0 & \boxed{*} & & & & \\
\vdots & 0 & \ddots & & & * \\
 & \vdots & \ddots & \boxed{*} & & \\
 & & & 0 & & \\
 & & & \vdots & & * \\
0 & 0 & \dots & 0 & &
\end{pmatrix} \leftarrow k+1
$$

$$
\uparrow \; j
$$

Set $k = k+1$.

End (column loop)

The end result of our algorithm is a row-echelon form **R** of **A**. On termination, the integer k is equal to the number of pivots that have been found during the row reduction of **A**. If we want to find the reduced row-echelon form of **A**, we need to sweep back from the last pivot in **R** back up and zero out the entries of each pivot column above the pivot. Finally, normalizing the pivots to be one by row scaling produces the reduced row-echelon form.

(Note : Columns that do not contain a pivot must be updated as required by the row reduction process. Their entries, however, generally cannot be zeroed out.)

The output of the preceding algorithm gives a **row echelon form** of **A** with the following simple properties:

Each row of a row echelon form (possibly) contains a number of leading zeros, possibly followed by a *first* nonzero entry $\boxed{}$ (the pivot) and arbitrary entries beyond that.

Zero rows can appear only as the bottom rows of a row echelon form.

For example, the matrices $\begin{pmatrix} \boxed{2} & 1 & -1 & 0 \\ 0 & 0 & \boxed{1} & 2 \end{pmatrix}$, $\begin{pmatrix} 0 & 0 & \boxed{-1} & 0 \\ 0 & 0 & 0 & \boxed{4} \end{pmatrix}$, and $\begin{pmatrix} 0 & \boxed{2} & 0 & 2 \\ 0 & 0 & \boxed{1} & 3 \\ 0 & 0 & 0 & 0 \end{pmatrix}$ are all in row-echelon form, but $\begin{pmatrix} 1 & -1 \\ 2 & 0 \end{pmatrix}$ is not.

The more stringent requirements for an upper triangular matrix to be in row echelon form are presented next.

(c) The Row-Echelon Form (REF) of a Matrix

A row echelon form can contain both zero rows and zero columns. While zero columns may appear in any position of a REF, zero rows may only appear at the bottom (i.e., below any pivot rows).

A matrix is in **row echelon form** if it satisfies the following three properties:

Zero row check: Any row consisting entirely of zeros is at the bottom of a REF below any row with a pivot.

Row check: Any lower row of a REF has its first nonzero entry (pivot) to the right of any pivot of a previous row.

Column check : If a column contains a nonzero entry in a lower position than any preceding column, then the last nonzero entry of this column (pivot) occurs precisely one position below the lowest nonzero position of all previous columns.

Any two of the preceding three checks suffice to establish that a matrix is in REF.

EXAMPLE 5

(a) $\begin{pmatrix} 0 & 1 & 0 & 1 & 2 \\ 1 & 0 & 0 & 1 & 1 \\ 0 & 0 & 1 & 0 & 1 \end{pmatrix}$ is not in row echelon form, since the leading nonzero entry in row 2 occurs before the pivot in row 1. It violates both the row and column checks. Swapping rows 1 and 2 will produce a REF.

(b) $\begin{pmatrix} \boxed{1} & 2 & 3 \\ 0 & 0 & 0 \\ 0 & 0 & 3 \end{pmatrix}$ is not in row-echelon form. Zero rows in REFs must always come last. Swap rows 2 and 3 to obtain a row-echelon form.

(c) $\begin{pmatrix} 0 & \boxed{1} & 2 & 3 \\ 0 & 0 & 0 & \boxed{-2} \\ 0 & 0 & 0 & 30 \end{pmatrix}$ is not in row-echelon form. The (3,4) entry 30 of the last row has to be eliminated by zeroing out below the second row pivot $\boxed{-2}$.

(d) The zero matrices $\mathbf{O}_{mn}$ and the identity matrices $\mathbf{I}_n$ are always in row-echelon form, having zero or n pivots, respectively. ◀

Equivalently, a matrix **R** is in row-echelon form if its nonzero entries have the form of an **upper staircase matrix**, where all steps have height equal to one, except possibly for a larger bottom step and a block of zero rows. The steps of a row-echelon form have variable-width landings between pivot columns. A typical upper staircase matrix looks like

$$\mathbf{R} = \begin{pmatrix} \boxed{*} & * & * & & & & * & & & * \\ 0 & 0 & \boxed{*} & & & & & & & \\ & & 0 & \boxed{*} & & & & * & & \\ \vdots & & & 0 & \boxed{*} & * & * & & & \\ \vdots & & & & 0 & 0 & 0 & \boxed{*} & * & * \\ & & & & & & & 0 & 0 & \boxed{*} \\ 0 & & & & & \cdots & & & & 0 \\ 0 & & & & & \cdots & & & & 0 \\ 0 & & & & & \cdots & & & & 0 \end{pmatrix}. \tag{2.4}$$

The zig–zag staircase profile line, drawn out for **R**, separates the zero lower left entries in **R** from the possibly nonzero entries $*$ in its upper right sector. The nonzero pivot entries marked by a box $\boxed{*}$ lie immediately to the right of each step of height one. The landing widths of **R** in Eq. (2.4) are 2, 1, 1, 3, 2, and 1 from the top row on down. All pivot steps have height one, while the bottom zero row step has height 3. Note that **R** in Eq. (2.4) satisfies the column check, since no subsequent column of **R** has a nonzero entry protruding further than one position below the lowest previous nonzero position. In Section 3.1, we shall see that the lengths of the pivot row landings of **R** determine whether a system of linear equations is uniquely solvable.

EXAMPLE 6 The staircase lines that separate the lower left zero entries from the upper right possibly nonzero entries of the matrices in Example 5 are as follows:

$$\begin{pmatrix} 0 & 1 & 0 & 1 & 2 \\ 1 & 0 & 0 & 1 & 1 \\ 0 & 0 & 1 & 0 & 1 \end{pmatrix}, \quad \begin{pmatrix} 1 & 2 & 3 \\ 0 & 0 & 0 \\ 0 & 0 & 3 \end{pmatrix}, \quad \begin{pmatrix} 0 & 1 & 2 & 3 \\ 0 & 0 & 0 & -2 \\ 0 & 0 & 0 & 30 \end{pmatrix}.$$

It is clear that none of these three matrices is in proper upper staircase form.

The matrix

$$\mathbf{R} = \begin{pmatrix} \boxed{2} & 0 & 1 & 2 & 3 & 4 \\ 0 & 0 & \boxed{3} & 1 & 1 & -1 \\ 0 & 0 & 0 & \boxed{4} & 0 & 7 \\ 0 & 0 & 0 & 0 & 0 & 0 \\ 0 & 0 & 0 & 0 & 0 & 0 \\ 0 & 0 & 0 & 0 & 0 & 0 \end{pmatrix},$$

however, is in upper staircase or row echelon form. ◀

In practice, we will have to reduce matrices to their REF repeatedly, economically, and correctly to be able to perform the tasks of linear algebra.

We strongly urge the students to learn the mechanics of this row reduction process well and become conscious of the **proper way and order of doing this**. For most of our purposes, obtaining a simple row echelon form will suffice. Throughout the hand computations with integer matrices, we urge students to try to compute REFs over the integers. This can be done by creating pivot rows via row operations that always have **unit pivots** equal to ± 1. This avoids having to deal with fractions and keeps the arithmetic challenges to a minimum. Numerically, this is not always wise (see Lecture 13), but for a first course and a basic understanding of the process, we think that this is a good practice.

A **caution** *about computational shortcuts:*

While the row echelon form reduction process using pencil and paper certainly is dull, repetitive, and prone to miscalculations, we believe that it cannot be replaced by learning a few calculator or keyboard strokes. In fact, we strongly advise against immediately using calculators or computers when learning the subject for the first time. Students should begin by computing low dimensional examples by hand.

Every course on linear algebra wants to foster the students' understanding of the subject. Any true understanding can only take place if we stand underneath and view the makings of our object of interest in detail. Every task outlined in Section 1.2 can be performed by reading up on a few software or calculator instructions and by punching in numbers and the appropriate commands. A day's work is all it takes to solve all computational problems in this book with the help of suitable software. Trying to understand what is going on takes a semester or more.

With this in mind, we urge our students (and instructors) to work through low-dimensional computations by hand first. This ensures an appropriate speed and leisure for comprehension. Then one can lose oneself in a problem, take wrong turns and—hopefully—find a correct path through the problem at hand and through the maze of matrices.

Hands–on time is important. No pain, no gain rules in the gym, as it does in the classroom.

(d) Properties of the Row Echelon and Reduced Row Echelon Forms

Before applying the row reduction process in subsequent chapters, we emphasize that both the REF and the RREF of a matrix $\mathbf{A}_{mn}$ with n columns has columns of two distinct types:

- the columns that contain a pivot $\square$, the **pivot columns**, and
- the columns that contain no pivot entry, called the **free columns**.

In Example 3, each column of $\mathbf{R}$ is a pivot column, while $\mathbf{R}$ in Eq. (2.4) has pivot columns in positions 1, 3, 4, 5, 8, and 10 and the columns 2, 6, 7, and 9 are free.

A row echelon form of a real or complex matrix $\mathbf{A}$ can be achieved in many ways. Different algorithms may use a different sequence of the three legitimate operations of Gaussian elimination than the one we choose. Hence, one should be prepared to accept different REFs as correct for the same matrix $\mathbf{A}$, or for the same augmented matrix $(\mathbf{A} \mid \mathbf{b})$ corresponding to a linear system $\mathbf{Ax} = \mathbf{b}$. In fact, if $\mathbf{R}$ is a REF with at least two pivot rows, we may add a nonzero multiple of the second row of $\mathbf{R}$ to its first row. This creates a different, but equivalent, REF for $\mathbf{A}$. Note, however, that the pivots appear in precisely the same columns in any REF $\mathbf{R}$ of $\mathbf{A}$. Thus, we conclude that

> The REF $\mathbf{R}$ of a real or complex matrix $\mathbf{A}_{mn}$ is generally not unique. However, the positioning of the pivot columns in a REF of $\mathbf{A}_{mn}$ is unique.

This is not disturbing, since any correctly computed REF $(\mathbf{R} \mid \tilde{\mathbf{b}})$ of $(\mathbf{A} \mid \mathbf{b})$ will be equivalent (in the sense of Gaussian row reduction equivalence) to any other. With regards to solving the related linear system $\mathbf{Ax} = \mathbf{b}$, the solutions to $\mathbf{Rx} = \tilde{\mathbf{b}}$ and $\mathbf{Ax} = \mathbf{b}$ must and will coincide for all correctly computed REFs $(\mathbf{R} \mid \tilde{\mathbf{b}})$ of $(\mathbf{A} \mid \mathbf{b})$.

What is unique about a REF **R** of a given matrix **A** is the number of pivots in **R** and their column location.

> The number of pivots in a REF **R** of a given real or complex matrix **A** is unique.
> This integer is called the **rank** of **A**.

The rank of a matrix **A** has the integer value $k \geq 0$ at the end of our earlier flow chart procedure. Clearly, k from the charted algorithm satisfies $k \leq \min\{m,\ n\}$ for any matrix $\mathbf{A}_{mn}$.

In Chapter 5, we shall learn of a dual definition for the rank of a matrix **A** in terms of its columns.

The reduced row echelon form, or **RREF** for short, in which all entries above each pivot are zeroed out as well and where all pivots are scaled to be one, finally is unique for each given real or complex matrix $\mathbf{A}_{mn}$.

> The RREF of a matrix **A** is unique.

Recall that in the RREF **R** of **A**, each pivot column is a multiple of some standard unit vector $\mathbf{e}_i$, while free columns may contain nonzero entries. These possibly nonzero free column entries of the RREF **R** are uniquely determined from **A**, once each pivot column has been row reduced to a unit vector. A proof of this is developed in Problems 5 and 6 of Section 5.1.P by using the ideas of Example 2 of Section 5.1.

EXAMPLE 7 Find all distinct REF and RREF patterns that are possible for 3×2 real matrices. Clearly, there are infinitely many REFs for such matrices; just rescale any nonzero row.

We order the possible zero and nonzero entry patterns of 3×2 REFs by their rank:

If rank $(\mathbf{A}) = 0$, then $\mathbf{A}_{3\times 2} = \mathbf{O}_{3\times 2}$, the zero matrix, which is in row-echelon and reduced row-echelon form.

If rank $(\mathbf{A}) = 1$, then any REF **R** of **A** has only one pivot, and hence only one nonzero row which occurs as the first row in **R**. The pivot entry can occur in the first, second, or third column. Thus, the possible patterns of such REFs of rank 1 are

$$\begin{pmatrix} \boxed{*} & * & * \\ 0 & 0 & 0 \end{pmatrix}, \quad \begin{pmatrix} 0 & \boxed{*} & * \\ 0 & 0 & 0 \end{pmatrix}, \quad \text{and} \quad \begin{pmatrix} 0 & 0 & \boxed{*} \\ 0 & 0 & 0 \end{pmatrix}.$$

Here $\boxed{*}$ denotes a real nonzero pivot entry, while $*$ signifies an arbitrary real number, including zero.

If rank $(\mathbf{A}) = 2$, then each of the two rows of a REF **R** of **A** contains a pivot and the following patterns may occur:

$$\begin{pmatrix} \boxed{*} & * & * \\ 0 & \boxed{*} & * \end{pmatrix}, \quad \begin{pmatrix} \boxed{*} & * & * \\ 0 & 0 & \boxed{*} \end{pmatrix}, \quad \text{and} \quad \begin{pmatrix} 0 & \boxed{*} & * \\ 0 & 0 & \boxed{*} \end{pmatrix}.$$

All together, each 3×2 matrix **A** has a REF with one of these seven patterns.

The possible rank 1 RREFs of **A** are

$$\begin{pmatrix} \boxed{1} & * & * \\ 0 & 0 & 0 \end{pmatrix}, \begin{pmatrix} 0 & \boxed{1} & * \\ 0 & 0 & 0 \end{pmatrix}, \text{ and } \begin{pmatrix} 0 & 0 & \boxed{1} \\ 0 & 0 & 0 \end{pmatrix}.$$

And all rank-2 RREFs of 3×2 matrices have one of the shapes

$$\begin{pmatrix} \boxed{1} & 0 & * \\ 0 & \boxed{1} & * \end{pmatrix}, \begin{pmatrix} \boxed{1} & * & 0 \\ 0 & 0 & \boxed{1} \end{pmatrix}, \text{ and } \begin{pmatrix} 0 & \boxed{1} & 0 \\ 0 & 0 & \boxed{1} \end{pmatrix},$$

since RREFs can contain only unit vectors $\mathbf{e}_i$ in their pivot columns. ◀

Note that the problem of computing the unique RREF of a real or complex matrix **A** correctly hinges on our ability to compute row reductions in absolute **precision**, without **rounding errors** or truncation errors, and, of course, without computational errors.

Once we compute approximately, such as with the help of a calculator or a computer, then these unavoidable errors will give (hopefully small) nonzero entries where we should expect zeros to appear via elimination in perfect arithmetic. This will make the rank of a real matrix **A**, defined as the number of nonzero pivots of any REF of **A**, very dubious to compute. This is particularly so, if **A** becomes sizable and has noninteger entries and we can no longer work with correct integer or fractional arithmetic. Hence, the following caveat applies:

> The rank of a matrix $\mathbf{A}_{mn}$ can in general not be reliably computed via row reduction, unless **A** is an integer matrix of small size.

In Chapter 12, we study the singular value decomposition of matrices $\mathbf{A}_{mn}$, which allows us to make much more reliable predictions of the numerical rank of any matrix **A** than can be made from row reduction.

Coming back to row reduction, for a given matrix **A** we say that the nonzero rows of a REF **R** of **A** are **row equivalent** to the original rows of **A**. One aim of Gaussian elimination and row reduction is to replace the row vectors of a matrix with a more revealing set of equivalent vectors that line up in echelon form and that carry the same information as **A**'s rows do with regards to linear equations.

We have introduced the backbone or workhorse of any modern elementary linear algebra course: the row reduction of matrices. There are many equivalent REFs for a real or complex matrix $\mathbf{A}_{mn}$. The RREF is, however, unique for each matrix.

2.1.P Problems

1. Is $\mathbf{A} = \begin{pmatrix} 0 & 7 & 3 & -3 & 5 & 3 \\ 0 & 0 & 1 & 4 & 0 & 0 \\ 0 & 0 & 0 & 3 & 1 & 10 \\ 1 & 0 & 0 & 0 & 4 & -2 \\ 0 & 0 & 0 & 1 & 0 & 2 \end{pmatrix}$ in row-echelon form? If not, put it into a REF.

2. Which of the following matrices are in row-echelon form, which are in reduced row-echelon form? (Draw the staircase diagrams.)

(a) $\begin{pmatrix} 1 & 2 & 3 & 0 & 7 & 8 \\ 0 & 1 & 0 & 0 & -9 & -5 \\ 0 & 0 & 3 & 6 & 9 & 0 \end{pmatrix}$.

(b) $\begin{pmatrix} 1 & 4 & 5 & 0 & 2 \\ 1 & 0 & 3 & 5 & -2 \\ 0 & 0 & 0 & 7 & 0 \\ 0 & 0 & 0 & 0 & 1 \end{pmatrix}$.

(c) $\begin{pmatrix} 0 & 0 & 3 \\ 0 & 0 & 1 \\ 0 & 0 & 0 \end{pmatrix}$.

(d) $\begin{pmatrix} 1 & 0 & 2 \\ 0 & 0 & 0 \\ 0 & 1 & 0 \end{pmatrix}$.

(e) $\begin{pmatrix} 0 & 0 & 1 & -3 \\ 1 & 0 & 0 & 0 \\ 0 & 2 & -1 & 7 \end{pmatrix}$.

(f) $\begin{pmatrix} 9 & 0 & 8 & -4 & -1 \\ 0 & 3 & 0 & 1 & -29 \\ 0 & 0 & 1 & 0 & 1 \end{pmatrix}$.

(g) Reduce two of the preceding matrices that are not in RREF to their RREFs.

3. Reduce $\mathbf{B} = \begin{pmatrix} 1 & 0 & 0 & 0 & 4 & -2 \\ 0 & 7 & 3 & -3 & 5 & 3 \\ 0 & 0 & 1 & 4 & 0 & 0 \\ 0 & 0 & 0 & 3 & 1 & 10 \\ 0 & 0 & 0 & 1 & 0 & 2 \end{pmatrix}$ to its RREF.

4. Reduce $\mathbf{B} = \begin{pmatrix} 1 & 1 & 11 & -1 & -1 \\ 0 & 4 & 2 & 0 & -7 \\ 0 & 0 & 0 & 2 & 1 \end{pmatrix}$ to its RREF.

5. Find a REF for

$$\mathbf{A} = \begin{pmatrix} -1 & -3 & 0 & 1 & 1 & -1 \\ 2 & 5 & -2 & 1 & 0 & 0 \\ 1 & 2 & 3 & 0 & -3 & 4 \end{pmatrix}.$$

Then find a different one from $\mathbf{A}$, if you can.

6. (a) How many pivots are there in the REF and the RREF of the **identity matrix**

$$\mathbf{I}_n = \begin{pmatrix} 1 & & 0 \\ & \ddots & \\ 0 & & 1 \end{pmatrix}_{nn} ?$$

(b) What is the rank of the identity matrix $\mathbf{I}_n$?

(c) Answer the same questions for the **zero matrix**

$$\mathbf{O}_{mn} = \begin{pmatrix} 0 & \dots & 0 \\ \vdots & & \vdots \\ 0 & \dots & 0 \end{pmatrix}_{mn}.$$

7. Find two 3×4 matrices $\mathbf{A}$ and $\mathbf{B}$, each of rank 2, so that , if possible,

(a) $\text{rank}(\mathbf{A} + \mathbf{B}) = 2$,

(b) $\text{rank}(\mathbf{A} + \mathbf{B}) = 0$,

(c) $\text{rank}(\mathbf{A} - \mathbf{B}) = 5$,

(d) $\text{rank}(\mathbf{A} + \mathbf{B}) = 1$,

(e) $\text{rank}(\mathbf{A} + \mathbf{B}) = 4$.

Explain why some of the foregoing requirements cannot be met by any 3×4 matrices $\mathbf{A}$ and $\mathbf{B}$.

8. Row reduce the matrices

$\mathbf{A} = \begin{pmatrix} 0 & 1 & 3 \\ 0 & 2 & -4 \end{pmatrix}$, $\mathbf{B} = \begin{pmatrix} 0 & 1 & 3 \\ 1 & 2 & 6 \end{pmatrix}$,

$\mathbf{C} = \begin{pmatrix} 0 & 0 & 0 & 0 \\ 0 & 0 & 0 & 0 \end{pmatrix}$, $\mathbf{D} = \begin{pmatrix} 0 \\ 0 \\ 1 \\ 14 \end{pmatrix}$,

$\mathbf{E} = \begin{pmatrix} 0 & 0 & 0 \\ 0 & 1 & -1 \\ 0 & 1 & 0 \\ 20 & 3 & 17 \end{pmatrix}$, and $\mathbf{F} = \begin{pmatrix} 0 \\ 4 \\ 0 \end{pmatrix}^T$ to their RREFs.

9. Row reduce the matrices to REF, and find their rank:

$$\begin{pmatrix} 1 & -1 & 2 \\ 0 & 1 & 4 \end{pmatrix}, \begin{pmatrix} 1 & 0 & -1 & 2 & -1 \\ 0 & 1 & 1 & -1 & 0 \\ 1 & 0 & 0 & 1 & 1 \end{pmatrix}.$$

10. Let $\mathbf{A}_{mn} = \begin{pmatrix} a & \dots & a \\ \vdots & & \vdots \\ a & \dots & a \end{pmatrix}$ for $a \in \mathbb{R}$.

(a) Find a REF of $\mathbf{A}$, depending on the value of $a \in \mathbb{R}$.

(b) Do the same for the RREF of $\mathbf{A}$.

(c) What is the rank of $\mathbf{A}$, depending on the value of $a \in \mathbb{R}$.

11. You have computed REF for an augmented matrix $(\mathbf{A} \mid \mathbf{b})$ as $\left(\begin{array}{ccc|c} 1 & 1 & 2 & 3 \\ 0 & 2 & -2 & 2 \\ 0 & 0 & 0 & 4 \end{array}\right)$, while your friend has obtained $\left(\begin{array}{ccc|c} 2 & -2 & 8 & 2 \\ 0 & -1 & 1 & -1 \\ 0 & 0 & 0 & -14 \end{array}\right)$ for the same problem.

(a) Explain what is going on.

(b) Write down at least <u>two</u> different explicit systems of linear equations

$$a_{i1}x_1 + \ldots + a_{i3}x_3 = b_i, \; i = 1, 2, 3,$$

with nonzero coefficients a_{ij} for all i, j that could have produced the first-mentioned REF.

12. (a) Row reduce both $\mathbf{A} = \begin{pmatrix} \boxed{1} & 2 & 3 \\ 0 & 2 & 2 \\ -1 & 4 & 1 \end{pmatrix}$ and $\mathbf{B} = \begin{pmatrix} \boxed{-1} & 4 & 1 \\ 0 & 2 & 2 \\ 1 & 2 & 3 \end{pmatrix}$ to a REF, starting with the given pivot in each case.

(b) Are the two computed REFs identical? Are they row equivalent?

(c) What are the RREFs of **A** and **B**? Why are they identical?

13. Find the rank of the matrix $\begin{pmatrix} 1 & 1 & 1 \\ 1 & 1 & 1 \\ 1 & 1 & 0 \end{pmatrix}$, a REF, and its RREF.

14. Fill in the blanks and prove:

(a) If $\mathbf{x}$ is a real vector, its matrix has rank one if and only if ...

(b) If $\mathbf{x}$ is a real vector, its matrix has rank zero if and only if ...

(c) Do the preceding answers depend on whether $\mathbf{x}$ is a row or column vector?

15. Let $n \geq 5$ and $\mathbf{A}_{mn}$ be a matrix of rank $n - 4$. Show that $n - m \leq 4$.

16. Write out all possible RREFs for

(a) 2×4 matrices **A**;

(b) 4×2 matrices **B**.

17. (a) What is the maximal possible rank of a 7×14 matrix?

(b) What is the minimal possible rank of a 7×14 matrix?

(c) What is the minimal possible rank of an $m \times n$ matrix?

(d) How about the maximal possible rank.

18. Find a 3×2 matrix **A** with a unique REF **R**, if possible.

19. (a) If the first row of the reduced REF **R** of a matrix **A** contains the vector $\begin{pmatrix} 0 & 1 & -2 & 0 & 4 & 0 \end{pmatrix}$, determine all ranks that **A** may possibly have. (Write out explicit RREFs **R** with this first row for each rank.)

(b) If the first row of the RREF **R** of a matrix $\mathbf{A}_{mn}$ contains the vector $\begin{pmatrix} 0 & 0 & \dots & 0 & 0 \end{pmatrix}$, what are the entries of **A**?

(c) How many different RREFs can have $\begin{pmatrix} 0 & \dots & 0 & 1 & 0 \end{pmatrix}$ as their first row?

(d) If the first pivot in a REF of **A** appears in position $k \geq 1$ of the first row, what are the entries in the first $k - 1$ columns of **A**?

(e) If the last row of a REF of $\mathbf{A}_{mn}$ is nonzero, what rank does **A** have?

20. Decide which values the entries in the positions marked by an asterisk $*$ may have so that the following matrices are in row echelon form and have the required rank:

(a) $\begin{pmatrix} 1 & 2 & * & * & * \\ 0 & * & * & * & 0 \\ 0 & 0 & 0 & 0 & * \\ 0 & 0 & 0 & 0 & 0 \end{pmatrix}$ with rank equal to 2.

(b) $\begin{pmatrix} 0 & * & * & * & * \\ 0 & * & 1 & * & * \\ 0 & 0 & 0 & * & * \\ 0 & 0 & 0 & 0 & 1 \end{pmatrix}$ with rank equal to 3.

(c) Repeat part (b) for a rank 4 RREF.

21. (a) Find the rank of the matrix

$$\mathbf{A} = \begin{pmatrix} 1 & 2 & 3 & 4 & 0 & -1 \\ -2 & -3 & -9 & -6 & -3 & 1 \\ 3 & 5 & 11 & 8 & 5 & -4 \\ 0 & -2 & 10 & 4 & 0 & 18 \end{pmatrix}.$$

(b) Find the rank of the matrix

$$\mathbf{B} = \begin{pmatrix} 1 & -2 & 3 & 0 \\ 2 & -3 & 5 & -2 \\ 3 & -9 & 11 & 10 \\ 4 & -6 & 8 & 4 \\ 0 & -3 & 5 & 0 \\ -1 & 1 & -4 & 18 \end{pmatrix}.$$

22. Find three different matrices **A**, **B**, and **C** with RREF

$$\mathbf{R} = \begin{pmatrix} 1 & 2 & 0 & 0 & 4 \\ 0 & 0 & 1 & 0 & -2 \\ 0 & 0 & 0 & 1 & 2 \\ 0 & 0 & 0 & 0 & 0 \end{pmatrix}.$$

23. Find the RREF of the matrix

$$\begin{pmatrix} -7 & 0 & 0 & -9 & -3 \\ 1 & -9 & 18 & -12 & 18 \\ -2 & 0 & 0 & -18 & 30 \\ 1 & 9 & -18 & -4 & 20 \end{pmatrix}.$$

24. (a) Assume that the ith column of the REF **R** consists entirely of zero entries. Describe all possible i^{th} columns of matrices **A** that row reduce to **R**.

(b) Assume that the ith column of the REF **R** contains nonzero entries. Describe all possible i^{th} columns of matrices **A** that row reduce to **R**.

25. Show that the row vectors $\mathbf{e}_1$ and $\mathbf{e}_2 \in \mathbb{R}^2$ are row equivalent to $(1, -1)$ and $(2, 1)$.

26. Are the two sets of vectors $\{(3, 0, 2), (1, -1, 4), (0, 3, -10)\} \subset \mathbb{R}^3$ and $\{(2, 1, 0), (1, 2, 0), (0, 2, 1)\}$ row equivalent?

27. Find all possible RREFs for the matrix $\mathbf{A} = \begin{pmatrix} 1 & 2 & -1 & 2 \\ 0 & 1 & 3 & 4 \\ 1 & -1 & a & b \end{pmatrix}$, depending on the values of a and $b \in \mathbb{R}$.

28. Repeat the previous problem for $\mathbf{B} = \begin{pmatrix} 1 & 1 & 1 \\ 2 & 2 & 2 \\ a & 1 & 1 \end{pmatrix}$.

29. Show that the matrices $\mathbf{A} = \begin{pmatrix} 1 & -2 & 3 \\ 2 & -5 & 2 \\ 1 & 2 & 19 \end{pmatrix}$ and $\mathbf{B} = \begin{pmatrix} 0 & 1 & 1 \\ 1 & 0 & -1 \\ 0 & 1 & 0 \end{pmatrix}$ are not row equivalent.

30. Find all RREFs $\mathbf{R} = \begin{pmatrix} \alpha & \beta \\ \gamma & \delta \end{pmatrix} \in \mathbb{R}^{2,2}$ with $\alpha + \beta + \gamma = \delta$.

31. Assume that $\mathbf{R} = \begin{pmatrix} \alpha & \beta \\ \gamma & \delta \end{pmatrix}$ is in row-echelon form. Show that if $\alpha = 0$, then $\alpha + \beta + \gamma \neq \delta$, unless $\mathbf{R} = \mathbf{O}_2$.

There is no need to overexercise row reductions here. Subsequent chapters are built on this algorithm. We will row reduce many a matrix before this course is over.

Teacher's Problem-Making Exercises

Lectures 1, 4, 5, and 13 are conceptual and noncomputational in nature. All other lectures contain exercises 'T $*$' in their problem sets $*$.1.P that show how to generate infinitely many standard problems for the chapter. These standard problems can be solved in integer or mostly integer arithmetic with the methods of the current chapter.

Many of the most useful achievements of linear algebra can be summarized as decoding information that is encoded in matrices and vectors. Among these are the solution of linear systems $\mathbf{Ax} = \mathbf{b}$ and finding eigenvalues and eigenvectors of a matrix, as well as singular values, least squares solutions etc.

In real life one starts with a given matrix problem and tries to decode the desired information using computer software. In school life, teachers must have a multitude of easily solvable problems at their disposal in order to train, drill, and test students in the fundamental skills. For this, knowing how to construct meaningful and easily hand-computable examples is essential.

In the area of matrices, it is easy to give ready recipes for easily decodable problems: In general, start with a nice, possibly integer result and encrypt it backwards into a more difficult, generic looking problem by reversing the decoding process that has just been taught.

T 2. To find matrices $\mathbf{B}_{mn}$ that row reduce nicely—preferably over the integers—and that have integer row echelon and/or integer reduced row echelon forms $\mathbf{R}_{mn}$, one starts with such an integer REF or RREF (with all pivots equal to one for the RREF, of course). Next pick a random $m \times m$ integer matrix $\mathbf{A}_{mm}$ and pre–multiply $\mathbf{A}_{mm}\mathbf{R}_{mn} =: \mathbf{B}_{mn}$ to obtain the matrix $\mathbf{B}$. If $\mathbf{A}$ is an invertible matrix, then $\mathbf{R}$ must be a REF or the RREF of $\mathbf{B}$. If $\mathbf{A}$ is singular, which happens quite rarely for random integer entries in $\mathbf{A}$, then a REF of $\mathbf{B}$ will be a REF for a collection—possibly a subcollection—of the rows of the original row echelon form **R**. Note that in both cases, the desired row reduced form of **B** can be found with integer elimination steps from **B**.

Hence, for an easy row reduction problem, specify **B** (= **AR**) and assign as an exercise to find **B**'s REF or RREF **R**.

EXAMPLE (a) For $\mathbf{R}_{3\times4} = \begin{pmatrix} 0 & \boxed{1} & 2 & 3 \\ 0 & 0 & \boxed{1} & -1 \\ 0 & 0 & 0 & 0 \end{pmatrix}$ and $\mathbf{A}_{3\times3} = \begin{pmatrix} 1 & 2 & -1 \\ 0 & 1 & 2 \\ -1 & 1 & 0 \end{pmatrix}$, we obtain $\mathbf{B}_{3\times4} := \mathbf{AR} = \begin{pmatrix} 0 & 1 & 4 & 1 \\ 0 & 0 & 1 & -1 \\ 0 & -1 & -1 & -4 \end{pmatrix}$. **R** is a REF of **B**, and $\begin{pmatrix} 0 & 1 & 0 & 5 \\ 0 & 0 & 1 & -1 \\ 0 & 0 & 0 & 0 \end{pmatrix}$ is the RREF of **B** and of **R**. Note that **A** is nonsingular.

(b) If $\mathbf{C} = \begin{pmatrix} 1 & 2 & -1 \\ 2 & 4 & -2 \\ 0 & 1 & -1 \end{pmatrix}$ then $\mathbf{D} := \mathbf{CR} = \begin{pmatrix} 0 & 1 & 4 & 1 \\ 0 & 2 & 8 & 2 \\ 0 & 0 & 1 & -1 \end{pmatrix}$ has the same RREF as **B**. Here **C** is singular.

(c) If one chooses the left factor **A** to be integer lower triangular with unit diagonal entries ± 1, one gains a clear preview of the respective elimination process for $\mathbf{B} := \mathbf{AR}$ and any integer upper staircase matrix **R**. ◀

2.2 Applications (MATLAB)

We explore MATLAB *with an eye on reducing matrices to row echelon form.*

The three operations of Gaussian elimination are easily expressed in MATLAB: For example, the command

```
A(2,:) = 4*A(2,:)
```

multiplies the second row `A(2,:)` of a given matrix **A** by 4, while the command

```
A = A([3 2 1 4],:)
```

swaps the first and third row of **A** since the chosen row index vector `[3 2 1 4]` indicates this permutation of the natural row order `[1 2 3 4]` in **A**. Finally,

```
A(2,:) = A(2,:) - A(2,1)/A(1,1) * A(1,:)
```

replaces the second row $\mathbf{a}_2 = \mathtt{A(2,:)}$ of **A** with $\mathbf{a}_2 - \frac{a_{21}}{a_{11}} \cdot \mathbf{a}_1$ in a typical elimination step, provided that $a_{11} \neq 0$. Note that in MATLAB, a colon : placed into an index position of a matrix selects all respective entries (i.e., a complete row or column of that matrix).

EXAMPLE 8 We row reduce

$$\mathbf{A} = \begin{pmatrix} 3 & 2 & 1 & 4 \\ 2 & 0 & -1 & 7 \\ 1 & 4 & -1 & 0 \\ 0 & 1 & 0 & 1 \end{pmatrix}$$

via MATLAB based Gaussian elimination as follows:

INPUT A:

```
>> A = [3 2 1 4; 2 0 -1 7; 1 4 -1 0; 0 1 0 1]
A =
      3    2    1    4
      2    0   -1    7
      1    4   -1    0
      0    1    0    1
```

SWAP ROWS 1 and 3 (for a unit pivot to appear in the (1,1) position):

```
>> A = A([3 2 1 4],:)
A =
      1    4   -1    0
      2    0   -1    7
      3    2    1    4
      0    1    0    1
```

ELIMINATE the (2,1) and (3,1) entries:

```
>> A(2,:)  = A(2,:)-A(2,1)/A(1,1)*A(1,:)
A =
      1     4    -1    0
      0    -8     1    7
      3     2     1    4
      0     1     0    1
>> A(3,:)=A(3,:)-A(3,1)/A(1,1)*A(1,:)
A =
      1     4    -1    0
      0    -8     1    7
      0   -10     4    4
      0     1     0    1
```

SWAP ROWS 3 and 4 (for unit (2,2) pivot):

```
>> A = A([1 4 3 2],:)
A =
      1     4    -1    0
      0     1     0    1
      0   -10     4    4
      0    -8     1    7
```

ELIMINATE (3,2) and (3,4) entries:

```
>> A(3,:)  = A(3,:)-A(3,2)/A(2,2)*A(2,:)
A =
      1    4   -1    0
      0    1    0    1
      0    0    4   14
      0   -8    1    7
```

```
>> A(4,:) = A(4,:)-A(4,2)/A(2,2)*A(2,:)

A =

     1     4    -1     0
     0     1     0     1
     0     0     4    14
     0     0     1    15
```

SWAP last two rows (for unit (3,3) pivot):

```
>> A = A([1 2 4 3],:)

A =

     1     4    -1     0
     0     1     0     1
     0     0     1    15
     0     0     4    14
```

ELIMINATE (4,3) entry:

```
>> A(4,:) = A(4,:)-A(4,3)/A(3,3)*A(3,:)

A =

     1     4    -1      0
     0     1     0      1
     0     0     1     15
     0     0     0    -46
```

(This is a row-echelon form for A.)

NORMALIZE the last pivot:

```
>> A(4,:) = A(4,:)/A(4,4)

A =

     1     4    -1     0
     0     1     0     1
     0     0     1    15
     0     0     0     1
```

ELIMINATE from the bottom up, first for the last pivot column to find the RREF:

```
>> A(3,:) = A(3,:)-A(3,4)/A(4,4)*A(4,:)

A =

     1     4    -1     0
     0     1     0     1
     0     0     1     0
     0     0     0     1
>> A(2,:) = A(2,:)-A(2,4)/A(4,4)*A(4,:)

A =

     1     4    -1     0
     0     1     0     0
     0     0     1     0
     0     0     0     1
```

ELIMINATE nonzero entries above the pivot in column 3:

```
  >> A(1,:) = A(1,:)-A(1,3)/A(3,3)*A(3,:)
A =
        1    4    0    0
        0    1    0    0
        0    0    1    0
        0    0    0    1
```

ELIMINATE nonzero (1,2) entry:

```
  >> A(1,:) = A(1,:)-A(1,2)/A(2,2)*A(2,:)
A =
        1    0    0    0
        0    1    0    0
        0    0    1    0
        0    0    0    1
```

(This is the RREF of A.)

Note that in MATLAB, blank spaces around equal, $+$, $-$, or $*$ signs are not necessary. The foregoing printout includes some blanks for reading clarity only. ◀

The standard hand computation scheme of Gaussian elimination in Section 2.1 has just been mimicked in MATLAB. MATLAB has a built–in function for this purpose that reduces matrices to their reduced row echelon form. The command

`rref(A)`

reduces **A** to its RREF. It has numerical priorities that differ from our desire for convenient unit pivots in the hand scheme, and hence it computes the RREF of a matrix **A** generally in a different sequence of elementary row operations. One can monitor its steps with the MATLAB command

`rrefmovie(A)`

and by following the on–screen instructions.

Finally, the MATLAB command `rank(A)` provides an estimate of the rank of a matrix $\mathbf{A}_{mn}$.

2.2.P Problems

1. Use the MATLAB function `rrefmovie`, and observe what pivots are chosen and how the algorithm proceeds for the matrix $\mathbf{A} = \begin{pmatrix} 3 & 2 & 1 & 4 \\ 2 & 0 & -1 & 7 \\ 1 & 4 & -1 & 0 \\ 0 & 1 & 0 & 1 \end{pmatrix}$ of Example 4.

   Vary the `format` from `format short` to `format rat` and play the "rrefmovie" in both MATLAB formats. What can you observe and see about the MATLAB algorithm that obtains the RREF. How does it compare with our computations in Example 4?

2. Explore the MATLAB function `rank(A)` by typing `help rank` and by evaluating the rank of several random matrices $\mathbf{A}_{mn}$. (For this, look up `help rand` and `help randn` here.)

   (a) How often do you find a random matrix $\mathbf{A}_{mn}$ with rank $(\mathbf{A}) < \min(m, n)$?

   (b) Compare your computed rank $(\mathbf{A})$ with the number of pivots of the RREF of **A** obtained from `rref(A)`.

(c) Find a matrix **A** where the MATLAB rank function and the number of computed pivots disagrees. (This is a hard problem.)

3. If $\mathbf{A} = \begin{pmatrix} 1 & 3 & 5 & -1 & 0 \\ 0 & 3 & 2 & -2 & 0 \\ 1 & -1 & -1 & -1 & -1 \end{pmatrix}$ and

$\mathbf{B} = \begin{pmatrix} 0 & 0 & 1 & 3 & -2 \\ 1 & 2 & -3 & 4 & 1 \\ 0 & 1 & 3 & -4 & 0 \end{pmatrix}$, then

(a) $\begin{pmatrix} 1 & 0 & 0 & 39 & -17 \\ 0 & 1 & 0 & -13 & 6 \\ 0 & 0 & 1 & 3 & -2 \end{pmatrix}$ is the RREF of the matrix ___ .

(b) What is the RREF of $2\mathbf{B} - 4\mathbf{A}$?

4. Test the following assertions via MATLAB:

(a) Does RREF($\mathbf{A} + \mathbf{B}$) = RREF($\mathbf{A}$) + RREF($\mathbf{B}$) for all compatibly sized matrices **A** and **B**?

(b) Can you find matrices **A** and **B**, both of size $m \times n$, so that RREF($\mathbf{A} - \mathbf{B}$) = RREF($\mathbf{A}$) − RREF($\mathbf{B}$) or so that RREF($\mathbf{A} + \mathbf{B}$) = RREF($\mathbf{A}$) + RREF($\mathbf{B}$)?

5. In MATLAB, one can transpose a matrix $\mathbf{A}_{mn}$ consisting of m rows and n columns to the matrix $\mathbf{A}^T$ with n rows and m columns by using the command `A'`.

(a) Find the rank of

$\mathbf{A}_{3,5} = \begin{pmatrix} 1 & 2 & 3 & 4 & 5 \\ 2 & 3 & 4 & 5 & 6 \\ 3 & 4 & 5 & 6 & 7 \end{pmatrix}$ and that of $(\mathbf{A}^T)_{5,3}$ using `rref` .

(b) Generate various 3×5 matrices and compare their ranks with that of their transposes.

6. Repeat Problems 1 through 5 of Section 2.1.P with the help of MATLAB.

7. Run `rrefmovie(A)` for `A = hilb(10)`, the 10×10 Hilbert matrix **A** with entries $a_{ij} = \frac{1}{i+j}$.

8. Repeat the previous problem for `B = 10*eps*hilb(10)`. What happens to the rank? Repeat with `C = eps*hilb(10)/100000` and `D = C/1000`. Watch the last columns in each run. More specifically, notice the top last column entries just before the last column is row reduced to become $\mathbf{e}_{10}$.

Finally, rework the problem using `hilb(20)` in place of `hilb(10)` .

9. In MATLAB, one can "flip" a matrix $\mathbf{A}_{mn}$ left-to-right by using the command `fliplr(A)` and "flip" it up–down by the command `flipud(A)`.

Test whether the rank of a matrix can be affected by flipping it left-to-right, up–down, or left-to-right and up–down simultaneously. Use the MATLAB `rref` command.

2.R Review Problems

1. (a) Find a matrix $\mathbf{A}_{4\times3}$ that cannot be row reduced to reduced row echelon form.

(b) Do this for a matrix $\mathbf{B}_{3\times4}$.

2. Describe the output of a complete Gaussian reduction, i.e., the REF and the RREF of an augmented matrix $(\mathbf{A} \mid \mathbf{b})$. Give all defining properties of these forms.

3. Is it true that some matrices can be reduced to different reduced row echelon forms by choosing different sets of pivots?

4. Do the pivot column positions of a REF depend on the row interchanges in the elimination process of the given matrix **A**?

5. The rank of a matrix is well defined mathematically and well computable numerically for any real or complex matrix: true or false?

6. (a) If $\mathbf{A}_{3\times2}$ has rank 1 and $\mathbf{A} = \begin{pmatrix} 0 & * & 0 \\ 0 & 0 & * \end{pmatrix}$, what are the possible (1,2) and (2,3) starred entries of **A**?

(b) How about for $\mathbf{A}_{3\times2}$ of rank 2?

7. Which of the 21 distinct matrix pairs made up of any two of the seven following matrices are row equivalent under Gaussian elimination (Why are there 21 such pairs here)?

$$\mathbf{A} = \begin{pmatrix} 1 & 0 & 1 \\ 0 & 2 & 0 \\ 0 & 0 & 3 \end{pmatrix}, \mathbf{B} = \begin{pmatrix} 0 & 0 & 1 \\ 0 & 1 & 0 \\ 1 & 0 & 0 \end{pmatrix},$$

$$\mathbf{C} = \begin{pmatrix} 1 & 3 & 2 \\ 2 & 5 & -1 \\ 0 & 1 & 5 \end{pmatrix}, \mathbf{E} = \begin{pmatrix} 0 & 1 & 5 \\ 4 & 7 & 0 \\ 1 & 8 & 0 \end{pmatrix},$$

$$\mathbf{F} = \begin{pmatrix} 1 & 1 & -8 \\ 0 & -1 & -5 \\ 0 & 0 & 0 \end{pmatrix}, \mathbf{G} = \begin{pmatrix} 1 & 5 & 0 \\ 0 & 1 & 0 \\ 0 & 2 & 0 \end{pmatrix},$$

$$\text{and } \mathbf{H} = \begin{pmatrix} 2 & 2 & 0 \\ 4 & 7 & 0 \\ 1 & -8 & 0 \end{pmatrix}.$$

8. (a) What does the RREF of a 5×3 matrix look like if it contains 3 pivots.

(b) Repeat part (a) for a 4×4 matrix with 4 pivots.

(c) List all possible RREFs for a 5×3 matrix. (There are 26 possible patterns.)

9. (a) Find a and $b \in \mathbb{R}$, so that the two matrices $\begin{pmatrix} 1 & 2 & 3 \\ -4 & -2 & 0 \end{pmatrix}$ and $\begin{pmatrix} 1 & 0 & a \\ 0 & -1 & b \end{pmatrix}$ are row equivalent.

(b) Find a and $b \in \mathbb{R}$, so that the two matrices $\begin{pmatrix} 1 & 2 & 3 \\ -4 & -2 & 0 \end{pmatrix}$ and $\begin{pmatrix} 1 & 0 & a \\ 0 & -1 & b \end{pmatrix}$ are <u>not</u> row equivalent.

10. Determine whether the two matrices $\mathbf{A} = \begin{pmatrix} 1 & 2 & 3 \\ -4 & -2 & 0 \end{pmatrix}$ and $\mathbf{B} = \begin{pmatrix} 1 & 0 & -1 \\ 0 & 2 & 4 \end{pmatrix}$ are row equivalent.

11. Let $\mathbf{A} = \begin{pmatrix} \cos(\theta) & \sin(\theta) \\ -\sin(\theta) & \cos(\theta) \end{pmatrix}$.

Show that $\mathbf{A}$ is row equivalent to $\mathbf{I}_2$ for all values of θ.

12. Decide which of the following matrices are

(a) in row-echelon form, but not in reduced row-echelon form,

(b) in reduced row-echelon form,

(c) in neither echelon form:

$$\mathbf{A} = \begin{pmatrix} 2 & 1 & 0 \\ 0 & 0 & 1 \\ 0 & 0 & 0 \\ 0 & 0 & 0 \end{pmatrix}, \ \mathbf{B} = \begin{pmatrix} 1 & 2 & 1 \\ 0 & 1 & * \\ 0 & 0 & * \end{pmatrix},$$

$$\mathbf{C} = \begin{pmatrix} 1 & 0 & 1 & 0 \\ 0 & 1 & 0 & 2 \end{pmatrix},$$

$$\mathbf{D} = \begin{pmatrix} 1 & 1 & 0 & 0 \\ 0 & 0 & -1 & 0 \\ 0 & 1 & 0 & 0 \end{pmatrix},$$

$$\mathbf{E} = \begin{pmatrix} 0 & 1 & * & 0 \\ 0 & 0 & 0 & 1 \\ 0 & 0 & 0 & 0 \end{pmatrix}, \ \mathbf{F} = \begin{pmatrix} 0 & 0 \\ 0 & 0 \\ 0 & 0 \\ 0 & 0 \end{pmatrix}.$$

13. Row reduce $\begin{pmatrix} -1 & 1 & 1 \\ 2 & 1 & 0 \\ 0 & 1 & 1 \\ 1 & 0 & -1 \end{pmatrix}$ to its RREF.

14. Which columns of a REF of

$$\mathbf{A} = \begin{pmatrix} 0 & 1 & 0 & 2 \\ 0 & 2 & 1 & 1 \\ 0 & 1 & 1 & -1 \\ 0 & 3 & 1 & 3 \end{pmatrix} \text{ and}$$

$$\mathbf{B} = \begin{pmatrix} -1 & 0 & 1 \\ 0 & 2 & 0 \\ -1 & 0 & 1 \end{pmatrix} \text{ contain a pivot?}$$

15. (a) Row reduce the matrix $\begin{pmatrix} a & 0 & 0 & 0 \\ e & b & 0 & 0 \\ f & g & c & 0 \\ h & k & \ell & d \end{pmatrix}$ to its RREF, where $a, b, c,$ and $d \neq 0$.

(b) Row reduce the matrix $\begin{pmatrix} a & e & f & g \\ 0 & b & g & h \\ 0 & 0 & c & \ell \\ 0 & 0 & 0 & d \end{pmatrix}$ to its RREF for real constants $a, b, c,$ and $d \neq 0$.

Standard Tasks and Questions

1. What is a row echelon or the reduced row echelon form of a given matrix $\mathbf{A}_{mn}$?

2. Determine whether two given augmented row echelon forms $(\mathbf{R} \mid \mathbf{b})$ and $(\tilde{\mathbf{R}} \mid \tilde{\mathbf{b}})$ are equivalent (i.e., represent the same linear system of equations.)

(Check that the two linear systems $\mathbf{Rx} = \mathbf{b}$ and $\tilde{\mathbf{R}}\mathbf{x} = \tilde{\mathbf{b}}$ have the same solution, check that the two REFs can be transformed into one another legitimately by Gaussian eliminations, or use the RREFs of $(\mathbf{R} \mid \mathbf{b})$ and $(\tilde{\mathbf{R}} \mid \tilde{\mathbf{b}})$.)

3. Separate the zero and nonzero entry triangles of a row-reduced matrix by drawing the staircase line. Then assert whether the reduced matrix is in row echelon form.

4. What is the rank of a matrix $\mathbf{A}_{mn}$? How many pivot and how many free columns will every row-echelon form $\mathbf{R}$ of $\mathbf{A}$ have if $\mathbf{A}$ has rank k?

Subheadings of Lecture Two

Some sides missing

CHAPTER

3

Linear Equations

Gaussian row reduction helps us solve systems of linear equations.

3.1 Lecture Three

Solvability and Solutions of Linear Systems

We study general solvability conditions for systems of linear equations by using the pivot information that is embedded in a row echelon form of the augmented system matrix. Then we describe techniques for solving linear systems.

(a) Solvability of Systems of Linear Equations

Every system of linear equations of the form (2.2)

$$\begin{array}{ccccc} a_{11}x_1 & + \cdots + & a_{1n}x_n & = & b_1 \\ \vdots & & \vdots & & \vdots \\ a_{m1}x_1 & + \cdots + & a_{mn}x_n & = & b_m \end{array}$$

can be written as a matrix $\times$ vector equation $\mathbf{Ax} = \mathbf{b}$ with a known **system matrix** $\mathbf{A}_{mn}$ and a known **right-hand-side** $\mathbf{b} \in \mathbb{R}^m$ for which we want to find a **solution vector** $\mathbf{x} \in \mathbb{R}^n$.

Using the linear transformations point of view, we see that a linear system $\mathbf{Ax} = \mathbf{b}$ can be solved for $\mathbf{x}$ precisely when the right-hand-side vector $\mathbf{b}$ belongs to the set of **images**

$$\{\mathbf{y} \in \mathbb{R}^m \mid \mathbf{Ax} = \mathbf{y} \text{ for some } \mathbf{x} \in \mathbb{R}^n\}.$$

For low dimensions m and n, one can solve linear systems by using geometry. For example, if $n = 2$, each linear equation in two variables $x_1, x_2 \in \mathbb{R}$ represents a line in the plane. If m such equations are given then the linear system $\mathbf{A}_{m\times 2}\mathbf{x} = \mathbf{b}$ has a simultaneous solution $\mathbf{x} \in \mathbb{R}^2$ precisely when the corresponding m lines all pass though one common point $\mathbf{x}$ of the plane.

EXAMPLE 1 (a) Study the system of linear equations given by the three line equations

$$\begin{aligned}
&\text{Line 1(a)} \quad & x_1 + 2x_2 &= 3, \\
&\text{Line 2(a)} \quad & 2x_1 - x_2 &= 0, \\
&\text{Line 3(a)} \quad & 3x_1 + x_2 &= -1.
\end{aligned}$$

Figure 3-1 shows a plot of the lines.

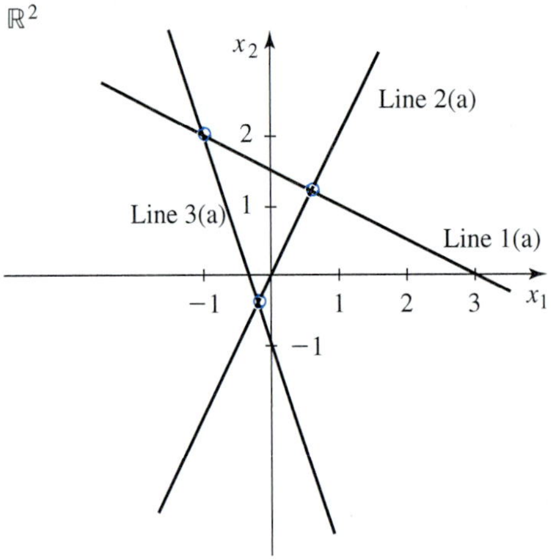

Figure 3-1

Any two of the lines have a point in common, but all three do not. Hence, this system of equations is unsolvable.

(b) Repeat part (a) for

$$\begin{aligned}
&\text{Line 1(b)} \quad & x_1 + 2x_2 &= 0, \\
&\text{Line 2(b)} \quad & 2x_1 - x_2 &= 0, \\
&\text{Line 3(b)} \quad & 3x_1 + x_2 &= 0.
\end{aligned}$$

The three lines are plotted in Figure 3-2.

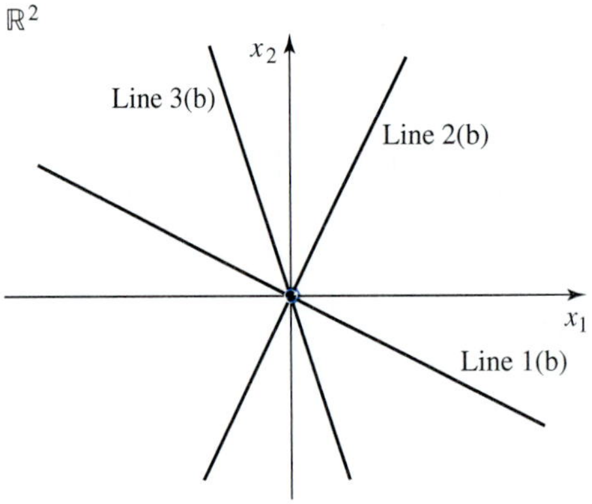

Figure 3-2

Hence, there is one point in $\mathbb{R}^n$ that lies on all three lines, namely, the origin $\mathbf{x} = \begin{pmatrix} 0 \\ 0 \end{pmatrix} \in \mathbb{R}^2$, which solves this system of equations.

(c) Repeat part (a) for

$$\begin{aligned} \text{Line 1(c)} \quad 2x_1 + x_2 &= 30, \\ \text{Line 2(c)} \quad 4x_1 + 2x_2 &= -20. \end{aligned}$$

The two lines are plotted in Figure 3-3.

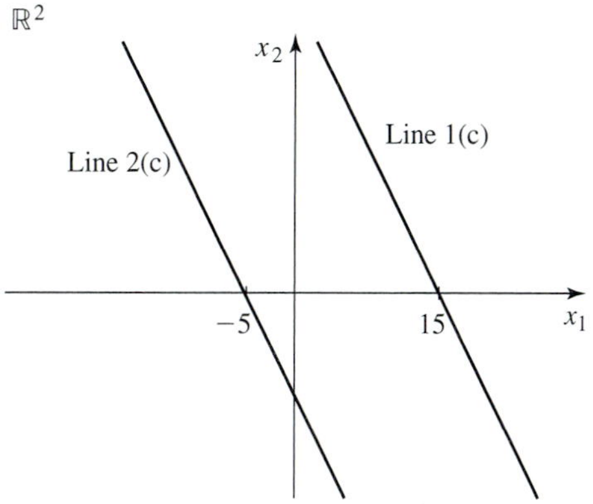

Figure 3-3

The two lines are parallel. Thus, they have no common point, and there is no solution for this set of equations.

(d) Consider the plot of the two lines

$$\begin{aligned} \text{Line 1(d)} \quad -x_1 + 2x_2 &= 4, \\ \text{Line 2(d)} \quad 2x_1 - 42x_2 &= -8. \end{aligned}$$

Their plot is shown in Figure 3-4.

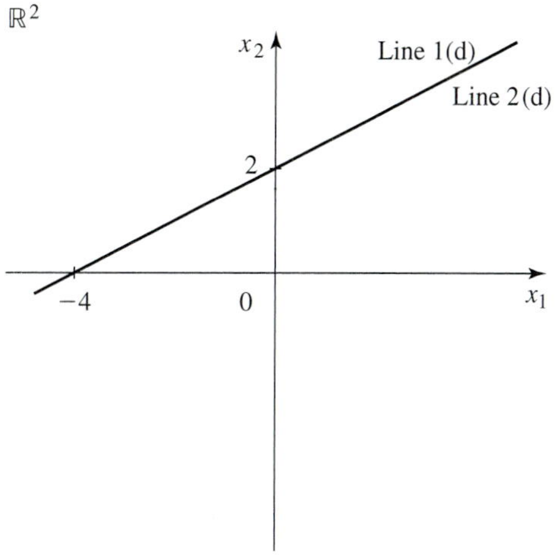

Figure 3-4

The two lines describe the same set of points in $\mathbb{R}^2$. By solving the first equation for x_2 in terms of $x := x_1 \in \mathbb{R}$, we see that any point of the form $\begin{pmatrix} x \\ \frac{x}{2}+2 \end{pmatrix} \in \mathbb{R}^2$ solves both equations. There are infinitely many solutions to this system of equations. ◀

The most useful tool for solving systems of linear equations is row reduction performed on the augmented matrix $(\mathbf{A} \mid \mathbf{b})$ in the form (2.1):

$$(\mathbf{A} \mid \mathbf{b}) = \left(\begin{array}{ccc|c} a_{11} & \cdots & a_{1n} & b_1 \\ \vdots & & \vdots & \vdots \\ a_{m1} & \cdots & a_{mn} & b_m \end{array}\right).$$

Through Gaussian elimination, we obtain a REF $(\mathbf{R} \mid \tilde{\mathbf{b}})$ that is row equivalent to the original system and that has the same solution as the original system. By examining the REF $(\mathbf{R} \mid \tilde{\mathbf{b}})$, we can determine whether a linear system $\mathbf{A}_{mn}\mathbf{x} = \mathbf{b}$ is solvable, and if it is, how many solutions it has. First we study solvability.

There are two cases:

1. For each **pivot row** $0, \cdots, 0, \boxed{*}, *, \cdots, *$ of $\mathbf{R}$ (with $\boxed{*} \neq 0$), there is at least one solution vector $\mathbf{x}$, so that

   $$\begin{pmatrix} 0 & \cdots & 0 & \boxed{*} & * & \cdots & * \end{pmatrix} \mathbf{x} = \tilde{b}_j,$$

   whatever the right-hand-side value $\tilde{b}_j \in \mathbb{R}$ may be.

   For example, for the pivot row $\begin{pmatrix} 0 & \boxed{2} & -1 & 1 \end{pmatrix}$ in R, the linear equation

   $$\begin{pmatrix} 0 & 2 & -1 & 1 \end{pmatrix} \mathbf{x} = \tilde{b}_j = 2$$

   is solvable with possible solution vectors $\mathbf{x} = \begin{pmatrix} 0 \\ 1 \\ 1 \\ 1 \end{pmatrix}$, $\mathbf{y} = \begin{pmatrix} 0 \\ 1 \\ 0 \\ 0 \end{pmatrix}$, $\mathbf{z} = \begin{pmatrix} 0 \\ 2 \\ 2 \\ 0 \end{pmatrix}$, or $\mathbf{w} = \begin{pmatrix} 0 \\ 0 \\ 0 \\ 2 \end{pmatrix}$.

2. For a **row without pivot** $0, \cdots, 0, \cdots, 0$ in R, made up entirely of zeros, the solvability of the whole linear system depends on whether the corresponding right-hand-side $\tilde{b}_j$ in $(\mathbf{R} \mid \tilde{\mathbf{b}})$ is zero.

   If $\tilde{b}_j$ is zero, then we can choose $\mathbf{x}$ arbitrarily, and the row equation $\mathbf{0} \cdot \mathbf{x} = 0$ is satisfied for $\mathbf{x}$. In fact, this trivial equation has no influence on the solution of the system.

   If $\tilde{b}_j$ is nonzero, however, we end up with the **inconsistency**

   $$\mathbf{0} \cdot \mathbf{x} = \tilde{b}_j \neq 0,$$

   which can never be solved for $\mathbf{x}$.

In case of an inconsistent row, note that the REF $(\mathbf{R} \mid \tilde{\mathbf{b}})$ of the augmented matrix $(\mathbf{A} \mid \mathbf{b})$ has one more pivot in its $(n+1)^{st}$ column $\tilde{\mathbf{b}}$ than the REF $\mathbf{R}$ of $\mathbf{A}$ does.

EXAMPLE 2 We repeat Example 1 using row reduction of the corresponding augmented matrices and decide upon solvability.

For part (a), we row reduce the augmented matrix $(\mathbf{A} \mid \mathbf{b})$:

x_1	x_2	$\mathbf{b}$	
1	2	3	
2	−1	0	$-2\ row_1$
3	1	−1	$-3\ row_1$
1	2	3	
0	−5	−6	
0	−5	−10	$-\ row_2$
1	2	0	
0	−5	0	*a REF* $(\mathbf{R}_1 \mid \tilde{\mathbf{b}})$ *of* $(\mathbf{A} \mid \mathbf{b})$
0	0	−4	

Since the REF $(\mathbf{R}_1 \mid \tilde{\mathbf{b}})$ contains the inconsistent third row $\left(0\ 0 \mid \boxed{-4}\right)$, the linear system is not solvable; that is, the three given lines have no common point of intersection.

For part (b), we solve $\mathbf{Bx} = \mathbf{0}$:

x_1	x_2	$\mathbf{0}$	
1	2	0	
2	−1	0	$-2\ row_1$
3	1	0	$-3\ row_1$
1	2	0	
0	−5	0	
0	−5	0	$-\ row_2$
1	2	0	
0	−5	0	*a REF* $(\mathbf{R}_2 \mid \mathbf{0})$ *of* $(\mathbf{B} \mid \mathbf{0})$
0	0	0	

There are no inconsistent rows. Using backsubstitution, we find that the solution $\mathbf{x}$ has the two components $x_2 = 0$ and $x_1 = 0$; that is, the three lines intersect at the origin.

In part (c), we attempt to solve $\mathbf{Cx} = \mathbf{c}$:

x_1	x_2	$\mathbf{c}$		
$\boxed{2}$	1	30		
$\boxed{4}$	2	−20	$-2\ row_1$	
$\boxed{2}$	1	30		*a REF* $(\mathbf{R}_3 \mid \tilde{\mathbf{c}})$ *of* $(\mathbf{C} \mid \mathbf{c})$
0	0	$\boxed{-80}$		

This system is unsolvable, since there is no $\mathbf{x} \in \mathbb{R}^2$ with $0x_1 + 0x_2 = -80$.

For part (d), we have

$\boxed{-1}$	2	4		
$\boxed{2}$	−4	−8	$+2\ row_1$	
$\boxed{-1}$	2	4		*a REF* $(\mathbf{R}_4 \mid \tilde{\mathbf{d}})$ *of* $(\mathbf{D} \mid \mathbf{d})$
0	0	0		

Since there are no inconsistent rows in $(\mathbf{R}_4 \mid \tilde{\mathbf{d}})$, this system is solvable. We will shortly learn about the number of its solutions. ◀

By using the notion of matrix **rank** that was defined in lecture two as the number of pivots in a REF of a given matrix $\mathbf{A}$, we express our findings as follows:

Theorem 3.1 A linear system of equations $\mathbf{Ax} = \mathbf{b}$ is **solvable** if and only if

$$\text{rank}\ \mathbf{A} = \text{rank}\ (\mathbf{A} \mid \mathbf{b}).$$ ◀

Note that in Example 2, only the second and fourth given linear systems are solvable, while the others are not according to Theorem 3.1. In Example 2 we have $\text{rank}(\mathbf{A}) = 2 < \text{rank}(\mathbf{A} \mid \mathbf{b}) = 3$, $\text{rank}\ (\mathbf{B} \mid \mathbf{0}) = \text{rank}(\mathbf{B}) = \text{rank}(\mathbf{R}_2) = \text{rank}(\mathbf{R}_2 \mid \mathbf{0}) = 2$, while $\text{rank}\ (\mathbf{C} \mid \mathbf{c}) = \text{rank}\ (\mathbf{R}_3 \mid \tilde{\mathbf{c}}) = 2 > \text{rank}\ (\mathbf{C}) = \text{rank}\ (\mathbf{R}_3) = 1$ and $\text{rank}\ (\mathbf{D} \mid \mathbf{d}) = \text{rank}\ (\mathbf{R}_4 \mid \tilde{\mathbf{d}}) = 1 = \text{rank}\ (\mathbf{D}) = \text{rank}\ (\mathbf{R}_4)$.

The solvability of a linear system $\mathbf{Ax} = \mathbf{b}$ depends on a row condition for the REF $(\mathbf{R} \mid \tilde{\mathbf{b}})$ of $(\mathbf{A} \mid \mathbf{b})$.

The REF $(\mathbf{R} \mid \tilde{\mathbf{b}})$ *must not contain an inconsistent row* $\begin{pmatrix} 0 & \dots & 0 & \mid & \boxed{\tilde{b}_j} \end{pmatrix}$ *with* $\tilde{b}_j \neq 0$.

In other words, each zero row in $\mathbf{R}$ must extend to a zero row in $(\mathbf{R} \mid \tilde{\mathbf{b}})$ for solvability.

Before actually solving specific linear systems, we deduce several elementary solvability results.

Lemma 1 If $\mathbf{Ax} = \mathbf{b}$ and $\mathbf{Ay} = \mathbf{b}$ for the same system matrix $\mathbf{A}$ and the same right-hand-side $\mathbf{b}$, then $\mathbf{A}(\mathbf{x} - \mathbf{y}) = \mathbf{0}$. ◀

Lemma 2 If $\mathbf{Az} = \mathbf{b}$ and $\mathbf{Aw} = \mathbf{0}$, then $\mathbf{A}(\mathbf{z} + \alpha\mathbf{w}) = \mathbf{b}$ for all scalars α. ◀

The proofs follow immediately from the linearity of **A**: For Lemma 1, we have $\mathbf{A}(\mathbf{x} - \mathbf{y}) = \mathbf{Ax} - \mathbf{Ay} = \mathbf{b} - \mathbf{b} = \mathbf{0}$, while $\mathbf{A}(\mathbf{z} + \alpha\mathbf{w}) = \mathbf{Az} + \alpha\mathbf{Aw} = \mathbf{b} + \mathbf{0} = \mathbf{b}$ in Lemma 2.

> Any two solutions **x** and **y** of a solvable linear system of equations $\mathbf{Ax} = \mathbf{b}$ differ by a vector that is mapped to zero by the linear transformation $\mathbf{x} \mapsto \mathbf{Ax}$.

The vectors **x** that a linear transformation **A** maps to zero (i.e., those **x** for which $\mathbf{Ax} = \mathbf{0}$) play a special role throughout linear algebra.

DEFINITION 1 The set of vectors $\mathbf{x} \in \mathbb{R}^n$ with $\mathbf{Ax} = \mathbf{0}$ for $\mathbf{A} \in \mathbb{R}^{m,n}$ is called the **kernel**, or **nullspace**, of **A**

$$\ker(\mathbf{A}) := \{\mathbf{x} \in \mathbb{R}^n \mid \mathbf{Ax} = \mathbf{0} \in \mathbb{R}^m\}.$$

Note that the nullspace or kernel of a matrix **A** is never empty since always $\mathbf{A0} = \mathbf{0}$. Thus, $\mathbf{0} \in \ker(\mathbf{A})$ for all **A**. Moreover, if **A** is $m \times n$, then $\ker(\mathbf{A}) \subset \mathbb{R}^n$ is part of **A**'s domain.

Lemma 3 The sum of any two vectors in $\ker(\mathbf{A})$ is again a vector in the kernel, and all scalar multiples of a vector in $\ker(\mathbf{A})$ lie in $\ker(\mathbf{A})$. That is, if **x** and $\mathbf{y} \in \ker(\mathbf{A})$, then $\alpha\mathbf{x} + \beta\mathbf{y} \in \ker(\mathbf{A})$ for all scalars α and β. ◀

The proof follows from linearity. (See Problem 27.)

Thus, the kernel of a matrix **A** is closed under the two vector-space operations of vector addition and scalar multiplication. We generalize this closure of matrix kernels and call a nonempty subset of $\mathbb{R}^n$ a **subspace** if it is closed under vector addition and scalar multiplication. Subspaces are studied in Chapter 4.

From Lemmas 1, 2, and 3, we gain a complete description of the **set of solutions x** for a linear system $\mathbf{Ax} = \mathbf{b}$ as long as $\mathbf{Ax} = \mathbf{b}$ is solvable in the sense of Theorem 3.1:

> $\{\mathbf{x} \mid \mathbf{Ax} = \mathbf{b}\}$ consists of one *particular solution* to $\mathbf{Ax} = \mathbf{b}$
> *plus* any vector in $\ker(\mathbf{A})$

In other words, the set of all solutions of a solvable linear system of equations $\mathbf{Ax} = \mathbf{b}$ consists of all sums of any **particular solution** $\mathbf{x}_{\text{part}}$ to $\mathbf{Ax} = \mathbf{b}$ and any **homogeneous solution** $\mathbf{x}_{\text{hom}}$ of the system $\mathbf{Ax} = \mathbf{0}$. If there are many vectors with $\mathbf{Ax} = \mathbf{0}$ then there are many solutions to $\mathbf{Ax} = \mathbf{b}$ provided the system is solvable. The homogeneous linear system $\mathbf{Ax} = \mathbf{0}$ can always be solved; $\mathbf{x} = \mathbf{0}$ will do.

If we row reduce a given matrix **A** to its REF **R**, we get a clear picture of the vectors $\mathbf{x}_{\text{hom}}$ in the kernel of **A**. For a solvable linear system $\mathbf{Ax} = \mathbf{b}$, the size of the kernel of **A** determines whether it has a unique solution or many solutions.

By definition, every REF $\mathbf{R}$ of $\mathbf{A}$ has as many pivot columns as the rank of $\mathbf{A}$. The columns of $\mathbf{R}$ without pivot are called **free columns**. When solving $\mathbf{Ax} = \mathbf{b}$, we can assign arbitrary scalar values α, β, γ, etc., to their corresponding variables, called the **free variables**. Subsequently, the **pivot variables** of a solvable linear system $\mathbf{Ax} = \mathbf{b}$ can be adjusted to solve the inhomogeneous system.

EXAMPLE 3 Describe all vectors in the kernel of the REF $\mathbf{R} = \begin{pmatrix} 0 & 1 & 2 & 0 & -1 \\ 0 & 0 & 0 & 2 & -1 \\ 0 & 0 & 0 & 0 & 2 \end{pmatrix}$.

$\mathbf{R}$ has three pivots and two free variables:

x_1	x_2	x_3	x_4	x_5	$\mathbf{0}$
0	$\boxed{1}$	2	0	−1	0
0	0	0	$\boxed{2}$	−1	0
0	0	0	0	$\boxed{2}$	0
↑		↑			
α		β			

The first and third columns of $\mathbf{R}$ are 'free'. Thus, we assign two arbitrary scalar values α and β to the free variables x_1 and x_3 and solve for the pivot variables x_2, x_4, and x_5 in terms of α and β. This we do by **backsubstitution** from the bottom up. Since the backsubstitution process needs to be mastered very well, we repeat it from Section 2.1.

The following diagram indicates the order of backsubstitution for the row reduced system $(\mathbf{R} \mid \mathbf{0})$ by the $\overset{\boxed{1}}{\longleftrightarrow}$ to $\overset{\boxed{3}}{\longleftrightarrow}$ symbols below. (Do $\boxed{1}$ first, $\boxed{2}$ next, then $\boxed{3}$, etc.

x_1	x_2	x_3	x_4	x_5	$\mathbf{0}$		
0	$\boxed{1}$	2	0	−1	0	$\overset{\boxed{3}}{\longleftrightarrow}$	$x_2 + 2x_3 - x_5 = 0 \Rightarrow x_2 = -2\beta$
0	0	0	$\boxed{2}$	−1	0	$\overset{\boxed{2}}{\longleftrightarrow}$	$2x_4 - x_5 = 0 \Rightarrow x_4 = 0$
0	0	0	0	$\boxed{2}$	0	$\overset{\boxed{1}}{\longleftrightarrow}$	$2x_5 = 0 \Rightarrow x_5 = 0$
↑		↑					
α		β					

Thus, $\mathbf{x} = \begin{pmatrix} x_1 \\ x_2 \\ x_3 \\ x_4 \\ x_5 \end{pmatrix} = \begin{pmatrix} \alpha \\ -2\beta \\ \beta \\ 0 \\ 0 \end{pmatrix}$ solves $\mathbf{Rx} = \mathbf{0}$ for any values of α and $\beta \in \mathbb{R}$, as can be easily checked.

We rewrite $\mathbf{x} = \alpha \begin{pmatrix} 1 \\ 0 \\ 0 \\ 0 \\ 0 \end{pmatrix} + \beta \begin{pmatrix} 0 \\ -2 \\ 1 \\ 0 \\ 0 \end{pmatrix}$ to express the two degrees of freedom for vectors in the kernel of $\mathbf{A}$. ◀

(b) Unique Solvability of Linear Systems

Solvable linear systems $\mathbf{Ax} = \mathbf{b}$ may or may not have **unique solutions** $\mathbf{x}$.

EXAMPLE 4 The linear system $\begin{matrix} 2x_1 - x_2 = 1 \\ 3x_2 = 6 \end{matrix}$ is solvable by Theorem 3.1, since the rank of $\begin{pmatrix} 2 & -1 \\ 0 & 3 \end{pmatrix}$ is two, which equals the rank of the augmented matrix $\left(\begin{array}{cc|c} 2 & -1 & 1 \\ 0 & 3 & 6 \end{array}\right)$. This system has the unique solution $\mathbf{x} = \begin{pmatrix} 1.5 \\ 2 \end{pmatrix}$ found by backsubstitution.

The system $\begin{matrix} 2x_1 - x_2 = 1 \\ -4x_1 + 2x_2 = -2 \end{matrix}$ is also solvable, since its augmented matrix $(\mathbf{A} \mid \mathbf{b})$ reduces to $\left(\begin{array}{cc|c} 2 & -1 & 1 \\ 0 & 0 & 0 \end{array}\right)$ with no inconsistent rows. All its solutions have the general form $\mathbf{z} = \begin{pmatrix} 1 + \alpha \\ 1 + 2\alpha \end{pmatrix} = \begin{pmatrix} 1 \\ 1 \end{pmatrix} + \alpha \begin{pmatrix} 1 \\ 2 \end{pmatrix}$ for any $\alpha \in \mathbb{R}$. Here $\begin{pmatrix} 1 \\ 1 \end{pmatrix}$ is a particular solution satisfying $\mathbf{A}\begin{pmatrix} 1 \\ 1 \end{pmatrix} = \begin{pmatrix} 1 \\ -2 \end{pmatrix}$, while $\begin{pmatrix} 1 \\ 2 \end{pmatrix}$ is a solution of the homogeneous system $\mathbf{Ax} = \mathbf{0}$. ◀

The unique solvability of a linear system $\mathbf{Ax} = \mathbf{b}$ is tied to the size of the kernel of $\mathbf{A}$. (See Lemma 2.)

Theorem 3.2 $\ker(\mathbf{A}) = \{\mathbf{0}\}$ if and only if any REF $\mathbf{R}$ of $\mathbf{A} \in \mathbb{R}^{m,n}$ has no free variables (i.e., if and only if rank $\mathbf{A} = n =$ the number of columns of $\mathbf{A}$). ◀

Proof If every column in a REF $\mathbf{R}$ of $\mathbf{A}$ has a pivot, then every pivot row landing in the upper staircase matrix $\mathbf{R}$ has length one. Hence, the augmented REF $(\mathbf{R} \mid \mathbf{0})$ is in specific upper staircase form without any free variables. It has the form

$$(\mathbf{R} \mid \mathbf{0}) = \left(\begin{array}{cccccc|c} \square & & & & & * & 0 \\ 0 & \square & & & & & 0 \\ & 0 & \square & & & & 0 \\ & & 0 & \square & & & 0 \\ & & & 0 & \square & & 0 \\ 0 & & & & 0 & \square & 0 \\ 0 & & & & & 0 & 0 \\ 0 & & & & & 0 & 0 \\ 0 & & & & & 0 & 0 \end{array}\right),$$

where the number of bottom zero rows depends on the problem. There may be none. Backsubstitution yields $\mathbf{x} = \mathbf{0}$ as the only solution to $\mathbf{Rx} = \mathbf{0}$. Thus, $\mathbf{x} = \mathbf{0}$ is the only solution of $\mathbf{Ax} = \mathbf{0}$ as well.

If, on the other hand rank $\mathbf{A} < n$, which is the number of columns in $\mathbf{A} \in \mathbb{R}^{m,n}$, then there are free variables in any REF $\mathbf{R}$ of $\mathbf{A}$, and these can be assigned any nonzero values. The values of the pivot variables can then be adjusted by backsubstitution to find a nonzero solution vector $\mathbf{x}$ with $\mathbf{Rx} = \mathbf{0}$ or $\mathbf{Ax} = \mathbf{0}$. Thus, $\mathbf{Ax} = \mathbf{0}$ has multiple solutions. ■

The solvability criterion in Theorem 3.1 is a row condition for the pivots in a REF $(\mathbf{R} \mid \tilde{\mathbf{b}})$ of $(\mathbf{A} \mid \mathbf{b})$, since no pivot row in $(\mathbf{R} \mid \tilde{\mathbf{b}})$ may contain its pivot in the $(n+1)^{\text{st}}$ position if $\mathbf{A}$ is $m \times n$.

In addition, the unique solvability of a linear system $\mathbf{Ax} = \mathbf{b}$ depends on a second condition, a column condition, for the REF of $(\mathbf{A} \mid \mathbf{b})$:

Theorem 3.3 A linear system $\mathbf{Ax} = \mathbf{b}$ is **uniquely solvable** if and only if

(a) $\mathbf{Ax} = \mathbf{b}$ is solvable.
(This is a row condition on $\mathbf{A}$ and $\mathbf{b}$: Each zero row of a REF $\mathbf{R}$ of $\mathbf{A}$ is also a zero row of the augmented REF $(\mathbf{R} \mid \tilde{\mathbf{b}})$.)
And
(b) every column of a REF $\mathbf{R}$ of $\mathbf{A}$ contains a pivot. ◀

Proof (Necessity): If $\mathbf{Ax} = \mathbf{b}$ is uniquely solvable, it is solvable, so condition (a) holds. If $\mathbf{Au} = \mathbf{b}$ and $\mathbf{z} \in \ker(\mathbf{A})$; that is, if $\mathbf{Az} = \mathbf{0}$, then by Lemma 2, we have $\mathbf{A}(\mathbf{u}+\mathbf{z}) = \mathbf{b}$ as well. Assuming unique solvability of $\mathbf{Ax} = \mathbf{b}$ makes $\mathbf{u}+\mathbf{z} = \mathbf{u}$, (i.e., $\mathbf{z} = \mathbf{0}$). Thus, $\ker(\mathbf{A}) = \{\mathbf{0}\}$ in this case. As $\ker(\mathbf{A}) = \{\mathbf{0}\}$, the only solution to $\mathbf{Ax} = \mathbf{0}$ is the zero vector. Hence, from Theorem 3.2, the REF $\mathbf{R}$ of $\mathbf{A}$ may not contain free variables, or alternatively, each column of $\mathbf{R}$ must contain a pivot, proving part (b).
(Sufficiency): Assume that conditions (a) and (b) hold for $\mathbf{A}$ and $\mathbf{b}$. If two vectors $\mathbf{u}$ and $\mathbf{v}$ solve the linear equation $\mathbf{Ax} = \mathbf{b}$ (i.e., if both $\mathbf{Au} = \mathbf{b}$ and $\mathbf{Av} = \mathbf{b}$), then from Lemma 1, we have $\mathbf{A}(\mathbf{u} - \mathbf{v}) = \mathbf{0}$ or $\mathbf{u} - \mathbf{v} \in \ker(\mathbf{A})$. Part (b) and Theorem 3.1 force $\ker(\mathbf{A}) = \{\mathbf{0}\}$. Consequently, $\mathbf{u} - \mathbf{v} = \mathbf{0}$. In other words, $\mathbf{u} = \mathbf{v}$ is the unique solution of the linear system $\mathbf{Ax} = \mathbf{b}$. ■

Note that the unique solvability of a linear system $\mathbf{Ax} = \mathbf{b}$ involves two conditions on the REF $(\mathbf{R} \mid \tilde{\mathbf{b}})$ of $(\mathbf{A} \mid \mathbf{b})$:

the solvability condition of Theorem 3.1 *and*

the trivial kernel column condition $\ker(\mathbf{R}) = \{\mathbf{0}\}$ for the REF $\mathbf{R}$ of $\mathbf{A}$ from Theorem 3.3(b).

The following summarizes our findings:

Fredholm Alternative
There are three mutually exclusive possibilities for a system of linear equations $\mathbf{Ax} = \mathbf{b}$: either

Ø. the linear system is unsolvable
1. the linear system has a unique solution, or
∞. there are infinitely many solutions for the linear system.

We illustrate this set of alternatives in the following example:

EXAMPLE 5 For which values of α and $\beta \in \mathbb{R}$ does the linear system

$$\begin{aligned} x_1 + x_2 + x_3 &= 30, \\ 3x_1 + x_2 - 2x_3 &= 20, \\ -x_1 + 3x_2 + \alpha x_3 &= \beta \end{aligned}$$

have

∅. no solution,
1. a unique solution, or
∞. infinitely many solutions?

Row reduction of the augmented system matrix $(\mathbf{A} \mid \mathbf{b})$ leads to the following:

$$\begin{array}{ccc|c|l} \boxed{1} & 1 & 1 & 30 & \\ \boxed{3} & 1 & -2 & 20 & -3\ row_1 \\ \boxed{-1} & 3 & \alpha & \beta & +\ row_1 \\ \hline \boxed{1} & 1 & 1 & 30 & \\ 0 & \boxed{-2} & -5 & -70 & \\ 0 & \boxed{4} & \alpha + 1 & \beta + 30 & +2\ row_2 \\ \hline \boxed{1} & 1 & 1 & 30 & \\ 0 & \boxed{-2} & -5 & -70 & \\ 0 & 0 & \alpha - 9 & \beta - 110 & \end{array}$$

If $\alpha = 9$, the system is unsolvable in case $\beta \neq 110$. Then the rank of the augmented matrix $(\mathbf{A} \mid \mathbf{b})$ is three, while the rank of $\mathbf{A}$ is only two.

The system $\mathbf{Ax} = \mathbf{b}$ is uniquely solvable if (a) it is solvable (i.e., $\alpha \neq 9$ or if $\alpha = 9$ and $\beta = 110$) and if (b) every column in a REF $\mathbf{R}$ of $\mathbf{A}$ has a pivot. Thus, the linear system is uniquely solvable for all $\alpha \neq 9$.

If $\alpha = 9$ and $\beta = 110$, then the system has infinitely many solutions, since the system is solvable according to Theorem 3.1 and it has a free column whose corresponding variable can be set arbitrarily. ◀

(c) Solving Systems of Linear Equations

We now indicate how to solve a system of linear equations $\mathbf{A}_{mn}\mathbf{x} = \mathbf{b}$ algorithmically for $\mathbf{x} \in \mathbb{R}^n$. The process itself will tell whether the given linear system is solvable and, if so, whether it has a unique solution.

Steps for solving a system of linear equations $\mathbf{A}_{mn}\mathbf{x} = \mathbf{b}$:

Step 1: Row reduce the augmented matrix $(\mathbf{A} \mid \mathbf{b})$ to a REF $(\mathbf{R} \mid \tilde{\mathbf{b}})$.
Here $\mathbf{R}$ must be at least in simple REF, but can be further reduced if desired.
If $\text{rank}(\mathbf{A}) < \text{rank}(\mathbf{A} \mid \mathbf{b})$ (i.e., if $\text{rank}(\mathbf{R}) < \text{rank}(\mathbf{R} \mid \tilde{\mathbf{b}})$), then the system cannot be solved. **STOP.**

Step 2: If rank(**R**) = **n** which is the number of columns of **A**, then ker(**A**) = {**0**}; that is, there are no solutions to $\mathbf{Ax} = \mathbf{0}$ other than $\mathbf{x} = \mathbf{0}$; $\mathbf{x}_{\text{hom}} = \mathbf{0}$. (Skip ahead to step 3.)
Otherwise, solve the **homogeneous system** $\mathbf{Rx} = \mathbf{0}$ and obtain all possible homogeneous solutions $\mathbf{x}_{\text{hom}}$. Their degree of freedom ℓ is found as the number of free columns, or free variables in the REF of **A**.
($\ell = n-$ rank(**A**).)

Step 3: Find **one particular solution** $\mathbf{x}_{\text{part}}$ of $\mathbf{Ax} = \mathbf{b}$ by, for example, setting all components of **x** that correspond to free variables (if any) in **R** equal to zero.
Thus, one $\mathbf{x}_{\text{part}}$ can be found by taking only the pivot columns of **R** (and leaving off the free columns) and by solving this (possibly shortened) linear system $\tilde{\mathbf{R}}\tilde{\mathbf{x}} = \tilde{\mathbf{b}}$ for the updated right-hand-side $\tilde{\mathbf{b}}$.

Step 4: Form the **general solution** $\mathbf{x}_{\text{gen}}$ to $\mathbf{Ax} = \mathbf{b}$ as

$$\mathbf{x}_{\text{gen}} \;=\; \mathbf{x}_{\text{part}} \;+\; \text{any linear combination of } \mathbf{x}_{\text{hom}}.$$

Step 5: Check your answer: Is $\mathbf{Ax}_{\text{gen}} = \mathbf{b}$?

Solving $\mathbf{Rx} = \tilde{\mathbf{b}}$ or $\mathbf{Rx} = \mathbf{0}$ in steps 2 and 3 is best done by backsubstitution from the last row upwards.

EXAMPLE 6 We start the solution process from the second step and assume that we are given the specific REF $(\mathbf{R} \mid \tilde{\mathbf{b}})$:

x_1	x_2	x_3	x_4	x_5	$\tilde{\mathbf{b}}$
[1]	–1	2	3	4	3
0	0	[2]	1	1	0
0	0	0	0	[4]	8
0	0	0	0	0	0
	↑ *free*		↑ *free*		
pivot		*pivot*		*pivot*	

This system is solvable since the only zero row in **R**, the fourth row, extends to a zero row of $(\mathbf{R} \mid \tilde{\mathbf{b}})$.

The homogeneous solution $\mathbf{x}_{\text{hom}}$ of the preceding system can be obtained by assigning different variables α and β to the two free variables associated with the free components x_2 and x_4. With yet unknown values for the pivot variables x_1, x_3, and x_5, we set

$$\mathbf{x}_{\text{hom}} = \begin{pmatrix} x_1 \\ \alpha \\ x_3 \\ \beta \\ x_5 \end{pmatrix}.$$

Now solve the homogeneous system $\mathbf{Rx} = \mathbf{0}$:

x_1	x_2	x_3	x_4	x_5	$\mathbf{0}$	
$\boxed{1}$	-1	2	3	4	0	↑
0	0	$\boxed{2}$	1	1	0	↑
0	0	0	0	$\boxed{4}$	0	↑ *by backsubstitution.*
0	0	0	0	0	0	
	↑ α		↑ β			

That is, solve for the pivot components x_5, x_3, and x_1 of $\mathbf{x}_{\text{hom}}$ in terms of α and β. The last nonzero equation $4x_5 = 0$ makes $x_5 = 0$. The next to last nonzero equation then becomes

$$2x_3 + \beta + 0 = 0, \quad \text{or} \quad x_3 = -\frac{\beta}{2}.$$

Finally, the first equation yields

$$x_1 - \alpha + 2\left(-\frac{\beta}{2}\right) + 3\beta + 0 = 0, \quad \text{or} \quad x_1 = \alpha - 2\beta.$$

Thus, $\mathbf{x}_{\text{hom}} = \begin{pmatrix} \alpha - 2\beta \\ \alpha \\ -\beta/2 \\ \beta \\ 0 \end{pmatrix}$ for any α and $\beta \in \mathbb{R}$. We rewrite

$$\mathbf{x}_{\text{hom}} = \alpha \begin{pmatrix} 1 \\ 1 \\ 0 \\ 0 \\ 0 \end{pmatrix} - \frac{\beta}{2} \begin{pmatrix} 4 \\ 0 \\ 1 \\ -2 \\ 0 \end{pmatrix}$$

as a linear combination of two generating vectors; that is, with $\widehat{\beta} := -\frac{\beta}{2}$, we have

$$\begin{aligned} \ker(\mathbf{A}) &= \{\mathbf{x} \mid \mathbf{A}\mathbf{x} = \mathbf{0}\} = \{\mathbf{x} \mid \mathbf{R}\mathbf{x} = \mathbf{0}\} \\ &= \left\{ \mathbf{x} = \alpha \begin{pmatrix} 1 \\ 1 \\ 0 \\ 0 \\ 0 \end{pmatrix} + \widehat{\beta} \begin{pmatrix} 4 \\ 0 \\ 1 \\ -2 \\ 0 \end{pmatrix} \,\middle|\, \alpha,\ \widehat{\beta} \in \mathbb{R} \right\}. \end{aligned}$$

The kernel of **A** is generated by the two vectors $\begin{pmatrix} 1 \\ 1 \\ 0 \\ 0 \\ 0 \end{pmatrix}$ and $\begin{pmatrix} 4 \\ 0 \\ 1 \\ -2 \\ 0 \end{pmatrix}$. This set of generating vectors is called a **basis** for the kernel of **A**. Bases are very important for linear spaces and will be studied in Chapter 5.

Note:

The components of the homogeneous solution $\mathbf{x}_{\text{hom}}$ *of a linear system always consist of either zero entries or variable sums such as* $2\alpha - \beta$, 3β, *or* $\gamma + 2\beta - \alpha$. *A real nonzero number such as* 2, -3 *or* 1.5 *can never occur as a component of* $\mathbf{x}_{hom}$ *other than as a factor for a variable* α, β, *etc.*

Next we try to find a particular solution for the row reduced system $\mathbf{Rx} = \tilde{\mathbf{b}}$. We can do so by setting all of the free variables, namely x_2 and x_4, equal to zero. That is, we simply omit the free columns 2 and 4 in $\mathbf{R}$ and solve the **shortened linear system** $\tilde{\mathbf{R}}\tilde{\mathbf{x}} = \tilde{\mathbf{b}}$ that contains only the pivot columns of $\mathbf{R}$ and $\tilde{\mathbf{b}}$.

x_1	x_3	x_5	$\tilde{\mathbf{b}}$			
[1]	2	4	3			↑
0	[2]	1	0		↑	
0	0	[4]	8	↑		
0	0	0	0			

This shortened system contains only the **pivot variables** x_1, x_3, and x_5. It is solved by backsubstitution: The last nonzero equation makes $x_5 = 2$. Then the previous equation becomes

$$2x_3 + 2 = 0, \text{ determining } x_3 = -1.$$

Finally, the top equation reads as

$$x_1 + 2(-1) + 4(2) = 3, \text{ or } x_1 = -3.$$

Thus, one particular solution of the inhomogeneous linear system $\mathbf{Ax} = \mathbf{b} \neq \mathbf{0}$ is

$\mathbf{x}_{\text{part}} = \begin{pmatrix} -3 \\ 0 \\ -1 \\ 0 \\ 2 \end{pmatrix}$, where we have carefully reinserted the zero entries in the free second and fourth position of $\mathbf{x}_{\text{part}}$.

Note:

A particular solution $\mathbf{x}_{\text{part}}$ *of a linear system can only contain specific real numbers in its components, but no variables or variable sums.*

Now synthesize the **general solution** $\mathbf{x}_{\text{gen}}$ of $\mathbf{Ax} = \mathbf{b}$ from its homogeneous and particular solutions $\mathbf{x}_{\text{hom}}$ and $\mathbf{x}_{\text{part}}$:

$$\begin{aligned} \mathbf{x}_{\text{gen}} &= \mathbf{x}_{\text{part}} + \mathbf{x}_{\text{hom}} \\ &= \begin{pmatrix} -3 \\ 0 \\ -1 \\ 0 \\ 2 \end{pmatrix} + \alpha \begin{pmatrix} 1 \\ 1 \\ 0 \\ 0 \\ 0 \end{pmatrix} + \widehat{\beta} \begin{pmatrix} 4 \\ 0 \\ 1 \\ -2 \\ 0 \end{pmatrix} \end{aligned}$$

for arbitrary constants α and $\widehat{\beta}$ $(= -\frac{\beta}{2}) \in \mathbb{R}$.

Equivalently, one can combine the process of finding $\mathbf{x}_{\text{hom}}$ and $\mathbf{x}_{\text{part}}$ for the row-reduced linear system $\mathbf{Rx} = \tilde{\mathbf{b}}$ by assigning variable entries α and β to the free variables x_2 and x_4 and by solving the complete system $\mathbf{Rx} = \tilde{\mathbf{b}}$ by backsubstitution all at once. Again note the backwards order of the steps in backsubstitution, as indicated by the $\boxed{1}, \ldots, \boxed{3}$ indicators in the following diagram:

x_1	x_2	x_3	x_4	x_5	$\tilde{\mathbf{b}}$		
$\boxed{1}$	-1	2	3	4	3	$\overset{\boxed{3}}{\longleftrightarrow}$	$x_1 - x_2 + 2x_3 + 3x_4 + 4x_5 = 3$ $\Rightarrow x_1 = -3 + \alpha - 2\beta$
0	0	$\boxed{2}$	1	1	0	$\overset{\boxed{2}}{\longleftrightarrow}$	$2x_3 + x_4 + x_5 = 0 \Rightarrow 2x_3 + \beta + 2 = 0$ $\Rightarrow x_3 = -\frac{\beta}{2} - 1$
0	0	0	0	$\boxed{4}$	8	$\overset{\boxed{1}}{\longleftrightarrow}$	$4x_5 = 8 \Rightarrow x_5 = 2$
0	0	0	0	0	0		
	$\uparrow$		$\uparrow$				
	α		β				

$$\text{Thus, } \mathbf{x}_{\text{gen}} = \begin{pmatrix} -3+\alpha-2\beta \\ \alpha \\ -\frac{\beta}{2}-1 \\ \beta \\ 2 \end{pmatrix} = \begin{pmatrix} -3 \\ 0 \\ -1 \\ 0 \\ 2 \end{pmatrix} + \alpha \begin{pmatrix} 1 \\ 1 \\ 0 \\ 0 \\ 0 \end{pmatrix} + \widehat{\beta} \begin{pmatrix} 4 \\ 0 \\ 1 \\ -2 \\ 0 \end{pmatrix} = \mathbf{x}_{\text{part}} +$$

$\mathbf{x}_{\text{hom}}$ for α and $\widehat{\beta} := -\frac{\beta}{2} \in \mathbb{R}$, just as before. ◀

Once a linear system $\mathbf{Ax} = \mathbf{b}$ *has been solved from its augmented REF* $(\mathbf{R} \mid \tilde{\mathbf{b}})$, *it is important to evaluate* $\mathbf{Ax}$ *for the computed solution* $\mathbf{x}$ *and to compare the result with* $\mathbf{b}$. *Hand arithmetic is fickle, and one has to* **check** *that* $\mathbf{Ax} = \mathbf{b}$ *is actually satisfied by* $\mathbf{x}$.

We state two further results:

Theorem 3.4 If the REF of a matrix $\mathbf{A}$ has ℓ free variables, then any solution of the homogeneous linear system $\mathbf{Ax} = \mathbf{0}$ has ℓ degrees of freedom. ◀

Remark:

If a linear system $\mathbf{Ax} = \mathbf{b}$ is solvable and if $\ker(\mathbf{A}) \neq \{\mathbf{0}\}$, then there are both infinitely many particular solutions $\mathbf{x}_{\text{part}}$ and infinitely many homogeneous solutions $\mathbf{x}_{\text{hom}}$ according to Lemmas 2 and 3, respectively.

Thus, when solving linear systems $\mathbf{Ax} = \mathbf{b}$ via row reduction, not only are there infinitely many ways to write an equivalent REF according to Chapter 2 (unless $\mathbf{A} = \mathbf{O}_{mn}$ or $\ker(\mathbf{A}) = \mathbb{R}^n$), but the description of the solutions $\mathbf{x}$ allows for infinite variations (except when $\ker(\mathbf{A}) = \{\mathbf{0}\}$).

One of the tasks of the next two chapters is to sort through this infinity and to understand and capture its essence.

To solve linear equations $\mathbf{Ax} = \mathbf{b}$ *from a REF* $(\mathbf{R} \mid \tilde{\mathbf{b}})$ *of the augmented matrix* $(\mathbf{A} \mid \mathbf{b})$ *is a craft that must be practiced and perfected when studying linear algebra.*

3.1.P Problems

1. Solve the system of linear equations

$$\begin{aligned} 3x_1 + x_2 &= 0, \\ x_1 - x_2 &= 2, \end{aligned}$$

and check your solution. Then solve the corresponding homogeneous linear system. Why are both systems uniquely solvable?

2. Decide whether the following statements are true or false for the system of linear equations $\begin{array}{l} 4x_1 \ +6x_2 = 14 \\ -3x_1 \ +7x_2 = 1 \end{array}$ without actually solving the system:

(a) The linear system has the solution $\mathbf{x} = \begin{pmatrix} -1 \\ 2 \end{pmatrix}$;

(b) The linear system has the solution $\mathbf{x} = \begin{pmatrix} 2 \\ -1 \end{pmatrix}$;

(c) The linear system has the solution $\mathbf{x} = \begin{pmatrix} 2 \\ 1 \end{pmatrix}$;

3. Solve $\begin{pmatrix} 3 & 6 & 0 \\ 6 & 1 & -11 \\ 2 & 0 & 35 \end{pmatrix} \mathbf{x} = \begin{pmatrix} 9 \\ 7 \\ 6 \end{pmatrix}$, and check your answer.

4. Find three different solution vectors $\mathbf{u}, \mathbf{v}, \mathbf{w} \in \mathbb{R}^2$ for the linear system $\mathbf{Ax} = \begin{pmatrix} 2 & 1 \\ 4 & 2 \end{pmatrix} \mathbf{x} = \begin{pmatrix} 30 \\ 60 \end{pmatrix} = \mathbf{b}$, and check that $\mathbf{Au} = \mathbf{Av} = \mathbf{Aw} = \mathbf{b}$ in each instance. What is the rank of $\mathbf{A}$?

5. Let $\mathbf{A} = \begin{pmatrix} 0 & -2 & 4 \\ 1 & 1 & 1 \\ 3 & 5 & -1 \\ 2 & 1 & 4 \end{pmatrix}$ and $\mathbf{r} = \begin{pmatrix} 3 \\ 1 \\ a \\ b \end{pmatrix}$, and define $\mathbf{B}$ to be the augmented matrix $\mathbf{B} = (\mathbf{A} \mid \mathbf{r})$.

(a) Write out the matrix $\mathbf{B}$, and determine the number of rows and columns of $\mathbf{B}$.

(b) What is the rank of $\mathbf{B}$ in dependence of a and $b \in \mathbb{R}$?

(c) For which a and b is the linear system $\mathbf{Ax} = \mathbf{r}$ solvable?

(d) Write out the solution to $\mathbf{Ax} = \mathbf{r}$ in parametric form, where a and b are chosen so that the system is solvable.

6. Solve the linear system of equations

$$\begin{aligned} 3w + a + 7y &= 6, \\ w + a + 4y &= 5, \\ 3 - w + y &= 5 \end{aligned}$$

for a, y, and $w \in \mathbb{R}$ using Gaussian elimination and backsubstitution. Check your answer. (*Hint*: To set up the problem, name the components of your unknown vector $\mathbf{x}$ first, by using the letters a, w, and y that appear in the problem. Then sort each equation according to the component order that you have chosen in $\mathbf{x}$.)

7. Solve the linear system

$$\begin{aligned} -f_1 + 3f_2 + 4f_3 &= 2, \\ f_1 + 2f_3 + 3f_2 &= 4, \\ 4f_2 \qquad + 4f_1 &= 4 \end{aligned}$$

for the f_i and check the solution. (Be careful with setting up the augmented matrix.)

8. Solve the linear system

$$\begin{pmatrix} 1 & 2 & 0 & -11 \\ 2 & 6 & -6 & 2 \\ 0 & 2 & -4 & -4 \\ -2 & 2 & -7 & -8 \end{pmatrix} \mathbf{x} = \begin{pmatrix} -11 \\ 18 \\ 6 \\ 7 \end{pmatrix},$$

and check your solution.

9. Solve the linear system

$$\begin{array}{rrrrrl} 2x_1 & +2x_2 & -x_3 & & +x_5 = & 2, \\ -x_1 & -x_2 & +2x_3 & -3x_4 & +x_5 = & -1, \\ x_1 & +x_2 & -2x_3 & & -x_5 = & -2, \\ & & x_3 & +x_4 & +x_5 = & 3 \end{array}$$

by row echelon form reduction of the augmented matrix. Mark the free and pivot variables. Give a formula for the general solution of this system, and check your answer for the original system.

10. (a) Find an echelon form for the matrix

$$A = \begin{pmatrix} 1 & 2 & 1 & 2 & 1 & 0 & 7 \\ 1 & 2 & 4 & -1 & 4 & 3 & 1 \\ 2 & 4 & 5 & 1 & 5 & -2 & 3 \\ 2 & 4 & 8 & -2 & 8 & 6 & 2 \\ 1 & 2 & 1 & 2 & 6 & 0 & 7 \end{pmatrix}.$$

Mark the pivot elements and the free variables.

(b) What is the rank of $\mathbf{A}$ in part (a).

(c) Is there a right-hand-side $\mathbf{b} \in \mathbb{R}^5$ for which $\mathbf{Ax} = \mathbf{b}$ is not solvable?

(d) How many degrees of freedom do the solutions of each solvable linear system $\mathbf{Ax} = \mathbf{b}$ have for the matrix $\mathbf{A}$ as a system matrix.

11. For which values of k is the linear system $\begin{pmatrix} 1 & 0 & 1 \\ 0 & 1 & k \\ 2k & 2k & 2k^2 \end{pmatrix} \mathbf{x} = \begin{pmatrix} 1 \\ -1 \\ k \end{pmatrix}$ solvable? For which k can the system not be solved. For which values of k can the system be solved uniquely? Solve the system for those k that allow a solution.

12. Find all values of k for which the linear system $\begin{pmatrix} 1 & 0 & 1 \\ 2 & -1 & k \\ 0 & -k & -1 \end{pmatrix} \mathbf{x} = \begin{pmatrix} k \\ 0 \\ 1 \end{pmatrix}$ is solvable and all k for which the system cannot be solved. For which k is the system uniquely solvable, if at all?

13. Decide whether the following linear system is solvable and solve it if possible:

$$\begin{array}{rcl} x_1 \quad +x_3 \quad +x_5 &=& 3, \\ x_2 \quad +x_4 + x_5 &=& 4, \\ x_1 + x_2 \quad\quad +x_5 &=& 2, \\ x_1 \quad +x_3 + x_4 \quad &=& -1, \\ x_1 + x_2 \quad -x_4 + x_5 &=& 2. \end{array}$$

14. Find the dimension and a basis of the space of solutions for the homogeneous system of linear equations

$$\begin{array}{rcl} x_1 + x_2 - x_3 - 2x_4 + x_5 &=& 0, \\ 2x_1 + 3x_2 - x_3 - 7x_4 + 4x_5 &=& 0, \\ 2x_1 + 4x_2 \quad - 10x_4 + 6x_5 &=& 0 \end{array}$$

(We have not formally defined "dimension" or "basis"; you can get an intuitive notion of these concepts by looking at Example 6; or you may skip ahead to Chapter 5.)

15. Let $T : \mathbb{R}^6 \longrightarrow \mathbb{R}^4$ be defined by

$$T(\mathbf{x}) = \begin{pmatrix} x_1 - x_3 + 2x_5 \\ -x_4 + x_5 \\ 0 \\ -x_6 \end{pmatrix}.$$

(a) Why is T a linear map?

(b) Find $T(\mathbf{e}_1)$, $T(\mathbf{e}_3)$, and $T(\mathbf{e}_5)$.

(c) Find the standard matrix representation $\mathbf{A}_\mathcal{E}$ of T. What is the rank of $\mathbf{A}$?

(d) Find two vectors $\mathbf{x} \neq \mathbf{y} \in \mathbb{R}^6$ with $T(\mathbf{x}) = T(\mathbf{y})$, if possible.

16. If $\mathbf{A} = \begin{pmatrix} 1 & 5 & 3 & -2 \\ 0 & 1 & 4 & 0 \\ 0 & 0 & 1 & -2 \\ 0 & 0 & 0 & 1 \end{pmatrix}$ and $\mathbf{b} = \begin{pmatrix} 34 \\ -22 \\ 0 \\ 1 \end{pmatrix}$, solve the following linear systems:

(a) $\mathbf{Ax} = \mathbf{b}$;

(b) $\mathbf{Ay} = \mathbf{y}$; and

(c) $\mathbf{Az} = -\mathbf{z} + 2\mathbf{b}$.

17. (a) Find the general solution of the linear system $\mathbf{Ax} = \mathbf{b}$ for $\mathbf{A} = \begin{pmatrix} 1 & 1 & 3 & 4 \\ 1 & 0 & 2 & 1 \\ 1 & 2 & 4 & 7 \\ 2 & 1 & 4 & 3 \end{pmatrix}$ and $\mathbf{b} = \begin{pmatrix} 1 \\ 2 \\ 0 \\ 0 \end{pmatrix}$.

Verify your answer.

(b) Find the general form of vectors in $\ker(\mathbf{A})$ for $\mathbf{A}$ in part (a).

What is $\text{rank}(\mathbf{A})$?

(c) Is the linear system $\mathbf{Ax} = \mathbf{c}$ uniquely solvable for any right-hand-side $\mathbf{c}$ and $\mathbf{A}$?

18. Give several examples of

(a) an unsolvable linear system of equations $\mathbf{Ax} = \mathbf{b}$, and

(b) a solvable, but not uniquely solvable, linear system $\mathbf{Ax} = \mathbf{b}$.

19. Is there a homogeneous linear system $\mathbf{A}_{15\times 7}\mathbf{x} = \mathbf{0}$ that cannot be solved for some particular system matrix $\mathbf{A}$?

20. What are the possibilities for a system of linear equations $\mathbf{A}_{mn}\mathbf{x}_n = \mathbf{b}_m$ as regards solvability, unique solvability, number of solutions etc?

State the theoretical facts and give examples with dense system matrices $\mathbf{A}_{mn}$ for $m, n \geq 3$ for each possible situation. (A matrix $\mathbf{A} = (a_{ij})$ is **dense** if $a_{ij} \neq 0$ for all indices i, j.) Then show that your chosen examples are indeed as claimed.

(*Hint*: There are three distinct cases.)

21. Try to solve the linear system

$$\begin{pmatrix} 4 & -1 & 0 & 0 \\ 1 & 4 & -1 & 0 \\ 0 & 1 & 5 & -1 \\ 0 & 0 & -1 & -4 \end{pmatrix} \mathbf{x} = \begin{pmatrix} 7 \\ 7 \\ -5 \\ -3 \end{pmatrix}.$$

(*Hint*: Swap the equations for a nice sequence of pivots.)

22. Let $\mathbf{A}_{mn} = \begin{pmatrix} a & \dots & a \\ \vdots & & \vdots \\ a & \dots & a \end{pmatrix}$ for $a \in \mathbb{R}$.

(a) For which right-hand-sides $\mathbf{b} \in \mathbb{R}^m$ can the linear system $\mathbf{Ax} = \mathbf{b}$ be solved?

(b) If $\mathbf{a} \neq 0$ and $\mathbf{b} = \beta \begin{pmatrix} 1 \\ \vdots \\ 1 \end{pmatrix} \in \mathbb{R}^m$ for $\beta \in \mathbb{R}$, show that every vector of the form

$\dfrac{\beta}{k \cdot a}(\mathbf{e}_{i_1} + \mathbf{e}_{i_2} + \ldots + \mathbf{e}_{i_k})$ solves $\mathbf{Ax} = \mathbf{b}$, where $1 \le k \le n$, $1 \le i_1 < \ldots < i_k \le n$ and $\mathbf{e}\ldots$ denotes a standard unit vector.

(c) How many different solution vectors are there in part (b)?

(*Hint*: For each fixed k, $1 \le k \le n$, find the number of such solution vectors.)

23. Let

$$T\begin{pmatrix} x_1 \\ x_2 \\ x_3 \\ x_4 \end{pmatrix} = \begin{pmatrix} x_1 - 2x_2 + x_1 + 4x_4 \\ \ln(e^{3x_3}) - 4\log_2(2^{(x_1 - x_2)}) \\ x_3 \cdot (\sin^2 x_2 + \cos^2 x_2) \end{pmatrix}.$$

(a) Why is T a linear mapping: $\mathbb{R}^4 \to \mathbb{R}^3$?

(b) Find the standard matrix representation $\mathbf{A}_{\mathcal{E}}$ for T.

(c) Find all vectors $\mathbf{x} \in \mathbb{R}^4$ with $T(\mathbf{x}) = \mathbf{0}$. (Use the matrix $\mathbf{A}_{\mathcal{E}}$ and solve $\mathbf{A}_{\mathcal{E}}\mathbf{x} = \mathbf{0}$.)

(d) Check your answer in part (c) for T as originally defined.

(*Hint*: Review log laws and trig identities.)

24. Determine whether the following systems of linear equations with the augmented matrix $(\mathbf{A} \mid \mathbf{b})$ have a solution, a unique solution, infinitely many solution, or are unsolvable from the given description:

(a) The augmented matrix $(\mathbf{A} \mid \mathbf{b})$ is square and its REF has a pivot in every row.

(b) $\text{rank}(\mathbf{A}) = \text{rank}(\mathbf{A} \mid \mathbf{b})$.

(c) $\mathbf{A}$ is 3×2 and $\mathbf{b} = \begin{pmatrix} 1 \\ -1 \\ 7 \end{pmatrix}$.

(d) $\mathbf{A}$ is 3×2 and $\mathbf{b} = \begin{pmatrix} 1 \\ 0 \\ 0 \end{pmatrix}$.

(e) $\mathbf{A}$ is 3×2 and $\mathbf{b} = \begin{pmatrix} 0 \\ 0 \\ 0 \end{pmatrix}$.

(f) $\mathbf{A}$ is 2×3 and $\mathbf{b} = \begin{pmatrix} 0 \\ 0 \end{pmatrix}$.

25. Solve the linear system of equations

$$\begin{aligned} x_1 + 2x_3 + x_4 - x_5 &= 4, \\ -2x_1 + x_2 - x_3 + 2x_4 + 3x_5 &= -1, \\ 2x_2 + 7x_3 + 9x_4 + x_5 &= 14, \\ 3x_1 - 2x_2 + 2x_3 - 3x_4 - 7x_5 &= -2 \end{aligned}$$

in the following steps:

(a) Determine the augmented matrix $(\mathbf{A} \mid \mathbf{b})$ for this linear system.

(b) Determine a REF $(\mathbf{R} \mid \widetilde{\mathbf{b}})$ for $(\mathbf{A} \mid \mathbf{b})$.

(c) Mark your pivot and free variables in $(\mathbf{R} \mid \widetilde{\mathbf{b}})$. Check for solvability.

(d) Use backsubstitution to find the general solution $\mathbf{x}_{\text{gen}}$ of the linear system.

(e) What is a particular solution $\mathbf{x}_{\text{part}}$ of the system? What is the homogeneous solution $\mathbf{x}_{\text{hom}}$?

(f) Check that for $\mathbf{x}_{\text{part}}$ and $\mathbf{x}_{\text{hom}}$ in part (e) we have $\mathbf{Ax}_{\text{part}} = \mathbf{b}$ and $\mathbf{Ax}_{\text{hom}} = \mathbf{0}$.

(g) Is $\mathbf{x} = \begin{pmatrix} 1 \\ 0 \\ 3 \\ -1 \\ 2 \end{pmatrix}$ a solution of the given linear system?

26. Solve the given linear systems graphically as in Example 1. Is either system uniquely solvable?

(a)
$$\begin{aligned} x_1 + 3x_2 &= 7, \\ 2x_1 - x_2 &= 0, \\ x_1 + 2x_2 &= 4. \end{aligned}$$

(b)
$$\begin{aligned} x_1 + 3x_2 &= 7, \\ 2x_1 - x_2 &= 0, \\ 2x_1 + 3x_2 &= 8. \end{aligned}$$

27. Prove Lemma 3 in detail.

28. What type of entries must always occur in

(a) a particular solution $\mathbf{x}_{\text{part}}$ and

(b) the general homogeneous solution $\mathbf{x}_{\text{hom}}$

of a solvable linear system?

29. Prove or disprove by giving a counterexample: If a system of linear equations $\mathbf{A}_{mn}\mathbf{x} = \mathbf{b}$ has more unknowns than equations ($n > m$), then the system is unsolvable or it has infinitely many solutions.

30. Prove the following or disprove by giving a counterexample: If a system of linear equations $\mathbf{A}_{mn}\mathbf{x} = \mathbf{b}$ has more equations than unknowns ($m > n$), then the system is uniquely solvable.

31. Solve the linear system

$$\begin{pmatrix} 1 & 2 & 1 & 1 & 0 \\ 2 & 3 & 1 & 2 & 1 \\ 0 & 1 & 2 & 2 & 2 \\ 2 & 3 & 1 & 4 & 5 \end{pmatrix} \mathbf{x} = \begin{pmatrix} 3 \\ 5 \\ 8 \\ 11 \end{pmatrix}.$$

Specifically, find the kernel of the system matrix **A**, as well as a particular solution of the linear system. Check your answers.

32. Prove the following or disprove by giving a counterexample:
If a system of linear equations $\mathbf{A}_{mn}\mathbf{x} = \mathbf{b}$ has the same number of equations as unknowns ($m = n$), then the system may not be solvable.

33. Solve $\begin{aligned} x - 2y + z &= -1, \\ 4x - 7y + z &= -1, \\ x + y + z &= 9, \end{aligned}$ and check your answer.

34. Draw the planar graph of each equation in the x–and y–coordinates, and determine geometrically and via Gaussian elimination whether the following linear systems are solvable:

(a) $x + 2y = 0$, $2x - y = 0$, and $3x + y = 0$.

(b) $x + 2y = 3$, $2x + 4y = 2$.

(c) $2x - 3y = 4$, $x - 2y = -2$, and $x - 5y = 6$.

35. Determine whether each of the following is true or false (include a proof or a counterexample):

(a) If the REF of a matrix **A** has free variables, then the corresponding linear system $\mathbf{A}\mathbf{x} = \mathbf{b}$ admits several solutions.

(b) If the REF of a matrix **A** has free variables, then the corresponding linear system $\mathbf{A}\mathbf{x} = \mathbf{b}$ may not be solvable.

(c) Complete the sentence: If the REF of a matrix **A** has free variables, then the corresponding linear system $\mathbf{A}\mathbf{x} = \mathbf{b}$ is solvable, provided that.

36. Let $T : \mathbb{R}^2 \to \mathbb{R}^3$ be a linear transformation with $T\begin{pmatrix} 1 \\ 0 \end{pmatrix} = \begin{pmatrix} 1 \\ -1 \\ 1 \end{pmatrix}$ and $T\begin{pmatrix} 1 \\ -1 \end{pmatrix} = \begin{pmatrix} 0 \\ 1 \\ 4 \end{pmatrix}$.

(a) Find the standard matrix representation $\mathbf{A}_{\mathcal{E}}$ of T. Check your answer and compute $\mathbf{A}_{\mathcal{E}}\begin{pmatrix} 2 \\ -1 \end{pmatrix}$.

(b) Find a point $\mathbf{x} = \begin{pmatrix} a \\ b \end{pmatrix} \in \mathbb{R}^2$ with $T(\mathbf{x}) = \begin{pmatrix} 2 \\ 0 \\ -1 \end{pmatrix}$ for T using part (a). Check your answer.

(c) Find a point $\mathbf{y} = \begin{pmatrix} c \\ d \end{pmatrix} \in \mathbb{R}^2$ with $T(\mathbf{y}) = \begin{pmatrix} 1 \\ 1 \\ 9 \end{pmatrix}$ for T using part (a). Check your answer.

37. Let $\mathbf{A} = \begin{pmatrix} 1 & 2 & 3 & 4 & 0 & -1 \\ -2 & -3 & -9 & -6 & -3 & 1 \\ 3 & 5 & 11 & 8 & 5 & -4 \\ 0 & -2 & 10 & 4 & 0 & 18 \end{pmatrix}$.

(a) Decide whether there are some right-hand-sides $\mathbf{b} \in \mathbb{R}^4$ for which the linear system $\mathbf{A}\mathbf{x} = \mathbf{b}$ is not solvable or whether every linear system $\mathbf{A}\mathbf{x} = \mathbf{b}$ with the specific matrix **A** is solvable.

(b) Assume that **b** is a right-hand-side vector for which the linear system $\mathbf{A}\mathbf{x} = \mathbf{b}$ is solvable. Decide whether there are always infinitely many solutions for the system or whether the solution **x** is unique for some right-hand sides **b**. (These parts requires a verbal answer with mathematical reasons.)

(c) Repeat parts (a) and (b) with the matrix
$$\mathbf{B} = \begin{pmatrix} 1 & -2 & 3 & 0 \\ 2 & -3 & 5 & -2 \\ 3 & -9 & 11 & 10 \\ 4 & -6 & 8 & 4 \\ 0 & -3 & 5 & 0 \\ -1 & 1 & -4 & 18 \end{pmatrix}.$$

38. Which of the two linear systems $\begin{pmatrix} 1 & 2 & 3 \\ 2 & 3 & 4 \end{pmatrix}\mathbf{x} = \mathbf{b}$ and $\begin{pmatrix} 1 & 2 \\ 2 & 3 \\ 3 & 4 \end{pmatrix}\mathbf{y} = \mathbf{c}$ can be solved for all right-hand sides $\mathbf{b} \in \mathbb{R}^2$ and $\mathbf{c} \in \mathbb{R}^3$?
If one of these system can be solved for a specific right-hand side, will the solution be unique?

39. Consider linear systems $\mathbf{A}\mathbf{x} = \mathbf{b}$ with the following sizes for the system matrix **A**:
(1) $\mathbf{A}_{4,5}$ (2) $\mathbf{A}_{4,2}$ (3) $\mathbf{A}_{5,5}$ (4) $\mathbf{A}_{3,4}$
(5) $\mathbf{A}_{4,12}$ (6) $\mathbf{A}_{8,5}$ (7) $\mathbf{A}_{6,4}$ (8) $\mathbf{A}_{1,1}$.

(a) Decide from **A**'s given size, which system $\mathbf{A}\mathbf{x} = \mathbf{b}$ may be solvable for all compatible right-hand sides **b**.

(b) Decide from **A**'s given size, which system $\mathbf{A}\mathbf{x} = \mathbf{b}$ may not be solvable for each compatible right-hand-side **b**.

(c) Decide from **A**'s given size which system $\mathbf{A}\mathbf{x} = \mathbf{b}$ may be uniquely solvable for all compatible right-hand sides **b**.

(d) Decide from **A**'s given size, which system $\mathbf{A}\mathbf{x} = \mathbf{b}$ may be uniquely solvable for some compatible right-hand sides **b**.

Teacher's Problem-Making Exercise

T 3. To come up with meaningful and easily computable problems involving linear equations, we may start with an integer system matrix $\mathbf{A} \in \mathbb{R}^{m,n}$ and an integer vector $\mathbf{x} \in \mathbb{R}^n$ and construct the integer right-hand-side $\mathbf{b}$ by evaluating $\mathbf{b} := \mathbf{A}\mathbf{x} \in \mathbb{R}^m$. The solution $\mathbf{x}$ of the linear system $\mathbf{A}\mathbf{x} = \mathbf{b}$ can then be found in integer arithmetic, provided that $\mathbf{A}$ can be row reduced over the integers. (See Problem T 2 and especially Example (c).)

EXAMPLE

(a) For $\mathbf{A} = \begin{pmatrix} 1 & -1 & 3 \\ 0 & 2 & 1 \\ 4 & 2 & -1 \\ 1 & 1 & 2 \end{pmatrix}$ and $\mathbf{x} = \begin{pmatrix} 1 \\ 2 \\ -3 \end{pmatrix}$ evaluate $\mathbf{b} = \mathbf{A}\mathbf{x}$. Then the given vector $\mathbf{x}$ is an integer solution of the linear system $\mathbf{A}\mathbf{x} = \mathbf{b}$.

(b) Construct a right-hand-side vector $\mathbf{c}$ for which the system $\mathbf{A}\mathbf{y} = \mathbf{c}$ is unsolvable for $\mathbf{A}$ as given in part (a). To be able to achieve this, we must have rank $(\mathbf{A}) < m$, which is the number of rows of $\mathbf{A}$. (See Theorem 3.1.) If this is so for $\mathbf{A}$, then all that needs to be done is to modify the computed right-hand-side $\mathbf{b} = \mathbf{A}\mathbf{x}$ slightly: A small random integer perturbation of $\mathbf{b}$ will most likely land outside the column space of $\mathbf{A}$ if rank$(\mathbf{A}_{mn}) < m$. (See the Solutions.) ◀

3.2 Applications

Circuits, Networks, Chemistry, and MATLAB

Linear equations and methods to solve them have been known since antiquity. Yet they have rested at the fringe of mathematics for many centuries until quite recently.

Certainly since classic Greek and Arab times and possibly before, there were little schoolbook ditties such as this in every mathematics primer:

> *As I went to the store to buy bread and candles, I was charged 2 silverlings for 3 candles and 1 bread today. Yesterday I paid 3 silverlings for 2 candles and 2 loafs of bread.*
>
> *Pray, what does the shop keeper charge for one candle and what for one loaf?*

Modern times have brought with them a new view of the world: the scientific, analytical viewpoint. With this, vast observational data of the physical world have been generated and beg to be interpreted. Astronomy set out several centuries ago to compute the orbits and periods of planets and stars. Geodesy helped map the new and old worlds. And so forth and so on. Nowadays land surveys via satellite will involve dozens of satellites and up to hundreds of thousands of geodesic points on earth to track positioning to high accuracy in the fractional-foot range. Serious global positioning systems (GPS) rely on solving millions of linear equations simultaneously, while consumer GPS gadgets work with just thousands of linear equations to achieve a position accuracy to within several yards. Solving such huge systems of linear equations has become a breadbasket for mathematicians, engineers, and programmers alike. It drives the modern technological age. Externally, this field is fueled by the huge amount of raw data that we and our technical gizmos collect and that have to be interpreted and subsequently compressed to give a graspable output for us. But there is also an intrinsic, natural drive to solve linear systems. Many of the basic laws of physics and science are best expressed in equation form. For example, in all network flows, the quantities entering a node must exit it, giving rise to one equation for each node. This basic network flow law is true in electricity and is called **Kirchhoff's first law**.

In an electric circuit, the sum of all inflowing and outflowing currents is zero at any junction according to Kirchhoff's first law, and the sum of all voltage drops along any closed loop of a circuit is zero by what is called **Kirchhoff's second law**.

EXAMPLE 7 Find the currents I_1, I_2, and I_3 in the branches of the circuit shown in Figure 3-5.

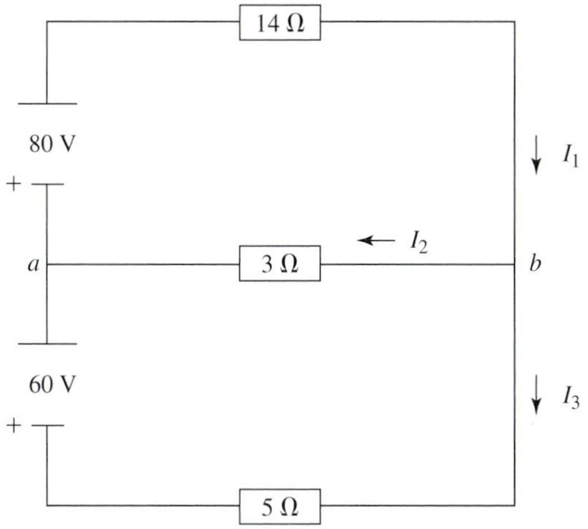

Figure 3-5

Here the 14 Ω box denotes a resistor of 14 Ohms, the symbol 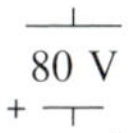

denotes an 80 Volt source, and the drawn-out lines show the electrical connections of this circuit.

Ohm's law $U = R \cdot I$ relates voltages U, resistances R, and currents I. By Ohm's law we know that the voltage drop U_1 associated with I_1 is $14I_1$, the one for I_2 is $U_2 = 3I_2$, and for I_3 it is $U_3 = 5I_3$. The total voltage drop in the upper loop associated with I_1, I_2, and the $80V$ source satisfies $14I_1 + 3I_2 = 80$, since $80V$ are supplied to this loop. The lower loop involving I_2, I_3, and the $60V$ source similarly satisfies $5I_3 - 3I_2 = 60$ by Kirchhoff's second law, where we acknowledge the reverse directions of the currents I_2 and I_3 by giving them opposite signs. If we additionally apply Kirchhoff's first law to the junctions at a or b in the figure, then we obtain a third equation $I_1 = I_2 + I_3$ that involves the three unknown currents I_j.

Thus, the unknown vector $\mathbf{x} = \begin{pmatrix} I_1 \\ I_2 \\ I_3 \end{pmatrix}$ of currents satisfies the following set of linear equations

$$\begin{aligned} 14I_1 + 3I_2 &= 80, \\ 5I_3 - 3I_2 &= 60, \\ I_1 - I_2 - I_3 &= 0, \end{aligned}$$

or, equivalently,

$$\begin{pmatrix} 14 & 3 & 0 \\ 0 & -3 & 5 \\ 1 & -1 & -1 \end{pmatrix} \mathbf{x} = \begin{pmatrix} 80 \\ 60 \\ 0 \end{pmatrix}.$$

If we swap equations 1 and 3 for ease with elimination, we obtain the equivalent augmented matrix

I_1	I_2	I_3			
$\boxed{1}$	–1	–1	0		
0	–3	5	60		
$\boxed{14}$	3	0	80	$-14\ row_1$	
$\boxed{1}$	–1	–1	0		
0	–3	5	60	$\cdot 6$	
0	17	14	80		
$\boxed{1}$	–1	–1	0		
0	–18	30	360	$+\ row_3$	*(to obtain a* ± 1 *pivot)*
0	17	14	80		
$\boxed{1}$	–1	–1	0		
0	$\boxed{-1}$	44	440		
0	$\boxed{17}$	14	80	$+17\ row_2$	
$\boxed{1}$	–1	–1	0	③ $\longleftrightarrow$	$I_1 = I_2 + I_3 \approx 6.46\ A$
0	$\boxed{-1}$	44	440	② $\longleftrightarrow$	$I_2 = -440 + 44I_3 \approx -3.46\ A$
0	0	$\boxed{762}$	7560	① $\longleftrightarrow$	$I_3 \approx 9.92\ A$

Thus, by backsubstituting, we get that the branch circuit currents are $I_3 = \dfrac{7560}{762} \approx 9.92\ amp$, $I_2 \approx -3.46\ amp$, and $I_1 \approx 6.46\ amp$. They solve this problem uniquely.

Knowing the voltage drops $U_i = R_i \cdot I_i$ from Ohm's law at each resistor and the current flows I_i now allows us to size the parts of this circuit: The 14Ω resistor drops $14 \cdot I_1 = 14 \cdot 6.46 = 90.4V$, for example. Thus, it requires a minimum wattage rating of $U_1 \cdot I_1 = 90.4 \cdot 6.46 = 584W$ and possibly at least twice that to insure the longevity of the circuit. Analogously, the 3Ω resistor should be able to dissipate at least 36 W and the 5Ω one at least 492 W, as can be readily checked. ◀

Kirchhoff's law likewise holds for traffic flow, water flow, etc. Electric circuits with their voltage drops are different from hydraulic flows and network flows that are "incompressible". Incompressible flows can only obey Kirchhoff's first law that "what enters a node must exit the node."

EXAMPLE 8 The following describes the simplified traffic flow in the area of Baltimore harbor, where the numbers represent fictitious data on the number of cars per hour.

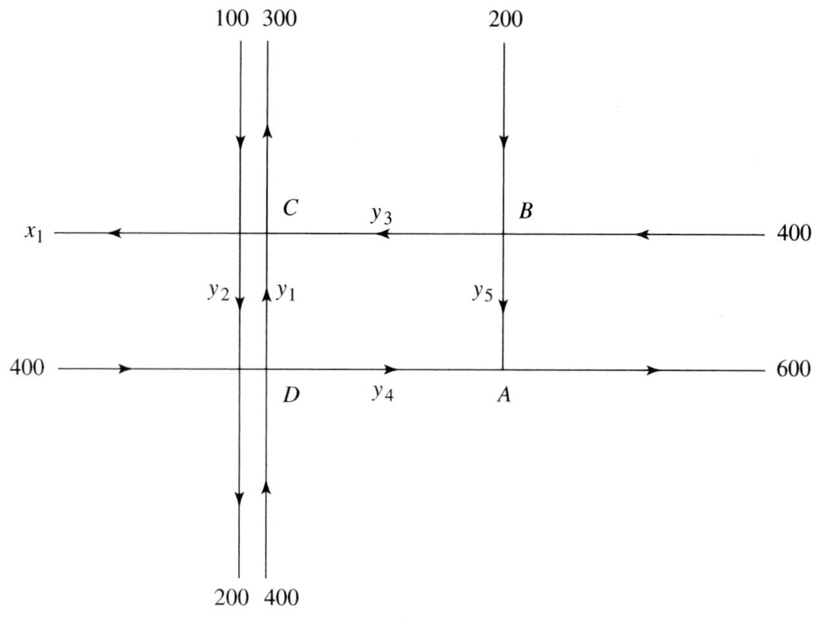

Figure 3-6

There are five unknown traffic flow sizes $y_1, \ldots, y_5$ in the diagram and the unknown number x_1 of cars leaving the area to the west. We use the inflow = outflow equations of Kirchhoff's first law at each of the four intersections A, B, C, and D and the total inflow = total outflow equation for the whole map area in order to obtain the following five linear equations for the six unknowns y_i, $i = 1, \ldots, 5$, and x_1.

The inflow = outflow equations at the four intersections are

$$\begin{aligned}
\text{at } A: &\quad y_4 + y_5 = 600 \\
\text{at } B: &\quad 200 + 400 = y_3 + y_5 \\
\text{at } C: &\quad y_1 + y_3 + 100 = y_2 + x_1 + 300 \\
\text{at } D: &\quad y_2 + 400 + 400 = y_1 + y_4 + 200,
\end{aligned}$$

while for the total flow, we have

$$100 + 200 + 400 + 400 + 400 = x_1 + 300 + 600 + 200.$$

This leads to the following system $\mathbf{Aw} = \mathbf{b}$ of five equations in our six unknown quantities that we combine in the vector $\mathbf{w} := \begin{pmatrix} y_1 \\ \vdots \\ y_5 \\ x_1 \end{pmatrix}$:

$$\begin{pmatrix} 0 & 0 & 0 & 1 & 1 & 0 \\ 0 & 0 & 1 & 0 & 1 & 0 \\ 1 & -1 & 1 & 0 & 0 & -1 \\ -1 & 1 & 0 & -1 & 0 & 0 \\ 0 & 0 & 0 & 0 & 0 & -1 \end{pmatrix} \mathbf{w} = \begin{pmatrix} 600 \\ 600 \\ 200 \\ -600 \\ -400 \end{pmatrix}.$$

If we reorder the individual equations judiciously, we are left to solve the following equivalent system:

y_1	y_2	y_3	y_4	y_5	x_1	**b**	
$\boxed{1}$	–1	1	0	0	–1	200	
$\boxed{-1}$	1	0	–1	0	0	–600	$+row_1$
0	0	1	0	1	0	600	
0	0	0	1	1	0	600	
0	0	0	0	0	–1	–400	
$\boxed{1}$	–1	1	0	0	–1	200	
0	0	$\boxed{1}$	–1	0	–1	–400	
0	0	$\boxed{1}$	0	1	0	600	$-row_2$
0	0	0	1	1	0	600	
0	0	0	0	0	–1	–400	
$\boxed{1}$	–1	1	0	0	–1	200	
0	0	$\boxed{1}$	–1	0	–1	–400	
0	0	0	$\boxed{1}$	1	1	1000	
0	0	0	$\boxed{1}$	1	0	600	$-row_3$
0	0	0	0	0	–1	–400	
$\boxed{1}$	–1	1	0	0	–1	200	
0	0	$\boxed{1}$	–1	0	–1	–400	
0	0	0	$\boxed{1}$	1	1	1000	
0	0	0	0	0	$\boxed{-1}$	–400	$\cdot(-1)$
0	0	0	0	0	$\boxed{-1}$	–400	$-row_4$
$\boxed{1}$	–1	1	0	0	–1	200	
0	0	$\boxed{1}$	–1	0	–1	–400	
0	0	0	$\boxed{1}$	1	1	1000	$(\mathbf{R} \mid \tilde{\mathbf{b}})$
0	0	0	0	0	$\boxed{1}$	400	
0	0	0	0	0	0	0	
	↑			↑			

We note that the system $\mathbf{Aw} = \mathbf{b}$ is solvable since $\text{rank}(\mathbf{A}) = \text{rank}(\mathbf{A} \mid \mathbf{b}) = 4$ from our row reduction. This is so because no pivot sticks out in the updated right-hand-side column $\tilde{\mathbf{b}}$ beyond the last pivot of $\mathbf{R}$. Further note that there are two free variables. The general solution has the form

$$\mathbf{x}_{\text{gen}} = \mathbf{x}_{\text{part}} + \mathbf{x}_{\text{hom}} = \begin{pmatrix} 0 \\ 0 \\ 600 \\ 600 \\ 0 \\ 400 \end{pmatrix} + \alpha \begin{pmatrix} 1 \\ 0 \\ -1 \\ -1 \\ 1 \\ 0 \end{pmatrix} + \beta \begin{pmatrix} 1 \\ 1 \\ 0 \\ 0 \\ 0 \\ 0 \end{pmatrix},$$

which should be verified by the students.

If the system had not been solvable, our collected data would have been suspect. Moreover, note that our traffic-flow numbers y_2 and y_5 act as free variables in the system, giving drivers options to choose their personal itinerary through the area. If this flow diagram had been represented by a uniquely solvable system (i.e., if its REF $\mathbf{R}$ had a pivot in every column according to Theorem 3.3), then the resulting traffic situation would have been unbearable: Every driver in the area would have to pick exactly the one route that would lead to the unique solution of the flow problem, leading to gridlock until everybody finally got it right (or flew off the handle).

There are real life benefits to linear systems that are not uniquely solvable. ◀

Next we study how to balance chemical reactions.

Chemical reactions are often described by two sets of chemical compounds and an arrow in between. For example, $H_2S + O_2 \to H_2O + SO_2$ describes the creation of sulfur dioxide SO_2 and water H_2O from H_2S and ozone O_2. However, the simple chemical formula is not balanced, since two oxygen atoms appear on its left and three on its right side. To properly balance chemical equations such as this, one needs to find positive integers α, β, γ, and δ, so that

$$\alpha H_2S + \beta O_2 = \gamma H_2O + \delta SO_2.$$

The number of atoms on the left- and the right-hand sides need to be in balance for each chemical element that occurs in the reaction equation.

If we identify $\mathbf{x} = \begin{pmatrix} x_1 \\ x_2 \\ x_3 \end{pmatrix}$ with the vector $\begin{pmatrix} \text{H} \\ \text{S} \\ \text{O} \end{pmatrix}$ for the three chemical elements hydrogen H, sulfur S, and oxygen O that are present, then this amounts to solving the following linear system

$$\alpha \begin{pmatrix} 2 \\ 1 \\ 0 \end{pmatrix} + \beta \begin{pmatrix} 0 \\ 0 \\ 2 \end{pmatrix} = \gamma \begin{pmatrix} 2 \\ 0 \\ 1 \end{pmatrix} + \delta \begin{pmatrix} 0 \\ 1 \\ 2 \end{pmatrix}$$

over the positive integers since H_2S is represented by the vector $\begin{pmatrix} 2 \\ 1 \\ 0 \end{pmatrix}$, O_2 by $\begin{pmatrix} 0 \\ 0 \\ 2 \end{pmatrix}$, etc.

EXAMPLE 9 (a) Balance the chemical reaction $H_2S + O_2 \to H_2O + SO_2$.
We rewrite the problem as

$$\begin{pmatrix} 2 & 0 & -2 & 0 \\ 1 & 0 & 0 & -1 \\ 0 & 2 & -1 & -2 \end{pmatrix} \begin{pmatrix} \alpha \\ \beta \\ \gamma \\ \delta \end{pmatrix} = \begin{pmatrix} 0 \\ 0 \\ 0 \end{pmatrix},$$

and solve a judiciously permuted set of the equations by row reduction:

$$\begin{array}{c|cccc|c|l}
S & \boxed{1} & 0 & 0 & -1 & 0 & \\
H & \boxed{2} & 0 & -2 & 0 & 0 & -2\ \mathit{row}_1 \\
O & 0 & 2 & -1 & -2 & 0 & \\
\hline
 & \boxed{1} & 0 & 0 & -1 & 0 & \\
 & 0 & 0 & -2 & 2 & 0 & \textit{swap rows 2 and 3} \\
 & 0 & 2 & -1 & -2 & 0 & \\
\hline
 & \boxed{1} & 0 & 0 & -1 & 0 & \longleftrightarrow \alpha = \delta \\
 & 0 & \boxed{2} & -1 & -2 & 0 & \longleftrightarrow 2\beta = \gamma + 2\delta \\
 & 0 & 0 & \boxed{-2} & 2 & 0 & \longleftrightarrow \gamma = \delta \\
\hline
 & \alpha & \beta & \gamma & \delta & &
\end{array}$$

For a positive integer solution, we must insure that $\gamma + 2\delta$ is even, which forces $\gamma = \delta$ to be even. After choosing $\delta = 2 = \gamma$, we obtain $\beta = 3$ and $\alpha = 2$. Thus, this chemical equation has the balanced form

$$2H_2S + 3O_2 \to 2H_2O + 2SO_2.$$

(b) To balance the chemical equation $CH_4 + O_2 \to C + H_2O$, we use the vector $\begin{pmatrix} \mathrm{C} \\ \mathrm{H} \\ \mathrm{O} \end{pmatrix}$ for the three elements carbon C, hydrogen H, and oxygen O that occur in the reaction. We solve $\alpha \begin{pmatrix} 1 \\ 4 \\ 0 \end{pmatrix} + \beta \begin{pmatrix} 0 \\ 0 \\ 2 \end{pmatrix} = \gamma \begin{pmatrix} 1 \\ 0 \\ 0 \end{pmatrix} + \delta \begin{pmatrix} 0 \\ 2 \\ 1 \end{pmatrix}$ over the positive integers as follows:

$$\begin{array}{c|cccc|c|l}
C & \boxed{1} & 0 & -1 & 0 & 0 & \\
H & \boxed{4} & 0 & 0 & -2 & 0 & -4\ \mathit{row}_1 \\
O & 0 & 2 & 0 & -1 & 0 & \\
\hline
 & \boxed{1} & 0 & -1 & 0 & 0 & \\
 & 0 & 0 & 4 & -2 & 0 & \textit{swap rows 2 and 3} \\
 & 0 & 2 & 0 & -1 & 0 & \\
\hline
 & \boxed{1} & 0 & -1 & 0 & 0 & \\
 & 0 & \boxed{2} & 0 & -1 & 0 & \\
 & 0 & 0 & \boxed{4} & -2 & 0 & \\
\hline
 & \alpha & \beta & \gamma & \delta & &
\end{array}$$

Thus, $\delta = 2$, $\gamma = 1$, $\beta = 1$, and $\alpha = 1$ balance this equation as follows:

$$CH_4 + O_2 \rightarrow C + 2H_2O.$$

To solve linear systems $\mathbf{Ax} = \mathbf{b}$ in MATLAB, we need to enter the system matrix $\mathbf{A}$ and the respective right-hand-side $\mathbf{b}$ and use the backslash command:

```
x = A\b
```

For $\mathbf{A}$ and $\mathbf{b}$ from Example 7, this command gives the unique solution

$$\mathbf{x} = \begin{pmatrix} 6.4567 \\ -3.4646 \\ 9.9213 \end{pmatrix}.$$

For Example 8, the MATLAB backslash command finds a particular solution

$$\mathbf{w}_{\text{part}} = \begin{pmatrix} 600 \\ 0 \\ 0 \\ 0 \\ 600 \\ 400 \end{pmatrix}.$$

Note that this particular solution differs from our hand computed one in Example 8. This is to be expected. (See the remark on the different implementations of Gaussian elimination at the end of Section 3.1.) All meaningful solutions to the traffic-flow problem must involve nonnegative integer entries in $\mathbf{w}$, since traffic laws frown upon backing up for whole city blocks, and we cannot accept fractional cars. For $\mathbf{A}$ and $\mathbf{b}$ in Example 8, MATLAB gives out the warning that the system matrix $\mathbf{A}_{5\times 6}$ is rank deficient of rank 4. If we enter the MATLAB command

```
null(A)
```

for the system matrix $\mathbf{A}$ of Example 8, we obtain two noninteger vectors $\mathbf{u}_1$ and $\mathbf{u}_2$ that generate the kernel of $\mathbf{A}$, namely,

$$\mathbf{u}_1 = \begin{pmatrix} 0.5951 \\ 0.7547 \\ 0.1596 \\ 0.1596 \\ -0.1596 \\ 0 \end{pmatrix} \text{ and } \mathbf{u}_2 = \begin{pmatrix} -0.4662 \\ 0.0440 \\ 0.5101 \\ 0.5101 \\ -0.5101 \\ 0 \end{pmatrix}.$$

Note that MATLAB computes in floating-point arithmetic by default. If we enter

```
format rat
```

it will put out the results as (approximate) rational numbers. Entering

```
format
```

toggles back to the default output format. For help, enter `help format` on the MATLAB command line.

To achieve integer homogeneous solutions for Example 8 in MATLAB, we modify the two computed vectors $\mathbf{u}_1$ and $\mathbf{u}_2 \in \ker(\mathbf{A})$. First we replace $\mathbf{u}_2$ by $\mathbf{u}_2 - \frac{0,5101}{0.1596}\mathbf{u}_1 \in \ker(\mathbf{A})$ and obtain $\begin{pmatrix} -2.3684 \\ -2.3684 \\ 0 \\ 0 \\ 0 \\ 0 \end{pmatrix}$. This we scale by its first coefficient and obtain one integer vector $\mathbf{w}_2 = \begin{pmatrix} 1 \\ 1 \\ 0 \\ 0 \\ 0 \\ 0 \end{pmatrix} \in \ker(A)$. Next we divide $\mathbf{u}_1$ by -0.1596 to obtain $\begin{pmatrix} -3.7287 \\ -4.7287 \\ -1 \\ -1 \\ 1 \\ 0 \end{pmatrix} \in \ker(\mathbf{A})$, which we further modify by adding $4.7287 \cdot \mathbf{w}_2$. The result is $\mathbf{w}_1 := \begin{pmatrix} 1 \\ 0 \\ -1 \\ -1 \\ 1 \\ 0 \end{pmatrix}$, which we know from Lemma 3 also belongs to the kernel of $\mathbf{A}$. Thus, $\mathbf{x}_{\text{hom}} = \alpha\mathbf{w}_1 + \beta\mathbf{w}_2$ describes the kernel of $\mathbf{A}$ in the same integer manner as our hand computations in Example 8 did.

For square matrices $\mathbf{A}_{nn}$ of arbitrary rank, MATLAB uses a partially pivoted Gauss elimination process to solve $\mathbf{Ax} = \mathbf{b}$ upon the command `A\b`; see Chapter 13. For nonsquare matrices $\mathbf{A}_{mn}$, MATLAB generally computes an orthogonal QR factorization of $\mathbf{A}$ (see Section 10.2), from which it finds the least-squares solution $\mathbf{x}_0$ with $\min_{\mathbf{x}} \|\mathbf{Ax} - \mathbf{b}\| = \|\mathbf{Ax}_0 - \mathbf{b}\|$ upon the command `A\b`. Here $\|\mathbf{y}\| := \sqrt{y_1^2 + \cdots + y_n^2}$ denotes the **norm** of a vector $\mathbf{y} \in \mathbb{R}^n$. For further details, see Chapter 12 and the MATLAB manual. For singular square system matrices $\mathbf{A}_{nn}$, please refer to the end of Section 12.3 to learn how to obtain the least-squares solution for $\mathbf{Ax} = \mathbf{b}$ in MATLAB.

We have shown how to interpret and solve electrical circuit, flow, and chemistry problems via linear equations.

3.2.P Problems

1. Set up a linear system of equations for the introductory shopkeeper problem, and determine the price of one bread and one candle.

2. Balance the following chemical equations:

   (a) $CH_4 + O_2 \rightarrow CO + H_2O$.

   (b) $(CH_4)_2 + (O_2)_3 \rightarrow CO + H_2O$.

3. Planes through the origin $\mathbf{0}$ in $\mathbb{R}^3$ can be described by their normal vector $\mathbf{a} \in \mathbb{R}^3$ as the kernel of $\mathbf{A} = \begin{pmatrix} - & \mathbf{a}^T & - \end{pmatrix}$.

   (a) Find three points on the plane through the origin that is normal to $(2,\ 4,\ -1)^T$.

   (b) Find the intersection of the two planes in $\mathbb{R}^3$ through the origin with normals $\mathbf{a}_1 = \begin{pmatrix} 2 & -2 & 1 \end{pmatrix}^T$ and $\mathbf{a}_2 = \begin{pmatrix} 0 & 1 & 2 \end{pmatrix}^T$. (*Hint*: use a kernel.)

   (c) Repeat part (b) with $\mathbf{a}_1 = \begin{pmatrix} 2 & -2 & 1 \end{pmatrix}^T$ and $\mathbf{a}_2 = \begin{pmatrix} -4 & 4 & -2 \end{pmatrix}^T$.

   (d) Find the intersection of three planes in $\mathbb{R}^3$ through the origin with respective normal vectors $\mathbf{a}_1 = \begin{pmatrix} 2 & -2 & 1 \end{pmatrix}^T$, $\mathbf{a}_2 = \begin{pmatrix} 0 & 1 & 2 \end{pmatrix}^T$, and $\mathbf{a}_3 = \begin{pmatrix} 1 & 1 & 1 \end{pmatrix}^T$.

4. (a) What is the equation of the plane in $\mathbb{R}^3$ with normal vector $\mathbf{a} = (1, -1, 1)^T$ through $\mathbf{x}_0 = (2, 0, 1)^T$?

   (b) Do the same for the normal vector $\mathbf{a} = (3, 2, -4)^T$ and the point $\mathbf{x}_0 = (1, -2, 3)^T$.

5. Find the point of intersection in $\mathbb{R}^3$ of the following three planes:

   normal to $(0, 2, -2)^T$ through $(2, 2, 0)^T$,
   normal to $(2, 3, -1)^T$ through $(0, 0, 0)^T$,
   normal to $(1, 0, -1)^T$ through $(1, 1, 1)^T$.

6. Balance the chemical equation
   $Zn + H_2SO_4 \rightarrow H_2S + ZnSO_4 + H_2O$.

7. Balance the following chemical equations:

   (a) $NaOH + CO_2 \rightarrow (Na)_2CO_3 + H_2O$.

   (b) $NaOH + (Fe)_2(SO_4)_3 \rightarrow$
   $\rightarrow (Na)_2SO_4 + Fe(OH)_3$.

   (c) $C_3H_8 + O_2 \rightarrow CO_2 + H_2O$.

8. Solve the electrical current problem for the following circuits:

   (a)

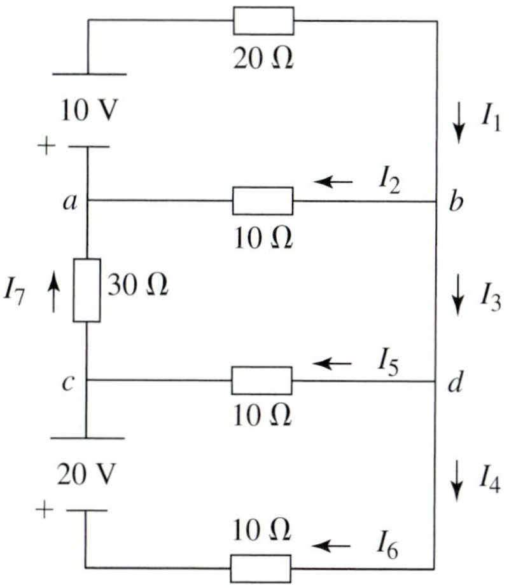

   (b)

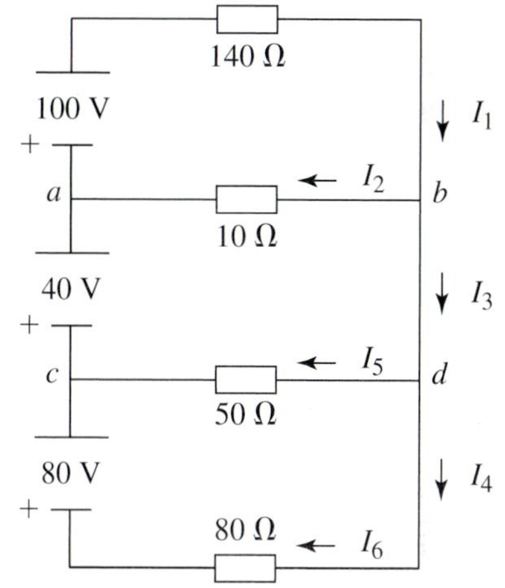

   by setting up the corresponding linear equations according to Kirchhoff's laws and by using both hand computations and MATLAB.

   Explain why the preceding circuit problems must have unique currents I_j in their branches. How does this physical necessity translate into the equivalent linear equations formulation of the problem?

   (c) Compute the minimal Watt ratings for each of the resistors in the circuits of parts (a) and (b).

9. Which traffic-flow numbers y_i are possible for the following roadside data:

(a)

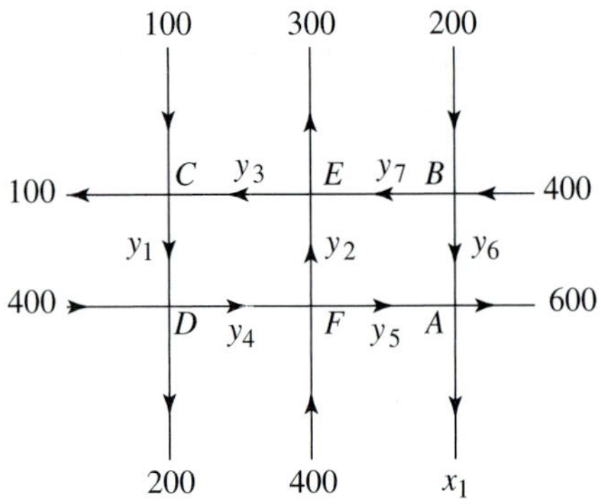

(b) Consider the altered traffic-flow problem with the road from F to E closed:

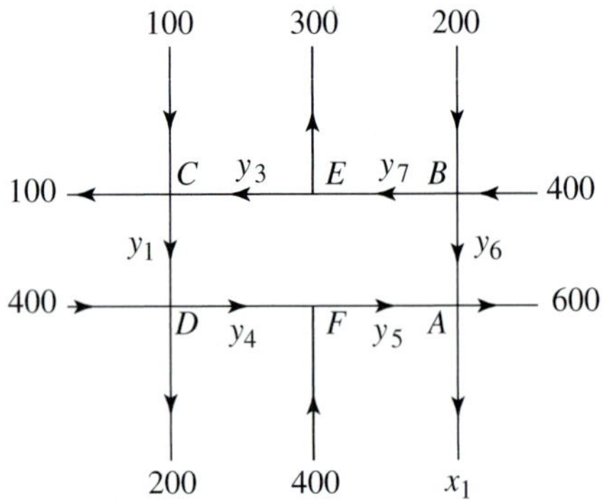

Check that the corresponding linear equations allow for free variables in both problems. This requirement is essential for a somewhat willful and smooth traffic flow.

10. For $\mathbf{A} = \begin{pmatrix} 1 & 2 & 3 \\ 2 & 3 & 4 \\ 5 & 6 & 7 \end{pmatrix}$, solve the linear systems

$$\mathbf{Ax} = \begin{pmatrix} 1 \\ 2 \\ 3 \end{pmatrix} \text{ and } \mathbf{Ay} = \begin{pmatrix} 6 \\ 9 \\ 12 \end{pmatrix} \text{ using MATLAB.}$$

Check the computed solutions. Find a basis of the kernel of **A** via MATLAB. Which of these two systems is solvable according to Theorem 3.1?

11. Repeat the previous problem with

$$\mathbf{B} = \begin{pmatrix} 1 & 2 & 3 & 4 \\ 0 & 1 & 2 & 3 \\ 4 & 3 & 2 & 1 \\ 3 & 2 & 1 & 0 \end{pmatrix} \text{ and the two right-hand sides}$$

$$\mathbf{b} = \begin{pmatrix} -2 \\ -2 \\ 2 \\ 2 \end{pmatrix} \text{ and } \mathbf{c} = \begin{pmatrix} 4 \\ 4 \\ 4 \\ -4 \end{pmatrix}.$$

12. Consider the two linear systems $\begin{pmatrix} 1 & 2 & 3 \\ 2 & 3 & 4 \end{pmatrix}$

$$\mathbf{x} = \begin{pmatrix} -2 \\ -2 \end{pmatrix} \text{ and } \begin{pmatrix} 1 & 2 \\ 2 & 3 \\ 3 & 4 \end{pmatrix} \mathbf{y} = \begin{pmatrix} -1 \\ 0 \\ 1 \end{pmatrix}$$

(a) Which of these linear systems can be solved in theory or in MATLAB? Why?

(b) Which of these systems is uniquely solvable?

(Refer to Problem 38 in Section 3.1.P.)

13. Solve the nonsensical problem from the introduction of Section 1.1:

Find $x \in \mathbb{R}$ with $2x = 6$ and $4x = -4$ by writing out the augmented system matrix $(\mathbf{A} \mid \mathbf{b})$ and using MATLAB's `A\b` command.

What is going on? (*Hint*: MATLAB treats this unsolvable problem as a **least-squares problem**; see Chapters 10 and 12.)

3.R Review Problems

1. Write out the definitions and concepts underlying the notions of :

   (a) A system of linear equations.

   (b) A solvable linear system.

   (c) An unsolvable linear system of equations.

   (d) A uniquely solvable linear system.

   (e) The two components of a general solution of a linear system.

   (f) The method and steps of solving linear systems of equations.

2. Describe all allowed steps of Gaussian elimination in full detail using an example of your own creation.

3. Use Gaussian reduction to show that the linear system

$$\begin{pmatrix} 1 & 0 & 3 \\ -1 & 1 & 2 \\ 2 & 0 & 6 \end{pmatrix} \mathbf{x} = \begin{pmatrix} 1 \\ 1 \\ k \end{pmatrix}$$

is solvable for certain values of k. For which k is the system uniquely solvable? Find the general solution.

4. What rank can the matrix $\mathbf{A} = \begin{pmatrix} -2 & 3 \\ \alpha & -9 \end{pmatrix}$ have, depending on the value of α.

5. Consider the following linear system $\mathbf{A}_{5\times 5}\mathbf{x} = \mathbf{b}$, where the right-hand-side $\mathbf{b}$ actually depends on a parameter β ($\mathbf{b} = \mathbf{b}(\beta)$) and only the third column entries of $\mathbf{A}$ are explicitly known:

$$\mathbf{A}_{5,5}\mathbf{x} = \begin{pmatrix} * & * & 0 & * & * \\ & & -3 & & \\ \vdots & \vdots & 0 & \vdots & \vdots \\ & & 9 & & \\ * & * & 0 & * & * \end{pmatrix} \mathbf{x} = \begin{pmatrix} 0 \\ -2 \\ 0 \\ \beta \\ 0 \end{pmatrix} = \mathbf{b}(\beta).$$

(a) Find all values of β for which the foregoing linear system is always solvable, no matter what values the unspecified entries marked by $*$ in $\mathbf{A}_{5,5}$ might have.

(b) Assume that β_1 is one of the values from part (a). Describe the set of all solutions $\mathbf{x}$ to $\mathbf{A}\mathbf{x} = \mathbf{b}(\beta_1)$ in terms of the four unspecified columns of $\mathbf{A}$.

(c) If β_2 does not belong to the values from part (a), try to specify values for the $*$ marked entries of $\mathbf{A}$, so that $\mathbf{A}\mathbf{x} = \mathbf{b}(\beta_2)$ is not solvable.

6. Find all values of $a \in \mathbb{R}$ for which the linear system $\begin{pmatrix} 2 & 0 & 1 \\ 1 & 2 & 0 \\ 0 & 1 & 2 \end{pmatrix} \mathbf{x} = \begin{pmatrix} 1 \\ 2 \\ a \end{pmatrix}$ is

(a) solvable.

(b) unsolvable.

In part (a), compute the solution $\mathbf{x}$ in terms of a and check your answer.

If either part (a), or part (b) cannot occur for the specific linear system, please quote the relevant theoretical facts.

7. Solve the linear system of equations

$$\begin{aligned} 2x + 3y^3 - 5\ln(z) &= 2, \\ -4x - 6y^3 - 14\ln(z) &= 8, \\ 6x + 9y^3 - 18\ln(z) &= 7.5, \end{aligned}$$

for x, y^3, and $\ln(z)$ first. Then find the values for the variables y and z. How many legitimate real solutions are there to this problem?

8. Let $\mathbf{R}_{m,n+1}$ be a REF for an augmented matrix $(\mathbf{A} \mid \mathbf{b})$ with $\mathbf{A} \in \mathbb{R}^{m,n}$ and $m, n \geq 1$. Determine whether each of the following is true or false:

(a) If $\text{rank}(\mathbf{R}) = 0$, then the linear system $\mathbf{A}\mathbf{x} = \mathbf{b}$ can be solved.

(b) If $\text{rank}(\mathbf{R}) = 0$, then the linear system $\mathbf{A}\mathbf{x} = \mathbf{b}$ cannot be uniquely solved.

(c) If $\text{rank}(\mathbf{R}) = 1$, then the linear system $\mathbf{A}\mathbf{x} = \mathbf{b}$ can be solved.

(d) If $\text{rank}(\mathbf{R}) = 1 = \text{rank}(\mathbf{A})$, then the linear system $\mathbf{A}\mathbf{x} = \mathbf{b}$ can be solved uniquely.

(e) If $\text{rank}(\mathbf{R}) = 1 = \text{rank}(\mathbf{A})$, then the linear system $\mathbf{A}\mathbf{x} = \mathbf{b}$ can be solved.

(f) If $m = n$ and $\text{rank}(\mathbf{R}) = n = \text{rank}(\mathbf{A})$, then the linear system $\mathbf{A}\mathbf{x} = \mathbf{b}$ can be uniquely solved.

(g) If $m = n$ and $\text{rank}(\mathbf{R}) = n$, then the linear system $\mathbf{A}\mathbf{x} = \mathbf{b}$ can be solved.

9. A linear system with more equations than unknowns is called an **overdetermined system**. Prove or disprove the following:

(a) There are overdetermined systems that cannot be solved.

(b) There are overdetermined systems that cannot be uniquely solved.

Justify your answers with an example if one of the statements above is true. If a statement is not true, state and prove what is true.

10. A linear system with more unknowns than equations is called an **underdetermined system**. Prove or disprove the following:

(a) There are underdetermined systems that can be solved.

(b) Every underdetermined system has at least one solution.

(c) There are underdetermined systems that can be uniquely solved.

(d) There are underdetermined systems with infinitely many solutions.

Justify your answers with an example or give a proof, as needed.

11. Let $\mathbf{A}_{4\times 6}$ have rank 4. Can every linear system $\mathbf{A}\mathbf{x} = \mathbf{b} \in \mathbb{R}^4$ be solved? Uniquely?

12. Assume that the REF of an augmented matrix $(\mathbf{A} \mid \mathbf{b})_{m\times(n+1)}$ has a pivot in its $(n+1)^{\text{st}}$ column. Is the linear system $\mathbf{Ax} = \mathbf{b}$ solvable for the given $\mathbf{b}$ and $\mathbf{A}$. Can it be uniquely solved?

13. Assume that the REF $\mathbf{R}$ of $\mathbf{A}_{mn}$ has a pivot in every row. Show that every linear equation $\mathbf{Ax} = \mathbf{b}$ can be solved for an arbitrary right hand side $\mathbf{b} \in \mathbb{R}^m$. What further condition on $\mathbf{A}$ will ensure unique solvability?

14. Prove or disprove for a matrix $\mathbf{A}_{mn}$, all of whose entries are rational numbers $\frac{a}{b}$ with integers a and $b \neq 0$:

    (a) There is an integer row echelon form for $\mathbf{A}$.

    (b) The homogeneous linear system $\mathbf{Ax} = \mathbf{0}$ has integer solutions.

    (c) If $\mathbf{b} \in \mathbb{R}^m$ has integer components, then $\mathbf{Ax} = \mathbf{b}$ has an integer solution, provided $\mathbf{A}$ has rational entries and the system is solvable.

15. Solve the linear system

$$\begin{aligned} 2x_1 - x_2 - x_3 &= 1 \\ 4x_1 - 2x_2 - 2x_3 &= 4 \end{aligned}$$

from its augmented matrix, if possible.

16. Let $\mathcal{P}_n$ denote the set of all real polynomials of degree at most n. Define $L : \mathcal{P}_n \to \mathcal{P}_n$ by $L(p) = p - 2p'$ where p' denotes the derivative.

    (a) Show that L is a linear transformation.

    (b) Which $p \in \mathcal{P}_n$ does L map to the zero function?

    (c) What is $L(3x^4 - 5x^3 + 17)$?

    If $L(p) = 7x - 3$, what is p?

    Given that $L(2q) = x^3 - 2x^2$, what is the polynomial q?

17. Let $\mathbf{A} = \begin{pmatrix} -1 & 4 & 2 & 1 \\ 1 & -4 & -4 & 2 \\ -2 & 8 & -1 & 10 \\ -1 & 4 & 2 & 4 \end{pmatrix}$ and $\mathbf{b} = \begin{pmatrix} -3 \\ 0 \\ -13 \\ 0 \end{pmatrix}$.

    (a) Why is the linear system $\mathbf{Ax} = \mathbf{b}$ solvable?

    (b) Find the general solution of the linear system $\mathbf{Ax} = \mathbf{b}$. (*Hint*: Use the work that is necessary to solve part (a).)

    (c) Check your answer.

18. (a) Show that not every vector $\mathbf{b} \in \mathbb{R}^3$ can be written as $\alpha \begin{pmatrix} 1 \\ -1 \\ 2 \end{pmatrix} + \beta \begin{pmatrix} 2 \\ 3 \\ -1 \end{pmatrix} + \gamma \begin{pmatrix} 1 \\ -6 \\ 4 \end{pmatrix}$ for some α, β, and $\gamma \in \mathbb{R}$.

    (b) Find four distinct vectors in $\mathbb{R}^3$ that can all be written as $\alpha \begin{pmatrix} 1 \\ -1 \\ 2 \end{pmatrix} + \beta \begin{pmatrix} 2 \\ 3 \\ -1 \end{pmatrix} + \gamma \begin{pmatrix} 1 \\ -6 \\ 7 \end{pmatrix}$.

    (c) Show that every vector in $\mathbb{R}^2$ can be written as $\alpha \begin{pmatrix} 2 \\ -3 \end{pmatrix} + \beta \begin{pmatrix} 7 \\ -3 \end{pmatrix}$ for some $\alpha, \beta \in \mathbb{R}$.

Standard Tasks and Questions

1. Represent a given explicit system of linear equations in matrix form $\mathbf{Ax} = \mathbf{b}$ and equivalently by the augmented matrix $(\mathbf{A} \mid \mathbf{b})$.

2. Is a given system of linear equations $\mathbf{Ax} = \mathbf{b}$ solvable? How about uniquely solvable? Is it unsolvable? How can one decide?

3. Why is a homogeneous system of linear equations $\mathbf{Ax} = \mathbf{0}$ always solvable?

   Define matrix and right hand side coefficients so that the linear systems $\mathbf{A}_{3\times2}\mathbf{x} = \mathbf{b}$ and $\mathbf{B}_{2\times3}\mathbf{x} = \mathbf{h}$ are both unsolvable.

4. Decide whether a given linear system of equations $\mathbf{Ax} = \mathbf{b}$ is solvable, and, if so, describe how to find all solutions $\mathbf{x}$ of $\mathbf{Ax} = \mathbf{b}$.

5. How many free variables does the solution of a given linear system $\mathbf{Ax} = \mathbf{b}$ allow? What determines their number?

6. Why and how do we use backsubstitution to solve a row-reduced linear system $\mathbf{Rx} = \tilde{\mathbf{b}}$?

7. What are the two components of the general solution of a system of linear equations $\mathbf{Ax} = \mathbf{b}$? How can they be obtained? What do they look like? How can the general solution $\mathbf{x}_{\text{gen}}$ be obtained in one step?

Subheadings of Lecture Three

Water in a dugout canoa

CHAPTER

4

Subspaces

We study subsets of $\mathbb{R}^n$ that are closed under the two linear vector space operations.

4.1 Lecture Four

The Image and Kernel of a Linear Transformation

Linearly closed subspaces of $\mathbb{R}^n$ can be described in two complementary ways.

(a) Two Generic Representations of a Subspace

If a matrix $\mathbf{A}_{mn} = \begin{pmatrix} | & & | \\ \mathbf{a}_1 & \dots & \mathbf{a}_n \\ | & & | \end{pmatrix}$ is given by its column vectors $\begin{pmatrix} | \\ \mathbf{a}_i \\ | \end{pmatrix} \in \mathbb{R}^m$, then the **image** of $\mathbb{R}^n$ under the linear transformation

$$\mathbf{x} \in \mathbb{R}^n \mapsto \mathbf{Ax} \in \mathbb{R}^m$$

is the set of all points of the form $\mathbf{Ax} \in \mathbb{R}^m$, namely,

$$\operatorname{im}(\mathbf{A}) := \{\mathbf{Ax} \mid \mathbf{x} \in \mathbb{R}^n\} = \left\{ \begin{pmatrix} | & & | \\ \mathbf{a}_1 & \dots & \mathbf{a}_n \\ | & & | \end{pmatrix} \begin{pmatrix} x_1 \\ \vdots \\ x_n \end{pmatrix} \middle| \, x_i \in \mathbb{R} \right\} =$$

$$= \left\{ x_1 \begin{pmatrix} | \\ \mathbf{a}_1 \\ | \end{pmatrix} + \dots + x_n \begin{pmatrix} | \\ \mathbf{a}_n \\ | \end{pmatrix} \middle| \, x_i \in \mathbb{R} \right\} \subset \mathbb{R}^m.$$

With this definition, recall from Chapter 3 that a linear system $\mathbf{Ax} = \mathbf{b}$ is solvable precisely when $\mathbf{b} \in \text{im}(\mathbf{A})$.

The image of $\mathbb{R}^n$ under $\mathbf{A}$ is the set of all **linear combinations** of the n columns $\begin{pmatrix} | \\ \mathbf{a}_i \\ | \end{pmatrix} \in \mathbb{R}^m$ of $\mathbf{A}_{mn}$. This image is a subset of $\mathbb{R}^m$ that is closed under the two linear operations of vector addition and scalar multiplication: If z and w lie in the image of $\mathbf{A}$, then $\mathbf{Au} = \mathbf{z}$ and $\mathbf{Av} = \mathbf{w}$ for two vectors $\mathbf{u}$ and $\mathbf{v}$ in $\mathbb{R}^n$. Consequently, the linear combination

$$\alpha\mathbf{z} + \beta\mathbf{w} = \alpha\mathbf{Au} + \beta\mathbf{Av} = \mathbf{A}(\alpha\mathbf{u} + \beta\mathbf{v})$$

belongs to $\text{im}(\mathbf{A})$ for all α, β, due to the linearity property (1.1) $f(\alpha\mathbf{x} + \beta\mathbf{y}) = \alpha f(\mathbf{x}) + \beta f(\mathbf{y})$ used for $\mathbf{A}$ in place of f, and $\mathbf{u}$ and $\mathbf{v}$ in place of $\mathbf{x}$ and $\mathbf{y}$.

The whole space $\mathbb{R}^m$ is closed under vector addition and scalar multiplication. Any linearly closed set of vectors is called a **vector space**. For the axiomatic definition of an abstract vector space, see Appendix C.

In this chapter, we study subsets of $\mathbb{R}^m$, such as $\text{Im}(\mathbf{A}_{mn}) \subset \mathbb{R}^m$, which are closed under the two vector space operations. Linearly closed subsets of $\mathbb{R}^m$ help us study linear algebra and linear transformations.

DEFINITION 1 A nonempty subset $U \subset \mathbb{R}^m$ is called a **subspace** of $\mathbb{R}^m$ if U is closed under both vector addition and scalar multiplication, i.e., if for every $\mathbf{u}, \mathbf{v} \in U$ and every $\alpha \in \mathbb{R}$ we have $\mathbf{u} + \mathbf{v} \in U$ and $\alpha\mathbf{u} \in U$.

EXAMPLE 1

(a) The singleton space $\{\mathbf{0}\} \subset \mathbb{R}^k$, containing only the origin, is a subspace of every k–space.

(b) The kernel $\ker(\mathbf{A})$ is a subspace of $\mathbb{R}^n$ for every $m \times n$ matrix $\mathbf{A}_{mn}$; see Lemma 3 in Section 3.1.

(c) $\mathbb{R}^k$ is a subspace of itself for every k.

Subsets of abstract vector spaces occur quite often in function spaces, such as polynomials; refer to Section 1.3 and Appendix C.

(d) The polynomials $\mathcal{P}_n$ of fixed degree less than or equal to n are a subspace of all functions $f : [a, b] \to \mathbb{R}$, since arbitrary linear combinations of nth-degree polynomials are n^{th}-degree polynomials.

(e) The polynomials p with $p(2) = 0$ form a subspace of $\mathcal{P}_n$, since if p and $q \in \mathcal{P}_n$ with $p(2) = 0$ and $q(2) = 0$, then $(p + q)(2) = p(2) + q(2) = 0 + 0 = 0$ and $(\alpha p)(2) = \alpha p(2) = \alpha \cdot 0 = 0$ for all $\alpha \in \mathbb{R}$.

(f) The polynomials $p \in \mathcal{P}_n$ with $p(1) = \beta \neq 0$ do not form a subspace, since for $q \in \mathcal{P}_n$ with $q(1) = \beta$, we have

$$(p + q)(1) = p(1) + q(1) = \beta + \beta = 2 \cdot \beta \neq \beta,$$

so that $p + q$ does not satisfy the defining condition for this subset of $\mathcal{P}_n$. ◀

In **set theory**, all subsets of a given set can be represented in one of two equivalent and intrinsic ways, namely,

- inclusively or
- exclusively.

That is, we define a subset by **inclusion** (by listing all of its members), or we define a subset by **exclusion** (by listing or describing all elements that do not belong to it).

We apply this dual representation of sets to linear algebra.

DEFINITION 2 The **span** of a set of vectors $\mathbf{u}_1, \ldots, \mathbf{u}_k \in \mathbb{R}^m$ is the subset

$$\text{span}\{\mathbf{u}_1, \ldots, \mathbf{u}_k\} := \left\{\mathbf{y} = \alpha_1\mathbf{u}_1 + \cdots + \alpha_k\mathbf{u}_k \in \mathbb{R}^m \mid \alpha_i \in \mathbb{R}\right\} \subset \mathbb{R}^m.$$

Lemma 1 The span of a set of vectors $\{\mathbf{u}_1, \ldots, \mathbf{u}_k\} \subset \mathbb{R}^m$ is a subspace. It consists of all **linear combinations**

$$\alpha_1\mathbf{u}_1 + \cdots + \alpha_k\mathbf{u}_k$$

of the given vectors $\mathbf{u}_i \in \mathbb{R}^m$ for arbitrary scalar coefficients $\alpha_i \in \mathbb{R}$. ◀

A proof of this Lemma should be developed in class by the students. *What needs to be proved here?*

With this result, the generic **inclusive subspace representation** of a subspace $U \subset \mathbb{R}^m$ becomes

$$U = \text{span}\{\mathbf{u}_1, \ldots, \mathbf{u}_k\} = \text{im}(\mathbf{U}_{mk}),$$

where the vectors $\mathbf{u}_1, \ldots, \mathbf{u}_k \in \mathbb{R}^m$ define membership in the subspace, $k \geq 1$, and $\mathbf{U} = \mathbf{U}_{mk} = \begin{pmatrix} | & & | \\ \mathbf{u}_1 & \ldots & \mathbf{u}_k \\ | & & | \end{pmatrix}$ is the $m \times k$ matrix with columns $\mathbf{u}_i$. Any vector $\mathbf{y} \in \mathbb{R}^m$ that can be obtained linearly from the $\mathbf{u}_i$ (i.e., as a linear combination of the $\mathbf{u}_i$), belongs to $\text{span}\{\mathbf{u}_1, \ldots, \mathbf{u}_k\}$. We can allow $k = 0$ here if we define the span of the empty set as the singleton subspace $U = \{\mathbf{0}\}$ in analogy to the empty sum $\sum_{i=1}^{0}$, which is standardly equated to zero. Since every vector $\mathbf{u} \in U = \text{span}\{\mathbf{u}_1, \ldots, \mathbf{u}_k\}$ is a **linear combination** of the given vectors $\mathbf{u}_1, \ldots, \mathbf{u}_k$, we say that the vectors $\{\mathbf{u}_i\}$ **span** or generate the subspace U.

Note that we use two words "span" here. One is a noun, **the span**. It refers to the subset of all linear combinations as in Definition 2. The other is the verb **to span**, meaning to generate a subspace. For example, the sentence *"The span of several vectors is the subspace that is spanned by all linear combinations of these vectors"* uses both the noun and the verb forms of "span".

Observe further that we denote the matrix $\mathbf{U} = \mathbf{U}_{mk} = \begin{pmatrix} | & & | \\ \mathbf{u}_1 & \dots & \mathbf{u}_k \\ | & & | \end{pmatrix}$ with the given vectors $\mathbf{u}_i \in \mathbb{R}^m$ as columns and the subspace $U = \text{span}\{\mathbf{u}_i\} = \text{im}(\mathbf{U}_{mk})$ by the same letter.

EXAMPLE 2 Does the vector $\mathbf{b} = \begin{pmatrix} 1 \\ 1 \\ -1 \end{pmatrix}$ lie in

$$\text{span}\left\{\mathbf{u}_1 = \begin{pmatrix} 1 \\ -1 \\ 3 \end{pmatrix}, \mathbf{u}_2 = \begin{pmatrix} -2 \\ 2 \\ -6 \end{pmatrix}, \mathbf{u}_3 = \begin{pmatrix} 1 \\ 0 \\ 4 \end{pmatrix}\right\} \subset \mathbb{R}^3?$$

To answer this question, we need to find out whether $\mathbf{b}$ is a linear combination of the three given vectors $\mathbf{u}_1$, $\mathbf{u}_2$, and $\mathbf{u}_3$. Interpreting the span of the vectors $\mathbf{u}_i$ as the image of the matrix $\mathbf{U} = \begin{pmatrix} | & | & | \\ \mathbf{u}_1 & \mathbf{u}_2 & \mathbf{u}_3 \\ | & | & | \end{pmatrix}$, we thus must check whether the linear system $\mathbf{Ux} = \mathbf{b}$ is solvable. If it is, then $\mathbf{b}$ belongs to the span; otherwise, it does not:

$$\begin{array}{ccc|c|l}
\boxed{1} & -2 & 1 & 1 & \\
\boxed{-1} & 2 & 0 & 1 & +\,row_1 \\
\boxed{3} & -6 & 4 & -1 & -3\,row_1 \\
\hline
\boxed{1} & -2 & 1 & 1 & \\
0 & 0 & \boxed{1} & 2 & \\
0 & 0 & \boxed{1} & -4 & -\,row_2 \\
\hline
\boxed{1} & -2 & 1 & 1 & \\
0 & 0 & \boxed{1} & 2 & \\
0 & 0 & 0 & \boxed{-6} &
\end{array}$$

The linear system is inconsistent according to Theorem 3.1, and thus the vector $\mathbf{b}$ does not lie in the span of the $\mathbf{u}_i$. ◀

We now turn to the second, the exclusive subset representation, and try to understand its meaning for linear algebra. An **exclusive representation of a subspace** $U \subset \mathbb{R}^n$ defines a subspace by excluding some vectors or directions $\mathbf{a}_i \in \mathbb{R}^n$ from belonging to U. In $\mathbb{R}^n$, it is most convenient to describe a subspace $U \subset \mathbb{R}^n$ exclusively by demanding that all vectors $\mathbf{u}$ of such a subspace be **orthogonal** to some excluded direction vectors $\mathbf{a}_i \in \mathbb{R}^n$. For a more detailed look at orthogonality, see Chapter 10.

DEFINITION 3 Two vectors $\mathbf{u}, \mathbf{v} \in \mathbb{R}^n$ are **orthogonal** if their dot product $\mathbf{u} \cdot \mathbf{v} = 0$. Orthogonality for vectors is denoted by the $\perp$ sign: $\mathbf{u} \perp \mathbf{v}$ if $\mathbf{u} \cdot \mathbf{v} = 0$.
A vector $\mathbf{a} \in \mathbb{R}^n$ is **orthogonal to a subspace** $U \subset \mathbb{R}^n$, denoted by $\mathbf{a} \perp U$, if $\mathbf{a} \cdot \mathbf{u} = 0$ for all $\mathbf{u} \in U$.
The set of all vectors that are orthogonal to every vector of a given subspace $U \subset \mathbb{R}^n$ is U's **orthogonal complement** in $\mathbb{R}^n$ denoted by

$$U^{\perp} := \{\mathbf{x} \in \mathbb{R}^n \mid \mathbf{x} \cdot \mathbf{u} = 0 \text{ for all } \mathbf{u} \in U\}.$$

EXAMPLE 3 (a) The vector $\begin{pmatrix} 1 \\ -1 \\ 2 \end{pmatrix}$ is orthogonal to each of the three vectors $\begin{pmatrix} 1 \\ 1 \\ 0 \end{pmatrix}$, $\begin{pmatrix} 1 \\ -1 \\ -1 \end{pmatrix}$ and $\begin{pmatrix} 0 \\ 0 \\ 0 \end{pmatrix} \in \mathbb{R}^3$; compute the three dot products.

(b) The zero vector of $\mathbb{R}^n$ is orthogonal to every vector in $\mathbb{R}^n$ since, $\mathbf{0} \cdot \mathbf{u} = 0$ for all vectors $\mathbf{u}$ (i.e., $\mathbf{0}^{\perp} = \mathbb{R}^n$ and $(\mathbb{R}^n)^{\perp} = \{\mathbf{0}\}$).

(c) For any subspace $U \subset \mathbb{R}^n$, the orthogonal complement $U^{\perp}$ is a subspace as well. If $\mathbf{x}$ and $\mathbf{y} \in U^{\perp}$, then $\mathbf{x} \cdot \mathbf{u} = 0 = \mathbf{y} \cdot \mathbf{u}$ for all $\mathbf{u} \in U$. Thus, $(\alpha\mathbf{x} + \beta\mathbf{y}) \cdot \mathbf{u} = \alpha\mathbf{x} \cdot \mathbf{u} + \beta\mathbf{y} \cdot \mathbf{u} = 0$ for all $\alpha, \beta \in \mathbb{R}$ and all $\mathbf{u} \in U$ (i.e., $\alpha\mathbf{x} + \beta\mathbf{y} \in U^{\perp}$ if $\mathbf{x}, \mathbf{y} \in U^{\perp}$). ◀

Observe that if a nonzero vector $\mathbf{a} \in \mathbb{R}^n$ is orthogonal to every vector of a subspace $U \subset \mathbb{R}^n$, then $\mathbf{a}$ cannot lie in U. This follows from the following orthogonality Lemma:

Lemma 2 The only vector $\mathbf{x} \in \mathbb{R}^n$ that is orthogonal to itself is the zero vector. ◀

Proof Note that $\mathbf{x} \cdot \mathbf{x} = \sum_i x_i^2 = 0$ if and only if each $x_i = 0$. ■

By definition of the dot product, we know that the set U of all vectors that are orthogonal to one given column vector $\mathbf{a}_1 \in \mathbb{R}^n$ is the set $U = \{\mathbf{u} \in \mathbb{R}^n \mid \mathbf{a}_1 \cdot \mathbf{u} = 0\}$. According to Chapter 3, the set U is identical to $\ker\begin{pmatrix} - & \mathbf{a}_1^T & - \end{pmatrix}$ where $\mathbf{a}_1^T$ denotes the row-vector transpose of the column $\mathbf{a}_1$. Using Definition 3, we observe that U can be expressed as $\text{span}\{\mathbf{a}_1\}^{\perp}$. Hence, if a subspace U excludes only one direction $\mathbf{a}_1$ with $\mathbf{a}_1 \perp U$, then U is the kernel or the nullspace of the $1 \times n$ matrix $\mathbf{A} = \begin{pmatrix} - & \mathbf{a}_1^T & - \end{pmatrix}$.

If U is represented exclusively by a set $\{\mathbf{a}_1, \ldots, \mathbf{a}_\ell\}$ of several vectors in $\mathbb{R}^n$ that are orthogonal to U and thus are excluded from U by Lemma 2, then

$U = \ker \begin{pmatrix} - & \mathbf{a}_1^T & - \\ & \vdots & \\ - & \mathbf{a}_\ell^T & - \end{pmatrix} = \text{span}\{\mathbf{a}_1, \ldots, \mathbf{a}_\ell\}^\perp$. If $\ell = 0$ here (i.e., if U excludes no vector in $\mathbb{R}^n$), then clearly $U = \mathbb{R}^n$.

Hence, we obtain

Theorem 4.1 **(Two generic representations of a subspace)** Every subspace U of $\mathbb{R}^n$ can be represented in two equivalent and complementary forms

- as a **span**

$$U = \text{span}\{\mathbf{u}_1, \ldots, \mathbf{u}_k\} = \text{im} \begin{pmatrix} | & & | \\ \mathbf{u}_1 & \ldots & \mathbf{u}_k \\ | & & | \end{pmatrix}$$

for certain vectors $\mathbf{u}_i \in U$, and

- as a **kernel**

$$U = \ker \begin{pmatrix} - & \mathbf{a}_1^T & - \\ & \vdots & \\ - & \mathbf{a}_\ell^T & - \end{pmatrix} = \text{span}\{\mathbf{a}_1, \ldots, \mathbf{a}_\ell\}^\perp = \{\mathbf{x}_{\text{hom}}\}$$

for certain vectors $\mathbf{a}_i \in \mathbb{R}^n$. Here $\mathbf{x}_{\text{hom}}$ denotes the solutions of the homogeneous linear system $\begin{pmatrix} - & \mathbf{a}_1^T & - \\ & \vdots & \\ - & \mathbf{a}_\ell^T & - \end{pmatrix} \mathbf{x} = \mathbf{0}$.

Conversely, every matrix kernel $\ker(\mathbf{A})$ and every matrix image $\text{im}(\mathbf{A})$ of an $m \times n$ matrix $\mathbf{A}$ is a subspace of $\mathbb{R}^n$ or $\mathbb{R}^m$, respectively. ◀

If $U = \text{im} \begin{pmatrix} | & & | \\ \mathbf{u}_1 & \ldots & \mathbf{u}_k \\ | & & | \end{pmatrix} = \text{span}\{\mathbf{u}_1, \ldots, \mathbf{u}_k\} \subset \mathbb{R}^n$ is an inclusively defined subspace, then the vectors $\mathbf{u}_i$ span all of U by Definition 2.

If, on the other hand, $U = \ker(A) = \text{span}\{\mathbf{a}_1, \ldots, \mathbf{a}_\ell\}^\perp \subset \mathbb{R}^n$ is defined exclusively by a matrix $\mathbf{A}_{\ell n}$, then the set of all homogeneous solutions $\mathbf{x}_{\text{hom}}$ to the linear system $\mathbf{Ax} = \mathbf{0}$ spans U.

EXAMPLE 4 Describe all subspaces U of $\mathbb{R}^3$ as images of matrix maps and as matrix kernels.

First note the restrictions that $U = \ker(\mathbf{A}) = \text{im}(\mathbf{B}) \subset \mathbb{R}^n$ poses on the matrices $\mathbf{A}$ and $\mathbf{B}$: $\mathbf{A}$ must be $m \times n$ and $\mathbf{B}$ $n \times \ell$ for arbitrary integers m and ℓ. Thus, in our case, $\mathbf{A}$ must be m by 3 and $\mathbf{B}$ 3 by ℓ, since $U \subset \mathbb{R}^3$.

To describe all possible REFs of matrices with a given number of rows and columns in Section 2.1, we ordered the answers by their rank (i.e., by the number of pivots involved).

To describe all subspaces of $\mathbb{R}^3$, we order them by their intrinsic size:

(0) The smallest subspace of $\mathbb{R}^3$ is the singleton space $U = \{\mathbf{0}\}$. Two complementary representations are

$$\text{im}(\mathbf{O}_3) = \{\mathbf{0}\} = \ker(\mathbf{I}_3)$$

for the 3×3 zero and identity matrices.

(1) If U allows movement in one nonzero direction $\mathbf{u}_1 \in \mathbb{R}^3$ only, then $U = \text{span}\{\mathbf{u}_1\} = \text{im}\begin{pmatrix} | & | & | & | \\ \mathbf{u}_1 & 3\mathbf{u}_1 & \mathbf{0} & -\mathbf{u}_1 \\ | & | & | & | \end{pmatrix}$ for example. Here the matrix $\mathbf{B}_{3\times 4} = \begin{pmatrix} | & | & | & | \\ \mathbf{u}_1 & 3\mathbf{u}_1 & \mathbf{0} & -\mathbf{u}_1 \\ | & | & | & | \end{pmatrix}$ with $\text{im}(\mathbf{B}) = U$ contains only multiples of the one generating vector $\mathbf{u}_1$ in its columns.
To express the same subspace U as a kernel $\ker(\mathbf{A})$, a REF of $\mathbf{A}_{k3}$ must have precisely one free variable. If $\mathbf{a}_1, \mathbf{a}_2 \in \mathbb{R}^3$ with $\mathbf{a}_i \perp \mathbf{u}_1$ and rank$\begin{pmatrix} - & \mathbf{a}_1^T & - \\ - & \mathbf{a}_2^T & - \end{pmatrix}$ $= 2$, then $\begin{pmatrix} - & \mathbf{a}_1^T & - \\ - & \mathbf{a}_2^T & - \end{pmatrix}\mathbf{u}_1 = \mathbf{0}$. Thus, $U = \ker\begin{pmatrix} - & \mathbf{a}_1^T & - \\ - & \mathbf{a}_2^T & - \end{pmatrix}$. Likewise, the matrix $\mathbf{A}_{3\times 3} = \begin{pmatrix} - & \mathbf{a}_1^T & - \\ - & 2\mathbf{a}_1^T & - \\ - & -3\mathbf{a}_2^T & - \end{pmatrix}$ has U as its kernel, since the repetition and scaling of the $\mathbf{a}_i^T$ in its rows does not affect orthogonality or the kernel.

(2) If a subspace U of $\mathbb{R}^3$ allows movement in more than one direction, but not in all three of $\mathbb{R}^3$, then $U = \text{span}\{\mathbf{u}_1, \mathbf{u}_2\} = \text{im}\begin{pmatrix} | & | \\ \mathbf{u}_1 & \mathbf{u}_2 \\ | & | \end{pmatrix}$, for example, where rank$\begin{pmatrix} | & | \\ \mathbf{u}_1 & \mathbf{u}_2 \\ | & | \end{pmatrix} = 2$. On the other hand, since the subspace U allows precisely two degrees of freedom, any matrix $\mathbf{A}$ with $\ker(\mathbf{A}) = \mathbf{U}$ must have two free variables; that is, if we choose a vector $\mathbf{0} \neq \mathbf{a}_1 \in \mathbb{R}^3$ with $\mathbf{a}_1 \perp \mathbf{u}_1$ and $\mathbf{a}_1 \perp \mathbf{u}_2$, then $U = \ker\begin{pmatrix} - & \mathbf{a}_1^T & - \end{pmatrix} = \ker\begin{pmatrix} - & \mathbf{a}_1^T & - \\ - & -3\mathbf{a}_1^T & - \end{pmatrix}$.

(3) Finally, if $U = \mathbb{R}^3$ is the maximal possible subspace of $\mathbb{R}^3$, then $U = \text{im}(\mathbf{I}_3)$ and $U = \ker(\mathbf{O}_3)$ are two complementary representations. Note the reversal of the generating matrices for U as an image and as a matrix kernel when compared with part (0). ◀

A subspace $\{\mathbf{0}\} \neq U \subset \mathbb{R}^n$ is generated by many different spanning sets. From chapter 2, we deduce the following:

Proposition: (a) A subspace $U = \text{span}\{\mathbf{a}_1, \ldots, \mathbf{a}_m\} \subset \mathbb{R}^n$ is spanned by the vectors $\{\mathbf{b}_1, \ldots, \mathbf{b}_j\}$ if and only if the nonzero rows of the RREFs of $\mathbf{A}_{mn} := \begin{pmatrix} - & \mathbf{a}_1^T & - \\ & \vdots & \\ - & \mathbf{a}_m^T & - \end{pmatrix}$ and $\mathbf{B}_{jn} := \begin{pmatrix} - & \mathbf{b}_1^T & - \\ & \vdots & \\ - & \mathbf{b}_j^T & - \end{pmatrix}$ are identical.

(b) The set of vectors $\{\mathbf{a}_1, \ldots, \mathbf{a}_k\} \subset \mathbb{R}^n$ spans all of $\mathbb{R}^n$ if and only the RREF of $\mathbf{A}_{kn} = \begin{pmatrix} - & \mathbf{a}_1^T & - \\ & \vdots & \\ - & \mathbf{a}_k^T & - \end{pmatrix}$ contains the identity matrix $\mathbf{I}_n$ in its first n rows. ◀

Proof If $\text{span}\{\mathbf{a}_i\} = \text{span}\{\mathbf{b}_\ell\}$, then each $\mathbf{a}_i$ is a linear combination of the $\mathbf{b}_\ell$ and vice versa. That is, the two sets of row vectors $\{\mathbf{a}_i^T\}$ and $\{\mathbf{b}_\ell^T\}$ are row equivalent in the sense of Gaussian elimination. Thus, their RREFs must agree, except possibly for the number of zero bottom rows. The converse implication that two row-equivalent sets of vectors span the same subspace is obvious.

Part (b) follows from part (a) and the fact that the standard unit vectors $\mathbf{e}_i \in \mathbb{R}^n$ span all of $\mathbb{R}^n$ and form $\mathbf{I}_n$ row– and columnwise. ■

EXAMPLE 5 (a) According to the Proposition, part (a), the subspace

$$U = \text{span}\{\mathbf{u}_1, \mathbf{u}_2, \mathbf{u}_3\} = \text{span}\left\{\begin{pmatrix} 1 \\ 2 \\ 0 \\ 1 \end{pmatrix}, \begin{pmatrix} 0 \\ 2 \\ 1 \\ 1 \end{pmatrix}, \begin{pmatrix} 0 \\ 0 \\ 1 \\ 0 \end{pmatrix}\right\} \subset \mathbb{R}^4$$

is also spanned by the four vectors

$$\{\mathbf{v}_1, \ldots, \mathbf{v}_4\} = \left\{\begin{pmatrix} 1 \\ 2 \\ 1 \\ 1 \end{pmatrix}, \begin{pmatrix} 1 \\ 4 \\ 1 \\ 2 \end{pmatrix}, \begin{pmatrix} 2 \\ 4 \\ 1 \\ 2 \end{pmatrix}, \begin{pmatrix} 0 \\ 2 \\ 2 \\ 1 \end{pmatrix}\right\},$$

since the RREFs of $\begin{pmatrix} - & \mathbf{u}_1^T & - \\ - & \mathbf{u}_2^T & - \\ - & \mathbf{u}_3^T & - \end{pmatrix}$ and $\begin{pmatrix} - & \mathbf{v}_1^T & - \\ & \vdots & \\ - & \mathbf{u}_4^T & - \end{pmatrix}$ are $\begin{pmatrix} 1 & 0 & 0 & 0 \\ 0 & 1 & 0 & 0.5 \\ 0 & 0 & 1 & 0 \end{pmatrix}$ and $\begin{pmatrix} 1 & 0 & 0 & 0 \\ 0 & 1 & 0 & 0.5 \\ 0 & 0 & 1 & 0 \\ 0 & 0 & 0 & 0 \end{pmatrix}$, respectively.

(b) The set of vectors $\left\{ \begin{pmatrix} 1 \\ 1 \\ 1 \end{pmatrix}, \begin{pmatrix} 0 \\ 1 \\ 1 \end{pmatrix}, \begin{pmatrix} 0 \\ 1 \\ -1 \end{pmatrix} \right\}$ spans $\mathbb{R}^3$ by the Proposition, part (b), since the associated REF contains a pivot in every row:

$$
\begin{array}{ccc|l}
\boxed{1} & 0 & 0 & \\
\boxed{1} & 1 & 1 & -\ row_1 \\
\boxed{1} & 1 & -1 & -\ row_1 \\
\hline
\boxed{1} & 0 & 0 & \\
0 & \boxed{1} & 1 & \\
0 & \boxed{1} & -1 & -\ row_2 \\
\hline
\boxed{1} & 0 & 0 & \\
0 & \boxed{1} & 1 & \\
0 & 0 & \boxed{-2} &
\end{array}
$$

Thus, it is row equivalent to the 3×3 identity matrix $\mathbf{I}_3$; just scale its last row and eliminate the (2, 3) nonzero entry. ◀

Many different matrices can have the same kernel. In particular, since every matrix kernel is uniquely determined by the RREF of the matrix according to Chapter 3, the uniqueness of the RREF of a matrix immediately establishes the following result:

Theorem 4.2 Two real matrices $\mathbf{A}_{mn}$ and $\mathbf{B}_{kn}$ have the same kernel in $\mathbb{R}^n$ if and only if the pivot rows of their RREFs coincide. ◀

(b) Translating between the Span and Kernel Representations of a Subspace

It is natural to ask how to translate between the two equivalent representations of one and the same subspace U.

In Section 3.1, we learned how to express the kernel of a matrix $\mathbf{A}_{mn}$ as the span of all homogeneous solutions $\mathbf{x}_{\text{hom}} \in \mathbb{R}^n$ to the homogeneous linear system $\mathbf{Ax} = \mathbf{0}$. Namely,

$$\ker(\mathbf{A}) = \{\mathbf{x}_{\text{hom}}\} = \text{span}\{\mathbf{u}_1, \ldots, \mathbf{u}_k\} \subset \mathbb{R}^n$$

if a REF of $\mathbf{A}$ has k free columns. Thus, $\ker(\mathbf{A}) = \text{im}(\mathbf{B}) = \text{span}\{\mathbf{b}_j\}_{j=1}^{\ell} \subset \mathbb{R}^n$, where $\mathbf{B}_{n\ell}$ is any matrix of rank k whose columns $\mathbf{b}_j \in \mathbb{R}^n$ are linear combinations of the kernel generating vectors $\mathbf{u}_i$.

Conversely, to express a span (i.e., a column space $\text{im}(\mathbf{B})$ with $\mathbf{B}$ $n \times \ell$) as a kernel $\ker(\mathbf{A})$ for $\mathbf{A}$ an $m \times n$ matrix we must make sure that each column $\mathbf{b}_j$ of $\mathbf{B}$ is mapped to zero under the map $\mathbf{b}_j \mapsto \mathbf{Ab}_j$. That is, each row of $\mathbf{A}$ must be orthogonal to each column of $\mathbf{B}$, and the number of free variables in a REF of $\mathbf{A}$ must be equal to the rank of $\mathbf{B}$. This problem thus involves solving the homogeneous linear system $\begin{pmatrix} - & \mathbf{b}_1^T & - \\ & \vdots & \\ - & \mathbf{b}_\ell^T & - \end{pmatrix} \mathbf{x} = \mathbf{0}$: Find a generating set $\{\mathbf{x}_{\text{hom}}\}$ for this kernel, and take m linear combinations thereof as the rows of $\mathbf{A}$ so that $\text{rank}(\mathbf{A}) = \dim(\ker(\mathbf{B}^T))$. Then $\ker(\mathbf{A}) = \text{im}(\mathbf{B})$, and, consequently, $\mathbf{AB}$ is the $m \times \ell$ zero matrix.

EXAMPLE 6 Express $\text{im}\begin{pmatrix} 2 & 1 & 3 \\ 1 & 2 & 3 \\ 0 & 1 & 1 \end{pmatrix}$ as a matrix kernel.

The image or column space of the given matrix $\mathbf{B}_{3\times 3} = \begin{pmatrix} 2 & 1 & 3 \\ 1 & 2 & 3 \\ 0 & 1 & 1 \end{pmatrix}$ is generated by the column vectors $\mathbf{b}_1 = \begin{pmatrix} 2 \\ 1 \\ 0 \end{pmatrix}$, $\mathbf{b}_2 = \begin{pmatrix} 1 \\ 2 \\ 1 \end{pmatrix}$, and $\mathbf{b}_3 = \begin{pmatrix} 3 \\ 3 \\ 1 \end{pmatrix}$.

We solve $\begin{pmatrix} - & \mathbf{b}_1^T & - \\ - & \mathbf{b}_2^T & - \\ - & \mathbf{b}_3^T & - \end{pmatrix} \mathbf{x} = \mathbf{0}$:

$$
\begin{array}{ccc|cl}
2 & 1 & 0 & 0 & \textit{swap rows 1 and 2} \\
1 & 2 & 1 & 0 & \\
3 & 3 & 1 & 0 & \\
\hline
\boxed{1} & 2 & 1 & & \\
\boxed{2} & 1 & 0 & 0 & -2\,row_1 \\
\boxed{3} & 3 & 1 & 0 & -3\,row_1 \\
\hline
\boxed{1} & 2 & 1 & 0 & \\
0 & \boxed{-3} & -2 & 0 & \\
0 & \boxed{-3} & -2 & 0 & -row_2 \\
\hline
\boxed{1} & 2 & 1 & 0 & \\
0 & \boxed{-3} & -2 & 0 & \\
0 & 0 & 0 & 0 &
\end{array}
$$

Thus, $\mathbf{x}_{\text{hom}} = \alpha \begin{pmatrix} 1 \\ -2 \\ 3 \end{pmatrix} =: \alpha \mathbf{u}$. And for $\mathbf{A}_{4\times 3}$ chosen as $\begin{pmatrix} - & \mathbf{u}^T & - \\ - & 2\mathbf{u}^T & - \\ - & \mathbf{0} & - \\ - & -3\mathbf{u}^T & - \end{pmatrix}$, for example, we have $\ker(\mathbf{A}) = \text{im}(\mathbf{B})$ as desired. We advise you to check that $\mathbf{AB} = \mathbf{O}_{4\times 3}$. ◀

We now describe a second method to obtain a kernel representation of a span and to express a given kernel as the image of a matrix transformation. This method is based on row reduction and will be further studied in Chapters 6 and 7. It is the nucleus of matrix inversion.

To express $U = \text{span}\{\mathbf{u}_1, \ldots, \mathbf{u}_k\} \subset \mathbb{R}^n$ as the kernel of a matrix $\mathbf{A}_{mn}$, we may proceed as follows:

The inclusive description span $\{\mathbf{u}_1, \ldots, \mathbf{u}_k\}$ builds a subspace, while the exclusive description $\ker(\mathbf{A}) = \text{span}\{\mathbf{v}_i\}^{\perp}$ delineates a subspace of $\mathbb{R}^n$ by using restrictions. Linear equations define the kernel. Their common solution forms it. One can find the restrictions for $\mathbf{y} \in \text{span}\{\mathbf{u}_i\}$ from the following scheme, whose original left side contains the vectors $\{\mathbf{u}_i\}$ as columns. To the right of these, we write the identity matrix $\mathbf{I}$ of order n whose columns are labelled as $y_1, \ldots, y_n$ for the n components of $\mathbf{y}$. Thus, the right-hand side of Eq. (4.1(a)) represents an arbitrary point $\mathbf{y} = \begin{pmatrix} y_1 \\ \vdots \\ y_n \end{pmatrix} \in \mathbb{R}^n$ as $\mathbf{I}_n\mathbf{y}$.

$$\begin{array}{ccc||ccc} & \mathbf{U} & & y_1 & \ldots & y_n \\ \hline | & & | & & & \\ \mathbf{u}_1 & \ldots & \mathbf{u}_k & & \mathbf{I}_n & \\ | & & | & & & \\ \hline \end{array} \tag{4.1(a)}$$

$$\downarrow \qquad\qquad \downarrow \qquad \text{by row reduction} \tag{4.1}$$

$$\begin{array}{ccc||ccc} & & & y_1 & \ldots & y_n \\ \hline \square & & * & & & \\ & \ddots & & & * & \\ 0 & & \square & & & \\ \hline 0 & \ldots & 0 & & & \\ \vdots & & \vdots & & \mathbf{A}_{mn} & \\ 0 & \ldots & 0 & & & \end{array} \tag{4.1(b)}$$

The scheme of Eq. (4.1) computes a row echelon form for the matrix $\mathbf{U} = \begin{pmatrix} | & & | \\ \mathbf{u}_1 & \ldots & \mathbf{u}_k \\ | & & | \end{pmatrix}$ and performs the same row operations simultaneously on the identity matrix on the right. This scheme has many applications. For example, it is a precursor to matrix inversion of Chapter 6 and to finding basis change matrices in Chapter 7.

The REF of $\mathbf{U}$, written out symbolically, appears on the left of the || divide in Eq. (4.1(b)). A vector $\mathbf{y} = \begin{pmatrix} y_1 \\ \vdots \\ y_n \end{pmatrix} = \sum_i y_i \mathbf{e}_i$ lies in the span of the $\{\mathbf{u}_i\}$ if and only if the trailing entries in $\mathbf{A}$ on the right-hand side below the horizontal line in Eq. (4.1(b)) satisfy $\mathbf{Ay} = \mathbf{0}$. This follows from Theorem 3.1 since the linear system $\mathbf{Ux} = \sum_i x_i \mathbf{u}_i = \mathbf{y}$ can only be solved for those right-hand sides $\mathbf{y}$ that do not increase the rank of $\mathbf{U}$ in the augmented matrix $(\mathbf{U} \mid \mathbf{y})$.

If the row reduction of $\mathbf{U}$ from Eq. (4.1(a)) to Eq. (4.1(b)) leads to j pivots, then the linear system (4.1(b)) contains $n-j$ zero bottom rows on the left of the $||$ divide. Hence, the matrix $\mathbf{A} = \mathbf{A}_{mn}$ on the right has $m := n-j$ rows and n columns. And the kernel of $\mathbf{A}$ satisfies $\ker(\mathbf{A}) = U = \operatorname{span}\{\mathbf{u}_1, \dots, \mathbf{u}_k\}$ by construction and Theorem 3.1.

EXAMPLE 7 For $\mathbf{u}_1 = \begin{pmatrix} 1 \\ 2 \\ -1 \\ 3 \end{pmatrix}$ and $\mathbf{u}_2 = \begin{pmatrix} 1 \\ 0 \\ 2 \\ -1 \end{pmatrix} \in \mathbb{R}^4$, express $U = \operatorname{span}\{\mathbf{u}_1, \mathbf{u}_2\}$ as the kernel of a matrix $\mathbf{A}_{r4}$. (The number r of its rows will emerge in the process.)

For this, we augment the matrix $\mathbf{U}_{4\times 2} = \begin{pmatrix} | & | \\ \mathbf{u}_1 & \mathbf{u}_2 \\ | & | \end{pmatrix}$ by $\mathbf{I}_4$ and row reduce the **multi-augmented matrix** $(\mathbf{U} \mid\mid \mathbf{I}_4)$ according to Eq. (4.1) to find the matrix $\mathbf{A}$ with $\ker(\mathbf{A}) = U$ in Eq. (4.1(b)):

	$\mathbf{u}_1$	$\mathbf{u}_2$	y_1	y_2	y_3	y_4	
	$\boxed{1}$	1	1	0	0	0	
$\mathbf{U}$ =	$\boxed{2}$	0	0	1	0	0	$-2\ row_1$
	$\boxed{-1}$	2	0	0	1	0	$+\ row_1$
	$\boxed{3}$	−1	0	0	0	1	$-3\ row_1$
	$\boxed{1}$	1	1	0	0	0	
	0	−2	−2	1	0	0	$+\ row_3$ *(for a 1 pivot)*
	0	3	1	0	1	0	
	0	−4	−3	0	0	1	
	$\boxed{1}$	1	1	0	0	0	
	0	$\boxed{1}$	−1	1	1	0	
	0	$\boxed{3}$	1	0	1	0	$-3\ row_2$
	0	$\boxed{-4}$	−3	0	0	1	$+4\ row_2$
	$\boxed{1}$	1	1	0	0	0	
	0	$\boxed{1}$	−1	1	1	0	
	0	0	4	−3	−2	0	
	0	0	−7	4	4	1	$= \mathbf{A}_{2\times 4}$

Students should check that $\mathbf{A}\mathbf{u}_1 = \mathbf{A}\begin{pmatrix} 1 \\ 2 \\ -1 \\ 3 \end{pmatrix} = \begin{pmatrix} 0 \\ 0 \end{pmatrix}$ and $\mathbf{A}\mathbf{u}_2 = \mathbf{A}\begin{pmatrix} 1 \\ 0 \\ 2 \\ -1 \end{pmatrix} = \begin{pmatrix} 0 \\ 0 \end{pmatrix}$ for the computed matrix $\mathbf{A}_{2\times 4} = \begin{pmatrix} 4 & -3 & -2 & 0 \\ -7 & 4 & 4 & 1 \end{pmatrix}$. Thus,

$$\operatorname{span}\left\{\begin{pmatrix}1\\2\\-1\\3\end{pmatrix},\begin{pmatrix}1\\0\\2\\-1\end{pmatrix}\right\} = U = \operatorname{im}\begin{pmatrix}1&1\\2&0\\-1&2\\3&-1\end{pmatrix} = \ker\begin{pmatrix}4&-3&-2&0\\-7&4&4&1\end{pmatrix}.$$

Note that in this scheme, the number of rows r in $\mathbf{A}_{rn}$ automatically comes out to be $r = n-$ the number of pivots of a REF for $\mathbf{U}$ with columns $\mathbf{u}_i$. Moreover, the constructed matrix $\mathbf{A}$ with $\ker(\mathbf{A}) = \mathbf{U}$ always has maximum rank, since at the start, we have rank $(\mathbf{U} \mid\mid \mathbf{I}_n) = n$ and row reduction does not alter rank.

The scheme of Eq. (4.1) can be reversed to express the kernel of a given matrix $\mathbf{A}_{mn}$ as the span of vectors. This gives an alternative method to the free variable backsubstitution method of Section 3.1 for solving homogeneous systems of linear equations.

Here are the steps to find a spanning set $\{\mathbf{u}_i\}$ of the kernel of a matrix $\mathbf{A}_{mn}$ using Eq. (4.1) in reverse:

First row reduce $\mathbf{A}_{mn}$ to a REF $\mathbf{R}_{mn}$. Drop all zero rows in $\mathbf{R}$, so that the resulting submatrix $\mathbf{R}^{(1)}_{kn}$ consists entirely of the pivot rows of $\mathbf{R}$. The number of rows in $\mathbf{R}^{(1)}$ equals rank $\mathbf{A} = k \leq m$.

$$\text{For example, } \mathbf{R}^{(1)} = \begin{pmatrix}\boxed{*}&*&*&*&*&*\\0&0&\boxed{*}&*&*&*\\&&0&\boxed{*}&*&*\\0&&&0&\boxed{*}&*\end{pmatrix} \text{ if}$$

$$\mathbf{R} = \begin{pmatrix}\boxed{*}&*&*&*&*&*\\0&0&\boxed{*}&*&*&*\\&&0&\boxed{*}&*&*\\0&&&0&\boxed{*}&*\\0&0&0&0&0&0\\0&0&0&0&0&0\end{pmatrix}.$$

Next for every free column in $\mathbf{R}^{(1)}$, perform the following two tasks starting from the front: If column i is free in $\mathbf{R}^{(1)}$, first insert the unit row vector $\mathbf{e}_i \in \mathbb{R}^n$ into $\mathbf{R}^{(1)}$ so that it becomes its ith row. At the same time, augment the updated matrix $\mathbf{R}^{(1)}$ by the unit column vector $\mathbf{e}_i \in \mathbb{R}^n$ on the right. Performing these two tasks for all free columns of $\mathbf{R}^{(1)}$ leads to a multi-augmented matrix $(\mathbf{R}^{(2)} \mid\mid \mathbf{e}_{i_1} \dots \mathbf{e}_{i_j})$. By construction $\mathbf{R}^{(2)}$ is $n \times n$ of rank n. It is in upper staircase form, and there are j unit vectors to the right of the $\mid\mid$ dividing line if $\mathbf{R}^{(1)}$ has j free columns.

For our sample REF $\mathbf{R}$, the multi-augmented matrix becomes

$$(\mathbf{R}^{(2)} \mid\mid \mathbf{e}_2, \mathbf{e}_6) = \left(\begin{array}{cccccc||cc}\boxed{*}&*&*&*&*&*&0&0\\0&1&0&0&0&0&1&0\\&0&\boxed{*}&*&*&*&0&0\\&&0&\boxed{*}&*&*&0&0\\&&&0&\boxed{*}&*&0&0\\0&&&&0&1&0&1\end{array}\right) \begin{array}{l}\\ \leftarrow \textit{inserted row}\\ \\ \\ \\ \leftarrow \textit{inserted row}\end{array}$$

with two augmented standard unit vector columns on the right, since $\mathbf{R}^{(1)}$ has two free columns in positions 2 and 6.

Now row reduce the multi-augmented matrix $(\mathbf{R}^{(2)} \,||\, \mathbf{e}_{i_1}, \ldots, \mathbf{e}_{i_j})$ to the form

$$\begin{array}{c|ccc} & \mathbf{u}_1 & \ldots & \mathbf{u}_j \\ \hline & * & & * \\ \mathbf{I}_n & \vdots & & \vdots \\ & * & & * \\ \hline \end{array}.$$

Then $\ker(\mathbf{A}) = \ker(\mathbf{R}) = \operatorname{span}\{\mathbf{u}_1, \ldots, \mathbf{u}_j\}$.

EXAMPLE 8 Express the kernel of the REF $\mathbf{R}_{4\times 5} = \begin{pmatrix} \boxed{1} & 1 & 0 & 1 & 1 \\ 0 & 0 & \boxed{1} & 2 & 0 \\ 0 & 0 & 0 & \boxed{2} & 1 \\ 0 & 0 & 0 & 0 & 0 \end{pmatrix}$ as a span.

Note that $\mathbf{R}$ contains one zero row and that its second and fifth columns are free. Drop the last zero row of $\mathbf{R}$ to obtain the 3×5 matrix $\mathbf{R}^{(1)}$. Insert copies of the row vectors $\mathbf{e}_2$ and $\mathbf{e}_5 \in \mathbb{R}^5$ between rows 1 and 2 and below row 4 of $\mathbf{R}^{(1)}$, respectively. At the same time, augment the updated matrix $\mathbf{R}^{(1)}$ with copies of the unit column vectors $\mathbf{e}_2$ and $\mathbf{e}_5$. Then the matrix

$$\mathbf{R}^{(2)} = \begin{pmatrix} 1 & 1 & 0 & 1 & 1 \\ 0 & 1 & 0 & 0 & 0 \\ 0 & 0 & 1 & 2 & 0 \\ 0 & 0 & 0 & 2 & 1 \\ 0 & 0 & 0 & 0 & 1 \end{pmatrix} \begin{array}{l} \\ \leftarrow \textit{inserted row} \\ \\ \\ \leftarrow \textit{inserted row} \end{array}$$

is row equivalent to $\mathbf{I}_5$. Next we row reduce the multi-augmented matrix $(\mathbf{R}^{(2)} \,||\, \mathbf{e}_2, \mathbf{e}_5)$ to $(\mathbf{I}_5 \,||\, \mathbf{u}_1, \mathbf{u}_2)$ using a backsubstitution-like upwards sweep:

$\mathbf{R}^{(2)}$					$\mathbf{e}_2$	$\mathbf{e}_5$	
$\boxed{1}$	1	0	1	$\boxed{1}$	0	0	$-\ row_5$
0	$\boxed{1}$	0	0	0	1	0	
0	0	$\boxed{1}$	2	0	0	0	
0	0	0	$\boxed{2}$	$\boxed{1}$	0	0	$-\ row_5$
0	0	0	0	$\boxed{1}$	0	1	
$\boxed{1}$	1	0	$\boxed{1}$	0	0	-1	$-1/2\ row_4$
0	$\boxed{1}$	0	0	0	1	0	
0	0	$\boxed{1}$	$\boxed{2}$	0	0	0	$-\ row_4$
0	0	0	$\boxed{2}$	0	0	-1	
0	0	0	0	$\boxed{1}$	0	1	
$\boxed{1}$	$\boxed{1}$	0	0	0	0	$-1/2$	$-\ row_2$
0	$\boxed{1}$	0	0	0	1	0	
0	0	$\boxed{1}$	0	0	0	1	
0	0	0	$\boxed{2}$	0	0	-1	$\div 2$
0	0	0	0	$\boxed{1}$	0	1	

$\mathbf{I}_5$	$\begin{array}{cc} -1 & -1/2 \\ 1 & 0 \\ 0 & 1 \\ 0 & -1/2 \\ 0 & 1 \end{array}$
	$\mathbf{u}_1$ $\quad$ $\mathbf{u}_2$

The two computed vectors $\mathbf{u}_1 = \begin{pmatrix} -1 \\ 1 \\ 0 \\ 0 \\ 0 \end{pmatrix}$ and $\mathbf{u}_2 = \begin{pmatrix} -1/2 \\ 0 \\ 1 \\ -1/2 \\ 1 \end{pmatrix}$ on the right side of the || divide span the kernel of $\mathbf{R}$. For convenience in writing, we replace $\mathbf{u}_2$ by the integer vector $2\mathbf{u}_2 = \begin{pmatrix} -1 \\ 0 \\ 2 \\ -1 \\ 2 \end{pmatrix}$ and obtain the integer spanning set $\{\mathbf{u}_1, 2\mathbf{u}_2\}$ for $\ker(\mathbf{A})$.

To check the work evaluate $\mathbf{R}\mathbf{u}_1$ and $\mathbf{R}(2\mathbf{u}_2)$; these two vectors should be zero. Also compare the result with finding $\mathbf{x}_{hom}$ by backsubstitution from $(\mathbf{R}_{4\times 5} \mid \mathbf{0})$. ◀

The process that we have used in Example 8 is the reverse, or backwards, process of Example 7.

(c) Vectors outside a Proper Subspace

In certain applications, such as for the matrix eigenvalue problem using vector iteration in Section 9.1, it is necessary to find nonzero vectors that do not lie in a given **proper subspace**. Here we call a subspace $U \subset \mathbb{R}^n$ **proper** if $U \neq \mathbb{R}^n$.

If a proper subspace U of $\mathbb{R}^n$ is given inclusively as $U = \text{span}\{\mathbf{u}_1, \dots, \mathbf{u}_k\} \neq \mathbb{R}^n$, then the exclusive subspace definition as a matrix kernel and the orthogonality Lemma 2 show that a nonzero vector $\mathbf{w} \in \ker\begin{pmatrix} - & \mathbf{u}_1^T & - \\ & \vdots & \\ - & \mathbf{u}_k^T & - \end{pmatrix}$ cannot belong to $\mathbf{U}$. For example, the vector $\mathbf{x} = \begin{pmatrix} 2 \\ 1 \\ 0 \end{pmatrix}$ lies in $\ker\begin{pmatrix} 1 & -2 & 0 \\ -2 & 4 & 3 \end{pmatrix}$, since $\begin{pmatrix} 1 & -2 & 0 \\ -2 & 4 & 3 \end{pmatrix}\begin{pmatrix} 2 \\ 1 \\ 0 \end{pmatrix} = \begin{pmatrix} 0 \\ 0 \end{pmatrix}$. And thus $\mathbf{x} \notin U = \text{span}\left\{\begin{pmatrix} 1 \\ -2 \\ 0 \end{pmatrix}, \begin{pmatrix} -2 \\ 4 \\ 3 \end{pmatrix}\right\}$. This can be checked by using Theorem 3.1 and Gaussian elimination on the augmented matrix $\left(\begin{array}{cc|c} 1 & -2 & 2 \\ -2 & 4 & 1 \\ 0 & 3 & 0 \end{array}\right)$: This system is unsolvable.

If, on the other hand, U is a subspace of $\mathbb{R}^n$ given exclusively as $U = \ker(\mathbf{A}) \neq \mathbb{R}^n$ for a matrix $\mathbf{A}_{mn} = (a_{ij})$ with a nonzero entry $a_{k\ell}$, then the standard unit vector $\mathbf{e}_\ell \notin \ker(\mathbf{A})$, since

$$\mathbf{A}\mathbf{e}_\ell = \begin{pmatrix} | & & | & & | \\ \mathbf{a}_1 & \dots & \mathbf{a}_\ell & \dots & \mathbf{a}_n \\ | & & | & & | \end{pmatrix} \mathbf{e}_\ell = \underset{\substack{\uparrow \\ \ell^{\text{th}} \text{ column of } A}}{\begin{pmatrix} * \\ * \\ a_{k\ell} \\ * \end{pmatrix}} \leftarrow k^{\text{th}} \text{ position} \quad \neq \mathbf{0} \in \mathbb{R}^n.$$

For example, for $\mathbf{A} = \begin{pmatrix} 0 & 0 & 0 & 7 \\ -2 & 0 & 0 & 0 \end{pmatrix}$, the two unit vectors $\mathbf{e}_1$ and $\mathbf{e}_4 \in \mathbb{R}^4$ do not belong to $\ker(\mathbf{A})$, since $\mathbf{A}\mathbf{e}_1 = \begin{pmatrix} | \\ \mathbf{a}_1 \\ | \end{pmatrix} = \begin{pmatrix} 0 \\ -2 \end{pmatrix} \neq \begin{pmatrix} 0 \\ 0 \end{pmatrix} \in \mathbb{R}^2$, while

$$\mathbf{A}\mathbf{e}_4 = \begin{pmatrix} | \\ \mathbf{a}_4 \\ | \end{pmatrix} = \begin{pmatrix} 7 \\ 0 \end{pmatrix} \neq \begin{pmatrix} 0 \\ 0 \end{pmatrix} \in \mathbb{R}^2.$$

Thus, for either of the two generic subspace representations, we can readily find vectors that do not belong to a given subspace $U \subset \mathbb{R}^n$, provided that $U \neq \mathbb{R}^n$.

Subspaces of $\mathbb{R}^n$ have been represented in two equivalent complementary forms.

4.1.P Problems

1. Check whether any of the following sets of vectors form a subspace. If a set forms a subspace, express it both as a matrix kernel and as a matrix image.

(a) $\left\{ \begin{pmatrix} x_1 \\ x_2 \end{pmatrix} \in \mathbb{R}^2 \mid x_1^2 + 2x_1x_2 + x_2^2 = 0 \right\}$;

(b) $\{\mathbf{x} \in \mathbb{R}^5 \mid x_1^3 - x_1(x_2 + x_1^2) + x_2(2 - x_1) = 0\}$;

(c) $\{\mathbf{x} \in \mathbb{R}^4 \mid x_1 = 2x_4,\ x_2 = -x_3\}$;

(d) $\{\mathbf{x} \in \mathbb{R}^3 \mid x_1 - x_2 + x_3 = 0,\ x_2 + 1 = 1,\ x_3 = 0\}$;

(e) $\left\{ \mathbf{x} \in \mathbb{R}^3 \mid x \perp \begin{pmatrix} 2 \\ 1 \\ -3 \end{pmatrix} \right\}$;

(f) $\left\{ \mathbf{x} = \alpha \begin{pmatrix} 1 \\ 0 \\ -1 \end{pmatrix} - \beta \begin{pmatrix} 1 \\ 0 \\ -1 \end{pmatrix} + \gamma \begin{pmatrix} 0 \\ 1 \\ 0 \end{pmatrix} \mid \alpha,\ \beta,\ \gamma \in \mathbb{R} \right\}$;

(g) $\{\mathbf{x} \in \mathbb{R}^4 \mid x_3 = 0\}$.

2. (a) Let $\mathbf{x} \neq \mathbf{0} \in \mathbb{R}^n$. Describe the line ℓ in $\mathbb{R}^n$ that contains $\mathbf{x}$ and the origin as a span.

(b) Show that the points on a line ℓ that contains the origin of $\mathbb{R}^n$ are closed under vector addition and scalar multiplication.

(c) Repeat part (b) for the singleton set $\{\mathbf{0}\} \subset \mathbb{R}^n$.

3. Find all vectors in $\mathbb{R}^4$ that are orthogonal to both $\begin{pmatrix} 1 \\ 2 \\ 0 \\ 1 \end{pmatrix}$ and $\begin{pmatrix} 0 \\ 1 \\ -1 \\ 1 \end{pmatrix} \in \mathbb{R}^4$. Show that these vectors form a subspace of $\mathbb{R}^4$.

4. (a) Show that the first quadrant $Q1 := \left\{ \begin{pmatrix} x_1 \\ x_2 \end{pmatrix} \in \mathbb{R}^2 \mid x_1 \geq 0,\ x_2 \geq 0 \right\}$ of the plane is closed under vector addition and multiplication by nonnegative scalars.

(b) Show the same for the third quadrant $Q3$ of $\mathbb{R}^2$.

(c) Show that $Q1 \cup Q3$ is closed under scalar multiplication by all real numbers.

(d) Is $Q1 \cup Q3$ a subspace of $\mathbb{R}^2$ or not?

(e) Are the quadrants $Q1$ or $Q3$ subspaces of $\mathbb{R}^2$?

(f) Draw pictures of all subspaces of $\mathbb{R}^2$.

5. Show that the set of positive real numbers $\mathbb{R}^+$ is closed under addition and multiplication by positive real numbers.

6. Express the singleton subspace $\{\mathbf{0}\} \subset \mathbb{R}^n$ as a span of at least one vector in $\mathbb{R}^n$.

7. (a) Find a matrix $\mathbf{A}_{k6}$, with $\ker(\mathbf{A}) = \text{span}\{\mathbf{e}_1\} \subset \mathbb{R}^6$.

(b) Do this for $\mathbf{B}_{\ell 4}$, with $\ker(\mathbf{B}) = \text{span}\{\mathbf{e}_3 - \mathbf{e}_4\} \subset \mathbb{R}^4$.

(c) Do this for $\mathbf{C}_{m5}$, with $\ker(\mathbf{C}) = \text{span}\{\mathbf{e}_1 + \mathbf{e}_5\} \subset \mathbb{R}^5$.

8. (a) Find a matrix $\mathbf{A}_{5,4}$, with $\ker(\mathbf{A}) = \{\mathbf{0}\} \subset \mathbb{R}^4$, if possible.

(b) Do this for $\mathbf{B}_{3,4}$, with $\ker(\mathbf{B}) = \{\mathbf{0}\} \subset \mathbb{R}^4$, if possible.

9. (a) Find a matrix $\mathbf{A}_{5,4}$, with $\ker(\mathbf{A}) = \mathbb{R}^4$, if possible.

(b) Do this for $\mathbf{B}_{3,4}$, with $\ker(\mathbf{B}) = \mathbb{R}^4$, if possible.

(c) Can you find a matrix $\mathbf{C}_{5,4}$, with $\ker(\mathbf{C}) = \mathbb{R}^5$?

10. (a) Find a spanning set for the kernel of

$$\mathbf{A} = \begin{pmatrix} 1 & 2 & 4 \\ 0 & 1 & 1 \\ 2 & 5 & 9 \end{pmatrix}.$$

(b) Do the same for $\mathbf{B} = \begin{pmatrix} 1 & 0 & 4 & 0 & 1 \\ 2 & 0 & 7 & 0 & 2 \end{pmatrix}$.

11. (a) Express $U = \text{span}\left\{\begin{pmatrix} 1 \\ 2 \\ -3 \\ -1 \end{pmatrix}\right\}$ as the kernel of three different matrices.

(b) What sizes can the matrices in part (a) possibly have?

12. (a) Express $U = \text{span}\left\{\begin{pmatrix} 1 \\ 2 \\ -3 \\ -1 \end{pmatrix}\right\}$ as the kernel of an upper triangular matrix $\mathbf{R}$ if possible.

(b) Do this for U and a lower triangular matrix $\mathbf{L}$.

13. Express the kernel of a matrix as a span for

(a) $\mathbf{A} = \begin{pmatrix} 2 \\ 1 \\ -5 \end{pmatrix}$.

(b) $\mathbf{B} = \begin{pmatrix} 1 & 4 & -5 & 0 \end{pmatrix}$.

(c) $\mathbf{C} = \begin{pmatrix} 1 & -3 & 2 \\ 2 & -7 & 1 \\ 1 & 0 & -1 \end{pmatrix}$.

(d) $\mathbf{D} = \begin{pmatrix} 0 & 0 & 1 & 3 & 5 \\ 0 & 0 & 1 & 2 & -1 \\ 0 & 1 & 0 & 0 & -1 \end{pmatrix}$.

(e) $\mathbf{H} = \begin{pmatrix} 1 & -1 & 2 & 3 & 4 & 3 \\ 0 & 0 & 2 & 1 & 1 & 0 \\ 0 & 0 & 0 & 0 & 4 & 8 \\ 0 & 0 & 0 & 0 & 0 & 0 \end{pmatrix}$.

14. Find all vectors in $\mathbb{R}^3$ that are orthogonal to both $\mathbf{u} = \begin{pmatrix} 1 \\ -1 \\ 2 \end{pmatrix}$ and $\mathbf{v} = \begin{pmatrix} -2 \\ 2 \\ -4 \end{pmatrix}$. Why do these vectors form a subspace?

15. Express the span of the following sets of vectors as the kernel of a matrix transformation.

(a) $U = \left\{\begin{pmatrix} 1 \\ 1 \\ 1 \end{pmatrix}, \begin{pmatrix} 2 \\ -2 \\ 2 \end{pmatrix}\right\}$.

(b) $V = \left\{\begin{pmatrix} 0 \\ 1 \\ 1 \\ 2 \end{pmatrix}, \begin{pmatrix} 2 \\ 0 \\ 2 \\ 0 \end{pmatrix}, \begin{pmatrix} 0 \\ 0 \\ 0 \\ 0 \end{pmatrix}\right\}$.

(c) $W = \left\{\begin{pmatrix} 0 \\ 0 \\ 1 \end{pmatrix}, \begin{pmatrix} 2 \\ -2 \\ 2 \end{pmatrix}, \begin{pmatrix} 1 \\ 0 \\ 1 \end{pmatrix}\right\}$.

16. Let $\mathbf{x}, \mathbf{y}, \mathbf{z}$ be three vectors in $\mathbb{R}^n$. Define $U = \text{span}\{\mathbf{x}, \mathbf{y}\}$. What connection is there between $\mathbf{z}$ lying in U and a nonhomogeneous system of linear equations? Explain.

17. Show that the polynomials $u_1(x) = 1 - x$, $u_2(x) = 1 + 2x$ and $u_3(x) = 3u_1(x)u_2(x)$ span the space of all polynomials $\mathcal{P}_2$ of degree less than or equal to 2.

18. Show that the set of all even positive integers is closed under addition and multiplication by positive integers. Do the even positive integers $\{2, 4, 6, \ldots\}$ form a subspace?

19. Let $U = \text{span}\left\{\mathbf{u} = \begin{pmatrix} 1 \\ 2 \\ -1 \end{pmatrix}\right\} \subset \mathbb{R}^3$ be a one-dimensional subspace of $\mathbb{R}^3$.

(a) Find a vector $\mathbf{v}$ so, that $\text{span}\{\mathbf{u}, \mathbf{v}\}$ is a plane in $\mathbb{R}^3$.

(b) Find a vector $\mathbf{w} \neq \mathbf{v}$ for $\mathbf{v}$ from part (a), so that $\text{span}\{\mathbf{u}, \mathbf{w}\} = \text{span}\{\mathbf{u}, \mathbf{v}\}$ is the same plane in $\mathbb{R}^3$.

(c) Find a vector $\mathbf{z}$ so that $\text{span}\{\mathbf{u},\ \mathbf{z}\} \neq \text{span}\{\mathbf{u},\ \mathbf{v}\}$ is a different plane in $\mathbb{R}^3$.

(d) Find a vector $\mathbf{x} \neq \mathbf{u}$, so that $\text{span}\{\mathbf{u},\ \mathbf{x}\} = U$.

20. Show that in Example 5 (a) each v_j belongs to U and that the linear system $\begin{pmatrix} | & | & | & | \\ \mathbf{v}_1 & \mathbf{v}_2 & \mathbf{v}_3 & \mathbf{v}_4 \\ | & | & | & | \end{pmatrix} \mathbf{x} = \begin{pmatrix} | \\ \mathbf{u}_i \\ | \end{pmatrix}$ can be solved for each right-hand-side $\mathbf{u}_i$. Compute the solution of each linear system.

21. (a) Draw three vectors $\mathbf{x}, \mathbf{y}, \mathbf{z}$ in the plane so that $\mathbf{x} \in \text{span}\{\mathbf{y}, \mathbf{z}\}$, $\mathbf{y} \in \text{span}\{\mathbf{x}, \mathbf{z}\}$ and $\mathbf{z} \in \text{span}\{\mathbf{x}, \mathbf{y}\}$.

(b) Can you draw three vectors $\mathbf{x}, \mathbf{y}, \mathbf{z}$ in the plane so that $\mathbf{x} \in \text{span}\{\mathbf{y}, \mathbf{z}\}$, $\mathbf{y} \in \text{span}\{\mathbf{x}, \mathbf{z}\}$ and $\mathbf{z} \notin \text{span}\{\mathbf{x}, \mathbf{y}\}$. Can anyone? Explain.

(c) Is the $\text{span}\{\mathbf{x}, \mathbf{y}\}$ always a two-dimensional space in $\mathbb{R}^n$ if $\mathbf{x}, \mathbf{y} \in \mathbb{R}^n$?

22. Show that the set of solutions $\{\mathbf{x} \mid \mathbf{A}\mathbf{x} = \mathbf{b}\}$ of a linear system of equations $\mathbf{A}\mathbf{x} = \mathbf{b}$ is not a subspace if $\mathbf{b} \neq \mathbf{0}$.

23. Use the method of Example 8 to find all homogeneous solutions of the linear system

$$\begin{pmatrix} 1 & 4 & 2 & 0 \\ 0 & 1 & 5 & -1 \\ 2 & 7 & -1 & 1 \end{pmatrix} \mathbf{x} = \begin{pmatrix} 0 \\ 0 \\ 0 \end{pmatrix}.$$

24. Show that $\ker(\mathbf{A}) = \ker(\mathbf{B})$ for $\mathbf{A} = \begin{pmatrix} 2 & -1 & 3 \\ -4 & 2 & -6 \\ 0 & 0 & 0 \end{pmatrix}$ and $\mathbf{B} = \begin{pmatrix} -1 & 0.5 & -1.5 \end{pmatrix}$.

25. (a) Find a vector $\mathbf{w} \notin \ker\begin{pmatrix} 1 & 2 & 3 \\ 0 & 1 & 1 \end{pmatrix}$, and check your answer.

(b) Find a vector $\mathbf{z} \notin \text{span}\left\{\begin{pmatrix} 1 \\ 1 \\ 1 \end{pmatrix}, \begin{pmatrix} 2 \\ -1 \\ 0 \end{pmatrix}\right\}$, and check your answer.

26. Show that for a matrix $\mathbf{A}_{mn} = \begin{pmatrix} | & & | \\ \mathbf{a}_1 & \dots & \mathbf{a}_n \\ | & & | \end{pmatrix}$, every vector in the kernel of $\mathbf{A}^T := \begin{pmatrix} - & \mathbf{a}_1^T & - \\ & \vdots & \\ - & \mathbf{a}_n^T & - \end{pmatrix}_{nm}$ is perpendicular to every vector in $\text{im}(\mathbf{A})$.

27. (a) List all subspaces of the real line $\mathbb{R}^1 = \mathbb{R}$.

(b) List all subspaces of the real plane $\mathbb{R}^2$. Express each subspace as a span and as a matrix kernel.

(c) List all subspaces $\mathbb{R}^3$. Express each subspace as a span.

28. Show that the matrices $\mathbf{A} = \begin{pmatrix} 1 & 2 & 3 \\ 2 & 1 & 0 \end{pmatrix}$ and $\mathbf{B} = \begin{pmatrix} 0 & 2 & 4 \\ 1 & -1 & -3 \\ 0 & -1 & -2 \end{pmatrix}$ have the same kernel.

29. Give a proof of Theorem 4.2.

30. (a) Show that $\text{im}\begin{pmatrix} 2 & -1 \\ 0 & 0 \end{pmatrix} = \text{im}\begin{pmatrix} 0 & 3 \\ 0 & 0 \end{pmatrix}$.

(b) Find two vectors $\mathbf{x}$ and $\mathbf{y} \in \mathbb{R}^2$ with $\begin{pmatrix} 2 & -1 \\ 0 & 0 \end{pmatrix} \mathbf{x} \neq \begin{pmatrix} 0 & 3 \\ 0 & 0 \end{pmatrix} \mathbf{x}$ and with $\begin{pmatrix} 2 & -1 \\ 0 & 0 \end{pmatrix} \mathbf{y} = \begin{pmatrix} 0 & 3 \\ 0 & 0 \end{pmatrix} \mathbf{y}$.

31. (a) Find a matrix $\mathbf{B}_{3\times 5}$ with $\text{im}(\mathbf{B}) = \ker(\mathbf{A}_{4\times 3})$ for

$$\mathbf{A} = \begin{pmatrix} 1 & 1 & 3 \\ -2 & 1 & 1 \\ 1 & 4 & 10 \\ 0 & 6 & 14 \end{pmatrix}.$$

(b) Compute $\mathbf{AB}$.

32. (a) Find a matrix $\mathbf{A}_{2\times 2}$ with $\ker(\mathbf{A}) = \text{im}(\mathbf{B}_{2\times 2})$ for $\mathbf{B} = \begin{pmatrix} 1 & 4 \\ -1 & -4 \end{pmatrix}$.

(b) Compute both $\mathbf{AB}$ and $\mathbf{BA}$.

33. Show that

(a) If $\mathbf{x} \in \ker(\mathbf{A})$ and $\mathbf{R}$ is an REF for $\mathbf{A}$, then $\mathbf{x} \in \ker(\mathbf{R})$.

(b) If $\mathbf{A}$ has the REF $\mathbf{R}$ and $\mathbf{x} \in \ker(\mathbf{R})$, then $\mathbf{A}\mathbf{x} = \mathbf{0}$.

34. Let $\mathbf{A}$ be an $n \times n$ matrix. Then

(a) Express the set $V = \{\mathbf{x} \mid \mathbf{A}\mathbf{x} = 2\mathbf{x}\} \subset \mathbb{R}^n$ as a matrix kernel $\ker(\mathbf{B}) = \{\mathbf{x} \mid \mathbf{B}\mathbf{x} = 0\}$ for an $n \times n$ matrix B.

(*Hint*: Make use of the fact that $\mathbf{x} = \mathbf{I}\mathbf{x}$ for the $n \times n$ identity matrix $\mathbf{I}$ once.)

(b) Show:

If $\mathbf{x} \in \mathbb{R}^n$ satisfies $\mathbf{A}\mathbf{x} = 2\mathbf{x}$ and $\mathbf{y} \in \mathbb{R}^n$ satisfies $\mathbf{A}\mathbf{y} = 2\mathbf{y}$, then

(i) $\mathbf{A}(\mathbf{x} + \mathbf{y}) = 2(\mathbf{x} + \mathbf{y})$ and

(ii) $\mathbf{A}(\alpha\mathbf{x}) = 2(\alpha\mathbf{x})$ for all real α.

(c) Explain verbally why the set $V = \{\mathbf{x} \mid \mathbf{A}\mathbf{x} = 2\mathbf{x}\}$ is a subspace of $\mathbb{R}^n$.

35. Determine whether each of the following is true or false:

(a) A nonzero column vector $\mathbf{a}_i \in \mathbb{R}^n$ of a matrix $\mathbf{A}_{nn}$ cannot lie in the kernel of $\mathbf{A}$.

(b) A nonzero column vector $\mathbf{a}_i \in \mathbb{R}^n$ of a matrix $\mathbf{A}_{nn}$ cannot lie in the kernel of $\mathbf{A}^T$.

36. (a) Show that the sets $V = \{\mathbf{x} \in \mathbb{R}^n \mid \mathbf{A}_{nn}\mathbf{x} = \mathbf{x}\}$ and $W = \{\mathbf{y} \in \mathbb{R}^n \mid \mathbf{A}_{nn}\mathbf{y} = -\mathbf{y}\}$ are both subspaces of $\mathbb{R}^n$.

(b) Find all solutions $\mathbf{x}$ of $\mathbf{A}\mathbf{x} = \mathbf{x}$ and all $\mathbf{y}$ of $\mathbf{A}\mathbf{y} = -\mathbf{y}$ for $\mathbf{A} = \begin{pmatrix} 0 & 1 \\ 1 & 0 \end{pmatrix}$.

37. Can one find two vectors $\mathbf{x} \neq \mathbf{y} \in \mathbb{R}^3$, so that $\mathbf{A}\mathbf{x} = \mathbf{x}$ and $\mathbf{A}\mathbf{y} = -\mathbf{y}$ for the matrix

$$\mathbf{A} = \begin{pmatrix} 3 & 1 & -1 \\ 2 & 0 & 1 \\ 1 & -1 & 1 \end{pmatrix}?$$

38. Does the vector $\mathbf{x} = \begin{pmatrix} 1 \\ 2 \\ -1 \\ 4 \end{pmatrix}$ belong to the kernel of

$$\mathbf{A} = \begin{pmatrix} 2 & 1 & 3 & 4 \\ 1 & 1 & -7 & 1 \\ 0 & 1 & 2 & 0 \end{pmatrix}?$$

39. A **line** ℓ in $\mathbb{R}^3$ is described parametrically by $\ell = \{\mathbf{x} = \mathbf{r}_0 + t\mathbf{r} \mid t \in \mathbb{R}\}$ for the direction vector $\mathbf{r}$ of ℓ and a specific point $\mathbf{r}_0$ on ℓ. Alternatively, two planes in $\mathbb{R}^3$ with nonparallel normal vectors $\mathbf{a}_1$ and $\mathbf{a}_2$ also describe a line in $\mathbb{R}^3$ as their intersection.

(a) For the line $\ell = \begin{pmatrix} 0 \\ 2 \\ -1 \end{pmatrix} + t\begin{pmatrix} 1 \\ -2 \\ 2 \end{pmatrix}$, find the normal equations $\mathbf{a}_i \cdot \mathbf{x} = b_i$ with $\mathbf{a}_i \in \mathbb{R}^3$, $b_i \in \mathbb{R}$ for $i = 1, 2$ of two planes whose intersection is ℓ.

(b) Repeat part (a) for $\ell = \begin{pmatrix} 2 \\ -1 \\ 1 \\ 1 \end{pmatrix} + t\begin{pmatrix} 1 \\ -1 \\ 2 \\ 1 \end{pmatrix} \subset \mathbb{R}^4$ and three hyperplanes $\mathbf{a}_i \cdot \mathbf{x} = b_i$ in $\mathbb{R}^4$.

40. (a) Find a parametric equation $\mathbf{x} = \mathbf{r}_0 + t\mathbf{r}$ with $\mathbf{x}$, $\mathbf{r}_0$, $\mathbf{r} \in \mathbb{R}^3$ and $t \in \mathbb{R}$ for the line of intersection of the two planes $2x_1 - x_2 + x_3 = 3$, $4x_1 + x_2 = 6$ in $\mathbb{R}^3$.

(b) Find a parametric equation for the line $\ell = \{\mathbf{r}_0 + t\mathbf{r} \mid t \in \mathbb{R}\}$ in $\mathbb{R}^2$ that is described by the normal line equation $2x_1 - 3x_2 = -1$.

4.2 Applications

Join and Intersection of Subspaces

We study subspaces that are derived from pairs of subspaces.

Any two subspaces U, $V \subset \mathbb{R}^n$ naturally generate two related subspaces, namely, their **join**

$$U + V := \{\mathbf{u} + \mathbf{v} \mid \mathbf{u} \in U, \mathbf{v} \in V\}$$

and their **intersection**

$$U \cap V := \{\mathbf{x} \in \mathbb{R}^n \mid \mathbf{x} \in U \ \underline{\text{and}}\ \mathbf{x} \in V\}.$$

Both the join $U + V$ and the intersection $U \cap V$ are subspaces of $\mathbb{R}^n$ if U and V are.

The join $U + V$ of U and V is no smaller subspace than U and V are (that is, both U and V are subspaces of $U + V$, or $U \subset U + V$ and $V \subset U + V$.)
On the other hand, the intersection $U \cap V$ of two subspaces is no larger subspace than either U or V (that is, $U \cap V \subset U$ and $U \cap V \subset V$.)

The following question arises naturally:

Given two subspaces U and V, either inclusively as spans of certain sets of vectors or exclusively as matrix kernels, how can one represent the subspaces $U + V$ and $U \cap V$ as spans or as kernels?

This question consists of six parts, two of which are easy to handle, while the remaining four require knowledge on how to write the span of vectors as a matrix kernel and vice versa (See Section 4.1).

First, the two easy cases for which just looking provides an instant proof.

Theorem 4.3 (a) If $U = \text{span}\{\mathbf{u}_1, \dots, \mathbf{u}_k\} \subset \mathbb{R}^n$ and $V = \text{span}\{\mathbf{v}_1, \dots, \mathbf{v}_\ell\} \subset \mathbb{R}^n$, then $U + V = \text{span}\{\mathbf{u}_1, \dots, \mathbf{u}_k, \mathbf{v}_1, \dots, \mathbf{v}_\ell\} \subset \mathbb{R}^n$. In words, the join of two subspaces defined as spans is the span of the union of their spanning sets.

(b) If $U = \ker(\mathbf{A}_{mn})$ and $V = \ker(\mathbf{B}_{\ell n})$, then $U \cap V = \ker\left(\begin{pmatrix} \mathbf{A} \\ \mathbf{B} \end{pmatrix}_{(m+\ell)\times n}\right)$ is the kernel of the block matrix comprised of $\mathbf{A}$ above $\mathbf{B}$. ◀

Students should think up and argue through their own proofs of Theorem 4.3.

For the remaining cases of writing the join of two kernels as a span for example, one has to convert the kernels to spans as detailed earlier and apply Theorem 4.3(a). Likewise, for representing the intersection of two spans, and for any mixed join or joint kernel of two subspaces given in different generic representations.

EXAMPLE 9 Represent the join $\ker\begin{pmatrix} 1 & 3 & -1 & 0 \end{pmatrix} + \ker\begin{pmatrix} 0 & 1 & -2 & 0 \\ 0 & 2 & 3 & 1 \end{pmatrix}$ as a span.

The first subspace $U := \ker\begin{pmatrix} 1 & 3 & -1 & 0 \end{pmatrix}$ is spanned by $\begin{pmatrix} -3 \\ 1 \\ 0 \\ 0 \end{pmatrix}, \begin{pmatrix} 1 \\ 0 \\ 1 \\ 0 \end{pmatrix}$,

and $\mathbf{e}_4 = \begin{pmatrix} 0 \\ 0 \\ 0 \\ 1 \end{pmatrix} \in \mathbb{R}^4$, according to Chapter 3. Simply evaluate $\begin{pmatrix} 1 & 3 & -1 & 0 \end{pmatrix}$ $\begin{pmatrix} 3 & 1 & 0 \\ -1 & 0 & 0 \\ 0 & 1 & 0 \\ 0 & 0 & 1 \end{pmatrix}$. A spanning set for the second subspace $V := \ker\begin{pmatrix} 0 & 1 & -2 & 0 \\ 0 & 2 & 3 & 1 \end{pmatrix}$ can be computed as follows using Chapter 3:
First row reduce the augmented matrix $(\mathbf{A} \mid \mathbf{0})$.

$$\begin{array}{cccc|c|l} 0 & \boxed{1} & -2 & 0 & 0 & \\ 0 & \boxed{2} & 3 & 1 & 0 & -2\ row_1 \\ \hline 0 & \boxed{1} & -2 & 0 & 0 & \\ 0 & 0 & \boxed{7} & 1 & 0 & \end{array}$$

Backsubstitution as in Chapter 3 yields $V = \text{span}\left\{\begin{pmatrix} 0 \\ 2 \\ 1 \\ -7 \end{pmatrix}, \begin{pmatrix} 1 \\ 0 \\ 0 \\ 0 \end{pmatrix}\right\}$.

Or use Section 4.1(b) and form the multiaugmented matrix

$$\left(\begin{array}{cccc||cc} 1 & 0 & 0 & 0 & 1 & 0 \\ 0 & 1 & -2 & 0 & 0 & 0 \\ 0 & 0 & 7 & 1 & 0 & 0 \\ 0 & 0 & 0 & 1 & 0 & 1 \end{array}\right) \begin{array}{l} \leftarrow \text{inserted row} \\ \\ \\ \leftarrow \text{inserted row} \end{array}$$

from the previously computed REF of **A**. Row reduction leads to the equivalent matrix

$$\left(\begin{array}{c||cc} & 1 & 0 \\ & 0 & -2/7 \\ \mathbf{I}_4 & & \\ & 0 & -1/7 \\ & 0 & 1 \end{array}\right).$$

Note that its last two columns give an equivalent spanning set for V; just scale its second member by -7 and reverse their order.

Thus, the join of the two kernels is $U+V = \text{span}\left\{\begin{pmatrix}-3\\1\\0\\0\end{pmatrix}, \begin{pmatrix}1\\0\\1\\0\end{pmatrix}, \begin{pmatrix}0\\0\\0\\1\end{pmatrix}, \begin{pmatrix}0\\2\\1\\-7\end{pmatrix}, \begin{pmatrix}1\\0\\0\\0\end{pmatrix}\right\}$. Note that $U+V = \mathbb{R}^4$. This follows from observing that $\mathbf{e}_1$ and $\mathbf{e}_4$ both occur in the spanning set of $U+V$. Then $\mathbf{e}_3 \in U+V$ follows, as well as $\mathbf{e}_2 \in U+V$ by, in turn, looking at the second vector $\mathbf{e}_1 + \mathbf{e}_3$ and the first vector $-3\mathbf{e}_1 + \mathbf{e}_2$ of the spanning set. Thus, $U+V$ contains the standard basis $\mathcal{E}$ of $\mathbb{R}^4$ (i.e., $U+V = \mathbb{R}^4$). ◀

EXAMPLE 10 Find a spanning set for $U \cap V$, where $U = \text{span}\left\{\begin{pmatrix}1\\2\\-1\\3\end{pmatrix}, \begin{pmatrix}1\\0\\2\\-1\end{pmatrix}\right\}$ and $V = \text{span}\left\{\begin{pmatrix}3\\2\\0\\1\end{pmatrix}, \begin{pmatrix}1\\0\\-1\\-1\end{pmatrix}\right\}$.

If we represent U and V as matrix kernels we can use Theorem 4.3(b) to express $U \cap V$ as the kernel of a larger sized matrix. Finding a spanning set of this kernel involves one more translation according to Section 4.1(b) or solving one homogeneous system of linear equations according to Section 3.1, which we shall do later.

We have represented U as $\ker\begin{pmatrix}4 & -3 & -2 & 0\\ -7 & 4 & 4 & 1\end{pmatrix}$ in Example 7 of Section 4.1.P. Likewise V can be presented as a kernel by for example using method (4.1) of Section 4.1(b):

$$\mathbf{V} = \begin{array}{cc||cccc} \mathbf{v}_1 & \mathbf{v}_2 & y_1 & y_2 & y_3 & y_4 \\ \hline 3 & 1 & 1 & 0 & 0 & 0 \\ 2 & 0 & 0 & 1 & 0 & 0 \\ 0 & -1 & 0 & 0 & 1 & 0 \\ 1 & -1 & 0 & 0 & 0 & 1 \end{array}$$

swap rows 1 and 4 for an easy pivot 1

$$
\begin{array}{rr||rrrr|l}
\boxed{1} & -1 & 0 & 0 & 0 & 1 & \\
\boxed{2} & 0 & 0 & 1 & 0 & 0 & -2\ \mathit{row}_1 \\
0 & -1 & 0 & 0 & 1 & 0 & \\
\boxed{3} & 1 & 1 & 0 & 0 & 0 & -3\ \mathit{row}_1 \\
\hline
\boxed{1} & -1 & 0 & 0 & 0 & 1 & \\
0 & 2 & 0 & 1 & 0 & -2 & \textit{swap rows 2 and 3 for the easy pivot } -1 \\
0 & -1 & 0 & 0 & 1 & 0 & \\
0 & 4 & 1 & 0 & 0 & -3 & \\
\hline
\boxed{1} & -1 & 0 & 0 & 0 & 1 & \\
0 & \boxed{-1} & 0 & 0 & 1 & 0 & \\
0 & \boxed{2} & 0 & 1 & 0 & -2 & +2\ \mathit{row}_2 \\
0 & \boxed{4} & 1 & 0 & 0 & -3 & +4\ \mathit{row}_2 \\
\hline
\boxed{1} & -1 & 0 & 0 & 0 & 1 & \\
0 & \boxed{-1} & 0 & 0 & 1 & 0 & \\
\cline{3-6}
0 & 0 & 0 & 1 & 2 & -2 & \\
0 & 0 & 1 & 0 & 4 & -3 & = \mathbf{B}
\end{array}
$$

Thus, $\mathbf{V} = \ker\begin{pmatrix} 0 & 1 & 2 & -2 \\ 1 & 0 & 4 & -3 \end{pmatrix}$ and

$$
U \cap V = \ker\begin{pmatrix} \mathbf{A} \\ \mathbf{B} \end{pmatrix} = \ker\begin{pmatrix} 4 & -3 & -2 & 0 \\ -7 & 4 & 4 & 1 \\ 0 & 1 & 2 & -2 \\ 1 & 0 & 4 & -3 \end{pmatrix}
$$

from Theorem 4.3(b). Next we compute a spanning set for this kernel: Reorder the rows of the preceding 4×4 matrix for a convenient pivot and solve the following equivalent homogeneous linear system:

$$
\begin{array}{rrrr|r|l}
\boxed{1} & 0 & 4 & -3 & 0 & \\
0 & 1 & 2 & -2 & 0 & \\
\boxed{4} & -3 & -2 & 0 & 0 & -4\ \mathit{row}_1 \\
\boxed{-7} & 4 & 4 & 1 & 0 & \\
\hline
\boxed{1} & 0 & 4 & -3 & 0 & \\
0 & \boxed{1} & 2 & -2 & 0 & \\
0 & \boxed{-3} & -18 & 12 & 0 & +3\ \mathit{row}_2 \\
0 & \boxed{4} & 32 & -20 & 0 & -4\ \mathit{row}_2 \\
\hline
\boxed{1} & 0 & 4 & -3 & 0 & \\
0 & \boxed{1} & 2 & -2 & 0 & \\
0 & 0 & \boxed{-12} & 6 & 0 & \\
0 & 0 & \boxed{24} & -12 & 0 & +2\ \mathit{row}_3 \\
\hline
\end{array}
$$

$$\begin{array}{cccc|c} \boxed{1} & 0 & 4 & -3 & 0 \\ 0 & \boxed{1} & 2 & -2 & 0 \\ 0 & 0 & \boxed{-12} & 6 & 0 \\ 0 & 0 & 0 & 0 & 0 \\ & & & \uparrow & \end{array}$$

Using backsubstitution we see that $\mathbf{x} = \begin{pmatrix} 2 \\ 2 \\ 1 \\ 2 \end{pmatrix}$ spans $U \cap V = \ker\begin{pmatrix} \mathbf{A} \\ \mathbf{B} \end{pmatrix}$.

Alternatively, using the method, of Eq. (4.1) of Section 4.1(b), we find that a row reduction of the multi–augmented matrix

$$\left(\begin{array}{cccc||c} 1 & 0 & 4 & -3 & 0 \\ 0 & 1 & 2 & -2 & 0 \\ 0 & 0 & -12 & 6 & 0 \\ 0 & 0 & 0 & 1 & 1 \end{array}\right) \quad \leftarrow \textit{inserted row}$$

to $(\mathbf{I}_4 \mid\mid \mathbf{y})$ yields $\mathbf{y} = \begin{pmatrix} 1 \\ 1 \\ 1/2 \\ 1 \end{pmatrix}$ on the right side. Scaling $\mathbf{y}$ by 2 shows that $\text{span}\{\mathbf{y}\} = \text{span}\{\mathbf{x}\} = U \cap V$.

Students should check that indeed $\mathbf{x} = 2\mathbf{y} = \mathbf{u}_1 + \mathbf{u}_2 = \mathbf{v}_1 - \mathbf{v}_2$ belongs to both U and V. ◀

We have learned how to find the join and intersection of any two subspaces.

4.2.P Problems

1. Use the subspace definition from Section 4.1 to prove that the join and the intersection of two subspaces of $\mathbb{R}^n$ are also subspaces.

2. (a) Draw two subspaces U and V in $\mathbb{R}^2$ whose union $U \cup V$ is not a subspace.

   (b) Draw two subspaces U and V in $\mathbb{R}^2$ whose union $U \cup V$ is a subspace of $\mathbb{R}^2$.

3. Express the join and the intersection of the two subspaces $U = \ker(1\ 2\ -1)$ and $V = \text{span}\left\{\begin{pmatrix} 1 \\ 0 \\ -1 \end{pmatrix}, \begin{pmatrix} 0 \\ 2 \\ 1 \end{pmatrix}\right\}$ of $\mathbb{R}^3$ as a span.

4. Express the join of $\ker(\mathbf{A})$ and $\ker(\mathbf{B})$ for $\mathbf{A} = \begin{pmatrix} 1 & 2 & 4 & -7 \\ 2 & 3 & 1 & 0 \end{pmatrix}$ and $\mathbf{B} = \begin{pmatrix} 3 & 5 & 5 & -7 \\ 1 & 1 & -3 & 7 \end{pmatrix}$ as a span.

5. (a) Find the join and the intersection of the two subspaces $U = \ker\begin{pmatrix} 0 & 1 & -1 & 0 \\ 2 & 1 & 0 & 1 \end{pmatrix}$ and $V = \text{span}\left\{\begin{pmatrix} 0 \\ 1 \\ -1 \\ 0 \end{pmatrix}, \begin{pmatrix} 2 \\ 1 \\ 0 \\ 1 \end{pmatrix}\right\}$ of $\mathbb{R}^4$.

   What are their dimensions?

   (b) Ditto for $U = \ker\begin{pmatrix} 0 & 1 & -1 & 0 \\ 2 & 1 & 0 & 1 \end{pmatrix}$ and $W = \text{span}\left\{\begin{pmatrix} -1 \\ 1 \\ 1 \\ 1 \end{pmatrix}, \begin{pmatrix} 1 \\ 0 \\ 0 \\ -2 \end{pmatrix}\right\}$.

6. Span the intersection and the join of the two subspaces $\ker(\mathbf{A})$ and $\ker(\mathbf{B})$ for $\mathbf{A} = \begin{pmatrix} 1 & 3 & 2 & 0 \\ 4 & 11 & 0 & 1 \\ -2 & 0 & 1 & 1 \end{pmatrix}$ and $\mathbf{B} = \begin{pmatrix} 0 & 11 & 2 & 3 \\ -1 & 3 & 3 & 1 \end{pmatrix}$.

7. Find a spanning set for the intersection and the join of the subspaces $U = \text{span}\left\{\begin{pmatrix} 1 \\ -1 \\ 0 \\ 2 \end{pmatrix}, \begin{pmatrix} 2 \\ 1 \\ 1 \\ 1 \end{pmatrix}\right\}$ and $V = \ker\begin{pmatrix} 1 & 0 & -1 & 0 \\ 0 & 2 & 0 & 1 \end{pmatrix}$.
What are the dimensions of U, V, $U + V$ and $U \cap V$?

8. How can one determine the intersection of a k–dimensional subspace U of $\mathbb{R}^n$ with a one-dimensional subspace $V = \text{span}\{\mathbf{v}\} \subset \mathbb{R}^n$ most quickly? (*Hint*: What intersections are possible?)

9. Let $U = \text{span}\left\{\begin{pmatrix} 1 \\ 2 \\ 0 \\ 1 \end{pmatrix}, \begin{pmatrix} -1 \\ 2 \\ 1 \\ 0 \end{pmatrix}\right\}$ and $V = \text{span}\left\{\begin{pmatrix} 3 \\ 1 \\ -1 \\ 1 \end{pmatrix}, \begin{pmatrix} 8 \\ -3 \\ -4 \\ 1 \end{pmatrix}\right\}$.

(a) Express U and V as matrix kernels. Check your answers.

(b) Express the intersection of U and V as a span.

10. True or false? Let U and V be two subspaces of $\mathbb{R}^n$. If $U + V = U \cap V$, then $U = V$.

11. Find the intersection and the join of the kernels and images for each of the following matrices:

$$\mathbf{A} = \begin{pmatrix} 0 & 1 & 0 \\ 0 & 0 & 1 \\ 0 & 0 & 0 \end{pmatrix}, \quad \mathbf{B} = \begin{pmatrix} 1 & 0 & 0 \\ 0 & 0 & 0 \\ 0 & 0 & 0 \end{pmatrix}, \quad \mathbf{C} = \begin{pmatrix} 1 & 2 \\ -1 & -2 \end{pmatrix}.$$

12. True or false? Let U and V be two subspaces of $\mathbb{R}^n$. Then $U \cup V = U + V$ if and only if $U \cup V$ is a subspace.

4.R Review Problems

1. Write out the definitions and concepts underlying the notions of

(a) A linear space.

(b) A subspace.

(c) A spanning set for a subspace.

(d) The kernel of a matrix A.

(e) The image associated with a matrix A.

2. (a) For $\mathbf{A} = \begin{pmatrix} 1 & -1 \\ 1 & -1 \end{pmatrix}$ show that $\ker(\mathbf{A}) = \text{im}(\mathbf{A}) \subset \mathbb{R}^2$.

(b) Try to find a 3×3 matrix $\mathbf{B}$ with the same image-equals-kernel property.

(c) Do this for a 4×4 matrix $\mathbf{C}$, if possible.

3. Let $\mathbf{a} = (a_1, \ldots, a_m) \in \mathbb{R}^m$ be a nonzero row vector. Show that the matrices $\mathbf{A}_1 = \begin{pmatrix} - & \mathbf{a} & - \end{pmatrix}_{1,m}$, $\mathbf{A}_2 = \begin{pmatrix} - & 2\mathbf{a} & - \\ - & -3\mathbf{a} & - \end{pmatrix}_{2,m}$, and $\mathbf{A}_3 = \begin{pmatrix} - & -\mathbf{a} & - \\ 0 & \cdots & 0 \\ - & 2\mathbf{a} & - \end{pmatrix}_{3,m}$ all have the same kernel. What is this kernel?

4. (a) Find a subspace of $\mathbb{R}^4$ that does not contain the vectors $\begin{pmatrix} 1 \\ 0 \\ -1 \\ 0 \end{pmatrix}$ and $\begin{pmatrix} 2 \\ 0 \\ 0 \\ 0 \end{pmatrix}$ and write it as a span and as a kernel.

(b) How many such subspaces are there in $\mathbb{R}^4$?

5. Write the subspace $U \subset \mathbb{R}^4$ spanned by $\begin{pmatrix} 1 \\ 0 \\ -1 \\ 0 \end{pmatrix}$ and $\begin{pmatrix} 2 \\ 0 \\ 0 \\ 0 \end{pmatrix}$ as a matrix kernel.

6. Prove that the union $U \cup V$ of two subspaces $U, V \subset \mathbb{R}^n$ is a subspace if and only if $U \subset V$ or $V \subset U$.

7. Express $-2\mathbf{e}_3 \in \mathbb{R}^3$ as a linear combination of the basis $\mathbf{u}_1 = \mathbf{e}_1$, $\mathbf{u}_2 = \mathbf{e}_1 - \mathbf{e}_2$ and $\mathbf{u}_3 = 2\mathbf{e}_1 - \mathbf{e}_2 + \mathbf{e}_3$ of $\mathbb{R}^3$.

8. Show that the set of right hand sides $\mathbf{b}$ for which the linear system of equations $\mathbf{Ax} = \mathbf{b}$ is solvable is a subspace of $\mathbb{R}^n$ for each matrix $\mathbf{A}_{nn}$. (Compare this with Problem 22 in section 4.1.P.)

9. What is the join $U + V$ of two subspaces U and V with $U \subset V$?

10. (a) Find a 2×2 matrix $\mathbf{A}$ with $\text{im}(\mathbf{A}) = \ker(\mathbf{A})$ and as many zero entries as possible. (Hint: There are several such matrices with only one nonzero entry.)

(b) Use the 2×2 matrix $\mathbf{A}$ from part (a) to construct a 4×4 matrix $\mathbf{B}$ with $\text{im}(\mathbf{B}) = \ker(\mathbf{B})$ and as many zero entries as possible.

11. Assume that $\ker(\mathbf{A}) = \text{im}(\mathbf{B})$ for two matrices $\mathbf{A}_{mn}$ and $\mathbf{B}_{n\ell}$.

(a) Evaluate the matrix product $\mathbf{AB}$.

(b) What can be said about the reverse order product $\mathbf{BA}$ provided that $\ell = m$?

12. Determine a spanning set for the subspace of all right hand sides b for which $\mathbf{Ax} = \mathbf{b}$ is solvable with $\mathbf{A} = \begin{pmatrix} 1 & 2 & 3 & -1 \\ 1 & 1 & 2 & -1 \\ 0 & 1 & -1 & 0 \end{pmatrix}$. (Hint: See Problem 8.)

13. Test whether the following sets are subspaces:

(a) $X = \left\{ \begin{pmatrix} a+b \\ a-b \end{pmatrix} \middle| a, b \in \mathbb{R} \right\} \subset \mathbb{R}^2$.

(b) $Y = \left\{ \begin{pmatrix} a+b-c \\ c-a \\ a+1-b \end{pmatrix} \middle| a, b, c \in \mathbb{R} \right\} \subset \mathbb{R}^3$.

(c) $Z = \left\{ \begin{pmatrix} a+2b+3c \\ b+c \\ -a+1-c \end{pmatrix} \middle| a, b, c \in \mathbb{R} \right\} \subset \mathbb{R}^3$.

14. Consider the entry sum $S(A) := a_{11} + a_{12} + a_{21} + a_{22}$ for each 2 by 2 matrix $\mathbf{A} = (a_{ij})$.
For which real constants c does the set

$$X = \{\mathbf{A} \in \mathbb{R}^{2,2} \mid S(\mathbf{A}) = c\}$$

form a subspace of $\mathbb{R}^{2,2}$?

15. Consider the set X of all 2×2 matrices $\mathbf{A} = (a_{ij})$ with $a_{11} = 3a_{22}$.
Show that X is a subspace of $\mathbb{R}^{2,2}$.

Standard Tasks and Questions

1. What kind of subsets of $\mathbb{R}^n$ are subspaces?
2. How can a subspace of $\mathbb{R}^n$ be represented?
3. Does a set of vectors $\{\mathbf{v}_1, \ldots, \mathbf{v}_\ell\}$ span a certain subspace?
4. Find all vectors that are orthogonal to a given set of vectors $\{\mathbf{a}_1, \ldots, \mathbf{a}_k\}$.
5. Express the kernel of a matrix as a span.
6. Express the span of several given vectors as a matrix kernel.

Subheadings of Lecture Four

Keeping above the muck

CHAPTER

5

Linear Dependence, Bases, and Dimension

This chapter uses the row-echelon-form reduction of a matrix to study how linear subspaces are generated.

5.1 Lecture Five, a Double Lecture

Minimal Spanning or Maximally Independent Sets of Vectors

We develop basic notions for describing and building subspaces.

(a) Linear Independence

A real matrix $\mathbf{A}_{mn}$ induces the linear transformation

$$\mathbf{A} : \mathbb{R}^n \longrightarrow \mathbb{R}^m$$

defined by the assignment $\mathbf{x} \mapsto \mathbf{Ax}$. The domain of $\mathbf{A}$ is all of $\mathbb{R}^n$, which is commonly described by the set

$$\{x_1\mathbf{e}_1 + \cdots + x_n\mathbf{e}_n \mid x_i \in \mathbb{R}\}$$

for the standard unit vector basis $\{\mathbf{e}_1, \mathbf{e}_2, \ldots, \mathbf{e}_n\} = \mathcal{E}$. Thus, $\mathbb{R}^n = \left\{ \begin{pmatrix} x_1 \\ \vdots \\ x_n \end{pmatrix} \middle| x_i \in \mathbb{R} \right\}$ has n degrees of freedom, one for each of its n components $x_1, \ldots, x_n \in \mathbb{R}$. In this sense, we say that $\mathbb{R}^n$ has **dimension** n.

The following question arises naturally:

For a given matrix $\mathbf{A}_{mn}$, what degree of freedom, what geometric dimension does its **image**

$$\operatorname{im}(\mathbf{A}) = \{\mathbf{y} \in \mathbb{R}^m \mid \mathbf{y} = \mathbf{A}\mathbf{x}\} = \left\{ x_1 \begin{pmatrix} | \\ \mathbf{a}_1 \\ | \end{pmatrix} + \cdots + x_n \begin{pmatrix} | \\ \mathbf{a}_n \\ | \end{pmatrix} \middle| x_i \in \mathbb{R} \right\}$$

have, where the vectors $\begin{pmatrix} | \\ \mathbf{a}_i \\ | \end{pmatrix} \in \mathbb{R}^m$ denote the columns of $\mathbf{A} = \begin{pmatrix} | & & | \\ \mathbf{a}_1 & \dots & \mathbf{a}_n \\ | & & | \end{pmatrix}$?

Recall from Section 4.1 that $\operatorname{im}(\mathbf{A})$ is the subspace of $\mathbb{R}^m$ spanned by the columns of $\mathbf{A}$. Thus, it has less than or equal to $m = \dim(\mathbb{R}^m)$ degrees of freedom.

EXAMPLE 1

(a) If $\mathbf{A} = \mathbf{O}_{mn}$, then $\operatorname{im}(\mathbf{A}) = \{\mathbf{0}\}$, the singleton space. Thus, there is no freedom to move in the image: $\mathbf{A}\mathbf{x} = \mathbf{0}$ for all $\mathbf{x} \in \mathbb{R}^n$.

(b) If $\mathbf{A}_{mm} = \mathbf{I}_m$, then $\operatorname{im}(\mathbf{A}) = \mathbb{R}^m$, the entire space. Thus, the freedom of movement in $\mathbf{A}$'s image or range space is as large as in $\mathbf{A}$'s domain. Actually, the domain and range are identical for the identity matrix $\mathbf{I}_m$.

(c) If $\mathbf{A}_{mn} = \begin{pmatrix} | & | & & | \\ \mathbf{a}_1 & \mathbf{0} & \cdots & \mathbf{0} \\ | & | & & | \end{pmatrix}$ for one vector $\mathbf{a}_1 \neq \mathbf{0} \in \mathbb{R}^m$, then $\operatorname{im}(\mathbf{A}) = \{\alpha \mathbf{a}_1 \mid \alpha \in \mathbb{R}\} \subset \mathbb{R}^m$. Thus, the image of $\mathbb{R}^n$ under $\mathbf{A}$ has only one degree of freedom along the line in the direction of $\mathbf{a}_1 \in \mathbb{R}^m$. ◀

This chapter works with $\mathbf{A}_{mn} = \begin{pmatrix} | & & | \\ \mathbf{a}_1 & \cdots & \mathbf{a}_n \\ | & & | \end{pmatrix}$ in column notation. To answer our question on the degree of freedom in $\operatorname{im}(\mathbf{A})$, we rely on our "helpers" from Chapters 2 and 3, namely, the row-echelon and reduced row-echelon forms of $\mathbf{A}$:

To find out about the size of $\operatorname{im}(\mathbf{A}) \subset \mathbb{R}^m$, we look at the effect that each column $\mathbf{a}_j$ of $\mathbf{A}$ has in adding to the span of the previous columns $\mathbf{a}_1, \dots, \mathbf{a}_{j-1}$ of $\mathbf{A}$. If $\mathbf{a}_j \in \operatorname{span}\{\mathbf{a}_1, \dots, \mathbf{a}_{j-1}\}$ for the jth column of $\mathbf{A}$, then $\mathbf{a}_j$ is a linear combination of the columns $\mathbf{a}_i$, $i \leq j{-}1$ that precede it. Therefore, in this case, any linear combination of the first j columns of $\mathbf{A}$ can be written as one of the first $j-1$ columns of $\mathbf{A}$. And $\mathbf{a}_j$ does not contribute a new direction to $\operatorname{im}(\mathbf{A})$. In case $\mathbf{a}_j \in \operatorname{span}\{\mathbf{a}_1, \dots, \mathbf{a}_{j-1}\}$, we recall from Chapters 2 and 3—Theorem 3.1 in particular—

that the rank of $\begin{pmatrix} | & & | \\ \mathbf{a}_1 & \dots & \mathbf{a}_j \\ | & & | \end{pmatrix}$ equals the rank of $\begin{pmatrix} | & & | \\ \mathbf{a}_1 & \dots & \mathbf{a}_{j-1} \\ | & & | \end{pmatrix}$ in this case. Thus, any REF of $\mathbf{A}$ must contain a free column (i.e., one without a pivot) in column position j if $\mathbf{a}_j \in \text{span}\{\mathbf{a}_1, \dots, \mathbf{a}_{j-1}\}$.

Therefore, any row reduced form of $\mathbf{A}$, such as a REF, gives us valuable information about the column space $\text{im}(\mathbf{A})$ of $\mathbf{A}$. In fact, there is a natural duality between the column space $\text{im}(\mathbf{A}) = \text{span}\,\{\mathbf{a}_1, \dots, \mathbf{a}_n\}$ and a REF of $\mathbf{A}$ that we shall exploit forthwith.

EXAMPLE 2 Let $\mathbf{A} = \begin{pmatrix} | & & | \\ \mathbf{a}_1 & \dots & \mathbf{a}_5 \\ | & & | \end{pmatrix} = \begin{pmatrix} 1 & 3 & 0 & 2 & 1 \\ -1 & -3 & 1 & -3 & 0 \\ 2 & 6 & -1 & 5 & 2 \end{pmatrix}$. To study the image space of the linear matrix transformation $\mathbf{x} \mapsto \mathbf{A}\mathbf{x}$, we row reduce $\mathbf{A}$:

$$
\begin{array}{r|ccccc|l}
\mathbf{A} = & \mathbf{a}_1 & \mathbf{a}_2 & \mathbf{a}_3 & \mathbf{a}_4 & \mathbf{a}_5 & \\
\hline
 & \boxed{1} & 3 & 0 & 2 & 1 & \\
 & \boxed{-1} & -3 & 1 & -3 & 0 & +\,row_1 \\
 & \boxed{2} & 6 & -1 & 5 & 2 & -2\,row_1 \\
\hline
 & \boxed{1} & 3 & 0 & 2 & 1 & \\
 & 0 & 0 & \boxed{1} & -1 & 1 & \\
 & 0 & 0 & \boxed{-1} & 1 & 0 & +\,row_2 \\
\hline
 & \boxed{1} & 3 & 0 & 2 & \boxed{1} & -\,row_3 \\
 & 0 & 0 & \boxed{1} & -1 & \boxed{1} & -\,row_3 \\
 & 0 & 0 & 0 & 0 & \boxed{1} & \\
\hline
 & \boxed{1} & 3 & 0 & 2 & 0 & \\
 & 0 & 0 & \boxed{1} & -1 & 0 & \quad = \mathbf{R} \\
 & 0 & 0 & 0 & 0 & \boxed{1} & \\
\hline
 & \boxed{1} & r_{12} & 0 & r_{14} & 0 & \\
 & 0 & 0 & \boxed{1} & r_{24} & 0 & \\
 & 0 & 0 & 0 & 0 & \boxed{1} & \\
 & & \uparrow & & \uparrow & &
\end{array}
$$

The foregoing RREF $\mathbf{R}$ of $\mathbf{A}$ has the particular shape $\mathbf{R} = \begin{pmatrix} \boxed{1} & r_{12} & 0 & r_{14} & 0 \\ 0 & 0 & \boxed{1} & r_{24} & 0 \\ 0 & 0 & 0 & 0 & \boxed{1} \end{pmatrix}$.

Each column $\begin{pmatrix} | \\ \mathbf{a}_i \\ | \end{pmatrix}$ of the original matrix $\mathbf{A}$ corresponds to either a pivot or a free column in its RREF $\mathbf{R}$. Specifically, the first, third, and fifth columns of $\mathbf{R}$ are pivot columns, while the columns two and four are free.

The RREF $\mathbf{R}$ of $\mathbf{A}$ expresses linear relations among the columns $\begin{pmatrix} | \\ \mathbf{a}_i \\ | \end{pmatrix}$ of $\mathbf{A}$:

For the first and second columns of $\mathbf{A}$, we have $\begin{pmatrix} 3 \\ -3 \\ 6 \end{pmatrix} = 3 \begin{pmatrix} 1 \\ -1 \\ 2 \end{pmatrix}$. This corresponds to

$$\begin{pmatrix} r_{12} \\ 0 \\ 0 \end{pmatrix} = r_{12} \begin{pmatrix} 1 \\ 0 \\ 0 \end{pmatrix} \tag{5.1(a)}$$

for the first two columns in the RREF $\mathbf{R}$. Similarly, for the first four columns of $\mathbf{A}$, we have $\begin{pmatrix} 2 \\ -3 \\ 5 \end{pmatrix} = 2 \begin{pmatrix} 1 \\ -1 \\ 2 \end{pmatrix} + (-1) \begin{pmatrix} 0 \\ 1 \\ -1 \end{pmatrix}$, or, equivalently,

$$\begin{pmatrix} r_{14} \\ r_{24} \\ 0 \end{pmatrix} = r_{14} \begin{pmatrix} 1 \\ 0 \\ 0 \end{pmatrix} + r_{24} \begin{pmatrix} 0 \\ 1 \\ 0 \end{pmatrix} \tag{5.1(b)}$$

for $\mathbf{R}$. This is no coincidence. Consider the two specific linear systems

$$\begin{pmatrix} | \\ \mathbf{a}_1 \\ | \end{pmatrix} x = \begin{pmatrix} | \\ \mathbf{a}_2 \\ | \end{pmatrix} \quad \text{and} \quad \begin{pmatrix} | & | \\ \mathbf{a}_1 & \mathbf{a}_3 \\ | & | \end{pmatrix} \begin{pmatrix} x_1 \\ x_2 \end{pmatrix} = \begin{pmatrix} | \\ \mathbf{a}_4 \\ | \end{pmatrix}$$

derived from $\mathbf{A}$. The RREF $\mathbf{R}$ of $\mathbf{A}$ indicates that both are solvable according to Theorem 3.1; hence, column $\mathbf{a}_2$ of $\mathbf{A}$ is a linear combination of $\mathbf{a}_1$, as is column $\mathbf{a}_4$ of $\mathbf{a}_1$ and $\mathbf{a}_3$.

> What do the relations (5.1(a)) and (5.1(b)) among the pivot and free columns of $\mathbf{A}$ mean for the image $\text{im}(\mathbf{A}) = \{\mathbf{Ax} \mid \mathbf{x} \in \mathbb{R}^n\} \subset \mathbb{R}^m$ of $\mathbf{A}$?

From our analysis, each free column of a REF $\mathbf{R}$ *of* $\mathbf{A}$ *is a certain linear combination of the preceding pivot columns. And this dependency translates verbatim to the columns of* $\mathbf{A}$.

In the RREF $\mathbf{R}$ of the given matrix $\mathbf{A}$, the second column is free and equals $r_{12} = 3$ times the first by Eq. (5.1(a)). This also holds for the columns of $\mathbf{A}$, namely, $\begin{pmatrix} | \\ \mathbf{a}_2 \\ | \end{pmatrix} = 3 \begin{pmatrix} | \\ \mathbf{a}_1 \\ | \end{pmatrix}$ by inspection. Similarly, $\mathbf{A}$'s fourth column $\begin{pmatrix} | \\ \mathbf{a}_4 \\ | \end{pmatrix} \in \text{span}\left\{ \begin{pmatrix} | \\ \mathbf{a}_1 \\ | \end{pmatrix}, \begin{pmatrix} | \\ \mathbf{a}_3 \\ | \end{pmatrix} \right\}$ by Eq. (5.1(b)). Thus, for our specific matrix $\mathbf{A}$, a generic point $\mathbf{y}$ in the image $\text{im}(\mathbf{A})$ has the form

$$\mathbf{y} = \mathbf{Ax} = \mathbf{A}\begin{pmatrix} x_1 \\ \vdots \\ x_5 \end{pmatrix} = x_1\begin{pmatrix} | \\ \mathbf{a}_1 \\ | \end{pmatrix} + x_2\begin{pmatrix} | \\ \mathbf{a}_2 \\ | \end{pmatrix} + \ldots + x_5\begin{pmatrix} | \\ \mathbf{a}_5 \\ | \end{pmatrix}$$

(by definition)

$$= x_1\begin{pmatrix} | \\ \mathbf{a}_1 \\ | \end{pmatrix} + 3x_2\begin{pmatrix} | \\ \mathbf{a}_1 \\ | \end{pmatrix} + x_3\begin{pmatrix} | \\ \mathbf{a}_3 \\ | \end{pmatrix} +$$

$$+ x_4\left(2\begin{pmatrix} | \\ \mathbf{a}_1 \\ | \end{pmatrix} - \begin{pmatrix} | \\ \mathbf{a}_3 \\ | \end{pmatrix}\right) + x_5\begin{pmatrix} | \\ \mathbf{a}_5 \\ | \end{pmatrix}$$

(using the dependencies $\mathbf{a}_2 = 3\mathbf{a}_1$ and $\mathbf{a}_4 = 2\mathbf{a}_1 - \mathbf{a}_3$)

$$= \alpha_1\begin{pmatrix} | \\ \mathbf{a}_1 \\ | \end{pmatrix} + \alpha_3\begin{pmatrix} | \\ \mathbf{a}_3 \\ | \end{pmatrix} + \alpha_5\begin{pmatrix} | \\ \mathbf{a}_5 \\ | \end{pmatrix}$$

(combining the three pivot column coefficients).

Here we have used the relations (5.1(a)) and (5.1(b)) for $\mathbf{R}$, and

$$\alpha_1 = x_1 + 3x_2 + 2x_4, \alpha_3 = x_3 - x_4, \text{ and } \alpha_5 = x_5 \in \mathbb{R}$$

are arbitrary parameters. Thus, the image of $\mathbb{R}^5$ under $\mathbf{A}$ has precisely three degrees of freedom. Every point of the image is a linear combination of the three column vectors $\begin{pmatrix} 1 \\ -1 \\ 2 \end{pmatrix}$, $\begin{pmatrix} 0 \\ 1 \\ -1 \end{pmatrix}$, and $\begin{pmatrix} 1 \\ 0 \\ 2 \end{pmatrix}$ of $\mathbf{A}$ that correspond to pivot columns in $\mathbf{R}$. Therefore, we say that the **dimension** of the image of $\mathbf{A}$ is three, or that

$$\dim(\text{im}(\mathbf{A})) = 3 (= \text{rank}(\mathbf{A})).$$

◀

This example leads to the definition of two complementary concepts for sets of vectors:

DEFINITION 1 **(Linear Dependence and Independence via Rank or Pivots)**

(a) A set of k vectors $\begin{pmatrix} | \\ \mathbf{a}_1 \\ | \end{pmatrix}, \ldots, \begin{pmatrix} | \\ \mathbf{a}_k \\ | \end{pmatrix} \in \mathbb{R}^m$ is **linearly independent** if a REF $\mathbf{R}$ of $\mathbf{A}_{mk} := \begin{pmatrix} | & & | \\ \mathbf{a}_1 & \ldots & \mathbf{a}_k \\ | & & | \end{pmatrix}$ has k pivots (i.e., if $\text{rank}\begin{pmatrix} | & & | \\ \mathbf{a}_1 & \ldots & \mathbf{a}_k \\ | & & | \end{pmatrix} = k =$ the number of vectors). In other words, every column of $\mathbf{R}_{mk}$ contains a pivot, or there are no free columns.

(b) A set of ℓ vectors $\begin{pmatrix} | \\ \mathbf{b}_1 \\ | \end{pmatrix}, \dots, \begin{pmatrix} | \\ \mathbf{b}_\ell \\ | \end{pmatrix} \in \mathbb{R}^m$ is **linearly dependent** if a REF **R** of $\mathbf{B}_{m\ell} := \begin{pmatrix} | & & | \\ \mathbf{b}_1 & \dots & \mathbf{b}_\ell \\ | & & | \end{pmatrix}$ has fewer than ℓ pivots, (i.e., if $\operatorname{rank}\begin{pmatrix} | & & | \\ \mathbf{b}_1 & \dots & \mathbf{b}_\ell \\ | & & | \end{pmatrix} < \ell =$ the number of vectors). In other words, not every column of $\mathbf{R}_{m\ell}$ contains a pivot, or there is at least one free column.

For example, the three columns $\begin{pmatrix} 1 \\ -1 \\ 2 \end{pmatrix}, \begin{pmatrix} 0 \\ 1 \\ -1 \end{pmatrix}$, and $\begin{pmatrix} 1 \\ 0 \\ 2 \end{pmatrix}$ of **A** in Example 2 are linearly independent, since they correspond to pivot columns in **R**. On the other hand, each of the column subsets

$$\left\{\begin{pmatrix} 1 \\ -1 \\ 2 \end{pmatrix}, \begin{pmatrix} 3 \\ -3 \\ 6 \end{pmatrix}\right\} \text{ and } \left\{\begin{pmatrix} 1 \\ -1 \\ 2 \end{pmatrix}, \begin{pmatrix} 3 \\ -3 \\ 6 \end{pmatrix}, \begin{pmatrix} 0 \\ 1 \\ -1 \end{pmatrix}, \begin{pmatrix} 2 \\ -3 \\ 5 \end{pmatrix}\right\}$$

of **A** is linearly dependent according to the RREF computed in Example 2.

The image of $\mathbb{R}^k$ under a matrix transform by $\mathbf{A}_{mk}$ and its associated REF **R** have allowed us to understand linear redundancy of column vectors in $\mathbb{R}^m$. Likewise, this dependence of column vectors can be interpreted via the kernel of the associated matrix **A**. For this, we use Section 3.1 and its criteria on the unique and nonunique solvability of homogeneous linear systems:
Every homogeneous linear system $\mathbf{Ax} = \mathbf{0}$ is solved by $\mathbf{x} = \mathbf{0}$. This trivial solution is the only solution of $\mathbf{Ax} = \mathbf{0}$ precisely when every column of a REF of **A** has a pivot. Example 2 shows that a set of columns $\{\mathbf{a}_i\}_{i=1}^k \subset \mathbb{R}^m$ is linearly independent if and only if the **column vector matrix** $\mathbf{A}_{mk} := \begin{pmatrix} | & & | \\ \mathbf{a}_1 & \dots & \mathbf{a}_k \\ | & & | \end{pmatrix}$ allows no free columns in its REF (i.e., if it has only the trivial kernel $\ker(\mathbf{A}) = \{\mathbf{0}\}$ according to Theorem 3.2).

Alternatively, a set of vectors $\{\mathbf{b}_j\}_{j=1}^\ell \subset \mathbb{R}^m$ is linearly dependent if their column matrix $\mathbf{B}_{m\ell} := \begin{pmatrix} | & & | \\ \mathbf{b}_1 & \cdots & \mathbf{b}_\ell \\ | & & | \end{pmatrix}$ has a nontrivial kernel (i.e., if $\ker(\mathbf{B}) \neq \{\mathbf{0}\}$).

Since the kernel of a matrix $\mathbf{A}_{mk}$ consists of all points $\begin{pmatrix} x_1 \\ \vdots \\ x_k \end{pmatrix} \in \mathbb{R}^k$, with $\mathbf{A}\mathbf{x} = x_1 \begin{pmatrix} | \\ \mathbf{a}_1 \\ | \end{pmatrix} + \cdots + x_k \begin{pmatrix} | \\ \mathbf{a}_k \\ | \end{pmatrix} = \mathbf{0} \in \mathbb{R}^m$, we obtain the following second equivalent definition of linear dependence and independence:

DEFINITION 2 **(Linear Dependence and Independence via the Homogeneous Linear Equation)**

(a) A set of k vectors $\begin{pmatrix} | \\ \mathbf{a}_1 \\ | \end{pmatrix}, \ldots, \begin{pmatrix} | \\ \mathbf{a}_k \\ | \end{pmatrix} \in \mathbb{R}^m$ is **linearly independent** if and only if $x_1 \begin{pmatrix} | \\ \mathbf{a}_1 \\ | \end{pmatrix} + \cdots + x_k \begin{pmatrix} | \\ \mathbf{a}_k \\ | \end{pmatrix} = \mathbf{0} \in \mathbb{R}^m$ implies $x_1 = \cdots = x_k = 0 \in \mathbb{R}$.

(b) A set of ℓ vectors $\begin{pmatrix} | \\ \mathbf{b}_1 \\ | \end{pmatrix}, \ldots, \begin{pmatrix} | \\ \mathbf{b}_\ell \\ | \end{pmatrix} \in \mathbb{R}^m$ is **linearly dependent** if and only if $x_1 \begin{pmatrix} | \\ \mathbf{b}_1 \\ | \end{pmatrix} + \cdots + x_\ell \begin{pmatrix} | \\ \mathbf{b}_\ell \\ | \end{pmatrix} = \mathbf{0}$ holds for some coefficients $x_j \in \mathbb{R}$, not all of which are equal to zero.

Since the homogeneous linear system $x_1 \begin{pmatrix} | \\ \mathbf{a}_1 \\ | \end{pmatrix} + \cdots + x_k \begin{pmatrix} | \\ \mathbf{a}_k \\ | \end{pmatrix} = \mathbf{0} \in \mathbb{R}^m$ is always solved by $\mathbf{x} = \mathbf{0}$, the preceding two criteria hinge on whether there is a nontrivial solution $\mathbf{x}$ for the homogeneous system $\mathbf{A}\mathbf{x} = \mathbf{0}$.

The two definitions of linear dependence and independence of vectors are equivalent, yet they serve different purposes: Definition 2 is mainly used in proofs, while Definition 1 works best in practical applications.

EXAMPLE 3

(a) Using Definition 2(b), the vectors $\mathbf{u}_1 = \begin{pmatrix} 1 \\ 0 \\ 1 \end{pmatrix}$, $\mathbf{u}_2 = \begin{pmatrix} 1 \\ 1 \\ 1 \end{pmatrix}$, and $\mathbf{u}_3 = \begin{pmatrix} 0 \\ 2 \\ 0 \end{pmatrix} \in \mathbb{R}^3$ are linearly dependent, since

$$2\mathbf{u}_1 - 2\mathbf{u}_2 + \mathbf{u}_3 = \begin{pmatrix} 2 \\ 0 \\ 2 \end{pmatrix} - \begin{pmatrix} 2 \\ 2 \\ 2 \end{pmatrix} + \begin{pmatrix} 0 \\ 2 \\ 0 \end{pmatrix} = \begin{pmatrix} 0 \\ 0 \\ 0 \end{pmatrix}.$$

(b) The set of vectors $\{\mathbf{e}_1, \mathbf{e}_1 + \mathbf{e}_2, \mathbf{e}_1 + \mathbf{e}_2 + \mathbf{e}_3\} \subset \mathbb{R}^4$ is linearly independent by Definition 2(a) since the equation

$$\alpha\mathbf{e}_1 + \beta(\mathbf{e}_1 + \mathbf{e}_2) + \gamma(\mathbf{e}_1 + \mathbf{e}_2 + \mathbf{e}_3) = \mathbf{0}$$

implies upon reordering that

$$(\alpha + \beta + \gamma)\mathbf{e}_1 + (\beta + \gamma)\mathbf{e}_2 + \gamma\mathbf{e}_3 = \begin{pmatrix} \alpha + \beta + \gamma \\ \beta + \gamma \\ \gamma \\ 0 \end{pmatrix} = \begin{pmatrix} 0 \\ 0 \\ 0 \\ 0 \end{pmatrix}.$$

Backsubstitution makes $\gamma = 0$, $\beta = 0$, and finally $\alpha = 0$. Hence, the only linear combination of the given vectors that is zero is the trivial one. ◀

EXAMPLE 4 We reconsider Example 3 in light of Definition 1 that relies on the rank of the column vector matrix:

(a) A REF $\mathbf{R}$ of the given column vector matrix $\mathbf{A}_{33} := \begin{pmatrix} | & | & | \\ \mathbf{u}_1 & \mathbf{u}_2 & \mathbf{u}_3 \\ | & | & | \end{pmatrix} = \begin{pmatrix} 1 & 1 & 0 \\ 0 & 1 & 2 \\ 1 & 1 & 0 \end{pmatrix}$ is computed as follows:

$$\begin{array}{ccc|l} \boxed{1} & 1 & 0 & \\ 0 & 1 & 2 & \\ \boxed{1} & 1 & 0 & -\,row_1 \\ \hline \boxed{1} & 1 & 0 & \\ 0 & \boxed{1} & 2 & = \mathbf{R} \\ 0 & 0 & 0 & \\ & & \uparrow & \end{array}$$

Clearly, the REF $\mathbf{R}$ of $\mathbf{A}$ has a free column, making the vectors $\mathbf{u}_i$ linearly dependent by part (b) of Definition 1. One can find the coefficients of the linear dependence from $\mathbf{R}$ by expressing the free third column $\begin{pmatrix} 0 \\ 2 \\ 0 \end{pmatrix}$ in $\mathbf{R}$ in terms of the pivot columns: $\begin{pmatrix} 0 \\ 2 \\ 0 \end{pmatrix} = 2\begin{pmatrix} 1 \\ 1 \\ 0 \end{pmatrix} - 2\begin{pmatrix} 1 \\ 0 \\ 0 \end{pmatrix}$. This column dependency in $\mathbf{R}$ translates verbatim to the column dependency $\mathbf{u}_3 = 2\mathbf{u}_2 - 2\mathbf{u}_1$ in $\mathbf{A}$, which agrees with the findings of Example 3(a).

(b) For the vectors $\{\mathbf{e}_1, \mathbf{e}_1 + \mathbf{e}_2, \mathbf{e}_1 + \mathbf{e}_2 + \mathbf{e}_3\}$ of Example 3(b), we note that their column matrix $\mathbf{B}_{43} := \begin{pmatrix} | & | & | \\ \mathbf{e}_1 & \mathbf{e}_1 + \mathbf{e}_2 & \mathbf{e}_1 + \mathbf{e}_2 + \mathbf{e}_3 \\ | & | & | \end{pmatrix} =$

$\begin{pmatrix} \boxed{1} & 1 & 1 \\ 0 & \boxed{1} & 1 \\ 0 & 0 & \boxed{1} \\ 0 & 0 & 0 \end{pmatrix}$ is already in row echelon form with three pivots. **B** has as many pivots as columns, and thus these vectors are linearly independent by part (a) of Definition 1. ◀

EXAMPLE 5 The linear dependence or independence of a set of vectors is a geometric condition.

(a) Let $\mathbf{u}_1 = \begin{pmatrix} 2 \\ 2 \end{pmatrix}$, $\mathbf{u}_2 = \begin{pmatrix} 1 \\ 4 \end{pmatrix}$, and $\mathbf{u}_3 = \begin{pmatrix} 1 \\ -2 \end{pmatrix} \in \mathbb{R}^2$ be given. For these vectors, there is a path from the origin $\begin{pmatrix} 0 \\ 0 \end{pmatrix}$ along <u>nonzero</u> multiples of the vectors $\mathbf{u}_i$ that leads back to the origin, as Figure 5-1 shows.

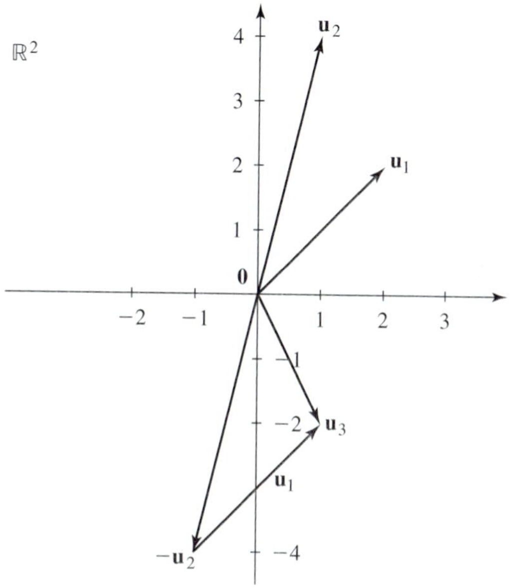

Figure 5-1

Namely $-\mathbf{u}_2 + \mathbf{u}_1 = \mathbf{u}_3$, or $\mathbf{u}_1 - \mathbf{u}_2 - \mathbf{u}_3 = \mathbf{0} \in \mathbb{R}^2$, which should be checked. Thus, the vectors $\{\mathbf{u}_1, \mathbf{u}_2, \mathbf{u}_3\}$ are linearly dependent according to Definition 2(b).

(b) However, any two of the vectors in part (a) are linearly independent, since there is no path from the origin that involves nonzero multiples of just two of the vectors $\mathbf{u}_1, \mathbf{u}_2, \mathbf{u}_3$ and leads back to $\mathbf{0} \in \mathbb{R}^2$. ◀

Next, we list several useful facts about linear dependence and linear independence of vectors:

Proposition 1

(a) Every single nonzero vector $\mathbf{x} \neq \mathbf{0} \in \mathbb{R}^m$ is linearly independent.
(b) Two vectors $\mathbf{x}$ and $\mathbf{y}$ in $\mathbb{R}^m$ are linearly dependent if and only if $\mathbf{x} = \alpha\mathbf{y}$ or $\mathbf{y} = \beta\mathbf{x}$ for some scalars $\alpha, \beta \in \mathbb{R}$.
(c) Any set of vectors that contains the zero vector is linearly dependent. In particular, the zero vector is linearly dependent.
(d) If a set of vectors is linearly dependent, then any set of vectors that contains this set is also linearly dependent.
(e) If a set of vectors is linearly independent, then any subset thereof is also linearly independent.
(f) Every set of more than m vectors in $\mathbb{R}^m$ is linearly dependent.
(g) If a linear system of equations $\mathbf{A}_{mn}\mathbf{x} = \mathbf{b}$ is solvable, then the set $\{\mathbf{a}_1, \ldots, \mathbf{a}_n, \mathbf{b}\}$ comprised of the column vectors $\mathbf{a}_i$ of $\mathbf{A}$ and the right hand side vector $\mathbf{b}$ is linearly dependent. (The converse is not true; that is, the $n+1$ columns of the augmented matrix $(\mathbf{A} \mid \mathbf{b})$ may be linearly dependent without the linear system being solvable.)
(h) A system of linear equations $\mathbf{A}_{mn}\mathbf{x} = \mathbf{b}$ is uniquely solvable if and only if the set $\{\mathbf{a}_1, \ldots, \mathbf{a}_n, \mathbf{b}\}$ is linearly dependent and the set $\{\mathbf{a}_1, \ldots, \mathbf{a}_n\}$ is linearly independent. ◀

Each student should try and reason through the preceding eight statements privately at first, using the two equivalent definition pairs for linear dependence and independence of vectors and Theorems 3.1, 3.2, and 3.3 when needed. Group discussions will only be fruitful afterwards, when they will be very useful to clear up any lingering misunderstandings.

(b) Minimal Spanning or Maximally Independent Sets of Vectors

The concept of linear independence helps us understand how to span subspaces. Subspaces were introduced in Chapter 4.

> This chapter's central question is about minimal spanning of subspaces of $\mathbb{R}^m$:
>
> How can one span a subspace U using the least number of generators?
>
> Or, equivalently, it concerns itself with maximally linearly independent sets of vectors:
>
> What is a maximal set of linearly independent vectors in a subspace U?

First we consider an inclusively represented subspace $U = \text{im}\begin{pmatrix} | & & | \\ \mathbf{u}_1 & \ldots & \mathbf{u}_k \\ | & & | \end{pmatrix}$.

As seen earlier, the linearly independent columns of $\mathbf{U}_{mk} = \begin{pmatrix} | & & | \\ \mathbf{u}_1 & \ldots & \mathbf{u}_k \\ | & & | \end{pmatrix}$ that correspond to the pivot columns in a REF of $\mathbf{U}$ are a minimal spanning set for U. Moreover, they are maximally linearly independent by Definition 2(a). We call such minimal spanning sets, or, equivalently, such maximally linearly independent sets, **bases** for U.

DEFINITION 3 **(Linearly Independent Spanning Set)** A **basis** of a subspace $\{\mathbf{0}\} \neq U \subset \mathbb{R}^m$ is a linearly independent and spanning set of vectors $\{\mathbf{u}_1, \ldots, \mathbf{u}_k\} \subset U$.

As the name indicates, the concept of a basis for a space is fundamental. It is particularly so from its quality as a maximally linearly independent set of vectors and, at the same time, as a minimal spanning set. Thus, a basis is an optimally small and expansive set of vectors.

The standard unit vectors $\mathbf{e}_1, \ldots, \mathbf{e}_m \in \mathbb{R}^m$, form a basis of $\mathbb{R}^m$, since span $\{\mathbf{e}_1, \ldots, \mathbf{e}_n\} = \text{im} \begin{pmatrix} 1 & & 0 \\ & \ddots & \\ 0 & & 1 \end{pmatrix} = \text{im}(\mathbf{I}_m) = \mathbb{R}^m$, and since the columns $\mathbf{e}_i$ of $\mathbf{I}_m$ are linearly independent according to Definition 1(a) or Definition 2(a).

Note that the identity matrix $\mathbf{I}_m$ is $m \times m$ and that $\mathbf{I}_m$ is equal to its own row echelon or reduced row echelon form with a pivot in every row and column. The columns $\mathbf{e}_i$ of $\mathbf{I}_m$ form the standard basis $\mathcal{E}$. This pivot property for the identity matrix $\mathbf{I}_m$ carries over to a more general setting.

Theorem 5.1 **(Basis Theorem)** A set of vectors $\{\mathbf{u}_i\}_{i=1}^k \subset \mathbb{R}^m$ is a basis of $\mathbb{R}^m$ if and only if $k = m$ and the REF of $\mathbf{U}_{mk} = \begin{pmatrix} | & & | \\ \mathbf{u}_1 & \cdots & \mathbf{u}_k \\ | & & | \end{pmatrix}$ has a pivot in every row and in every column. ◀

The proof follows from Definition 3 and Section 3.1. Students should develop it in class.

Note that for a basis $\{\mathbf{u}_1, \ldots, \mathbf{u}_k\} \subset \mathbb{R}^m$ of a subspace U, a REF of $\mathbf{U}_{mk} = \begin{pmatrix} | & & | \\ \mathbf{u}_1 & \ldots & \mathbf{u}_k \\ | & & | \end{pmatrix}$ must have a pivot in every column. Thus, Theorem 3.3 ensures the unique solvability of the linear system $\mathbf{Ux} = \mathbf{b}$ for every right-hand-side $\mathbf{b} \in U$. In light of Proposition 1(h), the Proposition in Section 4.1, and Chapter 3, a basis thus generates every vector $\mathbf{b}$ in its span uniquely as a linear combination. Hence, we have a second and equivalent definition of a basis.

DEFINITION 4 **(Unique Spanning Set)** A set of vectors $\{\mathbf{u}_i\}$ in $\mathbb{R}^m$ is called a **basis** of $\mathbb{R}^m$ if the $\mathbf{u}_i$ generate every vector in $\mathbb{R}^m$ uniquely.

A set of vectors $\{\mathbf{u}_i\}$ in a subspace $U \subset \mathbb{R}^m$ is called a **basis** of U, if the $\mathbf{u}_i \in U$ generate every vector in U uniquely.

We summarize the minimality of a basis as a spanning set for a subspace and its maximality as a linear independent set as follows.

Proposition 2 (a) If k vectors $\mathbf{x}_1, \ldots, \mathbf{x}_k$ span a subspace $U \subset \mathbb{R}^n$ and if $\mathbf{x}_i \neq \mathbf{z} \in U$ for all i, then the set $\{\mathbf{x}_1, \ldots, \mathbf{x}_k, \mathbf{z}\} \subset U$ is linearly dependent.

(b) If ℓ vectors $\mathbf{y}_1, \ldots, \mathbf{y}_\ell$ are linearly independent and lie in a subspace $U \subset \mathbb{R}^n$, then for any $i = 1, \ldots, \ell$, the set $\{\mathbf{y}_1, \ldots, \mathbf{y}_{i-1}, \mathbf{y}_{i+1}, \ldots, \mathbf{y}_\ell\}$ consisting of $\ell - 1$ vectors in U is not a spanning set of U. ◀

Finally, we describe an analogy with chemistry: The periodic table of chemical elements from hydrogen to the transuraniums forms a basis for all matter on earth and in space. These roughly 100 basic chemical elements are chemically independent in the sense that none of them can be obtained by a chemical reaction from the others.

Let us transpose this image to mathematics, namely, to a set of vectors $\mathbf{u}_i$ whose column vector matrix $\mathbf{A} = \begin{pmatrix} | & & | \\ \mathbf{u}_1 & \cdots & \mathbf{u}_k \\ | & & | \end{pmatrix}$ does <u>not</u> have a pivot in every column of its REF $\mathbf{R}$. In this case, $\mathbf{R}$ contains at least one free column and $\mathbf{A}\mathbf{x} = \mathbf{0}$ has multiple solutions; that is, the linear system

$$\alpha_1 \begin{pmatrix} | \\ \mathbf{u}_1 \\ | \end{pmatrix} + \cdots + \alpha_k \begin{pmatrix} | \\ \mathbf{u}_k \\ | \end{pmatrix} = \mathbf{0}$$

can be solved with coefficients α_i that are not all equal to zero by setting some of the free variables of a REF $\mathbf{R}$ of $\mathbf{A}$ equal to nonzero numbers and solving for the pivot variables in turn. (Refer to Sections 2.1 and 3.1 and to Definition 2).

Thus, some of the original vectors $\mathbf{u}_i$, namely, those that correspond to free variables in the REF of $\mathbf{A}$, can be expressed as a linear combination of the others; that is, they are not basic to the column space $U = \text{span}\{\mathbf{u}_1, \ldots, \mathbf{u}_k\}$ of $\mathbf{A}$. Their action, or direction, can be replicated by other vectors from the set $\{\mathbf{u}_1, \ldots, \mathbf{u}_k\}$, as we have seen in Example 2.

Our mathematical notions of basis and of linear dependence and independence thus are analogous to the notion of chemical independence. Linearly dependent vectors are not needed amongst the basis vectors that are essential, or minimal, to define a space or a subspace, just as H_2O is not needed as an element in the periodic table of chemistry, because it can be chemically derived from hydrogen H and oxygen O.

(c) Dimension

If $U \subset \mathbb{R}^m$ is a subspace, then according to Definitions 1 and 3, a set of vectors $\{\mathbf{u}_i\}_{i=1}^k$ in U is a **basis** of the subspace U if the vectors $\mathbf{u}_i$ span U <u>and</u> if the REF of $\mathbf{U}_{mk} = \begin{pmatrix} | & & | \\ \mathbf{u}_1 & \cdots & \mathbf{u}_k \\ | & & | \end{pmatrix}$ has a pivot in every column. We use this to define the **dimension** of a subspace.

DEFINITION 5 **(Dimension)** If the set of vectors $\{\mathbf{u}_1, \ldots, \mathbf{u}_k\}$ is a basis for a subspace $U \subset \mathbb{R}^m$, then we call k the **dimension** of U (i.e., $\dim(U) = k$).

In Section 5.2, we prove that the dimension of a subspace is unique (i.e., that every basis of a subspace $U \subset \mathbb{R}^m$ contains the same number of vectors).

We mention two additional facts for bases of $\mathbb{R}^m$:

Proposition 3 (a) Every set of m linearly independent vectors in $\mathbb{R}^m$ is a basis of $\mathbb{R}^m$.
(b) Every set of m vectors in $\mathbb{R}^m$ that spans all of $\mathbb{R}^m$ is a basis of $\mathbb{R}^m$. ◀

The proofs follow from counting pivots in the columns and rows of the associated REFs of the $m \times m$ column vector matrices in parts (a) and (b), respectively.

From Sections 2.1 and 3.1, we know that the degree of freedom in the homogeneous solutions $\{\mathbf{x}_{\text{hom}}\}$ of $\mathbf{Ax} = \mathbf{0}$ is equal to the number of free columns in a REF of $\mathbf{A}$. Since this degree of freedom is the dimension of $\mathbf{A}$'s kernel, we have

$$\dim(\ker(\mathbf{A})) = \text{the number of free columns of an REF of } \mathbf{A}.$$

Moreover, each column of a REF $\mathbf{R}$ of $\mathbf{A}$ is either a pivot column, contributing to its rank, or it is a free column, contributing to the dimension of the kernel of $\mathbf{A}$. Thus, we have proved the next theorem.

Theorem 5.2 **(Dimension Theorem)** Every matrix $\mathbf{A}_{mn}$ satisfies the following identities for each of its row echelon forms $\mathbf{R}$:

$$\begin{aligned}
n &= \text{the dimension of the domain of the map } \mathbf{x} \mapsto \mathbf{Ax} \\
&= \text{the number of columns of } \mathbf{A} \\
&= \text{the number of pivots in } \mathbf{R} + \text{the number of free variables in } \mathbf{R} \\
&= \operatorname{rank}(\mathbf{A}) + \dim(\ker(\mathbf{A})) \\
&= \dim(\operatorname{im}(\mathbf{A})) + \dim(\ker(\mathbf{A})).
\end{aligned}$$

◀

Our next result applies Theorem 5.2 to the kernel representation of subspace intersections in Theorem 4.3(b).

Corollary 1 If U and V are two subspaces of $\mathbb{R}^m$ of dimensions k and ℓ, respectively, then their intersection $U \cap V$ is a subspace of $\mathbb{R}^m$ whose dimension is at least $(k+\ell)-m$ and at most $\min\{k, \ell\}$. ◀

Proof According to Theorem 5.2 and Section 4.1, we can represent U as the kernel of a matrix $\mathbf{A}_{m-k,m}$ and V as $\ker(\mathbf{B}_{m-\ell,m})$ for two matrices $\mathbf{A}$ and $\mathbf{B}$ of rank $m-k$ and $m-\ell$, respectively. By Theorem 4.3(b), we have

$$U \cap V = \ker \begin{pmatrix} \mathbf{A} \\ \mathbf{B} \end{pmatrix}_{2m-(k+\ell),m}.$$

Thus, if the number of rows $2m-(k+\ell)$ of the block matrix $\begin{pmatrix} \mathbf{A} \\ \mathbf{B} \end{pmatrix}$ is less than m, then a REF of $\begin{pmatrix} \mathbf{A} \\ \mathbf{B} \end{pmatrix}$ will have at least $m-(2m-(k+\ell)) = k+\ell-m$ free variables, giving $U \cap V$ at least dimension $(k+\ell)-m$ as claimed. The result for the maximally possible dimension of $U \cap V$ follows from the fact that $U \cap V \subset U, V$. ■

Applying the Corollary to any two subspaces U and V in $\mathbb{R}^m$ with $\dim(U) = k$ and $\dim(V) = m - k + j$ for $j \geq 1$, we see that U and V must have a nonzero intersection in $\mathbb{R}^m$. For example this applies to two subspaces $U \subset \mathbb{R}^5$ of dimension four and $V \subset \mathbb{R}^5$ of dimension three: the intersection $U \cap V$ is at least a $(4+3)-5 = 2$ dimensional subspace of $\mathbb{R}^5$ and at most a three-dimensional subspace. However, the intersection of two subspaces in $\mathbb{R}^{10}$ of dimensions four and three, respectively, may be $\{\mathbf{0}\}$, since the dimension inequality $(4+3)-10 < 0$ offers no restriction in this case.

Finally, we mention a relation between the dimensions of two subspaces and that of their join and intersection without proof:

Theorem 5.3 **(Subspace Dimension Sum Theorem)** If U and $V \subset \mathbb{R}^n$ are two subspaces, then

$$\dim U + \dim V = \dim(U + V) + \dim(U \cap V).$$ ◀

(d) Finding Bases for Both Generic Subspace Representations

Next we look at the practicalities:

> How does one find a basis for the column space $U = \text{span}\{\mathbf{u}_1, \ldots, \mathbf{u}_k\}$ of a matrix $\mathbf{U}$, or for the nullspace $U = \ker(\mathbf{A})$ of a matrix $\mathbf{A}$?

In case of a column space $U = \text{span}\{\mathbf{u}_1, \ldots, \mathbf{u}_k\}$, we row reduce $\begin{pmatrix} | & & | \\ \mathbf{u}_1 & \cdots & \mathbf{u}_k \\ | & & | \end{pmatrix}$ and take those original columns $\mathbf{u}_j$ into a basis for U whose REF columns have a pivot. These $\mathbf{u}_j$ form a basis for U. The remaining $\mathbf{u}_i$, if any, are linearly dependent on these basis vectors for U; see Example 2. Moreover, the dimension of $U = \text{span}\{\mathbf{u}_1, \ldots, \mathbf{u}_k\}$ is equal to the number of vectors in a basis (i.e., to the number of pivots in a REF of $\begin{pmatrix} | & & | \\ \mathbf{u}_1 & \cdots & \mathbf{u}_k \\ | & & | \end{pmatrix}$).

For a basis of an exclusively represented subspace $U = \ker(\mathbf{A})$ and a matrix $\mathbf{A}_{mn}$ of rank k there must be k pivots in every REF of $\mathbf{A}$ and correspondingly $n-k$ free variables due to the Dimension Theorem 5.2. And $\mathbf{x}_{\text{hom}}$, the general homogeneous solution of $\mathbf{A}\mathbf{x} = \mathbf{0}$, depends on $n-k$ parameters; that is, $\ker(\mathbf{A})$ has a basis consisting of $n-k$ vectors. Consequently, $\ker(\mathbf{A})$ has dimension $n-k =$ the number of free variables of a REF of $\mathbf{A}$ according to Theorem 5.2. A basis of $\ker(\mathbf{A}) \subset \mathbb{R}^n$ can be found using Section 3.1.

EXAMPLE 6 Assume that a REF of an augmented homogeneous linear system $(\mathbf{A} \mid \mathbf{0})$ has the form

x_1	x_2	x_3	x_4	x_5	$\mathbf{b}$
0	$\boxed{4}$	1	0	−4	0
0	0	0	$\boxed{1}$	2	0 ,
0	0	0	0	0	0
↑		↑		↑	

with pivots and free variables as marked. Find a basis for $\ker(\mathbf{A})$.

As in Chapter 3 we assign arbitrary values $x_1 = \alpha$, $x_3 = \beta$, and $x_5 = \gamma$ to the three free variables and obtain

$$\mathbf{x}_{\text{hom}} = \begin{pmatrix} \alpha \\ -\frac{1}{4}\beta + \gamma \\ \beta \\ -2\gamma \\ \gamma \end{pmatrix} = \alpha \begin{pmatrix} 1 \\ 0 \\ 0 \\ 0 \\ 0 \end{pmatrix} + \frac{1}{4}\beta \begin{pmatrix} 0 \\ -1 \\ 4 \\ 0 \\ 0 \end{pmatrix} + \gamma \begin{pmatrix} 0 \\ 1 \\ 0 \\ -2 \\ 1 \end{pmatrix}$$

by backsubstitution. One basis for $U = \ker(\mathbf{A})$ is given by the set of generators $\begin{pmatrix} 1 \\ 0 \\ 0 \\ 0 \\ 0 \end{pmatrix}$, $\begin{pmatrix} 0 \\ -1 \\ 4 \\ 0 \\ 0 \end{pmatrix}$, and $\begin{pmatrix} 0 \\ 1 \\ 0 \\ -2 \\ 1 \end{pmatrix}$ of $\mathbf{x}_{\text{hom}}$, where we have taken pain to insure integer basis vectors for writing simplicity.

Clearly if a set of vectors $\{\mathbf{u}_1, \ldots, \mathbf{u}_k\}$ is a basis for any subspace U, then the scaled set $\{\alpha_1\mathbf{u}_1, \ldots, \alpha_k\mathbf{u}_k\}$ is also a basis for U as long as each $\alpha_i \neq 0$. (See Problem 30.) Thus, our scaling to convenient integer components is irrelevant to the process. ◀

We wrap up this section with the following conclusions:

> A **pivot in every row** of the REF of a matrix composed of a number of column vectors in $\mathbb{R}^m$ insures the **spanning condition**: The given column vectors span all of $\mathbb{R}^m$. (Note that a set of spanning vectors of a subspace $U \subset \mathbb{R}^m$ may or may not be linearly independent.)
>
> A **pivot in every column** of the REF of a number of vectors in $\mathbb{R}^m$ arranged as the columns of a matrix insures the unique representation by these vectors of every vector in their span; in other words it insures their **linear independence**. (Note that a set of linearly independent vectors in $\mathbb{R}^m$ may or may not span all of $\mathbb{R}^m$.)

If both of the above conditions are met simultaneously for a set of vectors, then the vectors form a basis. (That is, they are a minimal spanning or a maximally independent set.)

A Note on Mathematical Language:

Language reflects reality as reality reflects language.

It is important to use language correctly, especially mathematical language. Incorrect speaking or writing fosters wrong perceptions and improper images; that is, it hinders learning.
For example, the question *are there any pivots in a matrix* $\mathbf{A}_{mn}$? is wrongly phrased. A general matrix **A** has no pivots per se. We select one pivot in **A** to start the row reduction process. This process reveals more and more pivots and their positions. Thus, an unreduced matrix **A** has no pivots; they are revealed only after Gaussian elimination.
A REF **R** of **A**, however, contains as many pivots as **A**'s rank. Thus, only a REF contains pivots; they are the leading nonzero entry of each nonzero row of **R**. Hence, it is improper to say that

"... a matrix **A** has k pivots," or that
"... there is a pivot in every column of **A**."

Remember that there are no pivots in a general matrix, unless it is in row echelon form.
It is proper to say

"... the REF of **A** has k pivots,"
"... there is a pivot in every column of a REF of **A**,"
"... **A** has rank k," or
"... the k pivots of the RREF of **A** reveal **A** to have rank k," and so forth.

We have generalized the standard unit vector basis $\mathcal{E}$ to general vector space bases.

5.1.P Problems

1. Are the vectors $\begin{pmatrix} 1 \\ 2 \\ 1 \end{pmatrix}, \begin{pmatrix} 1 \\ 3 \\ 2 \end{pmatrix}, \begin{pmatrix} 0 \\ -1 \\ -1 \end{pmatrix}$, and $\begin{pmatrix} 0 \\ 0 \\ 1 \end{pmatrix} \in \mathbb{R}^3$

(a) linearly independent?

(b) spanning $\mathbb{R}^3$?

(c) a basis for $\mathbb{R}^3$?

2. Write out your own arguments for at least three of the statements of Proposition 1.

3. Answer the following questions with a YES or NO. In case of YES, give a simple example and verify that your example fits the situation. In case of NO, give your reason, please.

(a) Can four vectors in $\mathbb{R}^6$ span a five-dimensional subspace?

(b) Can six vectors in $\mathbb{R}^4$ span a five-dimensional subspace?

(c) Can six vectors in $\mathbb{R}^{10}$ span a four-dimensional subspace?

(d) Can three vectors in $\mathbb{R}^4$ be linearly dependent?

(e) Can four vectors in $\mathbb{R}^2$ be linearly independent?

(f) What are the possible ranks of a 5×17 matrix $\mathbf{A}$?

4. Decide by inspection which of the following sets of vectors are linearly independent and with are not. Which sets are a basis for their span?

(a) $\{\mathbf{e}_1 + \mathbf{e}_2, 2\mathbf{e}_2 + 2\mathbf{e}_3, \ldots, (n-1)\mathbf{e}_{n-1} + (n-1)\mathbf{e}_n\}$ for the standard unit vectors $\mathbf{e}_i \in \mathbb{R}^m$ and $n \leq m$.

(b) $\begin{pmatrix} 1 \\ 2 \\ 3 \\ 4 \end{pmatrix}, \begin{pmatrix} 0 \\ 0 \\ 0 \\ 0 \end{pmatrix}, \begin{pmatrix} -3 \\ 3 \\ 5 \\ -120 \end{pmatrix}.$

(c) $\begin{pmatrix} 1 \\ 2 \\ 3 \\ 4 \end{pmatrix}, \begin{pmatrix} 0 \\ 0 \\ 4 \\ 0 \end{pmatrix}, \begin{pmatrix} -3 \\ 3 \\ 5 \\ -120 \end{pmatrix}, \begin{pmatrix} 0 \\ 0 \\ 4 \\ 1 \end{pmatrix}$, and $\begin{pmatrix} 101 \\ -2 \\ 3 \\ 4 \end{pmatrix}.$

(d) $\begin{pmatrix} 3 \\ -9 \end{pmatrix}, \begin{pmatrix} -1 \\ 3 \end{pmatrix}.$

(e) $\begin{pmatrix} 0 \\ 0 \\ 0 \\ 0 \\ 0 \end{pmatrix}.$ (f) $\begin{pmatrix} 4 \\ 4 \\ 5 \end{pmatrix}, \begin{pmatrix} -1 \\ 6 \\ 0 \end{pmatrix}.$

(f) $\begin{pmatrix} 4 \\ 4 \end{pmatrix}, \begin{pmatrix} -1 \\ 6 \end{pmatrix}.$

5. (a) Let $\mathbf{u}_1 = \begin{pmatrix} 1 \\ 2 \\ 3 \end{pmatrix}, \mathbf{u}_2 = \begin{pmatrix} 0 \\ -1 \\ 3 \end{pmatrix}$, and $\mathbf{u}_3 = 2\mathbf{u}_2 - \mathbf{u}_1 = \begin{pmatrix} -1 \\ -4 \\ 3 \end{pmatrix}$. Find the RREF of the matrix $\mathbf{U} = \begin{pmatrix} | & & | \\ \mathbf{u}_1 & \ldots & \mathbf{u}_3 \\ | & & | \end{pmatrix}$. What does its last column tell you about the vectors $\mathbf{u}_i$?

(b) Let $\mathbf{w}_1 = \begin{pmatrix} 1 \\ 2 \\ 3 \\ 4 \end{pmatrix}, \mathbf{w}_2 = \begin{pmatrix} 0 \\ -1 \\ 3 \\ 2 \end{pmatrix}$, and $\mathbf{w}_3 = 2\mathbf{w}_1 - 3\mathbf{w}_2$. Show that the last column of the RREF of $\mathbf{W} = \begin{pmatrix} | & & | \\ \mathbf{w}_1 & \ldots & \mathbf{w}_3 \\ | & & | \end{pmatrix}$ is $\begin{pmatrix} 2 \\ -3 \\ 0 \\ 0 \end{pmatrix}.$

6. Generalize the ideas of the previous problem and of Example 1 to show that all entries of a reduced row echelon form of a matrix $\mathbf{A}$ are uniquely determined from the columns of $\mathbf{A}$.

7. (a) Find a basis for the span U of $\mathbf{u}_1 = \begin{pmatrix} 1 \\ 2 \\ -1 \\ 0 \end{pmatrix}, \mathbf{u}_2 = \begin{pmatrix} 2 \\ 5 \\ 0 \\ 1 \end{pmatrix}, \mathbf{u}_3 = \begin{pmatrix} -1 \\ -3 \\ -1 \\ -1 \end{pmatrix}$, and $\mathbf{u}_4 = \begin{pmatrix} 0 \\ 0 \\ 0 \\ 1 \end{pmatrix} \in \mathbb{R}^4.$

(b) Which of the standard unit vectors $\mathbf{e}_1, \mathbf{e}_2, \mathbf{e}_3$, or $\mathbf{e}_4 \in \mathbb{R}^4$ lie in U? (*Hint*: Recall Example 2 of Section 4.1; you will need to solve four systems of linear equations.)

8. Find a basis for the kernel of the matrix $\mathbf{A} = \begin{pmatrix} 0 & 1 & -1 & 1 \\ -1 & 0 & -1 & 0 \\ 2 & 5 & -3 & 0 \\ 1 & 2 & -1 & 0 \end{pmatrix}$. What is the dimension of the kernel of $\mathbf{A}$?

9. Prove or disprove each of the following:

(a) If four vectors $\mathbf{v}_i$ are given in $\mathbb{R}^m$ with $\mathbf{v}_2 = 2\mathbf{v}_1 - (-\mathbf{v}_2 + 2\mathbf{v}_1)$, then the vectors $\mathbf{v}_i$ are linearly dependent.

(b) If four vectors $\mathbf{v}_i$ are given in $\mathbb{R}^m$ with $\mathbf{v}_2 = \mathbf{v}_1 - \mathbf{v}_2$, then the $\mathbf{v}_i$ may be linearly independent.

(c) If three vectors $\mathbf{v}_i \in \mathbb{R}^5$ are linearly independent and $\mathbf{v}_4 \neq \mathbf{0} \in \mathbb{R}^5$, then the four vectors $\mathbf{v}_1, \mathbf{v}_2, \mathbf{v}_3, \mathbf{v}_4$ are also linearly independent.

(d) If the set of vectors $\{\mathbf{v}_i\}_{i=1}^k \subset \mathbb{R}^m$ is linearly dependent, then for any two vectors $\mathbf{v}_{k+1}, \mathbf{v}_{k+2} \in \mathbb{R}^m$ the larger set of vectors $\{\mathbf{v}_i\}_{i=1}^{k+2} \subset \mathbb{R}^m$ is linearly dependent.

(e) The columns of a matrix $\mathbf{A}$ are linearly dependent if $\mathbf{A}\mathbf{x} = \mathbf{0}$ has a unique solution. (What is the unique solution in this case?)

(f) The kernel of a matrix $\mathbf{A}$ is spanned by one nonzero vector if the REF of $\mathbf{A}$ has one pivot.

(g) The vectors in the kernel of a matrix $\mathbf{A}$ are always linearly independent.

(h) The column vectors of a matrix $\mathbf{A}_{mn}$ of rank n are linearly dependent.

(i) The column vectors of a matrix $\mathbf{A}_{mn}$ of rank m are linearly independent.

10. For which values of a are the three vectors $(1, a, 1)$, $(a, 1, 1)$, and $(1, 1, a) \in \mathbb{R}^3$ linearly dependent, and for which a are they linearly independent?

11. Define $\mathbf{u}_1 = \begin{pmatrix} -1 \\ 3 \\ 2 \end{pmatrix}$, $\mathbf{u}_2 = \begin{pmatrix} 1 \\ 1 \\ -1 \end{pmatrix}$, $\mathbf{u}_3 = \begin{pmatrix} -1 \\ -1 \\ -2 \end{pmatrix}$, and $\mathbf{u}_4 = \begin{pmatrix} 3 \\ -1 \\ -1 \end{pmatrix}$.

(a) Find a basis for $U = \text{span}\{\mathbf{u}_1, \mathbf{u}_2, \mathbf{u}_3, \mathbf{u}_4\}$.

(b) Does the linear system

$$\alpha\mathbf{u}_1 + \beta\mathbf{u}_2 + \gamma\mathbf{u}_3 + \delta\mathbf{u}_4 = \begin{pmatrix} 1 \\ -11 \\ -10 \end{pmatrix}$$

have a solution? (*Hint*: Rephrase this question as one about a matrix equation.)

(c) Show that $\mathbf{v} = \begin{pmatrix} 1 \\ -11 \\ -10 \end{pmatrix}$ belongs to U.

(d) Are the vectors $\mathbf{u}_1, \dots, \mathbf{u}_4$ linearly dependent or linearly independent?

(e) Are the vectors $\mathbf{v}, \mathbf{u}_1, \mathbf{u}_3$ linearly independent?

12. Find a minimal spanning set $M \neq \emptyset$ for a specific subspace U so that the vectors in M do not form a basis for U. (Note: there is only one such subspace $U \subset \mathbb{R}^m$.)

13. Find a basis for the kernel of $\mathbf{A} = \begin{pmatrix} 1 & 1 & 1 & 1 \\ 1 & 1 & 1 & 1 \\ 1 & 1 & 1 & 1 \\ 1 & 1 & 1 & 1 \end{pmatrix}$.

14. Let $\mathbf{A} = \begin{pmatrix} 1 & 2 & 4 & 1 \\ 2 & 4 & 8 & 2 \\ 3 & 6 & 2 & 0 \end{pmatrix}$.

(a) Find an integer basis for the column space of $\mathbf{A}$.

(b) Find an integer basis for $\mathbf{A}$'s kernel.

(c) Is the linear system $\mathbf{Ax} = \mathbf{0}$ solvable? Is it uniquely solvable?

(d) Is the linear system $\mathbf{Ax} = \mathbf{e}_i$ solvable for any $i = 1, \dots, 3$? Is it uniquely solvable for any $i = 1, \dots, 3$? Give your complete reasoning in each instance, please.

15. (a) Find the dimension of $U = \text{span}\left\{\begin{pmatrix} 1 \\ 1 \\ 0 \end{pmatrix}, \begin{pmatrix} 0 \\ 1 \\ 1 \end{pmatrix}, \begin{pmatrix} 1 \\ 2 \\ 1 \end{pmatrix}\right\} \subset \mathbb{R}^3$.

(b) Can you find a specific vector $\mathbf{b} \in \mathbb{R}^3$ that does not lie in U? Explain, please.

16. (a) Find a value for $d \in \mathbb{R}$, so that the vectors $\{\mathbf{u}_1, \dots, \mathbf{u}_4\} = \left\{\begin{pmatrix} 1 \\ 1 \\ 1 \\ 2 \end{pmatrix}, \begin{pmatrix} 4 \\ 1 \\ d \\ 3 \end{pmatrix}, \begin{pmatrix} 3 \\ 2 \\ 4 \\ 4 \end{pmatrix}, \begin{pmatrix} 1 \\ 0 \\ 2 \\ 1 \end{pmatrix}\right\} \subset \mathbb{R}^4$ are linearly independent. (*Hint*: reshuffle the vectors for the most convenient pivots.)

(b) If the $\{\mathbf{u}_i\}$ in part (a) are linearly independent for a specific choice of d, do they form a basis of $\mathbb{R}^4$ or not? Explain why!

(c) If the $\{\mathbf{u}_i\}$ in part (a) span $\mathbb{R}^4$ for a specific choice of d, can they be linearly dependent or not? Explain why!

17. Let $\mathbf{A} = \begin{pmatrix} -1 & * & 7 & * \\ -1 & 3 & 7 & * \\ -1 & * & 7 & * \\ 2 & * & ** & 3 \end{pmatrix}$ be a 4×4 matrix with partially prescribed real entries and partially unprescribed entries labelled as $*$ and $**$. (The entries marked by $*$ need not have the same value.)

(a) Write out <u>three</u> equivalent conditions for a square matrix to be singular.

(b) Find a value of the entry $**$ in position (4,3) of $\mathbf{A}$, so that the resulting matrix, to be <u>called $\mathbf{A}_{**}$ henceforth</u>, is singular, no matter what values the entries marked by $*$ may have.

(c) What ranks are possible for the matrix $\mathbf{A}_{**}$ from part (b)?

Please give explicit values to the entries of $\mathbf{A}$ originally marked by $*$ so that $\mathbf{A}_{**}$ achieves precisely each of these ranks.

18. (A hard problem with our means)

Let $\mathbf{u}_1 = \begin{pmatrix} 2 \\ -1 \\ 4 \end{pmatrix}$, $\mathbf{u}_2 = \begin{pmatrix} 1 \\ 2 \\ 0 \end{pmatrix}$, $\mathbf{u}_3 = \begin{pmatrix} 3 \\ -4 \\ 8 \end{pmatrix}$, $\mathbf{u}_4 = \begin{pmatrix} 1 \\ 1 \\ 1 \end{pmatrix}$

and $\mathbf{v}_1 = \begin{pmatrix} 1 \\ 0 \\ -1 \end{pmatrix}$, $\mathbf{v}_2 = \begin{pmatrix} 2 \\ 1 \\ 3 \end{pmatrix}$, $\mathbf{v}_3 = \begin{pmatrix} 0 \\ -1 \\ 1 \end{pmatrix}$, $\mathbf{v}_4 = \begin{pmatrix} 1 \\ 1 \\ 1 \end{pmatrix}$.

(a) Find a matrix **A** that describes the linear transformation $S : \mathbb{R}^3 \to \mathbb{R}^3$, defined by $S\mathbf{u}_2 = \mathbf{v}_2$, $S\mathbf{u}_3 = \mathbf{v}_3$, $S\mathbf{u}_4 = \mathbf{v}_4$. Check your answer.

(b) Find a matrix **B** that describes the linear transformation $T : \mathbb{R}^3 \to \mathbb{R}^3$, defined by $T\mathbf{u}_1 = \mathbf{v}_1$, $T\mathbf{u}_3 = \mathbf{v}_3$, $T\mathbf{u}_4 = \mathbf{v}_4$. Check your answer.

(c) Is the set $\{\mathbf{u}_1, \mathbf{u}_2, \mathbf{u}_3, \mathbf{u}_4\}$ linearly independent or dependent?

(d) Find one vector $\mathbf{u}_i$ that is a linear combination of the other $\mathbf{u}_j$, if this is possible. Give the explicit linear combination.

(e) Is the set $\{\mathbf{v}_1, \mathbf{v}_2, \mathbf{v}_3, \mathbf{v}_4\}$ linearly independent or dependent?

(f) Find one vector $\mathbf{v}_i$ that is a linear combination of the other $\mathbf{v}_j$, if this is possible. Give the explicit linear combination.

(g) What is the rank of **A** in part (a)? What about for **B** in part (b).

(h) For which right-hand sides **b** is the linear system $\mathbf{Ax} = \mathbf{b}$ solvable? How about uniquely solvable? What about $\mathbf{Bx} = \mathbf{b}$.

(i) Find a basis for the column space of the matrix **A** in (a).

(j) Find a basis for the row space of the matrix

$$\mathbf{V} = \begin{pmatrix} - & \mathbf{v}_1^T & - \\ - & \mathbf{v}_2^T & - \\ - & \mathbf{v}_3^T & - \\ - & \mathbf{v}_4^T & - \end{pmatrix}.$$

In each instance give, your answers and reasons in verbal form, and explain your thought process by quoting the underlying mathematical facts.

19. Determine whether the set of vectors $\left\{ \begin{pmatrix} 1 \\ 0 \\ 0 \\ 2 \end{pmatrix}, \begin{pmatrix} -2 \\ 1 \\ 0 \\ 1 \end{pmatrix}, \begin{pmatrix} 1 \\ -2 \\ 3 \\ -4 \end{pmatrix}, \begin{pmatrix} 3 \\ 1 \\ -4 \\ 9 \end{pmatrix} \right\}$ is linearly dependent. What is the dimension of their span in $\mathbb{R}^4$?

20. Let $\mathbf{x}, \mathbf{y}, \mathbf{z}$ be three vectors in $\mathbb{R}^m$. Define $U = \text{span}\{\mathbf{x}, \mathbf{y}\}$. Is there a way to answer the question whether the given set $\{\mathbf{x}, \mathbf{y}, \mathbf{z}\}$ is linearly dependent or independent by using a system of linear equations? Specify, please, if there is.

21. Show for any matrix $\mathbf{A}_{mn}$ that

$$n = \dim(\text{im}(\mathbf{A})) + \dim(\ker(\mathbf{A})).$$

22. Prove or disprove each of the following:

(a) The pivot rows of a REF **R** are linearly independent.

(b) The pivot columns of a REF **R** are linearly independent.

(c) The rows without pivots of a REF **R** are linearly dependent.

(d) The free columns of a REF **R** are linearly dependent.

23. Assume that the vectors $\mathbf{x}_1, \ldots, \mathbf{x}_n \in \mathbb{R}^m$ are linearly independent. Show that the three vectors $\mathbf{x} = \sum_{i=1}^{n} \mathbf{x}_i$, $\mathbf{x}_k$, and $\mathbf{x}_\ell$ are linearly independent if and only if $k \neq \ell$ and $m \geq 3$. (*Hint*: Use Definition 2.)

24. Consider a tetrahedron in $\mathbb{R}^3$ composed of four vertices **O**, **A**, **B**, and **C**, six edges of unit length and four equilateral triangles as faces. Assume that the tetrahedron rests on the x–y plane with corners at $\mathbf{O} = (0, 0, 0)$, $\mathbf{A} = (1, 0, 0)$, and $\mathbf{C} = (1/2, *, 0)$.

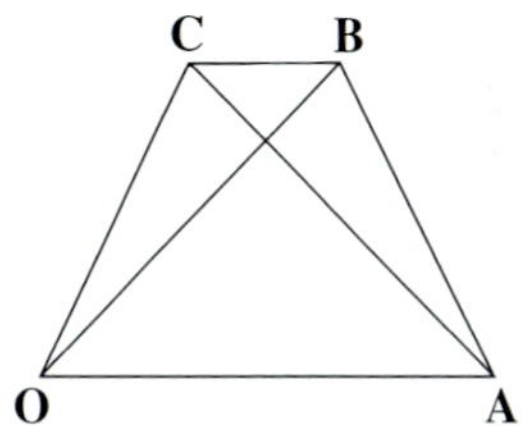

(a) Find the standard coordinates of **C** and of the fourth vertex **B**. (*Hint*: Redraw the figure and include the coordinate axes.)

(b) Show that the vectors $\overline{\mathbf{OA}}$, $\overline{\mathbf{OB}}$, and $\overline{\mathbf{OC}}$ are linearly independent.

25. Consider the three vectors a, b, and c that connect the origin $\mathbf{O} \in \mathbb{R}^3$ to two points **A** and **B** on the circumference of an upside down right regular cone with vertex at the origin, and to the center of the cone's base circle **C**, respectively.

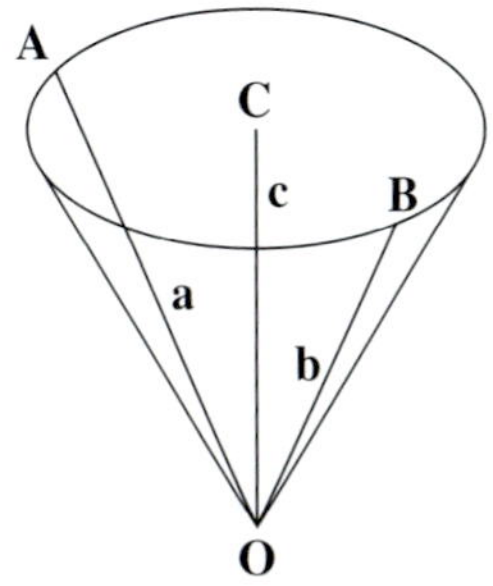

Show that the vectors **a**, **b**, and **c** are linearly dependent if and only if **A** and **B** either coincide or lie opposite each other on the base circle of this cone. (*Hint*: Include the coordinate axes in the plot.)

26. Show that if $P_d := \{(x, y, z) \mid ax + by + cz = d\}$ is a given plane in $\mathbb{R}^3$ with $d \neq 0$ for real numbers a, b, and c, then there exist three points **A**, **B**, **C**, on P_d for which the vectors $\overline{\mathbf{OA}}$, $\overline{\mathbf{OB}}$, and $\overline{\mathbf{OC}}$ are linearly independent.

27. Show that if $P_0 := \{(x, y, z) \mid ax + by + cz = 0\}$ is a given plane that contains the origin $\mathbf{0}$ in $\mathbb{R}^3$ for real numbers a, b, and c, not all of which are zero, then any three vectors connecting the origin $\mathbf{0}$ of $\mathbb{R}^3$ with any three points on P_0 are linearly dependent.

28. Let $U = \text{span}\{\mathbf{u}_i\}$ for $\mathbf{u}_1 = \begin{pmatrix} 1 \\ 2 \\ -2 \\ 0 \end{pmatrix}, \mathbf{u}_2 = \begin{pmatrix} 0 \\ 1 \\ 1 \\ -2 \end{pmatrix}$,

$\mathbf{u}_3 = \begin{pmatrix} -1 \\ 1 \\ 5 \\ -6 \end{pmatrix}$, and $\mathbf{u}_4 = \begin{pmatrix} 2 \\ 3 \\ -1 \\ 2 \end{pmatrix} \in \mathbb{R}^4$.

(a) Find a basis for U.

(b) Do either of the standard unit vectors $\mathbf{e}_1$ or $\mathbf{e}_2$ lie in U?

(c) What is the dimension of U?

(d) Which standard unit vector(s) $\mathbf{e}_i$ lie in U?

29. (An alternative approach to finding a basis:) For a given set of k vectors $\{\mathbf{r}_i\} \in \mathbb{R}^m$, we can form the $k \times m$ matrix $\mathbf{A} = \begin{pmatrix} - & \mathbf{r}_1^T & - \\ & \vdots & \\ - & \mathbf{r}_k^T & - \end{pmatrix}$ that contains the transposed given vectors as its rows. Row reduction of **A** will replace its rows with certain linear combinations thereof. When **A** has been reduced to a row echelon form **R**, then the transposed pivot rows of **R** will be linearly independent and will form a basis for $\text{span}\{\mathbf{r}_i\}$. (Proof?) Repeat Problems 7a(a), 11(a), and 15a(a) in this vein.

30. Prove that if the set of vectors $\{\mathbf{u}_1, \ldots, \mathbf{u}_k\}$ is a basis for a subspace $U \subset \mathbb{R}^m$, then the set $\{\alpha_1\mathbf{u}_1, \ldots, \alpha_k\mathbf{u}_k\}$ is another basis for U as long as each $\alpha_i \neq 0 \in \mathbb{R}$.

31. (*Mainly for math majors*) Prove at least the first five statements (a) – (e) of Proposition 1 by using

(a) Definition 1 of linear (in)dependence;

and separately by using

(b) Definition 2 of linear (in)dependence.

32. For a matrix $\mathbf{A}_{mn}$ and a vector $\mathbf{b} \in \mathbb{R}^m$ complete the following sentences and explain:

(a) If $\mathbf{Ax} = \mathbf{b}$ is solvable for every $\mathbf{b} \in \mathbb{R}^m$, then the rows of **A** are linearly . . . .

(b) If $\mathbf{Ax} = \mathbf{b}$ is unsolvable for at least one $\mathbf{b} \in \mathbb{R}^m$, then the rows of **A** are linearly . . . .

33. (a) Show that two subspaces U of dimensions 7 and V of dimension 5 in $\mathbb{R}^{10}$ must intersect in a subspace larger than $\{\mathbf{0}\}$. This subspace may have dimensions 5, 4, 3, or 2.

(b) Find 2 subspaces U and V of $\mathbb{R}^4$ for which $U \cap V$ is a one-dimensional subspace.

(c) Repeat part (b) so that $U \cap V = \{\mathbf{0}\}$.

(d) Repeat part (b) so that $U \cap V$ is two-dimensional.

34. (a) List all subspaces of the real plane $\mathbb{R}^2$. Express each subspace as a matrix kernel. (*Hint*: Use Chapters 3 and 4 and sort the subspaces by their dimension.)

(b) List all subspaces of real 3–space $\mathbb{R}^3$. Express each subspace as a matrix kernel. (*Hint*: Use Sections 3.1 and 4.1 and sort by dimension.)

35. Show that

(a) If the vectors $\mathbf{u}_1, \ldots, \mathbf{u}_k \in \mathbb{R}^m$ are linearly dependent, then for any matrix $\mathbf{A}_{\ell m}$ the vectors $\mathbf{Au}_1, \ldots, \mathbf{Au}_k \in \mathbb{R}^\ell$ are also linearly dependent.

(b) If the vectors $\mathbf{Bv}_1, \ldots, \mathbf{Bv}_j \in \mathbb{R}^\ell$ are linearly independent for a set of vectors $\{\mathbf{v}_i\} \in \mathbb{R}^m$ and a matrix $\mathbf{B} \in \mathbb{R}^{\ell,m}$, then the vectors $\mathbf{v}_1, \ldots, \mathbf{v}_j \in \mathbb{R}^m$ are also linearly independent.

36. Find a basis for $\ker\begin{pmatrix} 0 & 0 & 1 & 2 \\ 1 & -1 & 1 & 1 \\ 1 & -1 & 0 & -1 \\ 3 & 1 & 0 & 1 \\ 2 & -2 & 1 & 0 \end{pmatrix}$. What is its dimension?

37. Find the dimension of the subspace U that is spanned by the vectors

$$\begin{pmatrix} 0 \\ 0 \\ 1 \\ 2 \end{pmatrix}, \begin{pmatrix} 1 \\ -1 \\ 1 \\ 1 \end{pmatrix}, \begin{pmatrix} 1 \\ -1 \\ 0 \\ -1 \end{pmatrix}, \begin{pmatrix} 2 \\ -2 \\ 1 \\ 0 \end{pmatrix}, \begin{pmatrix} 3 \\ 1 \\ 0 \\ 1 \end{pmatrix}.$$

Find a basis for U from among the given vectors.

38. Show that if n is odd, then there is no $n \times n$ matrix $\mathbf{H}$ with $\text{im}(\mathbf{H}) = \ker(\mathbf{H})$. (*Hint*: Use the Dimension Theorem 5.2.) Then show for $n = 3$ that there are matrices $\mathbf{A}_{3,3}$ with $2\dim(\text{im}(\mathbf{A})) = \dim(\ker(\mathbf{A}))$ and matrices $\mathbf{B}_{3,3}$ with $\dim(\text{im}(\mathbf{B})) = 2\dim(\ker(\mathbf{B}))$.

39. For $\mathbf{A} = \begin{pmatrix} -7 & 9 \\ -6 & 8 \end{pmatrix}$, find a basis for $\mathbf{V} = \{\mathbf{x} \in \mathbb{R}^2 \mid \mathbf{A}\mathbf{x} = 2\mathbf{x}\} \subset \mathbb{R}^2$. Why is V a subspace?

40. Prove both parts of Proposition 2.

41. Let $\mathbf{A} = \begin{pmatrix} 1 & 0 & a & b & c & 0 \\ 0 & 0 & 1 & 0 & -2 & d \\ 0 & 0 & 0 & 0 & -1 & 4 \end{pmatrix}$ for arbitrary real constants $a, \ldots, d$.

(a) Show that the rows of $\mathbf{A}$ are linearly independent for all values of the constants.

(b) How many columns of $\mathbf{A}$ are linearly independent? Which ones are linearly independent?

(c) Find a basis for the column space of $\mathbf{A}$ that is independent of the values of the constants $a, \ldots, d$, if possible.

(d) Find a basis for the row space of $\mathbf{A}$ that is independent of the values of the constants $a, \ldots, d$, if possible.

42. (a) Show that for any two subspaces U and $V \subset \mathbb{R}^n$ we have $\dim(U + V) \geq \dim(U \cap V)$.

(b) Give an example of two subspaces in $\mathbb{R}^5$ with $\dim(U + V) = \dim(U \cap V)$.

43. If $\dim(U + V) = 7$ and $\dim(U \cap V) = 3$ for two subspaces U and V, what are the possible dimensions of U and V?

44. Repeat the previous problem if $\dim(U + V) = 9 = \dim(U \cap V)$.

5.2 Applications

Multiple Spanning Sets of One Subspace, MATLAB

We study different bases for one subspace.

In Section 5.1, we have shown that the columns of any rank m matrix $\mathbf{A}_{mm}$ form a basis of $\mathbb{R}^m$. Things get more complicated when we try to decide whether two sets of vectors in $\mathbb{R}^m$ span the same subspace $U \subset \mathbb{R}^m$.

If two sets of vectors $\{\mathbf{u}_i\}_{i=1}^k \subset \mathbb{R}^m$ and $\{\mathbf{v}_j\}_{j=1}^\ell \subset \mathbb{R}^m$ are given, then

$$U = \text{span}\{\mathbf{u}_i\} = \text{span}\{\mathbf{v}_j\} = V$$

are identical subspaces if and only if each $\mathbf{v}_j$ lies in $\text{span}\{\mathbf{u}_i\}$ <u>and</u> if each $\mathbf{u}_i \in \text{span}\{\mathbf{v}_j\}$. To settle the first inclusion $\mathbf{v}_j \in \text{span}\{\mathbf{u}_i\}$ for each j, we form the multi-augmented $m \times (k+\ell)$ matrix

$$\left(\begin{array}{ccc|ccc} | & & | & | & & | \\ \mathbf{u}_1 & \cdots & \mathbf{u}_k & \mathbf{v}_1 & \cdots & \mathbf{v}_\ell \\ | & & | & | & & | \end{array} \right)_{m,k+\ell}$$

and row reduce it to a REF $\mathbf{R}_{m,k+\ell}$.

Each vector $\mathbf{v}_j$ belongs to $\text{span}\{\mathbf{u}_i\}$ if and only if there are no pivot entries in $\mathbf{R}$ to the right of the vertical separator after column k. In this case, there is a solution to each of the linear systems $\sum_i \alpha_i \mathbf{u}_i = \mathbf{v}_j$ for $j = 1, \ldots, \ell$ by Theorem 3.1. Otherwise, at least one of the vectors $\mathbf{v}_j$ lies outside of $\text{span}\{\mathbf{u}_i\} = U$, and thus $V \neq U$.

Whether every $\mathbf{u}_i$ also lies inside $V = \text{span}\{\mathbf{v}_j\}$ hinges analogously on the same principle for the multiaugmented m by $(\ell+k)$ matrix

$$\left(\begin{array}{ccc|ccc} | & & | & | & & | \\ \mathbf{v}_1 & \cdots & \mathbf{v}_\ell & \mathbf{u}_1 & \cdots & \mathbf{u}_k \\ | & & | & | & & | \end{array}\right)_{m,\ell+k}$$

with reversed entry packages. Thus, we have

Theorem 5.4 Two sets of vectors $\{\mathbf{u}_i\}_{i=1}^k$ and $\left\{\mathbf{v}_j\right\}_{j=1}^\ell \subset \mathbb{R}^m$ span the same subspace if and only if the two multi-augmented $m \times (k+\ell)$ matrices

$$\left(\begin{array}{ccc|ccc} | & & | & | & & | \\ \mathbf{u}_1 & \cdots & \mathbf{u}_k & \mathbf{v}_1 & \cdots & \mathbf{v}_\ell \\ | & & | & | & & | \end{array}\right) \text{ and } \left(\begin{array}{ccc|ccc} | & & | & | & & | \\ \mathbf{v}_1 & \cdots & \mathbf{v}_\ell & \mathbf{u}_1 & \cdots & \mathbf{u}_k \\ | & & | & | & & | \end{array}\right)$$

have no pivots in their respective row echelon forms beyond their first k and first ℓ columns, respectively. ◀

Theorem 5.4 has very satisfying theoretical consequences.

Corollary 2 **Uniqueness of Dimension**

(a) If $\mathcal{U} = \{\mathbf{u}_1, \ldots, \mathbf{u}_k\}$ and $\mathcal{V} = \{\mathbf{v}_1, \ldots, \mathbf{v}_\ell\}$ are two bases for one subspace $W \subset \mathbb{R}^m$, then $\ell = k$.

(b) Let $W \subset \mathbb{R}^m$ be a subspace. Then its dimension $0 \leq k \leq m$ is uniquely defined. ◀

Proof Part (b) follows from part (a).

To prove part (a) we form the matrices $\mathbf{U}_{mk} := \begin{pmatrix} | & & | \\ \mathbf{u}_1 & \ldots & \mathbf{u}_k \\ | & & | \end{pmatrix}$ and $\mathbf{V}_{m\ell} := \begin{pmatrix} | & & | \\ \mathbf{v}_1 & \ldots & \mathbf{v}_\ell \\ | & & | \end{pmatrix}$. Since $\mathcal{U}$ and $\mathcal{V}$ are bases, the two sets of vectors $\{\mathbf{u}_i\}$ and $\left\{\mathbf{v}_j\right\}$ are both linearly independent. Hence, every REF of U has k pivots, while every REF of V has ℓ by Definition 1. By Theorem 5.4, a REF of $(\mathbf{U} \mid \mathbf{V})$ has no pivots beyond its first k columns; that is, the matrix $\mathbf{V}$ is row equivalent to a matrix with at most k nonzero rows. Hence, $\text{rank}(\mathbf{V}) = \ell \leq k$. The analogue holds for a REF of $(\mathbf{V} \mid \mathbf{U})$ (i.e., $\text{rank}(\mathbf{U}) = k \leq \ell$). This forces $\ell = k$. ■

EXAMPLE 7

(a) The two sets of vectors $\{\mathbf{u}_1, \mathbf{u}_2, \mathbf{u}_3\} = \left\{\begin{pmatrix} 1 \\ 0 \\ 2 \end{pmatrix}, \begin{pmatrix} 1 \\ 1 \\ 0 \end{pmatrix}, \begin{pmatrix} 3 \\ 2 \\ 2 \end{pmatrix}\right\}$ and $\{\mathbf{v}_1, \mathbf{v}_2, \mathbf{v}_3\} = \left\{\begin{pmatrix} 1 \\ 0 \\ 1 \end{pmatrix}, \begin{pmatrix} 2 \\ 1 \\ 0 \end{pmatrix}, \begin{pmatrix} 2 \\ 0 \\ 1 \end{pmatrix}\right\}$ do not span the same subspace of $\mathbb{R}^3$, since, the row reduction

$$(\mathbf{u}_1, \mathbf{u}_2, \mathbf{u}_3 \mid \mathbf{v}_1, \mathbf{v}_2, \mathbf{v}_3) = \begin{array}{ccc|ccc|l} \boxed{1} & 1 & 3 & 1 & 2 & 2 & \\ 0 & 1 & 2 & 0 & 1 & 0 & \\ \boxed{2} & 0 & 2 & 2 & 0 & 1 & -2\,row_1 \\ \hline \boxed{1} & 1 & 3 & 1 & 2 & 2 & \\ 0 & \boxed{1} & 2 & 0 & 1 & 0 & \\ 0 & \boxed{-2} & -4 & 0 & -4 & -3 & +2\,row_2 \\ \hline \boxed{1} & 1 & 3 & 1 & 2 & 2 & \\ 0 & \boxed{1} & 2 & 0 & 1 & 0 & \\ 0 & 0 & 0 & 0 & \boxed{-2} & -3 & \end{array}$$

exhibits a pivot beyond the vertical separator.

(b) The two sets of vectors $\{\mathbf{u}_1, \mathbf{u}_2\} = \left\{ \begin{pmatrix} 1 \\ 2 \end{pmatrix}, \begin{pmatrix} -2 \\ -4 \end{pmatrix} \right\}$ and $\{\mathbf{v}_1\} = \left\{ \begin{pmatrix} 5 \\ 10 \end{pmatrix} \right\}$ span the same subspace of $\mathbb{R}^2$, since both

$$(\mathbf{u}_1, \mathbf{u}_2 \mid \mathbf{v}_1) = \begin{array}{cc|c|l} \boxed{1} & -2 & 5 & \\ \boxed{2} & -4 & 10 & -2\,row_1 \\ \hline \boxed{1} & -2 & 5 & \\ 0 & 0 & 0 & \end{array}$$

and

$$(\mathbf{v}_1 \mid \mathbf{u}_1, \mathbf{u}_2) = \begin{array}{c|cc|l} \boxed{5} & 1 & -2 & \\ \boxed{10} & 2 & -4 & -2\,row_1 \\ \hline \boxed{5} & 1 & -2 & \\ 0 & 0 & 0 & \end{array}$$

do not have any pivots beyond their vertical separators, when row reduced. Note that for parts (a) and (b) both first mentioned sets of vectors $\{\mathbf{u}_i\}$ are linearly dependent, while the second sets $\{\mathbf{v}_i\}$ are by chance linearly independent. ◀

Bases of subspaces in $\mathbb{R}^m$ can be computed in MATLAB as follows.

If a subspace $U = \text{span}\{\mathbf{u}_i\}$ is given inclusively it is most expedient for computing purposes to enter the vectors $\mathbf{u}_i$ as row vectors in MATLAB and to use Problem 29 of Section 5.1.P: The RREF of $\begin{pmatrix} — & \mathbf{u}_1 & — \\ & \vdots & \\ — & \mathbf{u}_k & — \end{pmatrix}$ will contain a row vector basis for U in its pivot rows.

If a subspace $U = \ker(\mathbf{A})$ is given exclusively, then the MATLAB command

```
null(A)
```

finds a column vector basis for U from the QR decomposition of $\mathbf{A}$. (See Section 10.3.) The MATLAB command `null(A)` creates basis vectors of unit length, while `null(A,'r')` uses the RREF of $\mathbf{A}$ and finds a basis with rational components for $\ker(\mathbf{A})$.

We have learned how two spanning sets for one subspace can be compared.

5.2.P Problems

1. Show that the two sets of vectors $\{\mathbf{u}_i\} = \left\{\begin{pmatrix}1\\-1\\0\\2\end{pmatrix}, \begin{pmatrix}2\\1\\1\\1\end{pmatrix}\right\}$ and $\{\mathbf{v}_i\} = \left\{\begin{pmatrix}3\\0\\1\\3\end{pmatrix}, \begin{pmatrix}0\\3\\1\\-3\end{pmatrix}\right\}$ span the same subspace of $\mathbb{R}^4$.

2. Find bases for the intersection and the join of the subspaces U and V generated by $\left\{\begin{pmatrix}1\\-1\\0\\2\end{pmatrix}, \begin{pmatrix}2\\1\\1\\1\end{pmatrix}\right\}$ and by $\left\{\begin{pmatrix}3\\0\\1\\3\end{pmatrix}, \begin{pmatrix}1\\3\\1\\-3\end{pmatrix}\right\}$. What are their individual dimensions.

3. Let the vectors $\mathbf{x}_1, \dots, \mathbf{x}_4$ be linearly independent. Let $\mathbf{y}_1 = \mathbf{x}_1 + \mathbf{x}_2$, $\mathbf{y}_2 = \mathbf{x}_1 + \mathbf{x}_3$, and $\mathbf{y}_3 = \mathbf{x}_3 - \mathbf{x}_1$, as well as $\mathbf{z}_1 = \mathbf{x}_3 + \mathbf{x}_4$ and $\mathbf{z}_2 = \mathbf{x}_1 + 2\mathbf{x}_2$. Define $U = \operatorname{span}\{\mathbf{y}_i\}$ and $V = \operatorname{span}\{\mathbf{z}_j\}$. Find bases for the four subspaces U, V, $U \cap V$, and $U + V$.

4. Prove that the two sets of vectors $\left\{\begin{pmatrix}1\\1\\1\\0\end{pmatrix}, \begin{pmatrix}0\\1\\-1\\2\end{pmatrix}\right\}$ and $\left\{\begin{pmatrix}-1\\0\\-2\\2\end{pmatrix}, \begin{pmatrix}1\\2\\0\\2\end{pmatrix}\right\}$ span the same subspace of $\mathbb{R}^4$.

5. Show that the two matrices $\mathbf{A} = \begin{pmatrix}1 & 2 & 0 & 3\\ 2 & -1 & 4 & 0\\ 4 & 3 & 4 & 6\end{pmatrix}$ and $\mathbf{B} = \begin{pmatrix}3 & 1 & 4 & 3\\ 1 & -3 & 4 & -3\end{pmatrix}$ have the same kernel.

6. Verify Problem 5 using MATLAB.

7. Solve Problem 1 using Theorem 5.4 and MATLAB.

8. Solve Problem 13 of Section 4.1.P using MATLAB.

9. Find an integer basis for $\ker\begin{pmatrix}1 & 2 & 0 & 3\\ 2 & -1 & 4 & 0\\ 4 & 3 & 4 & 6\end{pmatrix}$ by using MATLAB.

5.R Review Problems

1. Write out the definitions and concepts underlying the notions of each of the following:

(a) Finding a basis for a space defined as a kernel of a linear map.

(b) A basis for a subspace.

(c) The definition of linear independent vectors in terms of a uniquely solvable linear system.

(d) Linearly dependent vectors.

2. Complete the sentences:

(a) If $\mathbf{A}$ has the REF $\mathbf{R}$, then the .... of $\mathbf{R}$ determine(s) a basis for the **row space** of $\mathbf{A}$, defined as the space spanned by the rows of $\mathbf{A}$.

(b) If $\mathbf{A}$ has the REF $\mathbf{R}$, then the .... of $\mathbf{R}$ determine(s) a basis for the column space of $\mathbf{A}$.

(c) If $\mathbf{A}$ has the REF $\mathbf{R}$, then the .... determine(s) a basis for the kernel of $\mathbf{A}$.

(d) If $\mathbf{A}$ has the row echelon form $\mathbf{R}$, then .... of $\mathbf{R}$ equal(s) .... of $\mathbf{A}$. (There are many fill-in possibilities here!)

3. (a) Can seven vectors in $\mathbb{R}^9$ be linearly independent?

(b) Can seven vectors in $\mathbb{R}^9$ span $\mathbb{R}^9$?

(c) Can seven vectors in $\mathbb{R}^9$ span a subspace $U \subset \mathbb{R}^9$ of dimension 6?

(d) Can seven vectors in $\mathbb{R}^9$ form a basis for $\mathbb{R}^9$?

(e) Answer the preceding four questions, but this time with nine given vectors in $\mathbb{R}^9$.

(f) How about for 12 given vectors in $\mathbb{R}^9$.

4. Fill in the blank: The zero vector is always linearly________ .

5. Fill in the blank: The two vectors $\mathbf{x}$ and $-3\mathbf{x}$ are always linearly________ provided $\mathbf{x} \neq \mathbf{0}$.

6. Fill in the blank: The three vectors $\mathbf{x}$, $a\mathbf{x}$ and $b\mathbf{x}$ are always linearly ________ , provided $\mathbf{x} \neq \mathbf{0}$, $a \neq b$ and $ab \neq 0$.

7. Find a basis for the space of vectors described as

$$\left\{ \mathbf{x} \in \mathbb{R}^4 \;\middle|\; \mathbf{x} = \begin{pmatrix} a & - & 2b & + & c \\ & & b & + & 2c \\ a & - & b & + & 3c \\ a & + & 3b & + & c \end{pmatrix} \right\}$$

for arbitrary real numbers a, b, and c.

8. Let $X = \{\mathbf{x}_i\}$ be a basis of $\mathbb{R}^m$. Consider the subset $X_j := \{\mathbf{x}_1, \ldots, \mathbf{x}_{j-1}, \mathbf{x}_{j+1}, \ldots, \mathbf{x}_m\} \subset X$ for any fixed $1 \leq j \leq m$.

(a) Show that X_j consists of linearly independent vectors.

(b) Determine all vectors $\mathbf{y} \in \mathbb{R}^m$, so that $X_j \cup \{\mathbf{y}\}$ is a basis of $\mathbb{R}^m$.

9. Show that the vectors $\mathbf{x} = \begin{pmatrix} a \\ b \end{pmatrix}$, $\mathbf{y} = \begin{pmatrix} c \\ d \end{pmatrix}$, and $\mathbf{z} = \begin{pmatrix} e \\ f \end{pmatrix}$ are linearly dependent for any choice of real numbers a, b, c, d, e, f.

10. Determine whether each of the following is true or false for two vectors $\mathbf{x} = \begin{pmatrix} a \\ b \\ c \end{pmatrix}$ and $\mathbf{y} = \begin{pmatrix} d \\ e \\ f \end{pmatrix} \in \mathbb{R}^3$:

(a) The vectors $\mathbf{x}$ and $\mathbf{y}$ are linearly independent for any choice of $a, \ldots, f \in \mathbb{R}$.

(b) The vectors $\mathbf{x}$ and $\mathbf{y}$ are linearly independent for some choice of $a, \ldots, f \in \mathbb{R}$.

(c) The vectors $\mathbf{x}$ and $\mathbf{y}$ are linearly dependent for some choice of $a, \ldots, f \in \mathbb{R}$.

(d) The vectors $\mathbf{x}$ and $\mathbf{y}$ are linearly dependent for any choice of $a, \ldots, f \in \mathbb{R}$.

11. Express $-2\mathbf{e}_3 \in \mathbb{R}^3$ as a linear combination of the basis $\mathbf{u}_1 = \mathbf{e}_1$, $\mathbf{u}_2 = \mathbf{e}_1 - \mathbf{e}_2$, and $\mathbf{u}_3 = 2\mathbf{e}_1 - \mathbf{e}_2 + \mathbf{e}_3$ of $\mathbb{R}^3$.

12. Prove or disprove for a matrix $\mathbf{A}_{mn}$:
If $\dim(\ker(\mathbf{A})) > \dim(\operatorname{im}(\mathbf{A}))$, then $n > m$.

13. (a) Show that if $\ker(\mathbf{A}) = \operatorname{im}(\mathbf{A})$ for a matrix $\mathbf{A}_{mn}$, then $\mathbf{A}$ is square (i.e., $m = n$ <u>and</u> n is even).

Find an example of a 2×2 matrix $\mathbf{A}$ with $\ker(\mathbf{A}) = \operatorname{im}(\mathbf{A})$.

(b) Repeat part (b) for a 4×4 and a 6×6 matrix.

14. If $f : \mathbb{R}^n \to \mathbb{R}^m$ is a function which maps k linearly independent vectors $\mathbf{x}_1, \ldots, \mathbf{x}_k \in \mathbb{R}^n$ to $f(\mathbf{x}_i) = \mathbf{0} \in \mathbb{R}^m$ for $i = 1, \ldots, k$, and if there are more than $n - k$ linearly independent vectors in the image $f(\mathbb{R}^n)$, then f cannot be ________ . (Explain, please!)
(*Hint*: Use the Dimension Theorem.)

15. (a) Does $\mathbf{w} = \begin{pmatrix} -1 \\ 12 \\ -47 \\ 23 \\ -5 \end{pmatrix}$ belong to the span of $\mathbf{u}_1 = \begin{pmatrix} -1 \\ 2 \\ -3 \\ 2 \\ -1 \end{pmatrix}$, $\mathbf{u}_2 = \begin{pmatrix} -2 \\ 5 \\ -11 \\ 4 \\ 0 \end{pmatrix}$, $\mathbf{u}_3 = \begin{pmatrix} 3 \\ -4 \\ -2 \\ -9 \\ 2 \end{pmatrix}$, and $\mathbf{u}_4 = \begin{pmatrix} -1 \\ -1 \\ 10 \\ -5 \\ 1 \end{pmatrix}$ in $\mathbb{R}^5$?

(b) Are the four vectors $\mathbf{u}_1, \ldots, \mathbf{u}_4$ linearly dependent or linearly independent? Are the five vectors $\mathbf{u}_1, \ldots, \mathbf{u}_4, \mathbf{w}$ linearly dependent or linearly independent?

(c) Find the coordinate vector $\mathbf{w}_U$ for $U = \{\mathbf{u}_1, \mathbf{u}_2, \mathbf{u}_3, \mathbf{u}_4\}$ if possible.

16. (a) Find the general solution of the linear system $\mathbf{A}\mathbf{x} = \mathbf{b}$ for $\mathbf{A} = \begin{pmatrix} 1 & 2 & 1 & -4 \\ 2 & 3 & 2 & -7 \\ 3 & -1 & -1 & -1 \\ 4 & 0 & -2 & -2 \\ 1 & 1 & 0 & -2 \end{pmatrix}$ and $\mathbf{b} = \begin{pmatrix} 1 \\ 2 \\ -5 \\ -8 \\ -1 \end{pmatrix}$. Check your answer.

What is the dimension of the kernel of $\mathbf{A}$? Find a basis for the image of $\mathbf{A}$.

(b) Can the linear system $\mathbf{A}\mathbf{x} = \mathbf{c}$ be solved for every right-hand-side $\mathbf{c} \in \mathbb{R}^5$ and the matrix $\mathbf{A}$ in part (a)?

(d) Find the maximal number of linearly independent rows of **A** by hand and by using MATLAB.

(e) Find the maximal number of linearly independent columns of **A** by hand and by using MATLAB.

17. Let $\mathbf{u}_1 = \begin{pmatrix} 1 \\ -1 \\ 0 \\ 2 \end{pmatrix}$, $\mathbf{u}_2 = \begin{pmatrix} -2 \\ 2 \\ 0 \\ -4 \end{pmatrix}$, $\mathbf{u}_3 = \begin{pmatrix} 0 \\ 2 \\ 2 \\ 1 \end{pmatrix}$, and $\mathbf{u}_4 = \begin{pmatrix} 2 \\ 0 \\ 2 \\ 5 \end{pmatrix}$, and define $U = \text{span}\{\mathbf{u}_1, \mathbf{u}_2, \mathbf{u}_3, \mathbf{u}_4\} \subset \mathbb{R}^4$.

(a) Show that $\mathbf{u}_4 \in \text{span}\{\mathbf{u}_1, \mathbf{u}_2, \mathbf{u}_3\}$.

(b) Use the work of part (a) to deduce that the set $\{\mathbf{u}_1, \mathbf{u}_3\}$ is a basis for U.

(c) Show that the vectors $\mathbf{u}_3$ and $\mathbf{u}_4$ also form a basis for U.

(d) What is the dimension of U?

18. For $\mathbf{A} = \begin{pmatrix} 1 & 1 & 1 & 1 \\ 1 & 1 & 1 & 1 \\ 1 & 1 & 2 & 1 \\ 1 & 1 & 1 & 4 \end{pmatrix}$ find :

(a) a basis for $\ker(\mathbf{A})$,

(b) a basis for $\text{im}(\mathbf{A})$,

(c) the dimensions of $\text{im}(\mathbf{A})$ and $\ker(\mathbf{A})$,

(d) three linearly independent rows,

(e) four linearly independent columns, if possible.

19. Let $\mathbf{A} = \begin{pmatrix} 1 & -2 & -1 & -2 \\ 3 & -5 & 1 & -3 \\ 1 & -3 & -6 & -7 \end{pmatrix}$ and $\mathbf{b} = \begin{pmatrix} 6 \\ 22 \\ 2 \end{pmatrix}$.

(a) Solve the linear system $\mathbf{Ax} = \mathbf{b}$, if possible.

(b) What is the rank of the augmented matrix $(\mathbf{A} \mid \mathbf{b})$? What is the dimension of the kernel of **A**?

(c) Find a linearly independent spanning set for the image of **A** and one for the kernel of **A**.

20. Let $\mathbf{A} = \begin{pmatrix} 1 & -2 & -1 & 1 & -1 \\ 1 & -3 & 2 & 1 & -4 \\ 3 & -4 & -8 & 1 & 8 \end{pmatrix}$.

(a) Show that $\text{im}(\mathbf{A}) = \mathbb{R}^3$.

(b) Find a basis for $\ker(\mathbf{A})$.

(c) What are the dimensions of $\text{im}(\mathbf{A})$ and of $\ker(\mathbf{A})$?

21. (a) Show that the three functions $\sin^2 x$, $\cos^2 x$ and $1 : \mathbb{R} \to \mathbb{R}$ are linearly dependent as functions.

(b) Show that the three functions $\sin(x)$, $\cos(x)$ and $1 : \mathbb{R} \to \mathbb{R}$ are linearly independent as functions.

22. Determine whether each of the following is true or false:

(a) If $\mathbf{x} \neq \mathbf{y} \in \mathbb{R}^2$, then the vectors **x** and **y** are linearly independent.

(b) If $\mathbf{x} \neq \mathbf{y} \in \mathbb{R}$, then the vectors **x** and **y** are linearly dependent.

(c) If $\mathbf{x} \neq -\mathbf{y}$ and $\mathbf{x} \neq \mathbf{y} \in \mathbb{R}^4$, then **x** and **y** are linearly independent.

23. (a) Can one find a plane p in $\mathbb{R}^3$ with the standard unit vectors $\mathbf{e}_1$, $\mathbf{e}_2$, and $\mathbf{e}_3 \in p$?

(b) Show that if p is a plane in $\mathbb{R}^3$ that contains the origin, then every three vectors in p are linearly dependent.

(c) How do parts (a) and (b) square with each other? Explain, please.

Standard Tasks and Questions

1. Is a given set of vectors $\{\mathbf{u}_1, \ldots, \mathbf{u}_k\}$ linearly independent or linearly dependent?
2. What is the dimension of $\text{span}\{\mathbf{u}_1, \ldots, \mathbf{u}_k\}$?
3. What is the dimension of $\ker(\mathbf{A})$? What is the dimension of $\text{im}(\mathbf{A})$?
4. Express one vector of a linearly dependent set of vectors as a linear combination of the other vectors.
5. Find a basis for the image $\text{im}(\mathbf{A})$; find a basis for $\ker(\mathbf{A})$ and a given matrix $\mathbf{A}_{mn}$.
6. What does the number of pivots in a REF **R** tell about the column vectors of a row equivalent matrix **A**? What does it tell about the kernel of **A**, what about the image?

Subheadings of Lecture Five

Composition of waves

CHAPTER

6

Composition of Linear Maps, Matrix Inverse, and Matrix Transpose

We study several linear maps operating in succession.

6.1 Lecture Six

Matrix products arise naturally from the composition of linear transformations. We apply matrix products to find inverse linear transformations and inverse matrices.

(a) The Composition of Linear Transformations

Assume that $T : \mathbb{R}^n \longrightarrow \mathbb{R}^m$ and $S : \mathbb{R}^m \longrightarrow \mathbb{R}^k$ are two linear transformations and that the range $\mathbb{R}^m$ of T equals the domain for S. In other words, we assume that the mappings T and S can be applied in sequence:

$$\mathbb{R}^n \xrightarrow{T} \mathbb{R}^m \xrightarrow{S} \mathbb{R}^k.$$

This sequence of maps defines the **composite map** $S \circ T : \mathbb{R}^n \to \mathbb{R}^k$ as

$$S \circ T(\mathbf{x}) := S(T(\mathbf{x})) \in \mathbb{R}^k$$

for each $\mathbf{x} \in \mathbb{R}^n$. As seen in Lemma 1 of Section 1.1, the composite map $S \circ T$ is linear if both S and T are linear.

Linear transformations $T : \mathbb{R}^n \to \mathbb{R}^m$ have been represented by their standard matrix representation $\mathbf{A}_{\mathcal{E}}$ since Chapter 1. For the composition of two linear maps, we now attempt to find the standard matrix representation $\mathbf{C}_{\mathcal{E}}$ of the **composite map** $C := S \circ T$.

If $\mathbf{A} = \mathbf{A}_{\mathcal{E}}$ is the standard matrix representation for the linear transformation $T : \mathbb{R}^n \to \mathbb{R}^m$, then we know from Section 1.1 that

$$\mathbf{A} = \begin{pmatrix} | & & | \\ T(\mathbf{e}_1) & \cdots & T(\mathbf{e}_n) \\ | & & | \end{pmatrix} = \begin{pmatrix} | & & | \\ \mathbf{a}_1 & \cdots & \mathbf{a}_n \\ | & & | \end{pmatrix}_{mn},$$

where $\mathbf{a}_i = T(\mathbf{e}_i) \in \mathbb{R}^m$ for the standard unit vectors $\mathbf{e}_1, , \ldots , \mathbf{e}_n \in \mathbb{R}^n$.

Likewise, for $\mathbf{B}$ representing $S : \mathbb{R}^m \to \mathbb{R}^k$, we have $\mathbf{B} = \begin{pmatrix} | & & | \\ S(\mathbf{e}_1) & \cdots & S(\mathbf{e}_m) \\ | & & | \end{pmatrix}_{km}$ for the standard unit vectors $\mathbf{e}_1, \ldots , \mathbf{e}_m$ of $\mathbb{R}^m$.

The standard matrix representation $\mathbf{C}_{\mathcal{E}}$ of the composite linear transformation $C = S \circ T : \mathbb{R}^n \to \mathbb{R}^k$ is formally defined as

$$\mathbf{C}_{\mathcal{E}} = \begin{pmatrix} | & & | \\ S \circ T(\mathbf{e}_1) & \cdots & S \circ T(\mathbf{e}_n) \\ | & & | \end{pmatrix}_{kn},$$

where $S \circ T(\mathbf{e}_i) := S(T(\mathbf{e}_i)) = S(\mathbf{a}_i) = \mathbf{B}\mathbf{a}_i$ by definition, $\mathbf{B}$ is the matrix representation of S, and the vectors $\mathbf{a}_i \in \mathbb{R}^m$ are the columns $T(\mathbf{e}_i)$ of $\mathbf{A}$. Thus,

$$\mathbf{C}_{\mathcal{E}} = \begin{pmatrix} | & & | \\ \mathbf{B}\mathbf{a}_1 & \cdots & \mathbf{B}\mathbf{a}_n \\ | & & | \end{pmatrix} = \mathbf{B} \begin{pmatrix} | & & | \\ \mathbf{a}_1 & \cdots & \mathbf{a}_n \\ | & & | \end{pmatrix} = \mathbf{BA}$$

is the **matrix product** of $\mathbf{B}_{km}$ and $\mathbf{A}_{mn}$, with $\mathbf{B}$ representing S and $\mathbf{A}$ representing T.

The **matrix product BA** is evaluated using dot products of rows of $\mathbf{B}$ with columns of $\mathbf{A}$ in the most obvious way: row $\mathbf{b}_j \in \mathbb{R}^m$ of $\mathbf{B}_{km}$ dots into column $\mathbf{a}_\ell \in \mathbb{R}^m$ of $\mathbf{A}_{mn}$ to give the scalar entry $c_{j\ell}$ in row j and column ℓ of $\mathbf{C}_{kn} = \mathbf{BA}$:

$$\mathbf{BA} = \begin{pmatrix} - & \mathbf{b}_j & - \end{pmatrix}_{km} \begin{pmatrix} | \\ \mathbf{a}_\ell \\ | \end{pmatrix}_{mn} = \begin{pmatrix} & | & \\ - & c_{j\ell} & - \\ & | & \end{pmatrix}_{kn} = \mathbf{C}.$$

For example, if $\mathbf{A}_{3,3} = \begin{pmatrix} 1 & 2 & -1 \\ 2 & 1 & 3 \\ 0 & 1 & -3 \end{pmatrix}$ and $\mathbf{B}_{3,2} = \begin{pmatrix} 1 & 3 \\ 2 & -1 \\ 4 & 0 \end{pmatrix}$, then the matrix product $\mathbf{AB} = \begin{pmatrix} 1 & 1 \\ 16 & 5 \\ -10 & -1 \end{pmatrix}_{3,2}$. Likewise, we can evaluate the following matrix product:

$$\begin{pmatrix} 1 & 2 & 4 & 3 \\ -2 & 1 & 0 & -5 \end{pmatrix}_{2,4} \begin{pmatrix} 2 & 1 & 0 \\ -1 & 4 & 1 \\ 1 & 2 & -1 \\ 3 & 4 & 2 \end{pmatrix}_{4,3} = \begin{pmatrix} 13 & 29 & 4 \\ -20 & -18 & -9 \end{pmatrix}_{2,3}.$$

However, the matrix product $\begin{pmatrix} 2 & 1 & -1 \\ 0 & 1 & -7 \end{pmatrix}_{2,3} \begin{pmatrix} 1 & 1 \\ 2 & -2 \end{pmatrix}_{2,2}$ cannot be formed, due to incompatible dimensions: The first matrix has rows in $\mathbb{R}^3$, while the columns of the second matrix are in $\mathbb{R}^2$.

By using matrix products we have obtained the following result:

> The **composition of two linear maps** $S \circ T$ is represented by the **matrix product BA** of the two individual matrix representations for their respective standard unit vector bases $\{\mathbf{e}_i\}$ of $\mathbb{R}^n$ or $\mathbb{R}^m$.
> That is, the matrix representation of the composition of two linear maps $S \circ T$ is the matrix product **BA** of their individual matrix representations **A** of T and **B** of S.

EXAMPLE 1 The functions $T(\mathbf{x}) = \begin{pmatrix} x_1 + 2x_2 \\ x_2 \end{pmatrix}$ and $S(\mathbf{z}) = \begin{pmatrix} z_1 \\ -z_1 + z_2 \end{pmatrix}$ are two linear maps defined for all $\mathbf{x}$ and $\mathbf{z} \in \mathbb{R}^2$. The matrix $\mathbf{A} = \begin{pmatrix} 1 & 2 \\ 0 & 1 \end{pmatrix}$ represents $T : \mathbb{R}^2 \to \mathbb{R}^2$, while $\mathbf{B} = \begin{pmatrix} 1 & 0 \\ -1 & 1 \end{pmatrix}$ represents $S : \mathbb{R}^2 \to \mathbb{R}^2$.

Now $S \circ T(\mathbf{x}) = S(T(\mathbf{x})) = S\begin{pmatrix} x_1 + 2x_2 \\ x_2 \end{pmatrix} = \begin{pmatrix} x_1 + 2x_2 \\ -x_1 - x_2 \end{pmatrix} : \mathbb{R}^2 \to \mathbb{R}^2$ by the definitions of S, T, and $S \circ T$. Thus, $S \circ T$ is represented by $\begin{pmatrix} 1 & 2 \\ -1 & -1 \end{pmatrix}$. Here the matrix product of the individual matrix representations for $S \circ T$ is $\mathbf{BA} = \begin{pmatrix} 1 & 0 \\ -1 & 1 \end{pmatrix}\begin{pmatrix} 1 & 2 \\ 0 & 1 \end{pmatrix} = \begin{pmatrix} 1 & 2 \\ -1 & -1 \end{pmatrix}$, which agrees with the matrix representation for $S \circ T$.

Since T and S both map $\mathbb{R}^2$ to $\mathbb{R}^2$, we can compose the two maps in reverse order; that is, we can form $T \circ S(\mathbf{z}) = T(S(\mathbf{z})) = T\begin{pmatrix} z_1 \\ -z_1 + z_2 \end{pmatrix} = \begin{pmatrix} -z_1 + 2z_2 \\ -z_1 + z_2 \end{pmatrix}$. Consequently, this composite linear mapping $T \circ S$ has the standard matrix representation $\begin{pmatrix} -1 & 2 \\ -1 & 1 \end{pmatrix}$, and the corresponding individual matrix representation product $\mathbf{AB} = \begin{pmatrix} 1 & 2 \\ 0 & 1 \end{pmatrix}\begin{pmatrix} 1 & 0 \\ -1 & 1 \end{pmatrix} = \begin{pmatrix} -1 & 2 \\ -1 & 1 \end{pmatrix}$ again agrees. ◀

Note that $S \circ T \neq T \circ S$ as mappings in the example since their matrix representations **BA** and **AB** differ. In general, matrix products **do not commute**, or $\mathbf{AB} \neq \mathbf{BA}$ for matrices.

Of course, if both **A** and **B** are 1×1 matrices, or scalars, then $\mathbf{AB} = \mathbf{BA}$, That is, 1×1 matrices always commute. Moreover, for any dimension n, there are matrix pairs $\mathbf{A}_{nn}$ and $\mathbf{B}_{nn}$ whose products commute. For example, $\mathbf{I}_n$, the identity matrix, commutes with every $n \times n$ matrix $\mathbf{A}$ because $\mathbf{A}_{nn}\mathbf{I}_n = \mathbf{A} = \mathbf{I}_n\mathbf{A}_{nn}$. Regarding products of nonsquare matrices, if $\mathbf{A}$ is $m \times n$ and $\mathbf{B}$ is $n \times m$, then both matrix products $\mathbf{AB} \in \mathbb{R}^{m,m}$ and $\mathbf{BA} \in \mathbb{R}^{n,n}$ can be formed. However, these two products can only be equal if $m = n$. But as Example 1 shows, even if $\mathbf{AB}$ and $\mathbf{BA}$ both have the same size $n \times n$, their products $\mathbf{AB}$ and $\mathbf{BA}$ can and will generally be different matrices, representing different linear transformations of $\mathbb{R}^n$.

In Problems 14 and 15 of Section 11.3.P, we deduce that whenever both matrix products $\mathbf{AB}$ and $\mathbf{BA}$ are defined for two given matrices $\mathbf{A}$ and $\mathbf{B}$, then these two matrix products have the same nonzero eigenvalues. More specifically, if $\mathbf{A}$ and $\mathbf{B}$ are both $n \times n$, as in our example for $n = 2$, then $\mathbf{AB}$ and $\mathbf{BA}$ are **similar** as linear transformations. **Matrix similarity** is introduced in Lecture 7, and it is studied throughout the second half of this book.

Caution: If for two matrices $\mathbf{A}$ and $\mathbf{B}$ both products $\mathbf{AB}$ and $\mathbf{BA}$ can be formed, then in general $\mathbf{AB} \neq \mathbf{BA}$.

If three mappings $f : X \to Y$, $g : Y \to Z$, and $h : Z \to W$ are given, then they can be composed in the order $h \circ g \circ f : X \to W$, since the appropriate ranges and domains match. Here we define $(h \circ g \circ f)(x) := h(g(f(x))) \in W$ for all $x \in X$. The composition of maps is **associative** in the following sense.

Theorem 6.1 If $f : X \to Y$, $g : Y \to Z$, and $h : Z \to W$ are functions and if $h \circ g \circ f : X \to W$ is defined by $(h \circ g \circ f)(x), := h(g(f(x))) \in W$ for all $x \in X$, then the triple composition of maps is associative, that bis,

$$h \circ g \circ f = h \circ (g \circ f) = (h \circ g) \circ f.$$ ◀

Proof We have $(h \circ (g \circ f))(x) = h((g \circ f)(x)) = h(g(f(x)))$ and $((h \circ g) \circ f)(x) = (h \circ g)(f(x)) = h(g(f(x)))$ for all $x \in X$ by definition. ■

Specifically, for three linear maps $T : \mathbb{R}^n \to \mathbb{R}^m$, $S : \mathbb{R}^m \to \mathbb{R}^k$, and $U : \mathbb{R}^k \to \mathbb{R}^\ell$ and their respective matrix representations $\mathbf{A}_{mn}$, $\mathbf{B}_{km}$, and $\mathbf{C}_{\ell k}$, we have:

Corollary 1 Matrix multiplication is associative. That is, for three matrices $\mathbf{A}_{mn}$, $\mathbf{B}_{km}$, and $\mathbf{C}_{\ell k}$, the **associative property**

$$(\mathbf{ABC})_{\ell n} := \mathbf{A}(\mathbf{BC}) = (\mathbf{AB})\mathbf{C}$$

holds for their multiplication. ◀

(b) The Inverse of a Linear Transformation

Some linear transformations can be inverted, while others cannot.

We now study the **inverse of a linear transformation** (i.e., when does it exist, and how can it be described and found).

Let $T : \mathbb{R}^n \to \mathbb{R}^m$ be a linear map with the standard matrix representation $\mathbf{A}_{mn}$. Our questions are as follows:

When can we find an **inverse map**

$$T^{-1} : \mathbb{R}^m \to \mathbb{R}^n,$$

so that

$$T^{-1}(T(\mathbf{x})) = \mathbf{x},$$

or so that $T^{-1} \circ T$ acts as the **identity transform** that maps every $\mathbf{x} \in \mathbb{R}^n$ to itself?

If a linear mapping $T : \mathbb{R}^n \to \mathbb{R}^m$ is invertible, is the inverse mapping $T^{-1} : \mathbb{R}^m \to \mathbb{R}^n$ also linear? If this is true, what is the associated standard matrix representation $\mathbf{A}_{nm}^{-1}$ of T^{-1}?

Will T likewise be the inverse of T^{-1}; that is, will

$$T(T^{-1}(\mathbf{y})) = \mathbf{y}$$

be the identity map for all $\mathbf{y} \in \mathbb{R}^m$?

To study these questions, we assume for the moment that $T : \mathbb{R}^n \to \mathbb{R}^m$ is linear and invertible and that T^{-1} is linear. The last assumption will be proved to be true for linear invertible maps in Lemma 1.

Under these assumptions, we make use of the standard matrix representations $\mathbf{A}$ and $\mathbf{A}^{-1}$ for the linear maps T and T^{-1}. As shown earlier, the composite linear map $T \circ T^{-1} : \mathbb{R}^m \to \mathbb{R}^m$ has the matrix representation $\mathbf{C} = \mathbf{A}\mathbf{A}^{-1}$, while $T^{-1} \circ T : \mathbb{R}^n \to \mathbb{R}^n$ is represented by the matrix product $\mathbf{A}^{-1}\mathbf{A}$.

Since $T \circ T^{-1}$ is assumed to be the identity map, $T(T^{-1}(\mathbf{e}_i)) = \mathbf{e}_i$ for every standard unit vector $\mathbf{e}_i$ of $\mathbb{R}^m$. Thus, the standard matrix representation $\mathbf{C}$ of $T \circ T^{-1}$ is

$$\mathbf{C}\left(= \mathbf{A}\mathbf{A}^{-1}\right) = \begin{pmatrix} | & & | \\ \mathbf{C}(\mathbf{e}_1) & \cdots & \mathbf{C}(\mathbf{e}_m) \\ | & & | \end{pmatrix} = \mathbf{I}_m, \qquad (6.1)$$

according to Chapter 1, since $\mathbf{C}(\mathbf{e}_i) = \mathbf{e}_i$. Hence, $\mathbf{A}_{mn}\mathbf{A}_{nm}^{-1} = \mathbf{I}_m$. To find $\mathbf{A}^{-1}$ from $\mathbf{A}$, we note that the columns $\mathbf{x}_i$ of

$$\mathbf{A}^{-1} = \begin{pmatrix} | & & | \\ \mathbf{A}^{-1}(\mathbf{e}_1) & \cdots & \mathbf{A}^{-1}(\mathbf{e}_m) \\ | & & | \end{pmatrix}_{nm} = \begin{pmatrix} | & & | \\ \mathbf{x}_1 & \ldots & \mathbf{x}_m \\ | & & | \end{pmatrix}$$

are mapped to the standard unit vectors $\mathbf{e}_i \in \mathbb{R}^m$ under $\mathbf{A}$, since $\mathbf{A}\mathbf{A}^{-1} = \mathbf{I}_m$. Thus, with $\mathbf{x}_i = \mathbf{A}^{-1}\mathbf{e}_i$, we have

$$\mathbf{A}\mathbf{x}_1 = \mathbf{A}(\mathbf{A}^{-1}(\mathbf{e}_1)) = \mathbf{I}\mathbf{e}_1 = \mathbf{e}_1, \mathbf{A}\mathbf{x}_2 = \mathbf{A}(\mathbf{A}^{-1}(\mathbf{e}_2)) = \mathbf{I}\mathbf{e}_2 = \mathbf{e}_2, \quad \text{etc}, \qquad (6.2)$$

for each $\mathbf{e}_i \in \mathbb{R}^m$, according to Eq. (6.1). The unknown columns $\mathbf{x}_i = \mathbf{A}^{-1}(\mathbf{e}_i) \in \mathbb{R}^n$ of the inverse matrix $\mathbf{A}^{-1}$ can be found from Eq. (6.2) by solving $\mathbf{A}\mathbf{x}_i = \mathbf{e}_i$ for each $i = 1, \ldots, m$. To find all $\mathbf{x}_i$ for $i = 1, \ldots, m$, we need to solve m linear systems with one system matrix $\mathbf{A}$ and m different right-hand sides $\mathbf{e}_i \in \mathbb{R}^m$. If $\mathbf{A}$ is invertible, we can do so in one complete row reduction sweep for the **multi-augmented** $m \times (n+m)$ **matrix**

$$(\mathbf{A}_{mn} \mid \mathbf{I}_m)$$

to its RREF

$$(\mathbf{I}_m \mid \mathbf{A}_{nm}^{-1}).$$

The row reduction process from $(\mathbf{A} \mid \mathbf{I})$ to its RREF $(\mathbf{I} \mid \mathbf{A}^{-1})$ not only gives us the matrix $\mathbf{A}^{-1}$, but it affords us the following insights:

- By comparing the sizes that are involved, we see that it appears necessary that $m = n$ for the inversion of a matrix $\mathbf{A}_{mn}$.
- Secondly, by the uniqueness property of the RREF, the inverse of an invertible linear map must be unique.
- Finally, since row reductions can be reversed, if $\mathbf{A}^{-1}$ is the inverse of $\mathbf{A}$, then $\mathbf{A}$ is the inverse of $\mathbf{A}^{-1}$ (i.e., $\mathbf{A}^{-1}\mathbf{A} = \mathbf{I} = \mathbf{A}\mathbf{A}^{-1}$).

In this section, we establish the first observation, while the latter two will be proved rigorously in Section 6.2

A function $f : X \to Y$ is called **invertible** if it has an inverse $f^{-1} : Y \to X$ with $f^{-1}(f(x)) = x$ for all $x \in X$. The following classifies invertible functions.

Theorem 6.2 A function $f : X \to Y$ is **invertible** if and only if

(a) f is **one–to–one**. (That is, if $f(x) = f(y)$, then $x = y$.)
In other words, every image $f(x)$ has a unique preimage x;
and

(b) f is **onto**. (That is, for every $y \in Y$ in the range of f, there is an $x \in X$ in the domain with $f(x) = y$.)
In other words, every point in Y is an image under f. ◀

The first requirement makes sure that f^{-1} is a function (i.e., that its images, being the pre–images x of $f(x)$, are uniquely defined). The second requirement insures that the inverse function f^{-1} is defined on its complete domain Y, which is the range of f.

If we specialize the two requirements of Theorem 6.2 from an arbitrary map to a linear transformation

$$T : \mathbb{R}^n \longrightarrow \mathbb{R}^m,$$

we obtain the following slightly amended requirements for the invertibility of T:

(a) *linear*: T is one-to-one if and only if $T(\mathbf{x}) = \mathbf{y}$ is uniquely solvable for every $\mathbf{y} \in T(\mathbb{R}^n)$, or
if and only if every column in the REF of $\mathbf{A}_{mn}$ has a pivot, or
if and only if n is the number of pivots in a REF of $\mathbf{A}_{mn}$; all by Theorem 3.2.

(b) *linear*: T is onto if and only if $T(\mathbb{R}^n) = \mathbb{R}^m$, or
if and only if $T\mathbf{x} = \mathbf{b}$ is solvable for every $\mathbf{b} \in \mathbb{R}^m$, or

if and only if every row in an REF of the standard matrix representation $\mathbf{A}_\mathcal{E}$ of T has a pivot, or

if and only if is the number of pivots in a REF of $\mathbf{A}_{mn}$; all by Theorem 3.1.

Combining these results, we obtain the following practical criterion:

Corollary 2 A linear transformation $T : \mathbb{R}^n \to \mathbb{R}^m$ is **invertible** if and only if its standard matrix representation $\mathbf{A}_\mathcal{E}$ is square and the REF of $\mathbf{A}_\mathcal{E}$ has $n = m$ pivots. ◀

Next we establish that the inverse map T^{-1} is linear if T is invertible and linear.

Lemma 1 If $T : \mathbb{R}^n \to \mathbb{R}^n$ is an invertible linear transformation, then the inverse mapping $T^{-1} : \mathbb{R}^n \to \mathbb{R}^n$, defined by the identity $T^{-1}(T(\mathbf{x})) = \mathbf{x}$ for all $\mathbf{x} \in \mathbb{R}^n$, is linear as well. ◀

Proof Let $\mathbf{w}$ and $\mathbf{z}$ be two arbitrary vectors in $\mathbb{R}^n$. Since T is invertible (i.e., onto and one-to-one), there are two unique vectors $\mathbf{x}$ and $\mathbf{y} \in \mathbb{R}^n$ with $T(\mathbf{x}) = \mathbf{w}$ and $T(\mathbf{y}) = \mathbf{z}$. Translated to T^{-1}, these equations become $T^{-1}(\mathbf{w}) = T^{-1}(T(\mathbf{x})) = \mathbf{x}$ and $T^{-1}(\mathbf{z}) = \mathbf{y}$. By the linearity of T, we have $T(\alpha\mathbf{x}) = \alpha T(\mathbf{x}) = \alpha\mathbf{w}$ and $T(\alpha\mathbf{x} + \beta\mathbf{y}) = \alpha T(\mathbf{x}) + \beta T(\mathbf{y}) = \alpha\mathbf{w} + \beta\mathbf{z}$ for all scalars α and β. Thus, T^{-1} is linear, because

$$T^{-1}(\alpha\mathbf{w} + \beta\mathbf{z}) = T^{-1}(T(\alpha\mathbf{x} + \beta\mathbf{y})) = \alpha\mathbf{x} + \beta\mathbf{y} = \alpha T^{-1}(\mathbf{w}) + \beta T^{-1}(\mathbf{z}). \quad \blacksquare$$

DEFINITION 1 A matrix $\mathbf{A}_{nn}$ is **nonsingular** if $\mathbf{A}$ is invertible. A matrix that is not invertible is called **singular**.

EXAMPLE 2

(a) Nonsquare matrices do not have inverses; hence they are always singular.

(b) Square matrices $\mathbf{A}_{nn}$ with n pivots are called **nonsingular**, or **invertible**, because the transformation of $\mathbf{x} \mapsto \mathbf{A}\mathbf{x}$ can be inverted by applying $\mathbf{A}^{-1}$ to a point $\mathbf{A}\mathbf{x}$ in its image and mapping $\mathbf{A}\mathbf{x}$ to $\mathbf{A}^{-1}(\mathbf{A}\mathbf{x}) = \mathbf{x}$. That is, each vector $\mathbf{x} \in \mathbb{R}^n$ can be retrieved from knowing its image $\mathbf{A}\mathbf{x}$. In other words, the square matrices $\mathbf{A}_{nn}$ of rank n are precisely the matrices that can be inverted. ◀

The following example presents a computational scheme that is based on Eq. (6.2) and that helps decide whether a square matrix $\mathbf{A}$ is invertible. It finds the matrix inverse $\mathbf{A}^{-1}$ if $\mathbf{A}$ is invertible. Otherwise, it stops.

EXAMPLE 3

(a) Does the inverse of $\mathbf{A} = \begin{pmatrix} 1 & 2 & 0 \\ 4 & -1 & 9 \\ -1 & 3 & -5 \end{pmatrix}$ exist? If so, what does it look like?

We set up the multi-augmented matrix $(\mathbf{A} \mid \mathbf{I}_3)$ and row reduce the left side $\mathbf{A}$ to its reduced row echelon form, working with row reductions all the way across the six columns of $(\mathbf{A} \mid \mathbf{I}_3)$. If we succeed to reduce $\mathbf{A}$ to $\mathbf{I}_3$, then $\mathbf{A}$ is invertible, and its inverse appears on the right of the RREF $(\mathbf{I}_3 \mid \mathbf{A}^{-1})$ of $(\mathbf{A} \mid \mathbf{I}_3)$.

A			$\mathbf{I}_3$			
1	2	0	1	0	0	
4	−1	9	0	1	0	$-4 \cdot row_1$
−1	3	−5	0	0	1	$+row_1$
1	2	0	1	0	0	
0	−9	9	−4	1	0	
0	5	−5	−1	0	1	$+5/9 \cdot row_2$
1	2	0	1	0	0	
0	−9	9	−4	1	0	
0	0	0	*	*	*	

The REF of $\mathbf{A}$ has only two pivots, so that its RREF cannot be equal to $\mathbf{I}_3$ with three pivots. Thus, by Corollary 2, the matrix $\mathbf{A}$ is not invertible.

(b) Is the matrix $\mathbf{B} = \begin{pmatrix} 1 & 2 & 0 \\ 4 & -1 & 9 \\ -1 & 3 & 0 \end{pmatrix}$ invertible? If so, find $\mathbf{B}^{-1}$.

We again set up a multiaugmented matrix $(\mathbf{B} \mid \mathbf{I}_3)$ and row reduce $(\mathbf{B} \mid \mathbf{I}_3)$ to its RREF. If the identity matrix $\mathbf{I}_3$ appears on the left side of the RREF, then $\mathbf{B}$ is invertible, according to Corollary 2.

B			$\mathbf{I}_3$			
1	2	0	1	0	0	
4	−1	9	0	1	0	$-4 \cdot row_1$
−1	3	0	0	0	1	$+row_1$
1	2	0	1	0	0	
0	−9	9	−4	1	0	
0	5	0	1	0	1	$+5/9 \cdot row_2$
1	2	0	1	0	0	
0	−9	9	−4	1	0	"*REF of* $(\mathbf{B} \mid \mathbf{I})$"
0	0	5	−11/9	5/9	1	

This REF has three pivots, and we can proceed to its RREF with $\mathbf{I}_3$ on the left by using Gaussian elimination from the last pivot back and up. By Corollary 2, the matrix $\mathbf{B}$ is invertible. Its inverse $\mathbf{B}^{-1}$ will appear on the right of the RREF $(\mathbf{I}_3 \mid \mathbf{B}^{-1})$ of $(\mathbf{B} \mid \mathbf{I}_3)$. We continue from the computed REF:

$\boxed{1}$	2	0	1	0	0	
0	$\boxed{-9}$	9	−4	1	0	$\div(-9)$
0	0	$\boxed{5}$	−11/9	5/9	1	$\div 5$
$\boxed{1}$	2	0	1	0	0	
0	$\boxed{1}$	−1	4/9	−1/9	0	$+\ row_3$
0	0	$\boxed{1}$	−11/45	1/9	1/5	
$\boxed{1}$	2	0	1	0	0	$-\ 2 \cdot row_2$
0	$\boxed{1}$	0	1/5	0	1/5	
0	0	$\boxed{1}$	−11/45	1/9	1/5	
$\boxed{1}$	0	0	3/5	0	−2/5	
0	$\boxed{1}$	0	1/5	0	1/5	*RREF of* $(\mathbf{B} \mid \mathbf{I})$
0	0	$\boxed{1}$	−11/45	1/9	1/5	
	$\mathbf{I}_3$			$\mathbf{B}^{-1}$		

By factoring out the common denominator, we obtain $\mathbf{B}^{-1} = \frac{1}{45}\begin{pmatrix} 27 & 0 & -18 \\ 9 & 0 & 9 \\ -11 & 5 & 9 \end{pmatrix}$. Now calculate the matrix products $\mathbf{B}\mathbf{B}^{-1}$ and $\mathbf{B}^{-1}\mathbf{B}$ to see that they indeed are both equal to $\mathbf{I}_3$. ◀

The inverse $\mathbf{A}^{-1}$ of a matrix can be used to express the solution of nonsingular linear systems $\mathbf{A}\mathbf{x} = \mathbf{b}$ as $\mathbf{x} = \mathbf{A}^{-1}\mathbf{b}$. Note, however, that to find $\mathbf{A}^{-1}$ from the multi-augmented matrix $(\mathbf{A} \mid \mathbf{I}_3)$ requires significantly more work than solving $\mathbf{A}\mathbf{x} = \mathbf{b}$ by row reduction of the singly augmented matrix $(\mathbf{A} \mid \mathbf{b})$. To evaluate $\mathbf{x}$ as $\mathbf{x} = \mathbf{A}^{-1}\mathbf{b}$ requires even further work. Thus, computing the matrix inverse in order to solve a system of linear equations is wasteful and cannot be recommended.

(c) The Transpose of a Matrix

As an application and for later use, we study the **transpose** of a matrix $\mathbf{A}_{mn} = (a_{ij})$ now; recall the definition of vector transposition from row to column notation and vice versa in the Introduction.

DEFINITION 2 Let $\mathbf{A}_{mn} = (a_{ij})$ be an $m \times n$ matrix with a_{ij} denoting the matrix entry in row i and column j. The **transpose** $\mathbf{A}^T$ of $\mathbf{A}$ is the $n \times m$ matrix with the entry a_{ji} of $\mathbf{A}$ appearing in row i and column j for each i and j.

If $\mathbf{A} = \begin{pmatrix} 1 & 2 & 3 \\ 0 & -1 & 4 \end{pmatrix}$, then $\mathbf{A}^T = \begin{pmatrix} 1 & 0 \\ 2 & -1 \\ 3 & 4 \end{pmatrix}$; that is, $\mathbf{A}$'s rows appear as columns in $\mathbf{A}^T$ and vice versa. Since transposing a matrix simply transforms its rows into columns, forming the double transpose $(\mathbf{A}^T)^T$ results in a double transposition and leads back to the original matrix **A**. In other words, the transpose of the transpose of **A** is **A** itself: $(\mathbf{A}^T)^T = \mathbf{A}$.

Since matrices do not always commute, we need to be cautious when transposing and inverting matrix products.

Theorem 6.3 If the sizes of two matrices **A** and **B** match so that the product **AB** can be formed, then the matrix product $\mathbf{B}^T\mathbf{A}^T$ can also be formed, and

$$\mathbf{B}^T\mathbf{A}^T = (\mathbf{AB})^T.$$

Proof If $\mathbf{A}_{mn} = \begin{pmatrix} - & \mathbf{a}_1 & - \\ & \vdots & \\ - & \mathbf{a}_m & - \end{pmatrix}$ and $\mathbf{B}_{nk} = \begin{pmatrix} | & & | \\ \mathbf{b}_1 & \cdots & \mathbf{b}_k \\ | & & | \end{pmatrix}$ are denoted by their rows and columns, respectively, then the transpose of their product $(\mathbf{A}_{mn}\mathbf{B}_{nk})_{mk}$ is defined as

$$\left(\begin{pmatrix} - & \mathbf{a}_1 & - \\ & \vdots & \\ - & \mathbf{a}_m & - \end{pmatrix}\begin{pmatrix} | & & | \\ \mathbf{b}_1 & \cdots & \mathbf{b}_k \\ | & & | \end{pmatrix}\right)^T = \left(\ \mathbf{a}_i\mathbf{b}_j\ \right)^T = \left(\ \mathbf{a}_j\mathbf{b}_i\ \right),$$

which is composed of row × column vector products. On the other hand,

$$(\mathbf{B}^T)_{kn}(\mathbf{A}^T)_{nm} = \begin{pmatrix} - & \mathbf{b}_1^T & - \\ & \vdots & \\ - & \mathbf{b}_k^T & - \end{pmatrix}\begin{pmatrix} | & & | \\ \mathbf{a}_1^T & \cdots & \mathbf{a}_m^T \\ | & & | \end{pmatrix}$$
$$= \left(\ \mathbf{b}_i^T\ \mathbf{a}_j^T\ \right) = \left(\ \mathbf{a}_j\ \mathbf{b}_i\ \right) = (\mathbf{AB})^T$$

is defined and contains the same entries, due to the properties of the row× column vector or dot product. ■

Theorem 6.4 If **A** and **B** are two invertible $n \times n$ matrices, then their product **AB** is invertible, and

$$(\mathbf{AB})^{-1} = \mathbf{B}^{-1}\mathbf{A}^{-1}.$$

Proof Observe that

$$(\mathbf{B}^{-1}\mathbf{A}^{-1})(\mathbf{AB}) = \mathbf{B}^{-1}\mathbf{A}^{-1}\mathbf{AB} = \mathbf{B}^{-1}\mathbf{I}_n\mathbf{B} = \mathbf{B}^{-1}\mathbf{B} = \mathbf{I}_n \quad \text{and}$$
$$(\mathbf{AB})(\mathbf{B}^{-1}\mathbf{A}^{-1}) = \mathbf{ABB}^{-1}\mathbf{A}^{-1} = \mathbf{AI}_n\mathbf{A}^{-1} = \mathbf{A}^{-1}\mathbf{A} = \mathbf{I}_n.$$ ■

The last two theorems show that inverting matrices and transposing matrices both have a similar structure when taking matrix products. The two operations behave in an **anti-commutative** way for products. Note, however, that the matrix transpose is defined and most easily computed for every matrix, while matrix inversion is restricted to certain square matrices (see Corollary 2) and takes considerable effort.

Finally, we have the following theorem:

Theorem 6.5 If $\mathbf{A}$ is a invertible $n \times n$ matrix, then its transpose $\mathbf{A}^T$ is also invertible, and

$$\left(\mathbf{A}^T\right)^{-1} = \left(\mathbf{A}^{-1}\right)^T.$$

Proof We have

$$\left(\mathbf{A}^{-1}\right)^T \mathbf{A}^T = \left(\mathbf{A}\mathbf{A}^{-1}\right)^T = \mathbf{I}_n$$

from Theorem 6.3. Thus, $\left(\mathbf{A}^T\right)^{-1} = \left(\mathbf{A}^{-1}\right)^T$ by definition. ∎

The composition of linear transformations can be represented by the product of their individual matrix representations. Invertibility of matrices and a method for finding the matrix inverse have been discussed.

6.1.P Problems

1. (a) If $\mathbf{A}$ is an $m \times n$ matrix and if $\mathbf{B}$ is $n \times k$, what is the size of the product matrix $\mathbf{AB}$?

(b) Compute the matrix products $\mathbf{AB}$ and $\mathbf{BA}$ for

$$\mathbf{A} = \begin{pmatrix} 1 & 3 & -7 & 2 \\ 2 & -1 & -3 & 0 \\ 4 & 9 & -3 & -7 \\ 1 & 0 & 0 & 2 \end{pmatrix} \text{ and}$$

$$\mathbf{B} = \begin{pmatrix} 0 & 0 & 1 & 0 \\ 2 & 1 & 0 & 0 \\ 3 & 0 & -4 & 1 \\ -1 & 5 & 0 & 2 \end{pmatrix} \text{ by hand.}$$

(c) Evaluate the matrix products $\mathbf{I}_m\mathbf{A}_{mn}$ and $\mathbf{O}_m\mathbf{A}_{mn}$ for an arbitrary $m \times n$ matrix $\mathbf{A}_{mn}$ and the identity and zero matrices $\mathbf{I}_m$ and $\mathbf{O}_m$.

(d) Evaluate the two matrix products $\mathbf{A}_{mn}\mathbf{I}_n$ and $\mathbf{A}_{mn}\mathbf{O}_n$ for an arbitrary $m \times n$ matrix $\mathbf{A}_{mn}$.

2. (a) Find two matrices $\mathbf{A}_{mn} \neq \mathbf{O}_{mn}$ and $\mathbf{B}_{nk} \neq \mathbf{O}_{nk}$ whose product $\mathbf{AB}_{m\times k}$ is the zero matrix.

(b) Find a matrix $\mathbf{A}_{nn} \neq \mathbf{O}_n$ with $\mathbf{A}^2 = \mathbf{AA} = \mathbf{O}_n$.

3. (a) Find the inverse of the matrix

$$\mathbf{A} = \begin{pmatrix} 1 & 4 & 3 \\ -1 & -2 & 0 \\ 2 & 2 & 3 \end{pmatrix},$$

if possible.

(b) Solve the linear system $\mathbf{Ax} = \begin{pmatrix} 12 \\ -12 \\ 8 \end{pmatrix}$ with $\mathbf{A}$ from part (a).

4. (a) Let the two mappings $A_\pm : \mathbb{R}^3 \to \mathbb{R}^3$ be defined by the assignments

$$\begin{pmatrix} x_1 \\ x_2 \\ x_3 \end{pmatrix} \mapsto \begin{pmatrix} x_1 \\ x_2 \\ x_3 \end{pmatrix} \pm 3 \begin{pmatrix} x_3 \\ x_2 \\ x_1 \end{pmatrix}.$$

Show that both A_+ and A_- are linear and invertible. Find both standard matrix representations and their inverse matrices, if possible.

(b) Define two mappings $B_\pm$ by

$$\begin{pmatrix} x_1 \\ x_2 \\ x_3 \end{pmatrix} \mapsto \begin{pmatrix} x_1 \\ x_2 \\ x_3 \end{pmatrix} \pm \begin{pmatrix} x_3 \\ x_2 \\ x_1 \end{pmatrix}.$$

Which of the two mappings B_+ or B_- is invertible. Find the standard two matrix representations for B_+ and B_-, decide whether these are invertible, and find the inverse, when it exists.

5. (a) For $\mathbf{E} = \begin{pmatrix} 0 & 0 & 1 \\ 0 & 1 & 0 \\ 1 & 0 & 0 \end{pmatrix}$, evaluate the matrix product $(\mathbf{I}_3 + \mathbf{E}) \cdot (\mathbf{I}_3 - \mathbf{E})$.

(b) If E is any $n \times n$ matrix with $\mathbf{E}^2 = \mathbf{I}_n$, what is $(\mathbf{I}_n + \mathbf{E})\ (\mathbf{I}_n - \mathbf{E})$? What is $(\mathbf{I}_n - \mathbf{E})\ (\mathbf{I}_n + \mathbf{E})$?

(c) What is $\mathbf{E}^k$ for $\mathbf{E}$ from part (a) and all integers k (positive, zero, or negative)?

6. Find the inverse of a general 2×2 matrix $\mathbf{A} = \begin{pmatrix} a & b \\ c & d \end{pmatrix}$ with arbitrary entries $a,\ b,\ c$, and $d \in \mathbb{R}$, if the inverse exists. Give an explicit formula for $\mathbf{A}_{2,2}^{-1}$ when it exists, together with a condition for its existence.

7. Find the inverse of $\mathbf{A} = \begin{pmatrix} 1 & 0 & 1 \\ 0 & 2 & 1 \\ 1 & 1 & 1 \end{pmatrix}$ and verify that $\mathbf{A}^{-1}\mathbf{A} = \mathbf{I}_3 = \mathbf{A}\mathbf{A}^{-1}$.

8. Explain why the inverse of the matrix $\mathbf{A} = \begin{pmatrix} 1 & -1 & 1 \\ 2 & 1 & 2 \\ 0 & 1 & -1 \end{pmatrix}$ exists and find $\mathbf{A}^{-1}$. Verify that $\mathbf{A}\mathbf{A}^{-1} = \mathbf{I}_3$.

9. Find the third column of $\mathbf{A}^{-1}$, where $\mathbf{A} = \begin{pmatrix} 1 & 2 & -4 \\ 0 & 2 & 3 \\ 2 & 3 & -1 \end{pmatrix}$, without actually computing any other columns of $\mathbf{A}^{-1}$.

10. Prove each of the following or give a counterexample to disprove it:

(a) If the columns of a square matrix $\mathbf{A}_{nn}$ form a basis for $\mathbb{R}^n$ then the matrix is invertible.

(b) If the rows of a matrix $\mathbf{A}_{mn}$ form a basis for $\mathbb{R}^n$, then the matrix is invertible.

(c) If the rows of a matrix $\mathbf{A}_{nn}$ do not form a basis for $\mathbb{R}^n$, then the columns of the matrix may still form a basis of $\mathbb{R}^n$.

11. Prove that $\mathbf{XA} = \mathbf{O}_{kn}$ for an invertible matrix $\mathbf{A}_{nn}$, then $\mathbf{X} = \mathbf{O}_{kn}$.

12. If $\mathbf{A}^{-1} = \mathbf{B}$ and $\mathbf{B}^T = \mathbf{C}^{-1}$, express the inverse of $\mathbf{A}^T\mathbf{B}$ in terms of $\mathbf{C}$.

13. What is the inverse of the product matrix $\mathbf{A}^T\mathbf{B}^{-1}(\mathbf{C}^T)^{-1}$? Check your answer.

14. Do the matrices $\mathbf{A} = \begin{pmatrix} 1 & -3 & 0 \\ 0 & 3 & 1 \\ 2 & -1 & 2 \end{pmatrix}$ and

$$\mathbf{B} = \begin{pmatrix} 1 & -2 & 1 & 3 \\ 0 & 1 & -2 & 1 \\ 0 & 0 & 3 & -4 \\ 2 & -1 & -4 & 9 \end{pmatrix}$$ have inverses? If they do, what are the inverses?

15. Find all values $d \in \mathbb{R}$ for which the matrix

$$\mathbf{A} = \begin{pmatrix} 1 & -1 & 2 & d \\ 0 & 4 & -2 & 0 \\ 2 & 0 & -d & 1 \\ 1 & 1 & 1 & 1 \end{pmatrix}$$ is nonsingular.

16. Prove or give a counterexample to disprove each of the following:

(a) If the linear system $\mathbf{Ax} = \mathbf{b}$ is solvable for every right-hand side $\mathbf{b}$, then $\mathbf{A}$ is invertible.

(b) If $\mathbf{A}_{nn}$ is square and the linear system $\mathbf{Ax} = \mathbf{b}$ can be solved for every right-hand side $\mathbf{b}$, then $\mathbf{A}$ is invertible.

(c) If the linear system $\mathbf{Ax} = \mathbf{b}$ has several solutions for one right-hand side $\mathbf{b}$, then $\mathbf{A}$ may be invertible.

(d) If the linear system $\mathbf{Ax} = \mathbf{b}$ has a unique solutions for one specific right-hand side $\mathbf{b}$, then $\mathbf{A}$ is invertible.

(e) If $\mathbf{A}_{nn}$ is square and the linear system $\mathbf{Ax} = \mathbf{b}$ has a unique solutions for one specific right-hand side $\mathbf{b}$, then $\mathbf{A}$ is invertible.

17. (a) Let $\mathbf{A}_{pq}$ and $\mathbf{B}_{rs}$ be two matrices for which both products $\mathbf{AB}$ and $\mathbf{BA}$ are defined. What does this imply about their sizes?

(b) Find two (different) 2×2 matrix pairs $\mathbf{A},\ \mathbf{B}$ and $\mathbf{C},\ \mathbf{D}$ where both

(1) $\mathbf{AB} = \mathbf{BA}$, and

(2) $\mathbf{CD} \neq \mathbf{DC}$.

18. (a) Show that $\begin{pmatrix} 1 & 2 & 2.5 \\ 0 & 2 & 1 \end{pmatrix} \begin{pmatrix} -2 & -2.5 \\ -1 & 0 \\ 2 & 1 \end{pmatrix} = \begin{pmatrix} 1 & 0 \\ 0 & 1 \end{pmatrix}$.

(b) Is $\begin{pmatrix} 1 & 2 & 2.5 \\ 0 & 2 & 1 \end{pmatrix}$ the inverse of $\begin{pmatrix} -2 & -2.5 \\ -1 & 0 \\ 2 & 1 \end{pmatrix}$? Why or why not?

(c) Compute $\begin{pmatrix} -2 & -2.5 \\ -1 & 0 \\ 2 & 1 \end{pmatrix} \begin{pmatrix} 1 & 2 & 2.5 \\ 0 & 2 & 1 \end{pmatrix}$.

(d) Repeat this problem with a 1×4 and a 4×1 matrix of your choice that is, find two vectors $\mathbf{x},\ \mathbf{y} \in \mathbb{R}^4$ with $\mathbf{x}^T\mathbf{y} = 1$, and compute $\mathbf{x}\mathbf{y}^T$.

19. Compute $\mathbf{A}^3$ for $\mathbf{A} = \begin{pmatrix} 0 & a & b \\ -a & 0 & c \\ -b & -c & 0 \end{pmatrix}$.

20. Compute the nth power of the matrices $\mathbf{A} = \begin{pmatrix} 0 & 1 \\ -1 & 0 \end{pmatrix}$, $\mathbf{B} = \begin{pmatrix} -1 & 1 \\ -1 & 0 \end{pmatrix}$, $\mathbf{C} = \begin{pmatrix} 0 & 1 \\ 1 & 0 \end{pmatrix}$, and $\mathbf{D} = \begin{pmatrix} 1 & 1 \\ 0 & 2 \end{pmatrix}$ for $n = 0, \pm 1, \pm 2, \dots$. (Here $\mathbf{A}^{-k} := (\mathbf{A}^{-1})^k$.)

21. Repeat Problems 18(a) and (b) from Section 5.1.P using matrix inverses.

22. Let $\mathbf{R} = \begin{pmatrix} r_{11} & & * \\ & \ddots & \\ 0 & & r_{nn} \end{pmatrix}$ be upper triangular.

(a) Show that $\mathbf{R}$ is invertible if and only if each diagonal entry $r_{ii} \neq 0$.

(b) Prove that if $\mathbf{R}$ is nonsingular, then

$$\mathbf{R}^{-1} = \begin{pmatrix} \frac{1}{r_{11}} & & * \\ & \ddots & \\ 0 & & \frac{1}{r_{nn}} \end{pmatrix}$$ is also upper triangular.

(c) What can be said about the inverses of lower triangular $n \times n$ matrices?

(d) Is the identity matrix $\mathbf{I}_n$ lower triangular, upper triangular, neither, or both?

23. Evaluate

(a) $\begin{pmatrix} a & d & g \\ b & e & h \\ c & f & i \end{pmatrix}^{-1} \begin{pmatrix} a \\ b \\ c \end{pmatrix}$, and

(b) $\begin{pmatrix} a & d & g \\ b & e & h \\ c & f & i \end{pmatrix}^{-1} \begin{pmatrix} 2d \\ 2e \\ 2f \end{pmatrix}$.

for arbitrary real constants $a, b, \dots, h, i$ that make the 3×3 matrix invertible.

24. If $\mathbf{A}_{nn}$ and $\mathbf{B}_{nn}$ both have rank n, show that their products $\mathbf{AB}$ and $\mathbf{BA}$ also have rank n.

25. (a) Find the inverse of the matrix

$$\mathbf{B} = \begin{pmatrix} 1 & 4 & 3 \\ -1 & -1 & 0 \\ 2 & 2 & 3 \end{pmatrix}.$$

(b) Solve the linear system $\mathbf{Bx} = \begin{pmatrix} 12 \\ -12 \\ 8 \end{pmatrix}$ with $\mathbf{B}$ from part (a).

26. If $\mathbf{A}_{nn}\mathbf{x} = \mathbf{0}$ holds for a nonzero vector $\mathbf{x} \in \mathbb{R}^n$, show that $\mathbf{A}$ cannot be inverted.

(*Hint*: The linear map $\mathbf{x} \to \mathbf{Ax}$ is not one-to-one.)

27. Repeat Problem 19 of Section 1.1.P in light of matrix inverses: Given a linear transformation T, where $T\begin{pmatrix} 4 \\ 1 \end{pmatrix} = \begin{pmatrix} -3 \\ 5 \end{pmatrix}$ and $T\begin{pmatrix} 5 \\ -1 \end{pmatrix} = \begin{pmatrix} -3 \\ 1 \end{pmatrix}$, the standard matrix $\mathbf{A}_\mathcal{E}$ for T satisfies $\mathbf{A}_\mathcal{E}\begin{pmatrix} 4 & 5 \\ 1 & -1 \end{pmatrix} = \begin{pmatrix} -3 & -3 \\ 5 & 1 \end{pmatrix}$. Solve this matrix equation for $\mathbf{A}_\mathcal{E}$. (Be careful: matrix products do not generally commute.)

28. Solve the matrix equation

$$\begin{pmatrix} 3 & 2 \\ 1 & 0 \end{pmatrix} \mathbf{X}_{2\times 3} \begin{pmatrix} 1 & -1 & 1 \\ 2 & 1 & 0 \\ 0 & 0 & 2 \end{pmatrix} = \begin{pmatrix} 1 & 2 & 1 \\ 4 & -1 & 1 \end{pmatrix}$$

for $\mathbf{X}$.

29. Solve the matrix equation $\mathbf{X} \cdot \mathbf{A} = \mathbf{O}_{?\times ?}$ for $\mathbf{X}_{5\times 2}$ and $\mathbf{A}_{2\times 3} = \begin{pmatrix} 1 & 0 & -1 \\ 2 & 2 & 1 \end{pmatrix}$:

(a) What is the size of the zero matrix here?

(b) How many unknown values does $\mathbf{X} = (x_{ij})$ contain?

(c) Write out an explicit system of linear equations involving the unknown entries of $\mathbf{X}$. How many equations are there for the x_{ij}?

(d) Solve the linear system in part (c).

(e) Do the same to find a matrix $\mathbf{Y}_{?\times ?} \neq \mathbf{O}$ with $\mathbf{A}_{2,3}\mathbf{Y} = \mathbf{O}_{?\times 4}$.

30. Show that if $\mathbf{A}_{mn}\mathbf{B}_{nm} = \mathbf{I}_m$, the identity matrix, then the columns of $\mathbf{B}$ are linearly independent.

(*Hint*: See Problem 35 of Section 5.1.P.)

31. Let $\{\mathbf{x}, \mathbf{y}\}$ be a basis for a subspace U of $\mathbb{R}^n$ and assume that $\mathbf{u} = a\mathbf{x} + b\mathbf{y}$ and $\mathbf{v} = c\mathbf{x} + d\mathbf{y} \in U$ for real coefficients a, b, c, d. Show that the two vectors $\mathbf{u}$ and $\mathbf{v}$ are a basis of U if $ad - bc \neq 0$. (*Hint*: Use Problem 6 and show linear independence first.)

32. (a) If $\mathbf{A}_{nn}$ contains the zero vector as one of its columns, can $\mathbf{A}$ be invertible? Why or why not?

(b) If $\mathbf{A}_{nn} = \begin{pmatrix} - & \mathbf{r}_1^T & - \\ & \vdots & \\ - & \mathbf{r}_n^T & - \end{pmatrix}$ contains two proportional rows (i.e., if $\mathbf{r}_j = \alpha \mathbf{r}_k$ for $j \neq k$ and $\alpha \in \mathbb{R}$), can $\mathbf{A}$ be invertible? Why or why not?

33. For $\mathbf{I}_n$, the $n \times n$ identity matrix, and any two matrices $\mathbf{A}_{mn}$ and $\mathbf{B}_{nk}$, show that both $\mathbf{IB} = \mathbf{B}$ and $\mathbf{AI} = \mathbf{A}$ hold.

34. If $\mathbf{A}$, $\mathbf{B}$, and $\mathbf{C}$ are matrices for which all triple products below can be formed, which of them will coincide?

A(**BC**), (**CB**)**A**, **B**(**AC**), (**BC**)**A**, (**AB**)**C**, (**AB**)**A**, **A**(**AB**), **C**(**AB**), **B**(**CA**), (**BA**)**C**, and **C**(**BA**).

35. Let $\mathbf{A} = \begin{pmatrix} 1 & 2 \\ 0 & 3 \end{pmatrix}$, $\mathbf{B} = \begin{pmatrix} -1 & 1 \\ 1 & 0 \end{pmatrix}$, $\mathbf{C} = \begin{pmatrix} 0 & 0 \\ 1 & 0 \end{pmatrix}$, and $\mathbf{E} = \begin{pmatrix} 0 & 1 \\ 1 & 0 \end{pmatrix}$.

(a) Compute the matrix products **AB**, **BA**, **AE**, **EA**, **BC**, **CB**, **BE**, **EB**, **AC**, **CA**, **CE**, and **EC**.

Which of the six matrix pairs that can be selected from **A**, **B**, **C**, and **E** commute?

(b) Form the triple matrix products **A**(**BE**), (**AB**)**E**, **C**(**AB**), and (**CA**)**B**. Which of these matrix triple products are associative?

36. (a) Evaluate the matrix product

$$\begin{pmatrix} 1 & -2 & 1 \\ 3 & -1 & -1 \\ -2 & 0 & 1 \end{pmatrix}\begin{pmatrix} 1 & -2 & -3 \\ 1 & -3 & -4 \\ 2 & -4 & -5 \end{pmatrix}.$$

(b) Find the inverse of $\begin{pmatrix} 1 & -2 & -3 \\ 1 & -3 & -4 \\ 2 & -4 & -5 \end{pmatrix}$, if it exists.

(c) Find the inverse of $\begin{pmatrix} 1 & 3 & -2 \\ -2 & -1 & 0 \\ 1 & -1 & 1 \end{pmatrix}$, if it exists.

37. (a) Show that every matrix product $\begin{pmatrix} 0 & 1 \\ 0 & 2 \end{pmatrix}$ a $\begin{pmatrix} x_{11} & x_{12} \\ x_{21} & x_{22} \end{pmatrix}$ has the form $\begin{pmatrix} x_{21} & x_{22} \\ 2x_{21} & 2x_{22} \end{pmatrix}$.

(b) Use part (a) to show that for $\mathbf{A} = \begin{pmatrix} 0 & 1 \\ 0 & 2 \end{pmatrix}$ there exists no 2×2 matrix **X** with $\mathbf{AX} = \mathbf{I}_2$.

(c) Can a matrix **Y** exist with $\mathbf{YA} = \mathbf{I}_2$ for A from part (b)?

38. (a) Find two nonsingular matrices **A** and $\mathbf{B} \in \mathbb{R}^{2,2}$ with $(\mathbf{AB})^{-1} = \mathbf{A}^{-1}\mathbf{B}^{-1}$, if possible.

(b) Find two nonsingular matrices **A** and $\mathbf{B} \in \mathbb{R}^{2,2}$ with $(\mathbf{AB})^{-1} \neq \mathbf{A}^{-1}\mathbf{B}^{-1}$, if possible.

39. (a) Find two invertible matrices **A** and $\mathbf{B} \in \mathbb{R}^{2,2}$ for which $\mathbf{A} + \mathbf{B}$ is also invertible.

(b) Find two matrices as in part (a), so that in addition $(\mathbf{A} + \mathbf{B})^{-1} \neq \mathbf{A}^{-1} + \mathbf{B}^{-1}$.

(c) Find two matrices as in part (a), so that $(\mathbf{A} + \mathbf{B})^{-1} = \mathbf{A}^{-1} + \mathbf{B}^{-1}$, if possible.

(d) Are there any 1×1 real matrices x and y, (i.e., real numbers) for which

$$(x + y)^{-1} = x^{-1} + y^{-1}?$$

40. If $\mathbf{A}^2 + 3\mathbf{A} - 2\mathbf{I}_n = \mathbf{O}_n$ holds for a square matrix **A**, show that $\mathbf{A}^{-1} = \frac{1}{2}(\mathbf{A} + 3\mathbf{I})$.

41. Let D be differentiation on $\mathcal{P}_3$, the space of all polynomials of degree less than or equal to three, and let $S := \{p \in \mathcal{P}_3 \mid p(0) = 0\}$.

(a) Show that $D : \mathcal{P}_3 \to \mathcal{P}_2$ is onto, but not one–to–one.

(b) Show that $D : S \to \mathcal{P}_3$ is one–to–one, but not onto.

42. Prove that $((\mathbf{A}^T)^{-1})^T = \mathbf{A}^{-1}$ holds for all invertible matrices **A**.

Teacher's Problem-Making Exercise

T 6. We outline how to find integer matrices whose inverses are likewise integer matrices. Such matrices are called **unimodular**.

If **A** and **B** are two integer $n \times n$ triangular matrices, one of which is upper triangular and the other lower triangular, then the two matrix products **AB** and **BA** are both integer $n \times n$ matrices. Moreover, if **A** and **B** are both nonsingular, then both matrix products **AB** and **BA** are nonsingular, and they will generally be **dense matrices**; that is, they will have nonzero entries in most positions. To insure integer inverses for both **AB** and **BA**, all that is needed is to ensure that the determinant of **AB** (and consequently that of **BA**) is ± 1. Hence, both **A** and **B** should be taken as triangular matrices (of opposite 'shape' for dense matrix products) all of whose diagonal entries are equal to ± 1.

EXAMPLE Let $\mathbf{A} = \begin{pmatrix} 1 & 7 & 2 \\ 0 & -1 & -2 \\ 0 & 0 & 1 \end{pmatrix}$ and $\mathbf{B} = \begin{pmatrix} 1 & 0 & 0 \\ 2 & 1 & 0 \\ -1 & 0 & 1 \end{pmatrix}$ be our matrix pair consisting of a unit upper and a unit lower triangular integer matrix.

Then $\mathbf{AB} = \begin{pmatrix} 13 & 7 & 2 \\ 0 & -1 & -2 \\ -1 & 0 & 1 \end{pmatrix}$ and $\mathbf{BA} = \begin{pmatrix} 1 & 7 & 2 \\ 2 & 13 & 2 \\ -1 & -7 & -1 \end{pmatrix}$ both have integer inverses that can be computed over the integers. ◀

6.2 Theory

Gauss Elimination Matrix Products, the Uniqueness of the Inverse, and Block Matrix Products

This section's aim is threefold:

(a) The Gaussian elimination process of a matrix $\mathbf{A}$ is recognized as a sequence of certain compositions of linear maps, or, in other words, as a succession of certain matrix multiplications performed on $\mathbf{A}$.

(b) We study the uniqueness of the matrix inverse and the set of all invertible matrices of a given size.

(c) We introduce block matrices whose entries are matrices themselves and study their multiplication, inverses, etc.

(a) Gaussian Elimination as a Matrix Product

Earlier chapters have shown that Gaussian elimination is a linear process involving certain modifications to the row vectors of a given matrix. For linear equations $\mathbf{Ax} = \mathbf{b}$ this process starts with the augmented matrix $(\mathbf{A} \mid \mathbf{b})$ and arrives in finitely many elimination steps at a REF $(\mathbf{R} \mid \tilde{\mathbf{b}})$. The linear system $\mathbf{Rx} = \tilde{\mathbf{b}}$ is equivalent to the original system of equations $\mathbf{Ax} = \mathbf{b}$ in the sense of solvability and having the same solution. In Gaussian elimination, the system matrix $\mathbf{A}$ in $(\mathbf{A} \mid \mathbf{b})$ is transformed step by step using one of three row operations (see Section 2.1):

(1) A multiple of one row is added to another,

(2) A row is multiplied by a nonzero constant, and

(3) Any two rows are interchanged.

We express each of these Gaussian row modifications as a matrix multiplication from the left on $\mathbf{A}$:

If $\mathbf{A}$ is given row–wise as $\mathbf{A}_{mn} = \begin{pmatrix} - & \mathbf{a}_1 & - \\ & \vdots & \\ - & \mathbf{a}_m & - \end{pmatrix}$, then:

(1) Multiplying $\mathbf{A}_{mn}$ from the left by

$$j \rightarrow \begin{pmatrix} 1 & & & 0 \\ & \ddots & & \\ & c_{jk} & \ddots & \\ 0 & & & 1 \end{pmatrix}_{mm} \quad \text{for any pair of indices } j \neq k$$

$$\uparrow$$
$$k$$

performs the addition of $c_{jk} \cdot \text{row}_k$ to row_j of $\mathbf{A}$. For example, for $j > k$,

$$\begin{pmatrix} 1 & & & & 0 \\ & \ddots & & & \\ & & \ddots & & \\ & & & \ddots & \\ & c_{jk} & & & \ddots \\ 0 & & & & 1 \end{pmatrix}_{mm} \mathbf{A}_{mn} = \begin{pmatrix} 1 & & & & 0 \\ & \ddots & & & \\ & & \ddots & & \\ & & & \ddots & \\ & c_{jk} & & & \ddots \\ 0 & & & & 1 \end{pmatrix} \cdot$$

$$\cdot \begin{pmatrix} - & \mathbf{a}_1 & - \\ & \vdots & \\ - & \mathbf{a}_k & - \\ & \vdots & \\ - & \mathbf{a}_j & - \\ & \vdots & \\ - & \mathbf{a}_m & - \end{pmatrix} = \begin{pmatrix} - & \mathbf{a}_1 & - \\ & \vdots & \\ - & \mathbf{a}_k & - \\ & \vdots & \\ - & c_{jk}\mathbf{a}_k + \mathbf{a}_j & - \\ & \vdots & \\ - & \mathbf{a}_m & - \end{pmatrix}_{mn} \begin{matrix} \\ \\ \\ \\ \leftarrow j \\ \\ \\ \end{matrix};$$

(2) Left-side multiplication of $\mathbf{A}_{mn}$ by

$$\begin{pmatrix} 1 & & & & & 0 \\ & \ddots & & & & \\ & & 1 & & & \\ & & & c & & \\ & & & & 1 & \\ & & & & & \ddots \\ 0 & & & & & 1 \end{pmatrix}_{mm} \leftarrow k$$

replaces the row $\mathbf{a}_k$ of $\mathbf{A}$ by $c \cdot \mathbf{a}_k$ in the product.

(3) Multiplying $\mathbf{A}_{mn}$ on the left by

$$\begin{pmatrix}
1 & & & & & & & & & 0 \\
& \ddots & & & & & & & & \\
& & 1 & & & & & & & \\
& & & 0 & & & 1 & & & \\
& & & & 1 & & & & & \\
& & & & & \ddots & & & & \\
& & & & & & 1 & & & \\
& & & 1 & & & & 0 & & \\
& & & & & & & & 1 & \\
& & & & & & & & & \ddots \\
0 & & & & & & & & & & 1
\end{pmatrix}_{mm}
\begin{matrix} \\ \\ \\ \leftarrow k \\ \\ \\ \\ \leftarrow j \\ \\ \\ \\ \end{matrix}$$

$$\begin{matrix} \uparrow & & \uparrow \\ k & & j \end{matrix}$$

for $j \neq k$ swaps the rows $\mathbf{a}_k$ and $\mathbf{a}_j$ in $\mathbf{A}_{mn}$.

The three left-sided multiplier matrices of Gaussian elimination have a special name.

DEFINITION 3 The square matrices

$$\mathbf{E}(c_{jk}) := \begin{pmatrix} 1 & & 0 \\ & \ddots & \\ c_{jk} & & \ddots \\ & & & 1 \end{pmatrix} \quad \text{for } j \neq k\,, \tag{6.3}$$

$$\mathbf{E}_k(c) := \begin{pmatrix} 1 & & & & & 0 \\ & \ddots & & & & \\ & & 1 & & & \\ & & & c & & \\ & & & & 1 & \\ & & & & & \ddots \\ 0 & & & & & & 1 \end{pmatrix} \quad \text{and} \tag{6.4}$$

$$\mathbf{E}_{jk} := \begin{pmatrix}
1 & & & & & & & & & 0 \\
& \ddots & & & & & & & & \\
& & 1 & & & & & & & \\
& & & 0 & & & 1 & & & \\
& & & & 1 & & & & & \\
& & & & & \ddots & & & & \\
& & & & & & 1 & & & \\
& & & 1 & & & & 0 & & \\
& & & & & & & & 1 & \\
& & & & & & & & & \ddots \\
0 & & & & & & & & & & 1
\end{pmatrix} \quad \text{for } j \neq k \tag{6.5}$$

are called **elementary elimination matrices**, or **elimination matrices** for short.

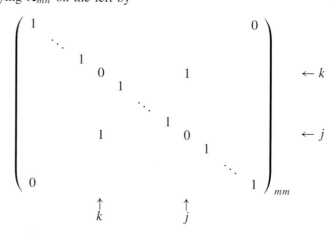

For a matrix $\mathbf{A}_{mn}$, a REF $\mathbf{R}$ can be obtained by multiplying $\mathbf{A}$ on the left by a finite sequence of elimination matrices $\mathbf{E}(c_{jk})$ with $j > k$, $\mathbf{E}_k(c)$, and $\mathbf{E}_{jk}$. In this process, the elimination matrices of type $\mathbf{E}(c_{jk})$ are formed for indices $j > k$, since they are used to zero a nonzero entry in position (j, k) below the pivot entry $(j > k)$ of column k. The RREF $\widehat{\mathbf{R}}$ of $\mathbf{R}$ (and of $\mathbf{A}$) is then obtained by applying further elimination matrices $\mathbf{E}(c_{jk})$ for $j < k$ that zero the entries above each pivot and by using elimination matrices of type $\mathbf{E}_k(c)$ for scaling the pivots to be equal to one. For a nonsingular square matrix $\mathbf{A}$ and its RREF $\mathbf{I}$, the row reduction process thus becomes a sequence of left multiplications with elimination matrices.

EXAMPLE 4 Let $\mathbf{A} = \begin{pmatrix} 2 & 1 & 0 \\ 4 & 0 & 1 \\ 6 & 2 & 1 \end{pmatrix}$. Then

$$\begin{pmatrix} 1 & 0 & 0 \\ -2 & 1 & 0 \\ 0 & 0 & 1 \end{pmatrix} \mathbf{A} = \begin{pmatrix} 2 & 1 & 0 \\ 0 & -2 & 1 \\ 6 & 2 & 1 \end{pmatrix},$$

$$\begin{pmatrix} 1 & 0 & 0 \\ 0 & 1 & 0 \\ -3 & 0 & 1 \end{pmatrix} \begin{pmatrix} 2 & 1 & 0 \\ 0 & -2 & 1 \\ 6 & 2 & 1 \end{pmatrix} = \begin{pmatrix} 2 & 1 & 0 \\ 0 & -2 & 1 \\ 0 & -1 & 1 \end{pmatrix},$$

and

$$\begin{pmatrix} 1 & 0 & 0 \\ 0 & 1 & 0 \\ 0 & -\frac{1}{2} & 1 \end{pmatrix} \begin{pmatrix} 2 & 1 & 0 \\ 0 & -2 & 1 \\ 0 & -1 & 1 \end{pmatrix} = \begin{pmatrix} 2 & 1 & 0 \\ 0 & -2 & 1 \\ 0 & 0 & \frac{1}{2} \end{pmatrix} = \mathbf{R}$$

is in row echelon form. The matrix product

$$\begin{pmatrix} 1 & 0 & 0 \\ 0 & 1 & 0 \\ 0 & -\frac{1}{2} & 1 \end{pmatrix} \begin{pmatrix} 1 & 0 & 0 \\ 0 & 1 & 0 \\ -3 & 0 & 1 \end{pmatrix} \begin{pmatrix} 1 & 0 & 0 \\ -2 & 1 & 0 \\ 0 & 0 & 1 \end{pmatrix} \mathbf{A} = \mathbf{R}$$

describes the row echelon form reduction of $\mathbf{A}$ via the three elementary elimination matrices $\mathbf{E}_1 = \begin{pmatrix} 1 & 0 & 0 \\ -2 & 1 & 0 \\ 0 & 0 & 1 \end{pmatrix}$, $\mathbf{E}_2 = \begin{pmatrix} 1 & 0 & 0 \\ 0 & 1 & 0 \\ -3 & 0 & 1 \end{pmatrix}$, and $\mathbf{E}_3 = \begin{pmatrix} 1 & 0 & 0 \\ 0 & 1 & 0 \\ 0 & -\frac{1}{2} & 1 \end{pmatrix}$.

Since $\mathbf{R}$ has three pivots, $\mathbf{A}$ is row equivalent to $\mathbf{I}_3$ by applying five further elimination matrix products in succession to $\mathbf{R}$ from the left as follows: $\begin{pmatrix} 1 & 0 & 0 \\ 0 & 1 & 0 \\ 0 & 0 & \frac{1}{2} \end{pmatrix} \mathbf{R} = \begin{pmatrix} 2 & 1 & 0 \\ 0 & -2 & 1 \\ 0 & 0 & 1 \end{pmatrix}$. Multiplying the result by the elimination matrices $\mathbf{E}_5 = \begin{pmatrix} 1 & 0 & 0 \\ 0 & 1 & -1 \\ 0 & 0 & 1 \end{pmatrix}$, $\mathbf{E}_6 = \begin{pmatrix} 1 & 0 & 0 \\ 0 & -\frac{1}{2} & 0 \\ 0 & 0 & 1 \end{pmatrix}$, $\mathbf{E}_7 = \begin{pmatrix} 1 & -1 & 0 \\ 0 & 1 & 0 \\ 0 & 0 & 1 \end{pmatrix}$, and $\mathbf{E}_8 = \begin{pmatrix} \frac{1}{2} & 0 & 0 \\ 0 & 1 & 0 \\ 0 & 0 & 1 \end{pmatrix}$ in sequence on the left reduces $\mathbf{A}$ to the identity matrix $\mathbf{I}_3$. This should be verified by the students. Thus, $\mathbf{E}_8 \cdots \mathbf{E}_1 \mathbf{A} = \mathbf{I}_3$. ◀

(b) Uniqueness of the Matrix Inverse

This subsection uses an alternative approach to matrix inverses.

For a given $n \times n$ matrix $\mathbf{A}$, define a vector mapping by assigning the image $\mathbf{Ax} \in \mathbb{R}^n$ to each vector $\mathbf{x} \in \mathbb{R}^n$. It is natural to call such a mapping $\mathbf{x} \mapsto \mathbf{Ax}$ **invertible** if its action can be reversed. In that case, there is an $n \times n$ matrix $\mathbf{B}$ such that

$$\mathbf{BAx} = \mathbf{x} = \mathbf{Ix}.$$

Here $\mathbf{B}$ retrieves $\mathbf{x}$ from its image $\mathbf{Ax}$ for all $\mathbf{x} \in \mathbb{R}^n$.

In this context, we define $\mathbf{B}$ to be an **inverse matrix** of $\mathbf{A}$ if $\mathbf{BA} = \mathbf{I}$, the identity matrix, and $\mathbf{A}$ to be **invertible**, or **nonsingular**, if such a matrix $\mathbf{B}$ exists for $\mathbf{A}$.

This is somewhat converse to starting with linear transformations when developing the earlier Sections 1.1 and 6.1.

This approach to matrix inversion is solely based on the notions of a matrix and of matrix products. What needs to be settled first is

- whether such an inverse matrix $\mathbf{B}$ is "the" inverse (i.e. uniqueness),
- whether this inverse $\mathbf{B}$ acting on the left of $\mathbf{A}_{nn}$ is also the right inverse of $\mathbf{A}$ (i.e., whether $\mathbf{BA} = \mathbf{I}_n$ implies $\mathbf{AB} = \mathbf{I}_n$),
- and how the invertible $n \times n$ matrices relate to each other.

All of this shall be done rigorously in the following four theorems:

Using Gaussian elimination matrices of the form of Eqs. (6.3), (6.4), and (6.5), we prove the following technical result with the help of a Lemma.

Theorem 6.6 Every invertible matrix $\mathbf{A}_{nn}$ is the product of elementary elimination matrices. ◀

Lemma 2 Each elimination matrix of Definition 3 is invertible. The inverse of an elimination matrix is again an elimination matrix of the same type.

In particular,

$$(\mathbf{E}(c_{jk}))^{-1} = \mathbf{E}(-c_{jk}),$$

for $j \neq k$; that is, adding c_{jk} times row k to row j of a matrix and then subtracting c_{jk} times row k from the altered j^{th} row, gives the original matrix back. Similarly,

$$(\mathbf{E}_k(c))^{-1} = \mathbf{E}_k(1/c) \quad \text{for all } c \neq 0;$$

i.e., scaling row k of a matrix by $c \neq 0$ and then scaling it back by the reciprocal $1/c$ again gives the original matrix. And lastly

$$(\mathbf{E}_{jk})^{-1} = \mathbf{E}_{jk} \quad \text{for all } j \text{ and } k,$$

that is, a row swap of rows j and k is reversed by a second identical swap.

In each case, $\mathbf{E}_{\dots}^{-1}\mathbf{E}_{\dots} = \mathbf{I}_n = \mathbf{E}_{\dots}\mathbf{E}_{\dots}^{-1}$; that is, the elimination matrix inverses are inverses both from the left and the right. ◀

The last commutativity assertion of the Lemma should be proved by the students.

Proof (Theorem 6.6)
As seen earlier, reducing an invertible matrix $\mathbf{A} = \mathbf{A}_{nn}$ to its RREF $\mathbf{I}_n$ can be achieved by multiplying $\mathbf{A}$ from the left by a finite number M of elimination matrices. Thus, if $\mathbf{A}$ is invertible, then

$$\mathbf{E}_M \cdots \mathbf{E}_1\mathbf{A} = \mathbf{I} \quad \text{for some integer } M.$$

Each of the elimination matrices $\mathbf{E}_j$ is invertible by Lemma 2. Thus, we can "peel off" the factors $\mathbf{E}_i$ in the last identity, being conscious of the order. We multiply by the inverse matrices $\mathbf{E}_i^{-1}$ on the left, starting with $\mathbf{E}_M$, $\mathbf{E}_{M-1}$, etc., backwards to obtain

$$\mathbf{E}_1^{-1} \cdots \mathbf{E}_M^{-1}\mathbf{E}_M \cdots \mathbf{E}_1\mathbf{A} = \mathbf{E}_1^{-1} \cdots \mathbf{E}_M^{-1}\mathbf{I},$$
$$\text{or } \mathbf{A} = \mathbf{E}_1^{-1} \cdots \mathbf{E}_M^{-1}.$$

This equation expresses $\mathbf{A}$ as a product of inverses of elimination matrices. By Lemma 2, these are elimination matrices in their own right. ■

Theorem 6.7 For an invertible matrix $\mathbf{A}_{nn}$, the left matrix inverse $\mathbf{B}$ with $\mathbf{BA} = \mathbf{I}_n$ is unique. ◀

Proof If two matrices $\mathbf{B}$ and $\mathbf{C}$ are both inverses of $\mathbf{A}$ (i.e., if $\mathbf{BA} = \mathbf{I}$ and $\mathbf{CA} = \mathbf{I}$), then

$$(\mathbf{B} - \mathbf{C})\mathbf{A} = \mathbf{O}_n \,. \tag{6.6}$$

From Theorem 6.6, we know that $\mathbf{A} = \mathbf{E}_N \cdots \mathbf{E}_1$ for certain elimination matrices $\mathbf{E}_i$. Let $\mathbf{E}_i^{-1}$ be the left (and right) inverse of $\mathbf{E}_i$ for each i according to Lemma 2, and set $\mathbf{D} = \mathbf{E}_1^{-1} \cdots \mathbf{E}_N^{-1}$. Then

$$\mathbf{AD} = \mathbf{E}_N \cdots \mathbf{E}_1\mathbf{E}_1^{-1} \cdots \mathbf{E}_N^{-1} = \mathbf{I}_n.$$

Thus, from Eq. (6.6),

$$\mathbf{B} - \mathbf{C} = (\mathbf{B} - \mathbf{C})\mathbf{I} = (\mathbf{B} - \mathbf{C})\mathbf{AD} = \mathbf{OD} = \mathbf{O}_n,$$

the $n \times n$ zero matrix, proving that $\mathbf{B} = \mathbf{C}$. ■

An alternative proof of Theorem 6.7 based on the concept of linear transformations is obtained as follows: If $\mathbf{BA} = \mathbf{I}$ and $\mathbf{CA} = \mathbf{I}$, then

$$(\mathbf{B} - \mathbf{C})\mathbf{Ax} = \mathbf{BAx} - \mathbf{CAx} = \mathbf{Ix} - \mathbf{Ix} = \mathbf{0} = \mathbf{Ox}$$

for all $\mathbf{x} \in \mathbb{R}^n$. Since the invertible matrix $\mathbf{A}$ is nonsingular, we have $\text{im}(\mathbf{A}) = \{\mathbf{Ax} \mid \mathbf{x} \in \mathbb{R}^n\} = \mathbb{R}^n$ according to Sections 5.1 and 6.1. Hence $\ker(\mathbf{B} - \mathbf{C}) = \text{im}(\mathbf{A}) = \mathbb{R}^n$ (i.e., $\mathbf{B} - \mathbf{C} = \mathbf{O}_n$, or $\mathbf{B} = \mathbf{C}$).

We note that for a nonsingular matrix $\mathbf{A}_{nn}$, the Gaussian elimination matrix equation $\mathbf{E}_M \cdots \mathbf{E}_1\mathbf{A} = \mathbf{I}$ and Theorem 6.7 together imply that the product $\mathbf{E}_M \cdots \mathbf{E}_1$ is the unique left inverse of $\mathbf{A}$. This observation reinforces the computational scheme for obtaining $\mathbf{A}^{-1}$ from $(\mathbf{A} \mid \mathbf{I})$ in Section 6.1. Note that the individual factors $\mathbf{E}_j$ and the number of them used in representing a given nonsingular matrix $\mathbf{A}$ as a product of elimination matrices can vary. Their product, however, is the unique matrix inverse $\mathbf{A}^{-1}$ of $\mathbf{A}$. (That is, it is always the same, no matter in which order and fashion the Gauss reduction of $\mathbf{A}$ to the identity matrix $\mathbf{I}_n$ has been achieved).

EXAMPLE 5 The nonsingular matrix $\mathbf{A} = \begin{pmatrix} 2 & 0 \\ 1 & 1 \end{pmatrix}$ can be row reduced to $\mathbf{I}_2$ using elimination matrices from the left in several ways.

For example, by scaling its first row by 1/2 to become (1 0) and subtracting the new first row from the second row of **A** transforms **A** to **I**. Expressed via Gaussian elimination matrices, this becomes

$$\begin{pmatrix} 1 & 0 \\ -1 & 1 \end{pmatrix}\begin{pmatrix} \frac{1}{2} & 0 \\ 0 & 1 \end{pmatrix}\begin{pmatrix} 2 & 0 \\ 1 & 1 \end{pmatrix} = \begin{pmatrix} 1 & 0 \\ 0 & 1 \end{pmatrix} = \mathbf{I}_2.$$

Alternatively, we can swap the two rows of **A** first, eliminate the (2,1) entry, scale the second row by – 1/2, and finally eliminate the (1,2) entry to obtain

$$\begin{pmatrix} 1 & -1 \\ 0 & 1 \end{pmatrix}\begin{pmatrix} 1 & 0 \\ 0 & -\frac{1}{2} \end{pmatrix}\begin{pmatrix} 1 & 0 \\ -2 & 1 \end{pmatrix}\begin{pmatrix} 0 & 1 \\ 1 & 0 \end{pmatrix}\begin{pmatrix} 2 & 0 \\ 1 & 1 \end{pmatrix} = \mathbf{I}_2.$$

Thus, two different elementary matrix products express $\mathbf{A}^{-1} =$

$$\begin{pmatrix} 1 & -1 \\ 0 & 1 \end{pmatrix}\begin{pmatrix} 1 & 0 \\ 0 & -\frac{1}{2} \end{pmatrix}\begin{pmatrix} 1 & 0 \\ -2 & 1 \end{pmatrix}\begin{pmatrix} 0 & 1 \\ 1 & 0 \end{pmatrix} = \begin{pmatrix} 1 & 0 \\ -1 & 1 \end{pmatrix}\begin{pmatrix} \frac{1}{2} & 0 \\ 0 & 1 \end{pmatrix}$$
$$= \begin{pmatrix} \frac{1}{2} & 0 \\ -\frac{1}{2} & 1 \end{pmatrix}.$$

(All of the matrix identities above should be checked by the students.) ◀

Finally, we have.

Theorem 6.8 Let $\mathbf{A}_{nn}$ be invertible. Then the unique left inverse **B** of **A** is also the unique right inverse of **A**. Namely, if $\mathbf{A}^{-1}\mathbf{A} = \mathbf{I}$, then $\mathbf{A}\mathbf{A}^{-1} = \mathbf{I}$ as well and vice versa. ◀

Proof As seen before, for an invertible matrix **A**, there are elimination matrices $\mathbf{E}_i$ with $\mathbf{E}_M \cdots \mathbf{E}_1\mathbf{A} = \mathbf{I}_n$. Assume that **B** is the unique left inverse of **A** by Theorem 6.7. Then $\mathbf{B} = \mathbf{E}_M \cdots \mathbf{E}_1$, and $\mathbf{E}_M \cdots \mathbf{E}_1\mathbf{A} = \mathbf{BA} = \mathbf{I}$. Consequently, $\mathbf{A} = \mathbf{E}_1^{-1} \cdots \mathbf{E}_M^{-1}$. Now look at the reverse-order product **AB**. We have

$$\mathbf{AB} = \mathbf{A}\mathbf{E}_M \cdots \mathbf{E}_1 = \mathbf{E}_1^{-1} \cdots \mathbf{E}_M^{-1}\mathbf{E}_M \cdots \mathbf{E}_1 = \mathbf{I}_n$$

as well, making $\mathbf{B} = \mathbf{E}_M \cdots \mathbf{E}_1$ also a right inverse.

Finally, if **C** is any right inverse of **A** with $\mathbf{AC} = \mathbf{I}$, then

$$\mathbf{B} = \mathbf{BI} = \mathbf{B}(\mathbf{AC}) = (\mathbf{BA})\mathbf{C} = \mathbf{IC} = \mathbf{C},$$

since $\mathbf{BA} = \mathbf{I}$. Hence an invertible matrix **A** admits precisely one inverse **B** that acts equally as a left and a right inverse, or,

$$\mathbf{BA} = \mathbf{I} = \mathbf{AB} \quad \text{if } \mathbf{B} = \mathbf{A}^{-1}.$$ ■

Theorem 6.8 is a theorem about matrix **commutativity**. It states that an invertible matrix **A** and its inverse $\mathbf{A}^{-1}$ always commute.

We are ready to describe some fundamental algebraic properties of the set of invertible $n \times n$ matrices.

Corollary The invertible $n \times n$ matrices $\mathbf{A}$ form a multiplicative group. Its unit element is the identity matrix $\mathbf{I}_n$.

Here we use the standard definition of a group.

DEFINITION 4 A nonempty set G with a **binary operation** $* : G \times G \to G$ is called a **group** $(G, *)$ if

(1) The binary operation $* : G \times G \to G$ is defined for all pairs of elements f and $g \in G$. In other words, the result $f * g$ of the binary operation again belongs to G for all $f,\ g \in G$.

(2) G contains a **unit element** e for which $e * g = g$ and $g * e = g$ for all $g \in G$.

(3) For every element $g \in G$ there is an **inverse element** $g \in G$ with $g * g = e = g * g$.

(4) The binary operation $* : G \times G \to G$ is **associative**; that is, for all elements $f,\ g$, and h in G, we have $(f * g) * h = f * (g * h) \in G$.

Note that the $n \times n$ matrices are associative under matrix multiplication, due to Corollary 1 of Section 6.1.

EXAMPLE 6

(a) The real numbers form a group under addition, with zero as the unit element.

(b) The nonzero real numbers form a different group under multiplication; their unit element is the number one.

(c) The nonsingular upper and the nonsingular lower triangular matrices each form a matrix group, as will be shown towards the end of this section.

(d) The integers $\mathbb{N} = \{\dots, -2, -1, 0, 1, 2, \dots\}$ form a group with addition as the binary operation. However, they do not form a group under multiplication, since the number zero has no multiplicative inverse.

Starting from the intuitive notion of an inverse mapping for the linear mapping that is induced by a nonsingular matrix, we have been able to classify the algebraic structure of the set of all invertible matrices of a fixed square size as being a group. Groups of matrices, functions, and operators play an important role in higher level mathematics.

Theorem 6.6 and Lemma 2 have been useful in understanding invertible matrices. Further, they allow us to express every $m \times n$ matrix $\mathbf{A}_{mn}$ as a product of three matrix factors.

Theorem 6.9 For every matrix $\mathbf{A}_{mn}$, there exist a permutation matrix $\mathbf{P}_{mm}$, a lower triangular matrix $\mathbf{L}_{mm}$, and an upper triangular matrix $\mathbf{R}_{mn}$, so that

$$\mathbf{A} = \mathbf{PLR}.$$

Specifically, the matrix $\mathbf{R}$ is a REF of $\mathbf{A}$.

A **permutation matrix P** is a square $n \times n$ matrix that contains $n^2 - n$ zero entries and n entries of one. Moreover, the nonzero entries of $\mathbf{P}$ are arranged so that each row and column contains precisely one entry of one. For example, the matrices $\begin{pmatrix} 0 & 1 \\ 1 & 0 \end{pmatrix}$, $\begin{pmatrix} 0 & 0 & 1 \\ 1 & 0 & 0 \\ 0 & 1 & 0 \end{pmatrix}$, and $\begin{pmatrix} 1 & 0 & 0 & 0 \\ 0 & 0 & 1 & 0 \\ 0 & 1 & 0 & 0 \\ 0 & 0 & 0 & 1 \end{pmatrix}$ are permutation matrices. However, the matrices $\begin{pmatrix} 1 & 0 & 0 \\ 0 & 1 & 0 \\ 1 & 0 & 0 \end{pmatrix}$, $\begin{pmatrix} 0 & 1 \\ -1 & 0 \end{pmatrix}$, and $\begin{pmatrix} 1 & 1 \\ 0 & 1 \end{pmatrix}$ are not. Multiplying a matrix $\mathbf{A}$ by a compatibly sized permutation matrix $\mathbf{P}$ on the left rearranges the rows of $\mathbf{A}$ in $\mathbf{PA}$. Likewise, the reverse-order matrix product $\mathbf{AP}$ changes the order of the columns of $\mathbf{A}$. Note that the transpose $\mathbf{P}^T$ of a permutation matrix is also a permutation matrix. It satisfies $\mathbf{P}^T\mathbf{P} = \mathbf{I}$ since $\mathbf{P}^T$ expresses the inverse permutations that are expressed by $\mathbf{P}$ itself. Thus, $\mathbf{P}^T = \mathbf{P}^{-1}$ for a permutation matrix.

The multiplicative decomposition of $\mathbf{A}$ as $\mathbf{PLR}$, or of $\mathbf{P}^T A$ as $\mathbf{LR}$ in Theorem 6.9, is called the **LR factorization** of the matrix $\mathbf{A}$. Note the symbolism here: $\mathbf{P}$ is a *permutation matrix*, $\mathbf{L}$ stands for a *lower triangular matrix*, while $\mathbf{R}$ stands for the *right factor*. $\mathbf{R}$ is a matrix in *row echelon form*,—it is in upper staircase form with special restrictions. (See the row and column conditions for row echelon forms in Section 2.1.)

To understand Theorem 6.9, consider a sequence of elementary Gaussian elimination matrices $\mathbf{E}_1, \ldots, \mathbf{E}_j$ that eliminates the first column of a matrix $\mathbf{A}_{mn}$:

$$\mathbf{E}_j \cdots \mathbf{E}_1 A = \begin{pmatrix} * & * & \ldots & * \\ 0 & * & \ldots & * \\ \vdots & \vdots & & \vdots \\ 0 & * & \ldots & * \end{pmatrix}. \tag{6.7}$$

The first elimination matrix $\mathbf{E}_1$ may be a swap matrix of the type of Eq. (6.5). When we eliminate the second column of the right hand matrix in (6.7) using further elimination matrices $\mathbf{E}_{j+1}, \ldots, \mathbf{E}_k$, we obtain

$$\mathbf{E}_k \cdots \mathbf{E}_{j+1}(\mathbf{E}_j \cdots \mathbf{E}_1\mathbf{A}) = \mathbf{E}_k \cdots \mathbf{E}_{j+1} \begin{pmatrix} * & * & \ldots & * \\ 0 & * & \ldots & * \\ \vdots & \vdots & & \vdots \\ 0 & * & \ldots & * \end{pmatrix}$$

$$= \begin{pmatrix} * & * & * & \ldots & * \\ 0 & * & * & \ldots & * \\ \vdots & 0 & * & \ldots & * \\ \vdots & \vdots & \vdots & & \vdots \\ 0 & 0 & * & \ldots & * \end{pmatrix}.$$

If in this process, $\mathbf{E}_{j+1}$ is a swap matrix of the form of Eq. (6.5), then its row swap operation could have been executed initially on $\mathbf{A}$, and the subsequent swap during the row reduction of column 2 would have been avoided. In this manner, all swaps that are encountered in the row reduction of **A** can be performed initially on **A** via one initial permutation matrix **P**. However, note that if we start out with **PA** instead of with **A**, then the elimination matrices that are needed to zero out the first two columns below their diagonal entries in (6.7) would be different from the elimination matrices in the sequence $\mathbf{E}_1, \ldots, \mathbf{E}_j, \mathbf{E}_{j+1}, \ldots, \mathbf{E}_k$.

Summarizing, we can always pull out the swap matrices of a complete row echelon form reduction of $\mathbf{A}_{mn}$ to the front. If we modify the original matrix **A** accordingly by the initial **permutation matrix**

$$\mathbf{P} = \prod \text{swap matrices of the complete row reduction process,}$$

then the matrix **PA** can be row reduced without any swaps. That is,

$$\widetilde{\mathbf{E}_M} \cdots \widetilde{\mathbf{E}_1}\mathbf{PA} = \mathbf{R}.$$

Here $\mathbf{R}_{mn}$ is in row echelon form, and each $\widetilde{\mathbf{E}_\ell}$ is a Gaussian elimination matrix of type (6.3) or (6.4). Thus, $\mathbf{PA} = \mathbf{LR}$ for the lower triangular matrix $\mathbf{L} := \widetilde{\mathbf{E}_1}^{-1} \cdots \widetilde{\mathbf{E}_M}^{-1}$. Consequently, $\mathbf{A} = \mathbf{P}^T\mathbf{LR}$ for the permutation matrix $\mathbf{P}^T$. This establishes Theorem 6.9. Note that **L** is solely the product of lower triangular elimination matrices of types (6.3) and (6.4), and thus it is lower triangular itself.

We study the **LR** factorization again in Section 10.2, when we investigate the orthogonal QR factorization of matrices $\mathbf{A}_{mn}$. Matrix factorizations are very important.

We mention one further result on arbitrary row manipulations of a multi-augmented matrix $(\mathbf{A}_{mn} \mid \mathbf{I}_n)$.

Corollary 3 If a matrix $(\mathbf{A}_{mn} \mid \mathbf{I}_n)$ is row reduced to $(\mathbf{S}_{mn} \mid \mathbf{T}_{mm})$ by using $m \times m$ Gaussian elimination matrices $\mathbf{E}_1, \ldots, \mathbf{E}_k$ in sequence, then

(a) **T** is nonsingular and equal to the product $\mathbf{E}_k \cdots \mathbf{E}_1$ of the elimination matrices that were used in the reduction process, and
(b) $\mathbf{S}_{mn} = \mathbf{T}_{mm}\mathbf{A}_{mn}$. ◀

This follows readily, since **A** is transformed to $\mathbf{S} = \mathbf{E}_k \cdots \mathbf{E}_1\mathbf{A}$, while simultaneously $\mathbf{T} = \mathbf{E}_k \cdots \mathbf{E}_1\mathbf{I} = \mathbf{E}_k \cdots \mathbf{E}_1$.

Consequently, if for $\mathbf{A}_{nn}$, we can achieve $\mathbf{S} = \mathbf{I}$ on the left via row reduction, then $\mathbf{T} = \mathbf{E}_k \cdots \mathbf{E}_1 = \mathbf{A}^{-1}$ appears in the right half of the row-reduced matrix of the corollary.

(c) Matrix Products and Block Matrix Products

A matrix product **AB** can be viewed in several ways. By writing $\mathbf{A} = \begin{pmatrix} | & & | \\ \mathbf{a}_1 & \ldots & \mathbf{a}_n \\ | & & | \end{pmatrix}_{mn}$ columnwise, the matrix product **AB** describes a matrix that contains certain linear combinations of the columns $\mathbf{a}_i$ of **A**. Namely,

$$\mathbf{AB} = \begin{pmatrix} | & & | \\ \mathbf{a}_1 & \cdots & \mathbf{a}_n \\ | & & | \end{pmatrix} \begin{pmatrix} b_{11} & \cdots & b_{1k} \\ \vdots & & \vdots \\ b_{n_1} & \cdots & b_{nk} \end{pmatrix}$$

$$= \begin{pmatrix} | & & | \\ \sum_{i=1}^{n} b_{i1} \begin{pmatrix} | \\ \mathbf{a}_i \\ | \end{pmatrix} & \cdots & \sum_{i=1}^{n} b_{ik} \begin{pmatrix} | \\ \mathbf{a}_i \\ | \end{pmatrix} \\ | & & | \end{pmatrix}.$$

In other words, each column of the matrix product **AB** lies in the span of the columns of its first factor **A**, or col{**AB**} ∈ span{col(**A**)}.

Alternatively, by writing $\mathbf{B} = \begin{pmatrix} - & \mathbf{b}_1 & - \\ & \vdots & \\ - & \mathbf{b}_n & - \end{pmatrix}_{nk}$ row–wise, we find that the matrix product **AB** contains certain linear combinations of the rows $\mathbf{b}_j$ of **B**, namely,

$$\mathbf{AB} = \begin{pmatrix} a_{11} & \cdots & a_{1n} \\ \vdots & & \vdots \\ a_{m1} & \cdots & a_{mn} \end{pmatrix} \begin{pmatrix} - & \mathbf{b}_1 & - \\ & \vdots & \\ - & \mathbf{b}_n & - \end{pmatrix} = \begin{pmatrix} \sum_{j=1}^{n} a_{1j}(-\mathbf{b}_j-) \\ \vdots \\ \sum_{j=1}^{n} a_{mj}(-\mathbf{b}_j-) \end{pmatrix}.$$

In other words, each row of the matrix product **AB** lies in the span of the rows of its second factor **B**, or row{**AB**} ∈ span{row(**B**)}.

Thus, matrix multiplication of a given matrix **A** by a matrix **B** on the right leads to the matrix **AB** whose column vectors are linear combinations of **A**'s original columns. Analogously, the same matrix product **AB** describes a matrix whose rows are linear combinations of **B**'s original rows.

The foregoing exposition gives us a **dual view** of matrix multiplication **A** times **B**. It depends on which object we focus on: either on the columns of the first matrix factor **A** or on the rows of the second one **B**.

Duality occurs frequently in mathematics. Whenever it occurs, it is an invitation to understand related objects and their meaning from multiple viewpoints.

As an application, we study the product of two upper triangular square matrices $\mathbf{R}_{nn}$ and $\mathbf{S}_{nn}$:

$$\mathbf{RS} = \begin{pmatrix} r_{11} & \cdots & r_{1n} \\ & \ddots & \vdots \\ 0 & & r_{nn} \end{pmatrix} \begin{pmatrix} s_{11} & \cdots & s_{1n} \\ & \ddots & \vdots \\ 0 & & s_{nn} \end{pmatrix} = \begin{pmatrix} r_{11}s_{11} & & * \\ & \ddots & \\ 0 & & r_{nn}s_{nn} \end{pmatrix}.$$

This product is upper triangular. If **R** is nonsingular, then on its diagonal there are n pivots r_{ii} with $r_{ii} \neq 0$. Hence, **R** is row equivalent to the identity matrix $\mathbf{I}_n$, best seen by using upper triangular Gauss elimination matrices of the two types

$\mathbf{E}(c_{jk})$ for $j < k$ and $\mathbf{E}_k(c)$. According to Section 6.1, the product of these upper triangular elimination matrices is equal to $\mathbf{R}^{-1}$. Combining the two results, we see that $\mathbf{R}^{-1}$ has upper triangular form as well.

Hence, we have proved the following:

Proposition: The upper triangular invertible matrices $\mathbf{R} \in \mathbb{R}^{n,n}$ form a matrix group. The same holds for the invertible lower triangular matrices $\mathbf{L}_{nn}$. Their unit element is the identity matrix $\mathbf{I}_n$. ◀

Matrices can be partitioned into **block matrices**. For example,

$$\mathbf{U}_{mn} = \left(\begin{array}{ccc|ccc} u_{11} & \dots & u_{1s} & u_{1,s+1} & \dots & u_{1n} \\ \vdots & & \vdots & \vdots & & \vdots \\ u_{r1} & \dots & u_{rs} & u_{r,s+1} & \dots & u_{rn} \\ \hline u_{r+1,1} & \dots & u_{r+1,s} & u_{r+1,,s+1} & \dots & u_{r+1,n} \\ \vdots & & \vdots & \vdots & & \vdots \\ u_{m1} & \dots & u_{ms} & u_{m,s+1} & \dots & u_{mn} \end{array}\right) = \begin{pmatrix} \mathbf{A} & \mathbf{B} \\ \mathbf{C} & \mathbf{D} \end{pmatrix}$$

gives $\mathbf{U}$ a block-matrix structure with two diagonal blocks $\mathbf{A}_{rs}$ and $\mathbf{D}_{m-r,n-s}$ and two off–diagonal blocks $\mathbf{B}_{r,n-s}$ and $\mathbf{C}_{m-r,s}$. In a block matrix, the matrix entries are matrices themselves.

For simplicity, we only consider block matrices whose diagonal blocks are square. For the matrix $\mathbf{U}_{mn}$, this means that $r = s$, and, consequently, $m = n$. We operate with such block matrices as we would with ordinary scalar entry matrices. Specifically for conformally partitioned block matrices $\mathbf{W} = (\mathbf{W}_{ij})$ and $\mathbf{Z} = (\mathbf{Z}_{jk})$ with ℓ square diagonal blocks $\mathbf{W}_{ii}$ and $\mathbf{Z}_{ii}$ of the same respective size, the product $\mathbf{WZ}$ is evaluated by performing **block-matrix dot products**

$$\mathbf{WZ} = \left(\sum_{j=1}^{\ell} \mathbf{W}_{ij}\mathbf{Z}_{jk}\right) \tag{6.8}$$

of the block rows $(\mathbf{W}_{i1}, \ \dots \ , \mathbf{W}_{i\ell})$ of $\mathbf{W}$ and the block columns $\begin{pmatrix} \mathbf{Z}_{1k} \\ \vdots \\ \mathbf{Z}_{\ell k} \end{pmatrix}$ of $\mathbf{Z}$.

Formula (6.8) coincides with the ordinary matrix product if each $\mathbf{W}_{ij}$, and $\mathbf{Z}_{jk}$ is 1×1 (i.e., if both $\mathbf{Z}$ and $\mathbf{W}$ are considered as being ordinary matrices).

EXAMPLE 7 We consider 2×2 ordinary matrices and 2×2 block matrices.

(a) For ordinary 2×2 lower triangular matrices, we have

$$\begin{pmatrix} 1 & 0 \\ a & 1 \end{pmatrix}\begin{pmatrix} 1 & 0 \\ b & 1 \end{pmatrix} = \begin{pmatrix} 1 & 0 \\ a+b & 1 \end{pmatrix} = \begin{pmatrix} 1 & 0 \\ b & 1 \end{pmatrix}\begin{pmatrix} 1 & 0 \\ a & 1 \end{pmatrix};$$

that is, these matrices commute for all values of $a, \ b \in \mathbb{R}$.

The same is true for 2×2 block lower triangular matrices of compatible block structure:

$$\begin{pmatrix} \mathbf{I}_k & \mathbf{0} \\ \mathbf{A}_{\ell k} & \mathbf{I}_\ell \end{pmatrix}\begin{pmatrix} \mathbf{I}_k & \mathbf{0} \\ \mathbf{B}_{\ell k} & \mathbf{I}_\ell \end{pmatrix} = \begin{pmatrix} \mathbf{I}_k & \mathbf{0} \\ \mathbf{A}+\mathbf{B} & \mathbf{I}_\ell \end{pmatrix} = \begin{pmatrix} \mathbf{I}_k & \mathbf{0} \\ \mathbf{B}_{\ell k} & \mathbf{I}_\ell \end{pmatrix}\begin{pmatrix} \mathbf{I}_k & \mathbf{0} \\ \mathbf{A}_{\ell k} & \mathbf{I}_\ell \end{pmatrix}.$$

(b) The inverse of $\begin{pmatrix} 1 & 0 \\ a & 1 \end{pmatrix} \in \mathbb{R}^{2,2}$ is $\begin{pmatrix} 1 & 0 \\ -a & 1 \end{pmatrix}$ by part (a). Likewise, the inverse of the block matrix $\begin{pmatrix} \mathbf{I}_k & \mathbf{0} \\ \mathbf{A}_{\ell k} & \mathbf{I}_\ell \end{pmatrix}$ is $\begin{pmatrix} \mathbf{I}_k & \mathbf{0} \\ -\mathbf{A}_{\ell k} & \mathbf{I}_\ell \end{pmatrix}$. This should be checked. ◀

6.2.P Problems

1. Verify the three statements of Lemma 2.

2. The matrix

$$\mathbf{G}_i = \begin{pmatrix} 1 & & & & & 0 \\ & \ddots & & & & \\ & & 1 & & & \\ & & a_{i+1,i} & 1 & & \\ & & \vdots & & \ddots & \\ 0 & & a_{n,i} & 0 & & 1 \end{pmatrix}$$

is called a composite Gaussian elimination matrix obtained by combining several matrices of type (6.3).

(a) Which elementary Gaussian elimination matrices are combined in $\mathbf{G}_i$? In what order must they be multiplied?

(b) Show that

$$\mathbf{G}_i^{-1} = \begin{pmatrix} 1 & & & & & 0 \\ & \ddots & & & & \\ & & 1 & & & \\ & & -a_{i+1,i} & 1 & & \\ & & \vdots & & \ddots & \\ 0 & & -a_{n,i} & 0 & & 1 \end{pmatrix}.$$

3. (a) Find a product of elementary Gaussian elimination matrices $\mathbf{G}$ that maps the vector $\begin{pmatrix} 4 \\ -1 \\ 2 \\ 3 \end{pmatrix}$ to $\begin{pmatrix} 4 \\ -1 \\ 0 \\ 0 \end{pmatrix}$.

(b) Find a second Gaussian elimination matrix product that does the same.

4. Prove or disprove the following: For any elementary elimination matrix $\mathbf{E}$, the linear system of equations $\mathbf{Ex} = \mathbf{b}$ is solvable for all right-hand sides $\mathbf{b}$. Can every such linear system be solved uniquely with all system matrices $\mathbf{E}$ of the form (6.3), (6.4), or (6.5)? Why?

5. Prove that for all scalars α and β and all square matrices $\mathbf{A}$, we have

$$(\mathbf{A} + \alpha\mathbf{I})(\mathbf{A} - \beta\mathbf{I}) = (\mathbf{A} - \beta\mathbf{I})(\mathbf{A} + \alpha\mathbf{I}).$$

6. Prove or disprove that if $(\mathbf{A} - 2\mathbf{I})(\mathbf{A} + 3\mathbf{I})\mathbf{x} = \mathbf{0}$ for a square matrix $\mathbf{A}$ and $\mathbf{x} \neq \mathbf{0}$, then $\mathbf{x} \in \ker(\mathbf{A} - 2\mathbf{I})$ or $\mathbf{x} \in \ker(\mathbf{A} + 3\mathbf{I})$.

What about the converse?

7. Given that $\mathbf{A}^{-1}\begin{pmatrix} 8 \\ 2 \\ 4 \end{pmatrix} = \begin{pmatrix} 2 \\ 0 \\ 0 \end{pmatrix}$, $\mathbf{A}^{-1}\begin{pmatrix} 4 \\ -1 \\ 0 \end{pmatrix} = \begin{pmatrix} 0 \\ -1 \\ 0 \end{pmatrix}$, and $\mathbf{A}\begin{pmatrix} 0 \\ 0 \\ 1 \end{pmatrix} = \begin{pmatrix} 1 \\ 2 \\ 0 \end{pmatrix}$ for a nonsingular matrix $\mathbf{A}_{3\times 3}$, find the matrix $\mathbf{A}$ and its inverse $\mathbf{A}^{-1}$. Check your answers for the given data.

8. (a) Solve $\begin{pmatrix} 1 & 3 & -1 \\ 7 & 0 & 1 \\ 1 & 0 & 1 \end{pmatrix}\mathbf{x} = \begin{pmatrix} 3 \\ 8 \\ 2 \end{pmatrix}$ using Gaussian elimination and backsubstitution; write out the corresponding elementary elimination matrices for each row reduction step and check your answer.

(b) Does the inverse $\mathbf{A}^{-1}$ of the preceding system matrix $\mathbf{A}$ exist? If it exists, calculate it.

(c) Evaluate $\mathbf{A}^{-1}\begin{pmatrix} 3 \\ 8 \\ 2 \end{pmatrix}$.

(d) Express both $\mathbf{A}$ and $\mathbf{A}^{-1}$ as the product of elimination matrices and an upper staircase matrix.

9. Let $\mathbf{D}_{mm}$ be a **diagonal matrix** (i.e., a matrix whose off–diagonal entries are all zero). Show that the matrix $\mathbf{D}_{mm}\mathbf{B}_{mn}$ describes a scaling of the rows of $\mathbf{B}$. What does the product $\mathbf{A}_{nm}\mathbf{D}_{mm}$ do to the rows of $\mathbf{A}$?

10. Let $\mathbf{R} = \begin{pmatrix} r_{11} & & r_{1n} \\ & \ddots & \\ 0 & & r_{nn} \end{pmatrix}$ be upper triangular. Describe

(a) the rows of $\mathbf{RB}_{nn}$, and

(b) the columns of $\mathbf{B}_{nn}\mathbf{R}$

for an arbitrary matrix $\mathbf{B}$.

Stump basis

(d) For which values of $c \in \mathbb{R}$ does the vector $\mathbf{u} = \begin{pmatrix} 1 \\ c \\ -1+c \end{pmatrix}$ lie in the image of T?

8. If $\mathbf{B} = \alpha_0 \mathbf{I}_n + \alpha_1 \mathbf{A} + \cdots + \alpha_k \mathbf{A}^k$ for real coefficients α_ℓ and a real $n \times n$ matrix $\mathbf{A}$, show that $\mathbf{AB} = \mathbf{BA}$ holds.

9. (a) Find two invertible matrices $\mathbf{A}, \mathbf{B} \in \mathbb{R}^{2,2}$ with $\mathbf{A}^{-1}\mathbf{BA} = \mathbf{B}$.

(b) Find two invertible matrices $\mathbf{A}, \mathbf{B} \in \mathbb{R}^{2,2}$ with $\mathbf{A}^{-1}\mathbf{BA} \neq \mathbf{B}$.

10. Under which conditions does the matrix equation $(\mathbf{A}+\mathbf{B})^2 = \mathbf{A}^2 + 2\mathbf{AB} + \mathbf{B}^2$ hold for two square matrices $\mathbf{A}$ and $\mathbf{B}$ of the same size?

Standard Questions

1. What are the size restrictions on two matrices $\mathbf{A}_{mn}$ and $\mathbf{B}_{k\ell}$, so that at least one of the matrix products $\mathbf{AB}$ or $\mathbf{BA}$ can be formed?

2. Which size matrices $\mathbf{A}_{mn}$ are not invertible?

3. When does a matrix have an inverse?

4. How can the matrix inverse $\mathbf{A}^{-1}$ be computed from $\mathbf{A}_{nn}$?

5. What properties does the inverse $\mathbf{A}^{-1}$ of a matrix $\mathbf{A}$ have?

6. When can a square matrix not be inverted?

7. What is the inverse (or transpose) of a product of square matrices?

Subheadings of Lecture Six

Basic Equations

p. 171	$\mathbf{AA}^{-1} = \mathbf{A}^{-1}\mathbf{A} = \mathbf{I}$	(matrix inverse)
p. 176	$(\mathbf{AB})^T = \mathbf{B}^T\mathbf{A}^T$	(transpose of matrix product)
p. 176	$(\mathbf{AB})^{-1} = \mathbf{B}^{-1}\mathbf{A}^{-1}$	(inverse of matrix product)
p. 177	$(\mathbf{A}^T)^{-1} = (\mathbf{A}^{-1})^T$	(inverse of matrix transpose)

Stump basis

6.R Review Problems

1. (a) Define the image of a matrix transformation $\mathbf{A} : \mathbb{R}^n \to \mathbb{R}^m$: im($\mathbf{A}$) := {. . . .

(b) Define the nullspace (or kernel) of a matrix transformation $\mathbf{A} : \mathbb{R}^n \to \mathbb{R}^m$: ker($\mathbf{A}$) := {. . . .

(c) If $\mathbf{A}$ is a square matrix and $\mathbf{A}^2 = \mathbf{O}_n$, show that $\text{im}(\mathbf{A}) \subset \ker(\mathbf{A})$.

(d) Find a 2×2 matrix $\mathbf{A}$ with $\text{im}(\mathbf{A}) \not\subset \ker(\mathbf{A})$.

(e) Evaluate A^2 for the matrix $\mathbf{A}$ found in part (d). Can $\mathbf{A}^2 = \mathbf{O}_2$?

(f) What is the relation between the image and the column space of a matrix $\mathbf{A}$?

2. (a) Show that if $ad - bc = 0$, then the linear system $\begin{pmatrix} a & b \\ c & d \end{pmatrix} \mathbf{x} = \begin{pmatrix} 0 \\ 0 \end{pmatrix}$ is not uniquely solvable.

(b) How does part (a) imply that in this case the system matrix $\mathbf{A}$ cannot be inverted.

(c) Write out three 2×2 matrices that are invertible and three that cannot be inverted.

3. Write out the definitions and concepts underlying the following notions and processes:

(a) The composition of two linear maps $T : \mathbb{R}^k \to \mathbb{R}^\ell$ and $S : \mathbb{R}^m \to \mathbb{R}^n$: What about the exponents $k,\ \ell,\ m,\ n$? What is $S \circ T$? And what is $T \circ S$? When is whichever defined, and when is a composition of maps linear?

(b) The matrix representation of a composition of linear maps.

(c) What is the inverse of a mapping, of a matrix? what size must a matrix have to have a chance to be invertible?

(d) How to invert a matrix.

4. Prove or disprove each of the following, or fill in the blanks:

(a) If $\mathbf{A}^T$ is invertible, then $\mathbf{A}^T$ is nonsingular.

(b) If $\mathbf{A}^T$ is not invertible, then $\mathbf{A}$ is singular.

(c) Assume that the RREF of a matrix $\mathbf{A}_{nn}$ has four pivots. Then $\mathbf{A}$ is invertible if and only if ________ .

(d) If the matrix product $\mathbf{BC}$ is invertible, then both $\mathbf{B}$ and $\mathbf{C}$ are invertible.

(e) If the matrix product $\mathbf{BC}$ is invertible and both $\mathbf{B}$ and $\mathbf{C}$ are square, then $\mathbf{B}$ and $\mathbf{C}$ are both nonsingular.

(f) A matrix $\mathbf{A}_{4,4}$ is invertible if and only if its columns are ________ in $\mathbb{R}^4$.

(g) If there is a right-hand-side vector $\mathbf{b}$ for which the linear system $\mathbf{Ax} = \mathbf{b}$ is unsolvable, then $\mathbf{A}$ may or may not be invertible.

(h) The matrix powers $\mathbf{A}^k$ are invertible for all integers $\mathbf{k}$ if and only if $\mathbf{A}$ is invertible.

5. Let $\mathbf{A} = \begin{pmatrix} 5 & -3 & -2 \\ 2 & -1 & 0 \\ 1 & -1 & -1 \end{pmatrix}$.

(a) Find $\mathbf{A}^{-1}$ and check your answer.

(b) Find the matrix $(\mathbf{A}^T)^{-1}$. (Think!)

(c) Solve $\mathbf{A}^T\mathbf{x} = \begin{pmatrix} 1 \\ -2 \\ 2 \end{pmatrix} = \mathbf{b}$.

6. Determine whether each of the following is true or false given that the matrix product $\mathbf{AB}$ is defined for two matrices $\mathbf{A}$ and $\mathbf{B}$:

(a) $\mathbf{AB}$ is a matrix whose columns are linear combinations of the columns of $\mathbf{A}$.

(b) $\mathbf{AB}$ is a matrix whose rows are linear combinations of the rows of $\mathbf{A}$.

(c) $\mathbf{BA}$ is a matrix whose rows are linear combinations of the rows of $\mathbf{A}$.

(d) $\mathbf{BAB}$ is a matrix whose columns are linear combinations of the columns of $\mathbf{B}$.

7. Let the linear transformation $T : \mathbb{R}^3 \to \mathbb{R}^3$ be given by the following point mappings

$$T\begin{pmatrix} 1 \\ -2 \\ 1 \end{pmatrix} = \begin{pmatrix} 2 \\ 0 \\ 1 \end{pmatrix},\quad T\begin{pmatrix} 1 \\ -3 \\ 3 \end{pmatrix} = \begin{pmatrix} 0 \\ 1 \\ 4 \end{pmatrix},$$

and $T\begin{pmatrix} 2 \\ -5 \\ 5 \end{pmatrix} = \begin{pmatrix} 1 \\ 0 \\ -1 \end{pmatrix}$.

(a) Write out a matrix equation that relates the standard matrix representation $\mathbf{A}_{\mathcal{E}}$ of T to the given vector mappings.

(b) What is $\mathbf{A}_{\mathcal{E}}$?

(c) How does T map the vector $\mathbf{x} = \begin{pmatrix} x_1 \\ x_2 \\ x_3 \end{pmatrix} \in \mathbb{R}^3$?

Find $T\begin{pmatrix} a \\ 2 \\ b \end{pmatrix}$.

These formats display the entries of **x** with all of the 16 digits that are carried by MATLAB in its floating-point computations. There will be slight differences in the trailing digits of the two computed solution vectors **x**, especially if **A** is nearly singular. The differences occur because of rounding errors. `A\b` computes the solution vector from the RREF of $(\mathbf{A} \mid \mathbf{b})$, while the inverse, `inv(A)`, is computed from the RREF of **A** alone.

Numerical algorithms and computer arithmetic often combine to violate fundamental mathematical rules, such as Theorem 6.5. Problem 6 deals with MATLAB's failure to invert triangular matrices as triangular ones in clear defiance of the theoretically and mathematically correct Proposition of Section 6.2. Students should develop a healthy critical attitude towards the reliability of even the best computed data.

Calculators and computers are wonderful, but not perfect, tools.

6.3.P Problems

1. Solve the linear system of equations $\mathbf{H}_n\mathbf{x} = \mathbf{e}_1 - \mathbf{e}_n$ for the **Hilbert matrix** $\mathbf{H} = \mathbf{H}_n = (h_{ij}) \in \mathbb{R}^{n,n}$ with $h_{ij} = \dfrac{1}{i+j}$ for $n = 3, 4, 5, 6, 7, 8$ by using the two MATLAB commands `x1 = H\b` and `C = inv(H); x2 = C*b` with $\mathbf{b} = \mathbf{e}_1 - \mathbf{e}_n$. Compare your computed results for increasing dimensions n by using `format long e`. The Hilbert matrix of size $n \times n$ can be generated in MATLAB via the command `H = hilb(n)`.

2. For what dimensions n can MATLAB invert the Hilbert matrix? How reliable is the MATLAB matrix inversion algorithm for $\mathbf{H}_n$ and increasing dimensions n? (Check the computed answers by evaluating $\mathbf{H}_n^{-1}\mathbf{H}_n$ and $\mathbf{H}_n\mathbf{H}_n^{-1}$ in MATLAB.)

3. Repeat Problem 1 for the singular matrix

$$\mathbf{E}_{nn} = \begin{pmatrix} 1 & \dots & 1 \\ \vdots & & \vdots \\ 1 & \dots & 1 \end{pmatrix}$$ and **b** as before. Observe what happens for various dimensions n.

4. Compare some of your hand-computed matrix inverses from Section 6.1.P with the MATLAB computed inverses.

5. Generate several 20×20 random matrices **A** in MATLAB, compute their inverses via `inv(A)`, and form $\mathbf{I}_{20} - \mathbf{A} * \mathbf{A}^{-1}$. Display this matrix in `format long e` to see what is going on.

6. Compute the inverses of several lower and upper triangular matrices in MATLAB:

(a) Generate upper triangular matrices **R** with the MATLAB commands:

```
n=5; R=diag(randn(1,n));
for i=1:n, R=R+diag(randn(1,n-i),i);
end
or
n=5; R=diag(randn(1,n));
for i=1:n,
R=R+10^i*diag(randn(1,n-i),i);
end.
```

(b) Create lower triangular matrices **L** by using the preceding command sequence and transposing `L=R'`.

(c) Investigate the inverses of lower and upper triangular matrices using the commands `inv(L)` or `inv(R)`: are the inverses lower and upper triangular matrices as they were proven to be theoretically? Use `format long` or `format long e` and watch the row echelon movie `rrefmovie(L)` as well.

(d) Can you find examples of upper triangular matrices **R** for which `inv(R')` $\neq$ `inv(R)'` in MATLAB, violating Theorem 6.5?

(e) Why is the computed inverse of an upper triangular matrix **R** again upper triangular in MATLAB?

(f) Why is this not so for lower triangular matrices **L** in MATLAB?

(b) The inverse of $\begin{pmatrix} 1 & 0 \\ a & 1 \end{pmatrix} \in \mathbb{R}^{2,2}$ is $\begin{pmatrix} 1 & 0 \\ -a & 1 \end{pmatrix}$ by part (a). Likewise, the inverse of the block matrix $\begin{pmatrix} \mathbf{I}_k & \mathbf{0} \\ \mathbf{A}_{\ell k} & \mathbf{I}_\ell \end{pmatrix}$ is $\begin{pmatrix} \mathbf{I}_k & \mathbf{0} \\ -\mathbf{A}_{\ell k} & \mathbf{I}_\ell \end{pmatrix}$. This should be checked. ◀

6.2.P Problems

1. Verify the three statements of Lemma 2.

2. The matrix

$$\mathbf{G}_i = \begin{pmatrix} 1 & & & & & 0 \\ & \ddots & & & & \\ & & 1 & & & \\ & & a_{i+1,i} & 1 & & \\ & & \vdots & & \ddots & \\ 0 & & a_{n,i} & 0 & & 1 \end{pmatrix}$$

is called a composite Gaussian elimination matrix obtained by combining several matrices of type (6.3).

(a) Which elementary Gaussian elimination matrices are combined in $\mathbf{G}_i$? In what order must they be multiplied?

(b) Show that

$$\mathbf{G}_i^{-1} = \begin{pmatrix} 1 & & & & & 0 \\ & \ddots & & & & \\ & & 1 & & & \\ & & -a_{i+1,i} & 1 & & \\ & & \vdots & & \ddots & \\ 0 & & -a_{n,i} & 0 & & 1 \end{pmatrix}.$$

3. (a) Find a product of elementary Gaussian elimination matrices $\mathbf{G}$ that maps the vector $\begin{pmatrix} 4 \\ -1 \\ 2 \\ 3 \end{pmatrix}$ to $\begin{pmatrix} 4 \\ -1 \\ 0 \\ 0 \end{pmatrix}$.

(b) Find a second Gaussian elimination matrix product that does the same.

4. Prove or disprove the following: For any elementary elimination matrix $\mathbf{E}$, the linear system of equations $\mathbf{E}\mathbf{x} = \mathbf{b}$ is solvable for all right-hand sides $\mathbf{b}$. Can every such linear system be solved uniquely with all system matrices $\mathbf{E}$ of the form (6.3), (6.4), or (6.5)? Why?

5. Prove that for all scalars α and β and all square matrices $\mathbf{A}$, we have

$$(\mathbf{A} + \alpha\mathbf{I})(\mathbf{A} - \beta\mathbf{I}) = (\mathbf{A} - \beta\mathbf{I})(\mathbf{A} + \alpha\mathbf{I}).$$

6. Prove or disprove that if $(\mathbf{A} - 2\mathbf{I})(\mathbf{A} + 3\mathbf{I})\mathbf{x} = \mathbf{0}$ for a square matrix $\mathbf{A}$ and $\mathbf{x} \neq \mathbf{0}$, then $\mathbf{x} \in \ker(\mathbf{A} - 2\mathbf{I})$ or $\mathbf{x} \in \ker(\mathbf{A} + 3\mathbf{I})$.

What about the converse?

7. Given that $\mathbf{A}^{-1}\begin{pmatrix} 8 \\ 2 \\ 4 \end{pmatrix} = \begin{pmatrix} 2 \\ 0 \\ 0 \end{pmatrix}$, $\mathbf{A}^{-1}\begin{pmatrix} 4 \\ -1 \\ 0 \end{pmatrix} = \begin{pmatrix} 0 \\ -1 \\ 0 \end{pmatrix}$, and $\mathbf{A}\begin{pmatrix} 0 \\ 0 \\ 1 \end{pmatrix} = \begin{pmatrix} 1 \\ 2 \\ 0 \end{pmatrix}$ for a nonsingular matrix $\mathbf{A}_{3\times 3}$, find the matrix $\mathbf{A}$ and its inverse $\mathbf{A}^{-1}$. Check your answers for the given data.

8. (a) Solve $\begin{pmatrix} 1 & 3 & -1 \\ 7 & 0 & 1 \\ 1 & 0 & 1 \end{pmatrix}\mathbf{x} = \begin{pmatrix} 3 \\ 8 \\ 2 \end{pmatrix}$ using Gaussian elimination and backsubstitution; write out the corresponding elementary elimination matrices for each row reduction step and check your answer.

(b) Does the inverse $\mathbf{A}^{-1}$ of the preceding system matrix $\mathbf{A}$ exist? If it exists, calculate it.

(c) Evaluate $\mathbf{A}^{-1}\begin{pmatrix} 3 \\ 8 \\ 2 \end{pmatrix}$.

(d) Express both $\mathbf{A}$ and $\mathbf{A}^{-1}$ as the product of elimination matrices and an upper staircase matrix.

9. Let $\mathbf{D}_{mm}$ be a **diagonal matrix** (i.e., a matrix whose off–diagonal entries are all zero). Show that the matrix $\mathbf{D}_{mm}\mathbf{B}_{mn}$ describes a scaling of the rows of $\mathbf{B}$. What does the product $\mathbf{A}_{nm}\mathbf{D}_{mm}$ do to the rows of $\mathbf{A}$?

10. Let $\mathbf{R} = \begin{pmatrix} r_{11} & & r_{1n} \\ & \ddots & \\ 0 & & r_{nn} \end{pmatrix}$ be upper triangular. Describe

(a) the rows of $\mathbf{R}\mathbf{B}_{nn}$, and

(b) the columns of $\mathbf{B}_{nn}\mathbf{R}$

for an arbitrary matrix $\mathbf{B}$.

11. Find the inverse of the block matrix $\mathbf{A} = \begin{pmatrix} \mathbf{A}_{11} & \mathbf{A}_{12} \\ \mathbf{0} & \mathbf{A}_{22} \end{pmatrix}$ if the diagonal blocks $\mathbf{A}_{ii}$ of $\mathbf{A}$ are both square and invertible.

12. For $\mathbf{A}_{mn}$ and $\mathbf{B}_{nk}$, the matrix product $\mathbf{AB}$ has m rows and k columns.

(a) Describe the individual entries of $\mathbf{AB}$.

(b) Describe each entry when $n = 1$: For example, what are the entries of $\begin{pmatrix} 1 \\ -2 \end{pmatrix} \begin{pmatrix} -1 & 4 & 0 & 12 \end{pmatrix}$ or $\begin{pmatrix} a_1 \\ a_2 \\ a_3 \end{pmatrix} \begin{pmatrix} b_1 & b_2 \end{pmatrix}$?

13. (a) Show that $\mathbf{P}^T = \mathbf{P} = \mathbf{P}^{-1}$ for any swap matrix $\mathbf{P} = \mathbf{E}_{jk}$ of type (6.5).

(b) Show that the inverse of a general **permutation matrix** $\mathbf{P}_{nn} := \prod \mathbf{E}_{ji}$ for swap matrices $\mathbf{E}_{ji}$ of type (6.5) is its transpose $\mathbf{P}^T$.

(Matrices $\mathbf{A}_{nn}$ with $\mathbf{A}^T A = \mathbf{I}_n$, such as permutation matrices, are called **orthogonal** and will be studied in Chapters 10 and 11.)

(*Hint*: Study how $\mathbf{P}$ maps an arbitrary vector $\mathbf{x} \in \mathbb{R}^n$, then how $\mathbf{P}^T$ maps $\mathbf{Px}$.)

14. Show that the transpose of an elementary elimination matrix $\mathbf{E}(c_{jk})$ of type (6.3) is an elimination matrix.

15. Find the inverse of the block matrix

$$\mathbf{A} = \left(\begin{array}{ccc|cc} 0 & 0 & 1 & 0 & 0 \\ 0 & 2 & 0 & 0 & 0 \\ 1 & 0 & 0 & 0 & 0 \\ \hline 0 & 0 & 0 & 0 & 1 \\ 0 & 0 & 0 & -1 & 0 \end{array}\right).$$

(*Hint*: You may work on the diagonal blocks only.)

16. Show that a block triangular matrix of the form $\mathbf{L} = \begin{pmatrix} \mathbf{A} & \mathbf{O} \\ \mathbf{B} & \mathbf{C} \end{pmatrix}$ with $\mathbf{A} \in \mathbb{R}^{m,m}$ and $\mathbf{C} \in \mathbb{R}^{k,k}$ is invertible if and only if both $\mathbf{A}$ and $\mathbf{C}$ are invertible.

6.3 Applications

MATLAB

In MATLAB matrices are multiplied by the * command

```
A * B
```

Here $\mathbf{A}$ and $\mathbf{B}$ must be two matrices that are compatible for multiplication. Find out what happens in MATLAB when one attempts to multiply two matrices that cannot be multiplied.
The transpose $\mathbf{B} = \mathbf{A}^T$ of a real matrix $\mathbf{A}$ is formed by the ' command

```
B = A'
```

while the inverse $\mathbf{C} = \mathbf{A}^{-1}$ of a nonsingular matrix $\mathbf{A}$ is computed in MATLAB via the command

```
C = inv(A)
```

It is quite instructive to solve a linear system $\mathbf{Ax} = \mathbf{b}$ with a nonsingular system matrix $\mathbf{A}$ in two ways: First find the solution vector $\mathbf{x}$ by using the command

```
x = A\b
```

and then evaluate $\mathbf{x} = \mathbf{A}^{-1}\mathbf{b}$ in MATLAB by invoking

```
x = inv(A)*b
```

Look at the entries of the two computed copies of $\mathbf{x}$ with great scrutiny by using the commands

```
format long   or   format long e
```

CHAPTER

7

Coordinate Vectors, Basis Change

We study how to describe position in one vector space with respect to different bases.

7.1 Lecture Seven

Matrix Representations with Respect to General Bases

As the standard unit vector basis has been extended to general bases, so can the standard matrix representation of a linear transformation via basis change similarity.

(a) Coordinate Vectors

A basis of a linear space helps us describe positions in the space. For example, in the standard basis $\mathcal{E} = \{\mathbf{e}_1, \mathbf{e}_2, \mathbf{e}_3\}$ of $\mathbb{R}^3$, the vector $\mathbf{x} = \begin{pmatrix} 1 \\ 2 \\ -1 \end{pmatrix} \in \mathbb{R}^3$ lies one unit along the first **coordinate axis** (i.e., along the first unit vector $\mathbf{e}_1 = \begin{pmatrix} 1 \\ 0 \\ 0 \end{pmatrix} \in \mathbb{R}^3$), two units along the second, and one negative unit (i.e., one unit in the opposite direction) along the third unit vector $\mathbf{e}_3$ of $\mathbb{R}^3$. (See Fig. 7-1.)

In terms of the standard unit vector basis $\mathcal{E} = \{\mathbf{e}_1, \mathbf{e}_2, \mathbf{e}_3\}$ of $\mathbb{R}^3$, we say that this vector $\mathbf{x}$ has the **coordinates** $\mathbf{x}_{\mathcal{E}} = \begin{pmatrix} 1 \\ 2 \\ -1 \end{pmatrix} = \mathbf{e}_1 + 2\mathbf{e}_2 - \mathbf{e}_3$ with respect to $\mathcal{E}$. We take the coefficients 1, 2, and –1 in the linear combination $\mathbf{x} = \mathbf{e}_1 + 2\mathbf{e}_2 - \mathbf{e}_3$ as the $\mathcal{E}$–coordinates of $\mathbf{x}$. This defines the **coordinate vector** $\mathbf{x}_{\mathcal{E}}$ of $\mathbf{x} \in \mathbb{R}^3$ with

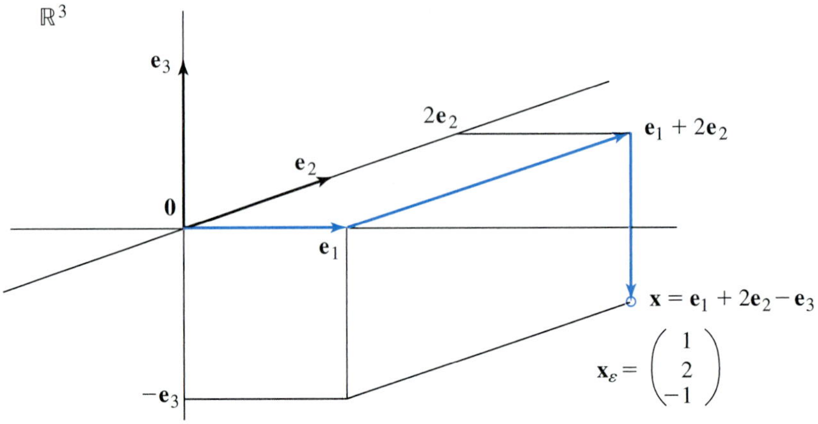

Figure 7-1

respect to the $\mathcal{E}$ basis. The $\mathcal{E}$–coordinate vector $\mathbf{x}_{\mathcal{E}}$ is unique for each $\mathbf{x} \in \mathbb{R}^3$, since $\mathcal{E}$ is a basis of $\mathbb{R}^3$.

According to Definition 4 of Section 5.1, a set of vectors $\{\mathbf{u}_1, \ldots, \mathbf{u}_n\}$ in $\mathbb{R}^n$ is a basis for $\mathbb{R}^n$ if each point $\mathbf{x} \in \mathbb{R}^n$ is a unique linear combination of the $\mathbf{u}_i$. Analogous to $\mathbb{R}^3$ and the $\mathcal{E}$–coordinate vector $\mathbf{x}_{\mathcal{E}}$ for the standard basis, the vector $\mathbf{x} \in \mathbb{R}^n$ has the **$\mathcal{U}$–coordinate vector**

$$\mathbf{x}_{\mathcal{U}} = \begin{pmatrix} \alpha_1 \\ \vdots \\ \alpha_n \end{pmatrix}, \qquad \text{if} \qquad \mathbf{x} = \alpha_1 \mathbf{u}_1 + \cdots + \alpha_u \mathbf{u}_n, \tag{7.1}$$

is the unique linear combination of the vectors $\mathbf{u}_i$ in the basis $\mathcal{U} = \{\mathbf{u}_1, \ldots, \mathbf{u}_n\}$ of $\mathbb{R}^n$. Thus, the assignment of a point $\mathbf{x} \in \mathbb{R}^n$ to its $\mathcal{U}$–coordinate vector $\mathbf{x}_{\mathcal{U}}$

$$\mathbf{x} \mapsto \mathbf{x}_{\mathcal{U}} \ : \ \mathbb{R}^n \to \mathbb{R}^n$$

is a function for each fixed basis $\mathcal{U}$ of $\mathbb{R}^n$. This function is linear, as we shall soon learn.

EXAMPLE 1 The two vectors $\mathbf{u}_1 = \begin{pmatrix} 1 \\ -1 \end{pmatrix}$ and $\mathbf{u}_2 = \begin{pmatrix} 1 \\ 2 \end{pmatrix}$ form a basis $\mathcal{U} = \{\mathbf{u}_1, \mathbf{u}_2\}$ for $\mathbb{R}^2$. (Why?) Find the $\mathcal{U}$–coordinate vector of an arbitrary point $\mathbf{w} = \begin{pmatrix} a \\ b \end{pmatrix} \in \mathbb{R}^2$.

We need to compute the $\mathcal{U}$–coordinates $\mathbf{w}_{\mathcal{U}} = \begin{pmatrix} \alpha_1 \\ \alpha_2 \end{pmatrix}$ of $\mathbf{w} = \begin{pmatrix} a \\ b \end{pmatrix} \in \mathbb{R}^2$, that is, we need to find scalars α_1 and α_2 with $\mathbf{w} = \alpha_1 \mathbf{u}_1 + \alpha_2 \mathbf{u}_2$. This is a linear equations problem: Solve $\begin{pmatrix} | & | \\ \mathbf{u}_1 & \mathbf{u}_2 \\ | & | \end{pmatrix} \begin{pmatrix} \alpha_1 \\ \alpha_2 \end{pmatrix} = \begin{pmatrix} a \\ b \end{pmatrix} = \mathbf{w}$ via row reduction of the augmented matrix:

U		w	
$\boxed{1}$	1	a	
$\boxed{-1}$	2	b	$+\ row_1$
$\boxed{1}$	1	a	
0	$\boxed{3}$	$a+b$	$\div 3$
$\boxed{1}$	$\boxed{1}$	a	$-\ row_2$
0	$\boxed{1}$	$\frac{a+b}{3}$	
$\boxed{1}$	0	$\frac{2a-b}{3}$	
0	$\boxed{1}$	$\frac{a+b}{3}$	

Thus, the $\mathcal{U}$–coordinate vector of $\mathbf{w}$ is given by $\mathbf{w}_{\mathcal{U}} = \begin{pmatrix} \frac{2a-b}{3} \\ \frac{a+b}{3} \end{pmatrix}$ for any $\mathbf{w} = \mathbf{w}_{\mathcal{E}} = \begin{pmatrix} a \\ b \end{pmatrix} \in \mathbb{R}^2$. By expressing

$$\mathbf{w}_{\mathcal{U}} = \begin{pmatrix} \frac{2a-b}{3} \\ \frac{a+b}{3} \end{pmatrix} = \frac{a}{3}\begin{pmatrix} 2 \\ 1 \end{pmatrix} + \frac{b}{3}\begin{pmatrix} -1 \\ 1 \end{pmatrix} = \frac{1}{3}\begin{pmatrix} 2 & -1 \\ 1 & 1 \end{pmatrix}\begin{pmatrix} a \\ b \end{pmatrix}$$

as a matrix $\times$ vector product, we see that the $\mathcal{E}$–coordinate vector $\mathbf{w}_{\mathcal{E}} = \begin{pmatrix} a \\ b \end{pmatrix}$ is mapped to its $\mathcal{U}$–coordinate vector $\mathbf{w}_{\mathcal{U}}$ by the linear transformation that has the matrix $\times$ vector form: $\mathbf{w}_{\mathcal{U}} = \frac{1}{3}\begin{pmatrix} 2 & -1 \\ 1 & 1 \end{pmatrix}\mathbf{w}_{\mathcal{E}}$. Here $\frac{1}{3}\begin{pmatrix} 2 & -1 \\ 1 & 1 \end{pmatrix}$ acts as the **basis change matrix**. This matrix transforms arbitrary $\mathcal{E}$–coordinate vectors $\mathbf{w} = \mathbf{w}_{\mathcal{E}}$ to their $\mathcal{U}$–coordinate vectors $\mathbf{w}_{\mathcal{U}}$. ◀

(b) The Basis Change Matrix

Assigning the $\mathcal{U}$–coordinate vector $\mathbf{w}_{\mathcal{U}}$ to a vector $\mathbf{w} = \mathbf{w}_{\mathcal{E}} \in \mathbb{R}^n$ in standard representation is a linear mapping $\mathbf{T}$ from $\mathcal{E}$–coordinate vectors to $\mathcal{U}$–coordinate ones:

If $\mathbf{w} = \mathbf{w}_{\mathcal{E}} = (a_1, \ldots, a_n)$, $\mathbf{v} = \mathbf{v}_{\mathcal{E}} = (b_1, \ldots, b_n)$, $T(\mathbf{w}_{\mathcal{E}}) = \mathbf{w}_{\mathcal{U}} = (\alpha_1, \ldots, \alpha_n)$, and $T(\mathbf{v}_{\mathcal{E}}) = \mathbf{v}_{\mathcal{U}} = (\beta_1, \ldots, \beta_n)$, then $\mathbf{w} = \sum_i \alpha_i \mathbf{u}_i$ and $\mathbf{v} = \sum_i \beta_i \mathbf{u}_i$ by Definition (7.1). These equations imply that

$$\begin{aligned} T(\mathbf{w}_{\mathcal{E}} + \mathbf{v}_{\mathcal{E}}) &= (\mathbf{w} + \mathbf{v})_{\mathcal{U}} = (\alpha_1 + \beta_1, \ldots, \alpha_n + \beta_n) = \mathbf{w}_{\mathcal{U}} + \mathbf{v}_{\mathcal{U}} \\ &= T(\mathbf{w}_{\mathcal{E}}) + T(\mathbf{v}_{\mathcal{E}}) \quad \text{and} \\ T(\gamma \mathbf{w}_{\mathcal{E}}) &= (\gamma \mathbf{w})_{\mathcal{U}} = (\gamma\alpha_1, \ldots, \gamma\alpha_n) = \gamma T(\mathbf{w}_{\mathcal{E}}). \end{aligned}$$

Thus, T satisfies the linearity conditions (1.2) for functions: the image of a linear combination of vectors under the coordinate change map T is equal to the same linear combination of the images of each individual ingredient vector.

Note that any basis of $\mathbb{R}^n$ should be as good as any other. The standard $\mathcal{E}$ basis was convenient at the start, but it usually hides much of the information that is embedded in a linear transformation $\mathbf{A} : \mathbb{R}^n \to \mathbb{R}^n$, as we shall see in Chapters 9, 11, 12, and 14.

To uncover the intrinsic information contained in each linear transformation, we study how to translate from one coordinate vector description $\mathbf{w}_{\mathcal{U}}$ of a point $\mathbf{w} = \mathbf{w}_{\mathcal{E}} \in \mathbb{R}^n$ to another $\mathbf{w}_{\mathcal{V}}$ for two different bases $\mathcal{U}$ and $\mathcal{V}$ of $\mathbb{R}^n$.

Question: Let $\mathcal{U} = \{\mathbf{u}_i\}$ and $\mathcal{V} = \{\mathbf{v}_i\}$ be two bases of $\mathbb{R}^n$, and let $\mathbf{w}_{\mathcal{U}}$ be the coordinate vector of a point $\mathbf{w}$ with respect to $\mathcal{U}$. What is its coordinate vectors $\mathbf{w}_{\mathcal{V}}$ for the other basis $\mathcal{V}$?
More generally, what is the basis change matrix from $\mathcal{U}$–coordinate vectors $\mathbf{w}_{\mathcal{U}}$ to $\mathcal{V}$–coordinate vectors $\mathbf{w}_{\mathcal{V}} \in \mathbb{R}^n$?

Assume that $\mathbf{x} \in \mathbb{R}^n$ has the $\mathcal{U}$–coordinate vector $\mathbf{x}_{\mathcal{U}} = \begin{pmatrix} \alpha_1 \\ \vdots \\ \alpha_n \end{pmatrix}$. According to Eq. (7.1), we have

$$\mathbf{x} = \alpha_1 \mathbf{u}_1 + \alpha_2 \mathbf{u}_2 + \cdots + \alpha_n \mathbf{u}_n = \begin{pmatrix} | & & | \\ \mathbf{u}_1 & \dots & \mathbf{u}_n \\ | & & | \end{pmatrix} \begin{pmatrix} \alpha_1 \\ \vdots \\ \alpha_n \end{pmatrix},$$

or $\mathbf{x} = \mathbf{U}\mathbf{x}_{\mathcal{U}}$. Likewise, $\mathbf{x} = \mathbf{V}\mathbf{x}_{\mathcal{V}}$. Here $\mathbf{U}$ and $\mathbf{V}$ are nonsingular—and hence invertible matrices whose columns contain the respective basis vectors $\mathbf{u}_i$ and $\mathbf{v}_i$ of $\mathcal{U}$ and $\mathcal{V}$. Thus, the two coordinate vectors $\mathbf{x}_{\mathcal{U}}$ and $\mathbf{x}_{\mathcal{V}}$ are related to the same point $\mathbf{x} = \mathbf{x}_{\mathcal{E}} \in \mathbb{R}^n$ as follows:

$$\mathbf{x} = \mathbf{x}_{\mathcal{E}} = \mathbf{U}\mathbf{x}_{\mathcal{U}} \quad \underline{\text{and}} \quad \mathbf{x} = \mathbf{x}_{\mathcal{E}} = \mathbf{V}\mathbf{x}_{\mathcal{V}}.$$

In particular, the standard basis $\mathcal{E} = \{\mathbf{e}_1, \dots, \mathbf{e}_n\}$ is represented by the identity matrix $\mathbf{I}_n$ with the standard vectors $\mathbf{e}_i$ as columns. Thus, our convention of equating a point $\mathbf{x} \in \mathbb{R}^n$ with its $\mathcal{E}$–coordinate vector $\mathbf{x}_{\mathcal{E}}$ is justified.

As both expressions $\mathbf{U}\mathbf{x}_{\mathcal{U}}$ and $\mathbf{V}\mathbf{x}_{\mathcal{V}}$ are equal to $\mathbf{x}$, they are equal to each other, or

$$\boxed{\mathbf{U}\mathbf{x}_{\mathcal{U}} = \mathbf{V}\mathbf{x}_{\mathcal{V}} \ (= \mathbf{x})}\,. \tag{7.2}$$

This is a **matrix equation** relating the two bases in the columns of $\mathbf{U}$ and $\mathbf{V}$ and the two coordinate vectors $\mathbf{x}_{\mathcal{U}}$ and $\mathbf{x}_{\mathcal{V}}$ of one point $\mathbf{x} = \mathbf{x}_{\mathcal{E}} \in \mathbb{R}^n$.

We answer our guiding question by solving for either $\mathbf{x}_{\mathcal{U}}$ or $\mathbf{x}_{\mathcal{V}}$ in Eq. (7.2). However, to do so, we need to be careful.

Matrix equations can only be solved correctly by using steps that are appropriate for matrix algebra. Here are some rules that must be observed when trying to simplify matrix equations:

(i) When multiplying a matrix equation by another matrix, only a matrix of properly matching size can be used. (See Sections 1.1 and 6.1.)
(ii) Only square nonsingular matrices have inverses. (See Section 6.1.)
(iii) The order of multiplication of two matrices is important, since matrices do not generally commute. (See the caution in Section 6.1.)

Specifically regarding Eq. (7.2), we can express $\mathbf{x}_{\mathcal{U}}$ in terms of $\mathbf{x}_{\mathcal{V}}$ by solving the matrix equation $\mathbf{U}\mathbf{x}_{\mathcal{U}} = \mathbf{V}\mathbf{x}_{\mathcal{V}}$ for $\mathbf{x}_{\mathcal{U}}$. Multiplying this equation on the left by $\mathbf{U}^{-1}$ finds $\mathbf{x}_{\mathcal{U}}$:

$$\mathbf{U}^{-1}\mathbf{U}\mathbf{x}_{\mathcal{U}} = \mathbf{I}\mathbf{x}_{\mathcal{U}} = \mathbf{x}_{\mathcal{U}} = \mathbf{U}^{-1}\mathbf{V}\mathbf{x}_{\mathcal{V}}.$$

On the other hand, multiplying Eq. (7.2) by $\mathbf{V}^{-1}$ on the left gives us $\mathbf{x}_{\mathcal{V}}$ in terms of $\mathbf{x}_{\mathcal{U}}$:

$$\mathbf{V}^{-1}\mathbf{U}\mathbf{x}_{\mathcal{U}} = \mathbf{V}^{-1}\mathbf{V}\mathbf{x}_{\mathcal{V}} = \mathbf{I}\mathbf{x}_{\mathcal{V}} = \mathbf{x}_{\mathcal{V}}.$$

That is,

$$\boxed{\mathbf{x}_{\mathcal{U}} = \mathbf{U}^{-1}\mathbf{V}\mathbf{x}_{\mathcal{V}}} \quad \text{and} \quad \boxed{\mathbf{x}_{\mathcal{V}} = \mathbf{V}^{-1}\mathbf{U}\mathbf{x}_{\mathcal{U}}}\,. \tag{7.3}$$

The product matrix $\mathbf{U}^{-1}\mathbf{V} =: \mathbf{X}_{\mathcal{U}\leftarrow\mathcal{V}}$ is called the **basis change matrix** from the basis $\mathcal{V}$ to the basis $\mathcal{U}$ of $\mathbb{R}^n$. And

$$\mathbf{X}_{\mathcal{V}\leftarrow\mathcal{U}} = \mathbf{V}^{-1}\mathbf{U}$$

is the basis change matrix from $\mathcal{U}$–coordinates to $\mathcal{V}$–coordinates. Their composition

$$\mathbf{X}_{\mathcal{V}\leftarrow\mathcal{U}}\,\mathbf{X}_{\mathcal{U}\leftarrow\mathcal{V}} = \mathbf{V}^{-1}\mathbf{U}\,\mathbf{U}^{-1}\mathbf{V} = \mathbf{I}_n$$

is the identity transformation, as is the reverse order composition

$$\mathbf{X}_{\mathcal{U}\leftarrow\mathcal{V}}\,\mathbf{X}_{\mathcal{V}\leftarrow\mathcal{U}} = \mathbf{U}^{-1}\mathbf{V}\,\mathbf{V}^{-1}U = \mathbf{I}_n.$$

Thus, the two basis change matrices $\mathbf{X}_{\mathcal{V}\leftarrow\mathcal{U}}$ and $\mathbf{X}_{\mathcal{U}\leftarrow\mathcal{V}}$ are mutually inverse linear transformations. They are analogous to the two parts of a dictionary. The matrix $\mathbf{X}_{\mathcal{U}\leftarrow\mathcal{V}}$ takes $\mathcal{V}$–coordinate vectors, or $\mathcal{V}$–words, to $\mathcal{U}$–coordinate vectors, or to $\mathcal{U}$–words, while $\mathbf{X}_{\mathcal{V}\leftarrow\mathcal{U}}$ translates $\mathcal{U}$–coordinate vectors into those for $\mathcal{V}$.

EXAMPLE 2 Let us revisit Example 1. From Eq. (7.3), $\mathbf{w}_{\mathcal{U}} = \mathbf{X}_{\mathcal{U}\leftarrow\mathcal{E}}\mathbf{w}_{\mathcal{E}} = \mathbf{U}^{-1}\mathbf{I}\mathbf{w}_{\mathcal{E}} = \mathbf{U}^{-1}\mathbf{w}_{\mathcal{E}}$ and $\mathbf{w} = \mathbf{w}_{\mathcal{E}} = \mathbf{X}_{\mathcal{E}\leftarrow\mathcal{U}}\mathbf{w}_{\mathcal{U}} = \mathbf{I}^{-1}\mathbf{U}\mathbf{w}_{\mathcal{U}} = \mathbf{U}\mathbf{w}_{\mathcal{U}}$ for $\mathbf{U} = \begin{pmatrix} | & | \\ \mathbf{u}_1 & \mathbf{u}_2 \\ | & | \end{pmatrix} = \begin{pmatrix} 1 & 1 \\ -1 & 2 \end{pmatrix}$.

The matrix inverse $\mathbf{U}^{-1} = \frac{1}{3}\begin{pmatrix} 2 & -1 \\ 1 & 1 \end{pmatrix}$ can be found by using the 2×2 matrix inversion formula of Problem 6 in Section 6.1.P ◀

The product formulation (7.3) of the basis change matrix $\mathbf{X}_{\mathcal{V}\leftarrow\mathcal{U}} = \mathbf{V}^{-1}\mathbf{U}$ reminds us that $\mathbf{X}_{\mathcal{V}\leftarrow\mathcal{U}}$ maps $\mathcal{U}$–coordinate vectors $\mathbf{x}_{\mathcal{U}}$ to $\mathcal{V}$–coordinate vectors $\mathbf{x}_{\mathcal{V}}$ in two stages: First $\mathbf{U}\mathbf{x}_{\mathcal{U}} = \mathbf{x}_{\mathcal{E}} = \mathbf{x}$ becomes an $\mathcal{E}$–vector, and then $\mathbf{V}^{-1}\mathbf{x}_{\mathcal{E}} = \left(\mathbf{V}^{-1}(\mathbf{U}\mathbf{x}_{\mathcal{U}})_{\mathcal{E}}\right)_{\mathcal{V}} = \mathbf{x}_{\mathcal{V}}$ becomes the $\mathcal{V}$–coordinate vector for $\mathbf{x}$. Thus, the product representation of $\mathbf{X}_{\mathcal{V}\leftarrow\mathcal{U}}$ as $\mathbf{V}^{-1}\mathbf{U}$ represents the composition of two transformations on $\mathbf{x}_{\mathcal{U}}$. The first one, by $\mathbf{U}$, translates the $\mathcal{U}$–coordinate vector $\mathbf{x}_{\mathcal{U}}$ to its standard basis representation $\mathbf{x}_{\mathcal{E}} = \mathbf{U}\mathbf{x}_{\mathcal{U}}$. Then $\mathbf{V}^{-1}$ takes this $\mathcal{E}$–coordinate vector $(\mathbf{U}\mathbf{x}_{\mathcal{U}})_{\mathcal{E}}$ of $\mathbf{x}_{\mathcal{U}}$ to its $\mathcal{V}$–coordinate vector $\mathbf{x}_{\mathcal{V}} = \mathbf{V}^{-1}\mathbf{U}\mathbf{x}_{\mathcal{U}}$:

$$\mathbf{x}_{\mathcal{U}} \;\mapsto\; \mathbf{U}\mathbf{x}_{\mathcal{U}} = \mathbf{x}_{\mathcal{E}} \;\mapsto\; \mathbf{V}^{-1}\mathbf{x}_{\mathcal{E}} = \mathbf{x}_{\mathcal{V}}$$

Let us explore this composition of maps as an analogue to translating between languages: $\mathbf{X}_{\mathcal{V}\leftarrow\mathcal{U}}$ translates $\mathcal{U}$–words (short for $\mathcal{U}$–coordinate vectors) $\mathbf{w}_{\mathcal{U}}$ to $\mathcal{V}$–words $\mathbf{w}_{\mathcal{V}}$. If this chapter were dealing with foreign languages, then the transformations of Ukrainian words $\mathbf{w}_{\mathcal{U}}$, say, to their Vietnamese counterparts $\mathbf{w}_{\mathcal{V}}$ could be achieved with the help of a Ukrainian-to-Vietnamese **dictionary**.

How could one translate from Ukrainian to Vietnamese, or from $\mathcal{U}$–coordinate vectors to $\mathcal{V}$–coordinate ones if an appropriate Ukrainian-to-Vietnamese dictionary $\mathbf{X}_{\mathcal{V}\leftarrow\mathcal{U}}$ were not handy?

In real life, it might be easier to find an English-speaking Ukrainian and an English-speaking Vietnamese person than find someone who is familiar with both Ukrainian and Vietnamese. Hence formula (7.3) suggests to use the $\mathcal{E}$–coordinate vector $\mathbf{w}_{\mathcal{E}}$ as an intermediary, or interpreter, when trying to translate a $\mathcal{U}$–word $\mathbf{w}_{\mathcal{U}}$ to a $\mathcal{V}$–word $\mathbf{w}_{\mathcal{V}}$, just as translators might do for languages:

$$\mathbf{w}_{\mathcal{V}} \underset{\text{def.}}{=} \mathbf{X}_{\mathcal{V}\leftarrow\mathcal{U}}\mathbf{w}_{\mathcal{U}} = \mathbf{X}_{\mathcal{V}\leftarrow\mathcal{E}}\,\mathbf{X}_{\mathcal{E}\leftarrow\mathcal{U}}\mathbf{w}_{\mathcal{U}}\ .$$

Here we have split the coordinate transformation from $\mathcal{U}$ to $\mathcal{V}$ into two parts: first from $\mathcal{U}$ to $\mathcal{E}$ operating on the $\mathcal{U}$–word $\mathbf{w}_{\mathcal{U}}$ via $\mathbf{X}_{\mathcal{E}\leftarrow\mathcal{U}}$, followed by the basis change transformation from $\mathcal{E}$ to $\mathcal{V}$ operating on the resulting $\mathcal{E}$–word $\mathbf{X}_{\mathcal{E}\leftarrow\mathcal{U}}\mathbf{w}_{\mathcal{U}}$.

Since $\mathbf{X}_{\mathcal{E}\leftarrow\mathcal{U}} = \mathbf{I}^{-1}\mathbf{U} = \mathbf{U}$ and $\mathbf{X}_{\mathcal{V}\leftarrow\mathcal{E}} = \mathbf{V}^{-1}\mathbf{I} = \mathbf{V}^{-1}$ according to Eq. (7.3), we once more see that

$$\mathbf{X}_{\mathcal{V}\leftarrow\mathcal{U}} = \mathbf{X}_{\mathcal{V}\leftarrow\mathcal{E}}\,\mathbf{X}_{\mathcal{E}\leftarrow\mathcal{U}} = \mathbf{V}^{-1}\mathbf{U}.$$

Note that the arrowed subscripts of a basis change matrix $\mathbf{X}_{..\,\leftarrow\,..}$ indicate which coordinate system is used as input (on the right end of the arrow) and which coordinate system as output (on the left end).

Let us evaluate a typical basis change matrix $\mathbf{W} := \mathbf{V}^{-1}\mathbf{U} = \mathbf{X}_{\mathcal{V}\leftarrow\mathcal{U}}$ now.

If $\mathbf{W} = \mathbf{V}^{-1}\mathbf{U}$, then $\mathbf{V}\mathbf{W} = \mathbf{U}$ after left multiplication of the equation by $\mathbf{V}$. Thus,

$$\mathbf{V}\begin{pmatrix} | & & | \\ \mathbf{w}_1 & \dots & \mathbf{w}_n \\ | & & | \end{pmatrix} = \begin{pmatrix} | & & | \\ \mathbf{u}_1 & \dots & \mathbf{u}_n \\ | & & | \end{pmatrix} = \mathbf{U}.$$

Here the matrix $\mathbf{V}$ and the column vectors $\mathbf{u}_i$ in $\mathbf{U}$ are known, while the column vectors $\mathbf{w}_i$ of $\mathbf{W} = \mathbf{V}^{-1}\mathbf{U} = \mathbf{X}_{\mathcal{V}\leftarrow\mathcal{U}}$ are unknown. This is a system of linear equations $\mathbf{V}\mathbf{w}_i = \mathbf{u}_i$ for each of the unknown columns $\mathbf{w}_i$ of $\mathbf{W}$ and $i = 1, \dots, n$. To find $\mathbf{W}$, we need to solve n systems of linear equations with the same system matrix $\mathbf{V}$. As in Section 6.1, we can combine our efforts here into one **multi-augmented matrix** $(\mathbf{V} \mid \mathbf{U})_{n \times 2n}$ and row reduce the n columns of $\mathbf{V}$ to their RREF $\mathbf{I}_n$, while simultaneously updating the right half of the augmented matrix; see Corollary 3 in Section 6.2. Then we obtain $\mathbf{W} = \mathbf{V}^{-1}\mathbf{U}$ in the last n columns (i.e., $(\mathbf{I}_n \mid \mathbf{V}^{-1}\mathbf{U}) = (\mathbf{I}_n \mid \mathbf{W})$).

EXAMPLE 3   What are the basis change matrices (a) from the basis $\mathcal{U} = \left\{ \begin{pmatrix} 1 \\ -1 \end{pmatrix}, \begin{pmatrix} 1 \\ 2 \end{pmatrix} \right\}$

to the basis $\mathcal{V} = \left\{ \begin{pmatrix} 1 \\ 1 \end{pmatrix}, \begin{pmatrix} 2 \\ 3 \end{pmatrix} \right\}$ of $\mathbb{R}^2$, and (b) back from $\mathcal{V}$ to $\mathcal{U}$?

We have $\mathbf{X}_{\mathcal{V} \leftarrow \mathcal{U}} = \mathbf{V}^{-1}\mathbf{U} = \begin{pmatrix} 1 & 2 \\ 1 & 3 \end{pmatrix}^{-1} \begin{pmatrix} 1 & 1 \\ -1 & 2 \end{pmatrix}$ and $\mathbf{X}_{\mathcal{V} \leftarrow \mathcal{U}}$ is obtained

by row reducing the multi-augmented matrix $(\mathbf{V} \mid \mathbf{U})_{2 \times 4}$ to $(\mathbf{I}_2 \mid \mathbf{V}^{-1}\mathbf{U})$:

V		U		
☐1	2	1	1	
☐1	3	−1	2	$-\,row_1$
☐1	☐2	1	1	$-2\,row_2$
0	☐1	−2	1	
$\mathbf{I}_2$		5	−1	$= \mathbf{V}^{-1}\mathbf{U} = \mathbf{X}_{\mathcal{V} \leftarrow \mathcal{U}}$
		−2	1	

The reverse direction basis change matrix $\mathbf{X}_{\mathcal{U} \leftarrow \mathcal{V}}$ is given by

$$(\mathbf{V}^{-1}\mathbf{U})^{-1} = \mathbf{U}^{-1}\mathbf{V} = (\mathbf{X}_{\mathcal{V} \leftarrow \mathcal{U}})^{-1} = \tfrac{1}{3} \begin{pmatrix} 1 & 1 \\ 2 & 5 \end{pmatrix}. \qquad \blacktriangleleft$$

To compute the basis change matrix between two bases $\mathcal{U}$ and $\mathcal{V}$ of $\mathbb{R}^n$, a row reduction scheme has to be carried out across each complete row of length $2n$ of the multi-augmented matrix $(\mathbf{V} \mid \mathbf{U})$ as indicated in the following scheme:

$$
\begin{array}{c}
\mathbf{X}_{\mathcal{V} \leftarrow \mathcal{U}} \\
\diagup | \diagdown \\
\begin{array}{c|c}
\mathbf{V} & \mathbf{U} \\
\vdots & \vdots \\
\downarrow & \downarrow \qquad \textit{by row reduction} \\
\vdots & \vdots \\
\mathbf{I}_n & \mathbf{V}^{-1}\mathbf{U} \quad = \mathbf{X}_{\mathcal{V} \leftarrow \mathcal{U}}
\end{array}
\end{array}
$$

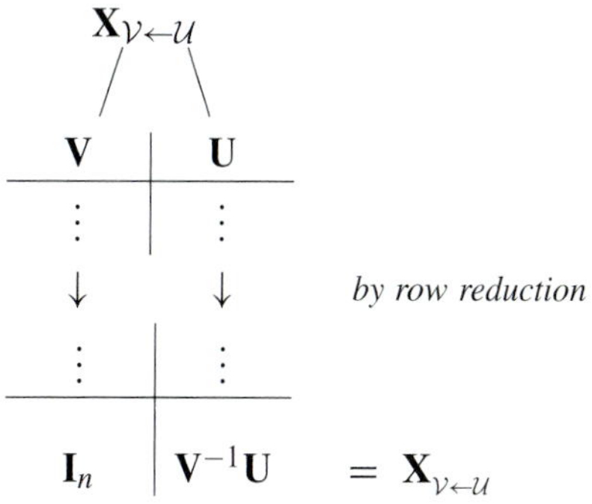

In this scheme, the matrices $\mathbf{U}$ and $\mathbf{V}$ contain the basis vectors of the two bases $\mathcal{U}$ and $\mathcal{V}$ as columns. The two bases are written down in the same order as they appear in the subscript of $\mathbf{X}_{\mathcal{V}\leftarrow\mathcal{U}}$. The $n \times 2n$ matrix $(\mathbf{V} \mid \mathbf{U})$ is row reduced to yield $(\mathbf{I}_n \mid \mathbf{V}^{-1}\mathbf{U})$ with $\mathbf{V}^{-1}\mathbf{U} = \mathbf{X}_{\mathcal{V}\leftarrow\mathcal{U}}$.

(c) Matrix Representations with Respect to General Bases

Why are we interested in basis changes?

The purpose of learning to express vectors in $\mathbb{R}^n$ with respect to arbitrary bases and to translate between coordinate vectors for different bases lies in a more intimate and more profound understanding of linear transformations.

From Section 1.1, the standard matrix representation

$$\mathbf{A}_{nn} = \begin{pmatrix} | & & | \\ T(\mathbf{e}_1) & \cdots & T(\mathbf{e}_n) \\ | & & | \end{pmatrix} = \mathbf{A}_{\mathcal{E}}$$

of a linear transformation $T : \mathbb{R}^n \to \mathbb{R}^n$ maps standard $\mathcal{E}$–coordinate vectors $\mathbf{x}$ $(= \mathbf{x}_{\mathcal{E}})$ to $\mathcal{E}$-vectors $\mathbf{Ax}$, so that $T(\mathbf{x}) = \mathbf{Ax}$. What happens to the matrix representation of T if we desire instead to map $\mathcal{U}$–coordinate vectors to $\mathcal{U}$–coordinate vectors by matrix $\times$ vector multiplication in the same way as the underlying linear transformation $T : \mathbb{R}^n \to \mathbb{R}^n$ does? How does the matrix representation of T change when using a general basis $\mathcal{U}$ to describe position in $\mathbb{R}^n$? What does the matrix representation $\mathbf{A}_{\mathcal{U}}$ of T look like for $\mathcal{U}$? How is $\mathbf{A}_{\mathcal{U}}$ related to $\mathbf{A}_{\mathcal{E}}$?

These are pressing questions that we can answer.

What is $\mathbf{A}_{\mathcal{U}}$, the **matrix representation of T with respect to the basis $\mathcal{U}$**? Note that

$$T\mathbf{x} = \underset{\substack{\mathcal{E}\text{–basis} \\ \text{description of } T\mathbf{x}}}{\mathbf{A}_{\mathcal{E}}\mathbf{x}_{\mathcal{E}}} \;\widehat{=}\; \left(\mathbf{X}_{\mathcal{U}\leftarrow\mathcal{E}}\left(\mathbf{A}_{\mathcal{E}}\mathbf{X}_{\mathcal{E}\leftarrow\mathcal{U}}\mathbf{x}_{\mathcal{U}}\right)_{\mathcal{E}\text{-vector}}\right)_{\mathcal{U}\text{-vector}} := \underset{\substack{\mathcal{U}\text{–basis} \\ \text{description of } T\mathbf{x}}}{\mathbf{A}_{\mathcal{U}}\mathbf{x}_{\mathcal{U}}},$$

since $\mathbf{x}_{\mathcal{E}} = \mathbf{X}_{\mathcal{E}\leftarrow\mathcal{U}}\mathbf{x}_{\mathcal{U}}$ and $\mathbf{X}_{\mathcal{U}\leftarrow\mathcal{E}}$ transforms the $\mathcal{E}$–coordinate vector $\mathbf{A}_{\mathcal{E}}\mathbf{X}_{\mathcal{E}\leftarrow\mathcal{U}}\mathbf{x}_{\mathcal{U}}$ of $T\mathbf{x}$ into a $\mathcal{U}$–coordinate vector (i.e., to $\mathbf{A}_{\mathcal{U}}\mathbf{x}_{\mathcal{U}}$).

While the actual entries of the coordinate vectors $(\mathbf{A}_{\mathcal{E}}\mathbf{x}_{\mathcal{E}})_{\mathcal{E}}$ and $(\mathbf{A}_{\mathcal{U}}\mathbf{x}_{\mathcal{U}})_{\mathcal{U}}$ are different, the equivalence symbol $\widehat{=}$ indicates that $\mathbf{A}_{\mathcal{E}}\mathbf{x}_{\mathcal{E}}$ and $\mathbf{A}_{\mathcal{U}}\mathbf{x}_{\mathcal{U}}$ designate the same point in $\mathbb{R}^n$, but expressed with respect to the different bases $\mathcal{E}$ and $\mathcal{U}$.

Thus, the matrix $\mathbf{A}_{\mathcal{U}}$, which maps $\mathcal{U}$–coordinate vectors $\mathbf{x}_{\mathcal{U}}$ to $\mathcal{U}$–vectors $\mathbf{A}_{\mathcal{U}}\mathbf{x}_{\mathcal{U}}$ in the same way as T does, is given by

$$\boxed{\mathbf{A}_{\mathcal{U}} = \mathbf{X}_{\mathcal{U}\leftarrow\mathcal{E}}\,\mathbf{A}_{\mathcal{E}}\,\mathbf{X}_{\mathcal{E}\leftarrow\mathcal{U}} = \mathbf{U}^{-1}\mathbf{A}_{\mathcal{E}}\mathbf{U}} \tag{7.4}$$

from formula (7.3). Transforming a square matrix $\mathbf{A}$ by multiplying $\mathbf{A}$ on the left and right simultaneously by mutually inverse matrices $\mathbf{U}^{-1}$ and $\mathbf{U}$ in $\mathbf{U}^{-1}\mathbf{AU}$ is called a **similarity transform** of the matrix $\mathbf{A}$. The matrices $\mathbf{A}$ and $\mathbf{B} := \mathbf{U}^{-1}\mathbf{AU}$ are called **similar**.

The **algebraic matrix equation** $\mathbf{A}_\mathcal{U} = \mathbf{U}^{-1}\mathbf{A}_\mathcal{E}\mathbf{U}$ relates two matrix representations of one linear transformation T for two bases. This equation will be studied throughout the remainder of the book. Therefore, we want to take another, more pictorial look at it:

On the left side of the equal sign, $\mathbf{A}_\mathcal{U}$ maps $\mathcal{U}$–coordinate vectors to $\mathcal{U}$–coordinate vectors as the underlying linear transformation T prescribes. The matrix product $\mathbf{U}^{-1}\mathbf{A}_\mathcal{E}\mathbf{U}$ on the right side of the equation describes the same vector mapping, but split into three stages: First the matrix product $\mathbf{U}^{-1}\mathbf{A}_\mathcal{E}\mathbf{U}$ maps any $\mathcal{U}$ vector $\mathbf{x}_\mathcal{U}$ to its $\mathcal{E}$–coordinate vector $\mathbf{U}\mathbf{x}_\mathcal{U}$. Then it transforms this standard $\mathcal{E}$ vector to its standard $\mathcal{E}$ vector image $\mathbf{A}_\mathcal{E}\mathbf{U}\mathbf{x}_\mathcal{U}$, as prescribed by T. Finally, it converts the resulting $\mathcal{E}$–coordinate vector back to $\mathcal{U}$-coordinates in forming $\mathbf{U}^{-1}\mathbf{A}_\mathcal{E}\mathbf{U}\mathbf{x}_\mathcal{U}$. With this understanding, we can draw a pictorial description of the algebraic matrix equation $\mathbf{A}_\mathcal{U} = \mathbf{U}^{-1}\mathbf{A}_\mathcal{E}\mathbf{U}$ as follows:

$$
\begin{array}{ccc}
\mathcal{U} \text{ vector} & & \mathcal{U} \text{ vector} \\
\mathbf{x}_\mathcal{U} & \xrightarrow{\ \mathbf{A}_\mathcal{U}\ } & \mathbf{A}_\mathcal{U}\mathbf{x}_\mathcal{U} = \mathbf{U}^{-1}\mathbf{A}_\mathcal{E}\mathbf{U}\mathbf{x}_\mathcal{U} \\
\downarrow \mathbf{U} & & \uparrow \mathbf{U}^{-1} \\
\mathbf{U}\mathbf{x}_\mathcal{U} = \mathbf{x}_\mathcal{E} & \xrightarrow{\ \mathbf{A}_\mathcal{E}\ } & \mathbf{A}_\mathcal{E}\mathbf{x}_\mathcal{E} = \mathbf{A}_\mathcal{E}\mathbf{U}\mathbf{x}_\mathcal{U} \\
\mathcal{E} \text{ vector} & & \mathcal{E} \text{ vector}
\end{array}
$$

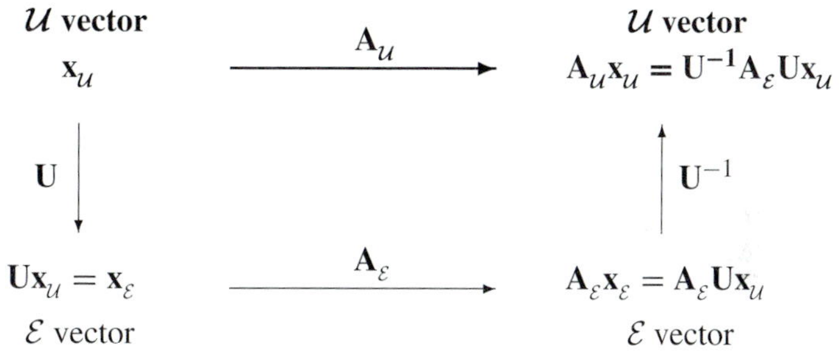

Any **equation** is two sided, and thus it can be read both from left to right and from right to left, giving rise to slightly different meanings. For example, the equations $2 \cdot 2 = 4$ and $(a - b)(a + b) = a^2 - b^2$ show us how to combine and simplify the two left side expressions. On the other hand, these equations with their sides reversed, namely, $4 = 2 \cdot 2$ and $a^2 - b^2 = (a - b)(a + b)$, show how to factor the integer 4 and any expression that is the difference of two squares.

In the preceding diagram, the matrix $\mathbf{A}_\mathcal{U}$ was expressed as a composition of three maps. We can multiply the given equation $\mathbf{A}_\mathcal{U} = \mathbf{U}^{-1}\mathbf{A}_\mathcal{E}\mathbf{U}$ by $\mathbf{U}$ on the left and by $\mathbf{U}^{-1}$ on the right to obtain the expression $\mathbf{U}\mathbf{A}_\mathcal{U}\mathbf{U}^{-1}$ for $\mathbf{A}_\mathcal{E}$. This equivalent equation $\mathbf{A}_\mathcal{E} = \mathbf{U}\mathbf{A}_\mathcal{U}\mathbf{U}^{-1}$ decomposes the standard matrix representation $\mathbf{A}_\mathcal{E}$ of T into the product of three maps in a slightly different fashion as follows:

$$
\begin{array}{ccc}
\mathcal{E} \text{ vector} & & \mathcal{E} \text{ vector} \\
\mathbf{x}_\mathcal{E} & \xrightarrow{\ \mathbf{A}_\mathcal{E}\ } & \mathbf{A}_\mathcal{E}\mathbf{x}_\mathcal{E} = \mathbf{U}\mathbf{A}_\mathcal{U}\mathbf{U}^{-1}\mathbf{x}_\mathcal{E} \\
\downarrow \mathbf{U}^{-1} & & \uparrow \mathbf{U} \\
\mathbf{U}^{-1}\mathbf{x}_\mathcal{E} = \mathbf{x}_\mathcal{U} & \xrightarrow{\ \mathbf{A}_\mathcal{U}\ } & \mathbf{A}_\mathcal{U}\mathbf{x}_\mathcal{U} = \mathbf{A}_\mathcal{U}\mathbf{U}^{-1}\mathbf{x}_\mathcal{E} \\
\mathcal{U} \text{ vector} & & \mathcal{U} \text{ vector}
\end{array}
$$

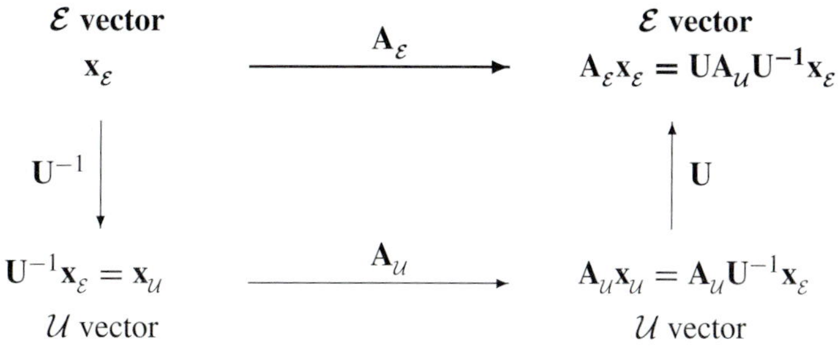

This diagram gives us another equivalent look at basis dependent matrix representations.

The standard matrix representation of a linear transformation $T : \mathbb{R}^n \to \mathbb{R}^n$ has been described as

$$\mathbf{A}_{\mathcal{E}} = \begin{pmatrix} | & & | \\ T\mathbf{e}_1 & \dots & T\mathbf{e}_n \\ | & & | \end{pmatrix} = \begin{pmatrix} | & & | \\ (T\mathbf{e}_1)_{\mathcal{E}} & \dots & (T\mathbf{e}_n)_{\mathcal{E}} \\ | & & | \end{pmatrix}$$

in Section 1.1. There we have emphasized that for the standard basis $\mathcal{E} = \{\mathbf{e}_1, \dots, \mathbf{e}_n\}$, the columns of $\mathbf{A}_{\mathcal{E}}$ are $T\mathbf{e}_i = (T\mathbf{e}_i)_{\mathcal{E}}$ for each $i = 1, \dots, n$.

As *one basis is as good as any other*, one should be able to express T for an arbitrary basis $\mathcal{U}$ of $\mathbb{R}^n$ by its $\mathcal{U}$ basis matrix representation

$$\mathbf{A}_{\mathcal{U}} = \begin{pmatrix} | & & | \\ (T\mathbf{u}_1)_{\mathcal{U}} & \dots & (T\mathbf{u}_n)_{\mathcal{U}} \\ | & & | \end{pmatrix}$$

by simply replacing each $\mathbf{e}_i$ with $\mathbf{u}_i$ and $\mathcal{E}$ with $\mathcal{U}$ in the original formula for $\mathbf{A}_{\mathcal{E}}$. This is, in fact, the case.

Proposition: If $T : \mathbb{R}^n \to \mathbb{R}^n$ is a linear transformation and $\mathcal{U} = \{\mathbf{u}_1, \dots, \mathbf{u}_n\} \subset \mathbb{R}^n$ is a basis for $\mathbb{R}^n$, then

$$\mathbf{A}_{\mathcal{U}} = \begin{pmatrix} | & & | \\ (T\mathbf{u}_1)_{\mathcal{U}} & \dots & (T\mathbf{u}_n)_{\mathcal{U}} \\ | & & | \end{pmatrix}$$

is the matrix representation of T with respect to the basis $\mathcal{U}$. ◀

Proof From Section 1.1, we have $T\mathbf{u}_i = \mathbf{A}_{\mathcal{E}}\mathbf{u}_i = \mathbf{A}_{\mathcal{E}} \begin{pmatrix} | & & | \\ \mathbf{u}_1 & \dots & \mathbf{u}_n \\ | & & | \end{pmatrix} \mathbf{e}_i$ for each $i = 1, \dots, n$, since $\begin{pmatrix} | \\ \mathbf{u}_i \\ | \end{pmatrix} = \begin{pmatrix} | & & | \\ \mathbf{u}_1 & \dots & \mathbf{u}_n \\ | & & | \end{pmatrix} \mathbf{e}_i$. Thus, $(T\mathbf{u}_i)_{\mathcal{E}} = \mathbf{A}_{\mathcal{E}}\mathbf{U}\mathbf{e}_i$ for the matrix $\mathbf{U} = \begin{pmatrix} | & & | \\ \mathbf{u}_1 & \dots & \mathbf{u}_n \\ | & & | \end{pmatrix}$. Recall that the $\mathcal{U}$–coordinate vector $\mathbf{b}_{\mathcal{U}}$ of any vector $\mathbf{b} = \mathbf{b}_{\mathcal{E}} \in \mathbb{R}^n$ is given by $\mathbf{b}_{\mathcal{U}} = \mathbf{U}^{-1}\mathbf{b}_{\mathcal{E}}$. Applied to $\mathbf{b} = (T\mathbf{u}_i)_{\mathcal{E}}$, this yields

$$(T\mathbf{u}_i)_{\mathcal{U}} = \mathbf{U}^{-1}(T\mathbf{u}_i)_{\mathcal{E}} = \mathbf{U}^{-1}\mathbf{A}_{\mathcal{E}}\mathbf{U}\mathbf{e}_i.$$

Thus, the matrix $\mathbf{A}_{\mathcal{U}} = \begin{pmatrix} | & & | \\ (T\mathbf{u}_1)_{\mathcal{U}} & \dots & (T\mathbf{u}_n)_{\mathcal{U}} \\ | & & | \end{pmatrix}$ has the same columns as $\mathbf{U}^{-1}\mathbf{A}_{\mathcal{E}}\mathbf{U}$. ■

One important goal of linear algebra (other than solving linear equations) is to find revealing matrix representations $\mathbf{A}_{\mathcal{U}}$ of a given linear transformation

$T : \mathbb{R}^n \to \mathbb{R}^n$ with respect to special bases $\mathcal{U}$ of $\mathbb{R}^n$. Specifically, we are looking for matrix representations $\mathbf{A}_\mathcal{U}$ of the linear transformation T that have "nice properties." According to Eq. (7.4), this involves matrix similarities $\mathbf{U}^{-1}\mathbf{A}\mathbf{U}$ of the standard matrix representation $\mathbf{A} = \mathbf{A}_\mathcal{E}$ for T. We want to find a specific basis $\mathcal{U}$ for T and $\mathbb{R}^n$ that reveals T's properties most clearly and openly in $\mathbf{A}_\mathcal{U} = \mathbf{U}^{-1}\mathbf{A}_\mathcal{E}\mathbf{U}$.

EXAMPLE 4 Assume that $\mathbf{A}$ is a 2×2 matrix with $\mathbf{Ax} = 2\mathbf{x}$ and $\mathbf{Ay} = -\mathbf{y}$ for two linearly independent vectors $\mathbf{x}, \mathbf{y} \in \mathbb{R}^2$. Then

$$\mathbf{A}\begin{pmatrix} | & | \\ \mathbf{x} & \mathbf{y} \\ | & | \end{pmatrix} = \begin{pmatrix} | & | \\ \mathbf{Ax} & \mathbf{Ay} \\ | & | \end{pmatrix} = \begin{pmatrix} | & | \\ 2\mathbf{x} & -\mathbf{y} \\ | & | \end{pmatrix} = \begin{pmatrix} | & | \\ \mathbf{x} & \mathbf{y} \\ | & | \end{pmatrix}\begin{pmatrix} 2 & 0 \\ 0 & -1 \end{pmatrix}.$$

As $\mathbf{x}$ and $\mathbf{y}$ are assumed to be linearly independent, the 2×2 matrix $\mathbf{U} = \begin{pmatrix} | & | \\ \mathbf{x} & \mathbf{y} \\ | & | \end{pmatrix}$ can be inverted; its REF has two pivots. The matrix equation $\mathbf{AU} = \mathbf{U}\begin{pmatrix} 2 & 0 \\ 0 & -1 \end{pmatrix}$ is equivalent to the similarity transform equation

$$\mathbf{U}^{-1}\mathbf{AU} = \begin{pmatrix} 2 & 0 \\ 0 & -1 \end{pmatrix} = \mathbf{A}_\mathcal{U}$$

for $\mathbf{A}$, This similarity gives a **diagonal matrix representation** $\mathbf{A}_\mathcal{U}$ of the underlying linear transformation with respect to the specific basis $\mathcal{U} = \{\mathbf{x}, \mathbf{y}\}$ of $\mathbb{R}^2$. ◀

The essential assumptions in Example 4 that allow us to represent the given linear transformation by the "nice" diagonal matrix $\mathbf{A}_\mathcal{U}$ for the basis $\mathcal{U} = \{\mathbf{x}, \mathbf{y}\}$ of $\mathbb{R}^2$ are the two matrix–vector equations

$$\mathbf{Ax} = 2\mathbf{x} \quad \text{and} \quad \mathbf{Ay} = -\mathbf{y}$$

that hold for $\mathbf{A}$.

In general, if $\mathbf{A}$ is an $n \times n$ matrix and $\mathbf{Ax} = \lambda\mathbf{x}$ for a vector $\mathbf{x} \neq \mathbf{0} \in \mathbb{R}^n$ or $\mathbb{C}^n$ and a scalar $\lambda \in \mathbb{R}$ or $\mathbb{C}$, then we say that $\mathbf{A}$ has the **eigenvalue** λ for the **eigenvector** $\mathbf{x}$.

Note that the equation $\mathbf{Ax} = \lambda\mathbf{x}$ is a **nonlinear equation** for the given matrix $\mathbf{A}_{nn}$, since the unknown quantities $\mathbf{x}$ and λ occur in product form; that is, the unknown eigenvalue λ is multiplied by the unknown eigenvector $\mathbf{x}$ on the right-hand side.

The notion of a basis change has led us to the nonlinear matrix eigenvalue/eigenvector equation which we shall study in depth in the subsequent chapters.

7.1.P Problems

1. If $\mathbf{x}_{\mathcal{U}} = \begin{pmatrix} 2 \\ -1 \\ 3 \end{pmatrix}$ is given and $\mathbf{X}_{\mathcal{U}\leftarrow\mathcal{V}} = \begin{pmatrix} 1 & 0 & 2 \\ -1 & 1 & 0 \\ 0 & 1 & 1 \end{pmatrix}$ is the basis change matrix from the basis $\mathcal{V}$ to the basis $\mathcal{U}$ of $\mathbb{R}^3$, what is the formula for the $\mathcal{V}$–coordinate vector for $\mathbf{x}$?
Compute $\mathbf{x}_{\mathcal{V}}$ explicitly.

2. Find the coordinate vector $\mathbf{x}_{\mathcal{B}}$ for $\mathbf{x} = \mathbf{x}_{\mathcal{E}} = \begin{pmatrix} -2 & 3 & -1 \end{pmatrix}^T$ and the basis $\mathcal{B} = \left\{ \begin{pmatrix} -2 \\ 1 \\ 0 \end{pmatrix}, \begin{pmatrix} 2 \\ -4 \\ 1 \end{pmatrix}, \begin{pmatrix} 4 \\ -1 \\ 2 \end{pmatrix} \right\}$ of $\mathbb{R}^3$.

3. Find the coordinate vector $\mathbf{x}_{\mathcal{E}}$ if $\mathbf{x}_{\mathcal{B}} = \begin{pmatrix} -2 & 3 & -1 \end{pmatrix}^T$ for the basis $\mathcal{B} = \left\{ \begin{pmatrix} -2 \\ 1 \\ 0 \end{pmatrix}, \begin{pmatrix} 2 \\ -4 \\ 1 \end{pmatrix}, \begin{pmatrix} 4 \\ -1 \\ 2 \end{pmatrix} \right\}$ of $\mathbb{R}^3$.

4. How are the two, previous problems related? What is the relation between the coordinate vectors $\mathbf{x}_{\mathcal{B}}$ and $\mathbf{x}_{\mathcal{E}}$?

5. Show that if $\mathbf{w}_{\mathcal{U}} = \mathbf{0} \in \mathbb{R}^n$ for one basis $\mathcal{U}$ of $\mathbb{R}^n$, then $\mathbf{w}_{\mathcal{V}} = \mathbf{0}$ for every basis $\mathcal{V}$ of $\mathbb{R}^n$. In particular, show that $\mathbf{v}_{\mathcal{U}} = \mathbf{0}$ if and only if $\mathbf{v} = \mathbf{0}$ is the origin of $\mathbb{R}^n$.

6. Prove that every basis change matrix $\mathbf{X}_{\mathcal{V}\leftarrow\mathcal{U}}$ between two bases $\mathcal{U}$ and $\mathcal{V}$ of $\mathbb{R}^n$ is nonsingular.

7. (a) Find the basis change matrix $\mathbf{X}_{\mathcal{V}\leftarrow\mathcal{U}}$ for the two bases

$$\mathcal{U} = \left\{ \begin{pmatrix} 1 \\ -2 \\ -1 \end{pmatrix}, \begin{pmatrix} 0 \\ -2 \\ 1 \end{pmatrix}, \begin{pmatrix} 1 \\ 1 \\ 0 \end{pmatrix} \right\}$$

and

$$\mathcal{V} = \left\{ \begin{pmatrix} 1 \\ 0 \\ -1 \end{pmatrix}, \begin{pmatrix} 1 \\ 2 \\ 1 \end{pmatrix}, \begin{pmatrix} 0 \\ -1 \\ 1 \end{pmatrix} \right\}$$

of $\mathbb{R}^3$.

(b) If a point $\mathbf{x} \in \mathbb{R}^3$ has the $\mathcal{U}$–coordinate vector $\mathbf{x}_{\mathcal{U}} = \begin{pmatrix} -12 \\ 7 \\ 2 \end{pmatrix}$, what are $\mathbf{x} = \mathbf{x}_{\mathcal{E}}$ and $\mathbf{x}_{\mathcal{V}}$ for $\mathcal{U}$ and $\mathcal{V}$ from part (a)?

8. The standard basis $\{\mathbf{e}_1, \mathbf{e}_2\}$ of $\mathbb{R}^2$ is rotated counterclockwise by 45° to obtain a new basis $\mathcal{U}$ of $\mathbb{R}^2$.

(a) What are the $\mathcal{E}$–basis coordinates of the basis vectors in $\mathcal{U}$?

(b) Write out the basis-change matrices $\mathbf{X}_{\mathcal{E}\leftarrow\mathcal{U}}$ and $\mathbf{X}_{\mathcal{U}\leftarrow\mathcal{E}}$.

(c) Verify that $\mathbf{X}_{\mathcal{E}\leftarrow\mathcal{U}}\,\mathbf{X}_{\mathcal{U}\leftarrow\mathcal{E}} = \mathbf{I}_2 = \mathbf{X}_{\mathcal{U}\leftarrow\mathcal{E}}\,\mathbf{X}_{\mathcal{E}\leftarrow\mathcal{U}}$.

(d) Represent the matrix $\mathbf{A} = \mathbf{A}_{\mathcal{E}} = \begin{pmatrix} 3 & 1 \\ 1 & 3 \end{pmatrix}$ with respect to the basis $\mathcal{U}$.

(e) Do this again, but for the identity matrix $\mathbf{B}_{\mathcal{E}} = \mathbf{I}_2$ and the zero matrix $\mathbf{C}_{\mathcal{E}} = \mathbf{O}_2$.

9. (a) Write out two matrix equations that relate the two coordinate vectors $\mathbf{w}_{\mathcal{V}}$ and $\mathbf{w}_{\mathcal{U}}$ with respect to two bases $\mathcal{U}$ and $\mathcal{V}$ of $\mathbb{R}^n$.

(b) Find the coordinate vector $\mathbf{w}_{\mathcal{U}}$ of $\mathbf{w} = \mathbf{w}_{\mathcal{E}} = \begin{pmatrix} 1 & 0 & 3 & -1 \end{pmatrix}^T$ with respect to the basis

$$\mathcal{U} = \left\{ \begin{pmatrix} 1 \\ 0 \\ 0 \\ 0 \end{pmatrix}, \begin{pmatrix} 1 \\ 1 \\ 0 \\ 0 \end{pmatrix}, \begin{pmatrix} 1 \\ 1 \\ 1 \\ 0 \end{pmatrix}, \begin{pmatrix} 1 \\ 1 \\ 1 \\ 1 \end{pmatrix} \right\} \text{ of } \mathbb{R}^4.$$

(c) If $\mathbf{z}_{\mathcal{U}} = \begin{pmatrix} -1 \\ 2 \\ -2 \\ 3 \end{pmatrix}$, where does $\mathbf{z}$ lie in the standard coordinate system $\mathcal{E}$ of $\mathbb{R}^4$?

(d) Find the basis change matrices $\mathbf{X}_{\mathcal{E}\leftarrow\mathcal{U}}$ and $\mathbf{X}_{\mathcal{U}\leftarrow\mathcal{E}}$ for $\mathcal{U}$ from part (b).

10. Draw the vectors $\mathbf{A}\mathbf{e}_1$ and $\mathbf{A}\mathbf{e}_2$ that result from transforming the standard basis $\mathbf{e}_1$ and $\mathbf{e}_2$ of $\mathbb{R}^2$ by matrix multiplication with $\mathbf{A} = \frac{1}{2}\begin{pmatrix} \sqrt{3} & -1 \\ 1 & \sqrt{3} \end{pmatrix}$. Show that $\{\mathbf{A}\mathbf{e}_i\}$ is a basis for $\mathbb{R}^2$. What does this basis change $\{\mathbf{e}_i\} \to \{\mathbf{A}\mathbf{e}_i\}$ actually do? Describe the transformation. What is the inverse transformation? What does it do? What is its matrix?

11. Assume that two sets of basis vectors for one subspace are given as $\mathcal{U} = \{\mathbf{u}_i\}_{i=1}^4$ and as $\mathcal{V} = \{\mathbf{v}_i\}_{i=1}^{\ell}$, and assume that they are related as follows:

$\mathbf{u}_1 = 2\mathbf{v}_3 - \mathbf{v}_4 + \mathbf{v}_1, \ \mathbf{u}_2 = \mathbf{v}_2 + \mathbf{v}_1,$

$\mathbf{u}_3 = -\mathbf{v}_3, \ \mathbf{u}_4 = \mathbf{v}_4 - \mathbf{u}_4$

(a) What is the integer ℓ used in $\mathcal{V}$?

(b) What is the basis change matrix $\mathbf{X}_{\mathcal{U}\leftarrow\mathcal{V}}$?

(c) What is the basis change matrix $\mathbf{X}_{\mathcal{U}\leftarrow\mathcal{U}}$?

(d) What is the basis change matrix $\mathbf{X}_{\mathcal{V}\leftarrow\mathcal{U}}$?

(e) Show that $\mathbf{X}_{\mathcal{U}\leftarrow\mathcal{U}} = \mathbf{X}_{\mathcal{V}\leftarrow\mathcal{V}}$.

(f) Find the $\mathcal{U}$–coordinate vector of $\mathbf{w}_{\mathcal{U}} = \begin{pmatrix} -1 \\ 2 \\ 3 \\ -1 \end{pmatrix}$.

(g) Find the $\mathcal{V}$–coordinate vector of $\mathbf{w}_{\mathcal{U}} = \begin{pmatrix} -1 \\ 2 \\ 3 \\ -1 \end{pmatrix}$.

12. If the basis change matrix between $\mathcal{U}$–coordinates and $\mathcal{V}$–coordinates is $\mathbf{X}_{\mathcal{V}\leftarrow\mathcal{U}} = \begin{pmatrix} 1 & 2 & 5 \\ -2 & 4 & 1 \\ 0 & 1 & -3 \end{pmatrix}$, which of the following are true, and which are false give reasons and derive the answers:

(a) $\dim(U) = 4$;

(b) $\dim(V) = 3 = \dim(U)$;

(c) $\dim(V) = 3$ and $\dim(U) < 3$.

(d) What is $\mathbf{w}_{\mathcal{U}}$ for $\mathbf{w}_{\mathcal{V}} = \begin{pmatrix} 1 \\ -1 \\ 1 \end{pmatrix}$?

(e) What is $\mathbf{w}_{\mathcal{V}}$ for $\mathbf{w}_{\mathcal{U}} = -\begin{pmatrix} 1 \\ -1 \\ 1 \end{pmatrix}$?

(f) What is $\mathbf{w}_{\mathcal{V}}$ for $\mathbf{w}_{\mathcal{U}} = -\begin{pmatrix} 0 \\ 0 \\ 0 \end{pmatrix}$?

13. Let $\mathcal{U} = \{\mathbf{u}_1, \dots, \mathbf{u}_k\}$ be a basis, and let $\mathcal{V} = \{\mathbf{u}_k, \dots, \mathbf{u}_1\}$ be the identical basis except that the basis vectors of $\mathcal{U}$ appear in reverse order.

(a) If $k = 4$ and $\mathbf{x}_{\mathcal{U}} = \begin{pmatrix} 1 & 3 & -4 & 7 \end{pmatrix}^T$ find $\mathbf{x}_{\mathcal{V}}$.

(b) What is the basis change matrix from $\mathcal{V}$–vectors to $\mathcal{U}$–vectors for arbitrary dimensions k?

(c) Show that for all k we have $\mathbf{X}_{\mathcal{V}\leftarrow\mathcal{U}} = \mathbf{X}_{\mathcal{U}\leftarrow\mathcal{V}}$.

14. (a) Find the coordinate vectors of the polynomials $u_1(x) = 1 - x$, $u_2(x) = 1 + 2x$, and $u_3(x) = 3u_1(x)u_2(x)$ with respect to the standard polynomial basis $\mathcal{E} = \{\mathbf{e}_1, \mathbf{e}_2, \mathbf{e}_3\} = \{1, x, x^2\}$ of $\mathcal{P}_{\in}$.

(b) Row reduce the matrix of coordinate vectors for the u_i in part (a) to conclude that the $\{u_i\}$ form a basis for $\mathcal{P}_2$.

(c) What are the two corresponding basis change matrices $\mathbf{X}_{\mathcal{E}\leftarrow\mathcal{U}}$ and $\mathbf{X}_{\mathcal{U}\leftarrow\mathcal{E}}$?

15. Solve the following matrix equations for $\mathbf{X}_{nn}$, where we assume all matrices to be $n \times n$ and nonsingular:

(a) $\mathbf{C} = \mathbf{UXAU}^{-1}$;

(b) $\mathbf{B} = \mathbf{A}^T\mathbf{AXB} - \mathbf{A}^2\mathbf{XB}$.

16. (a) For $T : \mathbb{R}^2 \to \mathbb{R}^2$ defined by $T\begin{pmatrix} x_1 \\ x_2 \end{pmatrix} = \frac{1}{2}\begin{pmatrix} x_1 + 3x_2 \\ 3x_1 + x_2 \end{pmatrix}$ show that there are two linearly independent vectors $\mathbf{u}, \ \mathbf{v} \in \mathbb{R}^2$ with $T(\mathbf{u}) = 2\mathbf{u}$ and $T(\mathbf{v}) = -\mathbf{v}$. That is, this linear transformation models Example 4.

(b) What is the standard matrix representation $\mathbf{A}_{\mathcal{E}}$ of T with respect to the unit vector basis $\mathcal{E}$.

(c) What is the matrix representation of T with respect to the basis $\mathcal{U} = \{\mathbf{u}, \ \mathbf{v}\}$ of $\mathbb{R}^2$ from part (a)?

17. (a) If $\begin{pmatrix} | & | & | \\ \mathbf{u}_1 & \mathbf{u}_2 & \mathbf{u}_3 \\ | & | & | \end{pmatrix} =$

$$\begin{pmatrix} | & | & | \\ \mathbf{v}_1 & \mathbf{v}_2 & \mathbf{v}_3 \\ | & | & | \end{pmatrix}\begin{pmatrix} 1 & -1 & 1 \\ 0 & 1 & 2 \\ 1 & 1 & 1 \end{pmatrix},$$

what is the basis change matrix $\mathbf{X}_{\mathcal{V}\leftarrow\mathcal{U}}$ for the two basis $\mathcal{U} = \{\mathbf{u}_i\}$ and $\mathcal{V} = \{\mathbf{v}_i\}$ of $\mathbb{R}^3$?

(b) What is $\mathbf{X}_{\mathcal{U}\leftarrow\mathcal{V}}$?

(c) What is $\mathbf{X}_{\mathcal{U}\leftarrow\mathcal{E}}$?

18. (a) If $\mathbf{U} = \mathbf{VB}$ for three nonsingular $n \times n$ matrices, find $\mathbf{U}^{-1}\mathbf{V}$ and $\mathbf{V}^{-1}\mathbf{U}$.

(b) If $\mathbf{U} = \mathbf{AV}$ find $\mathbf{U}^{-1}\mathbf{V}$ and $\mathbf{V}^{-1}\mathbf{U}$, if possible.

19. Let $\mathbf{A} = \begin{pmatrix} -7 & 9 \\ -6 & 8 \end{pmatrix}$.

(a) Try to find a nonzero vector $\mathbf{x} \in \mathbb{R}^2$ with $\mathbf{Ax} = 2\mathbf{x}$.

(b) Try to find a vector $\mathbf{y} \neq \mathbf{0} \in \mathbb{R}^2$ with $\mathbf{Ay} = -\mathbf{y}$.

(c) Try to find a vector $\mathbf{z} \neq \mathbf{0} \in \mathbb{R}^2$ with $\mathbf{Az} = \mathbf{z}$.

20. Assume that $\mathbf{A} = \begin{pmatrix} 1 & 1 & 0 \\ 2 & 1 & -1 \\ 3 & 2 & 2 \end{pmatrix}$ is the basis change matrix $\mathbf{X}_{\mathcal{V}\leftarrow\mathcal{U}}$ for the basis $\mathcal{U} = \left\{\begin{pmatrix} 1 \\ 0 \\ 2 \end{pmatrix}, \begin{pmatrix} -1 \\ 1 \\ 1 \end{pmatrix}, \begin{pmatrix} 0 \\ 1 \\ 2 \end{pmatrix}\right\}$ of $\mathbb{R}^3$.

(a) Write out the formula for a matrix $\mathbf{V}$ whose columns hold the basis vectors of $\mathcal{V}$ in terms of the given $\mathbf{A}$ and $\mathcal{U}$.

(b) Compute the basis vectors $\mathbf{v}_i$ in $\mathcal{V}$.

(c) If $\mathbf{x}_{\mathcal{U}} = \begin{pmatrix} 1 \\ -1 \\ 1 \end{pmatrix}$, find the point $\mathbf{x} = \mathbf{x}_{\mathcal{E}} \in \mathbb{R}^3$ and its $\mathcal{V}$–coordinate vector $\mathbf{x}_{\mathcal{V}}$. How can you check your answer?

21. (a) Show that the vectors $\begin{pmatrix} 1 \\ 1 \\ 1 \end{pmatrix}, \begin{pmatrix} 1 \\ 1 \\ 2 \end{pmatrix}$, and $\begin{pmatrix} 1 \\ 2 \\ 3 \end{pmatrix} \in \mathbb{R}^3$ are linearly independent.

(b) Show that the three vectors given in part (a) span $\mathbb{R}^3$.

(c) What is the coordinate vector $\mathbf{x}_{\mathcal{B}}$ of $\mathbf{x} = \mathbf{x}_{\mathcal{E}} = \begin{pmatrix} -3 \\ 2 \\ 4 \end{pmatrix} \in \mathbb{R}^3$ with respect to the basis $\mathcal{B} = \left\{ \begin{pmatrix} 1 \\ 1 \\ 1 \end{pmatrix}, \begin{pmatrix} 1 \\ 1 \\ 2 \end{pmatrix}, \begin{pmatrix} 1 \\ 2 \\ 3 \end{pmatrix} \right\}$ of $\mathbb{R}^3$?

(d) Represent the linear transformation $T(\mathbf{x}) = \begin{pmatrix} x_2 - 2x_1 - x_3 \\ x_3 - 2x_2 \\ 2x_3 + x_1 \end{pmatrix} : \mathbb{R}^3 \to \mathbb{R}^3$ with respect to the basis $\mathcal{B}$ of $\mathbb{R}^3$ in part (c).

22. Find the coordinate vector $\mathbf{x}_{\mathcal{B}}$ of the point $\mathbf{x} = \mathbf{x}_{\mathcal{E}} = \begin{pmatrix} 1 \\ 1 \\ 4 \end{pmatrix}$ with respect to the basis $\mathcal{B} = \left\{ \begin{pmatrix} 2 \\ -2 \\ 4 \end{pmatrix}, \begin{pmatrix} 1 \\ 1 \\ 1 \end{pmatrix}, \begin{pmatrix} 2 \\ 0 \\ 1 \end{pmatrix} \right\}$ of $\mathbb{R}^3$.

23. Represent the linear transformation $T(\mathbf{x}) = \begin{pmatrix} x_2 - 2x_1 - x_3 \\ x_3 - 2x_2 \\ 2x_3 + x_1 \end{pmatrix}$ with respect to the basis $\mathcal{B} = \left\{ \begin{pmatrix} 1 \\ 0 \\ 1 \end{pmatrix}, \begin{pmatrix} 1 \\ 1 \\ 2 \end{pmatrix}, \begin{pmatrix} 1 \\ 0 \\ 3 \end{pmatrix} \right\}$ of $\mathbb{R}^3$.

24. Let $\mathbf{X}_{\mathcal{V} \leftarrow \mathcal{U}} = \begin{pmatrix} 1 & 1 & 2 \\ -2 & -3 & 1 \\ 1 & 2 & -2 \end{pmatrix}$.

(a) Which of the three matrices $\mathbf{A} = \begin{pmatrix} -5 & -7 & -4 \\ 3 & 4 & -2 \\ 1 & 1 & 1 \end{pmatrix}$, $\mathbf{B} = \begin{pmatrix} -4 & -6 & -7 \\ 2 & 3 & 4 \\ 1 & 1 & 1 \end{pmatrix}$, or $\mathbf{C} = \begin{pmatrix} -4 & -6 & -7 \\ 3 & 4 & 5 \\ 1 & 1 & 1 \end{pmatrix}$ is the basis change matrix from $\mathcal{V}$ to $\mathcal{U}$–coordinate vectors?

(b) If $\mathbf{w}_{\mathcal{V}} = \begin{pmatrix} 1 \\ 2 \\ -1 \end{pmatrix}$, find $\mathbf{w}_{\mathcal{U}}$.

25. Let the linear transformation T from $\mathbb{R}^2$ to $\mathbb{R}^2$ be defined by $T\begin{pmatrix} x_1 \\ x_2 \end{pmatrix} = \begin{pmatrix} x_1 + x_2 \\ 2x_2 \end{pmatrix}$, and assume that $\mathbf{a} = \begin{pmatrix} 1 \\ -3 \end{pmatrix} \in \mathbb{R}^2$.

(a) Find the standard matrix representation $\mathbf{A}_{\mathcal{E}}$ of T.

(b) Show that $\mathbf{u}_1 = \begin{pmatrix} -1 \\ 3 \end{pmatrix}$ and $\mathbf{u}_2 = \begin{pmatrix} 3 \\ 1 \end{pmatrix}$ form a basis for $\mathbb{R}^2$.

(c) Find the $\mathcal{E}$–coordinate vector of $T(\mathbf{a})$.

(d) Find the matrix representation $\mathbf{A}_{\mathcal{U}}$ of T with respect to the basis $\mathcal{U} = \{\mathbf{u}_1, \mathbf{u}_2\}$ of $\mathbb{R}^2$.

(e) Find the $\mathcal{U}$–coordinate vectors of $\mathbf{a}$ and of $T(\mathbf{a})$ by using part (d).

(f) Change the $\mathcal{U}$–coordinate vector $(T(\mathbf{a}))_{\mathcal{U}}$ from part (d) to its $\mathcal{E}$–coordinate vector $(T(\mathbf{a}))_{\mathcal{E}}$ and compare the result with that of part (c).

26. Let $\mathcal{U} = \{a\mathbf{e}_1,\ b\mathbf{e}_2\}$ be the basis of $\mathbb{R}^2$ that results from the standard basis $\mathcal{E}$ by stretching for $a, b \neq 0$.

(a) Find the basis change matrices $\mathbf{X}_{\mathcal{E} \leftarrow \mathcal{U}}$ and $\mathbf{X}_{\mathcal{U} \leftarrow \mathcal{E}}$.

(b) Express the points $\begin{pmatrix} x \\ y \end{pmatrix}$ on the **ellipse** $\dfrac{x^2}{a^2} + \dfrac{y^2}{b^2} = 1$ via a matrix $\mathbf{Q}_{\mathcal{E}}$, the **quadratic form** matrix of the ellipse, for which $\frac{x^2}{a^2} + \frac{\mathbf{y}^2}{b^2} = (x\ y)\ \mathbf{Q}_{\mathcal{E}} \begin{pmatrix} x \\ y \end{pmatrix} = 1$.

(c) Describe the ellipse $\frac{x^2}{a^2} + \frac{\mathbf{y}^2}{b^2} = 1$, expressed in $\mathcal{E}$–coordinates, in terms of the $\mathcal{U}$-coordinates. What curve is it in $\mathcal{U}$–coordinate space?

(d) Describe the circle $x^2 + y^2 = 1$, expressed in $\mathcal{E}$–coordinates, in terms of the $\mathcal{U}$-coordinates. What is its shape in $\mathcal{U}$–coordinate space?

27. Prove that a set of vectors $\{\mathbf{v}_1, \ldots, \mathbf{v}_k\} \subset \mathbb{R}^n$ is linearly dependent if and only if the set of coordinate vectors $\left\{(\mathbf{v}_1)_{\mathcal{U}}, \ldots, (\mathbf{v}_k)_{\mathcal{U}}\right\}$ is linearly dependent for any basis $\mathcal{U}$ of $\mathbb{R}^n$.

28. Prove that if $\mathbf{x} = \mathbf{x}_{\mathcal{E}} = \alpha_1 \mathbf{v}_1 + \ldots + \alpha_k \mathbf{v}_k \in \mathbb{R}^n$ for any set of vectors $\mathbf{v}_i = \mathbf{v}_{i_{\mathcal{E}}}$ and any $\alpha_i \in \mathbb{R}$, then for the coordinate vectors $\mathbf{x}_{\mathcal{U}}$, $\mathbf{v}_{1_{\mathcal{U}}}, \ldots, \mathbf{v}_{k_{\mathcal{U}}}$ with respect to $\mathcal{U}$, we have $\mathbf{x}_{\mathcal{U}} = \alpha_1 \mathbf{v}_{1_{\mathcal{U}}} + \ldots + \alpha_k \mathbf{v}_{k_{\mathcal{U}}}$.

29. Compute the coordinate vectors of $\mathbf{x} = \mathbf{x}_{\mathcal{E}} = \begin{pmatrix} 1 \\ -3 \end{pmatrix}$ for the following bases of $\mathbb{R}^2$:

(a) $\mathcal{B} = \left\{ \begin{pmatrix} 2 \\ -1 \end{pmatrix}, \begin{pmatrix} 1 \\ -1 \end{pmatrix} \right\}$,

(b) $\mathcal{C} = \left\{ \begin{pmatrix} 0 \\ -1 \end{pmatrix}, \begin{pmatrix} 2 \\ 2 \end{pmatrix} \right\}$,

(c) $\mathcal{D} = \left\{ \begin{pmatrix} 1 \\ 4 \end{pmatrix}, \begin{pmatrix} 4 \\ 1 \end{pmatrix} \right\}$.

30. Find the basis change matrix from $\mathcal{U} = \left\{ \begin{pmatrix} 1 \\ 2 \end{pmatrix}, \begin{pmatrix} 2 \\ -1 \end{pmatrix} \right\}$ to $\mathcal{V} = \left\{ \begin{pmatrix} 3 \\ -1 \end{pmatrix}, \begin{pmatrix} 1 \\ 0 \end{pmatrix} \right\}$.

31. Find the coordinate vector $\mathbf{x}_{\mathcal{U}}$ of the point $\mathbf{x} = (\pi,\ 1,\ e^{-1})^T \in \mathbb{R}^3$ with respect to the basis $\mathcal{U} = \{\mathbf{e}_3,\ \mathbf{e}_2,\ -\mathbf{e}_1\}$ of $\mathbb{R}^3$. ($e = 2.71\ldots$)

32. Compute the basis change matrix from the basis $\mathcal{U} = \left\{ \begin{pmatrix} 0 \\ 2 \end{pmatrix}, \begin{pmatrix} 2 \\ 0 \end{pmatrix} \right\}$ to the standard basis $\mathcal{E}$ of $\mathbb{R}^2$.

Teacher's Problem-Making Exercise

T 7. For ease with hand computations, we need to be able to find two bases $\mathcal{V}$ and $\mathcal{U}$ of $\mathbb{R}^n$, composed of integer vectors whose basis change matrices $\mathbf{X}_{\mathcal{V}\leftarrow\mathcal{U}}$ and $\mathbf{X}_{\mathcal{U}\leftarrow\mathcal{V}}$ are both integer matrices. For this, we can use the method of constructing unimodular matrices $\mathbf{V}$ and $\mathbf{U}$ (whose columns are the respective basis vectors) as detailed in Problem T 6.

EXAMPLE For $\mathbb{R}^3$, let us construct $\mathbf{V} = \mathbf{AB}$ from $\mathbf{A} = \begin{pmatrix} 1 & 2 & -1 \\ 0 & 1 & 1 \\ 0 & 0 & -1 \end{pmatrix}$ and $\mathbf{B} = \begin{pmatrix} 1 & 0 & 0 \\ 1 & -1 & 0 \\ 0 & 1 & -1 \end{pmatrix}$ according to Problem T 6. Then the basis $\mathcal{V}$ is composed of the columns of $\mathbf{V} = \mathbf{AB}$, or

$$\mathcal{V} = \left\{ \begin{pmatrix} 3 \\ 1 \\ 0 \end{pmatrix}, \begin{pmatrix} -3 \\ 1 \\ -1 \end{pmatrix}, \begin{pmatrix} 1 \\ -2 \\ 1 \end{pmatrix} \right\}.$$

If we take the columns of $\mathbf{U} = \mathbf{BA} = \begin{pmatrix} 1 & 2 & -1 \\ 1 & 1 & -3 \\ 0 & 1 & 3 \end{pmatrix}$ for our second basis $\mathcal{U}$ of $\mathbb{R}^3$, then $\mathbf{X}_{\mathcal{V}\leftarrow\mathcal{U}} = \mathbf{V}^{-1}\mathbf{U} = \mathbf{B}^{-1}\mathbf{A}^{-1}\mathbf{BA}$ and $\mathbf{X}_{\mathcal{U}\leftarrow\mathcal{V}} = \mathbf{U}^{-1}\mathbf{V} = \mathbf{A}^{-1}\mathbf{B}^{-1}\mathbf{AB}$ will both be integer basis-change matrices. ◀

7.2 Theory

Rank, Matrix Transpose

Let $\mathbf{U}$ and $\mathbf{V} \in \mathbb{R}^{n,n}$ be two nonsingular matrices whose columns contain the vectors of two bases $\{\mathbf{u}_i\}$ and $\{\mathbf{v}_j\}$ for $\mathbb{R}^n$. Then the basis change matrix $\mathbf{X}_{\mathcal{V}\leftarrow\mathcal{U}}$ satisfies

$$\mathbf{X}_{\mathcal{V}\leftarrow\mathcal{U}} = \mathbf{V}^{-1}\mathbf{U}, \quad \text{or} \quad \mathbf{V}\mathbf{X}_{\mathcal{V}\leftarrow\mathcal{U}} = \mathbf{U}.$$

Thus, if $\mathbf{X}_{nn}$ is any nonsingular matrix and $\mathbf{V}_{nn}$ describes a basis of $\mathbb{R}^n$ in its columns, so does the product matrix $\mathbf{U}_{nn} = \mathbf{VX}$.

This situation becomes a bit more complicated if $\mathbf{V} \in \mathbb{R}^{n,m}$ is a rectangular matrix with $m < n$, containing a basis $\{\mathbf{v}_i\}_{i=1}^m$ of a proper subspace $V \subset \mathbb{R}^n$ in its columns. We use the symbols V for the subspace and $\mathbf{V}$ for the matrix comprising the basis vectors as columns. Then any matrix $\mathbf{U} \in \mathbb{R}^{n,m}$ likewise contains a basis of column vectors for the same subspace V precisely when $\mathbf{U} = \mathbf{VX}$ for a nonsingular matrix $\mathbf{X}_{mm}$. This generalizes the $n \times n$ case and follows from the following Lemma:

Lemma 1 If $\mathbf{A} \in \mathbb{R}^{n,m}$ has rank k, then

$$\text{rank}(\mathbf{AX}) = k \;=\; \text{rank}(\mathbf{A}) \;=\; \text{rank}(\mathbf{YA})$$

for every nonsingular $m \times m$ matrix $\mathbf{X}$ and every nonsingular $n \times n$ matrix $\mathbf{Y}$. ◀

Proof To show that $\text{rank}(\mathbf{YA}) = \text{rank}(\mathbf{A})$ for every nonsingular matrix $\mathbf{Y}_{nn}$, assume that $\mathbf{Y} = \mathbf{E}_1 \cdots \mathbf{E}_M$ for certain Gaussian elimination matrices $\mathbf{E}_i$, depending on $\mathbf{Y}$, according to Section 6.2. Then

$$\mathbf{E}_M^{-1} \cdots \mathbf{E}_1^{-1}(\mathbf{YA}) = \mathbf{A}_{nm}$$

represents a certain Gaussian row reduction of the product matrix $\mathbf{YA}$. This shows that $\mathbf{YA}$ and $\mathbf{A}$ are row equivalent via Gaussian elimination. Thus, $\mathbf{YA}$ and $\mathbf{A}$ must have the same RREF. Therefore, their ranks are equal.

To show that $\text{rank}(\mathbf{AX}) = \text{rank}(\mathbf{A}) = k$ for any $m \times m$ nonsingular matrix $\mathbf{X}$ is more complicated:

If $\mathbf{E}_1, \ldots, \mathbf{E}_L$ are the elimination matrices that reduce $\mathbf{A}_{nm}$ to its RREF $\mathbf{R}$ with k pivots and zero rows below row k, then by using the row interpretation of matrix products from the end of Section 6.2, we see that

$$(\mathbf{E}_L \cdots \mathbf{E}_1\mathbf{A})\mathbf{X} = \mathbf{RX} = \begin{pmatrix} \boxed{1} & * & 0 & * & * & 0 & * & * \\ & & \boxed{1} & * & * & \vdots & & \\ & & & \ddots & & \vdots & & \vdots \\ & & & & \ddots & 0 & & \\ & & & & & \boxed{1} & * & * \\ 0 & & \cdots & & 0 & 0 & \cdots & 0 \\ \vdots & & & & \vdots & \vdots & & \vdots \\ 0 & & \cdots & & 0 & 0 & \cdots & 0 \end{pmatrix} \mathbf{X} =$$

$$\begin{matrix} \uparrow & & \uparrow & & & \uparrow \\ j_1 & & j_2 & \ldots & & j_k \end{matrix}$$

$$= \begin{pmatrix} \text{linear combination of rows of } \mathbf{X} \\ \vdots \\ \text{linear combination of rows of } \mathbf{X} \\ \mathbf{0} \\ \vdots \\ \mathbf{0} \end{pmatrix} \begin{matrix} \\ \\ \leftarrow k \\ \\ \\ \\ \end{matrix}$$

$$= \begin{pmatrix} - & \mathbf{w}_1 & - \\ & \vdots & \\ - & \mathbf{w}_k & - \\ & \mathbf{0} & \\ & \vdots & \\ & \mathbf{0} & \end{pmatrix}.$$

Now use the fact that a square matrix $\mathbf{X}$ is nonsingular if and only if the rows of $\mathbf{X}$ are linearly independent. Look at any specific linear combination $\mathbf{w}_p$ for $1 \le p \le k$ of the rows $\mathbf{x}_i$ of $\mathbf{X}$ in the matrix $\mathbf{E}_L \cdots \mathbf{E}_1\mathbf{AX}$. If $j_1, \ldots, j_k$ are the k pivot column indices for the RREF $\mathbf{R}$ for $\mathbf{A}$ of rank k (i.e., if the pivots in $\mathbf{E}_L \cdots \mathbf{E}_1\mathbf{A}$ appear in the columns numbered $j_1, \ldots, j_k$ as indicated), then the linear combination for $\mathbf{w}_p$ contains row $\mathbf{x}_{j_p}$ of $\mathbf{X}$ with the coefficient 1. And for $q \ne p$, no other $\mathbf{w}_q$ uses this row $\mathbf{x}_{j_p}$ in its makeup. This is due to the RREF having unit vectors $\mathbf{e}_p$ in its j_p^{th} column for each $p = 1, \ldots, k$. Thus, any linear combination of the $\mathbf{w}_j$ contains a term involving those pivot–indexed rows $\mathbf{x}_{j_i}$ of $\mathbf{X}$ at most once.

To test for linear independence of the $\mathbf{w}_p$, we use Definition 2 of Section 5.1 and assume that

$$\alpha_1\mathbf{w}_1 + \cdots + \alpha_k\mathbf{w}_k = \mathbf{0}$$

for some scalars α_i. As each $\mathbf{w}_p$ is a linear combination of the rows $\mathbf{x}_i$ of $\mathbf{X}$, we can rewrite this equation in terms of the $\mathbf{x}_i$ as

$$\sum_{i=1}^{k} \alpha_i \mathbf{x}_{\text{pivot row}_{j_i}} + \sum (\text{various } \alpha_{..} \text{ coefficient sums})\mathbf{x}_{\text{nonpivot rows}} = \mathbf{0}.$$

By assumption, the rows $\mathbf{x}_i$ of $\mathbf{X}$ are linearly independent, and hence all of the coefficients in the sum must vanish, making $\alpha_i = 0$ for all pivot rows $\mathbf{x}_{j_i}$. Thus, the vectors $\{\mathbf{w}_1, \ldots, \mathbf{w}_p\}$ are linearly independent, and the matrix product $\mathbf{AX}$ has rank k. ■

Therefore,

Theorem 7.1 If the columns of a matrix $\mathbf{V}_{nm} = \begin{pmatrix} | & & | \\ \mathbf{v}_1 & \cdots & \mathbf{v}_m \\ | & & | \end{pmatrix}$ form a basis for a subspace $V \subset \mathbb{R}^n$, then for any nonsingular $m \times m$ matrix $\mathbf{X}$, the columns of the matrix $\mathbf{U}_{nm} := \mathbf{VX}$ likewise form a basis for the same subspace V. ◀

DEFINITION 1 A nonsingular matrix $\mathbf{X}_{mm}$ is called the **basis change matrix $\mathbf{X}_{\mathcal{V}\leftarrow\mathcal{U}}$** between two bases $\mathcal{U}$ and $\mathcal{V}$ for the same subspace of dimension m of $\mathbb{R}^n$, if $\mathbf{U} = \mathbf{VX}$.

Note that we can express $\mathbf{X}_{\mathcal{V}\leftarrow\mathcal{U}} = \mathbf{V}^{-1}\mathbf{U}$ as a product of two nonsingular matrices only in the case that the two bases $\mathcal{U}$ and $\mathcal{V}$ are bases for all of $\mathbb{R}^n$. If the two sets V and $U \subset \mathbb{R}^n$ each contain $m < n$ basis vectors of the same subspace of $\mathbb{R}^n$, then the matrices made up of the basis vectors as columns are not square, and thus they cannot be inverted. Hence we need to use the following slightly more general basis change matrix equation in the general subspace case:

$$\boxed{\mathbf{U}_{nm} = \mathbf{V}_{nm}\ \mathbf{X}_{\mathcal{V}\leftarrow\mathcal{U}}.} \tag{7.5}$$

This equation relates the basis column matrices $\mathbf{U}_{nm}$ and $\mathbf{V}_{nm}$ for the subspace bases $\mathcal{U}$ and $\mathcal{V}$, respectively, and differs from the $\mathbb{R}^n$ basis change formula (7.3) $\mathbf{X}_{\mathcal{V}\leftarrow\mathcal{U}} = \mathbf{V}_{nn}^{-1}\mathbf{U}_{nn}$.

The next theorem states a reassuring fact about matrix rank.

Theorem 7.2 The rank of a matrix $\mathbf{A}$ and that of its transpose $\mathbf{A}^T$ are the same for any matrix $\mathbf{A} \in \mathbb{R}^{m,n}$. ◀

Proof A RREF $\mathbf{R}$ of $\mathbf{A}$ with unit pivots can be achieved by a certain number M of Gaussian elimination matrices according to Section 6.2; that is,

$$\mathbf{E}_M \cdots \mathbf{E}_1\mathbf{A} = \begin{pmatrix} \boxed{1} & * & & \cdots & & * \\ 0 & \ddots & & & & \\ \vdots & & \boxed{1} & * & \cdots & * \\ 0 & \cdots & & & \cdots & 0 \\ \vdots & & & & & \vdots \\ 0 & \cdots & & & \cdots & 0 \end{pmatrix} \begin{matrix} \\ \\ \leftarrow k = \mathbf{R} \\ \\ \\ \\ \end{matrix},$$

with $\mathbf{R}$ having precisely $k = \text{rank}(\mathbf{A})$ pivot rows and zero rows below. Note that the pivots $\boxed{1}$ do not have to lie on the main diagonal of $\mathbf{R}$ as we appear to have indicated. (We did so only for simplicity of printing.) To investigate the rank of $\mathbf{A}^T$, we form

$$\mathbf{R}^T = (\mathbf{E}_M \cdots \mathbf{E}_1\mathbf{A})^T = \mathbf{A}^T\mathbf{E}_1^T \cdots \mathbf{E}_M^T = \begin{pmatrix} \boxed{1} & 0 & & & \cdots & 0 \\ * & \ddots & & & & \\ \vdots & & \boxed{1} & 0 & \cdots & 0 \\ \vdots & & * & \vdots & & \vdots \\ \vdots & & \vdots & \vdots & & \vdots \\ * & \cdots & * & 0 & \cdots & 0 \end{pmatrix}$$
$$\begin{matrix} \uparrow & \qquad & \uparrow \\ 1 & & k \end{matrix}$$

by using Theorem 6.3 repeatedly. From Lemma 1, applied M times, we see that

$$\text{rank}(\mathbf{A}^T) = \text{rank}(\mathbf{A}^T\mathbf{E}_1^T \cdots \mathbf{E}_M^T) = \text{rank}(\mathbf{R}^T),$$

since each transposed elimination matrix $\mathbf{E}_i^T$ is nonsingular by Theorem 6.5 and Lemma 1 of Section 6.2. If we further row reduce the matrix $\mathbf{R}^T$ to its REF, this reveals its rank to be k, since each of the successive pivots $\boxed{1}$ in $\mathbf{R}^T$ zeros out every nonzero element $*$ in the lower triangle of $\mathbf{R}^T$. Thus,

$$\text{rank}(\mathbf{A}) = \text{rank}(\mathbf{A}^T) = \text{the number of pivots in a REF of } \mathbf{A} \ \ (\text{ or of } \mathbf{A}^T). \qquad \blacksquare$$

Theorem 7.2 has the following corollary that makes the use of orthogonality in the exclusive subspace definition of chapter 4 more understandable.

Corollary Let $U = \text{span}\{\mathbf{u}_1, \ldots, \mathbf{u}_k\} \subset \mathbb{R}^n$ and assume that the vector $\mathbf{a} \notin U$. Then there exists a vector $\mathbf{b} \in \mathbb{R}^n$ with $\mathbf{b} \perp U$. Moreover, $\mathbf{0} \neq \mathbf{b} \notin U$. ◀

Proof Since $\mathbf{a} \notin \text{span}\{\mathbf{u}_i\} = U$, the linear system $\begin{pmatrix} | & & | \\ \mathbf{u}_1 & \ldots & \mathbf{u}_k \\ | & & | \end{pmatrix} \mathbf{x} = \mathbf{a}$ is not solvable. Thus, by Theorem 3.1, we have $\text{rank}(\mathbf{U}) < \text{rank}(\mathbf{U} \mid \mathbf{a})$ for the matrix $\mathbf{U} = \begin{pmatrix} | & & | \\ \mathbf{u}_1 & \ldots & \mathbf{u}_k \\ | & & | \end{pmatrix}_{nk}$. Consequently, $\text{rank}(\mathbf{U}) < n =$ the number of rows in $\mathbf{U}$. Hence, in each REF of $\mathbf{U}$, there is one row without a pivot. Look at $\mathbf{U}^T = \begin{pmatrix} - & \mathbf{u}_1^T & - \\ & \vdots & \\ - & \mathbf{u}_k^T & - \end{pmatrix}_{kn}$. $\mathbf{U}^T$ has the same rank as $\mathbf{U}$ by Theorem 7.2. Thus, any REF $\mathbf{R}_{kn}$ of $\mathbf{U}^T$ also has fewer than n pivots, giving each such $\mathbf{R}$ at least one free column. In other words, $\ker(\mathbf{U}^T) \neq \{\mathbf{0}\}$. Take any nonzero vector $\mathbf{b} \in \ker(\mathbf{U}^T)$. Then $\mathbf{b} \perp U = \text{span}\{\mathbf{u}_i\}$ according to Chapter 4, and $\mathbf{b} \notin U$; otherwise, $\mathbf{b} \cdot \mathbf{b} = \sum_i b_i^2 = 0$. $\blacksquare$

7.2.P Problems

1. (a) What is the rank of the matrix product $\mathbf{AB}$ for $\mathbf{A} = \begin{pmatrix} 1 & 0 & 0 \\ 1 & 1 & 0 \\ 1 & 1 & 1 \end{pmatrix}$ and $\mathbf{B} = \begin{pmatrix} 1 & 1 & 1 \\ 0 & 1 & 1 \\ 0 & 0 & 1 \end{pmatrix}$?

(b) What is the rank of the reverse order matrix product $\mathbf{BA}$?

(c) What is the rank of $\mathbf{C} = \mathbf{A} \cdot \begin{pmatrix} 1 & 2 \\ 2 & 4 \\ 3 & 6 \end{pmatrix}$ and of $\mathbf{H} = \begin{pmatrix} 17 & 101 & -93 \end{pmatrix} \cdot \mathbf{B}$ for $\mathbf{A}$ and $\mathbf{B}$ as defined in part (a)?

(d) Do you have to compute any of the matrix products in (c) to find their ranks?

2. Compute the ranks of both $\mathbf{A}$ and $\mathbf{A}^T$ via two separate row reductions for $\mathbf{A} = \begin{pmatrix} 1 & 3 & 4 & 0 & -1 \\ 0 & 2 & 3 & -1 & 0 \\ 1 & 0 & 0 & -1 & 0 \end{pmatrix}$ and for $\mathbf{A}^T$ (Just to make sure).

3. Let $\mathbf{A}_{3,3} = \begin{pmatrix} 1 & -1 & -2 \\ 2 & 3 & -1 \\ 0 & 0 & 2 \end{pmatrix}$.

(a) Compute the rank of the 3×4 matrix with columns $\mathbf{x}$, $\mathbf{Ax}$, $\mathbf{A}^2\mathbf{x}$, and $\mathbf{A}^3\mathbf{x}$ for $\mathbf{x} = \begin{pmatrix} 2 \\ -2 \\ 0 \end{pmatrix}$ and for $\mathbf{x} = \begin{pmatrix} 0 \\ 0 \\ -2 \end{pmatrix}$.

(b) Compute the rank of the 3×3 matrix with rows $(\mathbf{Ax})^T$, $\mathbf{x}^T$, and $(\mathbf{A}^3x)^T$ for $\mathbf{x} = \begin{pmatrix} 2 \\ -2 \\ 0 \end{pmatrix}$ and $\mathbf{x} = \begin{pmatrix} 0 \\ 0 \\ -2 \end{pmatrix}$.

(c) Do you have to compute the matrix powers $\mathbf{A}^2$ and $\mathbf{A}^3$ to complete the above parts?

4. What is the rank of a row vector? What of a column vector? What is the rank of the zero vector?

5. Assume that the two sets of vectors $U = \left\{ \begin{pmatrix} 1 \\ 1 \\ 1 \\ 0 \end{pmatrix}, \begin{pmatrix} 0 \\ 1 \\ -1 \\ 2 \end{pmatrix} \right\}$ and $V = \left\{ \begin{pmatrix} -1 \\ 0 \\ -2 \\ 2 \end{pmatrix}, \begin{pmatrix} 1 \\ 2 \\ 0 \\ 2 \end{pmatrix} \right\}$ span the same subspace of $\mathbb{R}^4$.

(a) Find the basis change matrix $\mathbf{X} = \mathbf{X}_{\mathcal{U} \leftarrow \mathcal{V}}$ with $\mathbf{x}_{\mathcal{U}} = \mathbf{X}x_{\mathcal{V}}$.

(b) Find $\mathbf{x}_{\mathcal{U}}$ for $\mathbf{x}_{\mathcal{V}} = \begin{pmatrix} 2 \\ -3 \end{pmatrix}$.

(c) Find $\mathbf{x}_{\mathcal{V}}$ for the point with $\mathbf{x}_{\mathcal{U}} = \begin{pmatrix} -1 \\ 4 \end{pmatrix}$.

6. Why does the REF of a matrix $\mathbf{A}_{mn}$ have precisely as many linearly independent rows as it has linearly independent columns?

7. Let $\mathcal{U} = \{\mathbf{u}_1, \mathbf{u}_2\} = \left\{ \begin{pmatrix} 1 \\ -1 \\ 2 \end{pmatrix}, \begin{pmatrix} 2 \\ 0 \\ 3 \end{pmatrix} \right\}$ be a basis for a subspace of $\mathbb{R}^3$.

(a) Write out a formula for the $\mathcal{E}$–coordinate vector $\mathbf{x}_{\mathcal{E}}$ of any $\mathcal{U}$–coordinate vector $\mathbf{x}_{\mathcal{U}}$.

(b) If $\mathcal{V} = \{\mathbf{v}_1, \mathbf{v}_2\}$ is another basis for the same subspace with $\mathbf{x}_{\mathcal{U}} = \mathbf{X}\mathbf{x}_{\mathcal{V}} = \begin{pmatrix} 1 & -1 \\ 2 & 0 \end{pmatrix} \mathbf{x}_{\mathcal{V}}$ for all points in the subspace, what are the $\mathcal{E}$–coordinate vectors of the basis vectors $\mathbf{v}_i$ for $\mathcal{V}$?

8. Find a vector $\mathbf{b}$ that is perpendicular to the subspace

$$U = \text{span}\left\{ \begin{pmatrix} 1 \\ 2 \\ 3 \end{pmatrix}, \begin{pmatrix} 0 \\ 2 \\ 1 \end{pmatrix} \right\} \subset \mathbb{R}^3.$$

9. Find two linearly independent vectors $\mathbf{b}$ and $\mathbf{c} \in \mathbb{R}^4$ that are perpendicular to the subspace

$$U = \text{span}\left\{ \begin{pmatrix} 1 \\ -2 \\ 3 \\ -1 \end{pmatrix}, \begin{pmatrix} 0 \\ 2 \\ 2 \\ -2 \end{pmatrix} \right\} \subset \mathbb{R}^4, \text{ if possible.}$$

10. (a) Prove or disprove the following:

If $\mathbf{Ax} = \mathbf{b}$ is solvable for every right-hand-side $\mathbf{b}$, then the linear system $\mathbf{A}^T\mathbf{y} = \mathbf{c}$ with the transposed system matrix is also solvable for every right-hand-side $\mathbf{c}$.

(b) If (a) is generally false, what further condition on $\mathbf{A}$ will make (a) universally true?

11. Let $\mathbf{A} = \mathbf{A}^T$ and $\mathbf{B} = \mathbf{B}^T$. Under what condition is it true that $\mathbf{AB} = \mathbf{BA}$?

12. (a) Show that for $\mathbf{C} = \sum_{i=1}^{k} \mathbf{A}_i$ and $\mathbf{D} = \sum_{j=1}^{m} \mathbf{B}_j$ and n by n matrices throughout, we have $\mathbf{CD}^T = \sum_{j=1}^{m} \sum_{i=1}^{k} \mathbf{A}_i \mathbf{B}_j^T$.

(b) Give a similar formula for the matrix product $\mathbf{D}^T\mathbf{C}^T$.

7.3 Applications

Subspace Basis Change, Calculus

We study two applications, namely, to basis changes in subspaces and to elementary calculus.

(a) Basis Change in Subspaces

We use Theorem 7.1 to find the basis change matrix between two different bases of one subspace of dimension $m < n$ of $\mathbb{R}^n$. If the two bases are $\{\mathbf{u}_1, \ldots, \mathbf{u}_m\}$ and $\{\mathbf{v}_1, \ldots, \mathbf{v}_m\}$, respectively, we need to find an $m \times m$ nonsingular matrix $\mathbf{X} = \mathbf{X}_{\mathcal{V} \leftarrow \mathcal{U}}$ with

$$\mathbf{U} = \mathbf{VX}$$

according to Eq. (7.5), where both $\mathbf{U} \in \mathbb{R}^{n,m}$ and $\mathbf{V} \in \mathbb{R}^{n,m}$ are given by their m columns $\mathbf{u}_i$ and $\mathbf{v}_j$, respectively. To find $\mathbf{X}$, we row reduce $\mathbf{V}$ using Gaussian elimination matrices $\mathbf{E}_M, \ldots, \mathbf{E}_1$ and obtain its RREF $\mathbf{E}_M \cdots \mathbf{E}_1\mathbf{V} = \begin{pmatrix} \mathbf{I}_m \\ \mathbf{0} \end{pmatrix}_{nm}$ according to Chapter 6. If we simultaneously perform these row operations on $\mathbf{U}$, we obtain

$$\mathbf{E}_M \cdots \mathbf{E}_1\mathbf{U} = (\mathbf{E}_M \cdots \mathbf{E}_1\mathbf{V})\mathbf{X}_{mm} = \begin{pmatrix} \mathbf{I}_m \\ \mathbf{0} \end{pmatrix}_{nm} \mathbf{X}_{mm} = \begin{pmatrix} \mathbf{X} \\ \mathbf{0} \end{pmatrix}_{nm}.$$

Thus, the top $m \times m$ submatrix $\mathbf{X}$ of $\mathbf{E}_M \cdots \mathbf{E}_1\mathbf{U}$ contains the basis change matrix $\mathbf{X}_{\mathcal{V}\leftarrow\mathcal{U}}$, while the bottom $n-m$ rows of $\mathbf{E}_M \cdots \mathbf{E}_1\mathbf{U}$ contain only zeros.

EXAMPLE 5 The two sets of vectors $U = \left\{ \begin{pmatrix} 1 \\ 2 \\ -1 \end{pmatrix}, \begin{pmatrix} -1 \\ 1 \\ 1 \end{pmatrix} \right\}$ and $V = \left\{ \begin{pmatrix} 0 \\ 3 \\ 0 \end{pmatrix}, \begin{pmatrix} 1 \\ 5 \\ -1 \end{pmatrix} \right\}$ span the same two-dimensional subspace of $\mathbb{R}^3$. (This statement can be verified by using the procedure of Theorem 5.3 in Section 5.2.)

Find the 2×2 basis-change matrix $\mathbf{X}_{\mathcal{V}\leftarrow\mathcal{U}}$.

As indicated, we start out with the multi-augmented matrix $(\mathbf{V} \mid \mathbf{U})$ and row reduce $\mathbf{V}$ to its RREF while simultaneously updating $\mathbf{U}$:

V		**U**		
[1]	−1	0	1	
[2]	1	3	5	$-2\ row_1$
[−1]	1	0	−1	$+\ row_1$
[1]	−1	0	1	
0	[3]	3	3	$\div 3$
0	0	0	0	
[1]	[−1]	0	1	$+\ row_2$
0	[1]	1	1	
0	0	0	0	
1	0	1	2	
0	1	1	1	$= \mathbf{X}_{\mathcal{V}\leftarrow\mathcal{U}}$
0	0	0	0	

Thus, $\mathbf{X}_{\mathcal{V}\leftarrow\mathcal{U}} = \begin{pmatrix} 1 & 2 \\ 1 & 1 \end{pmatrix}_{2\times 2}$ is the basis-change matrix satisfying $\mathbf{U} = \mathbf{V}\mathbf{X}$, which should be checked. ◀

To perform these calculations in MATLAB, we enter the two basis matrices $\mathbf{U}$ and $\mathbf{V}$, each in $\mathbb{R}^{n,m}$, and form the composite matrix $\mathbf{H}$ of size $n \times 2m$:

```
H = [V,U]
```

By calling

```
Y = rref(H); X = Y(1:m,m+1:2m)
```

we obtain the $m \times m$ basis-change matrix $\mathbf{X} = \mathbf{X}_{\mathcal{V}\leftarrow\mathcal{U}}$, where $\mathbf{U} = \mathbf{VX}$ in the top right corner of the RREF $\mathbf{Y}$ of $(\mathbf{V} \mid \mathbf{U})$.

(b) Calculus Applications

The set of all real-valued functions $\mathcal{F} : \mathbb{R} \to \mathbb{R}$ forms a linear space under function addition $(f+g)(x) := f(x) + g(x)$ and scalar multiplication $(\alpha f)(x) = \alpha f(x)$. The set $\mathcal{F}$ is a vector space under these two operations; see Appendix C for more on abstract vector spaces. This space is not finite dimensional. This follows from the Fundamental Theorem of Algebra:

The functions $f_j(x) := x^j : \mathbb{R} \to \mathbb{R}$, $j = 0, 1, 2, \ldots$, are linearly independent in $\mathcal{F}$ for any j.

To understand why, assume that a certain finite linear combination of the polynomials f_j is the zero function on all of $\mathbb{R}$; that is, assume that

$$p(x) = \alpha_0 f_0(x) + \alpha_1 f_1(x) + \cdots + \alpha_m f_m(x) = O = O(x) = 0$$

for all x. Using the Fundamental Theorem of Algebra (see Theorem 9.1) on the polynomial $p(x)$ of degree at most m, we have $p(x) = \gamma(x - \beta_1)\cdots(x - \beta_m)$ for $\gamma \in \mathbb{R}$ and certain $\beta_j \in \mathbb{C}$. Clearly, $p(x)$ has precisely m zeros $\beta_j \in \mathbb{C}$. If $p(x) = O(x)$, the zero function in $\mathcal{F}$, then necessarily $\gamma = 0$ and subsequently each coefficient $\alpha_j = 0$ in p. This proves linear independence for the polynomials $\{f_0, f_1, \ldots, f_j\} = \{1, x, \ldots, x^j\} \subset \mathcal{F}$ for any number j according to Definition 2 of Section 9.1.

As $\mathcal{F}$ is infinite dimensional, we have to restrict ourselves for our purposes to finite-dimensional subspaces thereof. In a finite dimensional part U of $\mathcal{F}$, we can introduce the notion of a basis as before. If F is a linear operator $F : \mathcal{F} \to \mathcal{F}$ such that $FU \subset U$ for the chosen finite-dimensional subspace U of $\mathcal{F}$, we can study F as we would any linear transformation with respect to a finite basis of U.

For example, let us consider **differentiation**, denoted by D, of functions $f \in \mathcal{F}^1$, the differentiable functions in $\mathcal{F}$. Clearly differentiation is a linear process, since

$$D(f+g) = (f+g)' = f' + g' = Df + Dg$$

and

$$D(\alpha f) = (\alpha f)' = \alpha f' = \alpha Df$$

for all differentiable functions $f, g \in \mathcal{F}^1$ and for all α. We intend to find simple **invariant subspaces** $U \subset \mathcal{F}^1$ under D (i.e., subspaces U of $\mathcal{F}^1$ with $DU \subset U$) on which we can study differentiation and its inverse operation, namely, the antiderivative for the differentiation map D.

EXAMPLE 6[1] (a) Consider $U = \text{span}\{\sin x, \ \cos x\} \subset \mathcal{F}^1$. If we take $\sin x$ and $\cos x$ as the basic functions of this subspace U and work with the basis

$$\mathcal{E} := \{\mathbf{e}_1 = \sin x, \ \mathbf{e}_2 = \cos x\},$$

[1]This example follows J. W. Rogers, Applications of linear algebra in calculus, American Math. Monthly, vol. 104 (1997), p. 20–26.

then

$$De_1 = D\sin x = (\sin x)' = \cos x = 0\cdot \mathbf{e}_1 + 1\cdot \mathbf{e}_2 = \begin{pmatrix} 0 \\ 1 \end{pmatrix}_{\mathcal{E}},$$

$$De_2 = D\cos x = (\cos x)' = -\sin x = -1\cdot \mathbf{e}_1 + 0\cdot \mathbf{e}_2 = \begin{pmatrix} -1 \\ 0 \end{pmatrix}_{\mathcal{E}}.$$

Thus, with respect to our chosen $\mathcal{E}$–basis $\{\sin x, \cos x\}$ of U the derivative operator D is represented by the 2×2 matrix

$$\mathbf{D}_{\mathcal{E}} = \begin{pmatrix} | & | \\ D\mathbf{e}_1 & D\mathbf{e}_2 \\ | & | \end{pmatrix} = \begin{pmatrix} 0 & -1 \\ 1 & 0 \end{pmatrix}$$

according to Section 1.1. If $\mathbf{D}_{\mathcal{E}}$ represents differentiation on U, then $\mathbf{D}_{\mathcal{E}}^{-1}$, the inverse operator, represents antidifferentiation, or **integration**, on U with respect to the same basis according to the Fundamental Theorem of Calculus. We find $\mathbf{D}_{\mathcal{E}}^{-1}$ by the standard matrix inversion process of Section 6.1 using row reductions :

$\mathbf{D}_{\mathcal{E}}$		$\mathbf{I}$		
0	−1	1	0	*swap rows 1 and 2*
1	0	0	1	
[1]	0	0	1	
0	[−1]	1	0	$\cdot(-1)$
1	0	0	1	
0	1	−1	0	
$\mathbf{I}$		$\mathbf{D}_{\mathcal{E}}^{-1}$		

Thus, $\mathbf{D}_{\mathcal{E}}^{-1} = \begin{pmatrix} 0 & 1 \\ -1 & 0 \end{pmatrix}$ and

$$\mathbf{D}_{\mathcal{E}}^{-1}\mathbf{e}_1 = D^{-1}\sin x = \int \sin x \, dx = \begin{pmatrix} 0 & 1 \\ -1 & 0 \end{pmatrix}\begin{pmatrix} 1 \\ 0 \end{pmatrix} = \begin{pmatrix} 0 \\ -1 \end{pmatrix}_{\mathcal{E}}$$

$$= 0\cdot \mathbf{e}_1 + 1\cdot \mathbf{e}_2 = \cos x \ ,$$

while

$$\mathbf{D}_{\mathcal{E}}^{-1}\mathbf{e}_2 = D^{-1}\cos x = \int \cos x \, dx = \begin{pmatrix} 0 & 1 \\ -1 & 0 \end{pmatrix}\begin{pmatrix} 0 \\ 1 \end{pmatrix} = \begin{pmatrix} 1 \\ 0 \end{pmatrix}_{\mathcal{E}}$$

$$= 1\cdot \mathbf{e}_1 + 0\cdot \mathbf{e}_2 = \sin x \ ,$$

precisely as in calculus.
Note that we do not retrieve the constants of integration in our approach, since the constant functions $f(x) = \text{const}$ do not belong to our chosen subspace $U = \text{span}\{\sin x, \ \cos x\} \subset \mathcal{F}^1$.

(b) Next we try to find the antiderivative of the real-valued function

$$f(t) = t^2 e^t \in \mathcal{F}$$

using linear algebra.

Calculus solves this problem via integration by parts. To solve this problem via linear algebra, our first effort is to find an invariant subspace $U \subset \mathcal{F}$ under the differentiation operator D that contains the given function f. Since

$$\begin{aligned} Df &= f'(x) = t^2e^t + 2te^t, \\ D^2 f &= f''(t) = t^2e^t + 3te^t + 2e^t, \\ D^3 f &= f'''(t) = t^2e^t + 5te^t + 5e^t, \end{aligned}$$

etc., f and all its derivatives $Df, D^2 f, \ldots$ lie in the D–invariant subspace

$$U = \text{span}\left\{t^2e^t,\ te^t,\ e^t\right\}.$$

Taking $\mathbf{e}_1 := t^2e^t$, $\mathbf{e}_2 := te^t$, and $\mathbf{e}_3 := e^t$ as the standard $\mathcal{E}$ basis for this subspace U, we compute

$$D\mathbf{e}_1 = (t^2e^t)' = t^2e^t + 2te^t = \begin{pmatrix} 1 \\ 2 \\ 0 \end{pmatrix}_{\mathcal{E}}$$

with respect to the basis $\mathcal{E}$. Similarly,

$$D\mathbf{e}_2 = (te^t)' = te^t + e^t = \begin{pmatrix} 0 \\ 1 \\ 1 \end{pmatrix}_{\mathcal{E}} \quad \text{and} \quad D\mathbf{e}_3 = (e^t)' = e^t = \begin{pmatrix} 0 \\ 0 \\ 1 \end{pmatrix}_{\mathcal{E}}.$$

Thus, from Chapter 1, differentiation on U acts like matrix $\times$ vector multiplication on the $\mathcal{E}$–coordinate vectors of any $f \in U$ for the matrix

$$\mathbf{D}_{\mathcal{E}} = \begin{pmatrix} | & | & | \\ D\mathbf{e}_1 & D\mathbf{e}_2 & D\mathbf{e}_3 \\ | & | & | \end{pmatrix} = \begin{pmatrix} 1 & 0 & 0 \\ 2 & 1 & 0 \\ 0 & 1 & 1 \end{pmatrix}.$$

To integrate $\int t^2e^t dt$, we find the antiderivatives of the functions that span U by matrix inversion:

$\mathbf{D}_{\mathcal{E}}$			$\mathbf{I}$			
$\boxed{1}$	0	0	1	0	0	
$\boxed{2}$	1	0	0	1	0	$-2\,row_1$
0	1	1	0	0	1	
$\boxed{1}$	0	0	1	0	0	
0	$\boxed{1}$	0	-2	1	0	
0	$\boxed{1}$	1	0	0	1	$-\,row_2$
$\boxed{1}$	0	0	1	0	0	
0	$\boxed{1}$	0	-2	1	0	
0	0	$\boxed{1}$	2	-1	1	
$\mathbf{I}_3$			$\mathbf{D}_{\mathcal{E}}^{-1}$			

Thus, $\mathbf{D}_{\mathcal{E}}^{-1} = \begin{pmatrix} | & | & | \\ \int \mathbf{e}_1 dt & \int \mathbf{e}_2 dt & \int \mathbf{e}_3 dt \\ | & | & | \end{pmatrix} = \begin{pmatrix} 1 & 0 & 0 \\ -2 & 1 & 0 \\ 2 & -1 & 1 \end{pmatrix}$ represents integration in U. Specifically,

$$\int \mathbf{e}_1 dt = \int t^2 e^t dt = \begin{pmatrix} 1 \\ -2 \\ 1 \end{pmatrix}_{\mathcal{E}} = t^2 e^t - 2te^t + 2e^t,$$

$$\int \mathbf{e}_2 dt = \int te^t dt = \begin{pmatrix} 0 \\ 1 \\ -1 \end{pmatrix}_{\mathcal{E}} = te^t - e^t,$$

$$\int \mathbf{e}_3 dt = \int e^t dt = \begin{pmatrix} 0 \\ 0 \\ 1 \end{pmatrix}_{\mathcal{E}} = e^t$$

from scanning the columns of $\mathbf{D}_{\mathcal{E}}^{-1}$ and expressing them via the chosen basis $\mathcal{E} = \{t^2 e^t,\ te^t,\ e^t\}$ for U. As $U = \text{span}\{t^2 e^t,\ te^t,\ e^t\}$ does not contain the constant functions, our indefinite integrals again miss their constants of integration.

(c) Solve the differential equation (see the Introduction)

$$f' + 2f = \cos x$$

via linear algebra.
We restrict ourselves to $U = \text{span}\{\sin x,\ \cos x\}$ according to part (a), since U is invariant under differentiation, and the right-hand-side function $\cos x$ lies in U. The differential operator L on the left-hand-side of the differential equation is

$$L(f) := f' + 2f = (D + 2I)f \stackrel{!}{=} \cos x\ .$$

L is represented by the matrix $\mathbf{D}_{\mathcal{E}} + 2\mathbf{I}_2 = \begin{pmatrix} 0 & -1 \\ 1 & 0 \end{pmatrix} + 2\begin{pmatrix} 1 & 0 \\ 0 & 1 \end{pmatrix} = \begin{pmatrix} 2 & -1 \\ 1 & 2 \end{pmatrix}$ on U as shown in part (a). This matrix representation of L operates on the coordinate vector of any function that can be expressed with respect to the basis $\mathcal{E} = \{\sin x,\ \cos x\}$ of U. To solve the given differential equation $Lf = f' + 2f = \cos x$ for a function $f(x) = \alpha \sin x + \beta \cos x = \begin{pmatrix} \alpha \\ \beta \end{pmatrix}_{\mathcal{E}} \in U$, we solve the equivalent linear system

$$Lf = (\mathbf{D}_{\mathcal{E}} + 2\mathbf{I}_2) f = \begin{pmatrix} 2 & -1 \\ 1 & 2 \end{pmatrix} \begin{pmatrix} \alpha \\ \beta \end{pmatrix} = \begin{pmatrix} 0 \\ 1 \end{pmatrix} = \mathbf{b},$$

since $\cos x = \begin{pmatrix} 0 \\ 1 \end{pmatrix}_{\mathcal{E}} \in U$:

$\mathbf{L} = \mathbf{D}_{\mathcal{E}} + 2\mathbf{I}_2$		$\mathbf{b}$	
2	−1	0	*swap rows 1 and 2*
1	2	1	
$\boxed{1}$	2	1	
$\boxed{2}$	−1	0	$-2\ row_1$
$\boxed{1}$	2	1	
0	$\boxed{-5}$	−2	

Thus, $\begin{pmatrix} \alpha \\ \beta \end{pmatrix} = \begin{pmatrix} 0.2 \\ 0.4 \end{pmatrix}$ by backsubstitution, and the function $f(x) = 0.2\sin x + 0.4\cos x$ solves $f' + 2f = \cos x$. Students should verify that for this f, we have

$$L(f) = f'(x) + 2f(x) = \cos x, \quad \text{for all } x \in \mathbb{R}.$$ ◀

We revisit Example 6(c) in Section 11.3.

These examples show how prevalent linear processes and linear operators are throughout mathematics and how beneficial it is to understand the power of linear algebra.

7.3.P Problems

1. (a) Let the linear transformation $T : \mathbb{R}^3 \to \mathbb{R}^4$ be represented with respect to the standard unit-vector bases $\mathcal{E}_3$, $\mathcal{E}_4$, of $\mathbb{R}^3$ and $\mathbb{R}^4$, respectively, by the matrix $\mathbf{A}_{\mathcal{E}_4 \leftarrow \mathcal{E}_3} = \begin{pmatrix} 1 & 2 & 5 \\ -1 & -7 & 0 \\ 1 & 0 & -3 \\ 0 & 1 & -2 \end{pmatrix}$. Find the matrix representation $\mathbf{A}_{\mathcal{B} \leftarrow \mathcal{C}}$ of T with respect to the two new bases $\mathcal{C} = \left\{ \begin{pmatrix} 1 \\ 2 \\ 0 \end{pmatrix}, \begin{pmatrix} -1 \\ 2 \\ 8 \end{pmatrix}, \begin{pmatrix} 1 \\ 0 \\ 3 \end{pmatrix} \right\}$ of $\mathbb{R}^3$ and $\mathcal{B} =$

$$\left\{ \begin{pmatrix} 4 \\ 1 \\ 2 \\ 0 \end{pmatrix}, \begin{pmatrix} -2 \\ 3 \\ 2 \\ -1 \end{pmatrix}, \begin{pmatrix} 0 \\ -11 \\ 12 \\ 6 \end{pmatrix}, \begin{pmatrix} 1 \\ 1 \\ -1 \\ 4 \end{pmatrix} \right\}$$

of $\mathbb{R}^4$.

(b) What is the matrix equation that links the two representations $\mathbf{A}_{\mathcal{E}_4 \leftarrow \mathcal{E}_3}$ and $\mathbf{A}_{\mathcal{B} \leftarrow \mathcal{C}}$ of T?

2. Let L be the mapping between the spaces of real valued function $B = \text{span}\{1,\ t,\ \sin(2t),\ \cos(2t)\}$ and $C = \text{span}\{1,\ \sin(2t) + \cos(2t),\ \sin(2t) - \cos(2t)\}$ defined by setting $L(f) = f' + 2f'' \in C = \text{span}\{c_j\}$ for each $f \in B = \text{span}\{b_i\}$.

(a) Find $L(\mathbf{b}_i)$ for each $i = 1, \ldots, 4$.

(b) Find the coordinate vectors $(L(b_i))_{\mathcal{C}}$ for each i.

(c) Find the matrix representation $\mathbf{A}$ of L with respect to the two different bases $\mathcal{B}$ and $\mathcal{C}$. What is the size of $\mathbf{A}$?

(d) Find the rank of L. Which functions $f \in B$ does L map to zero?

3. (a) Show that the functions $g(x) = -3\sin(x) + 2\cos(x)$ and $h(x) = 2\sin(x)$ span the space $U = \text{span}\{\sin(x),\ \cos(x)\}$.

(b) Find the basis change matrices $\mathbf{X}_{\{g,h\} \leftarrow \mathcal{U}}$ and $\mathbf{X}_{\mathcal{U} \leftarrow \{g,h\}}$

(c) Express the function $k(x) = 7\sin(x) - 4\cos(x)$ with respect to the basis $\{g, h\}$ of U. Express the function $\ell(x) = 7g(x) - 4h(x)$ with respect to the basis $\mathcal{U}$.

(d) Express the derivative operator D on U with respect to the basis $\sin(x)$ and $\cos(x)$.

(e) Express the derivative operator D on U with respect to the basis $g(x)$ and $h(x)$ from part (a).

(f) Repeat parts (d) and (e) for the second derivative D^2.

(g) Verify that the matrix representations of D^2 in part (f) are the squares of the matrices in parts (d) and (e), respectively.

4. The set $\mathcal{U} = \{u_i\} = \{1 + 2x^2,\ x - x^2,\ 3 + 2x - x^2\}$ is a basis for $\mathcal{P}_2$. (can you prove this?)

(a) Find the coordinate vector of the polynomials $p(x) = 3x^2 - 2x + 3$ and $q(x) = x^2 - 1$ with respect to $\mathcal{U}$.

(b) If a polynomial is expressed as $w(x) = 3u_1 - u_2$, what are the coefficients a_j of w in the standard $\sum a_j x^j$ notation?

5. Consider the two sets of real-valued functions

$$U = \{1, \cos(x), \cos^2(x), \cos^3(x)\}$$

and

$$V = \{1, \cos(x), \cos(2x), \cos(3x)\}.$$

(a) Use the trigonometric identities $\cos(2x) = 2\cos^2(x) - 1$ and $\cos(3x) = 4\cos^3(x) - 3\cos(x)$ to show that U and V span the same subspace.

(b) What is the basis change matrix $\mathbf{X}_{\mathcal{V}\leftarrow\mathcal{U}}$?

(c) Calculate the antiderivative of $h(x) = 3\cos^3(x) - \cos^2(x) + 5\cos(x)$ by transforming the $\mathcal{U}$–coordinate vector for h to $\mathcal{V}$–coordinates and integrating there.

6. (a) Express the differential operator $D : \mathcal{P}_n \to \mathcal{P}_n$ operating on the polynomials of degree less than or equal to n by an $n{+}1 \times n{+}1$ matrix $\mathbf{D}_{\mathcal{E}}$ with respect to the standard polynomial basis $\mathbf{e}_j = x^j$ for $j = 0, 1, \ldots, n$ of $\mathcal{P}_n$.

(b) Express the second derivative operator $D^2 : \mathcal{P}_n \to \mathcal{P}_n$ by the matrix power $(\mathbf{D}_{\mathcal{E}})^2$ for $\mathbf{D}_{\mathcal{E}}$ from part (a).

(c) What is the matrix representation for the third derivative operator $\mathcal{D}^3 : \mathcal{P}_n \to \mathcal{P}_n$?

(d) How about for the nth and ($n{+}1^{\text{st}}$ derivatives D^n and D^{n+1}.

7. (a) Find $\int 5e^t - 3t^2e^t + 2te^t\,dt$ by using matrix inversion.

(b) Find $\int e^t t^3\,dt$ using matrices.

7.R Review Problems

1. Explain in full technical detail how to obtain the matrix representation of a linear transformation $\mathbf{A}_{\mathcal{X}\leftarrow\mathcal{Y}}$ with respect to a basis $\mathcal{X}$ for its range space and another basis $\mathcal{Y}$ of its domain from the standard representation $\mathbf{A}_{\mathcal{E}_{..}\leftarrow\mathcal{E}_{..}}$ that is given according to the two standard $\mathcal{E}_{...}$ bases of its range and domain.

2. If $\mathcal{U} = \{\mathbf{u}_i\}_{i=1}^k \in \mathbb{R}^n$ is a basis of a subspace, what are the $\mathcal{U}$–coordinate vectors of each $\mathbf{u}_i$? In which space do these coordinate vectors lie?

3. For each of the following, determine whether the statement(s) are true or false and give reasons for your answer.

(a) If $\mathbf{X}$ is the basis change matrix from a basis $\mathcal{U}$ to a basis $\mathcal{V}$, then

(i) $\mathbf{w}_{\mathcal{V}} = \mathbf{w}_{\mathcal{U}}^T\mathbf{X}^T$.

(ii) $\mathbf{w}_{\mathcal{V}}^T = \mathbf{w}_{\mathcal{U}}^T\mathbf{X}^T$.

(iii) $\mathbf{w}_{\mathcal{V}} = \mathbf{X}^T\mathbf{w}_{\mathcal{U}}$.

(iv) $\mathbf{w}_{\mathcal{V}} = \mathbf{X}\mathbf{w}_{\mathcal{U}}$.

(v) $\mathbf{w}_{\mathcal{U}} = \mathbf{X}^{-1}\mathbf{w}_{\mathcal{V}}$.

(vi) $\mathbf{w}_{\mathcal{U}} = \mathbf{X}\mathbf{w}_{\mathcal{V}}$.

(b) A basis change matrix $\mathbf{X}$ is always square.

(c) The basis change matrix from $\mathcal{V}$–vectors to $\mathcal{U}$–vectors is $\mathbf{X}_{\mathcal{V}\leftarrow\mathcal{U}}^T$.

(d) The basis-change matrix $\mathbf{X}_{\mathcal{V}\leftarrow\mathcal{U}}$ can be found from a row echelon reduction of the multiaugmented matrix $(\mathbf{U} \mid \mathbf{V}^T)$.

(e) The basis-change matrix $\mathbf{X}_{\mathcal{V}\leftarrow\mathcal{U}}$ can be found from the RREF of the multi-augmented matrix $(\mathbf{U} \mid \mathbf{V})$.

(f) The basis-change matrix $\mathbf{X}_{\mathcal{V}\leftarrow\mathcal{U}}$ can be found from a row echelon reduction of the multi-augmented matrix $(\mathbf{V} \mid \mathbf{U})$.

4. Explain how any nonsingular real matrix $\mathbf{X}_{nn}$ can be viewed as a basis change matrix for $\mathbb{R}^n$. What two bases are involved with one basis change matrix $\mathbf{X}$? What are the sets of vectors that form the two bases?

5. (a) Show that

$$\mathcal{U} = \left\{ \begin{pmatrix} 2 \\ 1 \\ 0 \end{pmatrix}, \begin{pmatrix} 1 \\ 0 \\ 2 \end{pmatrix}, \begin{pmatrix} 1 \\ 1 \\ 1 \end{pmatrix} \right\}$$

is a basis for $\mathbb{R}^3$.

(b) Assume that

$$\mathcal{V} = \left\{ \begin{pmatrix} 1 \\ 0 \\ 1 \end{pmatrix}, \begin{pmatrix} 1 \\ 1 \\ 0 \end{pmatrix}, \begin{pmatrix} 0 \\ 1 \\ 1 \end{pmatrix} \right\}$$

is another basis of $\mathbb{R}^3$. Find the basis change matrix $\mathbf{X}_{\mathcal{V}\leftarrow\mathcal{U}}$.

(c) If $\mathbf{x}_{\mathcal{U}} = \begin{pmatrix} -1 \\ 3 \\ 2 \end{pmatrix}$ is the $\mathcal{U}$–coordinate vector of a point x in $\mathbb{R}^3$, find the coordinate vector $\mathbf{x}_{\mathcal{V}}$ of $\mathbf{x}$ with respect to the basis $\mathcal{V}$.

(d) What is the basis change matrix $\mathbf{X}_{\mathcal{E}\leftarrow\mathcal{V}}$ where $\mathcal{E}$ denotes the standard unit vector basis $\{\mathbf{e}_1, \mathbf{e}_2, \mathbf{e}_3\}$ of $\mathbb{R}^3$.

6. Prove the following:

(a) If $\mathbf{X}_{kk}$ is a basis change matrix with columns $\mathbf{x}_i$ and $k > 3$, then the three vectors $\mathbf{x}_1$, $\mathbf{x}_2$, and $\mathbf{x}_k$ are linearly independent.

(b) Conversely, three vectors that are linearly dependent cannot appear as some columns of the same basis change matrix.

7. To find the $\mathcal{U}$–coordinate vector $\mathbf{x}_{\mathcal{U}}$ of a point $\mathbf{x} = \mathbf{x}_{\mathcal{E}} \in \mathbb{R}^n$, one can compute the inverse $\mathbf{U}^{-1}$ of the associated matrix $\mathbf{U}$ and evaluate $\mathbf{x}_{\mathcal{U}} = \mathbf{U}^{-1}\mathbf{x}$, or one can solve the linear system $\mathbf{U}\mathbf{x}_{\mathcal{U}} = \mathbf{x}$ for the unknown coordinate vector $\mathbf{x}_{\mathcal{U}}$. Which method is to be preferred?

8. (a) Show that $\mathbf{u}_1 = \begin{pmatrix} 1 \\ 2 \\ 1 \end{pmatrix}$, $\mathbf{u}_2 = \begin{pmatrix} 1 \\ 0 \\ 1 \end{pmatrix}$ and $\mathbf{u}_3 = \begin{pmatrix} 0 \\ 1 \\ 1 \end{pmatrix}$ form a basis $\mathcal{U}$ of $\mathbb{R}^3$.

(b) Find the matrix representation $\mathbf{A}_{\mathcal{U}}$ for the linear transformation T defined by $T(\alpha\mathbf{u}_1 + \beta\mathbf{u}_2 + \gamma\mathbf{u}_3) = (\alpha - \beta + \gamma)\mathbf{u}_1 - (\beta + \gamma)\mathbf{u}_2 + (\alpha - \beta)\mathbf{u}_3$.

(c) Find the standard matrix $\mathbf{A}_{\mathcal{E}}$ of T from $\mathbf{A}_{\mathcal{U}}$. What is the formula?

Standard Questions

1. What is the coordinate vector $\mathbf{x}_{\mathcal{U}}$ of a point $\mathbf{x}$ in n–space with respect to a basis $\mathcal{U}$ of $\mathbb{R}^n$?

2. How are two coordinate vectors $\mathbf{x}_{\mathcal{U}}$ and $\mathbf{x}_{\mathcal{V}}$ of one point $\mathbf{x} \in \mathbb{R}^n$ related for two different bases $\mathcal{U}$ and $\mathcal{V}$ of $\mathbb{R}^n$?

3. How can one find the basis change matrix from $\mathcal{U}$–coordinate vectors $\mathbf{x}_{\mathcal{U}}$ to $\mathcal{V}$–coordinate vectors $\mathbf{x}_{\mathcal{V}}$?

4. What is the matrix representation of a linear transformation $T : \mathbb{R}^n \to \mathbb{R}^n$ with respect to an arbitrary basis $\mathcal{U}$ of $\mathbb{R}^n$?

5. How does the theory of basis change extend to two bases of the same proper subspace of R^n?

Subheadings of Lecture Seven

Basic Equations

p. 202	$\mathbf{x} = \mathbf{x}_{\mathcal{E}} = \mathbf{U}\mathbf{x}_{\mathcal{U}} = \mathbf{V}\mathbf{x}_{\mathcal{V}}$	(coordinate vector)
p. 203	$\mathbf{X}_{\mathcal{V}\leftarrow\mathcal{U}} = \mathbf{V}^{-1}\mathbf{U}$	(basis change matrix)
p. 206	$\mathbf{A}_{\mathcal{U}} = \mathbf{U}^{-1}\mathbf{A}_{\mathcal{E}}\mathbf{U}$	(matrix representation for basis $\mathcal{U}$)
p. 209	$\mathbf{A}\mathbf{x} = \lambda\mathbf{x}\ ,\ \ \mathbf{x} \neq \mathbf{0}$	(eigenvalue/eigenvector equation)

The way it was

CHAPTER

8

Determinants, λ–Matrices; an optional chapter

Determinants precede matrices in the history of mathematics by several centuries.[1] They are, however, of limited import to linear transformations and linear spaces. Yet many client disciplines and mathematicians use them in their presentations. For this reason, we include a short introduction to determinants.

The material of this chapter is optional for understanding the subsequent chapters. In the remaining chapters, we treat the matrix eigenvalue problem independently of determinants, as well as with their help.

Instructors and students may follow either approach to eigenvalues (i.e., use vector iteration, explained in Section 9.1, or use determinants, explained in Section 9.1.D.

The determinant is a scalar function of a square matrix that we use to better understand linear transformations.

8.1 Lecture Eight

Laplace Expansion, Gaussian Elimination, and Properties

This lecture introduces the determinant, a basic, but complicated computational tool for square matrices.

Determinants are peculiar functions of square matrices. If $\mathbf{A}_{nn}$ is a square matrix with n^2 scalar entries from $\mathbb{R}$ or $\mathbb{C}$, or with algebraic expressions or functions as entries, then $\det(\mathbf{A})$ is a certain sum of products of these entries. For example,

[1]Determinants first appeared in 1683, both in Japan and in Europe, though the term "determinant" was not used until the early 1800s by both Gauss and Cauchy. The term "matrix" was first put in print by Cayley in 1858. General $n \times n$ matrices were used in Sylvester's papers from around 1880. Frobenius adopted the term only in 1896. The word for and the concept of 'matrices' finally became widely used in the first decades of the 20th century.

det(**A**) is real if **A** is real. In general, the determinant of a square matrix **A** is a complicated **multilinear function** of the entries of **A**.

Determinants, taken of square matrices, are useful for the several applications: They describe the volume of the parallelepiped in $\mathbb{R}^n$ spanned by the columns of $\mathbf{A} \in \mathbb{R}^{n,n}$. They give us Cramer's rule for solving small systems of linear equations, and they help explain the coefficients that occur in Gaussian elimination. They are useful to decide upon linear dependence and independence of vectors and functions, and, most important for the theory, they help us transform the matrix eigenvalue problem into one of polynomial roots.

The aim of this section is technical:

Given a matrix $\mathbf{A}_{nn}$, *compute its determinant* $\det(\mathbf{A})$ *efficiently.*

There are essentially two ways to evaluate determinants:

- by Laplace expansion
- by row reduction.

Both methods can and will be used in tandem to make speed.

(a) Laplace Expansion

In **Laplace expansion**, the determinant of an $n \times n$ matrix A is evaluated inductively.

If $n = 1$, then $\mathbf{A}_{11} = (\alpha)$ and $\det \mathbf{A} = \alpha$, where α is a number or a mathematical expression of any sort such as $\alpha = 2.3, \alpha = -1, \alpha = \sin x, \alpha = \sqrt{x} - \cos^2 x, \alpha = \lambda - 2$ etc. Thus, $\det(2x-3) = 2x-3$ and $\det(-5) = -5$.

If $n = 2$, then $\mathbf{A}_{22} = \begin{pmatrix} \alpha & \beta \\ \gamma & \delta \end{pmatrix}$ with $\alpha, \beta, \gamma, \delta$ numbers or algebraic expressions, and $\det \mathbf{A} = \alpha\delta - \beta\gamma$. For example, $\det \begin{pmatrix} 1 & 2 \\ -1 & 3 \end{pmatrix} = 3 + 2 = 5$, and $\det \begin{pmatrix} x & 1 \\ 2 & x \end{pmatrix} = x^2 - 2$.

For $n \geq 1$, we use the **chess pattern matrix**

$$\mathbf{P}_{nn} = \begin{pmatrix} + & - & + & \cdots & (-1)^{n+1} \\ - & + & - & & \\ + & - & + & & \\ - & + & & \ddots & \\ \vdots & & & & \\ (-1)^{n+1} & & & & + \end{pmatrix}_{nn} = \left(\; (-1)^{i+j} \; \right)_{i,j=1}^{n} .$$

This sign pattern matrix looks like a black-and-white chess or checkers board, with alternating signs ± 1 assigned to its n^2 positions. The $(1, 1)$ entry of **P** is always $+1$.

EXAMPLE 1 The chess pattern matrices for $n = 1, 2, 3$, and 4 are $(\mathbf{1})$, $\begin{pmatrix} \mathbf{1} & -1 \\ -1 & \mathbf{1} \end{pmatrix}$, $\begin{pmatrix} \mathbf{1} & -1 & \mathbf{1} \\ -1 & \mathbf{1} & -1 \\ \mathbf{1} & -1 & \mathbf{1} \end{pmatrix}$, and $\begin{pmatrix} \mathbf{1} & -1 & \mathbf{1} & -1 \\ -1 & \mathbf{1} & -1 & \mathbf{1} \\ \mathbf{1} & -1 & \mathbf{1} & -1 \\ -1 & \mathbf{1} & -1 & \mathbf{1} \end{pmatrix}$, respectively, where we have highlighted the alternating positive entries for visual effect. ◀

The determinant of a matrix $\mathbf{A}_{nn}$ of size $n \geq 3$ can be evaluated using the chess pattern matrix $\mathbf{P}_{nn}$ and Laplace expansion. Laplace expansion proceeds either along a chosen row $\mathbf{r}_j$ of $\mathbf{A}$, or along a column $\mathbf{c}_k$ of $\mathbf{A}$. For best efficiency, we generally expand $\det(\mathbf{A})$ along a row or column of $\mathbf{A}$ with maximally many simple entries, such as zero or one.

Determinants of square matrices via Laplace expansion

(0) Pick a row j ((or a column k)) of $\mathbf{A}_{nn} = \begin{pmatrix} a_{11} & \dots & a_{1n} \\ \vdots & & \vdots \\ a_{j-1,1} & \dots & a_{j-1,n} \\ \mathbf{a}_{j,1} & \dots & \mathbf{a}_{j,n} \\ a_{j+1,1} & \dots & a_{j+1,n} \\ \vdots & & \vdots \\ a_{n,1} & \dots & a_{n,n} \end{pmatrix} \leftarrow j.$

(1) (<u>Row version</u>)
For each column index $\ell = 1, \dots, n$, denote the $n{-}1 \times n{-}1$ matrix $\widehat{\mathbf{A}}_{(j,\ell)}$ with row j <u>and</u> column ℓ omitted by

column ℓ of $\mathbf{A}$ omitted
↓

$$\widehat{\mathbf{A}}_{(j,\ell)} = \left(\begin{array}{ccc|ccc} a_{11} & \cdots & a_{1\ell-1} & a_{1\ell+1} & \cdots & a_{1n} \\ \vdots & \vdots & \vdots & \vdots & & \\ a_{j-1,1} & \cdots & a_{j-1\ell-1} & a_{j-1\ell+1} & \cdots & a_{jn} \\ \hline a_{j+1,1} & \cdots & a_{j+1\ell-1} & a_{j+1\ell+1} & \cdots & a_{j+1,n} \\ \vdots & \vdots & \vdots & \vdots & & \\ a_{n1} & \cdots & a_{n\ell-1} & a_{n\ell+1} & \cdots & a_{nn} \end{array}\right)_{n-1,n-1} \leftarrow \text{row } j \text{ omitted.}$$

$\widehat{\mathbf{A}}_{(j,\ell)}$ is obtained from $\mathbf{A}$ by omitting row j and column ℓ from $\mathbf{A}$. The omissions are indicated by vertical and horizontal lines in the display of $\widehat{\mathbf{A}_{(j,\ell)}}$.

(2) Evaluate $\det \mathbf{A}$ as the signed sum of products:

$$\det \mathbf{A} = \sum_{\ell=1}^{n} \Big(\underbrace{(-1)^{j+\ell}}_{\text{chess pattern matrix sign}} \cdot \underbrace{a_{j\ell}}_{(j,\ell) \text{ entry of } \mathbf{A}} \cdot \underbrace{\det(\widehat{\mathbf{A}}_{(j,\ell)})}_{\text{smaller sized matrix determinant}} \Big)$$

(3) The n smaller sized determinants of each submatrix $\widehat{\mathbf{A}_{(j,\ell)}}$ are evaluated recursively by again using Laplace expansion: Evaluate their even smaller sized subdeterminants by expansion until the respective sizes drop to two or one. The low order determinants can be evaluated directly from the determinant formulas for $n = 1, 2$.

Analogously, one can evaluate det(**A**) by **column sum expansion** along any column k of **A** as

$$\det \mathbf{A} = \sum_{\ell=1}^{n} \left((-1)^{\ell+k} a_{\ell k} \det(\widehat{\mathbf{A}_{(\ell,k)}})\right).$$

EXAMPLE 2

(a) Evaluate $\det \begin{pmatrix} 1 & 2 & 3 \\ 2 & 0 & -1 \\ 1 & 1 & 1 \end{pmatrix}$ by Laplace expansion along column 2. The chess-pattern matrix for $n = 3$ is $\begin{pmatrix} 1 & \mathbf{-1} & 1 \\ -1 & \mathbf{1} & -1 \\ 1 & \mathbf{-1} & 1 \end{pmatrix}$, with its relevant column 2 highlighted. Thus, by expanding along column 2, we obtain

$$\det \begin{pmatrix} 1 & \mathbf{2} & 3 \\ 2 & \mathbf{0} & -1 \\ 1 & \mathbf{1} & 1 \end{pmatrix} = (-1) \cdot 2 \cdot \det \begin{pmatrix} 2 & -1 \\ 1 & 1 \end{pmatrix} + 0 - 1 \cdot \det \begin{pmatrix} 1 & 3 \\ 2 & -1 \end{pmatrix}$$
$$= -2(2+1) - (-1-6) = 1.$$

Here we have used the 2×2 determinant definition for evaluating the two 2×2 subdeterminants.

(b) Evaluate $\det \mathbf{A}$ for $\mathbf{A} = \begin{pmatrix} 4 & 1 & 0 & 2 \\ -1 & 3 & 5 & 1 \\ 0 & 1 & 4 & 0 \\ 2 & -1 & -1 & 1 \end{pmatrix} \in \mathbb{R}^{4,4}$ by Laplace expansion.

We pick row expansion along row 3, since this row has two entries equal to zero. The chess pattern matrix for $n = 4$ is $\begin{pmatrix} 1 & -1 & 1 & -1 \\ -1 & 1 & -1 & 1 \\ \mathbf{1} & \mathbf{-1} & \mathbf{1} & \mathbf{-1} \\ -1 & 1 & -1 & 1 \end{pmatrix}$, with row 3 highlighted. Thus, det(**A**) =

$$\underset{\substack{\uparrow \\ \text{chess} \\ \text{pattern} \\ \text{sign}}}{(-1)} \quad \underset{\substack{\uparrow \\ (3,2) \\ \text{entry} \\ \text{value}}}{(1)} \quad \det \underset{\substack{\uparrow \\ \text{leftover} \\ \text{matrix}}}{\begin{pmatrix} \mathbf{4} & \mathbf{0} & \mathbf{2} \\ -1 & 5 & 1 \\ 2 & -1 & 1 \end{pmatrix}} + \underset{\substack{\uparrow \\ \text{chess} \\ \text{pattern} \\ \text{sign}}}{(1)} \quad \underset{\substack{\uparrow \\ (3,3) \\ \text{entry}}}{(4)} \quad \det \underset{\substack{\uparrow \\ \text{leftover} \\ \text{matrix}}}{\begin{pmatrix} 4 & 1 & \mathbf{2} \\ -1 & 3 & \mathbf{1} \\ 2 & -1 & \mathbf{1} \end{pmatrix}}$$

$$= -\left(4\det\begin{pmatrix} 5 & 1 \\ -1 & 1 \end{pmatrix} + 2\det\begin{pmatrix} -1 & 5 \\ 2 & -1 \end{pmatrix}\right) \qquad +$$

(↑ further expansion of 3×3 *determinants along row 1, and along column 3 ↓)*

$$+4\left(2\det\begin{pmatrix} -1 & 3 \\ 2 & -1 \end{pmatrix} - 1\det\begin{pmatrix} 4 & 1 \\ 2 & -1 \end{pmatrix} + 1\det\begin{pmatrix} 4 & 1 \\ -1 & 3 \end{pmatrix}\right)$$

$$= -(4(5+1) + 2(1-10)) + 4(2(1-6) - (-4-2) + (12+1))$$

(by direct evaluation of the 2×2 *determinants)*

$$= -(4 \cdot 6 - 2 \cdot 9) + 4(-10 + 6 + 13)$$

$$= -6 + 4(9) = -6 + 36 = 30.$$

◀

The row (or column) expansion of determinants is a rather tedious job. As every determinantal expansion of one $n \times n$ determinant generally involves evaluating n determinants of order $n-1$, each of which in turn requires $n-1$ determinant evaluations of order $n-2$, and so forth, it follows that repeated row (or column) expansion to evaluate $\det(\mathbf{A})$ is an $n! = n \cdot (n-1) \cdots 2 \cdot 1$ process for an $n \times n$ dense matrix $\mathbf{A}$.

The subsequent Chapters 9 through 14 deal with matrix eigenproblems. These can be solved theoretically in part by evaluating the determinants of **λ-matrices** $\mathbf{A}(\lambda)$, whose entries depend on one variable λ.

EXAMPLE 3 Find $\det(\mathbf{A}(\lambda))$ for $\mathbf{A}(\lambda) = \begin{pmatrix} \lambda - 2 & 1 & -1 \\ 1 & \lambda + 4 & -1 \\ 1 & 1 & \lambda \end{pmatrix}_{3,3}$ and $\lambda \in \mathbb{R}$ or $\mathbb{C}$.

Expanding along row 3 of $\mathbf{A}(\lambda)$ for its relatively easy factors $a_{3,\ldots}$, we obtain

$$\det\begin{pmatrix} \lambda - 2 & 1 & -1 \\ 1 & \lambda + 4 & -1 \\ \mathbf{1} & \mathbf{1} & \boldsymbol{\lambda} \end{pmatrix} = \det\begin{pmatrix} 1 & -1 \\ \lambda + 4 & -1 \end{pmatrix} - \det\begin{pmatrix} \lambda - 2 & -1 \\ 1 & -1 \end{pmatrix}$$

$$+\lambda\det\begin{pmatrix} \lambda - 2 & 1 \\ 1 & \lambda + 4 \end{pmatrix}.$$

The determinantal 2×2 formula, applied three times, then yields

$$\begin{aligned}\det(\mathbf{A}(\lambda)) &= -1 + (\lambda + 4) + (\lambda - 2) - 1 + \lambda((\lambda - 2)(\lambda + 4) - 1) \\ &= 2\lambda + \lambda(\lambda^2 + 2\lambda - 9) \\ &= \lambda(\lambda^2 + 2\lambda - 7) = \lambda(\lambda - 1 - \sqrt{8})(\lambda - 1 + \sqrt{8})\end{aligned}$$

in completely factored form. ◀

(b) Determinants via Gaussian Elimination

As an alternative to Laplace expansion, determinants of square matrices can be evaluated by using **row reduction**.

The main idea fuelling this approach is to use Gaussian elimination on $\mathbf{A}_{nn}$ and reduce $\mathbf{A}$ to a sparser REF $\mathbf{R}$. $\mathbf{A}$'s determinant is then evaluated from $\mathbf{R}$. More specifically, it is the modified product of $\mathbf{R}$'s diagonal entries, namely,

$\det(\mathbf{A}) = \pm c \prod r_{ii}$ for a certain $+/-$ sign and a constant c that both derive from the elimination process. For real matrices, this process is much quicker than Laplace expansion. However, for evaluating determinants of λ–matrices, Gaussian elimination is generally very cumbersome and not well suited.

The three basic operations of Gaussian elimination affect the value of the determinant $\det(\mathbf{A})$ in a known manner. Here are the rules for determinants of row equivalent matrices.

The first two rules assert that the determinant is a **multilinear function** (i.e., one that is linear in each row $\mathbf{r}_\ell$ of $\mathbf{A}_{nn}$.

1. **Adding or subtracting** nonzero multiples of one row to or from another (by an elimination of type (6.3)) has no effect on the determinant:

$$\det \begin{pmatrix} - & \mathbf{r}_1 & - \\ & \vdots & \\ - & \mathbf{r}_k & - \\ & \vdots & \\ - & \mathbf{r}_j + \alpha \mathbf{r}_k & - \\ & \vdots & \\ - & \mathbf{r}_n & - \end{pmatrix} \begin{matrix} \\ \\ \leftarrow k \\ \\ \leftarrow j \\ \\ \\ \end{matrix} = \det \begin{pmatrix} - & \mathbf{r}_1 & - \\ & \vdots & \\ - & \mathbf{r}_k & - \\ & \vdots & \\ - & \mathbf{r}_j & - \\ & \vdots & \\ - & \mathbf{r}_n & - \end{pmatrix} \tag{8.1}$$

2. **Scaling a row** by $c \neq 0$ (in an elimination step of type (6.4)) changes the determinant to c times the original determinant:

$$\det \begin{pmatrix} - & \mathbf{r}_1 & - \\ & \vdots & \\ - & \mathbf{r}_{j-1} & - \\ - & c \cdot \mathbf{r}_j & - \\ - & \mathbf{r}_{j+1} & - \\ & \vdots & \\ - & \mathbf{r}_n & - \end{pmatrix} \leftarrow j \; = c \det \begin{pmatrix} - & \mathbf{r}_1 & - \\ & \vdots & \\ - & \mathbf{r}_j & - \\ & \vdots & \\ - & \mathbf{r}_n & - \end{pmatrix} \tag{8.2}$$

The third rule makes the determinant an **alternating function**.

3. **Swapping rows**: Each row swap (by an elimination of type (6.5)) changes the determinant to the negative value of the original determinant:

$$\det \begin{pmatrix} - & \mathbf{r}_1 & - \\ & \vdots & \\ - & \mathbf{r}_{k-1} & - \\ - & \mathbf{r}_j & - \\ - & \mathbf{r}_{k+1} & - \\ & \vdots & \\ - & \mathbf{r}_{j-1} & - \\ - & \mathbf{r}_k & - \\ - & \mathbf{r}_{j+1} & - \\ & \vdots & \\ - & \mathbf{r}_n & - \end{pmatrix} \begin{matrix} \\ \\ \\ \leftarrow k \\ \\ \\ \\ \leftarrow j \\ \\ \\ \\ \end{matrix} = -\det \begin{pmatrix} - & \mathbf{r}_1 & - \\ & \vdots & \\ - & \mathbf{r}_k & - \\ & \vdots & \\ - & \mathbf{r}_j & - \\ & \vdots & \\ - & \mathbf{r}_n & - \end{pmatrix}. \tag{8.3}$$

If the three rules (8.1), (8.2), and (8.3) are followed and the constant c, as well as the sign changes are monitored throughout the Gaussian elimination process of $\mathbf{A}$ to a REF $\mathbf{R}$, then the determinant of $\mathbf{A}_{nn}$ is $\pm c \prod r_{ii}$.

EXAMPLE 4 (a) Evaluate det $\mathbf{A}$ via row reduction for the matrix $\mathbf{A}$ from Example 2 (b):

4	1	0	2		*swap rows 1 and 2:*
−1	3	5	1		*first swap : sign change: −1*
0	1	4	0		
2	−1	−1	1		

$\boxed{-1}$	3	5	1		
$\boxed{4}$	1	0	2	$+4\ row_1$	
0	1	4	0		
$\boxed{2}$	−1	−1	1	$+2\ row_1$	

$\boxed{-1}$	3	5	1		
0	13	20	6		*swap rows 2, 3: second swap:*
0	1	4	0		*sign changes to +1*
0	5	9	3		

$\boxed{-1}$	3	5	1		
0	$\boxed{1}$	4	0		
0	$\boxed{13}$	20	6	$-13\ row_2$	
0	$\boxed{5}$	9	3	$-5\ row_2$	

$\boxed{-1}$	3	5	1		
0	$\boxed{1}$	4	0		
0	0	−32	6	$-3\ row_4$	*intermediate matrix* **B**
0	0	−11	3		

$\boxed{-1}$	3	5	1		
0	$\boxed{1}$	4	0		
0	0	$\boxed{1}$	−3		
0	0	$\boxed{-11}$	3	$+11\ row_3$	

$\boxed{-1}$	3	5	1		
0	$\boxed{1}$	4	0		
0	0	$\boxed{1}$	−3		*row echelon form* **R**
0	0	0	$\boxed{-30}$		

In this example, the determinant of $\mathbf{A}$ is the same as that of $\mathbf{R}$ according to the rules (8.1) through (8.3), since we have not scaled ($c = 1$), and swapped an even number of times during the process. The determinant of $\mathbf{R}$ can be most easily evaluated by Laplace expansion:

$$\det(\mathbf{A}) = (-1)^{\text{number of swaps}} \det(\mathbf{R}) = \det(\mathbf{R}) = \det\begin{pmatrix} \mathbf{-1} & 3 & 5 & 1 \\ \mathbf{0} & 1 & 4 & 0 \\ \mathbf{0} & 0 & 1 & -3 \\ \mathbf{0} & 0 & 0 & -30 \end{pmatrix} =$$

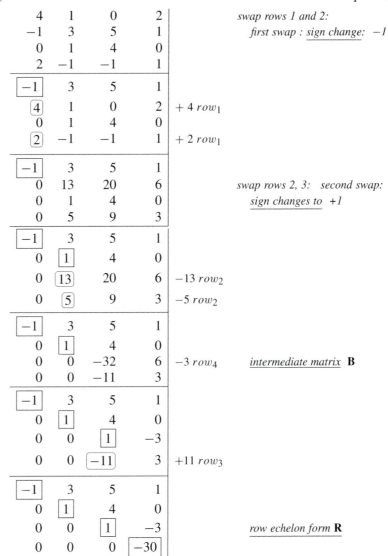

$$-1\det\begin{pmatrix}\mathbf{1} & 4 & 0\\ \mathbf{0} & 1 & -3\\ \mathbf{0} & 0 & -30\end{pmatrix} = -\det\begin{pmatrix}\mathbf{1} & -3\\ \mathbf{0} & -30\end{pmatrix} = 30 = \prod_{i=1}^{4} r_{ii}$$ by expanding first along the first column of $\mathbf{R}$ and analogously for each remainder matrix. Part 1 of the determinant Proposition below asserts that for upper triangular matrices $\mathbf{R}_{nn}$ we always have $\det(\mathbf{R}_{nn}) = \prod_{i=1}^{n} r_{ii}$.

It is sometimes more expedient to stop at an intermediate step of the Gaussian reduction and evaluate $\det(\mathbf{A})$ from an intermediate matrix such as $\mathbf{B}$. For our example, we have $\det(\mathbf{A}) = (-1)^{\text{number of swaps until reaching } \mathbf{B}_C} \cdot \det(\mathbf{B})$

$$= \det\begin{pmatrix}\mathbf{-1} & 3 & 5 & 1\\ \mathbf{0} & 1 & 4 & 0\\ \mathbf{0} & 0 & -32 & 6\\ \mathbf{0} & 0 & -11 & 3\end{pmatrix} \underset{\substack{\uparrow\\ \textit{expand}\\ \textit{along}\\ \textit{column 1}}}{=} -\det\begin{pmatrix}\mathbf{1} & 4 & 0\\ \mathbf{0} & -32 & 6\\ \mathbf{0} & -11 & 3\end{pmatrix} \underset{\substack{\uparrow\\ \textit{expand}\\ \textit{along new}\\ \textit{column 1}}}{=}$$

$$= -\det\begin{pmatrix}-32 & 6\\ -11 & 3\end{pmatrix} = -(-32\cdot 3 + 11\cdot 6) = 30,$$

where the final 2×2 determinant is evaluated directly from the 2×2 definition. Note that both results agree with that of Example 2(b).

(b) We use partial Gaussian elimination for the λ-matrix $\mathbf{A}(\lambda) = \begin{pmatrix}\lambda-2 & 1 & -1\\ 1 & \lambda+4 & -1\\ 1 & 1 & \lambda\end{pmatrix}$ of Example 3 to find $\det(\mathbf{A}(\lambda))$.

$\lambda-2$	1	-1	
$\boxed{1}$	$\lambda+4$	-1	*(Note the unusual position of the pivot.)*
$\boxed{1}$	1	λ	$-row_2$
$\lambda-2$	1	-1	
1	$\lambda+4$	-1	
0	$-\lambda-3$	$\lambda+1$	

To find the determinant of $\mathbf{A}(\lambda)$, we expand along the third row of the row-equivalent λ–matrix $\begin{pmatrix}\lambda-2 & 1 & -1\\ 1 & \lambda+4 & -1\\ 0 & -\lambda-3 & \lambda+1\end{pmatrix}$:

$$\det\mathbf{A}(\lambda) = \det\begin{pmatrix}\lambda-2 & 1 & -1\\ 1 & \lambda+4 & -1\\ \mathbf{0} & \mathbf{-\lambda-3} & \mathbf{\lambda+1}\end{pmatrix}$$

$$= -(-\lambda-3)\det\begin{pmatrix}\lambda-2 & -1\\ 1 & -1\end{pmatrix} + (\lambda+1)\det\begin{pmatrix}\lambda-2 & 1\\ 1 & \lambda+4\end{pmatrix}$$

$$= (\lambda+3)(-\lambda+2+1) + (\lambda+1)((\lambda-2)(\lambda+4)-1)$$
$$= \lambda^3 + 2\lambda^2 - 7\lambda = \lambda(\lambda^2 + 2\lambda - 7),$$

as before in Example 3. ◀

The best and most efficient method for evaluating determinants of low-order matrices $\mathbf{A}_{nn}$ or λ–matrices $\mathbf{A}(\lambda)_{nn}$ is to mix row reduction with Laplace expansion.

(c) Properties of Determinants

Here are some basic properties of determinants:

Proposition: **(Determinant Properties)**

1. $\det \begin{pmatrix} r_{11} & & * \\ & \ddots & \\ 0 & & r_{nn} \end{pmatrix} = r_{11} \cdot r_{22} \cdots r_{nn} = \prod_{i=1}^{n} r_{ii}$ for upper triangular matrices.
2. $\det(\mathbf{AB}) = \det \mathbf{A} \cdot \det \mathbf{B} = \det(\mathbf{BA})$ for any two square $n \times n$ matrices $\mathbf{A}$ and $\mathbf{B}$.
3. $\det(\mathbf{I}_n) = 1$ for every n.
4. $\det(\mathbf{A}^{-1}) = \dfrac{1}{\det \mathbf{A}}$ if $\det \mathbf{A} \neq 0$.
5. A matrix $\mathbf{A}$ is invertible if and only if $\det \mathbf{A} \neq 0$.
6. $\det(\mathbf{A}^T) = \det \mathbf{A}$.
7. $\det(\mathbf{A} + \mathbf{B}) \neq \det \mathbf{A} + \det \mathbf{B}$ in general.
8. $\det \mathbf{A} \neq 0$ if and only if every linear system $\mathbf{A}_{nn}\mathbf{x} = \mathbf{b}$ can be solved uniquely for $\mathbf{x}$ and every right-hand-side $\mathbf{b} \in \mathbb{R}^n$.
9. $\det(\mathbf{A}) \neq 0$ if and only if the columns of $\mathbf{A}_{nn}$ form a basis of $\mathbb{R}^n$.
10. If two rows (or columns) of $\mathbf{A}$ are identical, then $\det(\mathbf{A}) = 0$.
11. $\det(c\mathbf{A}) = (c)^n \det \mathbf{A}$ for every scalar c if $\mathbf{A}$ is $n \times n$.
12. If $\mathbf{A}$ is similar to $\mathbf{B}$ (i.e., if $\mathbf{X}^{-1}\mathbf{AX} = \mathbf{B}$), then $\det(\mathbf{A}) = \det(\mathbf{B})$.
13. If $\mathbf{A}$ is a block triangular matrix $\mathbf{A}_{nn} = \begin{pmatrix} \mathbf{B}_{kk} & \mathbf{C}_{k\ell} \\ 0 & \mathbf{D}_{\ell\ell} \end{pmatrix}$ with $k+\ell = n$, then $\det(\mathbf{A}) = \det(\mathbf{B}) \cdot \det(\mathbf{C})$. ◀

Property 12 follows from parts 2 and 4. It suggests the relevance of determinants when trying to solve the nonlinear matrix eigenvalue problem introduced in Section 7.1; see Section 9.1.D.

In light of the determinantal properties 4, 5, and 9, we add the following definition:

DEFINITION 1 An $n \times n$ matrix $\mathbf{A}$ is **nonsingular**, or **invertible**, if and only if $\det \mathbf{A} \neq 0$. A square matrix $\mathbf{A}_{nn}$ is **singular** if and only if $\det \mathbf{A} = 0$.

Determinants of square matrices can be evaluated in two possibly intertwined elementary ways.

8.1.P Problems

1. Find $\det\begin{pmatrix} 1 & 2 & 3 \\ 1 & 4 & 9 \\ 1 & 8 & 27 \end{pmatrix}$ in two completely different ways.

2. Find $\det\begin{pmatrix} 1 & 2 & 3 \\ 0 & 0 & 9 \\ 1 & -8 & 27 \end{pmatrix}$ in two completely different ways.

3. Ditto for $\det\begin{pmatrix} 2 & 1 & 1 \\ 1 & 2 & 1 \\ 1 & 1 & 2 \end{pmatrix}$.

4. Find all values $d \in \mathbb{R}$ for which the matrix $\mathbf{A} = \begin{pmatrix} 1 & -1 & 2 & d \\ 0 & 4 & -2 & 0 \\ 2 & 0 & -d & 1 \\ 1 & 1 & 1 & 1 \end{pmatrix}$ is singular.

5. Find $\det\begin{pmatrix} 1 & 1 & 1 \\ 2 & 4 & 9 \\ 3 & 27 & 81 \end{pmatrix}$.

6. Find the determinant of the λ–matrix $\mathbf{A} - \lambda\mathbf{I}$ for

$$\mathbf{A} = \begin{pmatrix} 4 & -1 & 0 & 0 \\ 1 & 4 & -1 & 0 \\ 0 & 1 & 5 & -1 \\ 0 & 0 & -1 & -4 \end{pmatrix}.$$

7. (a) Find the matrix representation $\mathbf{A}$ of the linear transformation $T : \mathbb{R}^3 \to \mathbb{R}^3$ defined as $T\begin{pmatrix} x_1 \\ x_2 \\ x_3 \end{pmatrix} = \begin{pmatrix} 2x_1 - x_3 \\ x_3 + x_1 + x_2 \\ x_2 \end{pmatrix}$ with respect to the basis $\mathcal{U} = \left\{\begin{pmatrix} 1 \\ 1 \\ -1 \end{pmatrix}, \begin{pmatrix} 0 \\ 1 \\ -1 \end{pmatrix}, \begin{pmatrix} 2 \\ 2 \\ 2 \end{pmatrix}\right\}$ of $\mathbb{R}^3$.

(b) Find all vectors $\mathbf{x}$ that T maps to zero.

(c) Find a basis for the column space of T.

(d) What is $\det \mathbf{A}_{\mathcal{E}}$?

(e) How does the determinant of a matrix representation for T change if we use a different basis for representing T? Give a proof.

8. Try to compute the determinants of $\mathbf{A} = \begin{pmatrix} -2 & 4 & 4 \\ 2 & -4 & -3 \\ 5 & -5 & 4 \end{pmatrix}$ and $\mathbf{B} = \begin{pmatrix} 1 & 3 & 3 & 3 \\ 3 & 1 & 3 & 3 \\ 3 & 3 & 1 & 3 \\ 3 & 3 & 3 & 1 \end{pmatrix}$ most efficiently. Explain your procedure.

9. Find the value of the determinant of the 6×6 matrix $\mathbf{A}$ with diagonal entries 3, upper and lower co–diagonal entries all equal to 1, and zeros elsewhere.

10. Evaluate $\det\begin{pmatrix} 1 & 1 & 1 \\ x & y & z \\ x^2 & y^2 & z^2 \end{pmatrix}$.

11. Evaluate $\det\begin{pmatrix} 0 & 0 & a \\ 0 & b & c \\ d & e & f \end{pmatrix}$ most efficiently.

12. If $\det(\mathbf{A}) = -5$ and $\mathbf{A}$ is $n \times n$, what is

(a) $\det(5\mathbf{A})$;

(b) $\det(3\mathbf{A}^{-1})$;

(c) $\det(\mathbf{A}^2)$;

(d) $-\det((2\mathbf{A})^3)$;

(e) $\det(\mathbf{A} - 6\mathbf{A})$?

13. Find all values of x for which $\det\begin{pmatrix} 2 & x^2 & x \\ -2 & 0 & 0 \\ 1 & 2 & 1 \end{pmatrix}$ is zero.

14. Find $\det\begin{pmatrix} 1 & 3 & -4 & 2 \\ 3 & 4 & -3 & 0 \\ 0 & 2 & -5 & 6 \\ 4 & 9 & -12 & 8 \end{pmatrix}$.

15. Determine whether $\mathbf{A} = \begin{pmatrix} 1 & 2 & -1 \\ 3 & 7 & -1 \\ -2 & -5 & 1 \end{pmatrix}$ is invertible.

16. Show: If $\mathbf{B} = \mathbf{X}^{-1}\mathbf{A}\mathbf{X}$ for a nonsingular matrix $\mathbf{X}_{nn}$, then $\det \mathbf{A} = \det \mathbf{B}$. (*Hint*: Use part 2 of the Proposition.)

17. Why is $\det\begin{pmatrix} 1 & 1 & 1 \\ x & y & z \\ y+z & x+z & x+y \end{pmatrix} = 0$ for all values x, y, and $z \in \mathbb{R}$?

18. Find all values of $x \in \mathbb{C}$ for which the matrix $\mathbf{A} = \begin{pmatrix} x-1 & 2 \\ -1 & x+1 \end{pmatrix}$ is invertible.

19. If for a matrix $\mathbf{A}_{nn}$ we know that $\det(\mathbf{A}) = 4$, what is the determinant of the enlarged matrix $\mathbf{B}_{n+1,n+1} = \begin{pmatrix} a & - & \mathbf{0} & - \\ * & & \mathbf{A} & \end{pmatrix}$ in terms of $a \in \mathbb{R}$ and $* \in \mathbb{R}^n$?

20. Evaluate $\det\begin{pmatrix} d & e & a \\ 0 & b & c \\ d & e & f \end{pmatrix} = bd(f-a)$ most efficiently.

21. (a) How many arithmetic operations does it take to evaluate the determinant of an $n \times n$ real matrix by

(i) repeated Laplace expansion

(ii) Gaussian elimination?

(b) Compare the amount of work in (a)(i) and (a)(ii) for $n = 2, 5, 10$, and 100.

22. Show that $\det(\mathbf{L}) = \prod_i \ell_{ii}$ for all lower triangular matrices $\mathbf{L} = \begin{pmatrix} \ell_{11} & & 0 \\ & \ddots & \\ * & & \ell_{nn} \end{pmatrix}$ by using parts 1 and 6 of the Proposition.

23. Evaluate $\det\begin{pmatrix} 2 & 0 & 0 & 0 \\ -35 & 0 & 1 & 0 \\ 0 & 1 & 1 & 1 \\ 734 & 0 & 2 & 4 \end{pmatrix}$.

24. Evaluate $\det\begin{pmatrix} 0 & 24 \\ 2 & -14 \end{pmatrix}$ and $\det\begin{pmatrix} 0 & 2 \\ 2 & 0 \end{pmatrix}$.

25. If $\det\begin{pmatrix} 1 & a & -1 \\ 2 & 4 & 0 \\ -1 & b & 1 \end{pmatrix} = 120$, what are all possible value for a and b.

26. Prove that if $\mathbf{A}_{nn}$ has a column or row of all zeros, then $\det(\mathbf{A}) = 0$.

27. Let $\mathbf{A} = \begin{pmatrix} -7 & 9 \\ -6 & 8 \end{pmatrix}$.

(a) Evaluate the function

$$f(x) = \det(\mathbf{A} - x\mathbf{I}_2).$$

(b) Find all real numbers x for which the matrix $\mathbf{A} - x\mathbf{I}_2$ is nonsingular.

(c) Find all real numbers x for which $\mathbf{A} - x\mathbf{I}_2$ is singular.

28. The matrix $\mathbf{A} = \begin{pmatrix} 1 & 2 & 4 & 1 \\ 2 & 3 & -1 & -2 \\ 1 & 0 & ? & -7 \\ -1 & -1 & ?? & 3 \end{pmatrix}$ has two illegible entries in positions (3,3) and (3,4). $\mathbf{A}$ is said to have determinant equal to -10. Can this possibly be so, or is this necessarily incorrect?

29. Find $\det(\mathbf{A})$ for $\mathbf{A} = \begin{pmatrix} 1 & 2 & 1 & 4 \\ 9 & 1 & 3 & 1 \\ 9 & 2 & 2 & 2 \\ 6 & 5 & 1 & 10 \end{pmatrix}$.

30. Show that $\det(\mathbf{A}^2) \geq 0$ for all real matrices $\mathbf{A}_{nn}$.

31. If $\mathbf{A}$ and $\mathbf{B}$ are two 5×5 matrices with $\det(\mathbf{A}) = -16$ and $\det(\mathbf{B}) = 2$, find

(a) $\det(-\mathbf{A}\mathbf{B}^{-1})$; (b) $\det(\mathbf{B}\mathbf{A}^2)$;

(c) $\det(\mathbf{B}^{-2}2\mathbf{A}^{-1}\mathbf{B}^2)$; (d) $\det(-\mathbf{B}^{-1}\mathbf{A}^{-1})$;

(e) $\det(\mathbf{A}^2\mathbf{B}^{-1})$; (f) $\det(2\mathbf{A} - \mathbf{B}^2)$.

32. Assume that for a given dimension n, the determinants of $\mathbf{A}$ and $-\mathbf{A}$ are the same for a real matrix $\mathbf{A}_{nn}$. Show that n is even or that $\mathbf{A}$ is singular.

33. Find

(a) $\det\begin{pmatrix} 0 & 2 & 0 \\ 0 & 0 & 3 \\ 1 & 0 & 0 \end{pmatrix}$ and $\det\begin{pmatrix} 0 & 2 & 0 \\ 0 & 0 & 3 \\ 1 & 2 & 3 \end{pmatrix}$.

(b) $\det\begin{pmatrix} 0 & 0 & 0 & 1/2 \\ 0 & 2 & 0 & 0 \\ -1 & 0 & 0 & 0 \\ 0 & 0 & 4 & 0 \end{pmatrix}$ and $\det\begin{pmatrix} 0 & 0 & 0 & 1/2 \\ 0 & 2 & 0 & 2 \\ -1 & 0 & 0 & -1 \\ 0 & 0 & 4 & 4 \end{pmatrix}$.

(c) $\det\begin{pmatrix} 0 & 1 & 0 & 1 \\ 3 & 0 & 0 & 3 \\ 0 & 0 & 2 & 2 \\ 3 & 1 & 2 & -1 \end{pmatrix}$.

(*Hint*: Find and exploit the pattern in the matrices of parts (a), (b), and (c).)

34. Let $\mathbf{A}_{7\times 7}$ be a block matrix of the form $\begin{pmatrix} \mathbf{B}_{3\times 3} & \mathbf{0} \\ \mathbf{D} & \mathbf{C}_{4\times 4} \end{pmatrix}$.

(a) What is the size of $\mathbf{D}$?

(b) If $\det(\mathbf{A}) = 20$ and $\det(\mathbf{B}) = -3$, what are the possible determinants for C?

35. Let $\mathbf{A} = \begin{pmatrix} \mathbf{B}_{kk} & \mathbf{D} \\ \mathbf{0} & \mathbf{C}_{kk} \end{pmatrix}$ with $\det(\mathbf{B}) = 14$ and a singular matrix $\mathbf{C}$.

(a) What are the possible determinants for $\mathbf{A}$?

(b) What is the size of $\mathbf{D}$?

(c) What determinants can $\mathbf{D}$ possibly have?

(d) Show that $\mathbf{A}$ is a singular matrix.

36. For which real values of $k \in \mathbb{R}$ is $\det\begin{pmatrix} 1 & 0 & k+1 \\ 0 & 2+k & 5 \\ 5 & 1 & 0 \end{pmatrix} = 0$?

37. Assume that $(\mathbf{A} + \mathbf{B})^2 = \mathbf{A}^2 + \mathbf{B}^2$ for two $n \times n$ matrices $\mathbf{A}$ and $\mathbf{B}$.

(a) Show that $\mathbf{AB} = -\mathbf{BA}$,

(b) Show that n must be even unless $\mathbf{A}$ or $\mathbf{B}$ are singular.

(c) Try to find two 2×2 matrices $\mathbf{A}$, $\mathbf{B}$ with $(\mathbf{A}+\mathbf{B})^2 = \mathbf{A}^2 + \mathbf{B}^2$.

(d) Try to find two nonsingular 2×2 matrices $\mathbf{A}$ and $\mathbf{B}$ with $(\mathbf{A}+\mathbf{B})^2 = \mathbf{A}^2 + \mathbf{B}^2$?

38. (a) Find two matrices $\mathbf{A}, \ \mathbf{B} \in \mathbb{R}^{2,2}$ for which $\det(\mathbf{A}+\mathbf{B}) \neq \det \mathbf{A} + \det \mathbf{B}$.

(b) Can you find two matrices $\mathbf{A}, \ \mathbf{B} \in \mathbb{R}^{2,2}$ with $\det(\mathbf{A}+\mathbf{B}) = \det \mathbf{A} + \det \mathbf{B}$. Explain.

39. (a) Assume that $\mathbf{A}$ is an $n \times n$ matrix with $\mathbf{A}\mathbf{A}^T = \mathbf{I}_n$, the identity matrix. What is the determinant of $\mathbf{A}$?

(*Hint*: Search for a usable part of the Proposition.)

(b) Find an $n \times n$ matrix with $\mathbf{A}\mathbf{A}^T = \mathbf{I}_n$.

(c) Repeat part (a) for $\mathbf{A}_{nn}$ with $\mathbf{A}\mathbf{A}^T = 3\mathbf{I}_n$.

40. Prove the determinantal product formula (i.e., part 2 of the proposition from earlier in this section) for 2×2 matrices: Show that $\det\left(\begin{pmatrix} a & b \\ c & d \end{pmatrix}\begin{pmatrix} \alpha & \beta \\ \gamma & \delta \end{pmatrix}\right) = \det\begin{pmatrix} a & b \\ c & d \end{pmatrix} \cdot \det\begin{pmatrix} \alpha & \beta \\ \gamma & \delta \end{pmatrix}$.

41. For $\mathbf{x} \in \mathbb{R}^n$ let $\mathbf{I}(\mathbf{x}, i)$ be the $n \times n$ matrix obtained from $\mathbf{I}_n$ by replacing its ith column $\mathbf{e}_i$ by $\mathbf{x}$.

(a) For $\mathbf{x} = \begin{pmatrix} -1 \\ 3 \end{pmatrix} \in \mathbb{R}^2$, write out the matrices $\mathbf{I}(\mathbf{x}, 1)$ and $\mathbf{I}(\mathbf{x}, 2)$.

(b) Repeat part (a) for $\mathbf{y} = \begin{pmatrix} 0 \\ 2 \\ -1 \end{pmatrix} \in \mathbb{R}^3$ and $\mathbf{I}(\mathbf{y}, 2)$ and $\mathbf{I}(\mathbf{y}, 3)$.

(c) Compute the determinant of $\mathbf{I}(\mathbf{x}, i)$ as a function of the entries of $\mathbf{x}$.

(*Hint*: Expand along row i of $\mathbf{I}(\mathbf{x}, i)$.)

42. For $\mathbf{b} \in \mathbb{R}^n$ and $\mathbf{A} \in \mathbb{R}^{n,n}$, let $\mathbf{A}(\mathbf{b}, i)$ be the $n \times n$ matrix obtained from $\mathbf{A}$ by making its ith column equal to $\mathbf{b}$, while leaving all other columns unchanged.

(a) Show that $\mathbf{A}\mathbf{I}(\mathbf{x}, i) = \mathbf{A}(\mathbf{b}, i)$ if and only if $\mathbf{A}\mathbf{x} = \mathbf{b}$.

(b) Prove **Cramer's rule** for a nonsingular matrix $\mathbf{A}_{nn}$: $\mathbf{A}\mathbf{x} = \mathbf{b}$ for $\mathbf{x} = \begin{pmatrix} x_1 \\ \vdots \\ x_n \end{pmatrix}$ if and only if

$$x_i = \frac{\det(\mathbf{A}(\mathbf{b}, i))}{\det(\mathbf{A})} \quad \text{for each } i = 1, \ldots, n.$$

(*Hint*: Use part (a) and Problem 41 (c).)

Teacher's Problem-Making Exercise

T 8. Our task is to construct low–dimensional matrices $\mathbf{A}_{nn}$ whose determinants can be computed easily by hand. Thus, we want to learn how to find integer matrices with small integer determinants.

Integer square matrix have integer determinants. We can generate random integer matrices with relatively small entries via the MATLAB call `A=floor(6*randn(3))` for size 3×3 for example. Unfortunately, for hand computations, the determinant of a small-dimensional matrix with small entries can become rather large.

As an alternative, we revisit problem T 6:

If $\mathbf{A}$ and $\mathbf{B}$ are both integer $n \times n$ triangular matrices, one of which is upper and the other lower triangular, then the two matrix products $\mathbf{AB}$ and $\mathbf{BA}$ are integer dense matrices. According to part 2 of the Proposition presented earlier, $\det(\mathbf{AB}) = \det(\mathbf{A}) \cdot \det(\mathbf{B}) = \det(\mathbf{BA})$. Differing from our design in problem T 6, we find that now both $\mathbf{A}$'s and $\mathbf{B}$'s diagonal entries can be arbitrary integers a_{ii} and b_{jj}. Then $\det(\mathbf{A}) = \prod a_{ii}$ and $\det(\mathbf{B}) = \prod b_{jj}$ from parts 1 and 6 of the Proposition, giving $\mathbf{AB}$ and $\mathbf{BA}$ the known integer determinant $\det(\mathbf{AB}) = \det(\mathbf{BA}) = \prod(a_{ii}b_{ii}) = \det(\mathbf{A}) \cdot \det(\mathbf{B})$.

EXAMPLE

(a) The determinant of the randomly generated matrix $\begin{pmatrix} -1 & 7 & -9 \\ -6 & 0 & 3 \\ -7 & 8 & 9 \end{pmatrix}$ is an unwieldy 687.

(b) Let $\mathbf{A} = \begin{pmatrix} 2 & 7 & 2 \\ 0 & -10 & -2 \\ 0 & 0 & 1 \end{pmatrix}$ and $\mathbf{B} = \begin{pmatrix} 1 & 0 & 0 \\ 2 & -3 & 0 \\ -1 & 0 & 2 \end{pmatrix}$ be our pair of integer upper and lower triangular matrices.

Then $\mathbf{AB} = \begin{pmatrix} 14 & -21 & 4 \\ 18 & 30 & -4 \\ -1 & 0 & 2 \end{pmatrix}$ and $\mathbf{BA} = \begin{pmatrix} 2 & 7 & 2 \\ 4 & 44 & 10 \\ -2 & -7 & 0 \end{pmatrix}$ both have determinants equal to $120 = \prod a_{ii} \prod b_{ii}$.

(c) If we place one or several zero entries on the diagonals of the triangular generators **A** or **B**, then the determinants of both **AB** and **BA** will be zero. This can be verified in integer arithmetic for **AB** and **BA** with $\mathbf{A} = \begin{pmatrix} 2 & 7 & 2 \\ 0 & 0 & -2 \\ 0 & 0 & 1 \end{pmatrix}$ and $\mathbf{B} = \begin{pmatrix} 1 & 0 & 0 \\ 2 & -3 & 0 \\ -1 & 0 & 2 \end{pmatrix}$, for example. ◀

8.2 Theory

Axiomatic Definition

The determinant function is uniquely defined by its alternating, multilinear, and normalization properties.

The determinant function "det" assigns a real number to a set of n vectors in $\mathbb{R}^n$, viewed as the columns or rows of a matrix $\mathbf{A}_{nn}$, or it assigns a single algebraic expression to an $n \times n$ matrix that contains algebraic expressions. The determinant function satisfies Eqs. (8.1), (8.2), and (8.3). In turn, a function $d : \mathbb{R}^{n,n} \to \mathbb{R}$ with these three properties is the determinant function if it is normalized, so that $d(\mathbf{I}_n) = 1$.

Equations (8.1) and (8.2) describe the **multilinear** behavior of the determinant: det(**A**) is linear in each of its rows, or

$$\det \begin{pmatrix} - & \mathbf{a}_1 & - \\ & \vdots & \\ - & \mathbf{a}_k + \alpha \mathbf{b} & - \\ & \vdots & \\ - & \mathbf{a}_n & - \end{pmatrix} = \det \begin{pmatrix} - & \mathbf{a}_1 & - \\ & \vdots & \\ - & \mathbf{a}_k & - \\ & \vdots & \\ - & \mathbf{a}_n & - \end{pmatrix} + \alpha \det \begin{pmatrix} - & \mathbf{a}_1 & - \\ & \vdots & \\ - & \mathbf{b} & - \\ & \vdots & \\ - & \mathbf{a}_n & - \end{pmatrix} \qquad (8.4)$$

for each k, all $\alpha \in \mathbb{R}$, and all row vectors $\mathbf{b} \in \mathbb{R}^n$. The third property (8.3) of determinants is called the **alternating property**, since the sign of a matrix determinant alternates with each swap of rows.

If a function d is defined for any set of n vectors in $\mathbb{R}^n$, and if it is alternating and multilinear in the sense of Eqs. (8.3) and (8.4), then the function $d : \mathbb{R}^{n,n} \to \mathbb{R}$ is the determinant function det : $\mathbb{R}^{n,n} \to \mathbb{R}$ if it is normalized, i.e., if

$$d(\mathbf{I}_n) = 1. \tag{8.5}$$

This last quality of the determinant function is called its **normalization property**.

Theorem 8.1 The determinant function det ($\mathbf{A}$) is the unique alternating, multilinear, and normalized function that assigns a real number det($\mathbf{A}$) to every real square matrix $\mathbf{A}_{nn}$.

Historically, the determinant function det($\mathbf{A}$) was first expressed by a multi term sum of $n!$ products of n entries each, taken from $\mathbf{A} = (a_{ij}) \in \mathbb{R}^{n,n}$:

$$\det \mathbf{A} \;=\; \sum_{\sigma \in \Upsilon_n} \epsilon_\sigma a_{1\sigma(1)} \cdots a_{n\sigma(n)}. \tag{8.6}$$

Here $\sum$ denotes summation, σ is a **permutation** of the numbers $1, \ldots, n$, and Υ_n is the set of all $n!$ permutations of the set of integers $\{1, 2, \ldots, n\}$. If $\sigma = [2, 3, 1, 4, 5, \ldots, n]$, for example, then σ reshuffles the first three integers 1,2,3 to appear in the order 2,3,1 and leaves the remaining integers $4, \ldots, n$ fixed. In Eq. (8.6), the symbol ϵ_σ denotes the **sign** ± 1 of the permutation σ induced by the number of swaps in σ. For our specific permutation σ, we have the following **crossing diagram**:

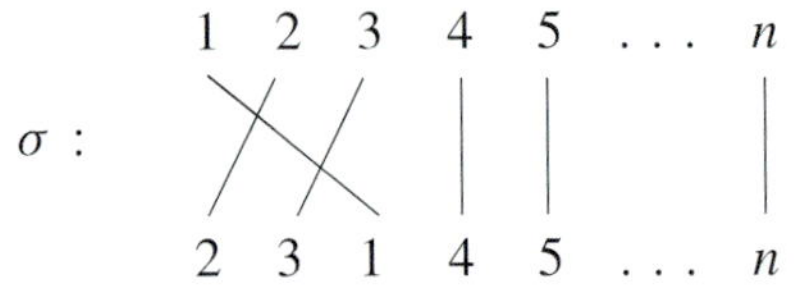

Thus, σ involves two swaps, as signified by the two line crossings that occur in the diagram. Hence its sign is $\epsilon_\sigma = (-1)^{\text{number of swaps}} = (-1)^2 = 1$. Note that each product term $a_{1\sigma(1)} \cdots a_{n\sigma(n)}$ in Eq. (8.6) involves n entries $a_{i\sigma(i)}$ of $\mathbf{A}$ for $i = 1, \ldots, n$ that lie in mutually exclusive rows and columns of $\mathbf{A}$, since for every permutation $\sigma \in \Upsilon_n$, we have $\sigma(i) \neq \sigma(j)$ for all $i \neq j$. For the specific permutation σ just presented, its term $\epsilon_\sigma a_{1\sigma(1)} \cdots a_{n\sigma(n)}$ in the determinantal sum expression (8.6) is equal to $1 \cdot a_{12}a_{23}a_{31}a_{44}a_{55} \cdots a_{nn}$.

If we use the historical definition (8.6) for the determinant, we can readily deduce part 2 of the Proposition in Section 8.1.

Lemma 1 For any two matrices $\mathbf{A}_{nn}$ and $\mathbf{B}_{nn}$, we have $\det(\mathbf{AB}) = \det(\mathbf{A}) \cdot \det(\mathbf{B}) = \det(\mathbf{BA})$.

Proof Write the second factor **B** of the matrix product of **AB** row–wise as in section 6.2:

$$\mathbf{AB} = \begin{pmatrix} a_{11} & \cdots & a_{1n} \\ \vdots & & \vdots \\ a_{n1} & \cdots & a_{nn} \end{pmatrix} \begin{pmatrix} - & \mathbf{b}_1 & - \\ & \vdots & \\ - & \mathbf{b}_n & - \end{pmatrix} = \begin{pmatrix} - & a_{11}\mathbf{b}_1 + \cdots + a_{1n}\mathbf{b}_n & - \\ & \vdots & \\ - & a_{n1}\mathbf{b}_1 + \cdots + a_{nn}\mathbf{b}_n & - \end{pmatrix}$$

By successive use of the determinant's row multilinearity (8.4), applied to the first row $a_{11}\mathbf{b}_1 + \cdots + a_{1n}\mathbf{b}_n$ of the product **AB** and so forth, we obtain

$$\det(\mathbf{AB}) = \det \begin{pmatrix} - & a_{11}\mathbf{b}_1 + \cdots + a_{1n}\mathbf{b}_n & - \\ & \vdots & \\ - & a_{n1}\mathbf{b}_1 + \cdots + a_{nn}\mathbf{b}_n & - \end{pmatrix}$$

$$= \sum_{i=1}^{n} a_{1i} \det \begin{pmatrix} - & \mathbf{b}_i & - \\ - & a_{21}\mathbf{b}_1 + \cdots + a_{2n}\mathbf{b}_n & - \\ & \vdots & \\ - & a_{n1}\mathbf{b}_1 + \cdots + a_{nn}\mathbf{b}_n & - \end{pmatrix}$$

$$= \sum_{i=1}^{n} \left(a_{1i} \sum_{j=1}^{n} \left(a_{2j} \det \begin{pmatrix} - & \mathbf{b}_i & - \\ - & \mathbf{b}_j & - \\ - & a_{31}\mathbf{b}_1 + \cdots + a_{3n}\mathbf{b}_n & - \\ & \vdots & \\ - & a_{n1}\mathbf{b}_1 + \cdots + a_{nn}\mathbf{b}_n & - \end{pmatrix} \right) \right) \qquad (*)$$

$$= \cdots$$

$$= \sum_{\sigma \in \Upsilon_n} a_{1\sigma(1)} \cdots a_{n\sigma(n)} \det \begin{pmatrix} - & \mathbf{b}_{\sigma(1)} & - \\ & \vdots & \\ - & \mathbf{b}_{\sigma(n)} & - \end{pmatrix},$$

where σ runs through all possible permutations of the integers $1, \ldots, n$. Note that all determinant terms in Eq. $(*)$ with repeated rows are zero, and thus these terms can be dropped from the sum, due to part 10 of the Proposition presented in Section 8.1.

With ϵ_σ denoting the $\pm$ sign induced by the number of swaps in the permutation σ, we have

$$\det \begin{pmatrix} - & \mathbf{b}_{\sigma(1)} & - \\ & \vdots & \\ - & \mathbf{b}_{\sigma(n)} & - \end{pmatrix} = \epsilon_\sigma \det \begin{pmatrix} - & \mathbf{b}_1 & - \\ & \vdots & \\ - & \mathbf{b}_n & - \end{pmatrix} = \epsilon_\sigma \det \mathbf{B}$$

from property (8.3).

Thus, $\det(\mathbf{AB}) = \sum_{\sigma \in \Upsilon_n} \epsilon_\sigma a_{1\sigma(1)} \cdots a_{n\sigma(n)} \det \mathbf{B} = \det \mathbf{A} \cdot \det \mathbf{B} = \det \mathbf{B} \cdot \det \mathbf{A} = \det(\mathbf{BA})$, proving Lemma 1 (or part 2 of the Proposition from Section 8.1). ■

8.2.P Problems

1. (a) Make a list of all permutations of $\{1,2,3\}$ and decide for which $\sigma \in \Upsilon_3$ we have $\epsilon_\sigma = -1$ and for which σ $\epsilon_\sigma = 1$.

(b) Repeat part (a) for $\{1,2,3,4\}$ and Υ_4.

(c) Write out the crossing diagram for the permutation $\{2,3,4,1\}$ and determine its sign.

2. Use the historical definition of the determinant function in order to evaluate

(a) $\det \begin{pmatrix} 0 & 0 & 0 & 2 \\ 0 & 0 & 3 & 0 \\ 0 & 4 & 0 & 0 \\ 5 & 0 & 0 & 0 \end{pmatrix}$.

(b) $\det \begin{pmatrix} 0 & 0 & 3 & 0 \\ 2 & 0 & 0 & 0 \\ 0 & 0 & 0 & -1 \\ 0 & 7 & 0 & 0 \end{pmatrix}$.

(c) $\det \begin{pmatrix} 1 & 0 & -1 \\ 0 & 4 & 0 \\ 1 & 0 & 7 \end{pmatrix}$.

(d) How many nonzero terms does the classical determinant formula for part (c) contain?

3. (a) Show that the equation (in the two variables x and y) of the line through the two points $\mathbf{a} = (a_1, a_2)$, $\mathbf{b} = (b_1, b_2) \in \mathbb{R}^2$ is given by $f(x,y) = \det \begin{pmatrix} x & y & 1 \\ a_1 & a_2 & 1 \\ b_1 & b_2 & 1 \end{pmatrix} = 0$. What is the slope of this line?

(b) What geometric object does the equation $g(x,y) = \det \begin{pmatrix} x & y & -17 \\ a_1 & a_2 & -17 \\ b_1 & b_2 & -17 \end{pmatrix} = 0$ describe?

(c) What is the relation of the line defined by $h(x,y) = \det \begin{pmatrix} -y & x & 1 \\ a_1 & a_2 & 1 \\ b_1 & b_2 & 1 \end{pmatrix} = 0$ to that of part (a) ? What points $(x,y) \in \mathbb{R}^2$ satisfy $h(x,y) = 0$?

(d) Generalize the foregoing determinantal equation of lines to describe the plane through three points $\mathbf{a} = (a_1, a_2, a_3)$, $\mathbf{b} = (b_1, b_2, b_3)$, and $\mathbf{c} = (c_1, c_2, c_3) \in \mathbb{R}^3$.

4. Show that if three points $\mathbf{a} = (a_1, a_2)$, $\mathbf{b} = (b_1, b_2)$, and $\mathbf{c} = (c_1, c_2) \in \mathbb{R}^2$ lie on the same line through the origin, then $\det \begin{pmatrix} a_1 & a_2 & 1 \\ b_1 & b_2 & 1 \\ c_1 & c_2 & 1 \end{pmatrix} = 0$. How about the converse implication?

5. Define $\mathbf{S}_n = \begin{pmatrix} 0 & & 2 \\ & \cdot^{\cdot^{\cdot}} & \\ 2 & & 0 \end{pmatrix} \in \mathbb{R}^{n,n}$, and let $\mathbf{A} \in \mathbb{R}^{n,n}$ be arbitrary.

(a) What is $\det(\mathbf{S}_n)$?

(b) Find $\det(\mathbf{S}_n\mathbf{A})$, $\det(\mathbf{S}_n^{-1}\mathbf{A})$, and $\det(\mathbf{A}^2\mathbf{S}_n^2)$ in terms of $\det(\mathbf{A})$.

6. Given that $\det(\mathbf{A}) = 14$, $\det(\mathbf{B}) = -2$, and $\det(\mathbf{C}) = 45$ for three $n \times n$ matrices $\mathbf{A}$, $\mathbf{B}$, and $\mathbf{C}$, what are

(a) $\det(\mathbf{ABC})$ if $n = 2$;

(b) $\det(\mathbf{ABC})$ if $n = 12$;

(c) $\det(\mathbf{AB}^T\mathbf{C}^{-1})$;

(d) $\det(\mathbf{A}^{-3}\mathbf{B}^{-1}\mathbf{C})$;

(e) $\det(\mathbf{C}^{-3}\mathbf{B}^{-1}\mathbf{A})$;

(f) $\det(\mathbf{A}^2\mathbf{B}^2\mathbf{C}^2)$?

7. If $\det(\mathbf{A}^2\mathbf{B}^{-1}) = 24$ and $\det(\mathbf{BC}^{-2}) = 2$ with $\det(\mathbf{C}) = 4$, what is $\det(\mathbf{A}^3)$, provided that each of the three matrices $\mathbf{A}$, $\mathbf{B}$, and $\mathbf{C}$ is 3 by 3?

8. (a) Show that if $\det(\mathbf{AB}) = 0$, then at least one of the matrices $\mathbf{A}$ and $\mathbf{B}$ is singular.

(b) Show that if both $\mathbf{A}$ and $\mathbf{B}$ are $n \times n$ with nonzero determinants, then $\det(\mathbf{AB}) \neq 0$.

9. If $\mathbf{A}_{nn}$ and its inverse $\mathbf{A}^{-1}$ both have only integer entries, what is the value of the determinant of $\mathbf{A}^{-1}$? Explain, please.

10. (a) Show that $\lim_{k\to\infty} \left|\det(\mathbf{A}^k)\right| = \infty$ if $|\det(\mathbf{A})| > 1$.

(b) Show that $\lim_{k\to\infty} \left|\det(\mathbf{A}^k)\right| = 0$ if $|\det(\mathbf{A})| < 1$.

(c) Show that $\lim_{k\to\infty} \left|\det(\mathbf{A}^k)\right| = 1$ if $|\det(\mathbf{A})| = 1$.

(d) If $\det(\mathbf{A}) = 1$, what is $\det(\mathbf{A}^k)$ for all exponents k?

(e) Repeat part (d) for $\mathbf{A}_{nn}$ with $\det(\mathbf{A}) = -1$.

11. Find all values of k for which $\det \begin{pmatrix} 2 & 1 & 1 \\ 1 & k+1 & 0 \\ -1 & 2 & 1 \end{pmatrix}$ is positive real.

12. Assume that $\det \begin{pmatrix} 1 & 2 & a \\ 2 & 2 & b \\ c & 0 & 0 \end{pmatrix} = 4$.

(a) Find det $\begin{pmatrix} 1 & 2 & a \\ 2 & 2 & b \\ c & 0 & 7 \end{pmatrix}$.

(b) Find a matrix **A** in the originally given form whose determinant is equal to 4; that is, specify the values of a, b, and c.

13. Show: If a row **r** of $\mathbf{A}_{nn}$ is replaced by $c\mathbf{r}$ for $c \neq 0 \in \mathbb{R}$ in $\tilde{\mathbf{A}}$, then $\det \tilde{\mathbf{A}} = c \cdot \det \mathbf{A}$.

14. Show that for two $n \times n$ matrices **A** and **B**, the two matrix products **AB** and **BA** are either both singular or both nonsingular.

8.3 Applications

Volume, Wronskian

Two natural applications for the determinant are the n–space volume function and the Wronskian.

The **volume function** vol_n assigns a nonnegative real number to each subset of $\mathbb{R}^n$. For n given vectors $\{\mathbf{a}_i\}_{i=1}^n \subset \mathbb{R}^n$ the n–dimensional box that the vectors $\mathbf{a}_i$ sustain is called a **parallelepiped**. This generalizes the concept of a parallelogram in 2–space. Every parallelepiped has a certain n–dimensional volume.

If $n = 1$ and $a \in \mathbb{R}$, then the one-dimensional volume of a is customarily called its **length**, or size:

$$\text{vol}_1(a) = |a| \quad = \quad |\det(a)|.$$

If $n = 2$ and **a** and $\mathbf{b} \in \mathbb{R}^2$, then the area of the parallelogram with sides **a** and **b** is measured by the "two dimensional volume," or **area function**

$$\text{vol}_2(\mathbf{a}, \mathbf{b}) := \left| \det \begin{pmatrix} | & | \\ \mathbf{a} & \mathbf{b} \\ | & | \end{pmatrix} \right|.$$

DEFINITION 2 For n vectors $\mathbf{a}_i \in \mathbb{R}^n, i = 1, \ldots, n$, we define the mathematical **volume** of their **parallelepiped** as

$$\text{vol}_n(\mathbf{a}_1, \ldots, \mathbf{a}_n) = \left| \det \begin{pmatrix} | & & | \\ \mathbf{a}_1 & \cdots & \mathbf{a}_n \\ | & & | \end{pmatrix} \right|.$$

By using the three defining properties of the determinant function in Theorem 8.1, we now deduce that the mathematical definition of an n–dimensional volume function as a determinant is actually correct and conforms with our intuitive perception of volume.

Note first that the intuitive volume function for the parallelepiped formed by a set of n vectors $\mathbf{a}_i \in \mathbb{R}^n$ is a linear function of each edge $\mathbf{a}_k$. For example, the intuitive volume function behaves analogously to Eq. (8.4) for the determinant:

$$\begin{aligned} \text{vol}_n(\mathbf{a}_1, \ldots, \mathbf{a}_k + \mathbf{y}, \ldots, \mathbf{a}_n) = {} & \text{vol}_n(\mathbf{a}_1, \ldots, \mathbf{a}_k, \ldots, \mathbf{a}_n) \\ & + \text{vol}_n(\mathbf{a}_1, \ldots, \mathbf{y}, \ldots, \mathbf{a}_n), \end{aligned}$$

since changing the kth edge $\mathbf{a}_k$ to $\mathbf{a}_k + \mathbf{y}$, while leaving all other edges fixed, means a certain elongation of the original solid. This adds volume to the parallelepiped, as indicated by the preceding equation. We visualize this in $\mathbb{R}^2$: The two thickly drawn parallelepipeds in Figure 8-1 are formed by $\mathbf{a}_1$ and $\mathbf{a}_k$, and by $\mathbf{a}_1$ and $\mathbf{y}$, respectively.

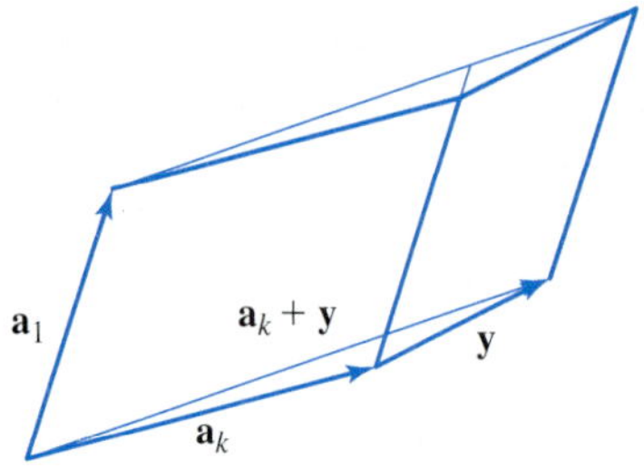

Figure 8-1

Together they have the same volume as the one formed by $\mathbf{a}_1$ and the modified kth edge $\mathbf{a}_k + \mathbf{y}$ drawn out thickly in Figure 8-2.

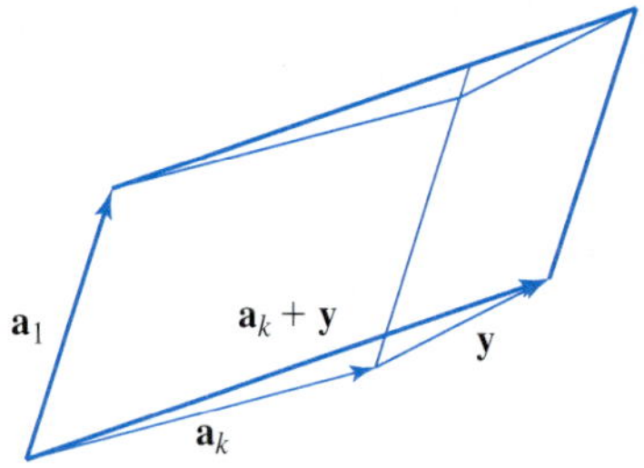

Figure 8-2

Next observe that the alternating property of determinants affects only the sign, and thus it is irrelevant to the volume function. A volume measure intuitively must satisfy $\text{vol}_n(S) \geq 0$ for all solids $S \subset \mathbb{R}^n$. Finally, for the unit cube that is spanned by the standard basis $\mathcal{E} = \mathbf{e}_1, \ldots, \mathbf{e}_n$, we have $\text{vol}_n(\mathbf{e}_1, \ldots, \mathbf{e}_n) = 1 = |\det(\mathbf{I}_n)|$; that is, the intuitive and mathematical volume functions agree for the unit cube.

Theorem 8.2 The intuitive volume of a parallelepiped in $\mathbb{R}^n$ with edges $\mathbf{a}_1, \ldots, \mathbf{a}_n \in \mathbb{R}^n$ is given by $\left| \det \begin{pmatrix} | & & | \\ \mathbf{a}_1 & \ldots & \mathbf{a}_n \\ | & & | \end{pmatrix} \right|$, the mathematical volume. ◀

If we transform a solid $S \subset \mathbb{R}^n$, be it a parallelepiped or of any other shape, via a linear transformation T, then its volume $\text{vol}_n(S)$ and the transformed volume $\text{vol}_n(T(S))$ of $T(S)$ are related via the determinant of the associated matrix representation $\mathbf{A}_{\mathcal{E}}$ of T.

Theorem 8.3 **(Volume Transform Theorem)** If $S \subset \mathbb{R}^n$ is a solid and $T : \mathbb{R}^n \to \mathbb{R}^n$ is a linear transformation with the standard matrix representation $\mathbf{A}_{\mathcal{E}}$, then

$$\text{vol}_n(T(S)) = |\det(\mathbf{A}_{\mathcal{E}})| \cdot \text{vol}_n(S). \tag{8.7}$$

◀

Thus, the determinant of a linear transformation T measures the stretching or shrinkage of the n–dimensional volume of objects under T. We include a proof for Theorem 8.3 in Section 11.3. It relies on properties of orthogonal matrices (see section 10.3), and on the polar decomposition of **A**.
Note: The volume transform formula (8.7) is independent of the chosen basis of $\mathbb{R}^n$, since a change of basis does not affect the determinant, due to part 12 of the Proposition presented in Section 8.1. This is so because $\det(\mathbf{A}_{\mathcal{E}}) = \det(\mathbf{U}^{-1}\mathbf{A}_{\mathcal{E}}\mathbf{U}) = \det(\mathbf{A}_{\mathcal{U}})$ according to Chapter 7.

Our second application of the determinant is to sets of functions $f : \mathbb{R} \to \mathbb{R}$. The concept and the definition of linear dependence and independence of vectors in $\mathbb{R}^n$, as expressed in Definition 2 of Section 5.1 for example, translate verbatim to the space of real valued functions

$$\mathcal{F} := \{f \mid f : \mathbb{R} \to \mathbb{R}\}.$$

When studying certain subspaces of $\mathcal{F}$ in detail in Section 7.3, we have called several sets of functions such as $\{\sin x, \cos x\}$ and $\{t^2e^t, te^t, e^t\}$ **linearly independent** without indicating a way to decide upon linear independence of functions in $\mathcal{F}$. A specific determinant function, called the **Wronskian**, will remedy our precarious situation.

DEFINITION 3 For n real-valued functions $f_1, \ldots, f_n \in \mathcal{F}$, each of which is $(n-1)$ times continuously differentiable on $\mathbb{R}$, we define

$$W(f_1, \ldots, f_n)(x) = \det \begin{pmatrix} f_1(x) & f_2(x) & \cdots & f_n(x) \\ f_1'(x) & f_2'(x) & \cdots & f_n'(x) \\ \vdots & \vdots & & \vdots \\ f_1^{(n-1)}(x) & f_2^{(n-1)}(x) & \cdots & f_n^{(n-1)}(x) \end{pmatrix}$$

to be the **Wronskian** of the set of functions $\{f_1, \ldots, f_n\}$. Here the symbols $f', \ldots, f^{(j)}$ denote the various higher order derivatives of a function $f : \mathbb{R} \to \mathbb{R}$.

Note that the Wronskian is a real-valued function in one variable $x \in \mathbb{R}$, which depends on n functions $f_1, \ldots, f_n : \mathbb{R} \to \mathbb{R}$ (i.e., $W : \mathcal{F}^n \to \mathcal{F}$).

The Wronskian determines whether a given set of functions $f_i : \mathbb{R} \to \mathbb{R}$ is linearly independent in $\mathcal{F}$.

Theorem 8.4 If $f_i : \mathbb{R} \to \mathbb{R}$ are $(n-1)$ times continuously differentiable functions defined on $\mathbb{R}$ for $i = 1, \ldots, n$, then the functions f_i are linearly independent if

$$W(f_1, \ldots, f_n)(x) \neq 0 \text{ for some } x \in \mathbb{R}. \quad ◀$$

The converse implication of Theorem 8.4 is generally not true; that is, if the Wronskian $W(f_1, \ldots, f_n)(x)$ for several functions $f_1, \ldots, f_n$ is zero for all x, then the f_i need not be linearly dependent.

Proof If the functions $f_i \in \mathcal{F}$ are linearly dependent, then according to Definition 2 of Section 5.1, there are real constants α_i, not all of which are zero, so that for all $x \in \mathbb{R}$

$$\alpha_1 f_1(x) + \alpha_2 f_2(x) + \cdots + \alpha_n f_n(x) = 0 = O(x)$$

is the zero function in $\mathcal{F}$. We can differentiate this identity up to order $n-1$ and obtain the following n identities for the f_i and their higher order derivatives:

$$\begin{pmatrix} f_1(x) & f_2(x) & \cdots & f_n(x) \\ f_1'(x) & f_2'(x) & \cdots & f_n'(x) \\ \vdots & & & \vdots \\ f_1^{(n-1)}(x) & f_2^{(n-1)}(x) & \cdots & f_n^{(n-1)}(x) \end{pmatrix} \begin{pmatrix} \alpha_1 \\ \vdots \\ \vdots \\ \alpha_n \end{pmatrix} = \mathbf{0} \in \mathbb{R}^n \qquad (8.8)$$

for the nonzero coefficient vector α. Thus, for all $x \in \mathbb{R}$, the system matrix in Eq. (8.8) is singular, or its determinant $W(f_1, \ldots, f_n)(x) = 0$ for all $x \in \mathbb{R}$ from the determinant Proposition of Section 8.1. ■

EXAMPLE 5 (a) The functions $\sin x$ and $\cos x : \mathbb{R} \to \mathbb{R}$ are linearly independent. To see this, look at

$$W(\sin, \cos)(x) = \det \begin{pmatrix} \sin x & \cos x \\ \cos x & -\sin x \end{pmatrix}$$
$$= -\sin^2 x - \cos^2 x = -1 \neq 0$$

for all x.

(b) The functions t^2e^t, te^t and $e^t : \mathbb{R} \to \mathbb{R}$ are linearly independent.

$$W(t^2e^t, te^t, e^t)(t) \overset{def.}{=} \det \begin{pmatrix} t^2e^t & te^t & e^t \\ 2te^t + t^2e^t & e^t + te^t & e^t \\ 2e^t + 4te^t + t^2e^t & 2e^t + te^t & e^t \end{pmatrix}$$

$$= (e^t)^3 \det \begin{pmatrix} t^2 & t & 1 \\ 2t + t^2 & 1 + t & 1 \\ 2 + 4t + t^2 & 2 + t & 1 \end{pmatrix}$$

(by extracting e^t from each row according to Eq. (8.2))

$$= e^{3t} \det \begin{pmatrix} t^2 & t & 1 \\ 2t & 1 & 0 \\ 2 + 4t & 2 & 0 \end{pmatrix}$$

(by subtracting row_1 from row_2 and from row_3)

$$= e^{3t} \cdot \det \begin{pmatrix} 2t & 1 \\ 2 + 4t & 2 \end{pmatrix} = e^{3t}(4t - 2 - 4t) = -2e^{3t} \neq 0.$$

(by Laplace expansion along column 3)

(c) The four polynomials $p(x) = x^2 - x$, $q(x) = x^2 + x$, $r(x) = x + 1$ and $s(x) = x^2 + 2x - 4 \in \mathcal{P}_2$ are linearly dependent according to Sections 1.3 and 5.1, since $\dim(\mathcal{P}_2) = 3$. Their Wronskian $W(p, q, r, s)(x) = \det \begin{pmatrix} x^2 - x & x^2 + x & x + 1 & x^2 + 2x - 4 \\ 2x - 1 & 2x + 1 & 1 & 2x + 2 \\ 2 & 2 & 0 & 2 \\ 0 & 0 & 0 & 0 \end{pmatrix}$ is equal to zero, since it contains a row of zeros.

(d) The first three polynomials p, q, and r in part (c) are linearly independent, since

$$\begin{aligned} W(p,q,r)(x) &= \det \begin{pmatrix} x^2-x & x^2+x & x+1 \\ 2x-1 & 2x+1 & 1 \\ 2 & 2 & 0 \end{pmatrix} \\ &= 2\det\begin{pmatrix} x^2+x & x+1 \\ 2x+1 & 1 \end{pmatrix} - 2\det\begin{pmatrix} x^2-x & x+1 \\ 2x-1 & 1 \end{pmatrix} \\ &= 2(x^2+x-(x+1)(2x+1)) - 2(x^2-x-(x+1)(2x-1)) \\ &= 2(2x-2(x+1)) = -4 \neq 0. \end{aligned}$$

Finally, we go to MATLAB. The determinant of an $n \times n$ real or complex matrix $\mathbf{A}$ can be evaluated with the command

```
det(A)
```

This command uses a Gaussian reduction on $\mathbf{A}$, counting swaps to adjust the sign, and the first part of our Proposition in Section 8.1 to obtain the numerical determinant of $\mathbf{A}$ as the properly signed product of the pivots of a REF $\mathbf{R}$ of $\mathbf{A}_{nn}$.

8.3.P Problems

1. Find the area of the triangle with vertices the origin, $\mathbf{v}_1 = \begin{pmatrix} 1 \\ 3 \end{pmatrix}$, and $\mathbf{v}_2 = \begin{pmatrix} -4 \\ 17 \end{pmatrix}$.

2. (a) If a solid $S \subset \mathbb{R}^n$ has zero volume in $\mathbb{R}^n$, what is the volume of $T(S)$ of any linear transformation $T : \mathbb{R}^n \to \mathbb{R}^n$?

(b) If $S \subset \mathbb{R}^n$ has volume 5 in $\mathbb{R}^n$, what is the volume of $T(S)$ and of $T^2(S) = T(T(S))$ if $T : \mathbb{R}^n \to \mathbb{R}^n$ is linear and the volume $T(U)$ of the unit cube U spanned by the standard basis vectors $\{\mathbf{e}_i\}$ in $\mathbb{R}^n$ is 7?

(c) Find a linear transformation $T : \mathbb{R}^n \to \mathbb{R}^n$ so that the volume of the transformed unit cube $T(U)$ is equal to 7.

(d) If $S \subset \mathbb{R}^n$ has volume 5 in $\mathbb{R}^n$, what is the volume of $T(S)$ and of $T^3(S)$ if $T : \mathbb{R}^n \to \mathbb{R}^n$ is linear with $\det(T) = -3$?

(e) If $T(S)$ has volume 24 for a linear map $T : \mathbb{R}^8 \to \mathbb{R}^8$ with $\det(T) = 3$, what is the volume of $S \subset \mathbb{R}^8$ and that of $T^{-1}(S)$? (Why can this T be inverted?)

3. Let $T : \mathbb{R}^2 \longrightarrow \mathbb{R}^2$ be a linear map that maps (4, 1) to $(-3, 5)$ and $(5, -1)$ to $(-3, 1)$. Find the area of $T(V)$ for a square $V \subset \mathbb{R}^2$ with side lengths equal to 3.

4. If $\mathbf{A} \in \mathbb{R}^{2,2}$ maps $\mathbf{e}_1$ to $3\mathbf{e}_1 - 2\mathbf{e}_2$ and $\mathbf{e}_2$ to $4\mathbf{e}_2 - 6\mathbf{e}_2$, what is the area of the image $\mathbf{A}(V)$ of a figure V with area 4 in $\mathbb{R}^2$?

5. Repeat Problem 4 if $\mathbf{B}$ maps $\mathbf{e}_1$ to $3\mathbf{e}_1 - 2\mathbf{e}_2$ and $\mathbf{e}_2$ to $\mathbf{e}_1 - 6\mathbf{e}_2$.

6. Show that the three functions e^x, $\cos(x)$, and $1 : \mathbb{R} \to \mathbb{R}$ are linearly independent.

7. Are the three functions 1, $\cos^2(x)$, and $\sin^2(x) : \mathbb{R} \to \mathbb{R}$ linearly independent?

8. Are the four functions 1, x, x^2, and $x^3 : \mathbb{R} \to \mathbb{R}$ linearly independent?

9. Determine whether the following sets of functions are linearly independent:

(a) $1 - x^2$ and $1 + x^2$;

(b) 1, $1 - 4x^2$, and $-x^2$;

(c) $f(x) = x + 1$, $g(x) = x - 1$, and $f(x)g(x)$;

(d) $x^{1/2}$ and $x^{5/2}$.

(e) $\sin(x + a)$ and $\cos(x)$ depending on a.

10. Evaluate determinants numerically via MATLAB: For this form, a number of 8×8 random matrices `A = randn(8); B = 17 * A;` . Compare the MATLAB output of `det(A)`, `det(B)/(17^8)`, `det(inv(A)) * det(A) -1`, `det(inv(B)) * det(B) -1` in `format long e`. What should the theoretical relations between these four quantities be? Repeat the exercise with larger sized matrices. What happens with these relations as the matrix size n becomes larger?

8.R Review Problems

1. Prove : If all columns or rows of a matrix $\mathbf{A}_{nn}$ add up to zero, then $\mathbf{A}$ has determinant zero.

2. If $\widetilde{\mathbf{B}}$ is derived from $\mathbf{A}_{nn}$ by writing the rows of $\mathbf{A}$ in reverse order (i.e., last row to first row), what is the determinant of $\widetilde{\mathbf{B}}$ in terms of the determinant of $\mathbf{A}$?

3. Compute the determinants of the following 100×100 matrices efficiently by hand:

(a) $$\begin{pmatrix} 1 & 2 & \dots & 100 \\ \vdots & \vdots & & \vdots \\ 1 & 2 & \dots & 100 \\ 10 & 10 & \dots & 10 \end{pmatrix}_{100\times 100}.$$

(b) $$\begin{pmatrix} 1 & \dots & 100 \\ \vdots & & \vdots \\ 1 & \dots & 100 \end{pmatrix}_{100\times 100}.$$

(c) $$\begin{pmatrix} 1 & \dots & 1 \\ \vdots & & \vdots \\ 1 & \dots & 1 \end{pmatrix}_{100\times 100}.$$

(d) $$\begin{pmatrix} 2 & 2 & 0 & \dots & 0 \\ 1 & 2 & 0 & & \\ 0 & \ddots & \ddots & \ddots & \\ & \ddots & \ddots & \ddots & 0 \\ 0 & & 0 & 1 & 2 \end{pmatrix}_{100\times 100}.$$

4. Repeat the above problem using MATLAB.

5. (a) Find both the determinant of $\mathbf{A} = \begin{pmatrix} 4 & -2 \\ -5 & 3 \end{pmatrix}$ and of the λ–matrix $\mathbf{A} - \lambda\mathbf{I}_2 = \begin{pmatrix} 4-\lambda & -2 \\ -5 & 3-\lambda \end{pmatrix}$.

(b) For which values of λ is the matrix $\mathbf{A} - \lambda\mathbf{I}_2$ singular?

6. Write out the historical formula (8.6) for the determinant of a 3×3 matrix $\mathbf{A}_{3,3} = (a_{ij})$ explicitly. (Use Problem 1(a) of 8.2.P)

7. How many terms are involved in evaluating the determinant of $\mathbf{A}_{nn}$ via formula (8.6) for arbitrary n?

8. Find the determinant of $\mathbf{A}^{-2} = \mathbf{A}^{-1}\mathbf{A}^{-1}$ for $\mathbf{A} = \begin{pmatrix} 2 & 0 & 0 \\ -12 & 1 & 0 \\ 13 & 78 & -3 \end{pmatrix}$.

9. Evaluate $\det \begin{pmatrix} 5 & 1 & 1 \\ 2 & 3 & 2 \\ 1 & 1 & 1 \end{pmatrix}$ by Laplace expansion along the last row.

10. If the vectors $\mathbf{x}_1, \dots, \mathbf{x}_4$ are known to be linearly independent, find all values of a, b, and c so that the vectors $\mathbf{y}_1 = a\mathbf{x}_2 + b\mathbf{x}_3$, $\mathbf{y}_2 = -a\mathbf{x}_1 + c\mathbf{x}_3$, and $\mathbf{y}_3 = -b\mathbf{x}_1 - c\mathbf{x}_2$ are also linearly independent.

11. For which values of a, b, and $c \in \mathbb{R}$ is the image of the matrix $\begin{pmatrix} 1 & 2 & a \\ 2 & 2 & b \\ c & 0 & 0 \end{pmatrix}$ different from $\mathbb{R}^3$? For which values of the constants are the columns of this matrix linearly independent?

12. Show that the matrix product $\mathbf{AB}$ of two $n \times n$ matrices is singular if and only if at least one of the matrices $\mathbf{A}$ and $\mathbf{B}$ is singular.

13. The *Matrices and Determinants* article at the history of math Web site mentioned in the Introduction has contained the following lines about one-third of the way down since its inception in 1996:

Rather remarkably the first appearance of a determinant in Europe appeared in exactly the same year 1683. In that year Leibniz wrote to de l'Hôpital. He explained that the system of equations

```
10 + 11x + 12y = 0
20 + 21x + 22y = 0
30 + 31x + 32y = 0
```

had a solution because

```
10.21.32 + 11.22.30 + 12.20.31
  = 10.22.31 + 11.20.32 + 12.21.30
```

which is exactly the condition that the coefficient matrix has determinant 0. Notice that here Leibniz is not using numerical coefficients but

> *two characters, the first marking in which equation it occurs, the second marking which letter it belongs to.*

Hence 21 *denotes what we might write as* a_{21}.

Note that Leibniz and Newton invented Calculus at about the same time.

What do you think of the preceding assertion on the solvability of this linear system and determinants? (*Hint*: rewrite the given system of linear equations in matrix form $\mathbf{Az} = \mathbf{b}$. What are $\mathbf{A}$ and $\mathbf{b}$ here; which $\mathbb{R}^n$ does $\mathbf{z}$ belong to?)

Do you think that Leibniz' letter to l'Hôpital is quoted correctly? Elaborate, using Chapters 3, 5, and 8 and the historical evidence.

Not everything published on the Web is correct; caveat lector.

Standard Questions and Tasks

1. Evaluate the determinant of a real matrix $\mathbf{A}_{nn}$.
2. Which matrices $\mathbf{A}_{nn}$ have zero determinants?
3. Evaluate the determinant of a λ–matrix $\mathbf{A}(\lambda)_{nn}$.
4. What is Laplace expansion? What does it achieve?
5. How can Gaussian elimination be used to evaluate determinants?
6. What is the determinant of a product of two or more $n \times n$ matrices?
7. How can we evaluate the volume of a solid and of a linearly transformed solid in n space?

Subheadings of Lecture Eight

Similar leaves

CHAPTER

9

Matrix Eigenvalues and Eigenvectors

Lecture nine comes doubly, or in two flavors:
The first version in Section 9.1 develops ways to find all eigenvalues and eigenvectors of a square matrix by using vector iteration.
The second version in Section 9.1.D does the same via determinants and relies on the optional determinant lecture eight.
Both approaches are equally valid. This author likes the determinant-free approach, but has alternated in his classes between the two. The two self–duplicating lectures[1] give instructors and readers a choice between two methods that find matrix eigenvalues and eigenvectors theoretically.

We study a linear space in light of a single linear transformation of itself.

9.1 Lecture Nine, Using Vector Iteration, a Double Lecture

Vanishing and Minimal Polynomial, Matrix Eigenanalysis, and Diagonalizable Matrices

This lecture does not rely on determinants or on Chapter 8.

We learn to extract intrinsic information of linear transformations and matrices, namely, their eigenvalues and eigenvectors. We recover this information by using the linear transformations themselves in a powerful way.

[1]There are kindred souls who like to use linear means for matrix eigenvalue problems rather than determinants. (See for example S. Axler, "Down with determinants!", *American Math. Monthly*, 103 (1995), p. 139–154.)

(a) The Definition of Eigenvalues and Eigenvectors

We continue where lecture seven has left off.

Given a real square matrix $\mathbf{A} = \mathbf{A}_\mathcal{E} \in \mathbb{R}^{n,n}$ that represents a linear transformation $T : \mathbb{R}^n \to \mathbb{R}^n$ with respect to the standard unit vector basis $\mathcal{E} = \{\mathbf{e}_1, \ldots, \mathbf{e}_n\}$ of $\mathbb{R}^n$, we want to find another basis $\mathcal{U} = \{\mathbf{u}_1, \ldots, \mathbf{u}_n\}$ of $\mathbb{R}^n$ (or $\mathbb{C}^n$) that gives us a simpler matrix representation $\mathbf{A}_\mathcal{U}$ of T. That is, we are looking for a basis $\mathcal{U}$ for which the matrix in Eq. (7.4) $\mathbf{A}_\mathcal{U} = \mathbf{U}^{-1}\mathbf{A}_\mathcal{E}\mathbf{U}$ is a simpler matrix than $\mathbf{A}_\mathcal{E}$. In order to do so, we need to extract **intrinsic geometric qualities** of the linear transformation T, or, equivalently, of its standard matrix representation $\mathbf{A}_\mathcal{E}$.

It is not apparent at this moment what exactly the notion of a "simpler" matrix means, or what the "intrinsic properties" of a linear transformation might be.

In light of Section 7.1, let us study specific **similarity transformations** $\mathbf{A}_\mathcal{U} = \mathbf{U}^{-1}\mathbf{A}_\mathcal{E}\mathbf{U}$ of the standard matrix representation $\mathbf{A}_\mathcal{E}$ of T. If $\mathbf{A}_\mathcal{U} = \mathbf{U}^{-1}\mathbf{A}_\mathcal{E}\mathbf{U}$, or, equivalently, if $\mathbf{A}_\mathcal{E}\mathbf{U} = \mathbf{U}\mathbf{A}_\mathcal{U}$ for a nonsingular matrix $\mathbf{U}$ and if $\mathbf{A}_\mathcal{U}$ is a **diagonal matrix**

$$\mathbf{A}_\mathcal{U} = \operatorname{diag}(\lambda_1, \ldots, \lambda_n) = \begin{pmatrix} \lambda_1 & & 0 \\ & \ddots & \\ 0 & & \lambda_n \end{pmatrix}$$

with $\lambda_i \in \mathbb{R}$ or $\mathbb{C}$, then $\mathbf{A}_\mathcal{U}$ is quite simple. In this case, we have

$$\mathbf{A}_\mathcal{E}\mathbf{U} = \mathbf{A}_\mathcal{E}\begin{pmatrix} | & & | \\ \mathbf{u}_1 & \cdots & \mathbf{u}_n \\ | & & | \end{pmatrix} = \begin{pmatrix} | & & | \\ \mathbf{u}_1 & \cdots & \mathbf{u}_n \\ | & & | \end{pmatrix}\begin{pmatrix} \lambda_1 & & 0 \\ & \ddots & \\ 0 & & \lambda_n \end{pmatrix} = \mathbf{U}\mathbf{A}_\mathcal{U},$$

or

$$\begin{pmatrix} | & & | \\ \mathbf{A}_\mathcal{E}\mathbf{u}_1 & \cdots & \mathbf{A}_\mathcal{E}\mathbf{u}_n \\ | & & | \end{pmatrix} = \begin{pmatrix} | & & | \\ \lambda_1\mathbf{u}_1 & \cdots & \lambda_n\mathbf{u}_n \\ | & & | \end{pmatrix}.$$

Thus, for $\mathbf{A}_\mathcal{U}$ to be most simply just diagonal, we must be able to find n linearly independent vectors $\mathbf{u}_i \in \mathbb{R}^n$ (or in $\mathbb{C}^n$) and scalars λ_i with $\mathbf{A}\mathbf{u}_i = \lambda_i\mathbf{u}_i$ for $i = 1, \ldots, n$. Here the set of columns $\{\mathbf{u}_1, \ldots, \mathbf{u}_n\}$ of $\mathbf{U}$ has to form a basis of $\mathbb{R}^n$ (or $\mathbb{C}^n$) for invertibility in Eq. (7.4) according to Section 6.1.

Consequently, if we want to find a simple and intrinsic representation $\mathbf{A}_\mathcal{U}$ of T (or of $\mathbf{A}_\mathcal{E}$), we must look for **eigenvalues** and **eigenvectors** of $\mathbf{A}_\mathcal{E}$.

DEFINITION 1 A scalar $\lambda \in \mathbb{R}$ (or $\mathbb{C}$) and a nonzero vector $\mathbf{u} \in \mathbb{R}^n$ (or $\mathbb{C}^n$) is called an **eigenvalue–eigenvector pair** for the square matrix $\mathbf{A}_{nn}$ if

$$\mathbf{A}\mathbf{u} = \lambda\mathbf{u}. \tag{9.1}$$

In this case, we say that $\mathbf{u} \neq \mathbf{0}$ is an **eigenvector** for the **eigenvalue** $\lambda \in \mathbb{R}$ (or $\mathbb{C}$) of $\mathbf{A}$.

To search for eigenvalue–eigenvector pairs of a given square matrix, $\mathbf{A} = \mathbf{A}_\mathcal{E}$ is a **nonlinear problem**. The unknown quantities λ and $\mathbf{u}$ occur in multiplied form on the right-hand side of Equation (9.1). In this nonlinear problem, our knowledge of how to solve linear equations $\mathbf{Ax} = \mathbf{b}$ that has helped us nicely in each chapter so far is of no use for finding matrix eigenvalues. It is a sad fact that starting an eigenanalysis of a square matrix $\mathbf{A}$ with a row reduction of $\mathbf{A}$ normally nets the student zero points for the eigenvalue problem. To illustrate, note that the RREF of all nonsingular square matrices is the same, namely, equal to $\mathbf{I}_n$. What information can be gleamed from $\mathbf{R} = \mathbf{I}_n$ that gives information about every nonsingular matrix $\mathbf{A}$, other than its rank? None! On the other hand, however, zero is an eigenvalue of each singular matrix, since, for a singular matrix $\mathbf{A}_{nn}$, we have $\ker(\mathbf{A}) \neq \{\mathbf{0}\}$; that is, there is a vector $\mathbf{x} \neq \mathbf{0}$ with $\mathbf{Ax} = \mathbf{0} = 0\mathbf{x}$. But no other eigenvalue of $\mathbf{A}$ can be ever found from a REF of $\mathbf{A}$.

To solve the matrix eigenvalue–eigenvector problem, we rewrite the eigenvalue–eigenvector equation (9.1) with $\mathbf{u} \neq \mathbf{0}$ as follows:

$$\begin{aligned}\mathbf{Au} = \lambda\mathbf{u} \quad &\Longleftrightarrow \quad \mathbf{Au} - \lambda\mathbf{u} = \mathbf{0} \quad \Longleftrightarrow \quad (\mathbf{A} - \lambda\mathbf{I})\mathbf{u} = \mathbf{0} \quad \text{for } \mathbf{u} \neq \mathbf{0} \\ &\Longleftrightarrow \quad \ker(\mathbf{A} - \lambda\mathbf{I}) \neq \{\mathbf{0}\} \quad \Longleftrightarrow \quad \mathbf{A} - \lambda\mathbf{I} \text{ is singular.}\end{aligned} \tag{9.2}$$

This relates the task of finding eigenvalues (and subsequently eigenvectors) of a matrix $\mathbf{A}_{nn}$ to finding singular matrices of the form $\mathbf{A} - \lambda\mathbf{I}$ for specific values of $\lambda \in \mathbb{R}$ or $\mathbb{C}$ that depend on $\mathbf{A}$. For an alternative approach using determinants see lecture 9.1.D.

Remark:

If for example, $\mathbf{R}$ is an upper triangular $n \times n$ matrix, then $\mathbf{R} - r_{ii}\mathbf{I}_n$ is singular for each diagonal entry r_{ii} of $\mathbf{R}$. That is, an upper triangular matrix has all of its eigenvalues on the diagonal. The same holds for lower triangular matrices $\mathbf{L}$ as well.

(b) Do Matrix Eigenvalues Exist?

The first question that we have to answer is whether eigenvalues and eigenvectors of a matrix $\mathbf{A}_{nn} \in \mathbb{R}^{n,n}$ must in fact exist. This can be answered by using the **Fundamental Theorem of Algebra** in conjunction with **vector iteration** (i.e., by forming the sequence of vectors $\mathbf{y}, \mathbf{Ay}, \mathbf{A}^2\mathbf{y}, \ldots, \mathbf{A}^j\mathbf{y}, \ldots \in \mathbb{R}^n$ for an arbitrary starting vector $\mathbf{y} \neq \mathbf{0} \in \mathbb{R}^n$) and looking for the roots of a related polynomial.

Theorem 9.1 **(Fundamental Theorem of Algebra)** Every polynomial

$$p(x) = a_m x^m + a_{m-1}x^{m-1} + \cdots + a_1 x + a_0$$

of degree $k \leq m$ with real or complex coefficients a_i has k **roots** $x_1, \ldots, x_k$ with $p(x_i) = 0$ in the complex plane $\mathbb{C}$ if repeated roots are counted with their multiplicity.

In other words, if $\deg(p) = k$ and $p(x) = a_k x^k + \cdots + a_1 x + a_0$ with $a_k \neq 0$, then $p(x) = a_k(x - x_1) \cdots (x - x_k)$ for its (possibly repeated) roots $x_i \in \mathbb{C}$. ◀

A polynomial p has **degree** k if k is the maximal integer for which the coefficient a_k is nonzero in the expression $p(x) = a_m x^m + a_{m-1}x^{m-1} + \cdots + a_k x^k + \cdots + a_1 x + a_0$. If the repeated roots of a polynomial p of degree k are collected (i.e., if

$$p(x) = a_k(x - x_1)^{n_1}(x - x_2)^{n_2} \cdots (x - x_j)^{n_j}$$

for distinct roots x_i with $n_i > 0$ for all i and $n_1 + n_2 + \ldots + n_j = k$), then we say that the root x_i of p has the **algebraic multiplicity** n_i for each $i = 1, \ldots, j$. For more on polynomials, see Appendix B.

Theorem 9.1 has the following consequence for matrices.

Theorem 9.2 Every square matrix $\mathbf{A} \in \mathbb{R}^{n,n}$ (or $\mathbb{C}^{n,n}$) has an eigenvalue in $\mathbb{C}$. ◀

Proof For $\mathbf{A} \in \mathbb{R}^{n,n}$ and an arbitrary vector $\mathbf{y} \neq \mathbf{0} \in \mathbb{R}^n$ look at the $n+1$ vectors

$$\mathbf{y}, \mathbf{Ay}, \mathbf{A}^2\mathbf{y}, \ldots, \mathbf{A}^{n-1}\mathbf{y}, \mathbf{A}^n\mathbf{y} \in \mathbb{R}^n.$$

As $n+1$ vectors in $\mathbb{R}^n$, these vectors are linearly dependent, due to Proposition 1 part (f) of Section 5.1. In other words, there must be $n+1$ coefficients $\alpha_i \in \mathbb{R}$, not all zero, so that

$$\alpha_0\mathbf{y} + \alpha_1\mathbf{Ay} + \cdots + \alpha_{n-1}\mathbf{A}^{n-1}\mathbf{y} + \alpha_n\mathbf{A}^n\mathbf{y} = \mathbf{0} \in \mathbb{R}^n.$$

Consequently, the $n \times n$ matrix

$$\mathbf{B} = \alpha_0\mathbf{I}_n + \alpha_1\mathbf{A} + \cdots + \alpha_{n-1}\mathbf{A}^{n-1} + \alpha_n\mathbf{A}^n$$

has a nontrivial kernel, since

$$\mathbf{By} = (\alpha_0\mathbf{I}_n + \alpha_1\mathbf{A} + \cdots + \alpha_{n-1}\mathbf{A}^{n-1} + \alpha_n\mathbf{A}^n)\mathbf{y} = \mathbf{0} \text{ for } \mathbf{y} \neq \mathbf{0}.$$

The matrix $\mathbf{B}$ is a polynomial in $\mathbf{A}$, or $\mathbf{B} = p(\mathbf{A})$ for the polynomial

$$p(x) = \alpha_0 + \alpha_1 x + \cdots + \alpha_{n-1}x^{n-1} + \alpha_n x^n.$$

Now use the Fundamental Theorem of Algebra: If $p(x)$ has degree $k \leq n$, then p factors into linear factors

$$p(x) = a_k(x - \lambda_1) \cdots (x - \lambda_k)$$

for some constants $\lambda_i \in \mathbb{C}$, the roots of p. Plugging in $\mathbf{A}$ in place of x, $\alpha_0\mathbf{I}_n = \alpha_0\mathbf{A}^0$ in place of the constant term $\alpha_0 = \alpha_0 x^0$ of p, and $\lambda_i\mathbf{I}$ for the scalar λ_i, we obtain

$$\mathbf{B} = p(\mathbf{A}) = \alpha_0\mathbf{I} + \alpha_1\mathbf{A} + \cdots + \alpha_{n-1}\mathbf{A}^{n-1} + \alpha_n\mathbf{A}^n = a_k(\mathbf{A} - \lambda_1\mathbf{I}) \cdots (\mathbf{A} - \lambda_k\mathbf{I})$$

with $\mathbf{By} = p(\mathbf{A})\mathbf{y} = \mathbf{0}$ for $\mathbf{y} \neq \mathbf{0}$.

Note that the matrix $\mathbf{B} = p(\mathbf{A}) \in \mathbb{R}^{n,n}$ is singular, since $\mathbf{0} \neq \mathbf{y} \in \ker(p(\mathbf{A}))$. If each of its matrix factors $\mathbf{A} - \lambda_i\mathbf{I}$ were nonsingular, so would be their product $\mathbf{B}$ from Lemma 1 of Section 7.2. Thus, we conclude that at least one of the factors $\mathbf{A} - \lambda_j\mathbf{I}$ is singular. This is precisely the eigenvalue condition (9.2) for $\mathbf{A}$. Hence the $n \times n$ matrix $\mathbf{A}$ has at least one eigenvalue $\lambda_j \in \mathbb{C}$. ■

According to Theorem 9.2 and the Fundamental Theorem of Algebra, matrix eigenvalue problems, even for real matrices, can generally only be solved over the complex field $\mathbb{C}$. Therefore, we have included an appendix on complex numbers and complex vectors, as well as one on finding roots of polynomials.

(c) The Vanishing Polynomial and the Minimal Polynomial

We now study the polynomial that occurs in the proof of Theorem 9.2 in more detail. For a matrix $\mathbf{A}_{nn}$ and a nonzero vector $\mathbf{y} \in \mathbb{R}^n$, assume that the polynomial $p_y(x) = \alpha_0 + \alpha_1 x + \cdots + \alpha_{m-1}x^{m-1} + \alpha_m x^m$ satisfies $p_y(\mathbf{A})\mathbf{y} = \mathbf{0} \in \mathbb{R}^n$ and has minimal degree. That is, assume that the coefficients α_i achieve the linear dependence of the sequence of vectors $\mathbf{y}, \mathbf{Ay}, \ldots, \mathbf{A}^m\mathbf{y}$ for the first time. Then, $\mathbf{A}^m\mathbf{y} \in \operatorname{span}\{\mathbf{y}, \mathbf{Ay}, \ldots, \mathbf{A}^{m-1}\mathbf{y}\}$, while the m vectors $\mathbf{y}, \mathbf{Ay}, \ldots, \mathbf{A}^{m-1}\mathbf{y}$ are linearly independent. Thus, the vector $\mathbf{A}^m\mathbf{y}$ is a linear combination of the lesser powers $\mathbf{A}^j\mathbf{y}$. Consequently, for a given matrix $\mathbf{A}$ and a vector $\mathbf{y} \neq \mathbf{0} \in \mathbb{R}^n$, we can always find a normalized polynomial $p_y(x)$ of minimal degree m with leading coefficient $\alpha_m = 1$ for which $p_y(\mathbf{A})\mathbf{y} = \mathbf{0} \in \mathbb{R}^n$.

DEFINITION 2 For a matrix $\mathbf{A}_{nn} \in \mathbb{R}^{n,n}$ (or $\mathbb{C}^{n,n}$) and a vector $\mathbf{y} \neq \mathbf{0} \in \mathbb{R}^n$, the monic polynomial p_y of minimal degree with $p_y(\mathbf{A})\mathbf{y} = \mathbf{0} \in \mathbb{R}^n$ is called the **vanishing polynomial** of $\mathbf{y}$ for $\mathbf{A}$. A polynomial $p(x)$ of degree k is called **monic** if its highest coefficient α_k equals one in the $\sum_{j=1}^{k} \alpha_j x^j$ power expansion of p.

For example, every eigenvector $\mathbf{v} \neq \mathbf{0}$ of a matrix $\mathbf{A}$ has a vanishing polynomial of degree one, since $\mathbf{Av} = \lambda\mathbf{v}$ implies $p_v(x) = x - \lambda$, because $p_v(\mathbf{A})\mathbf{v} = (\mathbf{A} - \lambda\mathbf{I})\mathbf{v} = \mathbf{0}$. Furthermore, the polynomials $p(x) = x^7 - 7x + 1$, $q(x) = 2 - 3x + 5x^3 + x^9$, and $r(x) = (x-1)(x+2)(x-5)^4$ are all monic, while $s(x) = 1 - x^2 + 3x$ and $t(x) = 2x^4 - 8$ are not.

If $p_y(x) = x^m + \cdots + \alpha_0$ is the vanishing polynomial for a nonzero vector $\mathbf{y} \in \mathbb{R}^n$ and $\mathbf{A}_{nn}$, and if p_y factors over $\mathbb{C}$ according to the fundamental theorem as $p_y(x) = (x - \lambda_1) \cdots (x - \lambda_m)$, then even more can be said about the individual matrix factors $\mathbf{A} - \lambda_i\mathbf{I}$ of $p_y(\mathbf{A}) = \prod_i (\mathbf{A} - \lambda_i\mathbf{I})$.

Corollary Let $p_y(x) = (x - \lambda_1) \cdots (x - \lambda_m)$ be the vanishing polynomial of a nonzero vector $\mathbf{y} \in \mathbb{R}^n$ and a matrix $\mathbf{A}_{nn}$ with $\lambda_i \in \mathbb{C}$. Then $\dim(\ker(\mathbf{A} - \lambda_i\mathbf{I})) \geq 1$ for each i. That is, each root $\lambda_i \in \mathbb{C}$ of the vanishing polynomial p_y of $\mathbf{y}$ and $\mathbf{A}$ is an eigenvalue of $\mathbf{A}$. ◀

Proof By definition $p_y(x)$ is the polynomial of minimal degree m for which

$$p_y(\mathbf{A})\mathbf{y} = (\mathbf{A} - \lambda_1\mathbf{I}) \cdots (\mathbf{A} - \lambda_m\mathbf{I})\mathbf{y} = \mathbf{0} \in \mathbb{R}^n.$$

Note that

$$(\mathbf{A} - \lambda\mathbf{I})(\mathbf{A} - \mu\mathbf{I}) = (\mathbf{A} - \mu\mathbf{I})(\mathbf{A} - \lambda\mathbf{I})$$

holds for all scalars λ and μ and any square matrix $\mathbf{A}$; that is, the order of the factors $(\mathbf{A} - \lambda_i\mathbf{I})$ of $p_y(\mathbf{A})$ does not matter. If $\ker(\mathbf{A} - \lambda_j\mathbf{I}) = \{\mathbf{0}\}$ for one j (i.e., if λ_j is not an eigenvalue of $\mathbf{A}$, then $\mathbf{A} - \lambda_j\mathbf{I}$ is invertible according to Chapter 6).

If one of the factors $\mathbf{A} - \lambda_j\mathbf{I}$ of $p_y(\mathbf{A})$ is invertible, then we have

$$(\mathbf{A} - \lambda_1\mathbf{I}) \cdots (\mathbf{A} - \lambda_{j-1}\mathbf{I})(\mathbf{A} - \lambda_{j+1}\mathbf{I}) \cdots (\mathbf{A} - \lambda_m\mathbf{I})\mathbf{y} = \mathbf{0}$$

as well. This follows by moving the invertible factor $\mathbf{A} - \lambda_j\mathbf{I}$ that is omitted to the front of the equation $p_y(\mathbf{A})\mathbf{y} = \mathbf{0}$ by using the commutativity of the matrix factors and by multiplying the resulting vector equation by the nonsingular matrix $(\mathbf{A} - \lambda_j \mathbf{I})^{-1}$ from the left. Note, however, that the new polynomial

$$q(x) = (x - \lambda_1) \cdots (x - \lambda_{j-1})(x - \lambda_{j+1}) \cdots (x - \lambda_m)$$

with $(x - \lambda_j)$ omitted is also monic. It has degree $m - 1$ and satisfies $q(\mathbf{A})\mathbf{y} = \mathbf{0}$ as well, contradicting minimality. Hence, every matrix factor of $p_y(\mathbf{A})$ of the form $(\mathbf{A} - \lambda_i \mathbf{I})$ must be singular, giving $\mathbf{A}$ the eigenvalues λ_i for each $i = 1, \ldots, m$. ∎

Conclusion: Every real or complex root of the vanishing polynomial p_y for a nonzero vector $\mathbf{y} \in \mathbb{R}^n$ and a matrix $\mathbf{A} \in \mathbb{R}^{n,n}$ is an eigenvalue of $\mathbf{A}$.

Our task of representing a square matrix $\mathbf{A}$ most simply consists of finding a basis $\mathcal{U}$ of $\mathbb{R}^n$ (or $\mathbb{C}^n$) that is made up of eigenvectors of $\mathbf{A}$, if possible. We set out on this quest by finding all eigenvalues of $\mathbf{A}$ from one or several vanishing polynomials. These we create for various nonzero vectors $\mathbf{y}_1, \mathbf{y}_2, \ldots \in \mathbb{R}^n$ until the associated eigenvectors span $\mathbb{R}^n$ (or $\mathbb{C}^n$), provided that this is possible for $\mathbf{A}$.

A randomly chosen vector $\mathbf{y} \in \mathbb{R}^n$ will usually have a vanishing polynomial of degree n for $\mathbf{A}$. For completeness, we include a more detailed description of what might happen if the vanishing polynomial $p_y(x)$ of a starting vector $\mathbf{y}$ does not have degree n.

If for example, the first choice $\mathbf{y}_1 \neq \mathbf{0} \in \mathbb{R}^n$ is by chance an eigenvector of $\mathbf{A}$, this is the best case scenario for the eigenanalysis of $\mathbf{A}$, since we have found one eigenvalue and eigenvector pair λ_1 and $\mathbf{y}_1$ of $\mathbf{A}$ instantly. However, this is the worst case scenario for the vanishing polynomial method of extracting <u>all</u> eigenvalues of $\mathbf{A}$: We have only found one such and need to continue:

First, we determine a basis of eigenvectors for the eigenvalue λ_1 of $\mathbf{A}$. If this eigenvector basis spans all of $\mathbb{R}^n$, then $\mathbf{A} = \lambda_1 \mathbf{I}$, or $\mathbf{A}$ is most trivially diagonal, which could have been noticed directly by inspection of $\mathbf{A}$. Otherwise, choose a vector $\mathbf{0} \neq \mathbf{y}_2 \in \mathbb{R}^n$ that is not an eigenvector of $\mathbf{A}$ for the eigenvalue λ_1 by using Section 4.1(c). Find its vanishing polynomial p_{y_2}. Compute the roots of $p_{y_2}(x)$ and find an eigenvector basis for each root of p_{y_2} that is different from the roots of p_{y_1}. Coalesce, or combine, the roots of p_{y_1} and p_{y_2} by taking only the distinct ones when forming $p_{(y_1,y_2)}(x)$ to be the least common multiple of p_{y_1} and p_{y_2}. For example, the least common multiple of $p_{y_1}(x) = (x-2)(x-3)^2$ and $p_{y_2}(x) = (x-2)^2(x-1)$, given in completely factored form, is the product of the highest powers of each distinct linear factor, namely, $p_{(y_1,y_2)}(x) = (x-2)^2(x-3)^2(x-1)$. In general, it is reasonable to expect that $p_{(y_1,y_2)}(\mathbf{A}) = \mathbf{O}_{nn}$.

If, however, we are unlucky again and $p_{(y_1,y_2)}(\mathbf{A}) \neq \mathbf{O}_{nn}$, then continue with a third vector $\mathbf{y}_3 \neq \mathbf{0}$ with $(p_{(y_1,y_2)}(\mathbf{A}))\mathbf{y}_3 \neq \mathbf{0}$: Find its vanishing polynomial p_{y_3}, the roots or eigenvalues, and the corresponding eigenspaces. Coalesce p_{y_3} with $p_{(y_1,y_2)}$ to obtain $p_{(y_1,y_2,y_3)}$ as the least common multiple of $p_{(y_1,y_2)}$ and p_{y_3}, etc.

To find vectors $\mathbf{y}_2, \mathbf{y}_3$, etc., that do not belong to a specified matrix kernel was the subject of Section 4.1(c). In fact, a randomly chosen integer vector $\mathbf{x}$ will most likely not belong to a given matrix kernel. One can check this quickly for $\mathbf{x}$ and adjust the vector, if it should by chance belong to the kernel in question.

Eventually, this process will come to an end: Then, $p_{(y_1,y_2,\ldots,y_j)}(\mathbf{A}) = \mathbf{O}_{nn}$, and either we have found n linearly independent eigenvectors $\mathbf{u}_i$ for $\mathbf{A}$ corresponding to the various roots of the vanishing polynomials p_{y_k}, or we have not. In the former case, $\mathbf{A}$ is diagonalizable; in the latter, $\mathbf{A}$ is not diagonalizable.

[*Note:* It is a very rare occurrence. (i.e., seemingly impossible) to choose an eigenvector $\mathbf{u}$ as the random starting vector for the vector iteration $\mathbf{u}, \mathbf{Au}, \mathbf{A}^2\mathbf{u}$, etc. Likewise, it is extremely rare to obtain a vanishing polynomial $p_y(x)$ for which $p_y(\mathbf{A})$ does not annihilate all of $\mathbb{R}^n$ (i.e., for which $p_y(\mathbf{A}) \neq \mathbf{O}_{nn}$). Thus, our previous elaborations are included mainly for those unlikely situations where one vanishing polynomial $p_y(x)$ will not tell the whole story of $\mathbf{A}$. Remember that for hand computations, we will almost never consider eigenproblems for matrices larger than 3×3.]

If $p_A(x) := p_{(y_1,y_2,\ldots,y_j)}(x)$ is a monic polynomial with leading coefficient equal to one and has minimal degree so that

$$\left(p_{(y_1,y_2,\ldots,y_j)}(\mathbf{A})\right)\mathbb{R}^n = \{\mathbf{0}\}, \text{ i.e., if } p_{(y_1,y_2,\ldots,y_j)}(\mathbf{A}) = \mathbf{O}_{nn},$$

then we call p_A the **minimal polynomial** of $\mathbf{A}$. If $p_A(x) = (x-\lambda_1)^{j_1}\cdots(x-\lambda_k)^{j_k}$ for the distinct eigenvalues $\lambda_1, \ldots, \lambda_k$ of $\mathbf{A}$, then the exponent $j_\ell \geq 1$ of $(x-\lambda_\ell)$ in $p_A(x)$ is called the **index** of the eigenvalue λ_ℓ of $\mathbf{A}$ for each $\ell = 1, \ldots, k$. The minimal polynomial $p_A(x)$ of a matrix $\mathbf{A}_{nn}$ has the following properties.

Proposition 1

(a) For an $n \times n$ matrix $\mathbf{A} \in \mathbb{R}^{n,n}$ (or $\mathbb{C}^{n,n}$), the minimal polynomial $p_A(x)$ is a polynomial with real (or complex) coefficients of degree less than or equal to n.

(b) The minimal polynomial $p_A(x)$ has real coefficients if $\mathbf{A} \in \mathbb{R}^{n,n}$. If $\mathbf{A} \in \mathbb{R}^{n,n}$, then $p_A(x)$ has real roots and possibly pairs of complex conjugate roots $\lambda = a + bi$ and $\overline{\lambda} = a - bi$ for $a, b \in \mathbb{R}$ and $i = \sqrt{-1} \in \mathbb{C}$.

(c) The roots of $p_A(x)$ lie in $\mathbb{C}$; they are the eigenvalues of $\mathbf{A}$.

(d) The minimum polynomial $p_A(x)$ is the least common multiple of the vanishing polynomials $p_{y_i}(x)$, $i = 1, \ldots, n$, for any basis $\{\mathbf{y}_1, \ldots, \mathbf{y}_n\}$ of $\mathbb{R}^n$.

(e) If the vanishing polynomial $p_y(x)$ of a vector $\mathbf{y} \neq \mathbf{0}$ and $\mathbf{A}$ has a repeated factor $(x-\lambda)^j$ with $j > 1$ for $\lambda \in \mathbb{R}$ (or $\mathbb{C}$), then $\mathbf{A}$ is not diagonalizable. In other words, if $\mathbf{A}$ has an eigenvalue of index greater than one, then $\mathbf{A}$ is not diagonalizable. ◀

For the minimum polynomial p_A, we have $p_A(\mathbf{A}) = \mathbf{O}_{nn}$, the zero matrix. Thus, we have proved the following theorem, considered to be one of the deep theorems of finite-dimensional vector spaces:

Theorem 9.3 **(Cayley–Hamilton Theorem)** Any $n+1$ successive powers $\mathbf{I}, \mathbf{A}, \mathbf{A}^2, \ldots, \mathbf{A}^n$ of a matrix $\mathbf{A} \in \mathbb{R}^{n,n}$ (or $\mathbb{C}^{n,n}$) are linearly dependent in the space of $n \times n$ matrices. That is, for each matrix $\mathbf{A}_{nn}$ there exists a monic polynomial $p(x)$ of degree less than or equal to n with $p(\mathbf{A}) = \mathbf{O}_{nn}$. ◀

The $n \times n$ matrices form a **vector space** (see Appendix C) in their own right: $n \times n$ real matrices can be added and scalar multiplied. There are n^2 entries in each $n \times n$ matrix $\mathbf{A}$. Hence, their space has dimension n^2. Using the familiar dimension argument on $\mathbb{R}^{n,n}$, for any $n \times n$ matrix $\mathbf{A}$ the n^2+1 matrix powers $\mathbf{I}, \mathbf{A}, \mathbf{A}^2, \ldots, \mathbf{A}^{n^2} \in \mathbb{R}^{n,n}$ must be linearly dependent in n^2 matrix space. The Cayley–Hamilton theorem cuts the maximal number of linearly independent powers $\mathbf{A}^j$ of every matrix $\mathbf{A} \in \mathbb{R}^{n,n}$ from n^2 down to n. This is quite an achievement.

(d) Finding Eigenvalues and Eigenspaces

We start with an example.

EXAMPLE 1 Find the eigenvalues and eigenvectors of $\mathbf{A} = \begin{pmatrix} 2 & 1 \\ -1 & 4 \end{pmatrix}$. To do so, let us look at the vector iteration $\mathbf{y}_1, \mathbf{A}\mathbf{y}_1, \mathbf{A}^2\mathbf{y}_1$ starting with $\mathbf{y}_1 = \begin{pmatrix} 1 \\ 1 \end{pmatrix}$.

We have $\mathbf{A}\mathbf{y}_1 = \begin{pmatrix} 2 & 1 \\ -1 & 4 \end{pmatrix}\begin{pmatrix} 1 \\ 1 \end{pmatrix} = \begin{pmatrix} 3 \\ 3 \end{pmatrix} = 3\mathbf{y}_1$. Thus, we have done the "nearly impossible" and picked an eigenvector for $\mathbf{A}$'s eigenvalue 3 as our starting vector $\mathbf{y}_1$ by chance. Let us find all eigenvectors for the eigenvalue $\lambda = 3$ of $\mathbf{A}$ next.

From the equivalence (9.2) following Definition 1, we know that an eigenvector $\mathbf{u} \neq \mathbf{0}$ for the eigenvalue 3 of A must satisfy $(\mathbf{A} - 3\mathbf{I})\mathbf{u} = \mathbf{0}$, or

$$\mathbf{u} \in \ker(\mathbf{A} - 3\mathbf{I}).$$

The row reduction for

$$\lambda = 3: \quad \mathbf{A} - 3\mathbf{I}: \quad \begin{array}{cc|cl} \boxed{-1} & 1 & 0 & \\ \boxed{-1} & 1 & 0 & -row_1 \\ \hline \boxed{-1} & 1 & 0 & \\ 0 & 0 & 0 & \\ & \uparrow & & \end{array}$$

shows that all eigenvectors for the eigenvalue $\lambda = 3$ of $\mathbf{A}$ are multiples of $\begin{pmatrix} 1 \\ 1 \end{pmatrix}$.

We are short of full degree 2 here, so we need to find another vector $\mathbf{y}_2 \neq \mathbf{0}$ that is not an eigenvector for the eigenvalue 3 of $\mathbf{A}$ and continue. If we choose $\mathbf{y}_2 = \begin{pmatrix} 1 \\ -1 \end{pmatrix}$, for example, then $\mathbf{A}\mathbf{y}_2 = \begin{pmatrix} 1 \\ -5 \end{pmatrix}$ and $\mathbf{A}^2\mathbf{y}_2 = \mathbf{A}(\mathbf{A}\mathbf{y}_2) = \begin{pmatrix} -3 \\ -21 \end{pmatrix}$. The three vectors $\mathbf{y}_2, \mathbf{A}\mathbf{y}_2, \mathbf{A}^2\mathbf{y}_2 \in \mathbb{R}^2$ are linearly dependent, since any three vectors in $\mathbb{R}^2$ must be linearly dependent. Their actual linear dependence becomes obvious after a row reduction of the vector iteration matrix to its RREF. This dependence then leads us to a vanishing polynomial for **A**:

A		$\mathbf{y}_2$	$\mathbf{A}\mathbf{y}_2$	$\mathbf{A}^2\mathbf{y}_2$	
2	1	$\boxed{1}$	1	−3	
−1	4	$\boxed{-1}$	−5	−21	$+ row_1$
		$\boxed{1}$	1	−3	
		0	$\boxed{-4}$	−24	$\div(-4)$
		$\boxed{1}$	$\boxed{1}$	−3	$- row_2$
		0	$\boxed{1}$	6	
		$\boxed{1}$	0	−9	
		0	$\boxed{1}$	6	

The RREF shows that $\mathbf{A}^2\mathbf{y}_2 = 6\mathbf{A}\mathbf{y}_2 - 9\mathbf{y}_2$ or $p_{y_2}(\mathbf{A}) = \mathbf{A}^2 - 6\mathbf{A} + 9\mathbf{I}$. Next, we factor $p_{y_2}(x) = x^2 - 6x + 9 = (x-3)^2$. This polynomial contains the vanishing polynomial for $\mathbf{y}_1$ and **A** as a factor. It has full degree ($n = 2$) and is normalized ($\alpha_n = 1$). Thus, it is the minimal polynomial p_A of **A**, due to Proposition 2(a). Clearly, $\lambda = 3$ is a double root of $p_A(x)$, and **A** has the double real eigenvalue 3.

Part (e) of Proposition 1 shows that **A** cannot be diagonalized. Let us explain why.

Since there is only one linearly independent eigenvector $\mathbf{u} = \begin{pmatrix} 1 \\ 1 \end{pmatrix}$ (and all of its nonzero multiples) for the double eigenvalue $\lambda = 3$ of **A**, there is something missing for **A**: We cannot find enough eigenvectors (two are needed) for this matrix $\mathbf{A} \in \mathbb{R}^{2,2}$ to diagonalize it. Hence, **A** is not diagonalizable. ◀

We define the space of eigenvectors or the **eigenspace** associated with a matrix eigenvalue next.

DEFINITION 3

(a) For a given matrix $\mathbf{A} \in \mathbb{R}^{n,n}$ (or $\mathbb{C}^{n,n}$) and each scalar $\lambda \in \mathbb{R}$ (or $\mathbb{C}$), we define the set

$$E(\lambda) = \ker(\mathbf{A} - \lambda\mathbf{I}) = \{\mathbf{x} \mid \mathbf{A}\mathbf{x} = \lambda\mathbf{x}\} \subset \mathbb{R}^n \text{ (or } \mathbb{C}^n).$$

(b) If λ is an eigenvalue of $\mathbf{A}_{nn}$, then

$$E(\lambda) = \ker(\mathbf{A} - \lambda\mathbf{I}) = \{\mathbf{x} \mid \mathbf{A}\mathbf{x} = \lambda\mathbf{x}\} \neq \{\mathbf{0}\}$$

is called the **eigenspace** for the eigenvalue λ of **A**.

(c) If λ is an eigenvalue of $\mathbf{A}_{nn}$, then the integer $\dim(E(\lambda)) \geq 1$ is called the **geometric multiplicity** of λ.

The set $E(\lambda)$ is a subspace of $\mathbb{R}^n$ (or $\mathbb{C}^n$) for any scalar $\lambda \in \mathbb{C}$, since it is defined as the kernel of a matrix. (See Section 4.1). Some simple properties of eigenvalues, eigenvectors, and the subspaces $E(\lambda)$ for a matrix $\mathbf{A}_{nn}$ are as follows:

Proposition 2 If $\mathbf{A}$ is an $n \times n$ real or complex matrix, then the following are true:

(a) $E(\lambda) = \{\mathbf{0}\} \Longleftrightarrow$ the scalar λ is not an eigenvalue of $\mathbf{A}$.
(b) $E(\lambda)$ is a subspace of $\mathbb{R}^n$ (or $\mathbb{C}^n$) for each scalar λ.
(c) If $\mathbf{x} \in E(\lambda)$ and $\mathbf{y} \in E(\mu)$ for two distinct eigenvalues $\lambda \neq \mu$ of $\mathbf{A}$ and corresponding eigenvectors $\mathbf{x} \neq \mathbf{0}$, $\mathbf{y} \neq \mathbf{0}$, then $\mathbf{x}$ and $\mathbf{y}$ are linearly independent.
(d) If A has the eigenvalue λ for the eigenvector $\mathbf{x} \neq \mathbf{0} \in \mathbb{R}^n$ (or $\mathbb{C}^n$), then the matrix $\mathbf{U}^{-1}\mathbf{A}\mathbf{U}$ has the same eigenvalue λ as $\mathbf{A}$ for the transformed eigenvector $\mathbf{U}^{-1}\mathbf{x}$ and any nonsingular matrix $\mathbf{U} \in \mathbb{R}^{n,n}$. ◀

Note particularly in part (d) that $\mathbf{A}_{\mathcal{U}} = \mathbf{U}^{-1}\mathbf{A}\mathbf{U}$ and $\mathbf{x}_{\mathcal{U}} = \mathbf{U}^{-1}\mathbf{x}_{\mathcal{E}}$ according to Section 7.1. That is, an eigenvector $\mathbf{x} = \mathbf{x}_{\mathcal{E}}$ of $\mathbf{A} = \mathbf{A}_{\mathcal{E}}$ is transformed to $\mathbf{x}_{\mathcal{U}}$ for $\mathbf{A}_{\mathcal{U}}$, while its eigenvalue does not change.

Proof Parts (a) and (b) are obvious from Definition 3.

For part (c), assume that an equation

$$(\text{E1}): \quad \alpha\mathbf{x} + \beta\mathbf{y} = \mathbf{0}$$

holds for some scalars α, β, where $\mathbf{x} \in E(\lambda)$ and $\mathbf{y} \in E(\mu)$ are two eigenvectors of $\mathbf{A}$. Then $\mathbf{A}(\alpha\mathbf{x} + \beta\mathbf{y}) = \mathbf{A}\mathbf{0} = \mathbf{0}$ by linearity. On the other hand, according to the eigenvalue–eigenvector equations for $\mathbf{A}$, λ, and μ, we have

$$\mathbf{A}(\alpha\mathbf{x} + \beta\mathbf{y}) = \alpha\mathbf{A}\mathbf{x} + \beta\mathbf{A}\mathbf{y} = \alpha\lambda\mathbf{x} + \beta\mu\mathbf{y}.$$

Thus, the equation

$$(\text{E2}): \quad \alpha\lambda\mathbf{x} + \beta\mu\mathbf{y} = \mathbf{0}$$

must hold for $\mathbf{x}$ and $\mathbf{y}$. We form the derived vector equation

$$\begin{aligned} \lambda \cdot (\text{E1}) - (\text{E2}): \quad & \lambda\alpha\mathbf{x} + \lambda\beta\mathbf{y} - \alpha\lambda\mathbf{x} - \beta\mu\mathbf{y} = \mathbf{0}, \\ & \beta\lambda\mathbf{y} - \beta\mu\mathbf{y} = \mathbf{0}, \text{ or} \\ & \beta(\lambda - \mu)\mathbf{y} = \mathbf{0}. \end{aligned}$$

This implies that $\beta = 0$, since $\lambda \neq \mu$ by assumption and $\mathbf{y} \neq \mathbf{0}$ for an eigenvector. Going back to (E1), we have $\alpha\mathbf{x} = \mathbf{0}$, or $\alpha = 0$ because $\mathbf{x} \neq \mathbf{0}$. Thus, $\mathbf{x}$ and $\mathbf{y}$ are linearly independent by Definition 2(a) of Section 5.1.

In part (d), assume that $\mathbf{A}\mathbf{x} = \lambda\mathbf{x}$ for $\mathbf{x} \neq \mathbf{0}$. Then $\mathbf{A}\mathbf{x} = \mathbf{A}\mathbf{U}\mathbf{U}^{-1}\mathbf{x} = \lambda\mathbf{x}$, since $\mathbf{U}\mathbf{U}^{-1} = \mathbf{I}$. Thus,

$$\mathbf{U}^{-1}\mathbf{A}\mathbf{U}(\mathbf{U}^{-1}\mathbf{x}) = \mathbf{U}^{-1}\lambda\mathbf{x} = \lambda\mathbf{U}^{-1}\mathbf{x}.$$

That is, the matrix $\mathbf{U}^{-1}\mathbf{A}\mathbf{U}$ has the eigenvalue λ for its eigenvector $\mathbf{U}^{-1}\mathbf{x} \neq \mathbf{0}$. ■

Let us gather all steps for a complete theoretical eigenvalue–eigenvector analysis of a given matrix $\mathbf{A}_{nn}$ via vector iteration $\mathbf{y}$, $\mathbf{A}\mathbf{y}$, $\mathbf{A}^2\mathbf{y}$, etc., now.

Steps for a complete eigen–analysis of a real matrix $\mathbf{A}_{nn}$
via vector iteration

1. For an arbitrarily chosen vector $\mathbf{y} \neq \mathbf{0} \in \mathbb{R}^n$ form the sequence of $n+1$ vectors
$$\mathbf{y}, \mathbf{Ay}, \mathbf{A}^2\mathbf{y} = \mathbf{A}(\mathbf{Ay}), \ldots, \mathbf{A}^n\mathbf{y} = \mathbf{A}(\mathbf{A}^{n-1}\mathbf{y}) \in \mathbb{R}^n.$$
(It is best to form these vectors one at a time by applying $\mathbf{A}$ to the most recently computed member of this sequence; do not form the matrix powers $\mathbf{A}^j$, for this will take the n-fold effort.)
2. Write out the matrix $\begin{pmatrix} | & | & | & & | \\ \mathbf{y} & \mathbf{Ay} & \mathbf{A}^2\mathbf{y} & \cdots & \mathbf{A}^n\mathbf{y} \\ | & | & | & & | \end{pmatrix}$ and row reduce it towards a REF $\mathbf{R}$. Stop the row reduction as soon as there is a free column.
3. Find the vanishing polynomial p_y for $\mathbf{y}$ and $\mathbf{A}$ from step 2 by terminating the row reduction when the first column of $\mathbf{R}$, say, column $j+1$, becomes free: The vector $\mathbf{A}^j\mathbf{y}$ in column $j+1$ is a linear combination of the previous j linearly independent columns $\mathbf{y}$, $\mathbf{Ay}$, $\ldots$, $\mathbf{A}^{j-1}\mathbf{y}$ according to Chapter 3 and Example 2 of Section 5.1.
To find the coefficients of this linear combination, continue to the RREF
$$\left(\begin{array}{ccc|c} & & & \alpha_0 \\ & \mathbf{I}_j & & \vdots \\ & & & \alpha_{j-1} \\ \hline 0 & \ldots & 0 & 0 \\ \vdots & & \vdots & \vdots \\ 0 & \ldots & 0 & 0 \end{array}\right)_{n,(j+1)} \quad \textit{involving the first } j+1 \textit{ columns}$$
$$\begin{pmatrix} | & | & & | & | \\ \mathbf{y} & \mathbf{Ay} & \cdots & \mathbf{A}^{j-1}\mathbf{y} & \mathbf{A}^j\mathbf{y} \\ | & | & & | & | \end{pmatrix}_{n,(j+1)}$$
only. The vanishing polynomial of $\mathbf{y}$ and $\mathbf{A}$ is
$$p_y(x) = x^j - \alpha_{j-1}x^{j-1} - \cdots - \alpha_1 x - \alpha_0.$$
It is a factor of the minimal polynomial $p_A(\lambda)$. Find the roots $\lambda_i \in \mathbb{C}$ of p_y; see Appendix B. Each root λ_i is an eigenvalue of $\mathbf{A}$, due to the corollary. If $p_y(x)$ has a repeated root (i.e., the index of one eigenvalue of $\mathbf{A}$ exceeds one.) STOP. $\mathbf{A}$ is not diagonalizable by Proposition 1(e).
(*Note:* with most choices of $\mathbf{y} \in \mathbb{R}^n$, the linear dependence of the iterated vectors $\mathbf{A}^k\mathbf{y}$ will involve the nth power vector $\mathbf{A}^n\mathbf{y}$. If $\mathbf{y}$ is an eigenvector for $\mathbf{A}$, then $\mathbf{y}$ and $\mathbf{Ay} = \lambda\mathbf{y}$ will, however, already be linearly dependent.)
4. Find a basis for the eigenspace $E(\lambda_i)$ of each root λ_i of $p_y(x)$ in step 3.

For each eigenvalue λ_i of $\mathbf{A}$, there may be more than one linearly independent eigenvector.

If you have found less than n linearly independent eigenvectors overall for all roots of p_y from step 3, go back to steps 2 and 3 with a vector $\mathbf{0} \neq \mathbf{v} \notin \ker(p_y(\mathbf{A}))$ and find its vanishing polynomial p_v for $\mathbf{A}$. Such a vector $\mathbf{v}$ can be found using Section 4.1(c), or randomly. The random choice method, however, requires checking that $\mathbf{v} \notin \ker(p_y(\mathbf{A}))$.

Then, repeat step 4 for $p_v(x)$, working only for the roots of p_v that are different from those of p_y. (If for the collected roots of the least common multiple $p_{(y,v)}(x)$ of p_y and p_v you again have found fewer than n linearly independent eigenvectors overall for $\mathbf{A}$, go back to steps 2 and 3 with a third vector $\mathbf{0} \neq \mathbf{w} \notin \ker(p_{(y,v)}(\mathbf{A}))$, etc.)

If, finally, $\ker p_{\dots}(\mathbf{A}) = \mathbb{R}^n$ and you have not been able to find n linearly independent eigenvectors for $\mathbf{A}$, then $\mathbf{A}$ is not diagonalizable: STOP (When looking for a basis for an eigenspace $E(\lambda_k)$, note that if the REF of the matrix $\mathbf{A} - \lambda_k \mathbf{I}$ does not have at least one free variable, then λ_k is **not** an eigenvalue of A; See Proposition 2(a). **Recheck** your work on steps 1 through 3. There is an **error**.)

5. If you were able to find n linearly independent eigenvectors $\mathbf{u}_i$ for $\mathbf{A}_{nn}$ in step 4 (possibly after some restarts), form the eigenvector basis $\mathcal{U} = \{\mathbf{u}_1, \dots, \mathbf{u}_n\}$ for $\mathbf{A}$ and the diagonal matrix representation $\mathbf{A}_{\mathcal{U}}$ of $\mathbf{A}$ with respect to the basis $\mathcal{U}$ as follows:

$$\mathbf{U} = \begin{pmatrix} | & & | \\ \mathbf{u}_1 & \cdots & \mathbf{u}_n \\ | & & | \end{pmatrix}, \quad \mathbf{A}_{\mathcal{U}} = \begin{pmatrix} \lambda_1 & & 0 \\ & \ddots & \\ 0 & & \lambda_n \end{pmatrix}.$$

(Note that the order of the eigenvectors $\mathbf{u}_i$ as columns of U determines the order of the eigenvalues λ_i of $\mathbf{A}$ that appear on the diagonal of $\mathbf{A}_{\mathcal{U}}$ and vice versa. You can swap the eigenvectors $\mathbf{u}_3$ and $\mathbf{u}_5$ in the columns of $\mathbf{U}$ only when simultaneously swapping λ_3 and λ_5 on the diagonal of $\mathbf{A}_{\mathcal{U}}$, for example.)

Now, $\boxed{\mathbf{AU} = \mathbf{UA}_{\mathcal{U}}}$. This matrix equation should be checked.

(This last matrix equation avoids having to deal with the matrix inverse $\mathbf{U}^{-1}$ of $\mathbf{U}$ and a matrix triple product in the more troublesome, but equivalent matrix similarity Equation (7.4) $\mathbf{A}_{\mathcal{U}} = \mathbf{U}^{-1}\mathbf{AU}$.)

EXAMPLE 2 Find the minimal polynomial for $\mathbf{A} = \begin{pmatrix} 3 & 3 \\ 2 & 4 \end{pmatrix}$ via vector iteration starting with $\mathbf{y} = \begin{pmatrix} 1 \\ -1 \end{pmatrix}$.

(1) We compute $\mathbf{Ay} = \begin{pmatrix} 0 \\ -2 \end{pmatrix}$ and $\mathbf{A}^2\mathbf{y} = \mathbf{A}(\mathbf{Ay}) = \mathbf{A}\begin{pmatrix} 0 \\ -2 \end{pmatrix} = \begin{pmatrix} -6 \\ -8 \end{pmatrix}$.

(2) Next, we form the 2×3 vector iteration matrix $\begin{pmatrix} | & | & | \\ \mathbf{y} & \mathbf{Ay} & \mathbf{A}^2\mathbf{y} \\ | & | & | \end{pmatrix}$ and row reduce it to its RREF:

A		y	Ay	A^2y	
3	3	[1]	0	−6	
2	4	[−1]	−2	−8	$+\ row_1$
		[1]	0	−6	
		0	[−2]	−14	$\div(-2)$
		[1]	0	−6	
		0	[1]	7	

(3) Note from the RREF that **y** and **Ay** are linearly independent, while $\mathbf{A}^2\mathbf{y}$ lies in the span of **y** and **Ay**. Specifically, $\mathbf{A}^2\mathbf{y} = -6\mathbf{y} + 7\mathbf{Ay}$, or

$$\mathbf{A}^2\mathbf{y} - 7\mathbf{Ay} + 6\mathbf{y} = (\mathbf{A}^2 - 7\mathbf{A} + 6\mathbf{I})\mathbf{y} = \mathbf{0},$$

making the vanishing polynomial $p_y(x) = x^2 - 7x + 6$. This polynomial is the minimal polynomial $p_A(x)$ of **A**, since it is monic of degree $2 = n$. ◀

Note that choosing a different starting vector for the vector iteration does not change the outcome for the minimal polynomial that is computed for **A** in this process. The minimal polynomial $p_A(x) = \prod_i (x - \lambda_i)^{j_i}$ with $\lambda_i \neq \lambda_k$ for $i \neq k$ is an intrinsic object associated with $\mathbf{A}_{nn}$ according to Proposition 2, since **A**'s eigenvalues λ_i are invariant under basis change. Each exponent j_i in p_A is called the **index** of the corresponding eigenvalue. The index of an eigenvalue is an intrinsic invariant of **A** and does not change for differing bases. Unfortunately, there is no time in an introductory course to explain why. Just note that a matrix $\mathbf{A} \in \mathbb{R}^{n,n}$ is diagonalizable if and only if its minimal polynomial has the form $p_A(x) = \prod_i (x - \lambda_i)$ (i.e., if and only if p_A factors into first degree factors $(x - \lambda_i)$ that do not repeat). See Proposition 1(e).

If we repeat Example 2 with the starting vector $\mathbf{v} = \begin{pmatrix} -1 \\ 2 \end{pmatrix}$, we obtain

A		v	Av	A^2v	
3	3	[−1]	3	27	
2	4	[2]	6	30	$+2\ row_1$
		[−1]	3	27	$\div(-1)$
		0	[12]	84	$\div(12)$
		[1]	0	−6	
		0	[1]	7	

That is, precisely the same linear relationship holds among $\mathbf{A}^2\mathbf{v}$, $\mathbf{A}\mathbf{v}$, and $\mathbf{v}$ as held between $\mathbf{A}^2\mathbf{y}$, $\mathbf{A}\mathbf{y}$, and $\mathbf{y}$ in Example 2.

EXAMPLE 3 Perform a complete eigenanalysis of $\mathbf{B} = \begin{pmatrix} 1 & 1 \\ -2 & 4 \end{pmatrix}$ and diagonalize $\mathbf{B}$ if possible by following the vector iteration procedure.

(1) –(3) Eigenvalues:

For $\mathbf{y} = \begin{pmatrix} 2 \\ 1 \end{pmatrix}$, we compute $\mathbf{B}\mathbf{y} = \begin{pmatrix} 3 \\ 0 \end{pmatrix}$ and $\mathbf{B}^2\mathbf{y} = \begin{pmatrix} 3 \\ -6 \end{pmatrix}$. Thus,

B		y	By	B^2y	
1	1	2	3	3	*swap rows 1 and 2*
−2	4	1	0	−6	
		[1]	0	−6	
		[2]	3	3	$-2\ row_1$
		[1]	0	−6	
		0	[3]	15	÷(3)
		[1]	0	−6	
		0	[1]	5	

and $\mathbf{B}^2\mathbf{y} = 5\mathbf{B}\mathbf{y} - 6\mathbf{y}$, or $\mathbf{p}_y(x) = p_B(x) = x^2 - 5x + 6 = (x-2)(x-3)$ by the degree argument used in Example 2. The eigenvalues of $\mathbf{B}$ are the roots 2 and 3 of $p_B(x)$.

4. Eigenvectors:

For $\lambda_1 = 2$:

$\mathbf{B} - 2\mathbf{I}$:

[−1]	1	0	
[−2]	2	0	$-2\ row_1$
[−1]	1	0	
0	0	0	
	↑		

Hence, $\mathbf{u}_1 = \begin{pmatrix} 1 \\ 1 \end{pmatrix}$ is a basis for the eigenspace $E(2)$ of $\mathbf{B}$.

(*Note:* The row reduction of $\mathbf{B} - 2\mathbf{I}$ must exhibit a free variable for the computed root $\lambda = 2$ of $p_B(\lambda)$ to be an eigenvalue of $\mathbf{B}$.)

For $\lambda_2 = 3$:

$$\mathbf{B} - 3\mathbf{I}: \quad \begin{array}{cc|c} \boxed{-2} & 1 & 0 \\ \boxed{-2} & 1 & 0 \\ \hline \boxed{-2} & 1 & 0 \\ 0 & 0 & 0 \\ & \uparrow & \end{array} \quad \begin{array}{l} \\ -row_1 \\ \\ \\ \\ \end{array}$$

Thus, $\mathbf{u}_2 = \begin{pmatrix} 1 \\ 2 \end{pmatrix}$ spans the eigenspace $E(3)$ of **B**.

5. There are two linearly independent eigenvectors $\mathbf{u}_1$ and $\mathbf{u}_2$ for the 2×2 matrix **B**, making **B** diagonalizable with respect to its eigenvector basis $\mathbf{u}_1$ and $\mathbf{u}_2$. This basis is assembled in the columns of

$$\mathbf{U} = \begin{pmatrix} | & | \\ \mathbf{u}_1 & \mathbf{u}_2 \\ | & | \end{pmatrix} = \begin{pmatrix} 1 & 1 \\ 1 & 2 \end{pmatrix}.$$

The diagonal representation $\mathbf{B}_{\mathcal{U}}$ of **B** with respect to this ordered basis is

$$\mathbf{B}_{\mathcal{U}} = \begin{pmatrix} 2 & 0 \\ 0 & 3 \end{pmatrix} (= \mathbf{U}^{-1}\mathbf{B}\mathbf{U}),$$

with **B**'s eigenvalues 2 and 3 appearing in the chosen order on the diagonal. The simple matrix identity $\mathbf{BU} = \mathbf{UB}_{\mathcal{U}}$ should be checked for the matrices **U**, $\mathbf{B}_{\mathcal{U}}$, and **B** just to be sure, rather than the equivalent, but more complicated, similarity equation $\mathbf{B}_{\mathcal{U}} = \mathbf{U}^{-1}\mathbf{B}_{\mathcal{E}}\mathbf{U}$. ◀

(e) Diagonalizable Matrices

Square matrices may or may not be diagonalizable as we have seen. When trying to diagonalize a matrix, we must keep in mind the following simple fact:

Theorem 9.4 A matrix $\mathbf{A}_{nn}$ is diagonalizable if and only if there are n linearly independent eigenvectors in $\mathbb{C}^n$ for **A**. ◀

This has the following consequence:

Corollary If $\mathbf{A}_{nn} \in \mathbb{R}^{n,n}$ (or $\mathbb{C}^{n,n}$) has n distinct eigenvalues, then **A** is diagonalizable over $\mathbb{C}$. In other words, if a matrix $\mathbf{A}_{nn}$ is not diagonalizable, then it must have repeated eigenvalues. ◀

The converse is false. Just think of the diagonal matrix $\mathbf{I}_n$ with its n-fold eigenvalue 1 on the diagonal.

> Remember that a matrix $\mathbf{A}_{nn}$ is not diagonalizable if a vanishing polynomial $p_y(x)$ of **A** has a repeated linear factor $(x-\lambda)^k$ for $k > 1$, (i.e., if one eigenvalue of **A** has an index greater than unity).
> Nondiagonalizable matrices **A** will be studied in Chapter 14.

We need to stress here that an eigenanalysis of a real matrix will normally lead to complex arithmetic. This is a clear consequence of the Fundamental Theorem of

Algebra. While the minimal polynomial of a real matrix must have real coefficients according to Proposition 1(b), most real coefficient polynomials will only split into first degree factors over the complex field $\mathbb{C}$. The matrices described earlier were prepared with care to stay in the reals when looking for the roots of a vanishing polynomial $p_y(x)$. Thus, an eigenanalysis of a general real matrix $\mathbf{A}$ is extremely difficult to do by hand, just consider the problem of evaluating $p_y(x)$ correctly from a vector iteration and its row reduction, and worse, that of factoring the polynomial $p_y(\lambda)$ for any sizable degree $n > 2$; see Appendix B for more details.

Clearly, we have reached the end of what we can do by hand, other than invent computers and design software for this purpose. (And this is just what mankind has begun to perfect, since the 1940s, 50s, and 60s.)

Finally, we want to emphasize that the possible diagonalization of any matrix hinges on the noncommutativity of the matrix product: $\mathbf{AB} \neq \mathbf{BA}$ for matrices in general, (See the caution in Section 6.1.)

If matrix products were or could be defined differently so that they were always to commute, then the $\mathbf{U}^{-1} \cdots \mathbf{U}$ left and right multiplications would cancel in the $\mathbf{A}_{\mathcal{U}} = \mathbf{U}^{-1}\mathbf{AU}$ diagonalization similarity (7.4), or in the matrix-simplification process, and we would almost never be able to see a diagonal representation $\mathbf{A}_{\mathcal{U}}$ of a square matrix $\mathbf{A} = \mathbf{A}_{\mathcal{E}}$. In other words, without the noncommutativity of matrix multiplication, a matrix diagonalizing eigenvector basis $\mathcal{U}$ would not exist except for those linear transformation that are already diagonal.

This is a universal fact worth contemplating: Complications, troubles, problems, and seemingly ill behavior somewhere often bear intrinsic fruit elsewhere in our human and mental endeavors.

We have learned how to use vector iteration and row reduction of specific matrices in order to solve nonlinear matrix eigenproblems in low dimensions.

9.1.P Problems

1. Which of the vectors $\mathbf{u}_1 = \begin{pmatrix} 1 \\ -1 \\ 1 \end{pmatrix}$, $\mathbf{u}_2 = \begin{pmatrix} 0 \\ 0 \\ 0 \end{pmatrix}$, $\mathbf{u}_3 = \begin{pmatrix} 2 \\ -1 \\ -1 \end{pmatrix}$, and $\mathbf{u}_4 = \begin{pmatrix} 1 \\ 2 \\ 0 \end{pmatrix}$ are eigenvectors for the matrix $\mathbf{A} = \begin{pmatrix} 6 & 7 & 7 \\ -7 & -8 & -7 \\ 7 & 7 & 6 \end{pmatrix}$. What are the corresponding eigenvalues, if any?

2. Find all eigenvalues and corresponding eigenvectors for the matrix $\mathbf{A} = \begin{pmatrix} 2 & 0 & 3 \\ 0 & -1 & 0 \\ 1 & 0 & 4 \end{pmatrix}$: Start with $\mathbf{y} = \begin{pmatrix} -1 \\ 1 \\ 1 \end{pmatrix}$ and find a vanishing polynomial $p_y(x)$ for $\mathbf{y}$ and $\mathbf{A}$, its roots, etc. Is $\mathbf{A}$ diagonalizable? Why and how?

3. Give an example of a real 2×2 matrix with no real eigenvalue.

4. Show that $\mathbf{A} = \begin{pmatrix} 1 & 0 \\ -2 & -2 \end{pmatrix}$ is diagonalizable. Find an eigenvector basis $\mathcal{U}$ for $\mathbf{A}$ and find $\mathbf{A}_{\mathcal{U}}$. Verify that $\mathbf{AU} = \mathbf{UA}_{\mathcal{U}}$.

5. Find the eigenvalues and corresponding eigenvectors for the matrices

(a) $\begin{pmatrix} 1 & 1 \\ 2 & 1 \end{pmatrix}$;

(b) $\begin{pmatrix} -1 & 8 \\ 1 & 1 \end{pmatrix}$;

(c) $\begin{pmatrix} 1 & 1 & 0 \\ 0 & 0 & 2 \\ 3 & 0 & -1 \end{pmatrix}$;

(d) $\begin{pmatrix} 1 & 2 & 3 \\ 0 & 0 & 0 \\ 0 & 0 & 2 \end{pmatrix}$;

(e) $\begin{pmatrix} 1 & 2 & 3 \\ 0 & 0 & 2 \\ 0 & 0 & 0 \end{pmatrix}$; and

(f) $\begin{pmatrix} 0 & 1 & 0 \\ 1 & 0 & -1 \\ 0 & 1 & 0 \end{pmatrix}$.

6. Consider $\mathbf{A} = \begin{pmatrix} 1 & -2 & 1 \\ -1 & -1 & -2 \\ 2 & 0 & 3 \end{pmatrix}$.

(a) Find the coefficients of the minimal polynomial of **A**.

(b) Factor the minimal polynomial into linear factors. (It has irrational roots.)

(c) Find the eigenvalues and eigenvectors of **A**.

(d) Find a diagonal representation of **A** if possible.

7. (a) Show that if $\mathbf{A} \in \mathbb{R}^{n,n}$ has the eigenvalue $\lambda \in \mathbb{R}$ for an eigenvector $\mathbf{x} \in \mathbb{R}^n$, then for any $\alpha \in \mathbb{R}$ the matrix $\mathbf{B} = \mathbf{A} + \alpha \mathbf{I}_n$ has the eigenvalue $\lambda + \alpha$. Why is **x** an eigenvector for **B** and its eigenvalue $\lambda + \alpha$?

(b) Show that 0 and 6 are eigenvalues of $\mathbf{A} = \begin{pmatrix} 4 & 8 \\ 1 & 2 \end{pmatrix}$. Is **A** diagonalizable? Which basis $\mathcal{U} = \{\mathbf{u}_1, \mathbf{u}_2\}$ will diagonalize **A**?

(c) Find all eigenvalues of $\mathbf{B} = \begin{pmatrix} 7 & 8 \\ 1 & 5 \end{pmatrix}$. Is **B** diagonalizable? How?

(d) Find the eigenvalues of the squared matrix $\mathbf{A}^2 = \mathbf{A} \cdot \mathbf{A}$ from part (b). (*Hint*: $\mathbf{A}^2\mathbf{x} = \mathbf{A}(\mathbf{A}\mathbf{x}) = \cdots \mathbf{x}$.)

8. Show that the linear transformation $T\begin{pmatrix} x \\ y \end{pmatrix} = \begin{pmatrix} y \\ -x \end{pmatrix}$ maps no real vector $\begin{pmatrix} x \\ y \end{pmatrix} \in \mathbb{R}^2$ to a real multiple of itself, except, of course, $\begin{pmatrix} x \\ y \end{pmatrix} = \begin{pmatrix} 0 \\ 0 \end{pmatrix}$. (Compare with Problem 22 of Section 9.R.)

9. Let $\mathbf{A} = \begin{pmatrix} 23 & 10 & -10 \\ -58 & -27 & 34 \\ -11 & -5 & 6 \end{pmatrix}$.

(a) Given that $\lambda = 3$ is an eigenvalue of **A**, find the corresponding eigenvector **x**.

(b) Choose an index i, $1 \le i \le 3$, for which the ith component x_i of **x** is nonzero. For the ith row $\mathbf{a}_i$ of **A** form $\mathbf{B}_0 = \mathbf{A} - \frac{1}{x_i}\mathbf{x}\mathbf{a}_i$. (*Note:* The matrix product $\mathbf{x}_{n1}\mathbf{a}_{i_{1n}}$ of a column vector $\mathbf{x} \in \mathbb{R}^n$ and a row vector $\mathbf{a}_i \in \mathbb{R}^n$ is an $n \times n$ matrix.)

(c) Show that $\mathbf{B}_0$ has the eigenvalue 0 and that every eigenvector **w** of B_0 has a zero in its ith position.

(d) Form $\mathbf{B}_1$ from $\mathbf{B}_0$ by dropping the ith row and the ith column of $\mathbf{B}_0$. Find the eigenvalues of the 2×2 matrix $\mathbf{B}_1$, starting with $\mathbf{y} = \begin{pmatrix} 1 \\ 1 \end{pmatrix}$, and compute $\mathbf{B}_1$'s eigenvectors.

(e) Verify as quickly as possible that the eigenvalues of $\mathbf{B}_1$ from part (c) are also eigenvalues of **A**, if **A** is diagonalizable and nonsingular. What are the corresponding eigenvectors?

(f) Is the originally given matrix **A** diagonalizable, and if so, how can **A** be diagonalized?

10. Which of the three matrices $\mathbf{A} = \begin{pmatrix} 2 & 0 & -1 \\ 0 & 3 & 0 \\ 0 & 0 & 2 \end{pmatrix}$, $\mathbf{B} = \begin{pmatrix} 2 & 1 & 0 \\ 0 & 3 & -2 \\ 0 & 0 & 2 \end{pmatrix}$, and $\mathbf{C} = \begin{pmatrix} 2 & 1 & 0 \\ 0 & 3 & 0 \\ 0 & 0 & 2 \end{pmatrix}$ are diagonalizable?

11. Let $\mathbf{A} = \begin{pmatrix} 1 & 2 \\ 2 & 1 \end{pmatrix}$.

(a) Show that $\mathbf{A}^2 - 2\mathbf{A} - 3\mathbf{I}_2 = \mathbf{O}_2$.

(b) Use part (a) to express $\mathbf{A}^{-1}$ as a polynomial in **A**.

(c) Find $\mathbf{A}^{-1}$ explicitly from part (b), and verify your result.

(d) Find all eigenvalues of **A**.

12. (a) If two matrices $\mathbf{A}_{m \times k}$ and $\mathbf{B}_{r \times s}$ are compatible for multiplication and if $\mathbf{T} = \mathbf{A} \cdot \mathbf{B}$, what sizes must **A** and **B** have? What is the size of **T**?

If $\mathbf{S} = \mathbf{B} \cdot \mathbf{A}$ can be formed, what sizes must **A** and **B** have? What size does the matrix **S** have?

(b) Given a row vector **u** and a column vector **v**, both of size n, what is the size of $\mathbf{u} \cdot \mathbf{v}$? What is the size of $\mathbf{v} \cdot \mathbf{u}$?

(c) Show that **v** is an eigenvector of $\mathbf{A} = \mathbf{v}\mathbf{u}$ if **v** is a nonzero column vector and **u** is a row vector, each of size n. What is the corresponding eigenvalue of **A**?

(d) Does **A** from part (c) have the eigenvalue zero? Find dim(ker(**A**)).

(e) Is it true that any square matrix made up as **A** in part (c) as a column times a row vector is diagonalizable? Why? How?

(f) Find all eigenvalues and eigenvectors of $\mathbf{B} = \begin{pmatrix} 1 & 2 & -3 \\ 2 & 4 & -6 \\ -3 & -6 & 9 \end{pmatrix}$, preferably by using the preceding parts.

13. Find a basis $\mathcal{U}$ of $\mathbb{R}^3$ with respect to which the linear transformation $T(\mathbf{x}) = \begin{pmatrix} 3x_1 + x_2 \\ x_2 \\ 4x_1 + 2x_2 + x_3 \end{pmatrix} : \mathbb{R}^3 \longrightarrow \mathbb{R}^3$ is represented in diagonal form $\mathbf{A}_{\mathcal{U}}$.
Perform a complete eigenanalysis for $\mathbf{A}_{\mathcal{E}}$. What is $\mathcal{U}$? What is $\mathbf{A}_{\mathcal{U}}$?

14. If a matrix $\mathbf{A}$ of arbitrary size $n \times n$ has only the two eigenvalues -1 and 4 and is diagonalizable, show that $\mathbf{A}^2 - 3\mathbf{A} = 4\mathbf{I}_n$. Why is $\mathbf{A}$ invertible? Express $\mathbf{A}^{-1}$ as a polynomial in $\mathbf{A}$.

15. Compute all eigenvalues and corresponding eigenvectors for $\mathbf{B} = \begin{pmatrix} 3 & -5 & 4 \\ 2 & -4 & 4 \\ 1 & -2 & 2 \end{pmatrix}$. Is $\mathbf{B}$ diagonalizable? Why or why not?

16. Find the eigenvalues of the matrix $\begin{pmatrix} 1 & 0 & 0 \\ -1 & 2 & -2 \\ -1 & 1 & -1 \end{pmatrix}$.
Is $\mathbf{A}$ diagonalizable?

17. Let $\mathbf{A} = \begin{pmatrix} 3 & -2 & 4 \\ -1 & 3 & -1 \\ -1 & 2 & -2 \end{pmatrix}$.

(a) Is 2 an eigenvalue of $\mathbf{A}$?

(b) Can you find a nonzero vector $\mathbf{x}$ with $\mathbf{A}\mathbf{x} = 2\mathbf{x}$?

(c) Evaluate $\mathbf{A}\mathbf{y}$ for $\mathbf{y} = \begin{pmatrix} 1 \\ -1 \\ 1 \end{pmatrix}$.

(d) Is $\mathbf{y} = \begin{pmatrix} 1 \\ -1 \\ 1 \end{pmatrix}$ an eigenvector of $\mathbf{A}$? Is $\mathbf{z} = \begin{pmatrix} -1 \\ 1 \\ -1 \end{pmatrix}$?

18. For $\mathbf{A} = \begin{pmatrix} a & b \\ c & d \end{pmatrix}$ with integer coefficients a, b, c, and d find conditions on the coefficients of $\mathbf{A}$, so that $\mathbf{A}$ has

(a) two real eigenvalues;

(b) two equal eigenvalues;

(c) two integer eigenvalues;

(d) two complex conjugate eigenvalues.

19. (a) Represent the matrix $\mathbf{A}_{\mathcal{E}} = \begin{pmatrix} 3 & 1 \\ 1 & 3 \end{pmatrix}$ with respect to the basis $\mathcal{U} = \left\{ \begin{pmatrix} 1 \\ -1 \end{pmatrix}, \begin{pmatrix} 2 \\ 2 \end{pmatrix} \right\}$ of $\mathbb{R}^2$.

(b) What are the eigenvalues of the matrix $\mathbf{A}_{\mathcal{E}}$; what are the corresponding eigenvectors?

20. Perform a complete eigenanalysis for the matrix

$$\mathbf{A} = \begin{pmatrix} 3 & -1 & -1 \\ -12 & 0 & 5 \\ 4 & -2 & -1 \end{pmatrix}.$$

Specifically, determine all eigenvalues and eigenvectors for $\mathbf{A}$. Determine whether $\mathbf{A}$ has a diagonal matrix representation $\mathbf{A}_{\mathcal{U}}$. If so, find an eigenvector basis $\mathcal{U}$ for $\mathbf{A}$.

21. Find the eigenvalues and corresponding eigenvectors for the matrix

$$\mathbf{A} = \begin{pmatrix} 0 & -1 & 0 \\ -1 & 2 & 1 \\ 0 & 1 & 4 \end{pmatrix}.$$

What is the matrix representation of $\mathbf{A}$ with respect to the eigenvector basis?

22. Show that if $\mathbf{x}$ is a real eigenvector of $\mathbf{A}_{nn} \in \mathbb{R}^{n,n}$ for a nonzero real eigenvalue λ, then $\mathbf{x}$ lies in the column space of $\mathbf{A}$. Is this result also true for the eigenvalue $\lambda = 0$?

23. Show that $\lambda = 0$ is an eigenvalue of $\mathbf{A}$ if and only if $\mathbf{A}$ is singular.

24. Let $\mathbf{A}_{nn} \in \mathbb{R}^{n,n}$ and assume that $\mathbf{X}_{nn}$ is nonsingular.

(a) Show that $(\mathbf{X}^{-1}\mathbf{A}\mathbf{X})^\ell = \mathbf{X}^{-1}\mathbf{A}^\ell\mathbf{X}$ for all positive integers $\ell \geq 0$.

(b) How can one use part (a) to evaluate the powers $\mathbf{A}^\ell$ of a diagonalizable matrix $\mathbf{A}$ efficiently?

25. Perform a complete eigenanalysis and check diagonalizability for

(a) $\mathbf{A} = 3\begin{pmatrix} 1 & 1 & 0 \\ 0 & 3 & -1 \\ 0 & 0 & -2 \end{pmatrix}$,

(b) $\mathbf{B} = \begin{pmatrix} -1 & 1 & 0 \\ 0 & 2 & -1 \\ 0 & 0 & -1 \end{pmatrix}$, and

(c) $\mathbf{C} = -\begin{pmatrix} 1 & 0 & 1 \\ 0 & 1 & 0 \\ 0 & 0 & -2 \end{pmatrix}$.

26. (a) Find the vanishing polynomial $p_u(x)$ for $\mathbf{u} = (0, 1, 1, 0)^T$ and $\mathbf{A}_{4,4}$, the diagonal matrix with the nonzero entries 1, 2, 1, and 2 on its diagonal. What is the minimal polynomial p_A of $\mathbf{A}$?

(b) What is the minimum polynomial of the identity matrix $\mathbf{I}_n$?

(c) What is the minimum polynomial of the zero matrix $\mathbf{O}_n$?

27. Let $\mathbf{A} = \begin{pmatrix} 0 & 2 \\ -8 & 0 \end{pmatrix} \in \mathbb{R}^{2,2}$.

(a) Find the eigenvalues of $\mathbf{A}$ over $\mathbb{C}$.

(b) Find the eigenvectors of $\mathbf{A}$ in $\mathbb{C}^2$.

28. Let $\mathbf{R} = \begin{pmatrix} 1 & 4 & 2 \\ 0 & -8 & 1 \\ 0 & 0 & 1 \end{pmatrix} \in \mathbb{R}^{3,3}$.

(a) Find the eigenvalues of $\mathbf{R}$.

(b) Is $\mathbf{R}$ diagonalizable?

29. (a) What is the matrix equation linking the standard matrix $\mathbf{A} = \mathbf{A}_{\mathcal{E}} \in \mathbb{R}^{n,n}$ of a linear transformation T to the matrix that represents the same linear transformation with respect to another basis $\mathcal{U}$ of $\mathbb{R}^n$?

(b) Find the matrix representation $\mathbf{A}_{\mathcal{U}}$ of $\mathbf{A} = \mathbf{A}_{\mathcal{E}} = \begin{pmatrix} 4 & -2 & 2 \\ 1 & 4 & -1 \\ -2 & 5 & -2 \end{pmatrix}$ for the basis $\mathcal{U} = \left\{ \begin{pmatrix} 1 \\ 0 \\ -2 \end{pmatrix}, \begin{pmatrix} -1 \\ 1 \\ 2 \end{pmatrix}, \begin{pmatrix} 0 \\ 1 \\ 1 \end{pmatrix} \right\}$.

(c) What are the eigenvalues of $\mathbf{A}$? Is $\mathbf{A}$ diagonalizable? Why or why not?

30. Perform a complete eigenanalysis on $\mathbf{A} = \begin{pmatrix} 1 & 3 \\ -3 & 1 \end{pmatrix} \in \mathbb{R}^{2,2}$. Show that $\mathbf{A}$ is diagonalizable over $\mathbb{C}$. Compute a basis $\mathcal{U}$ of $\mathbb{C}^2$ that represents $\mathbf{A}$ diagonally.

31. Repeat Example 2 with the starting vector $\mathbf{w} = \begin{pmatrix} 2 \\ -1 \end{pmatrix}$.

32. Decide whether $\mathbf{J} = \begin{pmatrix} 1 & 0 \\ 1 & 1 \end{pmatrix}$ is diagonalizable.

33. (a) Which of the vectors $\mathbf{u}_1 = \begin{pmatrix} 2 \\ 3 \end{pmatrix}$, $\mathbf{u}_2 = \begin{pmatrix} 1.5 \\ 1 \end{pmatrix}$, $\mathbf{u}_3 = \begin{pmatrix} 0 \\ 0 \end{pmatrix}$, $\mathbf{u}_4 = \begin{pmatrix} 1 \\ -1 \end{pmatrix}$, $\mathbf{u}_5 = \begin{pmatrix} 1 \\ -2 \end{pmatrix}$, $\mathbf{u}_6 = \begin{pmatrix} 3 \\ 2 \end{pmatrix}$, $\mathbf{u}_7 = \begin{pmatrix} -1 \\ -2 \end{pmatrix}$, and $\mathbf{u}_8 = \begin{pmatrix} 4 \\ -3 \end{pmatrix}$ are eigenvectors of the matrix $\mathbf{C} = \begin{pmatrix} 2 & -3 \\ -4 & 6 \end{pmatrix}$?

(b) Is the matrix $\mathbf{C}$ diagonalizable? What are its eigenvalues?

(c) Is the matrix $\mathbf{C}$ invertible? If so, what is $\mathbf{C}^{-1}$?

34. If we were to allow the zero vector in the matrix eigenvalue–eigenvector definition, how many eigenvalues would each matrix $\mathbf{A}_{nn}$ have, and what would they be?

35. For which $a \in \mathbb{R}$, does the matrix $\mathbf{A} = \begin{pmatrix} 1 & 1 & a \\ -1 & 0 & 2 \\ 2 & 1 & 4 \end{pmatrix}$ have the eigenvalue $\lambda = 2$, if at all?

36. Show that a block triangular matrix $\mathbf{U} = \begin{pmatrix} \mathbf{A} & \mathbf{O} \\ \mathbf{B} & \mathbf{C} \end{pmatrix}$ with $\mathbf{A} \in \mathbb{R}^{m,m}$ and $\mathbf{C} \in \mathbb{R}^{k,k}$ has the eigenvalues of $\mathbf{A}$ and $\mathbf{C}$ as its eigenvalues. (*Hint*: Refer to Problem 16 in Section 6.2.P.)

37. If $\mathbf{A}$ and $\mathbf{B} \in \mathbb{R}^{n,n}$ are both nonsingular, show that the matrix products $\mathbf{AB}$ and $\mathbf{BA}$ are similar. Conclude that for two nonsingular matrices the eigenvalues of $\mathbf{AB}$ and $\mathbf{BA}$ coincide.

38. Show that the two matrix products $\mathbf{BA}$ and $\mathbf{AB}$ have the same eigenvalues for any two $n \times n$ matrices $\mathbf{A}$ and $\mathbf{B}$.

39. Are $\mathbf{A} = \begin{pmatrix} 2 & 1 \\ 1 & 4 \end{pmatrix}$ and $\mathbf{B} = \begin{pmatrix} 6 & 7 \\ -1 & 0 \end{pmatrix}$ similar matrices?

40. Show that if two $n \times n$ matrices $\mathbf{A}$ and $\mathbf{B}$ are similar to the same matrix $\mathbf{C}_{nn}$, then there is a nonsingular matrix $\mathbf{W}_{nn}$ with $\mathbf{W}^{-1}\mathbf{AW} = \mathbf{B}$, that is, $\mathbf{A}$ and $\mathbf{B}$ are similar.

41. Investigate the validity of the matrix identity $(\mathbf{I} - \mathbf{A})^{-1} + (\mathbf{I} - \mathbf{A}^{-1})^{-1} = \mathbf{I}$ for nonsingular matrices $\mathbf{A}$ for which each of the inverses $(\mathbf{I} - \mathbf{A})^{-1}$ and $(\mathbf{I} - \mathbf{A}^{-1})^{-1}$ exists:

(a) Show that if $\mathbf{A}$ satisfies this matrix identity and if $\mathbf{X}$ is nonsingular, then $\mathbf{X}^{-1}\mathbf{AX}$ also satisfies the identity.

(b) Show that the matrix identity holds for all diagonalizable matrices $\mathbf{A}$ for which the various inverses can be formed.

(c) Check the truth of the identity for $\mathbf{A} = \begin{pmatrix} 2 & 1 \\ 0 & 1 \end{pmatrix}$.

Teacher's Problem Making Exercise (repeated in Section 9.1.D.P)

T 9. For ease with hand computations, one must learn to construct small integer matrices with integer eigenvalues and integer eigenvectors that may or may not be diagonalizable.

If $\mathbf{X}_{nn}$ is unimodular, then both $\mathbf{X}$ and $\mathbf{X}^{-1}$ are integer matrices. (See Problem T 6) If, moreover, $\mathbf{D}_{nn}$ is integer and diagonal, then the eigenvalues of $\mathbf{A} = \mathbf{X}\mathbf{D}\mathbf{X}^{-1}$ are the entries on the diagonal of $\mathbf{D}$, and its eigenvectors can be chosen as the columns of $\mathbf{X}$. Clearly $\mathbf{A}$ is diagonalizable.

If on the other hand, $\mathbf{J}_{nn}$ is an integer **Jordan block matrix** that is not diagonal, then $\mathbf{B} = \mathbf{X}\mathbf{J}\mathbf{X}^{-1}$ is an integer matrix that has the same eigenvalues as $\mathbf{J}$, some of which are repeated. The matrix $\mathbf{B}$ is not diagonalizable by construction, since $\mathbf{J}$ is not.

EXAMPLE We use the unimodular matrices $\mathbf{X} = \begin{pmatrix} 13 & 7 & 2 \\ 0 & -1 & -2 \\ -1 & 0 & 1 \end{pmatrix}$ and $\mathbf{Y} = \begin{pmatrix} 1 & 7 & 2 \\ 2 & 13 & 2 \\ -1 & -7 & -1 \end{pmatrix}$ from Problem T 6.

(a) For $\mathbf{D} = \begin{pmatrix} 1 & & \\ & -1 & \\ & & 1 \end{pmatrix}$, we know that $\mathbf{A}_1 = \mathbf{X}\mathbf{D}\mathbf{X}^{-1} = \begin{pmatrix} 29 & 210 & 364 \\ -4 & -29 & -52 \\ 0 & 0 & 1 \end{pmatrix}$ and $\mathbf{A}_2 = \mathbf{Y}\mathbf{D}\mathbf{Y}^{-1} = \begin{pmatrix} 1 & 14 & 28 \\ 0 & 27 & 52 \\ 0 & -14 & -27 \end{pmatrix}$ are both diagonalizable with eigenvalues 1, 1, −1, the diagonal entries of $\mathbf{D}$, and corresponding eigenvectors in the columns of $\mathbf{X}$ and $\mathbf{Y}$, respectively. Note that a hand computation of the eigendata for $\mathbf{A}_1$ looks more forbidding than for $\mathbf{A}_2$ in this example.

(b) To obtain nondiagonalizable integer matrices $\mathbf{A}$, let us work with $\mathbf{J} = \begin{pmatrix} 1 & 1 & \\ 0 & 1 & \\ & & -1 \end{pmatrix}$ and $\mathbf{Y}$. Then, $\mathbf{B} = \mathbf{Y}\mathbf{J}\mathbf{Y}^{-1} = \begin{pmatrix} -3 & -1 & -6 \\ -4 & -1 & -8 \\ 2 & 1 & 5 \end{pmatrix}$ and $\mathbf{B}$'s eigenvalues compute as 1 (double) and −1. However, the eigenvalue 1 of $\mathbf{B}$ has index 2, and there is an eigenvector deficiency for this double eigenvalue that makes $\mathbf{B}$ not diagonalizable. ◀

[*Note:* It is important to experiment with a number of examples so that one can find a matrix with integer eigenvalues and small integer entries for easy hand computations. MATLAB helps instructors to construct meaningful problems.]

9.1.D Lecture Nine, Using Determinants, a Double Lecture

Characteristic Polynomial, Matrix Eigenanalysis, and Diagonalizable Matrices

This lecture relies on Chapter 8 and determinants.

We learn to extract intrinsic information of linear transformations and matrices, namely, their eigenvalues and eigenvectors.

(a) The Definition of Eigenvalues and Eigenvectors

We continue where lecture seven has left off.

Given a real square matrix $\mathbf{A} = \mathbf{A}_{\mathcal{E}} \in \mathbb{R}^{n,n}$ that represents a linear transformation $T : \mathbb{R}^n \to \mathbb{R}^n$ with respect to the standard unit-vector basis $\mathcal{E} = \{\mathbf{e}_1, \ldots, \mathbf{e}_n\}$ of $\mathbb{R}^n$, we want to find another basis $\mathcal{U} = \{\mathbf{u}_1, \ldots, \mathbf{u}_n\}$ of $\mathbb{R}^n$ (or $\mathbb{C}^n$) that gives us a simpler matrix representation $\mathbf{A}_{\mathcal{U}}$ of T. That is, we are looking for a basis $\mathcal{U}$ for which the matrix in Eq. (7.4) $\mathbf{A}_{\mathcal{U}} = \mathbf{U}^{-1}\mathbf{A}_{\mathcal{E}}\mathbf{U}$ is a simpler matrix than $\mathbf{A}_{\mathcal{E}}$. In order to do so, we need to extract **intrinsic geometric qualities** of the linear transformation T, or, equivalently, of its standard matrix representation $\mathbf{A}_{\mathcal{E}}$.

It is not apparent at this moment what exactly the notion of a "simpler" matrix means, nor what the "intrinsic properties" of a linear transformation might be.

In light of Section 7.1, let us study specific **similarity transformations** $\mathbf{A}_{\mathcal{U}} = \mathbf{U}^{-1}\mathbf{A}_{\mathcal{E}}\mathbf{U}$ of the standard matrix representation $\mathbf{A}_{\mathcal{E}}$ of T. If $\mathbf{A}_{\mathcal{U}} = \mathbf{U}^{-1}\mathbf{A}_{\mathcal{E}}\mathbf{U}$, or, equivalently, if $\mathbf{A}_{\mathcal{E}}\mathbf{U} = \mathbf{U}\mathbf{A}_{\mathcal{U}}$ for a nonsingular matrix $\mathbf{U}$ <u>and</u> if $\mathbf{A}_{\mathcal{U}}$ is a **diagonal matrix**

$$\mathbf{A}_{\mathcal{U}} = \operatorname{diag}(\lambda_1, \ldots, \lambda_n) = \begin{pmatrix} \lambda_1 & & 0 \\ & \ddots & \\ 0 & & \lambda_n \end{pmatrix}$$

with $\lambda_i \in \mathbb{R}$ or $\mathbb{C}$, then $\mathbf{A}_{\mathcal{U}}$ is quite 'simple'. In this case, we have

$$\begin{aligned} \mathbf{A}_{\mathcal{E}}\mathbf{U} &= \mathbf{A}_{\mathcal{E}} \begin{pmatrix} | & & | \\ \mathbf{u}_1 & \cdots & \mathbf{u}_n \\ | & & | \end{pmatrix} \\ &= \begin{pmatrix} | & & | \\ \mathbf{u}_1 & \cdots & \mathbf{u}_n \\ | & & | \end{pmatrix} \begin{pmatrix} \lambda_1 & & 0 \\ & \ddots & \\ 0 & & \lambda_n \end{pmatrix} = \mathbf{U}\mathbf{A}_{\mathcal{U}} \end{aligned}$$

or

$$\begin{pmatrix} | & & | \\ \mathbf{A}_{\mathcal{E}}\mathbf{u}_1 & \cdots & \mathbf{A}_{\mathcal{E}}\mathbf{u}_n \\ | & & | \end{pmatrix} = \begin{pmatrix} | & & | \\ \lambda_1\mathbf{u}_1 & \cdots & \lambda_n\mathbf{u}_n \\ | & & | \end{pmatrix}.$$

Thus, for $\mathbf{A}_{\mathcal{U}}$ to be most simply just diagonal, we must be able to find n linearly independent vectors $\mathbf{u}_i \in \mathbb{R}^n$ (or in $\mathbb{C}^n$) and scalars λ_i with $\mathbf{A}\mathbf{u}_i = \lambda_i\mathbf{u}_i$ for $i = 1, \ldots, n$. Here, the set of columns $\{\mathbf{u}_1, \ldots, \mathbf{u}_n\}$ of U has to form a basis of $\mathbb{R}^n$ (or $\mathbb{C}^n$) for invertibility in Eq. (7.4) according to Section 6.1.

Consequently, if we want to find a simple and intrinsic representation $\mathbf{A}_{\mathcal{U}}$ of T (or of $\mathbf{A}_{\mathcal{E}}$), we must look for **eigenvalues** and **eigenvectors** of $\mathbf{A}_{\mathcal{E}}$.

DEFINITION 1 A scalar $\lambda \in \mathbb{R}$ (or $\mathbb{C}$) and a nonzero vector $\mathbf{u} \in \mathbb{R}^n$ (or $\mathbb{C}^n$) is called an **eigenvalue–eigenvector pair** for the square matrix $\mathbf{A}_{nn}$ if

$$\mathbf{Au} = \lambda\mathbf{u}. \tag{9.1}$$

In this case, we say that $\mathbf{u} \neq \mathbf{0}$ is an **eigenvector** for the **eigenvalue** $\lambda \in \mathbb{R}$ (or $\mathbb{C}$) of $\mathbf{A}$.

To search for eigenvalue–eigenvector pairs of a given square matrix $\mathbf{A} = \mathbf{A}_{\mathcal{E}}$ is a **nonlinear problem**. The unknown quantities λ and $\mathbf{u}$ occur in multiplied form on the right-hand side of Eq. (9.1). In this nonlinear problem, our knowledge of how to solve linear equations $\mathbf{Ax} = \mathbf{b}$ that has helped us nicely in each chapter so far is of no use for finding matrix eigenvalues. It is a sad fact that starting an eigenanalysis of a square matrix $\mathbf{A}$ with a row reduction of $\mathbf{A}$ normally nets the student zero points for the eigenvalue problem. To illustrate, note that the RREF of all nonsingular square matrices is the same, namely, equal to $\mathbf{I}_n$. What information can be gleamed from $\mathbf{R} = \mathbf{I}_n$ that gives information about every nonsingular matrix $\mathbf{A}$, other than its rank? None! On the other hand, however, zero is an eigenvalue of each singular matrix, since for a singular matrix $\mathbf{A}_{nn}$ we have $\ker(\mathbf{A}) \neq \{\mathbf{0}\}$; that is, there is a vector $\mathbf{x} \neq \mathbf{0}$ with $\mathbf{Ax} = \mathbf{0} = 0\mathbf{x}$. But no other eigenvalue of $\mathbf{A}$ can be ever found from a REF of $\mathbf{A}$.

To solve the matrix eigenproblem, we rewrite the **eigenvalue–eigenvector equation** (9.1) with $\mathbf{u} \neq \mathbf{0}$ as follows using the determinant function of Chapter 8:

$$\begin{aligned}
\mathbf{Au} = \lambda\mathbf{u} \iff\ & \mathbf{Au} - \lambda\mathbf{u} = \mathbf{0} && \iff \lambda\mathbf{u} - \mathbf{Au} = \mathbf{0} && \text{for } \mathbf{u} \neq \mathbf{0} \\
\iff\ & \lambda\mathbf{Iu} - \mathbf{Au} = \mathbf{0} && \iff (\lambda\mathbf{I} - \mathbf{A})\mathbf{u} = \mathbf{0} && \text{for some } \mathbf{u} \neq \mathbf{0} \\
\iff\ & \ker(\lambda\mathbf{I} - \mathbf{A}) \neq \{\mathbf{0}\} && \iff \lambda\mathbf{I} - \mathbf{A} && \text{is singular} \\
\iff\ & \det(\lambda\mathbf{I} - \mathbf{A}) = 0. && && (\textbf{characteristic equation})
\end{aligned} \tag{9.2.D}$$

This relates the task of finding eigenvalues (and subsequently eigenvectors) of a matrix $\mathbf{A}_{nn}$ to solving a determinantal equation for $\mathbf{A}$. For an alternative determinant-free approach, see Lecture 9.1.

Remark:

If, for example, $\mathbf{R}$ is an upper triangular $n \times n$ matrix, then $\mathbf{R} - r_{ii}\mathbf{I}_n$ is singular for each diagonal entry r_{ii} of $\mathbf{R}$. That is, an upper triangular matrix has all of its eigenvalues on the diagonal. The same holds for lower triangular matrices $\mathbf{L}$ as well.

(b) The Characteristic Polynomial of a Matrix

DEFINITION 2.D The determinantal expression $f_A(\lambda) = \det(\lambda\mathbf{I}_n - \mathbf{A})$ is called the **characteristic polynomial** of the matrix $\mathbf{A}_{nn}$.

The characteristic polynomial of a matrix $\mathbf{A}_{nn}$ has the following properties:

Proposition 1.D

(a) For a square matrix $\mathbf{A} \in \mathbb{R}^{n,n}$ (or $\mathbb{C}^{n,n}$), the characteristic polynomial $f_A(\lambda) = \det(\lambda\mathbf{I} - \mathbf{A})$ is a polynomial in λ of degree n.

(b) The characteristic polynomial $f_A(\lambda)$ has real coefficients if $\mathbf{A} \in \mathbb{R}^{n,n}$. If $\mathbf{A} \in \mathbb{R}^{n,n}$, then $f_A(x)$ has real roots and possibly pairs of complex conjugate roots $\lambda = a + bi$ and $\overline{\lambda} = a - bi$ for $a, b \in \mathbb{R}$ and $i = \sqrt{-1} \in \mathbb{C}$.

(c) The roots of $f_A(\lambda)$ lie in $\mathbb{C}$ and are the eigenvalues of $\mathbf{A}$.

(d) The characteristic polynomials $f_A(\lambda)$ and $f_{U^{-1}AU}(\lambda)$ are identical for every nonsingular matrix $\mathbf{U}_{nn}$ (i.e., $f_A(\lambda)$ is invariant under basis change). ◀

Proof Part (a) follows by Laplace expansion of $\det(\lambda\mathbf{I} - \mathbf{A})$. We prove part (d) as follows by using the determinant rules $\det(\mathbf{X})\det(\mathbf{X}^{-1}), = 1$ and $\det(\mathbf{A})\det(\mathbf{B}) = \det(\mathbf{AB})$ from the determinant Proposition of Section 8.1:

$$\begin{aligned} f_A(\lambda) &= \det(\lambda\mathbf{I} - \mathbf{A}) = \det(\mathbf{U}^{-1})\det(\lambda\mathbf{I} - \mathbf{A})\det(\mathbf{U}) \\ &= \det(\mathbf{U}^{-1}(\lambda\mathbf{I} - \mathbf{A})\mathbf{U}) = \det(\lambda\mathbf{U}^{-1}\mathbf{U} - \mathbf{U}^{-1}\mathbf{AU}) \\ &= \det(\lambda\mathbf{I} - \mathbf{U}^{-1}\mathbf{AU}) = f_{U^{-1}AU}(\lambda). \end{aligned}$$ ■

All of our further studies of the matrix simplification problem, or of the nonlinear matrix eigenvalue–eigenvector problem, hinge on the **Fundamental Theorem of Algebra**.

Theorem 9.1.D **(Fundamental Theorem of Algebra)** Every polynomial

$$p(x) = x^n + a_{n-1}x^{n-1} + \cdots + a_1 x + a_0$$

of degree n with real or complex coefficients a_i has n **roots** $x_1, \ldots, x_n$ with $p(x_i) = 0$ in the complex plane $\mathbb{C}$. Here repeated roots are counted with their multiplicity. In particular, $p(x) = (x - x_1)\cdots(x - x_n)$ for the (possibly repeated) roots $x_i \in \mathbb{C}$. ◀

If the repeated roots of a polynomial p of degree n are collected, or

$$p(x) = a_n(x - x_1)^{n_1}(x - x_2)^{n_2}\cdots(x - x_j)^{n_j}$$

for distinct roots x_i with $n_i > 0$ for all i and $n_1 + n_2 + \ldots + n_j = n$, then we say that the root x_i of p has the **algebraic multiplicity** n_i for each $i = 1, \ldots, j$.

For square matrices, we say analogously that an eigenvalue λ_i of $\mathbf{A}$ has the **algebraic multiplicity** n_i if the characteristic polynomial $f_A(\lambda) = \det(\lambda\mathbf{I} - \mathbf{A})$ factors as $f_A(\lambda) = (\lambda - \lambda_1)^{n_1}\cdots(\lambda - \lambda_k)^{n_k}$ over $\mathbb{C}$ for the distinct eigenvalues λ_i of A.

According to the definition of matrix eigenvalues as roots of a polynomial in Eq. (9.2.D) and to the Fundamental Theorem of Algebra, real matrix eigenvalue problems can generally only be solved over the complex field $\mathbb{C}$. Therefore, we have included an appendix on complex numbers and complex vectors, as well as one on finding roots of polynomials.

(c) Finding Eigenvalues and Eigenspaces

We start with an example.

EXAMPLE 1.D Find the eigenvalues of $\mathbf{A} = \begin{pmatrix} 2 & 1 \\ -1 & 4 \end{pmatrix}$. To do so, we look at the characteristic polynomial of $\mathbf{A}$:

$$\begin{aligned} f_A(\lambda) = \det(\lambda\mathbf{I} - \mathbf{A}) &= \det\left(\begin{pmatrix} \lambda & 0 \\ 0 & \lambda \end{pmatrix} - \begin{pmatrix} 2 & 1 \\ -1 & 4 \end{pmatrix}\right) \\ &= \det\begin{pmatrix} \lambda - 2 & -1 \\ 1 & \lambda - 4 \end{pmatrix} \\ &= (\lambda - 2)(\lambda - 4) + 1 = \lambda^2 - 6\lambda + 9 \\ &= (\lambda - 3)(\lambda - 3). \end{aligned}$$

Thus, $\lambda = 3$ is a double root of $f_A(\lambda)$, $\mathbf{A}$ has the double real eigenvalue 3, or for $\mathbf{A}$ the algebraic multiplicity of the eigenvalue $\lambda = 3$ is two. ◀

Having been able to find the intrinsic values (or eigenvalues) of the matrix $\mathbf{A}$ from its characteristic polynomial $f_A(\lambda)$, we are faced with finding the associated eigenvector(s) of $\mathbf{A}$ in order to name a specific basis $\mathcal{U}$ of $\mathbb{R}^n$ in which $\mathbf{A}_{\mathcal{U}} = \mathbf{U}^{-1}\mathbf{A}\mathbf{U}$ is in simple diagonal form, if this is possible. Finding the eigenvalues of a matrix is a nonlinear problem; however, finding the corresponding eigenvectors is a linear one.

EXAMPLE 2.D Find all eigenvectors for the double eigenvalue $\lambda = 3$ of $\mathbf{A} = \begin{pmatrix} 2 & 1 \\ -1 & 4 \end{pmatrix}$ from Example 1.D. From the equivalence (9.2.D) following Definition 1, we know that every eigenvector $\mathbf{u} \neq \mathbf{0}$ for the eigenvalue 3 of $\mathbf{A}$ must satisfy $(3\mathbf{I} - \mathbf{A})\mathbf{u} = \mathbf{0}$, or

$$\mathbf{u} \in \ker(3\mathbf{I} - \mathbf{A}) = \ker(\mathbf{A} - 3\mathbf{I}).$$

Hence, for $\lambda = 3$ we compute α REF:

$$\mathbf{A} - 3\mathbf{I}: \quad \begin{array}{cc|cl} \boxed{-1} & 1 & 0 & \\ \boxed{-1} & 1 & 0 & -row_1 \; . \\ \hline \boxed{-1} & 1 & 0 & \\ 0 & 0 & 0 & \\ & \uparrow & & \end{array}$$

There is one free variable in the REF of $\mathbf{A} - 3\mathbf{I}$. Therefore, there is one linearly independent eigenvector $\mathbf{u} = \begin{pmatrix} 1 \\ 1 \end{pmatrix}$ (and all of its nonzero multiples) for the double eigenvalue $\lambda = 3$ of $\mathbf{A}$.

Clearly, there is something amiss for $\mathbf{A}$. We cannot find enough eigenvectors (two are needed) for this matrix $\mathbf{A} \in \mathbb{R}^{2,2}$ to diagonalize it. Hence, $\mathbf{A}$ is not diagonalizable. ◀

Next, we define the space of eigenvectors or the **eigenspace** associated with a matrix eigenvalue.

DEFINITION 3.D (a) For a given matrix $\mathbf{A} \in \mathbb{R}^{n,n}$ (or $\mathbb{C}^{n,n}$) and each scalar $\lambda \in \mathbb{R}$ (or $\mathbb{C}$), we define the set

$$E(\lambda) := \ker(\mathbf{A} - \lambda\mathbf{I}) = \{\mathbf{x} \mid \mathbf{A}\mathbf{x} = \lambda\mathbf{x}\} \subset \mathbb{R}^n \text{ (or } \mathbb{C}^n).$$

(b) If λ is an eigenvalue of $\mathbf{A}_{nn}$, then

$$E(\lambda) = \ker(\mathbf{A} - \lambda\mathbf{I}) = \{\mathbf{x} \mid \mathbf{A}\mathbf{x} = \lambda\mathbf{x}\} \neq \{\mathbf{0}\}$$

is called the **eigenspace** for the eigenvalue λ of $\mathbf{A}$.

(c) If λ is an eigenvalue of $\mathbf{A}_{nn}$, then the integer $\dim(E(\lambda)) \geq 1$ is called the **geometric multiplicity** of λ.

The set $E(\lambda)$ is a subspace of $\mathbb{R}^n$ (or $\mathbb{C}^n$) for any scalar $\lambda \in \mathbb{C}$, since it is defined as the kernel of a matrix, see Section 4.1. Some simple properties of eigenvalues, eigenvectors, and the subspaces $E(\lambda)$ for a matrix $\mathbf{A}_{nn}$ are as follows:

Proposition 2.D Let $\mathbf{A}$ be an $n \times n$ real or complex matrix. Then, the following hold:

(a) $E(\lambda) = \{\mathbf{0}\} \iff$ the scalar λ is not an eigenvalue of $\mathbf{A}$.
(b) $E(\lambda)$ is a subspace of $\mathbb{R}^n$ (or $\mathbb{C}^n$) for each scalar λ.
(c) If $\mathbf{x} \in E(\lambda)$ and $\mathbf{y} \in E(\mu)$ for two distinct eigenvalues $\lambda \neq \mu$ of $\mathbf{A}$ and corresponding eigenvectors $\mathbf{x} \neq \mathbf{0}$, $\mathbf{y} \neq \mathbf{0}$, then $\mathbf{x}$ and $\mathbf{y}$ are linearly independent.
(d) If $\mathbf{A}$ has the eigenvalue λ for the eigenvector $\mathbf{x} \neq \mathbf{0} \in \mathbb{R}^n$ (or $\mathbb{C}^n$), then the matrix $\mathbf{U}^{-1}\mathbf{A}\mathbf{U}$ has the same eigenvalue λ as $\mathbf{A}$ for the transformed eigenvector $\mathbf{U}^{-1}\mathbf{x}$ and any nonsingular matrix $\mathbf{U} \in \mathbb{R}^{n,n}$. ◀

Note particularly in part (d) that $\mathbf{A}_{\mathcal{U}} = \mathbf{U}^{-1}\mathbf{A}\mathbf{U}$ and $\mathbf{x}_{\mathcal{U}} = \mathbf{U}^{-1}\mathbf{x}_{\mathcal{E}}$ according to Section 7.1. In other words, an eigenvector $\mathbf{x} = \mathbf{x}_{\mathcal{E}}$ of $\mathbf{A} = \mathbf{A}_{\mathcal{E}}$ is transformed to $\mathbf{x}_{\mathcal{U}}$ for $\mathbf{A}_{\mathcal{U}}$, while its eigenvalue does not change.

Proof Parts (a) and (b) are obvious from Definition 3.

For part (c), assume that an equation

$$(\text{E1}): \quad \alpha\mathbf{x} + \beta\mathbf{y} = \mathbf{0}$$

holds for some scalars α, β, where $\mathbf{x} \in E(\lambda)$ and $\mathbf{y} \in E(\mu)$ are two eigenvectors of $\mathbf{A}$. Then $\mathbf{A}(\alpha\mathbf{x} + \beta\mathbf{y}) = \mathbf{A}\mathbf{0} = \mathbf{0}$ by linearity. On the other hand, according to the eigenvalue–eigenvector equations for $\mathbf{A}$, λ, and μ, we have

$$\mathbf{A}(\alpha\mathbf{x} + \beta\mathbf{y}) = \alpha\mathbf{A}\mathbf{x} + \beta\mathbf{A}\mathbf{y} = \alpha\lambda\mathbf{x} + \beta\mu\mathbf{y}.$$

Thus, the equation

$$(\text{E2}); \quad \alpha\lambda\mathbf{x} + \beta\mu\mathbf{y} = \mathbf{0}$$

must hold for $\mathbf{x}$ and $\mathbf{y}$. We form the derived vector equation

$$\begin{aligned}\lambda \cdot (\text{E1}) - (\text{E2}): \quad & \lambda\alpha\mathbf{x} + \lambda\beta\mathbf{y} - \alpha\lambda\mathbf{x} - \beta\mu\mathbf{y} = \mathbf{0} \\ & \beta\lambda\mathbf{y} - \beta\mu\mathbf{y} = \mathbf{0} \\ & \beta(\lambda - \mu)\mathbf{y} = \mathbf{0},\end{aligned}$$

implying that $\beta = 0$, since $\lambda \neq \mu$ by assumption and $\mathbf{y} \neq \mathbf{0}$ for an eigenvector. Going back to (E1), we have $\alpha\mathbf{x} = \mathbf{0}$, or $\alpha = 0$, because $\mathbf{x} \neq \mathbf{0}$. Thus, $\mathbf{x}$ and $\mathbf{y}$ are linearly independent by Definition 2(a) of Section 5.1.

In part (d), assume that $\mathbf{Ax} = \lambda\mathbf{x}$ for $\mathbf{x} \neq \mathbf{0}$. Then $\mathbf{Ax} = \mathbf{AUU}^{-1}\mathbf{x} = \lambda\mathbf{x}$, since $\mathbf{UU}^{-1} = \mathbf{I}$. Thus

$$\mathbf{U}^{-1}\mathbf{AU}(\mathbf{U}^{-1}\mathbf{x}) = \mathbf{U}^{-1}\lambda\mathbf{x} = \lambda\mathbf{U}^{-1}\mathbf{x}.$$

That is, the matrix $\mathbf{U}^{-1}\mathbf{AU}$ has the eigenvalue λ for its eigenvector $\mathbf{U}^{-1}\mathbf{x} \neq \mathbf{0}$. ■

Let us gather all steps for a complete theoretical eigenvalue–eigenvector analysis of a given matrix $\mathbf{A}_{nn}$ via determinants now.

Steps for a complete eigenanalysis of a real matrix $\mathbf{A}_{nn}$ via determinants

1. Form the characteristic polynomial $f_{\mathbf{A}}(\lambda) = \det(\lambda\mathbf{I} - \mathbf{A})$ of degree n. (Keep the polynomial in factored form as much as possible).

   (One can **check** three coefficients of $f_{\mathbf{A}}(\lambda)$ quickly for accuracy: λ^n always has the coefficient one, λ^{n-1} must have the negative sum $-\sum_i a_{ii}$ of the diagonal entries a_{ii} of $\mathbf{A}$ as its coefficient, while the constant coefficient of $f_A(\lambda)$ is equal to $(-1)^n \det(\mathbf{A})$, all according to Theorem 9.5 in the next Section.)
2. Completely factor $f_A(\lambda)$ as $f_A(\lambda) = (\lambda - \lambda_1)^{n_1} \cdots (\lambda - \lambda_k)^{n_k}$. (In other words, find all distinct real or complex conjugate roots λ_i of $f_A(\lambda)$ and their multiplicities n_i. See Appendix B for details.)

   The roots of $f_A(\lambda)$, repeated as needed, are the eigenvalues of $\mathbf{A}$.
3. Find the corresponding eigenvectors u_i by solving $(\mathbf{A} - \lambda_i\mathbf{I})\mathbf{u}_i = \mathbf{0}$ for nonzero vectors $\mathbf{u}_i$ and each distinct $\lambda_i, i = 1, \ldots, k$ in turn. If there are less than n_i linearly independent eigenvectors for one eigenvalue λ_i, then $\mathbf{A}$ cannot be diagonalized.

   (When looking for a basis of an eigenspace $E(\lambda_j)$, note that if the row-echelon form of the matrix $\mathbf{A}-\lambda_j\mathbf{I}$ does not have at least one free variable, then $\ker(\mathbf{A} - \lambda_j\mathbf{I}) = \{\mathbf{0}\}$ and λ_j is **not** an eigenvalue of $\mathbf{A}$. See Proposition 2(a). **Recheck** parts 1 and 2. There is an **error** in parts 1 or 2, or in both.)
4. If you were able to find n linearly independent eigenvectors $\mathbf{u}_i$ for $\mathbf{A}_{nn}$ in part 3, form the eigenvector basis $\mathcal{U} = \{\mathbf{u}_1, \ldots, \mathbf{u}_n\}$ for $\mathbf{A}$ and the diagonal matrix representation $\mathbf{A}_{\mathcal{U}}$ of $\mathbf{A}$ with respect to the basis $\mathcal{U}$ as follows:

$$\mathbf{U} = \begin{pmatrix} | & & | \\ \mathbf{u}_1 & \cdots & \mathbf{u}_n \\ | & & | \end{pmatrix}, \quad \mathbf{A}_{\mathcal{U}} = \begin{pmatrix} \lambda_1 & & 0 \\ & \ddots & \\ 0 & & \lambda_n \end{pmatrix}.$$

   (Note that the order of the eigenvectors $\mathbf{u}_i$ as columns of $\mathbf{U}$ determines the order of the eigenvalues λ_i of $\mathbf{A}$ that appear on the diagonal of $\mathbf{A}_{\mathcal{U}}$ and vice versa. (You can swap the eigenvectors $\mathbf{u}_3$ and $\mathbf{u}_5$ in the columns of $\mathbf{U}$ only when simultaneously swapping λ_3 and λ_5 in $\mathbf{A}_{\mathcal{U}}$, for example.))

Now, $\boxed{\mathbf{AU} = \mathbf{UA}_{\mathcal{U}}}$. This matrix equation should be checked. (This last matrix equation avoids having to deal with the matrix inverse $\mathbf{U}^{-1}$ of $\mathbf{U}$ and matrix triple products in the more troublesome but equivalent matrix similarity equation (7.4) $\mathbf{A}_{\mathcal{U}} = \mathbf{U}^{-1}\mathbf{AU}$.)

EXAMPLE 3.D Perform a complete eigenanalysis of $\mathbf{B} = \begin{pmatrix} 1 & 1 \\ -2 & 4 \end{pmatrix}$ and diagonalize $\mathbf{B}$ if possible by following the procedure just described.

1. Characteristic polynomial of $\mathbf{B}$:

$$f_B(\lambda) = \det(\lambda\mathbf{I} - \mathbf{B}) = \det\begin{pmatrix} \lambda - 1 & -1 \\ 2 & \lambda - 4 \end{pmatrix} = (\lambda - 1)(\lambda - 4) + 2$$
$$= \lambda^2 - 5\lambda + 6.$$

2. Eigenvalues: $f_B(\lambda) = (\lambda - 2)(\lambda - 3)$, so that the eigenvalues of $\mathbf{B}$ are 2 and 3.
3. Eigenvectors: For $\lambda_1 = 2$:

$$\mathbf{B} - 2\mathbf{I}: \quad \begin{array}{cc|c} \boxed{-1} & 1 & 0 \\ \boxed{-2} & 2 & 0 \\ \hline \boxed{-1} & 1 & 0 \\ 0 & 0 & 0 \\ & \uparrow & \end{array} \quad \begin{array}{l} \\ -2\,row_1 \\ \\ \\ \\ \end{array}$$

Thus, $\mathbf{u}_1 = \begin{pmatrix} 1 \\ 1 \end{pmatrix}$ is a basis for the eigenspace $E(2)$ of $\mathbf{B}$.

(*Note:* that the row reduction of $\mathbf{B} - 2\mathbf{I}$ must exhibit a free variable for the computed root $\lambda = 2$ of $f_B(\lambda)$ to be an eigenvalue of $\mathbf{B}$.)

For $\lambda_2 = 3$,

$$\mathbf{B} - 3\mathbf{I}: \quad \begin{array}{cc|c} \boxed{-2} & 1 & 0 \\ \boxed{-2} & 1 & 0 \\ \hline \boxed{-2} & 1 & 0 \\ 0 & 0 & 0 \\ & \uparrow & \end{array} \quad \begin{array}{l} \\ -row_1 \\ \\ \\ \\ \end{array}$$

Thus, $\mathbf{u}_2 = \begin{pmatrix} 1 \\ 2 \end{pmatrix}$ spans the eigenspace $E(3)$ of $\mathbf{B}$.

4. There are two linearly independent eigenvectors $\mathbf{u}_1$ and $\mathbf{u}_2$ for the 2×2 matrix $\mathbf{B}$, making $\mathbf{B}$ diagonalizable with respect to its eigenvector basis $\mathbf{u}_1$ and $\mathbf{u}_2$. These we assemble in the columns of

$$\mathbf{U} = \begin{pmatrix} | & | \\ \mathbf{u}_1 & \mathbf{u}_2 \\ | & | \end{pmatrix} = \begin{pmatrix} 1 & 1 \\ 1 & 2 \end{pmatrix}.$$

The diagonal representation $\mathbf{B}_{\mathcal{U}}$ of $\mathbf{B}$ with respect to this ordered basis is

$$\mathbf{B}_{\mathcal{U}} = \begin{pmatrix} 2 & 0 \\ 0 & 3 \end{pmatrix} (= \mathbf{U}^{-1}\mathbf{B}\mathbf{U}),$$

with $\mathbf{B}$'s eigenvalues 2 and 3 appearing in the chosen order on the diagonal. The simple matrix identity $\mathbf{BU} = \mathbf{UB}_{\mathcal{U}}$ should be checked for the matrices $\mathbf{U}$, $\mathbf{B}_{\mathcal{U}}$, and $\mathbf{B}$ just to be sure, rather than the equivalent, but more complicated matrix-similarity equation $\mathbf{B}_{\mathcal{U}} = \mathbf{U}^{-1}\mathbf{BU}$. ◀

(d) Diagonalizable Matrices

Square matrices may or may not be diagonalizable as we have seen. When trying to diagonalize a matrix, we must keep in mind the following simple fact:

Theorem 9.2.D A matrix $\mathbf{A}_{nn}$ is diagonalizable if and only if there are n linearly independent eigenvectors in $\mathbb{C}^n$ for $\mathbf{A}$. ◀

This has the following consequence:

Corollary If $\mathbf{A}_{nn} \in \mathbb{R}^{n,n}$ (or $\mathbb{C}^{n,n}$) has n distinct eigenvalues, then $\mathbf{A}$ is diagonalizable over $\mathbb{C}$. In other words, if a matrix $\mathbf{A}_{nn}$ is not diagonalizable, then it must have repeated eigenvalues. ◀

The converse is false. Just think of the diagonal matrix $\mathbf{I}_n$ with its n-fold eigenvalue 1 on the diagonal.

> Remember that a matrix $\mathbf{A}_{nn}$ is not diagonalizable if the characteristic polynomial $f_A(\lambda)$ of $\mathbf{A}$ has a repeated factor $(\lambda - \lambda_i)^k$ for $k > 1$ and the corresponding eigenspace $E(\lambda_i)$ for $\mathbf{A}$ has dimension less than k. In other words, a square matrix $\mathbf{A}$ is not diagonalizable precisely when $\mathbf{A}$ has an eigenvalue with differing algebraic and geometric multiplicities.
>
> Nondiagonalizable matrices $\mathbf{A}$ will be studied in Chapter 14.

More on the two multiplicities of an eigenvalue is contained in Theorem 9.8.

We need to stress here that an eigenanalysis of a real matrix will normally lead us into complex arithmetic. This is a clear consequence of the Fundamental Theorem of Algebra. While the characteristic polynomial of a real matrix must have real coefficients according to Proposition 1.D(b), most real coefficient polynomials will only split into first degree factors over the complex field $\mathbb{C}$. The matrices described earlier were prepared with care to stay in the reals when looking for the roots of $f_A(\lambda)$. Thus, an eigenanalysis of a general real matrix $\mathbf{A}$ is extremely difficult to do by hand, just consider the problem of evaluating $f_A(\lambda)$ correctly, and worse, that of factoring the polynomial $f_A(\lambda)$ for any sizable degree $n > 2$; see Appendix B for more details.

Clearly, we have reached the end of what we can do by hand, other than invent computers and design software for this purpose. (And this is just what mankind has begun to perfect, since the 1940s, 50s, and 60s.)

Finally, we want to emphasize that the possible diagonalization of any matrix hinges on the noncommutativity of the matrix product: $\mathbf{AB} \neq \mathbf{BA}$ for matrices in general, see the caution in Section 6.1.

If matrix products were or could be defined differently so that they were always to commute, then the $\mathbf{U}^{-1}\cdots\mathbf{U}$ left and right multiplications would cancel in the $\mathbf{A}_{\mathcal{U}} = \mathbf{U}^{-1}\mathbf{AU}$ diagonalization similarity (7.4), or in the matrix-simplification process. And we would almost never be able to see a diagonal representation $\mathbf{A}_{\mathcal{U}}$ of a square matrix $\mathbf{A} = \mathbf{A}_{\mathcal{E}}$. In other words, without the noncommutativity of matrix multiplication, a matrix diagonalizing eigenvector basis $\mathcal{U}$ would not exist, except for those linear transformation that are already diagonal.

This is a universal fact worth contemplating: Complications, troubles, problems, and seemingly ill behavior somewhere often bear intrinsic fruit elsewhere in our human and mental endeavors.

We have learned how to use determinants and row reduction of specific matrices to solve nonlinear matrix eigenproblems in low dimensions.

9.1.D.P Problems

1. Which of the vectors $\mathbf{u}_1 = \begin{pmatrix} 1 \\ -1 \\ 1 \end{pmatrix}$, $\mathbf{u}_2 = \begin{pmatrix} 0 \\ 0 \\ 0 \end{pmatrix}$, $\mathbf{u}_3 = \begin{pmatrix} 2 \\ -1 \\ -1 \end{pmatrix}$, and $\mathbf{u}_4 = \begin{pmatrix} 1 \\ 2 \\ 0 \end{pmatrix}$ are eigenvectors for the matrix $\mathbf{A} = \begin{pmatrix} 6 & 7 & 7 \\ -7 & -8 & -7 \\ 7 & 7 & 6 \end{pmatrix}$. What are the corresponding eigenvalues, if any?

2. Find all eigenvalues and corresponding eigenvectors for the matrix $\mathbf{A} = \begin{pmatrix} 2 & 0 & 3 \\ 0 & -1 & 0 \\ 1 & 0 & 4 \end{pmatrix}$: Find the characteristic polynomial $f_A(x)$ of $\mathbf{A}$, its roots, etc. Is $\mathbf{A}$ diagonalizable? Why and how?

3. Give an example of a real 2×2 matrix with no real eigenvalue.

4. Show that $\mathbf{A} = \begin{pmatrix} 1 & 0 \\ -2 & -2 \end{pmatrix}$ is diagonalizable. Find an eigenvector basis $\mathcal{U}$ for $\mathbf{A}$ and $\mathbf{A}_{\mathcal{U}}$. Verify that $\mathbf{AU} = \mathbf{UA}_{\mathcal{U}}$.

5. Find the eigenvalues and corresponding eigenvectors for the matrices

(a) $\begin{pmatrix} 1 & 1 \\ 2 & 1 \end{pmatrix}$;

(b) $\begin{pmatrix} -1 & 8 \\ 1 & 1 \end{pmatrix}$;

(c) $\begin{pmatrix} 1 & 1 & 0 \\ 0 & 0 & 2 \\ 3 & 0 & -1 \end{pmatrix}$;

(d) $\begin{pmatrix} 1 & 2 & 3 \\ 0 & 0 & 0 \\ 0 & 0 & 2 \end{pmatrix}$;

(e) $\begin{pmatrix} 1 & 2 & 3 \\ 0 & 0 & 2 \\ 0 & 0 & 0 \end{pmatrix}$; and

(f) $\begin{pmatrix} 0 & 1 & 0 \\ 1 & 0 & -1 \\ 0 & 1 & 0 \end{pmatrix}$.

6. Consider $\mathbf{A} = \begin{pmatrix} 1 & -2 & 1 \\ -1 & -1 & -2 \\ 2 & 0 & 3 \end{pmatrix}$.

(a) Find the coefficients of the characteristic polynomial of $\mathbf{A}$.

(b) Factor the characteristic polynomial into linear factors. (It has irrational roots.)

(c) Find the eigenvalues and eigenvectors of $\mathbf{A}$.

(d) Find a diagonal representation of $\mathbf{A}$ if possible.

7. (a) Show that if $\mathbf{A} \in \mathbb{R}^{n,n}$ has the eigenvalue $\lambda \in \mathbb{R}$ for an eigenvector $\mathbf{x} \in \mathbb{R}^n$, then for any $\alpha \in \mathbb{R}$ the matrix $\mathbf{B} = \mathbf{A} + \alpha\mathbf{I}_n$ has the eigenvalue $\lambda + \alpha$. Why is $\mathbf{x}$ an eigenvector for $\mathbf{B}$ and its eigenvalue $\lambda + \alpha$?

(b) Show that 0 and 6 are eigenvalues of $\mathbf{A} = \begin{pmatrix} 4 & 8 \\ 1 & 2 \end{pmatrix}$. Is $\mathbf{A}$ diagonalizable? Which basis $\mathcal{U} = \{\mathbf{u}_1, \mathbf{u}_2\}$ will diagonalize $\mathbf{A}$?

(c) Find all eigenvalues of $\mathbf{B} = \begin{pmatrix} 7 & 8 \\ 1 & 5 \end{pmatrix}$. Is $\mathbf{B}$ diagonalizable? How?

(d) Find the eigenvalues of the squared matrix $\mathbf{A}^2 = \mathbf{A} \cdot \mathbf{A}$ from part (b). (*Hint*: $\mathbf{A}^2 x = \mathbf{A}(\mathbf{A}\mathbf{x}) = \cdots \mathbf{x}$.)

8. Show that the linear transformation $T\begin{pmatrix} x \\ y \end{pmatrix} = \begin{pmatrix} y \\ -x \end{pmatrix}$ maps no real vector $\begin{pmatrix} x \\ y \end{pmatrix} \in \mathbb{R}^2$ to a real multiple of itself, except, of course, $\begin{pmatrix} x \\ y \end{pmatrix} = \begin{pmatrix} 0 \\ 0 \end{pmatrix}$. (Compare with Problem 22 of Section 9.R.)

9. Let $\mathbf{A} = \begin{pmatrix} 23 & 10 & -10 \\ -58 & -27 & 34 \\ -11 & -5 & 6 \end{pmatrix}$.

(a) Given that $\lambda = 3$ is an eigenvalue of $\mathbf{A}$, find the corresponding eigenvector $\mathbf{x}$.

(b) Choose an index i, $1 \le i \le 3$, for which the ith component $\mathbf{x}_i$ of $\mathbf{x}$ is nonzero. For the ith row $\mathbf{a}_i$ of $\mathbf{A}$ form $\mathbf{B}_0 = \mathbf{A} - \frac{1}{x_i}\mathbf{x}\,\mathbf{a}_i$. (Note that the matrix product $\mathbf{x}_{n1}\,\mathbf{a}_{i_{1n}}$ of a column vector $\mathbf{x} \in \mathbb{R}^n$ and a row vector $\mathbf{a}_i \in \mathbb{R}^n$ is an $n \times n$ matrix.)

(c) Show that $\mathbf{B}_0$ has the eigenvalue 0 and that every eigenvector $\mathbf{w}$ of $\mathbf{B}_0$ has a zero in its ith component.

(d) Form $\mathbf{B}_1$ from $\mathbf{B}_0$ by dropping the ith row and the ith column of $\mathbf{B}_0$. Find the eigenvalues of the 2×2 matrix $\mathbf{B}_1$ and compute its eigenvectors.

(e) Verify as quickly as possible that the eigenvalues of $\mathbf{B}_1$ from part (c) are also eigenvalues of $\mathbf{A}$. What are the corresponding eigenvectors?

(f) Is $\mathbf{A}$ diagonalizable, and, if so, how can A be diagonalized?

10. Which of the three matrices $\mathbf{A} = \begin{pmatrix} 2 & 0 & -1 \\ 0 & 3 & 0 \\ 0 & 0 & 2 \end{pmatrix}$, $\mathbf{B} = \begin{pmatrix} 2 & 1 & 0 \\ 0 & 3 & -2 \\ 0 & 0 & 2 \end{pmatrix}$, and $\mathbf{C} = \begin{pmatrix} 2 & 1 & 0 \\ 0 & 3 & 0 \\ 0 & 0 & 2 \end{pmatrix}$ are diagonalizable?

11. Let $\mathbf{A} = \begin{pmatrix} 1 & 2 \\ 2 & 1 \end{pmatrix}$.

(a) Show that $\mathbf{A}^2 - 2\mathbf{A} - 3\mathbf{I}_2 = \mathbf{O}_2$.

(b) Use part (a) to express $\mathbf{A}^{-1}$ as a polynomial in $\mathbf{A}$.

(c) Find $\mathbf{A}^{-1}$ explicitly from part (b), and verify your result.

(d) Find the eigenvalues of $\mathbf{A}$.

12. (a) If two matrices $\mathbf{A}_{m \times k}$ and $\mathbf{B}_{r \times s}$ are compatible for multiplication and if $\mathbf{T} := \mathbf{A} \cdot \mathbf{B}$, what sizes must $\mathbf{A}$ and $\mathbf{B}$ have? What is the size of $\mathbf{T}$?
If $\mathbf{S} = \mathbf{B} \cdot \mathbf{A}$ can be formed, what sizes must $\mathbf{A}$ and $\mathbf{B}$ have? What size does the matrix $\mathbf{S}$ have?

(b) Given a row vector $\mathbf{u}$ and a column vector $\mathbf{v}$, both of size n, what is the size of $\mathbf{u} \cdot \mathbf{v}$? What is the size of $\mathbf{v} \cdot \mathbf{u}$?

(c) Show that $\mathbf{v}$ is an eigenvector of $\mathbf{A} = \mathbf{v}\mathbf{u}$ if $\mathbf{v}$ is a nonzero column vector and $\mathbf{u}$ is a row vector, each of size n. What is the corresponding eigenvalue of $\mathbf{A}$?

(d) Does $\mathbf{A}$ from part (c) have the eigenvalue zero? Find $\dim(\ker(\mathbf{A}))$.

(e) Is it true that any square matrix made up as $\mathbf{A}$ in part (c) as a column times a row vector is diagonalizable? Why? How?

(f) Find all eigenvalues and eigenvectors of $\mathbf{B} = \begin{pmatrix} 1 & 2 & -3 \\ 2 & 4 & -6 \\ -3 & -6 & 9 \end{pmatrix}$, preferably by using the preceding parts.

13. Find a basis $\mathcal{U}$ of $\mathbb{R}^3$ with respect to which the linear transformation $T(\mathbf{x}) = \begin{pmatrix} 3x_1 + x_2 \\ x_2 \\ 4x_1 + 2x_2 + x_3 \end{pmatrix} : \mathbb{R}^3 \longrightarrow \mathbb{R}^3$ is represented in diagonal form $\mathbf{A}_{\mathcal{U}}$.
Perform a complete eigenanalysis for $\mathbf{A}_{\mathcal{E}}$. What is $\mathcal{U}$? What is $\mathbf{A}_{\mathcal{U}}$?

14. If a matrix $\mathbf{A}$ of arbitrary size $n \times n$ has only the two eigenvalues -1 and 4 and is diagonalizable, show that $\mathbf{A}^2 - 3\mathbf{A} = 4\mathbf{I}_n$. Why is $\mathbf{A}$ invertible? Express $\mathbf{A}^{-1}$ as a polynomial in $\mathbf{A}$.

15. Compute all eigenvalues and corresponding eigenvectors for $\mathbf{B} = \begin{pmatrix} 3 & -5 & 4 \\ 2 & -4 & 4 \\ 1 & -2 & 2 \end{pmatrix}$. Is $\mathbf{B}$ diagonalizable? Why or why not?

16. Find the eigenvalues of the matrix $\begin{pmatrix} 1 & 0 & 0 \\ -1 & 2 & -2 \\ -1 & 1 & -1 \end{pmatrix}$. Is $\mathbf{A}$ diagonalizable?

17. Let $\mathbf{A} = \begin{pmatrix} 3 & -2 & 4 \\ -1 & 3 & -1 \\ -1 & 2 & -2 \end{pmatrix}$.

(a) Is 2 an eigenvalue of $\mathbf{A}$?

(b) Can you find a nonzero vector $\mathbf{x}$ such that $\mathbf{Ax} = 2\mathbf{x}$?

(c) Evaluate $\mathbf{Ay}$ for $\mathbf{y} = \begin{pmatrix} 1 \\ -1 \\ 1 \end{pmatrix}$.

(d) Is $\mathbf{y} = \begin{pmatrix} 1 \\ -1 \\ 1 \end{pmatrix}$ an eigenvector of $\mathbf{A}$? Is $\mathbf{z} = \begin{pmatrix} -1 \\ 1 \\ -1 \end{pmatrix}$?

18. For $\mathbf{A} = \begin{pmatrix} a & b \\ c & d \end{pmatrix}$ with integer coefficients a, b, c, and d, find conditions on the coefficients of $\mathbf{A}$ so that $\mathbf{A}$ has

(a) two real eigenvalues;

(b) two equal eigenvalues;

(c) two integer eigenvalues;

(d) two complex conjugate eigenvalues.

19. (a) Represent the matrix $\mathbf{A}_{\mathcal{E}} = \begin{pmatrix} 3 & 1 \\ 1 & 3 \end{pmatrix}$ with respect to the basis $\mathcal{U} = \left\{ \begin{pmatrix} 1 \\ -1 \end{pmatrix}, \begin{pmatrix} 2 \\ 2 \end{pmatrix} \right\}$ of $\mathbb{R}^2$.

(b) What are the eigenvalues of the matrix $\mathbf{A}_{\mathcal{E}}$; what are the corresponding eigenvectors?

20. Perform a complete eigenanalysis for the matrix

$$\mathbf{A} = \begin{pmatrix} 3 & -1 & -1 \\ -12 & 0 & 5 \\ 4 & -2 & -1 \end{pmatrix}.$$

Specifically, determine all eigenvalues and eigenvectors for $\mathbf{A}$. Determine whether $\mathbf{A}$ has a diagonal matrix representation $\mathbf{A}_{\mathcal{U}}$. If so, find an eigenvector basis $\mathcal{U}$ for $\mathbf{A}$.

21. Find the eigenvalues and corresponding eigenvectors for the matrix

$$\mathbf{A} = \begin{pmatrix} 0 & -1 & 0 \\ -1 & 2 & 1 \\ 0 & 1 & 4 \end{pmatrix}.$$

What is the matrix representation of $\mathbf{A}$ with respect to the eigenvector basis?

22. Show that if x is a real eigenvector of $\mathbf{A}_{nn} \in \mathbb{R}^{n,n}$ for a nonzero real eigenvalue λ, then $\mathbf{x}$ lies in the column space of $\mathbf{A}$. Is this result also true for the eigenvalue $\lambda = 0$?

23. Show that $\lambda = 0$ is an eigenvalue of $\mathbf{A}$ if and only if $\mathbf{A}$ is singular.

24. Let $\mathbf{A}_{nn} \in \mathbb{R}^{n,n}$, and assume that $\mathbf{X}_{nn}$ is nonsingular.

(a) Show that $(\mathbf{X}^{-1}\mathbf{AX})^{\ell} = \mathbf{X}^{-1}\mathbf{A}^{\ell}\mathbf{X}$ for all positive integers $\ell \geq 0$.

(b) How can one use part (a) to evaluate the powers $\mathbf{A}^{\ell}$ of a diagonalizable matrix $\mathbf{A}$ efficiently?

25. Perform a complete eigenanalysis and check diagonalizability for

(a) $\mathbf{A} = 3\begin{pmatrix} 1 & 1 & 0 \\ 0 & 3 & -1 \\ 0 & 0 & -2 \end{pmatrix}$,

(b) $\mathbf{B} = \begin{pmatrix} -1 & 1 & 0 \\ 0 & 2 & -1 \\ 0 & 0 & -1 \end{pmatrix}$, and

(c) $\mathbf{C} = -\begin{pmatrix} 1 & 0 & 1 \\ 0 & 1 & 0 \\ 0 & 0 & -2 \end{pmatrix}$.

26. (a) Find the characteristic polynomial for the 4×4 diagonal matrix $\mathbf{A}$ with the nonzero entries 1, 2, 1, and 2 on its diagonal.

What is the characteristic polynomial of the identity matrix $\mathbf{I}_n$?

(b) What is the characteristic polynomial of the zero matrix $\mathbf{O}_n$?

27. Let $\mathbf{A} = \begin{pmatrix} 0 & 2 \\ -8 & 0 \end{pmatrix} \in \mathbb{R}^{2,2}$.

(a) Find the eigenvalues of $\mathbf{A}$ over $\mathbb{C}$.

(b) Find the eigenvectors of $\mathbf{A}$ in $\mathbb{C}^2$.

28. Let $\mathbf{R} = \begin{pmatrix} 1 & 4 & 2 \\ 0 & -8 & 1 \\ 0 & 0 & 1 \end{pmatrix} \in \mathbb{R}^{3,3}$.

(a) Find the eigenvalues of $\mathbf{R}$.

(b) Is $\mathbf{R}$ diagonalizable?

29. (a) What is the matrix equation linking the standard matrix $\mathbf{A} = \mathbf{A}_{\mathcal{E}} \in \mathbb{R}^{n,n}$ of a linear transformation T to the matrix that represents the same linear transformation with respect to another basis $\mathcal{U}$ of $\mathbb{R}^n$?

(b) Find the matrix representation $\mathbf{A}_{\mathcal{U}}$ of $\mathbf{A} = \mathbf{A}_{\mathcal{E}} = \begin{pmatrix} 4 & -2 & 2 \\ 1 & 4 & -1 \\ -2 & 5 & -2 \end{pmatrix}$ for the basis $\mathcal{U} = \left\{ \begin{pmatrix} 1 \\ 0 \\ -2 \end{pmatrix}, \right.$

$\left(\begin{array}{c}-1\\1\\2\end{array}\right), \left(\begin{array}{c}0\\1\\1\end{array}\right)\Big\}$.

(c) What are the eigenvalues of **A**? Is **A** diagonalizable? Why or why not?

30. Perform a complete eigenanalysis on $\mathbf{A} = \begin{pmatrix} 1 & 3 \\ -3 & 1 \end{pmatrix} \in \mathbb{R}^{2,2}$. Show that **A** is diagonalizable over $\mathbb{C}$. Compute a basis $\mathcal{U}$ of $\mathbb{C}^2$ that represents **A** diagonally.

31. Find the characteristic polynomial of $\mathbf{A} = \begin{pmatrix} 3 & 3 \\ 2 & 4 \end{pmatrix}$. Does **A** have distinct eigenvalues?

32. Decide whether $\mathbf{J} = \begin{pmatrix} 1 & 0 \\ 1 & 1 \end{pmatrix}$ is diagonalizable.

33. (a) Which of the vectors $\mathbf{u}_1 = \begin{pmatrix} 2 \\ 3 \end{pmatrix}$, $\mathbf{u}_2 = \begin{pmatrix} 1.5 \\ 1 \end{pmatrix}$, $\mathbf{u}_3 = \begin{pmatrix} 0 \\ 0 \end{pmatrix}$, $\mathbf{u}_4 = \begin{pmatrix} 1 \\ -1 \end{pmatrix}$, $\mathbf{u}_5 = \begin{pmatrix} 1 \\ -2 \end{pmatrix}$, $\mathbf{u}_6 = \begin{pmatrix} 3 \\ 2 \end{pmatrix}$, $\mathbf{u}_7 = \begin{pmatrix} -1 \\ -2 \end{pmatrix}$, and $\mathbf{u}_8 = \begin{pmatrix} 4 \\ -3 \end{pmatrix}$ are eigenvectors of the matrix $\mathbf{C} = \begin{pmatrix} 2 & -3 \\ -4 & 6 \end{pmatrix}$? Is the matrix **C** diagonalizable? What are its eigenvalues?

(b) Is the matrix **C** invertible? If so, what is $\mathbf{C}^{-1}$?

34. If we were to allow the zero vector in the matrix eigenvalue–eigenvector definition, how many eigenvalues would each matrix $\mathbf{A}_{nn}$ have, and what would they be?

35. For which $a \in \mathbb{R}$ does the matrix $\mathbf{A} = \begin{pmatrix} 1 & 1 & a \\ -1 & 0 & 2 \\ 2 & 1 & 4 \end{pmatrix}$ have the eigenvalue $\lambda = 2$, if at all.

36. Show that a block triangular matrix $\mathbf{U} = \begin{pmatrix} \mathbf{A} & \mathbf{O} \\ \mathbf{B} & \mathbf{C} \end{pmatrix}$ with $\mathbf{A} \in \mathbb{R}^{m,m}$ and $\mathbf{C} \in \mathbb{R}^{k,k}$ has the eigenvalues of **A** and **C** as its eigenvalues. (*Hint*: Refer to Problem 16 in Section 6.2.P.)

37. If **A** and $\mathbf{B} \in \mathbb{R}^{n,n}$ are both nonsingular, show that the matrix products **AB** and **BA** are similar. Conclude that for two nonsingular matrices the eigenvalues of **AB** and **BA** coincide.

38. Show that the two matrix products **BA** and **AB** have the same eigenvalues for any two $n \times n$ matrices **A** and **B**.

39. Are $\mathbf{A} = \begin{pmatrix} 2 & 1 \\ 1 & 4 \end{pmatrix}$ and $\mathbf{B} = \begin{pmatrix} 6 & 7 \\ -1 & 0 \end{pmatrix}$ similar matrices?

40. Show that if two $n \times n$ matrices **A** and **B** are similar to the same matrix $\mathbf{C}_{nn}$, then there is a nonsingular matrix $\mathbf{W}_{nn}$ with $\mathbf{W}^{-1}\mathbf{A}\mathbf{W} = \mathbf{B}$. (That is, **A** and **B** are similar.)

41. Investigate the validity of the matrix identity $(\mathbf{I} - \mathbf{A})^{-1} + (\mathbf{I} - \mathbf{A}^{-1})^{-1} = \mathbf{I}$ for nonsingular matrices **A** for which each of the inverses $(\mathbf{I} - \mathbf{A})^{-1}$ and $(\mathbf{I} - \mathbf{A}^{-1})^{-1}$ exists:

(a) Show that if **A** satisfies this matrix identity and if **X** is nonsingular, then $\mathbf{X}^{-1}\mathbf{A}\mathbf{X}$ also satisfies the identity.

(b) Show that the matrix identity holds for all diagonalizable matrices **A** for which the various inverses can be formed.

(c) Check the truth of the identity for $\mathbf{A} = \begin{pmatrix} 2 & 1 \\ 0 & 1 \end{pmatrix}$.

Teacher's Problem Making Exercise (repeated from Section 9.1.P)

T 9. For ease with hand computations, one must learn to construct small integer matrices with integer eigenvalues and integer eigenvectors that may or may not be diagonalizable.

If $\mathbf{X}_{nn}$ is unimodular, then both **X** and $\mathbf{X}^{-1}$ are integer matrices, see Problem T 6. If, moreover, $\mathbf{D}_{nn}$ is integer and diagonal, then the eigenvalues of $\mathbf{A} := \mathbf{X}\mathbf{D}\mathbf{X}^{-1}$ are the entries on the diagonal of **D**, and its eigenvectors can be chosen as the columns of **X**. Clearly **A** is diagonalizable.

If, on the other hand, $\mathbf{J}_{nn}$ is an integer **Jordan block matrix** that is not diagonal, then $\mathbf{B} := \mathbf{X}\mathbf{J}\mathbf{X}^{-1}$ is an integer matrix that has the same eigenvalues as **J**, some of which are repeated. The matrix **B** is not diagonalizable by construction, since **J** is not.

EXAMPLE We use the unimodular matrices $\mathbf{X} = \begin{pmatrix} 13 & 7 & 2 \\ 0 & -1 & -2 \\ -1 & 0 & 1 \end{pmatrix}$ and $\mathbf{Y} = \begin{pmatrix} 1 & 7 & 2 \\ 2 & 13 & 2 \\ -1 & -7 & -1 \end{pmatrix}$ from Problem T 6.

(a) For $\mathbf{D} = \begin{pmatrix} 1 & & \\ & -1 & \\ & & 1 \end{pmatrix}$, we know that $\mathbf{A}_1 := \mathbf{X}\mathbf{D}\mathbf{X}^{-1} = \begin{pmatrix} 29 & 210 & 364 \\ -4 & -29 & -52 \\ 0 & 0 & 1 \end{pmatrix}$ and $\mathbf{A}_2 := \mathbf{Y}\mathbf{D}\mathbf{Y}^{-1} = \begin{pmatrix} 1 & 14 & 28 \\ 0 & 27 & 52 \\ 0 & -14 & -27 \end{pmatrix}$ are both diagonalizable with eigenvalues 1, 1, -1, the diagonal entries of $\mathbf{D}$, and the corresponding eigenvectors in the columns of $\mathbf{X}$ and $\mathbf{Y}$, respectively. Note that a hand computation of the eigendata for $\mathbf{A}_1$ looks more forbidding than for $\mathbf{A}_2$ in this example.

(b) To obtain nondiagonalizable integer matrices $\mathbf{A}$, let us work with $\mathbf{J} = \begin{pmatrix} 1 & 1 & \\ 0 & 1 & \\ & & -1 \end{pmatrix}$ and $\mathbf{Y}$. Then $\mathbf{B} := \mathbf{Y}\mathbf{J}\mathbf{Y}^{-1} = \begin{pmatrix} -3 & -1 & -6 \\ -4 & -1 & -8 \\ 2 & 1 & 5 \end{pmatrix}$ and $\mathbf{B}$'s eigenvalues compute as 1 (double) and –1. However, there is an eigenvector deficiency for the double eigenvalue 1 that makes $\mathbf{B}$ not diagonalizable.

[*Note:* It is important to experiment with a number of examples so that one can find a matrix with integer eigenvalues and small integer entries for easy hand computations. MATLAB helps instructors to construct meaningful problems.] ◀

9.2 Theory

Geometry, Vector Iteration, and Eigenvalue Functions

We study geometric and algebraic qualities of a matrix that are revealed by its eigenvalues and eigenvectors.

The term "eigen" in the words eigenvalue and eigenvectors comes from German where "eigen" means "belonging to" or "property of." Thus, from the language root of "eigen", an eigenanalysis of a matrix $\mathbf{A}_{nn}$ means to analyze, or to investigate all that is the property of $\mathbf{A}$ (i.e., $\mathbf{A}$'s intrinsic meaning or action as a mapping: $\mathbf{x} \mapsto \mathbf{A}\mathbf{x}$). If $\mathbf{A}$ represents a linear transformation $T : \mathbb{R}^n \to \mathbb{R}^n$, then an eigenanalysis of $\mathbf{A}$ will explain T in a fundamental, deep, and intrinsic way.

Reversing our approach from that of Lecture 1 now, we ask the question of what quality—pray—is fundamental or intrinsic for a specific linear transformation T? This sounds like a very demanding question that we attempt to answer now.

First, we look at the geometric information embedded in matrix eigenvalues and eigenvectors. For $n = 2$ and a diagonalizable square matrix $\mathbf{A} \in \mathbb{R}^{2,2}$, we plot the mapping of the eigenvectors $\mathbf{u}_1$ and $\mathbf{u}_2$ under the action of $\mathbf{A}$ similarly to how we have plotted the images $\mathbf{A}\mathbf{e}_1$ and $\mathbf{A}\mathbf{e}_2$ of the unit vectors $\mathbf{e}_1$ and $\mathbf{e}_2 \in \mathbb{R}^2$ under $\mathbf{x} \mapsto \mathbf{A}\mathbf{x}$ in Example 12 of Section 1.3. Here, we normalize $\mathbf{u}_1$ and $\mathbf{u}_2$ so that their maximal components are equal to ± 1. Thus, the eigenvectors $\mathbf{u}_i$ are scaled to touch an edge of the union of the two unit squares in the first or fourth quadrant of $\mathbb{R}^2$.

EXAMPLE 4 If $\mathbf{A} \in \mathbb{R}^{2,2}$ has two eigenvectors $\mathbf{u}_1$ and $\mathbf{u}_2$ for two real eigenvalues, then we can obtain a geometric understanding of the linear transformation $\mathbf{x} \mapsto \mathbf{Ax}$ by plotting the images $\mathbf{Au}_1 = \lambda_1\mathbf{u}_1$ and $\mathbf{Au}_2 = \lambda_2\mathbf{u}_2$. We use the same matrices as in Example 12 of Section 1.3.

(a) The matrix $\mathbf{A} = \begin{pmatrix} 2 & -1 \\ -2 & -2 \end{pmatrix}$ has the eigenvalues $\sqrt{6} \approx 2.45$ and $-\sqrt{6} \approx -2.45$ for the eigenvectors $\mathbf{u}_1 = \begin{pmatrix} 1 \\ -0.449 \end{pmatrix}$ and $\mathbf{u}_2 = \begin{pmatrix} 0.225 \\ 1 \end{pmatrix}$. These have been normalized so that their maximal components are equal to 1. See Figure 9-1.

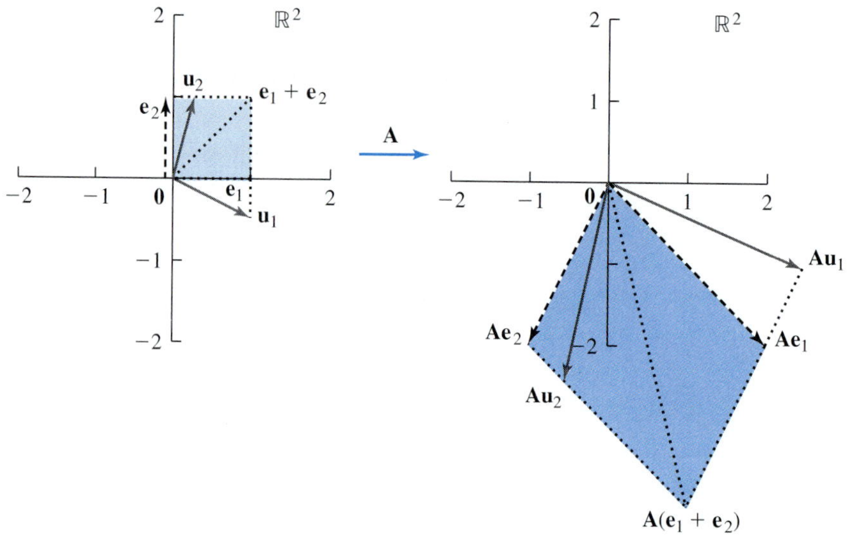

Figure 9-1

The vector $\mathbf{Au}_1 = \sqrt{6}\mathbf{u}_1$ has the same direction as $\mathbf{u}_1$ and is stretched by a factor around 2.45. Similarly, $\mathbf{Au}_2 = -\sqrt{6}\mathbf{u}_2$ lies on the same line through the origin as $\mathbf{u}_2$ does; however, it points in the opposite direction and is similarly elongated.

(b) The upper triangular matrix matrix $\mathbf{B} = \begin{pmatrix} 1 & 1 \\ 0 & 1.5 \end{pmatrix}$ exhibits its eigenvalues 1 and 1.5 on the diagonal according to Theorem 9.6(b). The corresponding eigenvectors are $\mathbf{u}_1 = \mathbf{e}_1$ and $\mathbf{u}_2 = \begin{pmatrix} 1 \\ 0.5 \end{pmatrix}$, leading to the graphs in Figure 9-2.
Again, note that the images $\mathbf{Bu}_1$ and $\mathbf{Bu}_2$ of the two eigenvectors lie on the same lines through the origin as the eigenvectors $\mathbf{u}_1$ and $\mathbf{u}_2$ do. The vectors

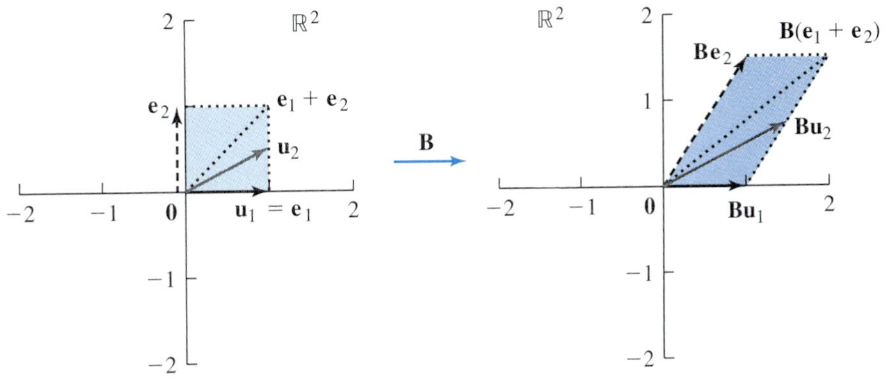

Figure 9-2

$\mathbf{Bu}_1$ and $\mathbf{u}_1$ are identical, while $\mathbf{Bu}_2 = 1.5\mathbf{u}_2$ is an elongated copy of $\mathbf{u}_2$. **B** represents a **shear**.

(c) If $b \neq 0 \in \mathbb{R}$, then the matrix $\mathbf{C} = \begin{pmatrix} a & -b \\ b & a \end{pmatrix}$ has two complex eigenvalues $a + bi$ and $a - bi \notin \mathbb{R}$. Thus, there are no real vectors whose directions are preserved under the linear mapping $\mathbf{x} \mapsto \mathbf{Cx}$. This is obvious from the nature of the mapping $\mathbf{x} \mapsto \mathbf{Cx}$ as a **dilated rotation**: No real vector—except for the zero vector—is left unaltered under the rotation and dilation induced by **C** if $b \neq 0$; see Figure 1-8 in Section 1.3.

(d) The matrix $\mathbf{D} = \begin{pmatrix} 2 & 0 \\ 0 & 0 \end{pmatrix}$ has the eigenvalues 2 and 0 with eigenvectors $\mathbf{e}_1$ and $\mathbf{e}_2$, respectively. Thus, Figure 1-9 in Chapter 1 already shows the mapping of **D**'s eigenvectors.

(e) The matrix $\mathbf{E} = \begin{pmatrix} 0 & 2 \\ 0 & 0 \end{pmatrix}$ has the double eigenvalue 0, but **E** is not diagonalizable, since the associated eigenspace $E(0) = \text{span}\{\mathbf{e}_1\}$ is only one-dimensional. Thus, our simple eigenanalysis for the matrix E is insufficient for a useful geometric interpretation. ◀

Next, we study repetitive processes involving a linear transformation $T : \mathbb{R}^n \to \mathbb{R}^n$.

Assume for simplicity that the given standard matrix $\mathbf{A}_{\mathcal{E}} \in \mathbb{R}^{n,n}$ for T has a basis of real eigenvectors $\{\mathbf{u}_i\}_{i=1}^n$ in $\mathbb{R}^n$ for real eigenvalues λ_i. Then, every vector $\mathbf{x} \in \mathbb{R}^n$ is the unique linear combination of the eigenvector basis $\{\mathbf{u}_1, \ldots, \mathbf{u}_n\}$ according to Definition 4 of Section 5.1. In other words,

$$\mathbf{x} = \alpha_1 \mathbf{u}_1 + \cdots + \alpha_n \mathbf{u}_n \qquad \text{with certain scalars } \alpha_i \in \mathbb{R}.$$

The matrix $\mathbf{A} = \mathbf{A}_{\mathcal{E}}$ represents the linear transformation $T : \mathbb{R}^n \to \mathbb{R}^n$ defined by $T(\mathbf{x}) := \mathbf{Ax}$. In terms of its n eigenvalue–eigenvector pairs λ_i and $\mathbf{u}_i$, we observe that

$$\begin{aligned}\mathbf{Ax} &= \mathbf{A}(\alpha_1 \mathbf{u}_1 + \cdots + \alpha_n \mathbf{u}_n) = \alpha_1 \mathbf{Au}_1 + \cdots + \alpha_n \mathbf{Au}_n \\ &= \alpha_1 \lambda_1 \mathbf{u}_1 + \cdots + \alpha_n \lambda_n \mathbf{u}_n\end{aligned}$$

for every $\mathbf{x} \in \mathbb{R}^n$. Similarly,

$$\mathbf{A}^2\mathbf{x} = \mathbf{A}(\mathbf{A}\mathbf{x}) = \alpha_1\lambda_1^2\mathbf{u}_1 + \cdots + \alpha_n\lambda_n^2\mathbf{u}_n, \quad \text{and consequently}$$
$$\mathbf{A}^k\mathbf{x} = \alpha_1\lambda_1^k\mathbf{u}_1 + \cdots + \alpha_n\lambda_n^k\mathbf{u}_n \quad \text{for all integers } k. \tag{9.3}$$

Let us interpret Equation (9.3) as k becomes large. For this, assume further that the eigenvalues λ_i of $\mathbf{A}$ are ordered, so that

$$\lambda_1 > |\lambda_2| \geq \cdots \geq |\lambda_n| \geq 0.$$

Factoring out λ_1^k on the right-hand side, we rewrite Eq. (9.3) as

$$\mathbf{A}^k\mathbf{x} = \lambda_1^k\left(\alpha_1\mathbf{u}_1 + \alpha_2\left(\frac{\lambda_2}{\lambda_1}\right)^k\mathbf{u}_2 + \cdots + \alpha_n\left(\frac{\lambda_n}{\lambda_1}\right)^k\mathbf{u}_n\right). \tag{9.4}$$

Since λ_1 is the **dominant eigenvalue** of $\mathbf{A}$ (i.e., the one of largest magnitude), we have $\left|\frac{\lambda_j}{\lambda_1}\right| < 1$ for each $j \geq 2$. Thus, the "essence of $\mathbf{A}$" appears in its k-fold application to an arbitrary starting vector $\mathbf{x} \in \mathbb{R}^n$:

Namely, if $\alpha_1 \neq 0$ in the $\mathcal{U}$–coordinate vector $\mathbf{x}_{\mathcal{U}} = (\alpha_1, \ldots, \alpha_n)^T$ of $\mathbf{x} \in \mathbb{R}^n$, then for large integers k, the normalized image

$$\frac{\mathbf{A}^k\mathbf{x}}{\lambda_1^k}$$

tends towards a vector in the direction of $\pm\mathbf{u}_1$ according to Eq. (9.4), or towards the direction of the dominant eigenvector $\mathbf{u}_1$ of $\mathbf{A}$, no matter what vector $\mathbf{x}$ with $\alpha_1 \neq 0$ was originally chosen. This is so, since for $j > 1$, the coefficient $\left(\frac{\lambda_j}{\lambda_1}\right)^k$ of each $\mathbf{u}_j$ tends to 0 as $k \to \infty$ in Eq. (9.4). Moreover, for sufficiently large powers k, the length of $\mathbf{A}^k\mathbf{x}$ changes approximately by the factor $|\lambda_1|$ from that of the previous power iterate $\mathbf{A}^{k-1}\mathbf{x}$.

These are two intrinsic qualities of the underlying linear transformation $\mathbf{x} \mapsto \mathbf{A}\mathbf{x}$ that an eigenanalysis of $\mathbf{A}$ will exhibit, namely, the dominant eigenvalue and eigenvector pair.

EXAMPLE 5 For $\mathbf{B} = \begin{pmatrix} 1 & 1 \\ -2 & 4 \end{pmatrix}$ of Example 3 or 3.D with its eigenvalues 3 and 2, we take $\mathbf{x} = \begin{pmatrix} 1 \\ -1 \end{pmatrix}$ as a starting vector. Then, $\mathbf{B}^{10}\mathbf{x} = \begin{pmatrix} -115026 \\ -233124 \end{pmatrix}$. Note that the next iterate $\mathbf{B}^{11}\mathbf{x} = \begin{pmatrix} -348150 \\ -702444 \end{pmatrix}$ increases in its components roughly by the factor of $\mathbf{B}$'s maximal eigenvalue $\lambda_{max} = 3$ when compared with $\mathbf{B}^{10}\mathbf{x}$. Moreover, $\mathbf{B}^{50}\mathbf{x} = \begin{pmatrix} -1.435796 \cdot 10^{24} \\ -2.871592 \cdot 10^{24} \end{pmatrix}$. Look at the coordinate ratio of approximately $1:2$ in $\mathbf{B}^j\mathbf{x}$: The iterated vectors tend towards a vector in the opposite direction of the eigenvector $\mathbf{u}_2 = \begin{pmatrix} 1 \\ 2 \end{pmatrix}$ that was computed in Examples 3 and 3.D. This is so because our specific starting vector $\mathbf{x} = 3\mathbf{u}_1 - 2\mathbf{u}_2$ with $\mathbf{x}_{\mathcal{U}} = \begin{pmatrix} 3 \\ -2 \end{pmatrix}$ uses the negative coordinate $\alpha_2 = -2 < 0$ for the dominant eigenvector $\mathbf{u}_2 = \begin{pmatrix} 1 \\ 2 \end{pmatrix}$ of $\mathbf{B}$. ◀

The coefficients of the characteristic polynomial f_A of $\mathbf{A}_{nn}$ have useful relations to the eigenvalues of $\mathbf{A}$, pointing to further intrinsic qualities of $\mathbf{A}$.

Theorem 9.5 The coefficients a_0 and a_{n-1} of the characteristic polynomial

$$f_A(x) = x^n + a_{n-1}x^{n-1} + \cdots + a_1x + a_0 = \prod_{i=1}^{n}(x - \lambda_i)$$

of a real or complex square matrix $\mathbf{A}_{nn}$ are related to the entries of $\mathbf{A}$ and to $\mathbf{A}$'s eigenvalues λ_i as follows:

(a) $-a_{n-1} = \sum_{i=1}^{n} \lambda_i = \sum_{i=1}^{n} a_{ii} =:$ trace($\mathbf{A}$), the **trace** of $\mathbf{A}$,
where the diagonal entries of $\mathbf{A}_{nn}$ are denoted by a_{ii}. (trace condition)

(b) $(-1)^n a_0 = \prod_{i=1}^{n} \lambda_i =:$ det($\mathbf{A}$), the **determinant** of $\mathbf{A}_{nn}$.
(determinant condition)

◀

Proof (via determinants)

(a) We see that $-\lambda_1 - \lambda_2 \cdots - \lambda_n = a_{n-1}$ by looking at the terms involving x^{n-1} in the product expansion of the characteristic polynomial $f_A(x) = \prod_i (x - \lambda_i)$. On the other hand, as

$$f_A(x) = \det(x\mathbf{I} - \mathbf{A}) = \det \begin{pmatrix} x - a_{11} & & * \\ & \ddots & \\ * & & x - a_{nn} \end{pmatrix},$$

the coefficient of x^{n-1} in f_A must be equal to $-\sum a_{ii}$. This follows by using the historical determinant definition (8.6).

(b) Observe that $f_A(0) = \det(0\mathbf{I} - \mathbf{A}) = (-1)^n \det \mathbf{A}$ from part 11 of the determinant Proposition in Section 8.1. On the other hand, $f_A(0) = a_0 = (-1)^n \prod_i \lambda_i$ from the polynomial expansion of $f_A(x)$.
(A proof of the trace condition (a) that does not rely on the characteristic polynomial or on determinants is given in Example 4 in Section 11.3.) ■

Part (b) of Theorem 9.5 can serve as an intrinsic or "linear transformation-based" definition of the **determinant function** $\det : \mathbb{R}^{n,n} \to \mathbb{R}$: Define the **determinant** of a matrix $\mathbf{A}_{nn}$ as

$$\det(\mathbf{A}) := \prod_i \lambda_i$$

for the eigenvalues λ_i of $\mathbf{A}$, counted repeatedly if repeated.

Our next result involves matrix transposition and describes the eigenvalues of triangular matrices explicitly.

Theorem 9.6 (a) For a real square matrix $\mathbf{A}_{nn}$, the eigenvalues of $\mathbf{A}$ and those of its transpose $\mathbf{A}^T$ coincide.

(b) The eigenvalues of a triangular matrix $\mathbf{T}_{nn}$ are the diagonal entries t_{ii} of $\mathbf{T}$.

(c) The set of eigenvalues of a triangular block matrix such as

$$\mathbf{A}_{nn} = \begin{pmatrix} \mathbf{A}_1 & & & * \\ \mathbf{O} & \mathbf{A}_2 & & \\ & & \ddots & \\ \mathbf{O} & & \mathbf{O} & \mathbf{A}_\ell \end{pmatrix}$$

with square diagonal blocks $\mathbf{A}_k$ of size n_k is the union of the eigenvalues of the diagonal blocks $\mathbf{A}_k$ for $k = 1, \ldots, \ell$. ◀

Proof (a) From Theorem 7.2, we know that $\text{rank}(\mathbf{A}) = \text{rank}(\mathbf{A}^T)$ for every matrix $\mathbf{A}$. By Eq. (9.2) or Eq. (9.2.D), we know that λ is an eigenvalue of $\mathbf{A}$ if and only if $\text{rank}(\mathbf{A} - \lambda\mathbf{I}) < n$. Combining the two, we see that $\text{rank}(\mathbf{A} - \lambda\mathbf{I}) < n$ if and only if $\text{rank}((\mathbf{A} - \lambda\mathbf{I})^T) = \text{rank}(\mathbf{A}^T - \lambda\mathbf{I}) < n$ (i.e., statement (a)).

(b) The rank of a lower triangular matrix $\mathbf{L}$ is the same as that of the upper triangular matrix $\mathbf{L}^T =: \mathbf{R}$ for any $\mathbf{L}$ by Theorem 7.2. For upper triangular matrices $\mathbf{R}_{nn}$, we observe that $\text{rank}(\mathbf{R} - r_{ii}\mathbf{I}) < n$ for each diagonal entry r_{ii} of $\mathbf{R}$, since the REF of $\mathbf{R} - r_{ii}\mathbf{I}$ has at least one zero in position (i, i), and thus it cannot contain n pivots. Hence, according to Eq. (9.2) or Eq. (9.2.D), each diagonal entry r_{ii} of $\mathbf{R}$ is an eigenvalue of $\mathbf{R}$.

(c) The matrix $\mathbf{A} - \lambda\mathbf{I}_n$ is singular precisely when one of $\mathbf{A}_k - \mathbf{I}_{n_k}$ is singular. ■

From Section 7.1 and the matrix similarity formula (7.4), it is apparent that two similar matrices $\mathbf{A}_{nn}$ and $\mathbf{B}_{nn} = \mathbf{X}^{-1}\mathbf{A}\mathbf{X}$ represent the same linear transformation of $\mathbb{R}^n$, but with respect to two different bases. Thus, the eigenstructures of similar matrices should be identical.

The following theorem establishes this result rigorously:

Theorem 9.7 If $\mathbf{X}$ is nonsingular $n \times n$ and if $\mathbf{A}_{nn}$ and $\mathbf{B}_{nn} = \mathbf{X}^{-1}\mathbf{A}\mathbf{X}$ are similar matrices, then the following hold:

(a) $\mathbf{A}$ and $\mathbf{B}$ have the same eigenvalues.

(b) If $\mathbf{A}\mathbf{u} = \lambda\mathbf{u}$ for $\lambda \in \mathbb{C}$ and $\mathbf{u} \neq \mathbf{0} \in \mathbb{C}^n$, then $\mathbf{B}\mathbf{X}^{-1}\mathbf{u} = \lambda\mathbf{X}^{-1}\mathbf{u}$, that is, if $\mathbf{A}$ has the eigenvector $\mathbf{u}$ for the eigenvalue λ, then $\mathbf{B}$ has the eigenvector $\mathbf{X}^{-1}\mathbf{u}$ for the same eigenvalue λ.

(c) If $\mathbf{A}$ is diagonalizable, then $\mathbf{B}$ is diagonalizable and vice versa. ◀

Proof (a) follows from (b).

(b) Since $\mathbf{A} = \mathbf{X}\mathbf{B}\mathbf{X}^{-1}$, we have $\mathbf{A}\mathbf{u} = \mathbf{X}\mathbf{B}\mathbf{X}^{-1}\mathbf{u} = \lambda\mathbf{u}$, or $\mathbf{B}\left(\mathbf{X}^{-1}\mathbf{u}\right) = \lambda\left(\mathbf{X}^{-1}\mathbf{u}\right)$.

(c) If $\mathbf{A}_{nn}$ is diagonalizable, there is a basis $\{\mathbf{u}_1, \ldots, \mathbf{u}_n\}$ of n eigenvectors in $\mathbb{C}^n$ for $\mathbf{A}$ according to Theorem 9.4 or Theorem 9.2.D. Consequently, the n eigenvectors $\mathbf{X}^{-1}\mathbf{u}_i$ of $\mathbf{B}$ from part (b) are also linearly independent. This is so, since $\mathbf{z} = \sum_{i=1}^{n} (\alpha_i \mathbf{X}^{-1}\mathbf{u}_i) = \mathbf{0}$ implies $\mathbf{X}^{-1}\sum_i(\alpha_i \mathbf{u}_i) = \mathbf{0}$ or $\sum_i \alpha_i \mathbf{u}_i = \mathbf{0}$, since $\mathbf{X}$ is nonsingular, or $\ker(\mathbf{X}) = \{\mathbf{0}\}$. As the n eigenvectors $\mathbf{u}_i$ of $\mathbf{A}$ are assumed linearly independent, the last identity and Definition 2 of Section 5.1 make every coefficient α_i of $\mathbf{z}$ equal to zero. Thus, the n eigenvectors $\mathbf{X}^{-1}\mathbf{u}_i$ of $\mathbf{B}$ are linearly independent, making $\mathbf{B}$ diagonalizable as well. ■

Our final theoretical result compares the index and the two different multiplicities of matrix eigenvalues. Recall their definitions:

For each matrix $\mathbf{A} \in \mathbb{R}^{n,n}$ the characteristic polynomial can be factored over $\mathbb{C}$ as $f_A(\lambda) = (\lambda - \lambda_1)^{n_1} \cdots (\lambda - \lambda_k)^{n_k}$ with distinct roots λ_j and $n_1 + \cdots + n_k = n$. In this notation, each eigenvalue λ_j of $\mathbf{A}$ has the **algebraic multiplicity** n_j. Likewise, the minimal polynomial p_A of $\mathbf{A}$ can be factored as $p_A(\lambda) = (\lambda - \lambda_1)^{\ell_1} \cdots (\lambda - \lambda_k)^{\ell_k}$ for the same distinct eigenvalues λ_j of $\mathbf{A}$, but with possibly different exponents ℓ_j. The integer ℓ_j is called the **index** of the eigenvalue λ_j.

How do the index and the algebraic multiplicity of an eigenvalue λ_j of $\mathbf{A}$ relate to its **geometric multiplicity**, defined as $\dim(E(\lambda_j)) = \dim(\ker(\mathbf{A} - \lambda_j \mathbf{I}))$?

Theorem 9.8 **(Eigenvalue Multiplicity)**

(a) The geometric multiplicity of each root λ_j of the characteristic polynomial $f_A(\lambda)$ does not exceed its algebraic multiplicity. In other words, for each eigenvalue λ_j of a matrix $\mathbf{A}_{nn}$, we have

$$\text{geom. mult. } (\lambda_j) \le \text{ alg. mult. } (\lambda_j), \text{ or}$$
$$\dim(E(\lambda_j)) \le n_j$$

if $f_A(\lambda) = \det(\lambda I - \mathbf{A}) = (\lambda - \lambda_1)^{n_1} \cdots (\lambda - \lambda_k)^{n_k}$ for distinct λ_j and $n_1 + \cdots + n_k = n$.

(b) Consequently, a matrix $\mathbf{A}_{nn}$ is diagonalizable if and only if

(1) geom. mult. $(\lambda_j) =$ alg. mult. (λ_j), or
(2) $\text{index}(\lambda_j) = 1$ for each eigenvalue λ_j of $\mathbf{A}$.

(c) The index of each matrix eigenvalue λ_j does not exceed its algebraic multiplicity n_j. ◀

Proof Part (c) follows directly, since the minimum polynomial $p_A(\lambda)$ of degree less than or equal to n divides the characteristic polynomial $f_A(\lambda)$ of degree n. For part (a), if μ is an eigenvalue of $\mathbf{A}$ with geometric multiplicity $m \ge 1$, then there are m linearly independent eigenvectors $\mathbf{u}_1, \ldots, \mathbf{u}_m$ satisfying

$$\mathbf{A}_{nn}\begin{pmatrix} | & & | \\ \mathbf{u}_1 & \ldots & \mathbf{u}_m \\ | & & | \end{pmatrix}_{nm} = \begin{pmatrix} | & & | \\ \mathbf{u}_1 & \ldots & \mathbf{u}_m \\ | & & | \end{pmatrix}_{nm} \begin{pmatrix} \mu & & 0 \\ & \ddots & \\ 0 & & \mu \end{pmatrix}_{mm}.$$ Since $\dim \ker \begin{pmatrix} - & \mathbf{u}_1^T & - \\ & \vdots & \\ - & \mathbf{u}_m^T & - \end{pmatrix}_{mn} = n - m$, we can find $n - m$ linearly independent

vectors $\mathbf{u}_{m+1}, \ldots, \mathbf{u}_n$ in $\mathbb{R}^n$ (or $\mathbb{C}^n$) that are orthogonal to the eigenspace $E(\lambda) = \text{span}\{\mathbf{u}_1, \ldots, \mathbf{u}_m\}$ according to the Dimension Theorem 5.2 of Section 5.1. Look at the matrix product $\mathbf{AU}$ with $\mathbf{U}_{nn} = \begin{pmatrix} | & & | & | & & | \\ \mathbf{u}_1 & \ldots & \mathbf{u}_m & \mathbf{u}_{m+1} & \ldots & \mathbf{u}_n \\ | & & | & | & & | \end{pmatrix}$:

$$\begin{aligned}
\mathbf{A}_{nn}\mathbf{U}_{nn} &= \mathbf{A}\begin{pmatrix} | & & | & | & & | \\ \mathbf{u}_1 & \ldots & \mathbf{u}_m & \mathbf{u}_{m+1} & \ldots & \mathbf{u}_n \\ | & & | & | & & | \end{pmatrix} \\
&= \begin{pmatrix} | & & | & | & & | \\ \mu\mathbf{u}_1 & \ldots & \mu\mathbf{u}_m & \mathbf{A}\mathbf{u}_{m+1} & \ldots & \mathbf{A}\mathbf{u}_n \\ | & & | & | & & | \end{pmatrix} \\
&= \begin{pmatrix} | & & | & | & & | \\ \mathbf{u}_1 & \ldots & \mathbf{u}_m & \mathbf{u}_{m+1} & \ldots & \mathbf{u}_n \\ | & & | & | & & | \end{pmatrix} \left(\begin{array}{ccc|c} \mu & & 0 & \\ & \ddots & & \mathbf{X} \\ 0 & & \mu & \\ \hline & \mathbf{0} & & \mathbf{Y} \end{array}\right),
\end{aligned}$$

where $\begin{pmatrix} | & & | \\ \mathbf{A}\mathbf{u}_{m+1} & \ldots & \mathbf{A}\mathbf{u}_n \\ | & & | \end{pmatrix} = \begin{pmatrix} | & & | \\ \mathbf{u}_1 & \ldots & \mathbf{u}_n \\ | & & | \end{pmatrix} \left(\begin{array}{c} \mathbf{X} \\ \hline \mathbf{Y} \end{array}\right)$ expresses the images $\mathbf{A}\mathbf{u}_{m+1}, \ldots, \mathbf{A}\mathbf{u}_n \in \mathbb{R}^n$ in terms of the basis $\{\mathbf{u}_1, \ldots, \mathbf{u}_n\}$. Since $\mathbf{U}$ is nonsingular, we have $\mathbf{U}^{-1}\mathbf{AU} = \left(\begin{array}{c|c} \mu\mathbf{I}_m & \mathbf{X} \\ \hline \mathbf{0} & \mathbf{Y} \end{array}\right)$. Consequently,

$$\begin{aligned}
\det(\lambda\mathbf{I} - \mathbf{A}) = f_A(\lambda) = f_{U^{-1}AU}(\lambda) &= \det(\lambda\mathbf{I} - \mathbf{U}^{-1}\mathbf{AU}) \\
&= \det\left(\begin{array}{c|c} (\lambda - \mu)\mathbf{I}_m & -\mathbf{X} \\ \hline \mathbf{0} & \lambda\mathbf{I}_{n-m} - \mathbf{Y} \end{array}\right) \\
&= (\lambda - \mu)^m \det(\lambda\mathbf{I}_{n-m} - \mathbf{Y})
\end{aligned}$$

according to Proposition 9.1.D part (d) of Section 9.1.D and the determinant Proposition, part 13, of Section 8.1.

Thus, if μ has geometric multiplicity m as an eigenvalue of $\mathbf{A}$, then it is at least an m–fold root of the characteristic equation of $\mathbf{A}$. Part (b)(1) is obvious from Sections 9.1 and 9.1.D. Part (b)(2) follows, since $p_A(\lambda) = p_{X^{-1}AX}(\lambda)$ for any nonsingular matrix $\mathbf{X}$ and since for diagonalizable matrices $\mathbf{B} = \mathbf{X}^{-1}\mathbf{AX}$, every eigenvalue index is equal to one. ■

The eigenvalues and eigenvectors of a square matrix determine its behaviour as a linear transformation.

9.2.P Problems

1. Let $\mathbf{A} = \begin{pmatrix} 4 & -1 & 0 & 0 \\ 1 & 4 & -1 & 0 \\ 0 & 1 & 5 & -1 \\ 0 & 0 & -1 & -4 \end{pmatrix}$.

(a) (Via vector iteration and Section 9.1) Find the vanishing polynomial of the vector $\mathbf{y} = \begin{pmatrix} 1 \\ -1 \\ 1 \\ 0 \end{pmatrix}$ for the matrix $\mathbf{A}$.

(b) (Via determinants and Section 9.1.D) Find the characteristic polynomial $\det(\lambda\mathbf{I} - \mathbf{A})$ for the matrix $\mathbf{A}$.

2. Consider $\mathbf{A} = \begin{pmatrix} 1 & -2 & -1 \\ -1 & 0 & -2 \\ 2 & 0 & 3 \end{pmatrix}$.

(a) (Via determinants) Find the coefficients of the characteristic polynomial of $\mathbf{A}$.

(b) For $\mathbf{v}_1 = \begin{pmatrix} 1 \\ 0 \\ 1 \end{pmatrix}$ compute $\mathbf{v}_2 = \mathbf{A}\mathbf{v}_1$, $\mathbf{v}_3 = \mathbf{A}^2\mathbf{v}_1$, and $\mathbf{v}_4 = \mathbf{A}^3\mathbf{v}_1$ most economically. (*Hint*: It is a bad idea to compute the powers of $\mathbf{A}$. Use $\mathbf{v}_3 = \mathbf{A}\mathbf{v}_2$, etc., instead.)

(c) Solve the linear system of equations
$$\begin{pmatrix} | & | & | \\ \mathbf{v}_1 & \mathbf{v}_2 & \mathbf{v}_3 \\ | & | & | \end{pmatrix} \begin{pmatrix} c_0 \\ c_1 \\ c_2 \end{pmatrix} = \begin{pmatrix} | \\ \mathbf{v}_4 \\ | \end{pmatrix}$$
with the $\mathbf{v}_i$ from part (b). Is $\mathcal{V} = \{\mathbf{v}_1, \mathbf{v}_2, \mathbf{v}_3\}$ a basis for $\mathbb{R}^3$? Why?

(d) Verify that $\mathbf{A}\begin{pmatrix} | & | & | \\ \mathbf{v}_1 & \mathbf{v}_2 & \mathbf{v}_3 \\ | & | & | \end{pmatrix} =$
$$\begin{pmatrix} | & | & | \\ \mathbf{v}_1 & \mathbf{v}_2 & \mathbf{v}_3 \\ | & | & | \end{pmatrix} \begin{pmatrix} 0 & 0 & c_0 \\ 1 & 0 & c_1 \\ 0 & 1 & c_2 \end{pmatrix}.$$

(e) Find the matrix representation of the linear transformation $\mathbf{x} \mapsto \mathbf{A}\mathbf{x}$ with respect to the basis $\mathcal{V} = \{\mathbf{v}_1, \mathbf{v}_2, \mathbf{v}_3\}$ as defined in part (b). (*Hint*: There is no need to compute anything here.)

3. Prove that if for two given matrices $\mathbf{A}_{nn}$ and $\mathbf{B}_{mm}$ with $n \geq m$ there is a matrix $\mathbf{X}_{nm}$ of rank m with $\mathbf{AX} = \mathbf{XB}$, then every eigenvalue of $\mathbf{B}$ is an eigenvalue of $\mathbf{A}$.

4. (a) Show that the two square matrices $\mathbf{A}$ and $\mathbf{X}^{-1}\mathbf{AX}$ have the same characteristic polynomial (when using determinants) or the same minimal polynomial (without using determinants) for any nonsingular $n \times n$ matrix $\mathbf{X}$.

Prove that all similar matrices have the same eigenvalues.

(b) Find two 2×2 matrices with identical eigenvalues that are not similar.

5. (a) If λ is an eigenvalue for the eigenvector $\mathbf{x}$ of a nonsingular matrix $\mathbf{A}$, show that λ^{-1} is an eigenvalue of $\mathbf{A}^{-1}$. What is the corresponding eigenvector?

(b) Show that $g(\lambda)$ is an eigenvalue of $g(\mathbf{A})$ for any polynomial g if λ is an eigenvalue of $\mathbf{A}$.

6. (a) If $\mathbf{A}_{nm} = \mathbf{u}\mathbf{v}^T$ for two column vectors $\mathbf{u}, \mathbf{v} \neq \mathbf{0}$ with $\mathbf{u} \in \mathbb{R}^n$, $\mathbf{v} \in \mathbb{R}^m$, show that rank $\mathbf{A} = 1$.

(b) If $\mathbf{u}, \mathbf{v} \in \mathbb{R}^n$ are nonzero column vectors, find all eigenvalues and eigenvectors of $\mathbf{A} = \mathbf{u}\mathbf{v}^T$. What is $\det \mathbf{A}$? (Careful, the result depends on n.) Under which condition is $\mathbf{A}$ diagonalizable?

7. Let $\mathbf{A} = \begin{pmatrix} -1 & * & 7 & * \\ -1 & 3 & 7 & * \\ -1 & * & 7 & * \\ 2 & * & -14 & 3 \end{pmatrix}$ be a 4×4 matrix with partially prescribed real entries and partially unprescribed entries labelled as $*$.

(a) Is $\mathbf{A}$ singular?

(b) If two eigenvalues of $\mathbf{A}$ (with its unspecified and unknown entries marked by $*$) are 5 and 3, find all eigenvalues of $\mathbf{A}$. (*Hint*: Think trace.)

(c) Is $\mathbf{A}$ from part (b) similar to a diagonal matrix $\mathbf{D}$? Which one? How many different ones?

(d) Find at least one eigenvector for $\mathbf{A}$. Which eigenvalue of $\mathbf{A}$ does this eigenvalue correspond to?

8. Let $\mathbf{A} = \begin{pmatrix} 2 & 3 & -47 & 7 \\ 0 & -1 & 22 & -14 \\ 0 & 0 & 1 & 12 \\ 0 & 0 & 0 & 11 \end{pmatrix}$.

(a) What are the eigenvalues of $\mathbf{A}$?

(b) Find the eigenvalues of $\mathbf{A}^{14}$.

(c) What direction will the vector $\mathbf{A}^{200}\mathbf{x}$ point to for $\mathbf{x} = \begin{pmatrix} 1 \\ 1 \\ 1 \\ 2 \end{pmatrix}$?

9. Let $\mathbf{A} = \begin{pmatrix} 5 & 4 \\ -2 & -4 \end{pmatrix}$.

(a) Find a matrix $\mathbf{X}$ so that $\mathbf{X}^{-1}\mathbf{A}\mathbf{X} = \begin{pmatrix} 4 & 0 \\ 0 & -3 \end{pmatrix}$.

(b) Find another matrix $\mathbf{Y} \neq \mathbf{X}$ with $\mathbf{Y}^{-1}\mathbf{A}\mathbf{Y} = \begin{pmatrix} 4 & 0 \\ 0 & -3 \end{pmatrix}$.

(c) What relation is there between $\mathbf{X}$ in (a) and $\mathbf{Y}$ in part (b)?

10. (a) If a matrix $\mathbf{A}$ has the eigenvalues 1, 3, and 15 without repeats, what is the minimal or the characteristic polynomial for $\mathbf{A}$?

(b) If $\mathbf{B}$ has the same eigenvalues as $\mathbf{A}$ above, except that 15 is repeated three times while 1 is a double eigenvalue, write out the possible minimal or the characteristic polynomial $p_{B^T}(x)$ or $f_{B^T}(x)$ of $\mathbf{B}^T$.

11. (With determinants) Assume that the characteristic polynomial of $\mathbf{A}$ is $f_A(x) = (x-3)^2(x-1)(x+3)(x^2-9)$.

(a) What is the size of $\mathbf{A}$?

(b) What are the eigenvalues of $\mathbf{A}$? What are the eigenvalue multiplicities? Which multiplicities?

(c) Must $\mathbf{A}$ be diagonalizable?

(d) What are the eigenvalues of $\mathbf{A}^2$? What are their multiplicities?

(e) Could $\mathbf{A}^2$ be diagonalizable?

(f) What is the characteristic polynomial of $\mathbf{A}^2$?

12. If $\mathbf{A}$ has only the eigenvalues $+1$ and -1 and is diagonalizable, show that $\mathbf{A}^3 = \mathbf{A}$. (*Hint*: What is $\mathbf{A}^2$ in this case?)

13. Let $\mathbf{A}(\alpha) := \mathbf{B} + \alpha\mathbf{I}_n$ for an $n \times n$ matrix $\mathbf{B}$.

(a) Show that if $\mathbf{B}$ has the eigenvalues $\beta_1, \ldots, \beta_n \in \mathbb{C}$, then the matrix $\mathbf{A}(\alpha)$ has the eigenvalues $\beta_1 + \alpha_1, \ldots, \beta_n + \alpha \in \mathbb{C}$ for all α. How are the eigenvectors of $\mathbf{B}$ and $\mathbf{A}(\alpha)$ related?

(b) Find all eigenvalues of $\mathbf{B} = \begin{pmatrix} 1 & \ldots & 1 \\ \vdots & & \vdots \\ 1 & \ldots & 1 \end{pmatrix}_{nn}$.

(*Hint*: Look at $\mathbf{B}$'s kernel.)

(c) What is the determinant of

$$\mathbf{C} = \begin{pmatrix} 0 & 1 & \ldots & 1 \\ 1 & 0 & \ddots & \vdots \\ \vdots & \ddots & \ddots & 1 \\ 1 & \ldots & 1 & 0 \end{pmatrix}_{nn} ?$$

(*Hint*: Use Theorem 9.5 and part (b).)

14. Show that if all eigenvalues λ_i of a square matrix $\mathbf{A}_{nn}$ satisfy $|\lambda_i| < 1$, then $\lim_{k\to\infty} \det\left(\mathbf{A}^k\right) = 0$.

15. (a) What are the eigenvalues of $\mathbf{L} = \begin{pmatrix} 12 & 0 & 0 \\ 8 & -3 & 0 \\ -4 & \pi & -17 \end{pmatrix}$?

(b) Are the two matrices $\mathbf{L}$ from part (a) and $\mathbf{R} = \begin{pmatrix} -17 & e & 3 \\ 0 & 12 & \pi^2 \\ 0 & 0 & -3 \end{pmatrix}$ similar or not?

16. (a) Assume that $\mathbf{A}_{nn}$ has rank k. Specify as many eigenvalues of $\mathbf{A}$ as you can from this information.

(b) If $\mathbf{B}$ has the k-fold eigenvalue 2, what is the rank of $\mathbf{B} - 2\mathbf{I}$?

17. Assume that the matrix $\mathbf{A}$ has the eigenvalue λ with algebraic multiplicity n_λ and geometric multiplicity $k_\lambda = \dim(E(\lambda))$. Show that for any $\alpha \in \mathbb{C}$, the matrix $\mathbf{A} + \alpha\mathbf{I}$ has the eigenvalue $\lambda + \alpha$ with the same algebraic multiplicity n_λ and with the same geometric multiplicity k_λ as $\mathbf{A}$ does for λ. (*Hint*: For the algebraic multiplicity part, look at the characteristic polynomials $f_{A+\alpha I}(t+\alpha)$ and $f_A(t)$.)

9.3 Applications

Stochastic Matrices, Systems of Linear DEs, and MATLAB

As the repeated application of a linear transformation T and forming iterated vectors $T^k\mathbf{x}$ reveals T's properties, we now analyze processes with repetition by using matrix models.

We begin with an example.

EXAMPLE 6 Four adjacent basins are filled with a certain chemical in different concentrations. This chemical permeates or seeps through the walls between the basins with the following percentages per unit of time (Figure 9-3):

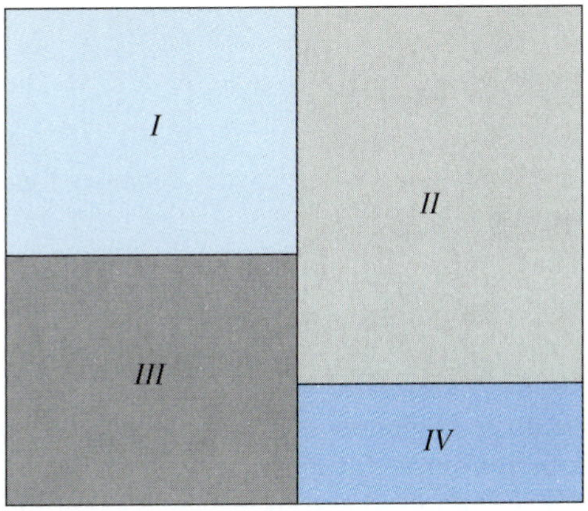

Figure 9-3

$$\begin{array}{llr} \text{Seepage :} & \text{basin } I \to II: & 20\% \\ & I \to III: & 5\% \\ & II \to I: & 10\% \\ & II \to III: & 30\% \\ & II \to IV: & 16\% \\ & III \to I & 12\% \\ & III \to II & 20\% \\ & III \to IV & 30\% \\ & IV \to II & 20\% \\ & IV \to III & 10\% \end{array}$$

Consequently, the four basins each retain the following percentages per unit of time:

$$\begin{array}{llr} \text{Retention :} & \text{basin } I: & 75\% \\ & II: & 44\% \\ & III: & 38\% \\ & IV: & 70\% \end{array}$$

If the original concentrations of the chemical are given by the percentages

$$\begin{array}{llr} \text{Concentration :} & \text{basin } I: & 60\% \\ & II: & 20\% \\ & III: & 80\% \\ & IV: & 10\%, \end{array}$$

what will the concentration be in each basin after 2, 5, 10, or 100 time periods?

Let $x_i \in \mathbb{R}$ denote the concentration level of the chemical in basin i and let $\mathbf{x} = \begin{pmatrix} x_1 \\ \vdots \\ x_4 \end{pmatrix}$ be the concentration level vector. From the given data, the original concentration level vector is $\mathbf{x} = \begin{pmatrix} 0.6 \\ 0.2 \\ 0.8 \\ 0.1 \end{pmatrix}$. During each time period, this state vector $\mathbf{x}$ changes according to the seepage figures in the following linear way from $\mathbf{x}$ to $\mathbf{Ax}$:

$$\mathbf{Ax} = \begin{pmatrix} a_{11} & a_{12} & a_{13} & a_{14} \\ \vdots & & & \vdots \\ a_{41} & a_{42} & a_{43} & a_{44} \end{pmatrix} \begin{pmatrix} x_1 \\ x_2 \\ x_3 \\ x_4 \end{pmatrix}$$

Here, the diagonal entries a_{11}, a_{22}, a_{33}, and a_{44} of $\mathbf{A}$ are the retention percentages of each basin, since, for example, $a_{11}x_1$ of x_1 will remain in basin I after one time period. The off–diagonal entries contain the seepage rates; that is,

> a_{12} indicates the percentage of seepage from basin $II \to I$, and, in general, a_{ij} denotes the percentage of seepage from basin j to basin i in one time period.

Thus, $\mathbf{A} = \begin{pmatrix} 0.75 & 0.1 & 0.12 & 0 \\ 0.2 & 0.44 & 0.2 & 0.2 \\ 0.05 & 0.3 & 0.38 & 0.1 \\ 0 & 0.16 & 0.3 & 0.7 \end{pmatrix}$ describes the **transition matrix** of the concentration-level vectors between consecutive time periods. Note that no entry of $\mathbf{A}$ is negative and that the sum of each column of $\mathbf{A}$ is equal to one. This is so because each column of $\mathbf{A}$ describes the retention and seepage percentages of one basin, and because the process preserves the overall amount of chemicals.

Nonnegative matrices all of whose column sums are equal to one are called **stochastic matrices**. They appear regularly in models of consumer product selections, voting behavior, or in cell-membrane transport problems like the one here, because these **processes** are **conservative** (i.e., they do not alter the overall quantities, such as the number of consumers, voters, or chemicals.)

For the given starting concentration vector $\mathbf{x} = \begin{pmatrix} 0.6 \\ 0.2 \\ 0.8 \\ 0.1 \end{pmatrix}$, we compute

$\mathbf{A}^2\mathbf{x} = \begin{pmatrix} 0.4839 \\ 0.3781 \\ 0.2844 \\ 0.3736 \end{pmatrix}$, $\mathbf{A}^5\mathbf{x} = \begin{pmatrix} 0.3742 \\ 0.3997 \\ 0.2925 \\ 0.4536 \end{pmatrix}$, $\mathbf{A}^{10}\mathbf{x} = \begin{pmatrix} 0.3196 \\ 0.4000 \\ 0.2994 \\ 0.5010 \end{pmatrix}$, and $\mathbf{A}^{100}\mathbf{x} = \begin{pmatrix} 0.3045 \\ 0.4000 \\ 0.3011 \\ 0.5144 \end{pmatrix}$. The last concentration vector signifies convergence of the concentration levels to their steady-state values somewhere between 30.11% in basin III and 51.44% in basin IV. ◀

Using Theorem 9.6, we prove the following result:

Theorem 9.9 Every stochastic $n \times n$ matrix $\mathbf{A}$ has the eigenvalue one.

Proof Stochastic matrices $\mathbf{A}_{nn}$ have constant column sums equal to one. Thus, the transpose of a stochastic matrix $\mathbf{A}$ has constant row sums equal to one. The matrix transpose has the same eigenvalues as the original matrix, due to Theorem 9.6(a), and $\mathbf{A}^T \begin{pmatrix} 1 \\ \vdots \\ 1 \end{pmatrix} = \begin{pmatrix} 1 \\ \vdots \\ 1 \end{pmatrix}$ by $\mathbf{A}^T$'s constancy of row sums equal to one. This is the stipulated eigenvalue–eigenvector equation for the eigenvalue $\lambda = 1$ of $\mathbf{A}^T$. ■

MATLAB computes matrix eigenvalues upon the command

```
eig(A)
```

or upon the extended MATLAB command

```
[U, D] = eig(A)
```

The former returns the real or complex eigenvalues of $\mathbf{A}$, while the latter returns the matrix $\mathbf{U}$ of eigenvector columns together with a diagonal matrix $\mathbf{D}$ containing the eigenvalues of $\mathbf{A}$ on its diagonal, both in conformal order. Thus, $\mathbf{A}_{\mathcal{U}} := \mathbf{D} = \mathbf{U}^{-1}\mathbf{A}\mathbf{U}$ is a diagonal matrix representation of $\mathbf{A}$ with respect to the computed eigenvector basis $\mathcal{U}$ of $\mathbb{R}^n$, or of $\mathbb{C}^n$, if $\mathbf{A}$ has complex eigenvalues.

To find an eigenvector basis for an eigenvalue a of $\mathbf{A}_{nn}$ in MATLAB, use the command `null(A - a*eye(n))`.

For Example 6, we find from MATLAB that $\mathbf{A}$'s dominant eigenvalue is one with corresponding eigenvector $\mathbf{u}_1 = -\begin{pmatrix} 0.3905 \\ 0.5130 \\ 0.3861 \\ 0.6597 \end{pmatrix}$. It takes but a moment to realize that the vector $\mathbf{A}^{100}\mathbf{x}$ computed by us earlier satisfies $\mathbf{A}^{100}\mathbf{x} = -0.7797 \cdot \mathbf{u}_1$. That is, the vector iteration $\{\mathbf{A}^j\mathbf{u}\}$ has converged to a multiple of the MATLAB computed dominant eigenvector $\mathbf{u}_1$ of $\mathbf{A}$ when $j = 100$.

Iterated vectors such as $\mathbf{A}^2\mathbf{x}, \mathbf{A}^5\mathbf{x}, \mathbf{A}^{10}\mathbf{x}$, and $\mathbf{A}^{100}\mathbf{x}$ can be computed in MATLAB via the command sequence

```
A^2*x, A^5*x, A^10*x, A^100*x
```

provided that $\mathbf{A}$ is a square matrix and $\mathbf{x}$ is a vector of compatible size. Note that MATLAB generally computes matrix powers according to the formula $\mathbf{A}^j := \mathbf{U}\mathbf{D}^j\mathbf{U}^{-1}$ from a diagonalizing similarity $\mathbf{U}^{-1}\mathbf{A}\mathbf{U} = \mathbf{D}$ of $\mathbf{A}$ using the internal MATLAB command `[U, D] = eig(A)`. This MATLAB process is somewhat harsh and a bit unreliable for obtaining accurate matrix powers if $\mathbf{A}$ is not diagonalizable.

Finding a vanishing polynomial for a vector $\mathbf{y} \neq \mathbf{0}$ and $\mathbf{A} \in \mathbb{R}^{n,n}$ as in Section 9.1 can be mimicked in MATLAB by forming the vector iteration matrix $\begin{pmatrix} | & | & | & \\ \mathbf{y} & \mathbf{Ay} & \mathbf{A}^2\mathbf{y} & \dots \\ | & | & | & \end{pmatrix}$ and computing its REF using the command

```
C = rref([y, A*y, A*A*y, A*A*A*y]) or
C = rref([y, A*y, A^2*y, A^3*y])
```

if $n = 3$, for example. From step 3 of the eigenanalysis procedure in Section 9.1, the negative coefficients of the vanishing polynomial $p_y(x)$ for the powers $x^{n-1}, \dots, x^1$, and x^0 appear (in reverse order) in the last column of the computed RREF $\mathbf{C}$. Thus the MATLAB row vector

```
py = [1, -fliplr(C(:,4)')]
```

will contain the coefficients of the vanishing polynomial p_y for $\mathbf{y} \neq \mathbf{0} \in \mathbb{R}^3$ in decreasing order. To check out what `fliplr` does to a row vector or to a matrix, enter `help fliplr`. To find the eigenvalues of $\mathbf{A}$ as in Section 9.1 via MATLAB, we can compute the roots of the vanishing polynomial p_y using the command

```
eigenvalues = roots(py)
```

The characteristic polynomial of a square matrix $\mathbf{A}$ can be computed in MATLAB via the command `poly(A)`, which outputs the polynomial coefficients in descending order. These are computed from $\mathbf{A}$'s eigenvalues internally.

Interestingly, in MATLAB the roots of a polynomial p are computed by using the QR matrix eigenvalue algorithm applied to the companion matrix for p. Computing polynomial roots as matrix eigenvalues is highly efficient and accurate; see Chapter 13.

EXAMPLE 7 Let $\mathbf{A} = \begin{pmatrix} 2 & 1 & 2 \\ -1 & 0 & 1 \\ 2 & 0 & -1 \end{pmatrix}$. MATLAB computes the vector iteration matrix $\mathbf{P} = \begin{pmatrix} | & | & | & | \\ \mathbf{y} & \mathbf{Ay} & \mathbf{A}^2\mathbf{y} & \mathbf{A}^3\mathbf{y} \\ | & | & | & | \end{pmatrix}$ for the starting vector $\mathbf{y} = \begin{pmatrix} 2 \\ 3 \\ 0 \end{pmatrix}$ and its reduced REF C as follows:

```
>> P=[y,A*y,A*A*y,A*A*A*y]
P =
       2      1      8     15
      -3     -2      3    -10
       0      4     -2     18
>> C=rref(P)
C =
       1      0      0      1
       0      1      0      5
       0      0      1      1
```

Next, we form the vanishing polynomial p_y for $\mathbf{y}$ and $\mathbf{A}$ from the last column of $\mathbf{C}$ and find its roots in MATLAB:

```
>> py=[1 -fliplr(C(:,4)')]
py =
        1    -1    -5    -1
>> roots(py)
ans =
         2.8662e+00
        -1.6554e+00
        -2.1076e-01
```

Of course, these roots agree with the more standard MATLAB evaluation of matrix eigenvalues using the `eig` command for **A**:

```
>> eig(A)
ans =
         2.8662e+00
        -1.6554e+00
        -2.1076e-01
```

The characteristic polynomial of **A** is computed via `poly(A)` in MATLAB as having the coefficients `ans = 1.0000 -1.0000 -5.0000 -1.0000`. This agrees with `py` above.

Note that to find the vanishing polynomial for a vector **y** and **A** or to find the minimal and characteristic polynomial of a matrix **A** is comparatively easy for low dimensions. Finding the roots of a third or higher degree polynomial is generally very hard unless one uses matrix methods. Just try to factor $p_y(x) = x^3 - x^2 - 5x - 1$ of Example 7 by hand. p_y has no rational roots. For hints on how to find out by hand whether an integer coefficient polynomial has some integer roots see Appendix B.

Next, we turn to analysis. For a differentiable function $f : \mathbb{R} \to \mathbb{R}$ in one variable t and a real constant c, the **differential equation**

$$f'(t) = cf(t) \qquad \text{for } t \in \mathbb{R} \tag{9.5}$$

is called a **linear differential equation**, because both of its sides are linear in the function f. Namely, if f and g are two functions that satisfy Eq. (9.5), then $f'(t) = cf(t)$ and $g'(t) = cg(t)$. Consequently, $\alpha f + \beta g$ solves Eq. (9.5) for all scalars α and β because

$$(\alpha f + \beta g)' = \alpha f' + \beta g' = \alpha cf + \beta cg = c(\alpha f + \beta g).$$

Thus, the space of solutions to (9.5),

$$\mathcal{S} := \{f \mid f : \mathbb{R} \to \mathbb{R} \text{ differentiable with } f'(t) = cf(t)\}$$

is a subspace of all one-variable real valued functions $f : \mathbb{R} \to \mathbb{R}$. Note that $\mathcal{S} \neq \emptyset$, since the zero function, defined by $O(t) := 0$ for all t satisfies Eq. (9.5). From calculus all solutions of Eq. (9.5) have the form $f(t) = ke^{ct}$ for an arbitrary constant k. The differential equation (9.5) is used in calculus to study the exponential function.

What happens when we generalize Eq. (9.5) to several dimensions? For this, we parametrize points $\mathbf{x} \in \mathbb{R}^n$ by a time parameter $t \in \mathbb{R}$. That is, we set

$$\mathbf{x} = \mathbf{x}(t) = \begin{pmatrix} x_1(t) \\ \vdots \\ x_n(t) \end{pmatrix} \in \mathbb{R}^n.$$

Then, the point $\mathbf{x}(t)$ describes a location or position in n–space that varies in time. As the tip of the nose of a walker or the end of a piece of chalk used at the blackboard does in three–dimensional space.

The **tangent**, or **velocity vector** of such a parametrized **location vector** $\mathbf{x}(t)$ is defined as the component-wise derivative $\mathbf{x}'(t) := \begin{pmatrix} x_1'(t) \\ \vdots \\ x_n'(t) \end{pmatrix}$. If $\mathbf{x}'(t)$ is proportional to $\mathbf{x}(t)$ as it was in Eq. (9.5) for one dimension, then instead of one constant c, we need to insert a constant matrix $\mathbf{A}_{nn}$ into the n–dimensional analog of Eq. (9.5). Thus, we obtain a **system of linear differential equations**

$$\mathbf{x}'(t) = \mathbf{A}\mathbf{x}(t) \qquad \text{for } t \in \mathbb{R} \text{ and a constant matrix } \mathbf{A} \in \mathbb{R}^{n,n}. \tag{9.6}$$

To solve such **systems of linear differential equations** for $\mathbf{x}$ and a given matrix $\mathbf{A}_{nn}$, we use our knowledge of the solution in one dimension and linear algebra. For simplicity, we assume that the matrix $\mathbf{A}_{nn}$ is diagonalizable. Let $\mathbf{U}_{nn}$ be a matrix with an eigenvector basis for $\mathbf{A}$ in its columns. Then $\mathbf{U}^{-1}\mathbf{A}\mathbf{U} = \mathbf{D} = \operatorname{diag}(\lambda_i)$ is diagonal, and $\mathbf{A} = \mathbf{U}\mathbf{D}\mathbf{U}^{-1}$. Thus, Eq. (9.6) is equivalent to

$$\mathbf{x}'(t) = \mathbf{U}\mathbf{D}\mathbf{U}^{-1}\mathbf{x}(t).$$

Multiplying on the left by $\mathbf{U}^{-1}$ leads to

$$\mathbf{U}^{-1}\mathbf{x}'(t) = \mathbf{D}\mathbf{U}^{-1}\mathbf{x}(t). \tag{9.7}$$

Now, use the following lemma.

Lemma If $\mathbf{B}$ is a constant $n \times n$ matrix, then the derivative $\big(\mathbf{B}\mathbf{x}(t)\big)'$ of $\mathbf{B}\mathbf{x}(t)$ with respect to t is equal to $\mathbf{B}\mathbf{x}'(t)$ for all differentiable functions $\mathbf{x} : \mathbb{R} \to \mathbb{R}^n$. ◀

Proof If $\mathbf{x}(t) = \begin{pmatrix} x_1(t) \\ \vdots \\ x_n(t) \end{pmatrix} \in \mathbb{R}^n$ and $\mathbf{B}_{nn} = (b_{ij}) \in \mathbb{R}^{n,n}$, then we have

$$\mathbf{B}\mathbf{x}(t) = \begin{pmatrix} \sum_{j=1}^{n} b_{1j}x_j(t) \\ \vdots \\ \sum_{j=1}^{n} b_{nj}x_j(t) \end{pmatrix} \in \mathbb{R}^n.$$

By differentiating the ith component function $(\mathbf{Bx}(t))_i$ of $\mathbf{Bx}(t)$, we obtain

$$\left(\mathbf{Bx}(t)\right)'_i = \left(\sum_{j=1}^{n} b_{ij}x_j(t)\right)' = \sum_{j=1}^{n} b_{ij}x'_j(t) = \left(\mathbf{Bx}'(t)\right)_i$$

for each $i = 1, \dots, n$ according to the rules of differentiation. ■

Next, set $\mathbf{v}(t) := \mathbf{U}^{-1}\mathbf{x}(t)$ in Eq. (9.7). According to the Lemma, we have

$$\mathbf{v}'(t) = \mathbf{Dv}(t). \tag{9.8}$$

In Eq. (9.8), every component function $v_i(t)$ of $\mathbf{v}(t) = \begin{pmatrix} v_1(t) \\ \vdots \\ v_n(t) \end{pmatrix}$ is "separated" from every other, since $v'_i(t) = \lambda_i v_i(t)$ for each i, due to the diagonal system matrix $\mathbf{D}$. Each of the n individual component-function differential equations $v'_i(t) = \lambda_i v_i(t)$ in Eq. (9.8) is in turn solved by $v_i(t) = k_i e^{\lambda_i t}$. This gives us the composite solution $\mathbf{v}(t) = \begin{pmatrix} k_1 e^{\lambda_1 t} \\ \vdots \\ k_n e^{\lambda_n t} \end{pmatrix}$ of Eq. (9.8) expressed in terms of the eigenvector basis $\mathcal{U}$ of $\mathbf{A}$. Converting back to the standard basis; that is, setting

$$\mathbf{x} = \mathbf{x}(t) = \mathbf{Uv}(t) = \mathbf{U}\begin{pmatrix} k_1 e^{\lambda_1 t} \\ \vdots \\ k_n e^{\lambda_n t} \end{pmatrix}$$

according to Section 7.1, finally solves the original problem (9.6).

EXAMPLE 8 Solve the linear system of differential equations $\mathbf{x}'(t) = \mathbf{Ax}(t)$ for $\mathbf{A} = \begin{pmatrix} 3 & 1 & 0 \\ 0 & 1 & 0 \\ 4 & 2 & 1 \end{pmatrix}$ from Problem 13 of Sections 9.1.P or 9.1.D.P.

The matrix $\mathbf{A}$ has the eigenvalues $\lambda_1 = 1$, $\lambda_2 = 1$, and $\lambda_3 = 3$. The corresponding eigenvectors are $\mathbf{u}_1 = \begin{pmatrix} 0 \\ 0 \\ 1 \end{pmatrix}$, $\mathbf{u}_2 = \begin{pmatrix} -1 \\ 2 \\ 0 \end{pmatrix}$, and $\mathbf{u}_3 = \begin{pmatrix} 1 \\ 0 \\ 2 \end{pmatrix}$, respectively. Thus, $\mathbf{A}$ is diagonalizable and the equivalent diagonal system of linear differential equations (9.8) has the solution $\mathbf{v}(t) = \begin{pmatrix} k_1 e^t \\ k_2 e^t \\ k_3 e^{3t} \end{pmatrix}$ in terms of the eigenvector basis $\mathcal{U} = \{\mathbf{u}_1, \mathbf{u}_2, \mathbf{u}_3\}$. Consequently, the general solution of $\mathbf{x}'(t) = \mathbf{Ax}(t)$ is

$$\mathbf{x}(t) = \mathbf{Uv}(t) = \begin{pmatrix} 0 & -1 & 1 \\ 0 & 2 & 0 \\ 1 & 0 & 2 \end{pmatrix}\begin{pmatrix} k_1 e^t \\ k_2 e^t \\ k_3 e^{3t} \end{pmatrix} = \begin{pmatrix} -k_2 e^t + k_3 e^{3t} \\ 2k_2 e^t \\ k_1 e^t + 2k_3 e^{3t} \end{pmatrix}$$

$\mathbf{U}_{2\times 2} = \begin{pmatrix} | & | \\ \mathbf{z} & \overline{\mathbf{z}} \\ | & | \end{pmatrix} = \begin{pmatrix} | & | \\ \mathbf{u}+i\mathbf{w} & \mathbf{u}-i\mathbf{w} \\ | & | \end{pmatrix}$ of $\mathbf{A}_{2\times 2}$, we obtain the complex-valued solution

$$\mathbf{x}(t) = \mathbf{U}\mathbf{v}(t) = \begin{pmatrix} | & | \\ \mathbf{u}+i\mathbf{w} & \mathbf{u}-i\mathbf{w} \\ | & | \end{pmatrix} \begin{pmatrix} k_1 e^{at}(\cos(bt) + i\sin(bt)) \\ k_2 e^{at}(\cos(bt) - i\sin(bt)) \end{pmatrix} \tag{9.9}$$

of the original differential equation (9.6). Here, we have written the complex exponential function $e^{\lambda t}$ as

$$e^{\lambda t} = e^{(a+bi)t} = e^{at}e^{ibt} = e^{at}(\cos(bt) + i\sin(bt))$$

using the well-known formula

$$e^{iy} = \cos(y) + i\sin(y)$$

for purely imaginary exponents.

But we are only interested in real trajectories $\mathbf{x}(t) \in \mathbb{R}^2$ amongst all solutions of the form (9.9). By multiplying out the matrix $\times$ vector product $\mathbf{U}\mathbf{v}(t)$ in Eq. (9.9), we obtain

$$\mathbf{x}(t) = e^{at}\Big((k_1+k_2)(\cos(bt)\mathbf{u} - \sin(bt)\mathbf{w}) + i(k_1-k_2)(\sin(bt)\mathbf{u} + \cos(bt)\mathbf{w})\Big).$$

Hence, $\mathbf{x}(t) \in \mathbb{R}^2$ if and only if $k_1 = k_2 \in \mathbb{R}$. With $k = 2k_1$, the real solutions of Eq. (9.6) then have the form

$$\mathbf{x}(t) = ke^{at}(\cos(bt)\mathbf{u} - \sin(bt)\mathbf{w}) \tag{9.10}$$

for arbitrary $k \in \mathbb{R}$, provided that $\lambda = a + bi \in \mathbb{C}$ for $b \neq 0$ and $\mathbf{A}\mathbf{z} = \lambda\mathbf{z}$ for $\mathbf{z} = \mathbf{u} + i\mathbf{w}$ with $\mathbf{u}, \mathbf{w} \in \mathbb{R}^2$.

As t varies in Eq. (9.10), observe the following phenomena: The vector part $\cos(bt)\mathbf{u} - \sin(bt)\mathbf{w}$ of $\mathbf{x}(t)$ describes a point on an ellipse in the real plane with conjugate radii $\mathbf{u}$ and $\mathbf{w}$. This is the oscillatory, or elliptic part of the solution. It is amplified or stretched by the real factor ke^{at} of $\mathbf{x}(t)$. If $a < 0$ in $\lambda = a + bi$, then the magnitude of $\mathbf{x}(t)$ decreases for increasing t If $a > 0$, the magnitude increases without bound, while for $a = 0$, the point $\mathbf{x}(t)$ stays on the ellipse defined by $\mathbf{u}$ and $\mathbf{w}$ for all times t. Thus, $\mathbf{x}(t)$ represents a spiralling movement in the plane, turning tighter or wider, or remaining forever on one ellipse, depending on the sign of the real part a of $\mathbf{A}$'s eigenvalue $\lambda \in \mathbb{C}$.

Moreover, the initial point $\mathbf{x}(0) = k\mathbf{u}$ lies on the line through the origin and $\mathbf{u} \in \mathbb{R}^2$. A small amount of time later, we have $\mathbf{x}\left(\frac{\pi}{2|b|}\right) = -\text{sign}(b)ke^{\frac{a\pi}{2|b|}}\mathbf{w}$. That is, $\mathbf{x}\left(\frac{\pi}{2|b|}\right)$ has moved onto the line through the origin and $\mathbf{w} \in \mathbb{R}^2$. This occurs towards $-\mathbf{w}$ if $b > 0$ and towards $\mathbf{w}$ otherwise.

To summarize for $\mathbf{A} \in \mathbb{R}^{2,2}$ with two complex conjugate eigenvalues $\lambda = a + bi, b \neq 0$ and $\overline{\lambda}$, and corresponding eigenvectors $\mathbf{z} = \mathbf{u} + i\mathbf{w}$ and $\overline{\mathbf{z}} = \mathbf{u} - i\mathbf{w}$, respectively: the sign of the real part a of λ determines whether the real solution $\mathbf{x}(t)$ of $\mathbf{x}'(t) = \mathbf{A}\mathbf{x}(t)$ spirals outwards (if $a > 0$) away from the origin, spirals inwards (if $a < 0$) towards the origin of $\mathbb{R}^2$, or circles (if $a = 0$) along an elliptical

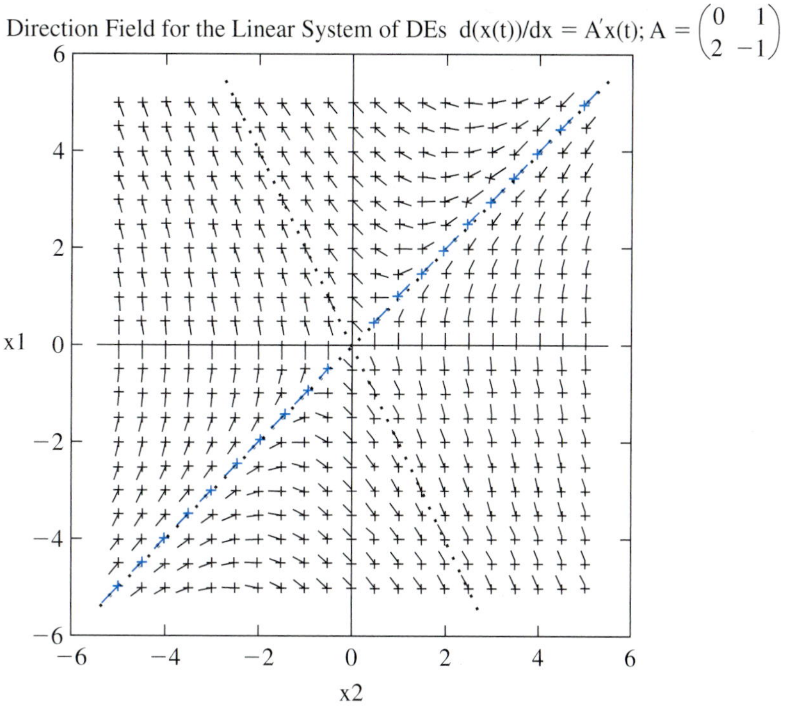

Figure 9-4

the eigenvalue-eigenvector equation $\mathbf{Az} = \lambda\mathbf{z}$ has only real quantities on its left-hand side, while its right-hand side is not real. If $\mathbf{A}_{nn}$ is real and $\mathbf{Az} = \lambda\mathbf{z}$ for $\lambda \in \mathbb{C}$ and $\mathbf{z} \in \mathbb{C}^n$, then by complex conjugation, we have $\overline{\mathbf{Az}} = \overline{\lambda\mathbf{z}}$, or $\mathbf{A}\overline{\mathbf{z}} = \overline{\lambda}\overline{\mathbf{z}}$, since the complex conjugate matrix $\overline{\mathbf{A}}$ is equal to $\mathbf{A}$ here. Thus, the complex eigenvalues of a real $n \times n$ matrix $\mathbf{A}$ come in complex-conjugate pairs λ and $\overline{\lambda} \in \mathbb{C}$, with corresponding complex-conjugate eigenvectors $\mathbf{z}$ and $\overline{\mathbf{z}} \in \mathbb{C}^n$.

Specifically, for $n = 2$ and a real matrix $\mathbf{A}_{2\times 2}$ with a pair of complex-conjugate eigenvalues $\lambda = a+bi$ and $\overline{\lambda} = a-bi$, the general solution of Eq. (9.8) has the form

$$v(t) = \begin{pmatrix} k_1 e^{\lambda t} \\ k_2 e^{\overline{\lambda} t} \end{pmatrix},$$

expressed in terms of the complex-eigenvector basis $\mathbf{z} = \mathbf{u} + i\mathbf{w}$ and $\overline{\mathbf{z}} = \mathbf{u} - i\mathbf{w}$ of $\mathbf{A}$. Here, $\mathbf{u}$ and $\mathbf{w} \in \mathbb{R}^n$ are the real and imaginary part vectors of the complex eigenvector $\mathbf{z}$, respectively. For example, if $\mathbf{z} = \begin{pmatrix} 2-i \\ i+3 \end{pmatrix} \in \mathbb{C}^2$, then $\mathbf{z} = \begin{pmatrix} 2 \\ 3 \end{pmatrix} + i\begin{pmatrix} -1 \\ 1 \end{pmatrix}$ and $\overline{\mathbf{z}} = \begin{pmatrix} 2+i \\ -i+3 \end{pmatrix}$. Using the eigenvector matrix

$\mathbf{U}_{2\times 2} = \begin{pmatrix} | & | \\ \mathbf{z} & \overline{\mathbf{z}} \\ | & | \end{pmatrix} = \begin{pmatrix} | & | \\ \mathbf{u}+i\mathbf{w} & \mathbf{u}-i\mathbf{w} \\ | & | \end{pmatrix}$ of $\mathbf{A}_{2\times 2}$, we obtain the complex-valued solution

$$\mathbf{x}(t) = \mathbf{U}\mathbf{v}(t) = \begin{pmatrix} | & | \\ \mathbf{u}+i\mathbf{w} & \mathbf{u}-i\mathbf{w} \\ | & | \end{pmatrix} \begin{pmatrix} k_1 e^{at}(\cos(bt) + i\sin(bt)) \\ k_2 e^{at}(\cos(bt) - i\sin(bt)) \end{pmatrix} \tag{9.9}$$

of the original differential equation (9.6). Here, we have written the complex exponential function $e^{\lambda t}$ as

$$e^{\lambda t} = e^{(a+bi)t} = e^{at}e^{ibt} = e^{at}(\cos(bt) + i\sin(bt))$$

using the well-known formula

$$e^{iy} = \cos(y) + i\sin(y)$$

for purely imaginary exponents.

But we are only interested in real trajectories $\mathbf{x}(t) \in \mathbb{R}^2$ amongst all solutions of the form (9.9). By multiplying out the matrix $\times$ vector product $\mathbf{U}\mathbf{v}(t)$ in Eq. (9.9), we obtain

$$\mathbf{x}(t) = e^{at}\Big((k_1 + k_2)(\cos(bt)\mathbf{u} - \sin(bt)\mathbf{w}) + i(k_1 - k_2)(\sin(bt)\mathbf{u} + \cos(bt)\mathbf{w})\Big).$$

Hence, $\mathbf{x}(t) \in \mathbb{R}^2$ if and only if $k_1 = k_2 \in \mathbb{R}$. With $k = 2k_1$, the real solutions of Eq. (9.6) then have the form

$$\mathbf{x}(t) = ke^{at}(\cos(bt)\mathbf{u} - \sin(bt)\mathbf{w}) \tag{9.10}$$

for arbitrary $k \in \mathbb{R}$, provided that $\lambda = a + bi \in \mathbb{C}$ for $b \neq 0$ and $\mathbf{Az} = \lambda\mathbf{z}$ for $\mathbf{z} = \mathbf{u} + i\mathbf{w}$ with $\mathbf{u}, \mathbf{w} \in \mathbb{R}^2$.

As t varies in Eq. (9.10), observe the following phenomena: The vector part $\cos(bt)\mathbf{u} - \sin(bt)\mathbf{w}$ of $\mathbf{x}(t)$ describes a point on an ellipse in the real plane with conjugate radii $\mathbf{u}$ and $\mathbf{w}$. This is the oscillatory, or elliptic part of the solution. It is amplified or stretched by the real factor ke^{at} of $\mathbf{x}(t)$. If $a < 0$ in $\lambda = a + bi$, then the magnitude of $\mathbf{x}(t)$ decreases for increasing t If $a > 0$, the magnitude increases without bound, while for $a = 0$, the point $\mathbf{x}(t)$ stays on the ellipse defined by $\mathbf{u}$ and $\mathbf{w}$ for all times t. Thus, $\mathbf{x}(t)$ represents a spiralling movement in the plane, turning tighter or wider, or remaining forever on one ellipse, depending on the sign of the real part a of $\mathbf{A}$'s eigenvalue $\lambda \in \mathbb{C}$.

Moreover, the initial point $\mathbf{x}(0) = k\mathbf{u}$ lies on the line through the origin and $\mathbf{u} \in \mathbb{R}^2$. A small amount of time later, we have $\mathbf{x}\left(\frac{\pi}{2|b|}\right) = -\text{sign}(b)ke^{\frac{a\pi}{2|b|}}\mathbf{w}$. That is, $\mathbf{x}\left(\frac{\pi}{2|b|}\right)$ has moved onto the line through the origin and $\mathbf{w} \in \mathbb{R}^2$. This occurs towards $-\mathbf{w}$ if $b > 0$ and towards $\mathbf{w}$ otherwise.

To summarize for $\mathbf{A} \in \mathbb{R}^{2,2}$ with two complex conjugate eigenvalues $\lambda = a + bi, b \neq 0$ and $\overline{\lambda}$, and corresponding eigenvectors $\mathbf{z} = \mathbf{u} + i\mathbf{w}$ and $\overline{\mathbf{z}} = \mathbf{u} - i\mathbf{w}$, respectively: the sign of the real part a of λ determines whether the real solution $\mathbf{x}(t)$ of $\mathbf{x}'(t) = \mathbf{Ax}(t)$ spirals outwards (if $a > 0$) away from the origin, spirals inwards (if $a < 0$) towards the origin of $\mathbb{R}^2$, or circles (if $a = 0$) along an elliptical

By differentiating the ith component function $(\mathbf{Bx}(t))_i$ of $\mathbf{Bx}(t)$, we obtain

$$\left(\mathbf{Bx}(t)\right)'_i = \left(\sum_{j=1}^{n} b_{ij}x_j(t)\right)' = \sum_{j=1}^{n} b_{ij}x'_j(t) = \left(\mathbf{Bx}'(t)\right)_i$$

for each $i = 1, \dots, n$ according to the rules of differentiation. ■

Next, set $\mathbf{v}(t) := \mathbf{U}^{-1}\mathbf{x}(t)$ in Eq. (9.7). According to the Lemma, we have

$$\mathbf{v}'(t) = \mathbf{Dv}(t). \tag{9.8}$$

In Eq. (9.8), every component function $v_i(t)$ of $\mathbf{v}(t) = \begin{pmatrix} v_1(t) \\ \vdots \\ v_n(t) \end{pmatrix}$ is "separated" from every other, since $v'_i(t) = \lambda_i v_i(t)$ for each i, due to the diagonal system matrix $\mathbf{D}$. Each of the n individual component-function differential equations $v'_i(t) = \lambda_i v_i(t)$ in Eq. (9.8) is in turn solved by $v_i(t) = k_i e^{\lambda_i t}$. This gives us the composite solution $\mathbf{v}(t) = \begin{pmatrix} k_1 e^{\lambda_1 t} \\ \vdots \\ k_n e^{\lambda_n t} \end{pmatrix}$ of Eq. (9.8) expressed in terms of the eigenvector basis $\mathcal{U}$ of $\mathbf{A}$. Converting back to the standard basis; that is, setting

$$\mathbf{x} = \mathbf{x}(t) = \mathbf{Uv}(t) = \mathbf{U}\begin{pmatrix} k_1 e^{\lambda_1 t} \\ \vdots \\ k_n e^{\lambda_n t} \end{pmatrix}$$

according to Section 7.1, finally solves the original problem (9.6).

EXAMPLE 8 Solve the linear system of differential equations $\mathbf{x}'(t) = \mathbf{Ax}(t)$ for $\mathbf{A} = \begin{pmatrix} 3 & 1 & 0 \\ 0 & 1 & 0 \\ 4 & 2 & 1 \end{pmatrix}$ from Problem 13 of Sections 9.1.P or 9.1.D.P.

The matrix $\mathbf{A}$ has the eigenvalues $\lambda_1 = 1$, $\lambda_2 = 1$, and $\lambda_3 = 3$. The corresponding eigenvectors are $\mathbf{u}_1 = \begin{pmatrix} 0 \\ 0 \\ 1 \end{pmatrix}$, $\mathbf{u}_2 = \begin{pmatrix} -1 \\ 2 \\ 0 \end{pmatrix}$, and $\mathbf{u}_3 = \begin{pmatrix} 1 \\ 0 \\ 2 \end{pmatrix}$, respectively. Thus, $\mathbf{A}$ is diagonalizable and the equivalent diagonal system of linear differential equations (9.8) has the solution $\mathbf{v}(t) = \begin{pmatrix} k_1 e^t \\ k_2 e^t \\ k_3 e^{3t} \end{pmatrix}$ in terms of the eigenvector basis $\mathcal{U} = \{\mathbf{u}_1, \mathbf{u}_2, \mathbf{u}_3\}$. Consequently, the general solution of $\mathbf{x}'(t) = \mathbf{Ax}(t)$ is

$$\mathbf{x}(t) = \mathbf{Uv}(t) = \begin{pmatrix} 0 & -1 & 1 \\ 0 & 2 & 0 \\ 1 & 0 & 2 \end{pmatrix}\begin{pmatrix} k_1 e^t \\ k_2 e^t \\ k_3 e^{3t} \end{pmatrix} = \begin{pmatrix} -k_2 e^t + k_3 e^{3t} \\ 2k_2 e^t \\ k_1 e^t + 2k_3 e^{3t} \end{pmatrix}$$

for arbitrary real constants k_i. Students should verify that the derivative $\mathbf{x}'(t)$ of the computed solution $\mathbf{x}(t)$ is indeed equal to $\mathbf{Ax}(t) = \begin{pmatrix} 3 & 1 & 0 \\ 0 & 1 & 0 \\ 4 & 2 & 1 \end{pmatrix} \mathbf{x}(t)$. ◀

Specifically, for $n = 2$, we now visualize the position vector $\mathbf{x}(t) \in \mathbb{R}^2$ together with its movement as prescribed by the differential equation (9.6) graphically. At every point $\mathbf{x}(t) \in \mathbb{R}^2$, the equation $\mathbf{x}'(t) = \mathbf{Ax}(t)$ specifies $\mathbf{x}'(t)$ (i.e., the direction of movement and its magnitude). We will study the **direction field** that is induced on $\mathbb{R}^2$ by the Eq. (9.6) $\mathbf{x}'(t) = \mathbf{Ax}(t)$ in dependence of the eigenstructure of the system matrix $\mathbf{A}_{2\times 2}$.

EXAMPLE 9 (Real eigenvalues) Take $\mathbf{A} = \begin{pmatrix} 0 & 1 \\ 2 & -1 \end{pmatrix}$ in Eq. (9.6). $\mathbf{A}$ has two real eigenvalues $\lambda_1 = 1$ and $\lambda_2 = -2$ for the eigenvectors $\mathbf{u}_1 = \begin{pmatrix} 1 \\ 1 \end{pmatrix}$ and $\mathbf{u}_2 = \begin{pmatrix} -1 \\ 2 \end{pmatrix}$, respectively. At each point $\mathbf{x} = \begin{pmatrix} x_1 \\ x_2 \end{pmatrix} \in \mathbb{R}^2$ with $-5 \le x_i \le 5$ for which $4x_i$ is an integer, we plot the direction of movement for the solution $\mathbf{x}(t)$ of $\mathbf{x}'(t) = \mathbf{Ax}(t)$ at the grid points marked by "+" using a "flag." For example, the flag $\overset{|}{+}$ indicates a movement to the north, or a southerly wind, while $-\!\!+$ indicates an east wind, or movement to the west from the grid point +.
In Figure 9-4, note that along the dotted one-dimensional eigenspace $E(1) = \text{span}\left\{\begin{pmatrix} 1 \\ 1 \end{pmatrix}\right\}$ of $\mathbf{A}$, all blue flags point away from the origin, while along the eigenspace $E(-2) = \text{span}\left\{\begin{pmatrix} -1 \\ 2 \end{pmatrix}\right\}$ they point towards it. In general, negative real eigenvalues indicate movement of $\mathbf{x}(t)$ towards the origin along the corresponding eigenspace, while positive eigenvalues cause outward movement on their eigenspaces. The movement of points not in an eigenspace is the resultant of the eigenmovements according to the point's position with respect to the various eigenspaces; all according to the parallelogram law of forces.

The two eigenspaces $E(1)$ and $E(-2)$ of $\mathbf{A}$ partition the plane into four sectors, each of which behaves uniformly, but differently, under the action of Eq. (9.6): Every trajectory $\mathbf{x}(t)$ stays in the sector in which it starts. For the given matrix $\mathbf{A}$, a trajectory in the sector that contains the positive x_1-axis crosses it going upwards in a clockwise rotation. In the sector with the negative x_1-axis, the trajectories turn downwards, also in a clockwise sense, while the two sectors with parts of the x_2-axis in them experience counterclockwise movement to the right or left, respectively. ◀

Next we investigate the solution of the differential equation (9.6) for a real $n \times n$ matrix $\mathbf{A}$ with a complex eigenvalue $\lambda = a + bi$. Here $a, b \in \mathbb{R}$ with $b \neq 0$. Such a matrix $\mathbf{A}$ must have a complex eigenvector $\mathbf{z} \in \mathbb{C}^n$ for λ, since otherwise arc. The sign of the imaginary part b of λ together with the real and imaginary part vectors $\mathbf{u}$ and $\mathbf{w}$ of the eigenvector $\mathbf{z} = \mathbf{u} + i\mathbf{w}$ determine the direction of the spiral rotation, namely from $\mathbf{w}$ to $\mathbf{u} \in \mathbb{R}^2$ if $b > 0$ and from $\mathbf{u}$ to $\mathbf{w}$ if $b < 0$.

EXAMPLE 10 (Complex eigenvalues)

(a) The matrix $\mathbf{A} = \begin{pmatrix} -1 & 1 \\ -4 & -1 \end{pmatrix}$ has the complex eigenvalue $\lambda = a + bi = -1 + 2i$ for the eigenvector $\mathbf{z} = \begin{pmatrix} 1 \\ 2i \end{pmatrix} = \begin{pmatrix} 1 \\ 0 \end{pmatrix} + i \begin{pmatrix} 0 \\ 2 \end{pmatrix} = \mathbf{u} + i\mathbf{w}$.

Thus, the motion prescribed by $\mathbf{x}'(t) = \mathbf{A}\mathbf{x}(t)$ on $\mathbb{R}^2$ is an ever-tightening spiral ($a < 0$) that rotates from $\mathbf{w} = 2\mathbf{e}_2$ to $\mathbf{u} = \mathbf{e}_1$ ($b > 0$) in a clockwise elliptical spiral with major axes $\mathbf{u} = \mathbf{e}_1$ and $\mathbf{w} = 2\mathbf{e}_2$, since $\mathbf{u} \perp \mathbf{w}$ here, see Figure 9-5.

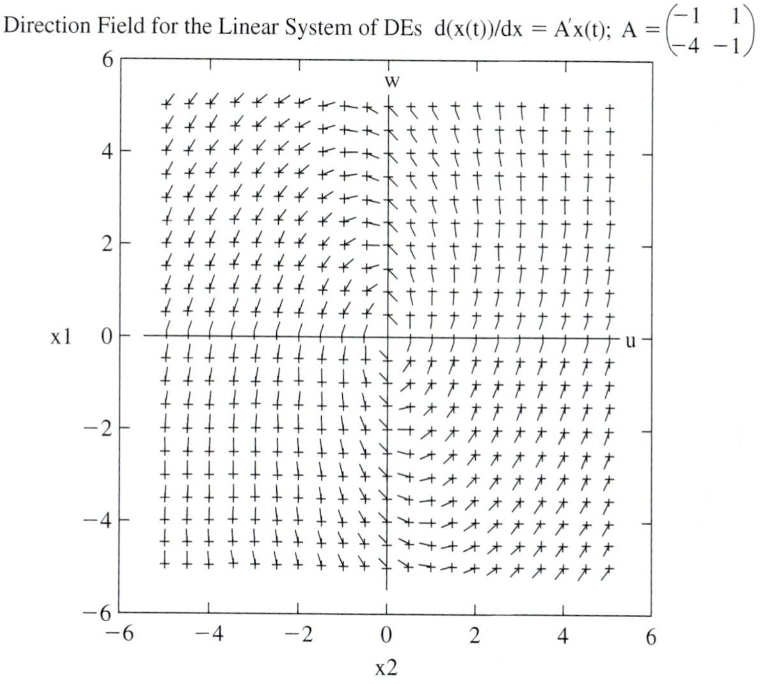

Figure 9-5

(b) The matrix $\mathbf{B} = \begin{pmatrix} 1 & -1 \\ 1 & 1 \end{pmatrix}$ creates the direction field plot in Figure 9-6. This is the pattern of an outward counterclockwise spiral for the solution $\mathbf{x}(t)$ to the differential equation. Hence, $\mathbf{B}$ must have two complex conjugate eigenvalues with positive real parts. In fact, $\lambda = 1+i$ is an eigenvalue for the eigenvector $\mathbf{z} = \begin{pmatrix} -1 \\ i \end{pmatrix} = -\mathbf{e}_1 + i\mathbf{e}_2 = \mathbf{u} + i\mathbf{w}$ of $\mathbf{B}$. And every complex

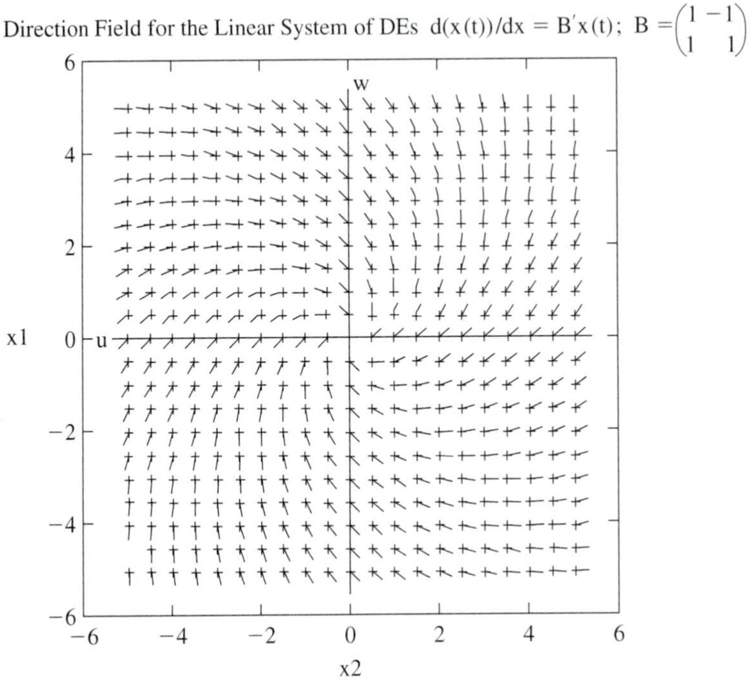

Figure 9-6

vector of the form $\mathbf{z} = \begin{pmatrix} c \\ s \end{pmatrix} + i \begin{pmatrix} s \\ -c \end{pmatrix}$ for $c^2 + s^2 = 1, c, s \in \mathbb{R}$ is an eigenvector of B for $\lambda = 1 + i$. This explains why the spirals in Figure 9-6 favor no direction: the underlying ellipse is a circle. ◀

To plot the direction field spirals for complex conjugate pairs of eigenvalues of real matrices $\mathbf{A}_{2\times 2}$ with their underlying ellipses most convincingly, we must determine and plot the major axes of the ellipse, rather than use the conjugate radii $\mathbf{u}$ and $\mathbf{w}$ that are embedded in the eigenvector $\mathbf{z} = \mathbf{u} + i\mathbf{w}$ with $\mathbf{u}, \mathbf{w} \in \mathbb{R}^2$. If $\mathbf{u} \perp \mathbf{w}$ for an eigenvector $\mathbf{z} = \mathbf{u} + i\mathbf{w}$, then $\mathbf{u}$ and $\mathbf{w}$ are the major axes of the spiralling ellipse, and we are done. This was the case in Examples 10(a) and (b). If $\mathbf{u} \not\perp \mathbf{w}$. (i.e., if $\mathbf{u}^T\mathbf{w} \neq 0$), then we modify the computed eigenvector $\mathbf{z} = \mathbf{u} + i\mathbf{w}$ for λ to become $\widehat{\mathbf{z}} = (c + i)(\mathbf{u} + i\mathbf{w}) \in E(\lambda)$ for a judiciously chosen real scalar c, so that for the modified eigenvector $\widehat{\mathbf{z}} = \widehat{\mathbf{u}} + i\widehat{\mathbf{w}}$ of $\mathbf{A}$, we have $\widehat{\mathbf{u}}, \widehat{\mathbf{w}} \in \mathbb{R}^2$ and $\widehat{\mathbf{u}} \perp \widehat{\mathbf{w}}$. To find c, observe that

$$\widehat{\mathbf{z}} = (c + i)(\mathbf{u} + i\mathbf{w}) = c\mathbf{u} - \mathbf{w} + i(\mathbf{u} + c\mathbf{w}) = \widehat{\mathbf{u}} + i\widehat{\mathbf{w}}$$

holds for two real orthogonal vectors $\widehat{\mathbf{u}}$ and $\widehat{\mathbf{w}}$ if and only if $\widehat{\mathbf{u}} = c\mathbf{u} - \mathbf{w} \perp \mathbf{u} + c\mathbf{w} = \widehat{\mathbf{w}}$. In other words, we need to solve the equation

$$(c\mathbf{u} - \mathbf{w})^T(\mathbf{u} + c\mathbf{w}) = 0$$

for $c \in \mathbb{R}$, if possible. Expansion leads to the quadratic polynomial

$$\mathbf{u}^T\mathbf{w}c^2 + (\mathbf{u}^T\mathbf{u} - \mathbf{w}^T\mathbf{w})c - \mathbf{u}^T\mathbf{w} = 0$$

or to

$$c^2 + \frac{\mathbf{u}^T\mathbf{u} - \mathbf{w}^T\mathbf{w}}{\mathbf{u}^T\mathbf{w}} - 1 = 0.$$

This equation has the real solution

$$c = \frac{\mathbf{w}^T\mathbf{w} - \mathbf{u}^T\mathbf{u}}{2\mathbf{u}^T\mathbf{w}} + \sqrt{\left(\frac{\mathbf{w}^T\mathbf{w} - \mathbf{u}^T\mathbf{u}}{2\mathbf{u}^T\mathbf{w}}\right)^2 + 1}.$$

Hence, for this choice of c, the eigenvector $\widehat{\mathbf{z}} = \widehat{\mathbf{u}} + i\widehat{\mathbf{w}}$ for $\mathbf{A}$ and λ with $\widehat{\mathbf{u}} = c\mathbf{u} - \mathbf{w}$ and $\widehat{\mathbf{w}} = \mathbf{u} + c\mathbf{w}$ has mutually orthogonal real and imaginary parts vectors, indicating the major axes of the spiralling ellipses. We use this in our MATLAB routine `dflDE.m` that plots direction fields. This routine is a self-contained `m.file`. † To use it, the file should be copied into the directory from which MATLAB is called. It can then be applied by entering `dflDE(A,a,m,e,title)` on the MATLAB command line. Here, the values for $\mathbf{A}$, a, m, e and 'title' are assumed prespecified. Calling `help dflDE` gives instructions on its use:

```
function  dflDE(A,a,m,e,title)

% This function m-file plots the direction field of a 2 by 2 linear system
% of DEs of the form  x'(t)=A x(t), where x is in R^2 and A is real 2 by 2.
%
% Inputs: 2 by 2 Matrix A; the real number a which gives the boundary of the
%         plot: -a <= x, y <= a  in uniform steps of size  h= (2a)/(m-1);
%         m = # nodes per line (square grid) (suggestion: set m = 9, 13 or 21)
%         If e = 1: we draw the eigenspaces for real eigenvalues; and for
%         complex eigenvalues we draw the major axes of the elliptical spirals
%         from an orthogonal set of real and imaginary eigenvector parts into
%         the plot. (e is optional)
%         If title = 0, we plot without a title line. The default is set to
%         title = 1.  (Specifying 'title' is also optional)
% Output: Direction field plot; input data for A.

if nargin < 4, e = 0; end, if nargin < 5, title = 1; end,
[n n] = size(A);                                                % initialize

if n ~= 2,  'A is not 2 by 2',  return, end;                     % data checks
if a <= 0
 'right interval endpoint is negative; bad choice; a > 0 only', return, end
hold off                                                         % clear screen

h = 2*a/(m-1); edge=(a+h)/a; x = -a:h:a; y = x; len = 0.6*h;  % initialize grid
plot([0,0],edge*[-a,a],'-'); axis(edge*[-a a -a a]), hold on; % hold images
axis('square'); axis auto; plot(edge*[-a,a],[0,0],'-');       % axis plotting
xlabel('x2'); ylabel('x1','Rotation', 0);                      % labels
if title == 1, if a == 5                                        % for a = 5 :
text(-1.32*a, 1.35*a,' Direction Field for the Linear System of DEs '),
text(0.05*a, 1.35*a,' d(x(t))/dx = A * x(t);  A =  ') % adjust name of matrix
```

† Download from `http://www.auburn.edu/~uhligfd/TLA/download`

```
text(0.85*a,1.39*a,[num2str(A(1,1)),'     ', num2str(A(1,2))]);
text(0.85*a,1.31*a,[num2str(A(2,1)),'     ', num2str(A(2,2))]); end
if a ~= 5                                                  % for a ~= 5 :
text(-1.35*a, 1.39*a,' Direction Field for the Linear System of DEs ')
text(0.06*a, 1.39*a,' d(x(t))/dx = A * x(t);  A =  ') % adjust name of matrix
text(0.88*a,1.43*a,[num2str(A(1,1)),'     ', num2str(A(1,2))]);
text(0.88*a,1.35*a,[num2str(A(2,1)),'     ', num2str(A(2,2))]); end, end

for i=1:m, plot(x(i)*ones(1,m),y,'+'); end                 % plot mesh points

for i=1:m                               % plot scaled tangent vectors
  for j=1:m
    t = A*[x(i);y(j)];                                 % tangent vector x'(t)
    if norm(t) ~= 0, t = len*t/norm(t); end            % normalized
    if norm(t) ~= 0
       plot([x(i);x(i)+t(1)],[y(j);y(j)+t(2)],'-');  end, end, end

% follow with eigenvector plots, if desired:

if e == 1
  [v d]=eig(A);
  if norm(d-d') == 0                                   % for real eigenvalues
    v1 = v(:,1); E1=1.1*a*v1/norm(v1,inf);
    plot([E1(1),-E1(1)],[E1(2),-E1(2)],':');
    v2 = v(:,2); E2=1.1*a*v2/norm(v2,inf);
    plot([E2(1),-E2(1)],[E2(2),-E2(2)],':');
  else                                                 % for complex eigenvalues
    vr = real(v(:,1)); vi = imag(v(:,1)); vrdotvi = vr'*vi;
    if vrdotvi ~= 0                                    % find principal axes
      whirl = -(vr'*vr-vi'*vi)/(2*vrdotvi);
      whirl = whirl + sqrt(whirl^2 + 1);
      VR = whirl*vr - vi; vi = whirl*vi + vr; vr = VR; end
    E1=1.1*a*vr/norm(vr,inf);
    plot([E1(1),-E1(1)],[E1(2),-E1(2)],':');
    text(1.05*E1(1),1.05*E1(2),'u');
    E2=1.1*a*vi/norm(vi,inf);
    plot([E2(1),-E2(1)],[E2(2),-E2(2)],':');
    text(1.05*E2(1),1.05*E2(2),'w');
    end , end
```

We have seen two different applications of matrix models whose behavior is entirely determined by matrix eigenvalues and eigenvectors.

9.3.P Problems

1. Let $\mathbf{A}_{nn}$ be a nonnegative matrix ($a_{ij} \geq 0$ for all i and j) with column sums $\sum_{i=1}^{n} a_{ij}$ equal to one for all $j = 1, \ldots, n$. Let $\mathbf{x} \in \mathbb{R}^n$ have nonnegative components x_i, so that $\sum_{i=1}^{n} x_i = 1$.

Show that all entries of the vector $\mathbf{Ax}$ are also nonnegative and that their sum equals one as well.

2. (a) Show that if a matrix $\mathbf{A}_{nn}$ has constant row sums, then $\mathbf{e} = (1, \ldots, 1)^T$, the column vector of all ones, is an eigenvector of $\mathbf{A}$. What is the corresponding eigenvalue?

(b) Show that $\mathbf{A} = \begin{pmatrix} 3 & 1 & 0 \\ 4 & 1 & -1 \\ 3 & 5 & -4 \end{pmatrix}$ has at least one positive real eigenvalue.

(c) Without explicitly computing the eigenvalues of $\mathbf{A}$ in part (b), deduce that $\mathbf{A}$ has either

(i) three real eigenvalues or

(ii) a pair of complex complex conjugate eigenvalues besides the positive one from part (b).

(d) Assuming that (c)(ii) holds for $\mathbf{A}$, find the value of the real part a of the complex conjugate eigenvalue pair $\lambda_{1,2} = a \pm b\, i$ of $\mathbf{A}$. (*Hint*: Do not use the characteristic polynomial; think "trace" instead.)

(e) Fill in the unspecified entries marked by an $*$ symbol of a matrix $\mathbf{B} = \begin{pmatrix} 3 & * & * \\ * & 1 & * \\ * & * & -4 \end{pmatrix}$ with nonzero entries in such a way that the resulting matrix will surely have three real eigenvalues. Please give two such examples.

3. Tonight there are two parties in town. If 80% of the attendants at party 1 stay there while 20% hop over to the other party in any one given hour; and if 60% attending party 2 do not partyhop, while 40% do every hour, what will the likely attendance be at the two parties after 4 hours if there are initially 100 guests at party 1 and 140 at party 20? If the parties lasted forever, would there finally be more guests at party 1 or at party 2?

4. Find the terminal distribution $\mathbf{x}_\infty$ of two substances in a permeable wall two–pond system $\mathbf{x}_{i+1} = \mathbf{A}\mathbf{x}_i$ with transition matrix $\mathbf{A} = \begin{pmatrix} 0.5 & 0.7 \\ 0.5 & 0.3 \end{pmatrix}$, provided $\mathbf{x}_0 = \begin{pmatrix} 0.4 \\ 0.6 \end{pmatrix}$.

5. Repeat the previous problem with $\mathbf{B} = \begin{pmatrix} 0.5 & 0.3 \\ 0.5 & 0.7 \end{pmatrix}$ and $\mathbf{x}_0$.

6. (a) Use MATLAB to compute the ranks of $\mathbf{A}$ and $\mathbf{B}$ in Problems 4 and 5.

(b) Find the ranks of $\mathbf{A}^{40}$ and $\mathbf{B}^{40}$ with MATLAB.

(c) Do MATLAB computations respect Theorem 6.4 or Lemma 1 of Section 7.1 on the rank of matrix products? Explain, please.

7. (a) Compute all eigenvalues and corresponding eigenvectors by hand for $\mathbf{A} = \begin{pmatrix} 0 & 1 & 0 \\ 1 & 0 & -1 \\ 0 & 1 & 0 \end{pmatrix}$ and $\mathbf{B} = \begin{pmatrix} 3 & -5 & 4 \\ 2 & -4 & 4 \\ 1 & -2 & 2 \end{pmatrix}$. Are $\mathbf{A}$ or $\mathbf{B}$ diagonalizable? Use the MATLAB function `eig` to find the numerical eigenvalues of $\mathbf{A}, \mathbf{A}^T, \mathbf{B}$, and $\mathbf{B}^T$. Compare your results! What do you find? (Use `format long e` here.)

(b) What is the only 3×3 diagonalizable matrix with eigenvalues 0, 0, and 0?

(c) What is the only 3×3 diagonalizable matrix with eigenvalues 1, 1, and 1?

8. Try to find matrices $\mathbf{A}$ of small size for which the MATLAB computed eigenvalues differ from those computed for $\mathbf{A}^T$ due to rounding errors. (Use `format long e`.)

9. Solve the following linear systems of differential equations:

(a) $\mathbf{x}'(t) = \begin{pmatrix} 3 & -1 \\ -6 & 4 \end{pmatrix} \mathbf{x}(t)$,

(b) $\mathbf{x}'(t) = \begin{pmatrix} 2 & 2 \\ -8 & 0 \end{pmatrix} \mathbf{x}(t)$.

10. Verify Example 8; that is, compute the derivative of the solution function $\mathbf{x}(t)$ and show that $\mathbf{x}'(t) = \mathbf{A}\mathbf{x}(t)$ for the matrix $\mathbf{A}$ as given.

11. Show that $\mathbf{x}(t) = \begin{pmatrix} k_1 e^t + 2k_3 e^{3t} \\ 2k_2 e^t \\ -k_2 e^t + k_3 e^{3t} \end{pmatrix}$ solves the system of differential equations $\mathbf{x}'(t) = \begin{pmatrix} 1 & 2 & 4 \\ 0 & 1 & 0 \\ 0 & 1 & 3 \end{pmatrix} \mathbf{x}(t)$.

12. Predict from eigenvalue theory what the direction field of $\mathbf{x}'(t) = \mathbf{A}\mathbf{x}(t)$ will look like for the system matrices

(a) $\mathbf{A} = \mathbf{I}_2$,

(b) $\mathbf{A} = -\mathbf{I}_2$,

(c) $\mathbf{A} = \begin{pmatrix} 0 & 1 \\ 1 & 0 \end{pmatrix}$,

(d) $\mathbf{A} = \begin{pmatrix} 0 & -1 \\ 1 & 0 \end{pmatrix}$,

(e) $\mathbf{A} = \mathbf{O}_2$, and

(f) $\mathbf{A} = \begin{pmatrix} 0 & 0 \\ 0 & 2 \end{pmatrix}$.

13. Plot the six direction fields for Problem 12 using the MATLAB m.file `df1DE.m`.

14. Repeat Problem 12 for the discrete iteration $\mathbf{x}^{(m)} = \mathbf{A}^m \mathbf{x}^{(0)}$ and $\mathbf{x}^{(0)} \in \mathbb{R}^2$ for all six matrices $\mathbf{A}$. Where and how does the initial point $\mathbf{x}^{(0)}$ move for large m? How does this iteration differ from the direction field movement induced by the differential equation $\mathbf{x}'(t) = \mathbf{A}\mathbf{x}(t)$ of Problems 12 and 13?

15. Use the direction field plot routine to determine approximate eigenvectors graphically for the matrices

(a) $\mathbf{A} = \begin{pmatrix} 1 & 2 \\ -1 & 3 \end{pmatrix}$,

(b) $\mathbf{B} = \begin{pmatrix} -4 & 1 \\ 1 & -1 \end{pmatrix}$,

(c) $\mathbf{C} = \begin{pmatrix} 2 & -5 \\ 5 & 2 \end{pmatrix}$, and

(d) $\mathbf{D} = \begin{pmatrix} 0 & 1 \\ 3 & -4 \end{pmatrix}$.

(e) What do the plots reveal about the corresponding eigenvalues for the four matrices?

16. Evaluate the real solution $\mathbf{x}(t)$ in Eq. (9.10) for $t = 0$ and $t = \frac{\pi}{2|b|}$ in order to determine the type of spiralling rotation of the solution $\mathbf{x}$ of $\mathbf{x}'(t) = \mathbf{A}\mathbf{x}(t)$ for a real 2×2 matrix $\mathbf{A}$ with the complex conjugate eigenvalues $\lambda = a \pm bi$.

9.R Review Problems

1. Complete the following statements and definitions:

(a) The eigenvalues and eigenvectors of a ________ matrix $\mathbf{A}$ are ________ and vectors that are related by ________ for $\mathbf{A}$.

(b) The ________ vector cannot be an eigenvector of any matrix.

(c) If $\mathbf{x}$ is an eigenvector for $\mathbf{A}$, then so is ________. (There are many correct fill–ins here.)

(d) The eigenvalue–eigenvector equation is ________ for ________ and ________.

(e) The eigenvalues of a matrix $\mathbf{A}$ satisfy the ________ for $\mathbf{A}$.

(f) If $\ker(\mathbf{A} - 3\mathbf{I}) = \{\mathbf{0}\}$, then 3 is ________ of $\mathbf{A}$.

2. Complete the following definitions:

(a) Two $n \times n$ matrices $\mathbf{A}$ and $\mathbf{B}$ are *similar*, if ....

(b) A square matrix $\mathbf{A}$ is *diagonalizable*, if ....

3. Suppose $\mathbf{A} \in \mathbb{R}^{3,3}$ is real and singular. If $\mathbf{A}$ has one positive eigenvalue and diagonal entries 1, 3, and -4 show that $\mathbf{A}$ must have three real eigenvalues. Must $\mathbf{A}$ have two positive eigenvalues? May it have two?

4. Let $\mathbf{A} = \begin{pmatrix} 1 & 4 \\ 5 & 2 \end{pmatrix}$.

(a) What are the eigenvalues and eigenvectors of $\mathbf{A}$? Is $\mathbf{A}$ diagonalizable?

(b) Answer the same questions for $\mathbf{B} = \mathbf{A} - 15\mathbf{I}_2$.

(c) Answer the same questions for $\mathbf{C} = (\mathbf{A} + 3\mathbf{I}_2)^2$.

5. (a) Find a matrix $\mathbf{A}$ so that the sum of two eigenvalues of $\mathbf{A}$ is an eigenvalue of $\mathbf{A}$ as well.

(b) Show that the sum of two eigenvalues of a matrix $\mathbf{A}$ is generally not an eigenvalue of $\mathbf{A}$.

(c) Are there analogous examples and statements for the sum or difference of matrix eigenvectors?

6. Is is true or false that if A is diagonalizable with eigenvalues 0 and 1 only, then $\mathbf{A}^2 = \mathbf{A}$.

7. If $\mathbf{A}_{nn}$ has the minimal (without determinants) or the characteristic (with determinants) polynomial $p_A(x) = x(x-2)(x+2)(x^2+16)$ or $f_A(x) = x(x-2)(x+2)(x^2+16)$, respectively, decide whether the following statements are true or false:

(a) $n = 4$.

(b) $\mathbf{A}$ may be real.

(c) $\mathbf{A}$ may not be diagonalizable.

(d) If $\mathbf{A}$ is real, then there is a matrix $\mathbf{X}$ in $\mathbb{R}^{55}$ for which $\mathbf{X}^{-1}\mathbf{A}\mathbf{X}$ is diagonal.

(e) $\mathbf{A}$ is singular.

(f) $\mathbf{A}$ has an eigenspace $E(2)$ of dimension __ .

8. Prove that for a nonzero eigenvalue λ and the corresponding eigenvector $\mathbf{x}$ of a matrix $\mathbf{A}_{nn}$, the vectors $\mathbf{A}\mathbf{x}$, $\mathbf{A}^2\mathbf{x}$ and $\mathbf{A}^3\mathbf{x}$ are also eigenvectors of $\mathbf{A}$. What are their corresponding eigenvalues?

9. Let $\mathbf{C} = \begin{pmatrix} 0 & & & & -a_0 \\ 1 & \ddots & & & -a_1 \\ 0 & \ddots & \ddots & & \vdots \\ & & \ddots & \ddots & \vdots \\ 0 & \dots & 0 & 1 & -a_{n-1} \end{pmatrix}_{nn}$ be a so-called **companion matrix**.

(a) (Using Section 9.1) Find the vanishing polynomial of $\mathbf{C}$ and the first-unit vector $\mathbf{e}_1$ via vector iteration

(b) (Using Section 9.1.D and Chapter 8) Find the characteristic polynomial of $\mathbf{C}$. (*Hint*: Be careful with the signs.)

10. Show the following:

(a) If $\mathbf{A}^2 = \mathbf{A}$ for a matrix $\mathbf{A} \in \mathbb{R}^{n,n}$, then $(\mathbf{I}_n - \mathbf{A})^2 = \mathbf{I}_n - \mathbf{A}$.

(b) If $(\mathbf{I}_n - \mathbf{A})^2 = \mathbf{I}_n - \mathbf{A}$ for a matrix $\mathbf{A} \in \mathbb{R}^{n,n}$, then $\mathbf{A}^2 = \mathbf{A}$ as well.

(c) If $\mathbf{A}^2 = \mathbf{A}$, then only 1 or 0 can be an eigenvalue of $\mathbf{A}$.

(d) Find 3×3 real matrices $\mathbf{A}$ with $\mathbf{A}^2 = \mathbf{A}$ that have

(i) only the eigenvalue 1,

(ii) only the eigenvalue 0, or

(iii) both the eigenvalue 1 and 0.

11. Show that $\mathbf{A} = \begin{pmatrix} 1 & 2 \\ -1 & 3 \end{pmatrix}$ has the complex eigenvalues $\lambda_1 = 2 + i$ and $\lambda_2 = 2 - i$ with corresponding eigenvectors $\mathbf{u}_1 = \begin{pmatrix} 2 \\ 1+i \end{pmatrix}$ and $\mathbf{u}_2 = \begin{pmatrix} 2 \\ 1-i \end{pmatrix}$.

12. Write down four 2×2 matrices that are not diagonalizable.

13. Decide which of the following 3×3 matrices are diagonalizable and which are not:

(a) $\mathbf{A} = \begin{pmatrix} 2 & 0 & 0 \\ -1 & 0 & -1 \\ 1 & 2 & 3 \end{pmatrix}$.

(b) $\mathbf{B} = \begin{pmatrix} 0 & 2 & -6 \\ 2 & 0 & 6 \\ 1 & -1 & 5 \end{pmatrix}$.

(c) $\mathbf{C} = \begin{pmatrix} 2 & 1 & 1 \\ -1 & -1 & -2 \\ 1 & 3 & 4 \end{pmatrix}$.

(d) $\mathbf{F} = \begin{pmatrix} 0 & 0 & 1 \\ 0 & 0 & 1 \\ 1 & 2 & 0 \end{pmatrix}$.

14. What can an REF $\mathbf{R}$ of a square matrix $\mathbf{A}_{nn}$ tell us about $\mathbf{A}$'s eigenvalues, if anything?

15. Consider the two matrix processes: $\mathbf{x}'(t) = \mathbf{A}\mathbf{x}(t)$ and $\mathbf{x}^{(m)} = \mathbf{A}^m\mathbf{x}^{(0)}$ for the same 2×2 real matrix $\mathbf{A}$. What properties of $\mathbf{A}$ govern the two processes? How do they differ, in which way are they the same?

16. Determine whether each of the following is true or false:

(a) If $\mathbf{A}$ is diagonalizable and has only one eigenvalue, then $\mathbf{A}$ is a diagonal matrix.

(b) If $\mathbf{A}$ has only the eigenvalue zero, then $\mathbf{A}$ is diagonalizable.

(c) If $\mathbf{A}$ has only the eigenvalue zero, then $\mathbf{A}$ is the zero matrix.

(d) If $\mathbf{A} \in \mathbb{R}^{n,n}$ and λ is one eigenvalue of $\mathbf{A}$, then $-\lambda$ is another.

(e) If $\mathbf{A} \in \mathbb{R}^{n,n}$ and λ is one eigenvalue of $\mathbf{A}$, then $\overline{\lambda}$ is another.

(f) If $\mathbf{A} \in \mathbb{R}^{n,n}$ and λ is one eigenvalue of $\mathbf{A}$, then $\frac{1}{\lambda}$ is another.

(g) If $\mathbf{A} \in \mathbb{R}^{n,n}$ and λ is one eigenvalue of $\mathbf{A}$, then $\lambda + \mu$ is an eigenvalue of $\mathbf{B} := \mathbf{A} - \mu\mathbf{I}$.

17. (a) Show that $\ker(\mathbf{A}) \neq \{\mathbf{0}\}$ for $\mathbf{A} = \begin{pmatrix} -1 & 2 & 1 \\ 3 & -6 & -3 \\ 1 & 0 & 1 \end{pmatrix}$.

(b) Deduce directly from part (a) that $\mathbf{B} = \begin{pmatrix} 1 & 2 & 1 \\ 3 & -4 & -3 \\ 1 & 0 & 3 \end{pmatrix}$ has the eigenvalue 2, while $\mathbf{C} = \begin{pmatrix} 0 & 2 & 1 \\ 3 & -5 & -3 \\ 1 & 0 & 2 \end{pmatrix}$ has the eigenvalue 1.

(c) Show that the matrices $\mathbf{B}$ and $\mathbf{C}$ of part (b) have an eigenvector in common.

(d) Do the matrices $\mathbf{A}$ and $\mathbf{B}$ commute? Do the matrices $\mathbf{B}$ and $\mathbf{C}$ commute?

18. (a) Decide whether the matrix $\mathbf{A} = \begin{pmatrix} 1 & 3 & 0 \\ 2 & 1 & 3 \\ 0 & 1 & 1 \end{pmatrix}$ is diagonalizable.

(b) What are the eigenvalues of $\mathbf{B} = \mathbf{A} + 2\mathbf{I} = \begin{pmatrix} 3 & 3 & 0 \\ 2 & 3 & 3 \\ 0 & 1 & 3 \end{pmatrix}$? Is $\mathbf{B}$ diagonalizable or invertible?

(c) If you can, write out several 3×3 matrices that have repeated eigenvalues, and are diagonalizable.

19. Let $\mathbf{A} = \begin{pmatrix} 1 & 1 & 0 & 0 \\ 0 & 1 & 1 & 0 \\ 0 & 0 & 1 & 1 \\ 0 & 0 & 0 & 1 \end{pmatrix}$ and

$\mathbf{B} = \begin{pmatrix} 1 & 1 & 0 & 0 \\ 0 & 1 & 0 & 0 \\ 0 & 1 & 1 & 1 \\ 0 & 0 & 0 & 1 \end{pmatrix}$.

(a) Show that $\mathbf{A}$ and $\mathbf{B}$ have the same eigenvalues.

(b) Show that neither $\mathbf{A}$ nor $\mathbf{B}$ is diagonalizable.

(c) Are $\mathbf{A}$ and $\mathbf{B}$ similar as matrices? (*Hint*: Consider Proposition 2, part (d), and Theorem 9.7(b).)

20. Let $S\begin{pmatrix} x_1 \\ x_2 \end{pmatrix} = \begin{pmatrix} x_1 + x_2 \\ 2x_1 - x_2 \end{pmatrix}$ and $T\begin{pmatrix} x_1 \\ x_2 \end{pmatrix} = \begin{pmatrix} -2x_1 + x_2 \\ -x_1 + 3x_2 \end{pmatrix}$ be two transformations of $\mathbb{R}^2$.

(a) Why are S and T linear transformations? What are their standard matrix representations $\mathbf{A}_{\mathcal{E}}$ and $\mathbf{B}_{\mathcal{E}}$?

(b) Find the standard matrix representations of $S \circ T$, $T \circ S$, T^2, and S^2. (Be careful: What does T^2 mean?)

(c) Is one of $T \circ S$ or $S \circ T$ diagonalizable over the reals? Are they both?

(d) Are T or S diagonalizable over the reals?

21. Consider the matrix $\mathbf{A} = \begin{pmatrix} 1 & -2 & -6 & 2 \\ -2 & 4 & -3 & 1 \\ -6 & -3 & -4 & 3 \\ 2 & 1 & 3 & 4 \end{pmatrix}$.

(a) Find the eigenvalues of **A**.

(b) Find bases for the eigenspaces of **A**.

(c) Is **A** diagonalizable? Why, why not?

22. Find the eigenvalues of $\mathbf{A} = \begin{pmatrix} 0 & 1 \\ 1 & 0 \end{pmatrix}$. Is **A** diagonalizable?

Standard Questions and Tasks

1. How can one easily decide whether a given vector $\mathbf{v} \in \mathbb{C}^n$ is an eigenvector of a matrix $\mathbf{A} \in \mathbb{R}^{n,n}$? what if $\mathbf{v} \in \mathbb{R}^n$?
2. How can one find all eigenvalues of a given square matrix **A**.
3. Using Section 9.1, find the minimal polynomial of a given matrix $\mathbf{A} \in \mathbb{R}^{n,n}$.

   Alternatively, using Section 9.1.D, find the characteristic polynomial of a matrix $\mathbf{A} \in \mathbb{R}^{n,n}$.
4. Which vectors are eigenvectors of a matrix **A**? What is an eigenspace of a matrix?
5. Find a basis for each eigenspace of a given square matrix **A**.
6. When is an $n \times n$ real matrix **A** diagonalizable?
7. How can one find a basis $\mathcal{U}$, if possible, that diagonalizes a given matrix $\mathbf{A} \in \mathbb{R}^{n,n}$.
8. Write out a matrix $\mathbf{A} \in \mathbb{R}^{3,3}$, element by element, that is not diagonalizable.

Subheadings of Lecture Nine

Basic Equations for Lecture Nine

p. 254 $\mathbf{A}_{\mathcal{U}} = \mathbf{U}^{-1}\mathbf{A}_{\mathcal{E}}\mathbf{U}$ or $\mathbf{A}_{\mathcal{E}}\mathbf{U} = \mathbf{U}\mathbf{A}_{\mathcal{U}}$ (matrix similarity)

p. 254 $\mathbf{A}\mathbf{u} = \lambda\mathbf{u},\ \mathbf{u} \neq \mathbf{0}$ (eigenvalue–eigenvector equation)

p. 261 $E(\lambda) = \ker(\mathbf{A} - \lambda\mathbf{I})$ (eigenspace)

Subheadings of Lecture Nine D

Basic Equations for Lecture Nine D

p. 273	$\mathbf{A}_{\mathcal{U}} = \mathbf{U}^{-1}\mathbf{A}_{\mathcal{E}}\mathbf{U}$ or $\mathbf{A}_{\mathcal{E}}\mathbf{U} = \mathbf{U}\mathbf{A}_{\mathcal{U}}$	(matrix similarity)
p. 274	$\mathbf{A}u = \lambda\mathbf{u},\ \mathbf{u} \neq \mathbf{0}$	(eigenvalue–eigenvector equation)
p. 274	$f_{\mathbf{A}}(\lambda) = \det(\lambda\mathbf{I} - \mathbf{A})$	(characteristic equation)
p. 277	$E_A(\lambda) = \ker(\mathbf{A} - \lambda\mathbf{I})$	(eigenspace)

Nothing perpendicular

CHAPTER

10

Orthogonal Bases and Orthogonal Matrices

Vector space bases can have additional qualities, such as orthogonality. Vector orthogonality prepares the study of special linear transformations.

10.1 Lecture Ten

Length, Orthogonality, and Orthonormal Bases

We study one highly desirable quality of the standard basis $\mathcal{E}$ and model other orthonormal bases after it.

The standard matrix representation $\mathbf{A}_{\mathcal{E}}$ for the standard basis $\mathcal{E} = \{\mathbf{e}_1, \ldots, \mathbf{e}_n\}$ of $\mathbb{R}^n$ from Section 1.1 has been extended to matrix representations $\mathbf{A}_{\mathcal{U}}$ for general bases $\mathcal{U}$ in Chapters 7 and 9. By definition, a basis is a linearly independent spanning set, or equivalently, it is a set of vectors with a unique coordinate vector for every point in its space. (See Section 5.1).

(a) Length and Orthogonality of Vectors

There is more to the standard basis $\mathcal{E}$ than being a basis of $\mathbb{R}^n$ with the linear independence and unique spanning properties—namely, each standard vector $\mathbf{e}_i$ of $\mathcal{E}$ has length one, and any two standard vectors $\mathbf{e}_i$ and $\mathbf{e}_j$ are mutually orthogonal if $i \neq j$.

DEFINITION 1

(a) The **length** of a column vector $\mathbf{x} \in \mathbb{R}^n$ is given by the value of its **vector norm**

$$\|\mathbf{x}\| := \sqrt{x_1^2 + \cdots + x_n^2} = \sqrt{\mathbf{x}^T\mathbf{x}} = \sqrt{\mathbf{x} \cdot \mathbf{x}} \in \mathbb{R}.$$

Here $\mathbf{x}^T\mathbf{x}$ describes the matrix product of the transposed row vector $\mathbf{x}^T$ with $\mathbf{x}$ itself, while $\mathbf{x}\cdot\mathbf{x}$ denotes the equivalent dot product of $\mathbf{x}$ with itself. (*Note:* $\mathbf{x}^T\mathbf{x} = \mathbf{x}\cdot\mathbf{x} = x_1^2 + \cdots x_n^2 \geq 0$ for all $\mathbf{x} \in \mathbb{R}^n$ as the sum of squares.)

(b) Any vector $\mathbf{x} \in \mathbb{R}^n$ of length one, (i.e., with $\|\mathbf{x}\| = 1$) is called a **unit vector**.

(c) Two vectors $\mathbf{x}$ and $\mathbf{y} \in \mathbb{R}^n$ are **orthogonal** if

$$\mathbf{x}^T\mathbf{y} = \mathbf{y}^T\mathbf{x} = \mathbf{x}\cdot\mathbf{y} = \mathbf{y}\cdot\mathbf{x} = \mathbf{0} \in \mathbb{R}.$$

This is denoted by the **orthogonal** or **perpendicular sign** $\perp$ as $\mathbf{x} \perp \mathbf{y}$.

(d) A set of vectors $\{\mathbf{u}_1, \ldots, \mathbf{u}_k\} \subset \mathbb{R}^n$ is **orthonormal** if each vector $\mathbf{u}_i$ has unit length and if each pair of vectors $\mathbf{u}_i$ and $\mathbf{u}_j$ with $i \neq j$ is orthogonal—that is, if

$$\mathbf{u}_\ell^T\mathbf{u}_\ell = \mathbf{u}_\ell\cdot\mathbf{u}_\ell = 1 \ \underline{\text{and}}\ \mathbf{u}_i^T\mathbf{u}_j = \mathbf{u}_i\cdot\mathbf{u}_j = 0$$

for all ℓ and all $i \neq j$.

If $\mathbf{x} \neq \mathbf{0} \in \mathbb{R}^n$ has length different from one, then

$$\mathbf{y} := \frac{\mathbf{x}}{\|\mathbf{x}\|}$$

is the **unit vector** parallel to $\mathbf{x}$, i.e., $\mathbf{y}$ has length 1.

Any two nonzero vectors $\mathbf{x}$ and $\mathbf{y}$ whose dot product $\mathbf{x}\cdot\mathbf{y}$ is zero sustain a 90° angle between them due to the standard definition of the **cosine** function,

$$\cos\angle(\mathbf{x}, \mathbf{y}) := \frac{\mathbf{x}\cdot\mathbf{y}}{\|\mathbf{x}\|\ \|\mathbf{y}\|},$$

for $\mathbf{x}, \mathbf{y} \neq \mathbf{0} \in \mathbb{R}^n$. Thus, two vectors $\mathbf{x}$ and $\mathbf{y} \in \mathbb{R}^n$ enclose a 90° angle if and only if $\cos\angle(\mathbf{x}, \mathbf{y}) = 0$. In other words, two vectors $\mathbf{x}$ and $\mathbf{y}$ are **perpendicular** to each other, or $\mathbf{x} \perp \mathbf{y}$, if and only if $\mathbf{x}\cdot\mathbf{y} = 0$. We will deduce the cosine definition from first principles in a later subsection.

The **vector norm** $\|..\| : \mathbb{R}^n \to \mathbb{R}$ satisfies the following four properties:

(1) $\|\mathbf{x}\| \geq 0$ for all $\mathbf{x} \in \mathbb{R}^n$, and $\|\mathbf{x}\| = 0 \in \mathbb{R}$ if and only if $\mathbf{x} = \mathbf{0} \in \mathbb{R}^n$.
(2) $\|\alpha\mathbf{x}\| = |\alpha|\|\mathbf{x}\|$ for all vectors $\mathbf{x} \in \mathbb{R}^n$ and all scalars $\alpha \in \mathbb{R}$.
(3) $|\mathbf{x}\cdot\mathbf{y}| \leq \|\mathbf{x}\|\ \|\mathbf{y}\|$ for all vectors $\mathbf{x}, \mathbf{y} \in \mathbb{R}^n$ (**Cauchy–Schwarz inequality**).
(4) $\|\mathbf{x}+\mathbf{y}\| \leq \|\mathbf{x}\| + \|\mathbf{y}\|$ for all vectors $\mathbf{x}$ and $\mathbf{y} \in \mathbb{R}^n$ (**triangle inequality**).

The first two vector norm properties follow directly from the definition. The Cauchy–Schwarz inequality is obviously true for $\mathbf{y} = \mathbf{0}$, and it can be proved by using property (1) for the vector $\mathbf{x} - \frac{\mathbf{x}\cdot\mathbf{y}}{\mathbf{y}\cdot\mathbf{y}}\mathbf{y}$ if $\mathbf{y} \neq \mathbf{0}$. The triangle inequality in (4) in turn follows from the Cauchy–Schwarz inequality. See Problems 23 and 26 in Section 10.1.P

[Incidentally, **norm inequalities** are generally easier to derive for the "squared inequality" such as $(\mathbf{x}\cdot\mathbf{y})^2 \leq \|\mathbf{x}\|\ \|\mathbf{y}\|^2$ and $\|\mathbf{x}+\mathbf{y}\|^2 \leq (\|\mathbf{x}\|+\|\mathbf{y}\|)^2$ in (3) and (4).]

Finally, we note that the Cauchy–Schwarz inequality makes the definition of the cosine function in $\mathbb{R}^n$ feasible, since it implies that

$$-1 \leq \cos\angle(\mathbf{x}, \mathbf{y}) = \frac{\mathbf{x}\cdot\mathbf{y}}{\|\mathbf{x}\|\ \|\mathbf{y}\|} \leq 1$$

for all $\mathbf{x}, \mathbf{y} \neq \mathbf{0}$.

(b) Orthonormal Sets of Vectors and Orthonormal Matrices

Sets of vectors that are mutually orthogonal or orthonormal play a special role for matrices.

Theorem 10.1 Every orthonormal set of vectors $\{\mathbf{u}_1, \ldots, \mathbf{u}_k\} \subset \mathbb{R}^n$ is linearly independent.

Proof We use Definition 2(a) of Section 5.1 to derive linear independence of the $\mathbf{u}_i$. Assume that $\mathbf{u} = \alpha_1\mathbf{u}_1 + \cdots + \alpha_k\mathbf{u}_k = \mathbf{0}$ for some scalars α_i. Taking the dot product of $\mathbf{u}$ with any given $\mathbf{u}_j$ yields

$$\begin{aligned}\mathbf{u}_j \cdot \mathbf{u} &= \mathbf{u}_j \cdot (\alpha_1\mathbf{u}_1 + \cdots + \alpha_k\mathbf{u}_k)\\ &= \alpha_1\mathbf{u}_j \cdot \mathbf{u}_1 + \cdots + \alpha_j\mathbf{u}_j \cdot \mathbf{u}_j + \cdots + \alpha_k\mathbf{u}_j \cdot \mathbf{u}_k\\ &= \alpha_j\mathbf{u}_j \cdot \mathbf{u}_j = \alpha_j,\end{aligned}$$

since $\mathbf{u}_j \cdot \mathbf{u}_i = 0$ for all $j \neq i$ and $\mathbf{u}_j \cdot \mathbf{u}_j = 1$ by definition. But $\mathbf{u}$ was assumed to be the zero vector, implying $\mathbf{u}_j \cdot \mathbf{u} = 0$. Thus, every one of $\mathbf{u}$'s coefficients α_j must be zero, completing the proof. ■

Consequently, any set of n mutually orthonormal vectors $\mathbf{u}_i \in \mathbb{R}^n$ automatically forms a basis of $\mathbb{R}^n$ according to Proposition 2 of Section 5.1. Such a basis of orthonormal vectors is called an **orthonormal basis**, or an **ONB** for short. The standard basis $\mathcal{E} = \{\mathbf{e}_1, \ldots, \mathbf{e}_n\}$ of Section 1.1 is an example of an ONB for $\mathbb{R}^n$. We now study orthonormal bases in detail. If $\mathcal{U} = \{\mathbf{u}_1, \ldots, \mathbf{u}_n\}$ is an ONB of $\mathbb{R}^n$, then

$$\mathbf{u}_i \cdot \mathbf{u}_j = \mathbf{u}_i^T\mathbf{u}_j = \delta_{ij} := \begin{cases} 1 & \text{if } i = j \\ 0 & \text{if } i \neq j \end{cases},$$

the **Kronecker δ function**. Hence, we have the following lemma and theorem:

Lemma 1 If $\mathbf{U}$ is the $n \times n$ matrix composed of an orthonormal basis of $\mathbb{R}^n$ as columns, then $\mathbf{U}^T\mathbf{U} = \mathbf{I}_n$. Consequently, $\mathbf{U}^{-1} = \mathbf{U}^T$ if $\mathbf{U}_{nn}$ has mutually orthonormal columns.

Theorem 10.2 Let $\mathcal{U} = \{\mathbf{u}_1, \ldots, \mathbf{u}_n\}$ is an ONB of $\mathbb{R}^n$. Then the $\mathcal{U}$–coordinate vector $\mathbf{x}_\mathcal{U}$ and the standard coordinate vectors $\mathbf{x}_\mathcal{E}$ of each vector $\mathbf{x} \in \mathbb{R}^n$ are related in the following way:

$$\mathbf{x}_\mathcal{U} = \mathbf{U}^T\mathbf{x}_\mathcal{E} = \begin{pmatrix} \mathbf{u}_1 \cdot \mathbf{x} \\ \vdots \\ \mathbf{u}_n \cdot \mathbf{x} \end{pmatrix} \quad \text{for } \mathbf{U} = \begin{pmatrix} | & & | \\ \mathbf{u}_1 & \cdots & \mathbf{u}_n \\ | & & | \end{pmatrix}.$$

For a quick understanding of the theorem, recall from Section 7.1 that

$$\mathbf{x}_\mathcal{U} = \mathbf{X}_{\mathcal{U}\leftarrow\mathcal{E}}\mathbf{x}_\mathcal{E} = \mathbf{U}^{-1}\mathbf{x}_\mathcal{E} = \mathbf{U}^T\mathbf{x}_\mathcal{E}$$

in this case, since $\mathbf{U}^T\mathbf{U} = \mathbf{I}$, or $\mathbf{U}^{-1} = \mathbf{U}^T$ from Lemma 1.

DEFINITION 2 Real square matrices $\mathbf{U}_{nn}$ that contain n mutually orthonormal real column vectors are called **orthogonal matrices**.

Complex square matrices that contain mutually orthonormal complex column vectors are called **unitary matrices**.

Lemma 2 Every real orthogonal matrix $\mathbf{U}_{nn}$ satisfies the matrix equation $\mathbf{U}^T\mathbf{U} = \mathbf{I}_n = \mathbf{U}\mathbf{U}^T$. Complex unitary matrices $\mathbf{U}$ satisfy $\mathbf{U}^*\mathbf{U} = \mathbf{I}_n = \mathbf{U}\mathbf{U}^*$, where the symbol $\mathbf{U}^*$ denotes the **complex conjugate transpose matrix** of $\mathbf{U}$ i.e., $\mathbf{U}^* := \overline{\mathbf{U}}^T = \overline{(\mathbf{U}^T)}$. ◀

Real orthogonal matrices are unitary as well, since $\mathbb{R} \subset \mathbb{C}$. The identity matrix $\mathbf{I}_n$ is the classic example of an orthogonal and of a unitary matrix, representing the standard basis $\mathcal{E}$ of both $\mathbb{R}^n$ and $\mathbb{C}^n$ in its columns.

Orthogonal and unitary matrices help us understand how prosperous it is to work with orthonormal bases $\mathcal{U}$ of $\mathbb{R}^n$. For example, coordinate changes are near trivial for orthonormal bases compared with the tedious inversion of a basis representing matrix $\mathbf{U}$ in the general basis case. (Refer to Section 7.1 and compare with Theorem 10.2.)

(c) Constructing Orthonormal Bases for Subspaces

Before studying orthonormal bases for a matrix representation of linear transformations, we must learn how to "rectify," "right," or "orthogonalize" a given set of vectors or a given basis.

> **Question:** How can one find an orthonormal basis for a given subspace U?

Assume for simplicity that the given subspace U is determined inclusively (i.e., that $U = \text{span}\{\mathbf{u}_1, \ldots, \mathbf{u}_k\}$; see Sections 3.1 or 4.1(b) for converting a matrix kernel to a span).

For a one–dimensional subspace $U = \text{span}\{\mathbf{u}_1\}$, we only need to normalize the given vector $\mathbf{u}_1 \neq \mathbf{0} \in U$ to obtain the ONB $\left\{\frac{\mathbf{u}_1}{\|\mathbf{u}_1\|}\right\}$ for U.

For a two–dimensional subspace $U = \text{span}\{\mathbf{u}_1, \mathbf{u}_2\}$, we may assume that the two given nonzero generators $\mathbf{u}_1$ and $\mathbf{u}_2$ of U are not orthogonal. (If they are orthogonal, we only need to normalize both in order to find the ONB $\left\{\frac{\mathbf{u}_1}{\|\mathbf{u}_1\|}, \frac{\mathbf{u}_2}{\|\mathbf{u}_2\|}\right\}$ for $U = \text{span}\{\mathbf{u}_1, \mathbf{u}_2\}$.) To solve this orthogonalization problem, we first construct two vectors $\mathbf{v}_1$ and $\mathbf{v}_2 \in U$ that are orthogonal (i.e., that satisfy $\mathbf{v}_1 \cdot \mathbf{v}_2 = 0$) and for which $U = \text{span}\{\mathbf{u}_1, \mathbf{u}_2\} = \text{span}\{\mathbf{v}_1, \mathbf{v}_2\}$.

Then we normalize the orthogonal set $\{\mathbf{v}_1, \mathbf{v}_2\} \subset U$ to obtain an ONB for U.

To start the orthogonalization process for $\mathbf{u}_1$ and $\mathbf{u}_2$, we set

$$\mathbf{v}_1 := \mathbf{u}_1 \neq \mathbf{0}.$$

Next, we construct a specific vector $\mathbf{v}_2 \in \text{span}\{\mathbf{u}_1, \mathbf{u}_2\} = U$ with $\mathbf{v}_2 \perp \mathbf{v}_1$. (See Figure 10-1).

To understand this process, it is important to realize that the vector $\mathbf{x}$ connecting a **starting point s** in space with an **endpoint e** always has the algebraic form

$$\mathbf{x} = \textit{endpoint} - \textit{starting point} = \mathbf{e} - \mathbf{s}.$$

This is so because the vector sum *starting point* $+ \mathbf{x} = \mathbf{s} + (\mathbf{e} - \mathbf{s}) = \mathbf{e}$ is equal to the endpoint $\mathbf{e}$, as depicted in Figure 10-2.

Applying this to Figure 10-1, we have

$$\mathbf{v}_2 := \textit{endpoint} - \textit{starting point} = \mathbf{u}_2 - \alpha \mathbf{v}_1$$

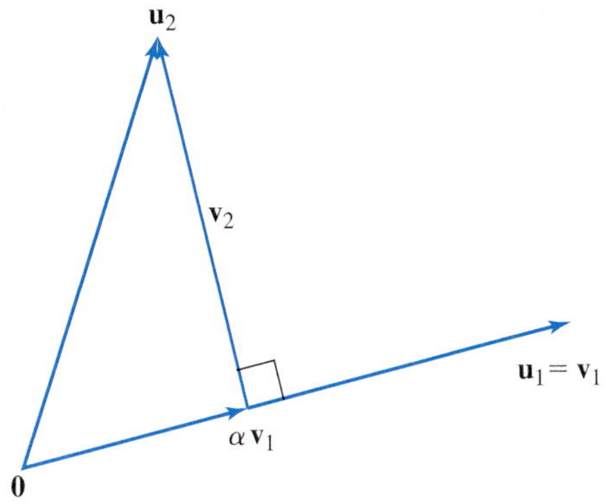

Figure 10-1

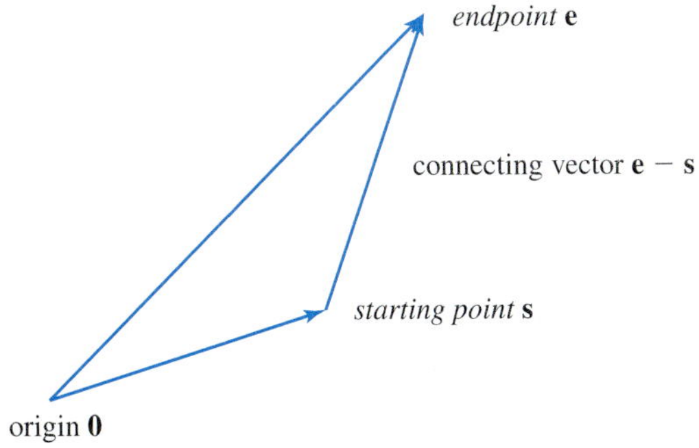

Figure 10-2

for a yet unknown scalar $\alpha \in \mathbb{R}$. For orthogonality, this scalar α must make $\mathbf{v}_2 \perp \mathbf{v}_1$, or

$$\mathbf{v}_2 \cdot \mathbf{v}_1 = (\mathbf{u}_2 - \alpha\mathbf{v}_1) \cdot \mathbf{v}_1 = \mathbf{u}_2 \cdot \mathbf{v}_1 - \alpha\mathbf{v}_1 \cdot \mathbf{v}_1 = 0.$$

Solving for α gives $\alpha = \dfrac{\mathbf{u}_2 \cdot \mathbf{v}_1}{\mathbf{v}_1 \cdot \mathbf{v}_1}$, since $\mathbf{v}_1 \cdot \mathbf{v}_1 = \|\mathbf{v}_1\|^2 \neq 0$.

Looking at Figure 10-1, we now observe that the **distance** of any point $\mathbf{u}_2$ from the line through the origin $\mathbf{0}$ and $\mathbf{u}_1 \neq \mathbf{0}$ is given by the length of the vector $\mathbf{v}_2 = \mathbf{u}_2 - \dfrac{\mathbf{u}_2 \cdot \mathbf{u}_1}{\mathbf{u}_1 \cdot \mathbf{u}_1}\mathbf{u}_1$.

Remark:

Knowing that $\alpha = \dfrac{\mathbf{u}_2 \cdot \mathbf{v}_1}{\mathbf{v}_1 \cdot \mathbf{v}_1}$ allows us to deduce that the definition of the **cosine** of the **angle** between any two nonzero vectors $\mathbf{x}, \mathbf{y} \in \mathbb{R}^n$, as $\cos\angle(\mathbf{x}, \mathbf{y}) := \dfrac{\mathbf{x} \cdot \mathbf{y}}{\|\mathbf{x}\|\ \|\mathbf{y}\|}$ agrees with the intuitive elementary one of $\dfrac{\text{“adjacent”}}{\text{“hypotenuse”}}$ in $\mathbb{R}^2$. For this, we relabel and redraw Figure 10-1 in terms of the two given vectors $\mathbf{x}$ and $\mathbf{y}$, producing Figure 10-3. If $\mathbf{x}$ and $\mathbf{y}$ are linearly independent vectors in $\mathbb{R}^n$, then in the plane $\text{span}\{\mathbf{x}, \mathbf{y}\} \subset \mathbb{R}^n$ we may rely on the standard measure of $\cos(\theta)$ as

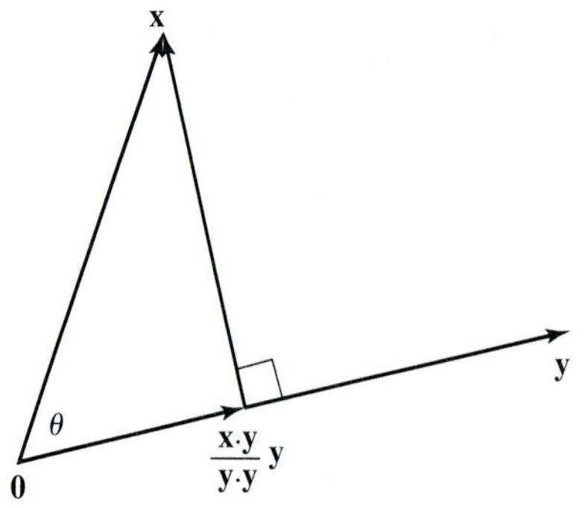

Figure 10-3

the ratio of the length of the signed adjacent leg to the length of the hypotenuse in the supporting right triangle. That is, in Figure 10-3, we have

$$\cos(\theta) = \frac{\text{'adjacent'}}{\text{'hypotenuse'}} = \frac{\pm\left\|\frac{\mathbf{x}\cdot\mathbf{y}}{\|\mathbf{y}\|^2}\mathbf{y}\right\|}{\|\mathbf{x}\|} = \frac{\mathbf{x}\cdot\mathbf{y}}{\|\mathbf{x}\|\ \|\mathbf{y}\|}\left\|\frac{\mathbf{y}}{\|\mathbf{y}\|}\right\| = \frac{\mathbf{x}\cdot\mathbf{y}}{\|\mathbf{x}\|\|\mathbf{y}\|},$$

since the vector $\frac{\mathbf{y}}{\|\mathbf{y}\|}$ has length 1. If $\mathbf{x}$ and $\mathbf{y}$ are linearly dependent, then $\mathbf{x} = \alpha\mathbf{y}$ (or $\alpha\mathbf{x} = \mathbf{y}$) for $\alpha \in \mathbb{R}$ and $\cos(\theta) = 1$ for the angle $\theta = 0°$ sustained between $\mathbf{x}$ and $\mathbf{y}$. If $\mathbf{x} = \alpha\mathbf{y}$, then $\frac{\mathbf{x}\cdot\mathbf{y}}{\|\mathbf{x}\|\ \|\mathbf{y}\|} = \frac{\alpha\|\mathbf{y}\|^2}{\alpha\|\mathbf{y}\|^2} = 1 = \cos(0)$, and similarly in case $\alpha\mathbf{x} = \mathbf{y}$. In other words, the cosine definition as the ratio of dot products and norms generalizes the elementary one from $\mathbb{R}^2$ to $\mathbb{R}^n$.

Let us return to the orthogonalization process: If two vectors $\mathbf{v}_1$ and $\mathbf{v}_2$ are orthogonal, then any scalar multiples of the two are also orthogonal. For two integer vectors $\mathbf{u}_1$ and $\mathbf{u}_2$, we will exploit this observation for ease with hand computations. Accordingly, we replace $\mathbf{v}_2 = \mathbf{u}_2 - \frac{\mathbf{u}_2 \cdot \mathbf{v}_1}{\mathbf{v}_1 \cdot \mathbf{v}_1}\mathbf{v}_1$ by its equivalent integer multiple

$$\|\mathbf{v}_1\|^2\mathbf{v}_2 = \mathbf{v}_1 \cdot \mathbf{v}_1\mathbf{u}_2 - \mathbf{u}_2 \cdot \mathbf{v}_1\mathbf{v}_1.$$

Thus, for $\dim(U) = 2$, we determine an ONB $\{\mathbf{w}_1, \mathbf{w}_2\}$ of $U = \text{span}\{\mathbf{u}_1, \mathbf{u}_2\} = \text{span}\{\mathbf{v}_1, \mathbf{v}_2\}$ in two steps as follows:

1. Set
$$\mathbf{v}_1 := \mathbf{u}_1 \neq \mathbf{0} \text{ and}$$
$$\mathbf{v}_2 := \mathbf{v}_1 \cdot \mathbf{v}_1\mathbf{u}_2 - \mathbf{u}_2 \cdot \mathbf{v}_1\mathbf{v}_1 \quad \textit{(orthogonalize over the integers).}$$
2. Set
$$\mathbf{w}_1 := \frac{\mathbf{v}_1}{\|\mathbf{v}_1\|} \text{ and}$$
$$\mathbf{w}_2 := \frac{\mathbf{v}_2}{\|\mathbf{v}_2\|} \text{ if } \mathbf{v}_2 \neq \mathbf{0} \quad \textit{(normalize over the reals).}$$

This process of orthogonalizing one vector with respect to another is generalized to an arbitrary number k of generators of $U = \text{span}\{\mathbf{u}_1, \dots, \mathbf{u}_k\}$ in the **Gram–Schmidt process**, which automatically determines the dimension of U by discarding linearly dependent spanning vectors $\mathbf{u}_j$ in U.

Gram–Schmidt Orthogonalization Process
Given: k nonzero vectors $\mathbf{u}_1, \ldots, \mathbf{u}_k \in \mathbb{R}^n$.

(a) **Orthogonalize** in $k-1$ levels:

First level : Orthogonalize $\mathbf{u}_2, \ldots, \mathbf{u}_k$ with respect to $\mathbf{u}_1$.

$\mathbf{v}_1 := \mathbf{u}_1;$

$\mathbf{v}_2 := \mathbf{v}_1 \cdot \mathbf{v}_1\mathbf{u}_2 - \mathbf{u}_2 \cdot \mathbf{v}_1\mathbf{v}_1;$

If $\mathbf{v}_2 = \mathbf{0}$, discard $\mathbf{v}_2$, and continue with $\mathbf{u}_3, \ldots, \mathbf{u}_k$. If the $\mathbf{u}_i$ are integer vectors, scale $\mathbf{v}_2$ to have mutually prime integer components.

$\mathbf{v}_3 := \mathbf{v}_1 \cdot \mathbf{v}_1\mathbf{u}_3 - \mathbf{u}_3 \cdot \mathbf{v}_1\mathbf{v}_1;$

If $\mathbf{v}_3 = \mathbf{0}$, discard $\mathbf{v}_3$, and continue with $\mathbf{u}_4, \ldots, \mathbf{u}_k$. If the $\mathbf{u}_i$ are integer vectors, scale $\mathbf{v}_3$ to have mutually prime integer components.

$\vdots$

$\mathbf{v}_k := \mathbf{v}_1 \cdot \mathbf{v}_1\mathbf{u}_k - \mathbf{u}_k \cdot \mathbf{v}_1\mathbf{v}_1;$

Discard or scale as warranted.

Now, span$\{\mathbf{v}_2, \ldots, \mathbf{v}_k\} \perp \mathbf{v}_1$. This should be checked i.e., the $k-1$ dot products $\mathbf{v}_1 \cdot \mathbf{v}_i$ should be zero for $i = 2, \ldots, k$.

Second level : Orthogonalize $\mathbf{v}_3, \ldots, \mathbf{v}_k$ with respect to $\mathbf{v}_2$.

$\mathbf{v}_3 := \mathbf{v}_2 \cdot \mathbf{v}_2\mathbf{v}_3 - \mathbf{v}_3 \cdot \mathbf{v}_2\mathbf{v}_2;$

Discard or scale as warranted.

$\vdots$

$\mathbf{v}_k := \mathbf{v}_2 \cdot \mathbf{v}_2\mathbf{v}_k - \mathbf{v}_k \cdot \mathbf{v}_2\mathbf{v}_2;$

Discard or scale as warranted.

Now, span$\{\mathbf{v}_3, \ldots, \mathbf{v}_k\} \perp \mathbf{v}_1, \mathbf{v}_2$. This should be checked by verifying that $\mathbf{v}_1 \cdot \mathbf{v}_i = 0$ and $\mathbf{v}_2 \cdot \mathbf{v}_i = 0$ for $i = 3, \ldots, k$.

$\vdots \qquad \vdots$

$(k-2)^{nd}$ level : Orthogonalize $\mathbf{v}_{k-1}$ and $\mathbf{v}_k$ with respect to $\mathbf{v}_{k-2}$.

$\mathbf{v}_{k-1} := \mathbf{v}_{k-2} \cdot \mathbf{v}_{k-2}\mathbf{v}_{k-1} - \mathbf{v}_{k-1} \cdot \mathbf{v}_{k-2}\mathbf{v}_{k-2};$

Discard or scale as warranted.

$\mathbf{v}_k := \mathbf{v}_{k-2} \cdot \mathbf{v}_{k-2}\mathbf{v}_k - \mathbf{v}_k \cdot \mathbf{v}_{k-2}\mathbf{v}_{k-2};$

Discard or scale as warranted.

Now, span$\{\mathbf{v}_{k-1}, \mathbf{v}_k\} \perp \mathbf{v}_1, \mathbf{v}_2, \ldots, \mathbf{v}_{k-2}$. This should be checked.

And finally:

$(k-1)^{st}$ level : Orthogonalize $\mathbf{v}_k$ with respect to $\mathbf{v}_{k-1}$.

$$\mathbf{v}_k \;:=\; \mathbf{v}_{k-1} \cdot \mathbf{v}_{k-1}\mathbf{v}_k - \mathbf{v}_k \cdot \mathbf{v}_{k-1}\mathbf{v}_{k-1};$$

Discard or scale as warranted.

Now $\mathbf{v}_k \perp \mathbf{v}_1, \mathbf{v}_2, \dots, \mathbf{v}_{k-2}, \mathbf{v}_{k-1}$. Check this.

The computed nonzero vectors $\{\mathbf{v}_i\}_{i=1}^{\ell}$ with $\ell \le k$ are mutually orthogonal, linearly independent, and span the subspace generated by $\mathbf{u}_1, \dots, \mathbf{u}_k$. (If all given vectors $\mathbf{u}_i$ have integer components, it is best for hand computations to "stretch" or "shrink" each intermediate vector $\mathbf{v}_j$ so that it has integer and relatively prime components. Orthogonality is not affected by rescaling the intermediate vectors $\mathbf{v}_j$.)

(b) **Normalize:**

$$\mathbf{w}_1 := \frac{\mathbf{v}_1}{\|\mathbf{v}_1\|}, \dots, \mathbf{w}_j := \frac{\mathbf{v}_j}{\|\mathbf{v}_j\|}, \dots, \mathbf{w}_\ell := \frac{\mathbf{v}_\ell}{\|\mathbf{v}_\ell\|}.$$

The $\{\mathbf{w}_i\}$ for $i = 1, \dots, \ell$ form an ONB for $U = \text{span}\{\mathbf{u}_1, \dots, \mathbf{u}_k\}$; $\dim U = \ell$.

The process of orthogonalizing k vectors $\mathbf{u}_1, \dots, \mathbf{u}_k \in \mathbb{R}^n$ in levels via Gram–Schmidt is analogous to row reducing the matrix $\mathbf{A} = \begin{pmatrix} - & \mathbf{u}_1 & - \\ & \vdots & \\ - & \mathbf{u}_k & - \end{pmatrix}$ via Gaussian elimination in the following sense:

For simplicity, we assume that the given row vectors $\mathbf{u}_i$ are linearly independent and that the complete row reduction of $\mathbf{A}$ to a REF R can proceed without any swaps. The first-level sweep of row reduction on A creates equivalent rows $\widetilde{\mathbf{u}}_j$ for $j = 2, \dots, k$ that lie in the coordinate plane span $\{\mathbf{e}_2, \dots, \mathbf{e}_n\} \subset \mathbb{R}^n$, since their first entries are all equal to zero. Geometrically, the row reduction of one row $\mathbf{u}_j$ via the pivot $u_{11} \neq 0$ of row $\mathbf{u}_1$ projects $\mathbf{u}_j$ onto span $\{\mathbf{e}_2, \dots, \mathbf{e}_n\}$ by using $\mathbf{u}_1$ and the elimination coefficient $-\frac{u_{j1}}{u_{11}}$. (See Chapter 2 and Figure 10-4.)

Having found $\widetilde{\mathbf{u}}_2 \in \text{span}\{\mathbf{u}_1, \mathbf{u}_2\}$ with first entry $\widetilde{u}_{21} = 0$ and second entry $\widetilde{u}_{22} \neq 0$ as assumed, the further row reduction of $\mathbf{A}$ proceeds by using $\widetilde{\mathbf{u}}_2$ to update each $\widetilde{\mathbf{u}}_j \in$ span $\{\mathbf{e}_2, \dots, \mathbf{e}_n\}$ to a vector that lies in the more restricted subspace span$\{\mathbf{e}_3, \dots, \mathbf{e}_n\}$ of $\mathbb{R}^n$ for $j = 3, \dots, k$, etc. In other words, to create a REF R for $\mathbf{A}$ via Gaussian elimination, at the first level we alter $n-1$ row vectors in $\mathbf{A}$ via $\mathbf{A}$'s first row $\mathbf{u}_1$, then at the second level we update the trailing updated $n-2$ rows of $\mathbf{A}$ via its second updated row, then we update the trailing $n-3$ rows via a third row, etc.

The Gram–Schmidt algorithm operates in levels in much the same way. Here, however, we update with an eye on orthogonality rather than zero leading coefficients.

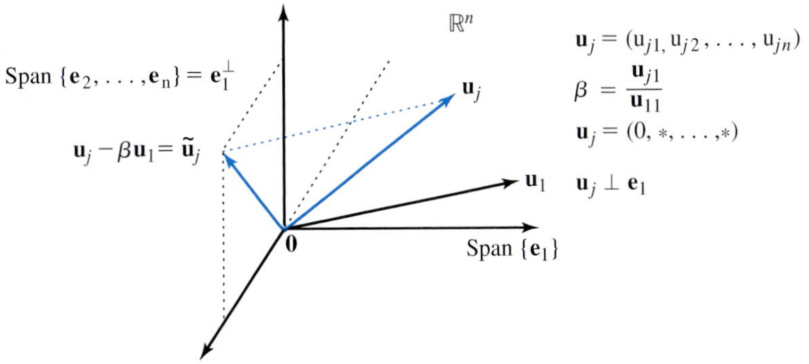

Figure 10-4

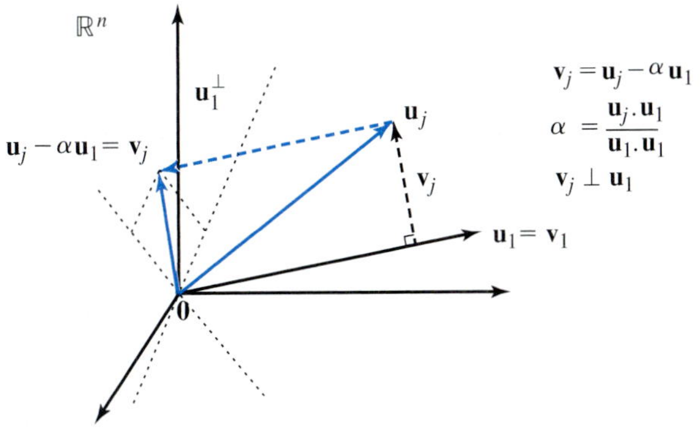

Figure 10-5

First, we replace each $\mathbf{u}_j$ for $j = 2, \ldots, k$ by a vector $\mathbf{v}_j \in \text{span}\{\mathbf{u}_1, \mathbf{u}_j\}$ that lies in the $n-1$-dimensional subspace $\mathbf{u}_1^{\perp} = \{\mathbf{v} \in \mathbb{R}^n \mid \mathbf{v} \cdot \mathbf{u}_1 = 0\}$, as depicted in Figure 10-5. Then we update the newly computed vectors $\mathbf{v}_3, \ldots, \mathbf{v}_n$ to lie in $\text{span}\{\mathbf{v}_2, \mathbf{v}_j\}$ <u>and</u> in $\mathbf{v}_2^{\perp}$ for $j = 3, \ldots, k$, etc., until we normalize.

The Gram–Schmidt process clearly has the same level mechanism of updating vectors, but it has a different objective than Gaussian elimination.

EXAMPLE 1 Find an ONB $\mathcal{W} = \{\mathbf{w}_j\}$ for $U = \text{span}\{\mathbf{u}_i\} = \text{span}\left\{\begin{pmatrix} 4 \\ 0 \\ 3 \end{pmatrix}, \begin{pmatrix} 1 \\ 1 \\ 0 \end{pmatrix}, \begin{pmatrix} 0 \\ 0 \\ -1 \end{pmatrix}\right\}$.

How far is $\mathbf{u}_2 = \begin{pmatrix} 1 \\ 1 \\ 0 \end{pmatrix}$ from the line connecting the origin and $\mathbf{u}_1 = \begin{pmatrix} 4 \\ 0 \\ 3 \end{pmatrix}$?

We are given $k = 3$ vectors and need $k - 1$, or two, levels of Gram–Schmidt. We start at the first level and orthogonalize both $\mathbf{u}_2$ and $\mathbf{u}_3$ with respect to $\mathbf{u}_1$:

$$\mathbf{v}_1 := \mathbf{u}_1 = \begin{pmatrix} 4 \\ 0 \\ 3 \end{pmatrix};$$

$$\mathbf{v}_2 := \mathbf{v}_1 \cdot \mathbf{v}_1 \mathbf{u}_2 - \mathbf{u}_2 \cdot \mathbf{v}_1 \mathbf{v}_1 = 25 \begin{pmatrix} 1 \\ 1 \\ 0 \end{pmatrix} - 4 \begin{pmatrix} 4 \\ 0 \\ 3 \end{pmatrix}$$

$$= \begin{pmatrix} 25 - 16 \\ 25 \\ -12 \end{pmatrix} = \begin{pmatrix} 9 \\ 25 \\ -12 \end{pmatrix} \perp \mathbf{v}_1;$$

$$\mathbf{v}_3 := \mathbf{v}_1 \cdot \mathbf{v}_1 \mathbf{u}_3 - \mathbf{u}_3 \cdot \mathbf{v}_1 \mathbf{v}_1 = 25 \begin{pmatrix} 0 \\ 0 \\ -1 \end{pmatrix} + 3 \begin{pmatrix} 4 \\ 0 \\ 3 \end{pmatrix}$$

$$= \begin{pmatrix} 12 \\ 0 \\ -25 + 9 \end{pmatrix} = \begin{pmatrix} 12 \\ 0 \\ -16 \end{pmatrix} \simeq \begin{pmatrix} 3 \\ 0 \\ -4 \end{pmatrix} =: \mathbf{v}_3 \perp \mathbf{v}_1.$$

Here, we have denoted the scaling of $\mathbf{v}_3$ by one-fourth symbolically using the symbol $\simeq$. Scaling vectors for our convenience has no influence on their orthogonality. That $\mathbf{v}_2$ and $\mathbf{v}_3$ are both orthogonal to $\mathbf{v}_1$ should be checked by evaluating the two dot products $\mathbf{v}_1 \cdot \mathbf{v}_2$ and $\mathbf{v}_1 \cdot \mathbf{v}_3$.

The required distance from $\begin{pmatrix} 1 \\ 1 \\ 0 \end{pmatrix}$ to the line through the origin and $\mathbf{u}_1$ is given by

$$\left\| \mathbf{u}_2 - \frac{\mathbf{u}_2 \cdot \mathbf{v}_1}{\mathbf{v}_1 \cdot \mathbf{v}_1} \mathbf{v}_1 \right\| = \frac{1}{\mathbf{v}_1 \cdot \mathbf{v}_1} \| \mathbf{v}_1 \cdot \mathbf{v}_1 \mathbf{u}_2 - \mathbf{u}_2 \cdot \mathbf{v}_1 \mathbf{v}_1 \|$$

$$= \frac{1}{25} \left\| \begin{pmatrix} 9 \\ 25 \\ -12 \end{pmatrix} \right\| = \frac{\sqrt{81 + 625 + 144}}{25} = \frac{\sqrt{850}}{25} \approx \frac{29.15}{25} \doteq 1.166.$$

We continue with second-level Gram–Schmidt and orthogonalize the newly computed vector $\mathbf{v}_3$ with respect to the new $\mathbf{v}_2$:

$$\mathbf{v}_3 = \mathbf{v}_2 \cdot \mathbf{v}_2 \mathbf{v}_3 - \mathbf{v}_3 \cdot \mathbf{v}_2 \mathbf{v}_2 = 850 \begin{pmatrix} 3 \\ 0 \\ -4 \end{pmatrix} - 75 \begin{pmatrix} 9 \\ 25 \\ -12 \end{pmatrix} \quad \text{(scale by 1/25)}$$

$$\simeq 34 \begin{pmatrix} 3 \\ 0 \\ -4 \end{pmatrix} - 3 \begin{pmatrix} 9 \\ 25 \\ -12 \end{pmatrix} = \begin{pmatrix} 102 - 27 \\ -75 \\ -136 + 36 \end{pmatrix} = \begin{pmatrix} 75 \\ -75 \\ -100 \end{pmatrix} \quad \text{(scale by 1/25)}$$

$$\simeq \begin{pmatrix} 3 \\ -3 \\ -4 \end{pmatrix} =: \mathbf{v}_3 \perp \mathbf{v}_1, \mathbf{v}_2.$$

Again, we need to check the orthogonality of the computed vector $\mathbf{v}_3$ to $\mathbf{v}_2$ and $\mathbf{v}_1$. (*Note:* For integer starting vectors $\mathbf{u}_i$ our Gram–Schmidt process avoids fractional arithmetic until its normalization stage.)

Normalizing the final vectors $\mathbf{v}_1, \mathbf{v}_2$, and $\mathbf{v}_3$ gives us the orthogonal set of vectors $\mathbf{w}_1, \mathbf{w}_2$, and $\mathbf{w}_3$ with $\text{span}\{\mathbf{w}_i\} = \text{span}\{\mathbf{v}_i\} = \text{span}\{\mathbf{u}_i\} = U$. The ONB $\mathcal{W} = \{\mathbf{w}_i\}_{i=1}^3$ of $\mathbb{R}^3$ is given by the three mutually orthogonal unit vectors

$$\mathbf{w}_1 = \frac{\mathbf{v}_1}{\|\mathbf{v}_1\|} = \frac{1}{5}\begin{pmatrix} 4 \\ 0 \\ 3 \end{pmatrix}, \mathbf{w}_2 = \frac{\mathbf{v}_2}{\|\mathbf{v}_2\|} = \frac{1}{\sqrt{850}}\begin{pmatrix} 9 \\ 25 \\ -12 \end{pmatrix}, \text{ and}$$

$$\mathbf{w}_3 = \frac{\mathbf{v}_3}{\|\mathbf{v}_3\|} = \frac{1}{\sqrt{34}}\begin{pmatrix} 3 \\ -3 \\ -4 \end{pmatrix}.$$

◀

We mention that orthogonalizing any but pre-cooked-up sets of vectors is a horrible task to do by hand. This chapter calls out for the use of a computer.

The next item concerns **basis extensions**: If $\{\mathbf{u}_1, \dots, \mathbf{u}_k\}$ are linearly independent vectors in $\mathbb{R}^n$ with $k < n$, it is often desirable to extend the set of $\mathbf{u}_i$ to a basis of all of $\mathbb{R}^n$. One can do so by following Chapter 4 and finding a basis for the nullspace of dimension $n - k$ for the matrix $\mathbf{B}_{kn} = \begin{pmatrix} - & \mathbf{u}_1^T & - \\ & \vdots & \\ - & \mathbf{u}_k^T & - \end{pmatrix}$. The kernel of this matrix $\mathbf{B}$ consists of all vectors in $\mathbb{R}^n$ that are orthogonal to the given vectors $\mathbf{u}_i$. Hence, we have the following method for obtaining an orthonormal basis extension:

Proposition 1 Let $\{\mathbf{u}_1, \dots, \mathbf{u}_k\}$ be an orthonormal set of vectors in $\mathbb{R}^n$ with $k < n$. Then this set of vectors can be extended to an ONB $\mathbf{u}_1, \dots, \mathbf{u}_k, \mathbf{w}_{k+1}, \dots, \mathbf{w}_n$ of $\mathbb{R}^n$ by choosing a spanning set $\mathbf{v}_{k+1}, \dots, \mathbf{v}_n$ of $\ker\begin{pmatrix} - & \mathbf{u}_1^T & - \\ & \vdots & \\ - & \mathbf{u}_k^T & - \end{pmatrix}$ and orthonormalizing this set via Gram–Schmidt to an ONB $\{\mathbf{w}_{k+1}, \dots, \mathbf{w}_n\}$ of the kernel. ◀

In later chapters, we often have to create an orthogonal matrix $\mathbf{U}_{nn}$ with prescribed orthonormal first columns $\mathbf{u}_1, \dots, \mathbf{u}_k$. We refer to this basis-extension process as **orthonormal fill–in**.

(d) Orthogonal Projections onto a Subspace

Let V be a subspace of $\mathbb{R}^n$ with the ONB $\mathcal{V} = \{\mathbf{v}_1, \dots, \mathbf{v}_k\}$. Then we can easily decide whether a vector $\mathbf{u} \in \mathbb{R}^n$ lies in V, what its distance is to V if it does not, and which vector $\mathbf{v} \in V$ is closest to $\mathbf{u}$. All this can be determined by using the geometry that underlies the Gram–Schmidt process.

Proposition 2 Assume that the subspace $V \subset \mathbb{R}^n$ has the ONB $\mathcal{V} = \{\mathbf{v}_1, \dots, \mathbf{v}_k\}$ and that $\mathbf{u} \in \mathbb{R}^n$. Then the following hold true:

(a) For the vector $\mathbf{v}$ defined as

$$\mathbf{v} := \mathbf{v}_1 \cdot \mathbf{u}\ \mathbf{v}_1 + \cdots + \mathbf{v}_k \cdot \mathbf{u}\ \mathbf{v}_k,$$

the vector $\mathbf{u} - \mathbf{v}$ is orthogonal to every vector $\mathbf{w} \in V$.

(b) The shortest **distance** between $\mathbf{u} \in \mathbb{R}^n$ and the subspace V is

$$\text{dist}(\mathbf{u}, V) := \|\mathbf{u} - \mathbf{v}\|,$$

with $\mathbf{v}$ defined as in part (a).

(c) The vector $\mathbf{v} \in V$ that is **closest** to $\mathbf{u} \in \mathbb{R}^n$ is

$$\mathbf{v} = \mathbf{v}_1 \cdot \mathbf{u}\ \mathbf{v}_1 + \cdots + \mathbf{v}_k \cdot \mathbf{u}\ \mathbf{v}_k$$

from part (a). Here, $\mathbf{v}$ is the **orthogonal projection** of the vector $\mathbf{u}$ onto the subspace V.

(d) The vector $\mathbf{u} \in \mathbb{R}^n$ lies in V if and only if

$$\mathbf{u} = \mathbf{v}_1 \cdot \mathbf{u}\,\mathbf{v}_1 + \cdots + \mathbf{v}_k \cdot \mathbf{u}\,\mathbf{v}_k,$$

or equivalently, if and only if $\text{dist}(\mathbf{u}, V) = 0$. ◀

For a set $\{\mathbf{v}_1, \ldots, \mathbf{v}_k\}$ of orthonormal vectors in $\mathbb{R}^n$, the assignment of a vector

$$\mathbf{u} \in \mathbb{R}^n \longmapsto \mathbf{v} = \mathbf{v}_1 \cdot \mathbf{u}\,\mathbf{v}_1 + \cdots + \mathbf{v}_k \cdot \mathbf{u}\,\mathbf{v}_k \in V$$

is a linear transformation that projects all of $\mathbb{R}^n$ orthogonally onto the subspace $V = \text{span}\{\mathbf{v}_1, \ldots, \mathbf{v}_k\} \subset \mathbb{R}^n$. The matrix representation $\mathbf{P}_V$ of this **orthogonal projection** is given by the formula

$$\mathbf{P}_V(\mathbf{u}) = \begin{pmatrix} | & & | \\ \mathbf{v}_1 & \cdots & \mathbf{v}_k \\ | & & | \end{pmatrix} \begin{pmatrix} - & \mathbf{v}_1^T & - \\ & \vdots & \\ - & \mathbf{v}_k^T & - \end{pmatrix} \mathbf{u} = \mathbf{v},$$

according to part (a) of Proposition 2. That is, $(\mathbf{P}_V)_{\mathcal{E}} = \mathbf{V}\mathbf{V}^T$ for $\mathbf{V} = \begin{pmatrix} | & & | \\ \mathbf{v}_1 & \cdots & \mathbf{v}_k \\ | & & | \end{pmatrix}_{nk}$. Note that $\mathbf{V}\mathbf{V}^T$ is an $n \times n$ matrix of rank k that differs from $\mathbf{I}_n$ if $k < n$. On the other hand, $\mathbf{V}^T\mathbf{V} = \mathbf{I}_k$ since the columns $\mathbf{v}_i$ of $\mathbf{V}$ form an ONB for V of dimension $k \leq n$. In fact, the matrix $\mathbf{P}_V$ of the orthogonal projection onto the subspace V is independent of the chosen ONB. (See Problem 34.)

The Gram–Schmidt process uses orthogonal projections repeatedly to find an ONB of an arbitrary subspace $\text{span}\{\mathbf{u}_1, \ldots, \mathbf{u}_k\} \subset \mathbb{R}^n$. (Refer to Figure 10-1.)

EXAMPLE 2 Let $V = \text{span}\left\{ \frac{1}{\sqrt{3}} \begin{pmatrix} 1 \\ 1 \\ 1 \end{pmatrix}, \frac{1}{\sqrt{2}} \begin{pmatrix} 1 \\ -1 \\ 0 \end{pmatrix} \right\} \subset \mathbb{R}^3$ and $\mathbf{u} = \begin{pmatrix} 1 \\ 2 \\ 3 \end{pmatrix} \in \mathbb{R}^3$.

(a) The orthogonal projection of **u** onto V with its given ONB $\mathcal{V}$ is the point

$$\mathbf{P}_V(\mathbf{u}) = \mathbf{V}\mathbf{V}^T\mathbf{u} = \begin{pmatrix} \frac{1}{\sqrt{3}} & \frac{1}{\sqrt{2}} \\ \frac{1}{\sqrt{3}} & -\frac{1}{\sqrt{2}} \\ \frac{1}{\sqrt{3}} & 0 \end{pmatrix} \begin{pmatrix} \frac{1}{\sqrt{3}} & \frac{1}{\sqrt{3}} & \frac{1}{\sqrt{3}} \\ \frac{1}{\sqrt{2}} & -\frac{1}{\sqrt{2}} & 0 \end{pmatrix} \begin{pmatrix} 1 \\ 2 \\ 3 \end{pmatrix}$$

$$= \begin{pmatrix} \frac{1}{\sqrt{3}} & \frac{1}{\sqrt{2}} \\ \frac{1}{\sqrt{3}} & -\frac{1}{\sqrt{2}} \\ \frac{1}{\sqrt{3}} & 0 \end{pmatrix} \begin{pmatrix} \frac{6}{\sqrt{3}} \\ -\frac{1}{\sqrt{2}} \end{pmatrix} = \begin{pmatrix} 1.5 \\ 2.5 \\ 2 \end{pmatrix}.$$

Thus, the vector $\mathbf{v} = \begin{pmatrix} 1.5 \\ 2.5 \\ 2 \end{pmatrix} \in V$ is the vector that is closest to the given $\mathbf{u} \in \mathbb{R}^3$. The distance from **u** to the subspace V is $\text{dist}(\mathbf{u}, V) = \|\mathbf{u} - \mathbf{v}\| = \left\| \begin{pmatrix} -0.5 \\ -0.5 \\ 1 \end{pmatrix} \right\| = \sqrt{1.5}$.

(b) Determine the third component of the vector $\mathbf{w} = \begin{pmatrix} 2 \\ 1 \\ a \end{pmatrix}$ so that $\mathbf{w} \in V$. For **w**, we compute $\mathbf{P}_V(\mathbf{w}) = \begin{pmatrix} \frac{3+a}{3} + \frac{1}{2} \\ \frac{3+a}{3} - \frac{1}{2} \\ \frac{3+a}{3} \end{pmatrix}$. Thus, $\mathbf{P}_V(\mathbf{w}) = \mathbf{w} = \begin{pmatrix} 2 \\ 1 \\ a \end{pmatrix}$ precisely for $a = \frac{3}{2}$. In other words, the line $\left\{ \begin{pmatrix} 2 \\ 1 \\ 0 \end{pmatrix} + a \begin{pmatrix} 0 \\ 0 \\ 1 \end{pmatrix} \right\} \subset \mathbb{R}^3$ intersects the given subspace V at $\begin{pmatrix} 2 \\ 1 \\ 1.5 \end{pmatrix}$.

(c) Using part (c) of Proposition 2 instead, we can find $\mathbf{P}_V(\mathbf{u})$ for $\mathbf{u} = \begin{pmatrix} 1 \\ 2 \\ 3 \end{pmatrix}$ more quickly as $\mathbf{P}_V(\mathbf{u}) := \mathbf{v} := \mathbf{v}_1 \cdot \mathbf{u}\ \mathbf{v}_1 + \mathbf{v}_2 \cdot \mathbf{u}\ \mathbf{v}_2 = \frac{6}{\sqrt{3}}\mathbf{v}_1 - \frac{1}{\sqrt{2}}\mathbf{v}_2 = \frac{6}{3}\begin{pmatrix} 1 \\ 1 \\ 1 \end{pmatrix} - \frac{1}{2}\begin{pmatrix} 1 \\ -1 \\ 0 \end{pmatrix} = \begin{pmatrix} 1.5 \\ 2.5 \\ 2 \end{pmatrix}$. ◀

Remark on our choice of algorithms:

For a given set of vectors $\mathbf{u}_1, \ldots, \mathbf{u}_k \in \mathbb{R}^n$, the Gram–Schmidt algorithm is most often described differently in elementary textbooks as follows:

$$\begin{aligned}
\mathbf{v}_1 &:= \mathbf{u}_1; \quad \text{if } \mathbf{v}_1 = \mathbf{0}, \text{ discard } \mathbf{u}_1 \text{ and restart the process with } \mathbf{u}_2. \\
\mathbf{v}_2 &:= \mathbf{u}_2 - \frac{\mathbf{u}_2 \cdot \mathbf{v}_1}{\mathbf{v}_1 \cdot \mathbf{v}_1}\mathbf{v}_1; \text{ if } \mathbf{v}_2 = \mathbf{0}, \text{ discard } \mathbf{u}_2 \\
&\qquad \text{and continue with } \mathbf{u}_3, \ldots, \mathbf{u}_k. \\
\mathbf{v}_3 &:= \mathbf{u}_3 - \frac{\mathbf{u}_3 \cdot \mathbf{v}_1}{\mathbf{v}_1 \cdot \mathbf{v}_1}\mathbf{v}_1 - \frac{\mathbf{u}_3 \cdot \mathbf{v}_2}{\mathbf{v}_2 \cdot \mathbf{v}_2}\mathbf{v}_2; \\
&\qquad \text{if } \mathbf{v}_3 = \mathbf{0}, \text{ discard } \mathbf{u}_3, \text{ and continue with } \mathbf{u}_4, \ldots, \mathbf{u}_k. \\
&\;\vdots \\
\mathbf{v}_j &:= \mathbf{u}_j - \frac{\mathbf{u}_j \cdot \mathbf{v}_1}{\mathbf{v}_1 \cdot \mathbf{v}_1}\mathbf{v}_1 - \cdots - \frac{\mathbf{u}_j \cdot \mathbf{v}_{j-1}}{\mathbf{v}_{j-1} \cdot \mathbf{v}_{j-1}}\mathbf{v}_{j-1}; \\
&\qquad \text{if } \mathbf{v}_j = \mathbf{0}, \text{ discard and proceed analogously.} \\
&\;\vdots
\end{aligned}$$

Lastly, normalize each $\mathbf{v}_j$ to $\mathbf{w}_j$ of unit length.

This variant is the **classical Gram–Schmidt algorithm**. It replaces each given vector $\mathbf{u}_j$ by $\mathbf{v}_j = \mathbf{u}_j - \mathbf{P}_{\text{span}\{\mathbf{v}_1,\ldots,\mathbf{v}_{j-1}\}}\mathbf{u}_j$ according to Proposition 2. In other words, it computes the vector connecting the projection of $\mathbf{u}_j$ onto the subspace span$\{\mathbf{v}_1, \ldots, \mathbf{v}_{j-1}\}$ and $\mathbf{u}_j$. The computed vector $\mathbf{v}_j$ is orthogonal to the subspace span$\{\mathbf{v}_1, \ldots, \mathbf{v}_{j-1}\}$ with its ONB $\mathcal{V}_{j-1} := \{\mathbf{v}_1, \ldots, \mathbf{v}_{j-1}\}$ by Proposition 2(a).

The classical Gram-Schmidt algorithm uses the same intermediate scalars $\mathbf{v}_j \cdot \mathbf{v}_j$ and $\mathbf{u}_j \cdot \mathbf{v}_i$, and the same ratios thereof, as the earlier **modified Gram–Schmidt algorithm** does. Only the order of the vector modifications is different. In theory, the two versions find the identical orthogonal vectors $\mathbf{v}_1, \ldots, \mathbf{v}_\ell$ and the same ONB $\{\mathbf{w}_1, \ldots, \mathbf{w}_\ell\}$ in exactly the same number of operations. However, the classical Gram–Schmidt algorithm is numerically unstable (i.e., even for moderate numbers of starting vectors $\mathbf{u}_1, \ldots, \mathbf{u}_k$, its computed output vectors $\mathbf{w}_j$ are, in general, not precisely orthogonal). On the other hand, the modified Gram–Schmidt algorithm always leads stably to mutually orthogonal vectors and to an ONB of span$\{\mathbf{u}_i\}$. It does so more simply by projecting only onto one-dimensional subspaces throughout.

Unfortunately, many standardly taught algorithms are **unstable**. The quadratic formula for solving quadratic equations is such an unstable algorithm. See Appendices A and B.

And so is **Cramer's rule** (see Problem 42 in Section 8.1.P), which expresses the individual components x_i of the solution vector $\mathbf{x}$ for a linear system $\mathbf{Ax} = \mathbf{b}$ with an invertible system matrix $\mathbf{A}$ in terms of determinantal quotients as follows:

$$x_i = \frac{\det\begin{pmatrix} | & & | & | & | & & | \\ \mathbf{a}_1 & \cdots & \mathbf{a}_{i-1} & \mathbf{b} & \mathbf{a}_{i+1} & \cdots & \mathbf{a}_n \\ | & & | & | & | & & | \end{pmatrix}}{\det(\mathbf{A})}.$$

Here, $\mathbf{A} = \begin{pmatrix} | & & | \\ \mathbf{a}_1 & \ldots & \mathbf{a}_n \\ | & & | \end{pmatrix}$ is written columnwise, and the right-hand side $\mathbf{b}$ is inserted in place of column $\mathbf{a}_i$ of $\mathbf{A}$ in the numerator to obtain x_i.

While some unstable methods, such as Cramer's rule, have theoretical uses, there is no reason to perpetuate teaching or learning an algorithm in an unstable version, especially if the stable form can be taught and learned just as easily. Even the most pedagogically perfect derivation of an unstable algorithm is just that: obsolete.

As teachers, we must measure our lesson contents analogously to the Hippocratic oath of doctors: `never to harm the student's grasp of a subject.`

An orthonormal basis can be easily constructed from a general basis.

10.1.P Problems

1. (a) Find one nonzero vector $\mathbf{x} \in \mathbb{R}^3$ that is orthogonal to $(1\ 1\ 1)$.

   (b) How many linearly independent such vectors are there in $\mathbb{R}^3$?

2. Find two linearly independent vectors $\mathbf{y}, \mathbf{z} \in \mathbb{R}^3$ that are orthogonal to $(1\ -1\ 1)$.

3. (a) Can you find three linearly independent vectors that are orthogonal to $(1\ -1\ -1)$? Why, or why not? How?

   (b) Repeat part (a) for the zero vector.

4. Find a vector $\mathbf{x} \in \mathbb{R}^4$ that is orthogonal to each of the rows of $\mathbf{A} = \begin{pmatrix} 1 & 7 & 0 & 2 \\ 2 & 2 & -3 & 4 \\ -1 & 1 & 2 & -2 \end{pmatrix}$.

5. Repeat Problem 1 of Section 8.3.P. Find the area of the triangle with vertices of the origin, $\mathbf{v}_1 = (1, 3)$, and $\mathbf{v}_2 = (-4, 17)$ as follows:

   (a) What is the distance of $\mathbf{v}_1$ to the origin? (This will be the length of the base of the triangle.)

   (b) What is the distance of $\mathbf{v}_2$ to the line through $\mathbf{v}_1$ and the origin?

   (c) Evaluate the triangular area by the formula $A = \frac{1}{2} base \times height$. Compare.

6. Find an ONB for the subspace $U = \text{span}$ $\left\{ \begin{pmatrix} 1 \\ 0 \\ 0 \\ 2 \end{pmatrix}, \begin{pmatrix} -2 \\ 1 \\ 0 \\ 1 \end{pmatrix}, \begin{pmatrix} 1 \\ -2 \\ 3 \\ -4 \end{pmatrix}, \begin{pmatrix} 3 \\ 1 \\ -4 \\ 9 \end{pmatrix} \right\}$. What is the dimension of $U \subset \mathbb{R}^4$?

7. Find an ONB for the column space of the matrix

$$\mathbf{U} = \begin{pmatrix} 1 & 0 & 3 \\ -1 & 1 & 2 \\ 2 & 0 & 6 \end{pmatrix}.$$

8. Show that the rows of an orthogonal matrix $\mathbf{U}$ are mutually orthonormal.

9. If $\mathbf{U} = \{\mathbf{u}_1, \ldots, \mathbf{u}_k\} \subset \mathbb{R}^n$ is an ONB and $\mathbf{x} = \sum_i \alpha_i \mathbf{u}_i \in U$, then $\|\mathbf{x}\| = \sqrt{\sum_i \alpha_i^2}$.

10. If $U = \{\mathbf{u}_1, \ldots, \mathbf{u}_k\} \subset \mathbb{R}^n$ is an ONB, then $\mathbf{x} = \sum_i ((\mathbf{u}_i \cdot \mathbf{x})\mathbf{u}_i)$ and $\|\mathbf{x}\| = \sqrt{\sum_i (\mathbf{u}_i \cdot \mathbf{x})^2}$ for each $\mathbf{x} \in U$.

11. Find an orthonormal basis for the space spanned by $\mathbf{u}_1 = \begin{pmatrix} 3 \\ 0 \\ -4 \\ 5 \end{pmatrix}$, $\mathbf{u}_2 = \begin{pmatrix} 8 \\ -3 \\ 1 \\ -4 \end{pmatrix}$, and $\mathbf{u}_3 = \begin{pmatrix} 1 \\ -5 \\ -3 \\ -3 \end{pmatrix}$.

12. Find an integer orthogonal basis for the span of $\mathbf{u}_1 = \begin{pmatrix} 1 \\ 0 \\ 1 \\ 0 \end{pmatrix}$ and $\mathbf{u}_2 = \begin{pmatrix} 0 \\ 1 \\ 1 \\ 1 \end{pmatrix}$.

13. Find the orthogonal projection $\alpha \mathbf{v}_1$ of the vector $\mathbf{v}_2 = \begin{pmatrix} 2 \\ -3 \end{pmatrix}$ onto the line through the origin and $\mathbf{v}_1 = \begin{pmatrix} 1 \\ 4 \end{pmatrix}$. (See Figure 10-1.)

14. (a) Find the orthogonal projection $\mathbf{P}_V(\mathbf{v}_3) = \alpha\mathbf{v}_1 + \beta\mathbf{v}_2$ of the vector $\mathbf{v}_3 = \begin{pmatrix} 2 \\ -3 \\ 1 \end{pmatrix}$ onto the plane V through the origin, $\mathbf{v}_1 = \begin{pmatrix} 1 \\ 4 \\ 0 \end{pmatrix}$, and $\mathbf{v}_2 = \begin{pmatrix} 1 \\ 1 \\ -1 \end{pmatrix} \in \mathbb{R}^3$.

(b) Repeat part (a) for the non-orthogonal projection $\widetilde{\mathbf{P}}_V(\mathbf{v}_3) :=$

$$= \begin{pmatrix} | & | \\ \mathbf{v}_1 & \mathbf{v}_2 \\ | & | \end{pmatrix} \begin{pmatrix} - & \mathbf{v}_1^T & - \\ - & \mathbf{v}_2^T & - \end{pmatrix} \mathbf{v}_3 \in$$

span $\{\mathbf{v}_1, \mathbf{v}_2\}$, and find the scalars α and β with $\widetilde{\mathbf{P}}_V(\mathbf{v}_3) = \alpha\mathbf{v}_1 + \beta\mathbf{v}_2$. Investigate whether $\widetilde{\mathbf{P}}_V(\mathbf{v}_3) \perp \mathbf{v}_1, \mathbf{v}_2$.

15. (a) Find the orthogonal projection $a = \alpha\mathbf{u}_1$ of the vector $\mathbf{u}_2 = \begin{pmatrix} 1 \\ -3 \end{pmatrix}$ onto the line through the origin and $\mathbf{u}_1 = \begin{pmatrix} 1 \\ -4 \end{pmatrix}$.

(b) Find a nonzero vector $\mathbf{u}$ that is orthogonal to $\mathbf{u}_1 = \begin{pmatrix} 1 \\ -4 \end{pmatrix}$.

(c) How far does $\mathbf{u}_2 = \begin{pmatrix} 1 \\ -3 \end{pmatrix}$ lie from the line through the origin and $\mathbf{u}_1 = \begin{pmatrix} 1 \\ -4 \end{pmatrix}$.

(d) Find the matrix $\mathbf{P}_{\mathcal{U}}$ of the orthogonal projection of $\mathbb{R}^2$ onto U, the span of $\mathbf{u}_1$.

(e) Find $\mathbf{P}_{\mathcal{U}}(\mathbf{u}_2)$ and compare with the result in (a).

16. (a) Give at least three different reasonings that the vectors $\mathbf{u}_1 = \begin{pmatrix} 1 \\ 2 \\ -1 \end{pmatrix}$, $\mathbf{u}_2 = \begin{pmatrix} 2 \\ -1 \\ 2 \end{pmatrix}$, $\mathbf{u}_3 = \begin{pmatrix} 0 \\ 0 \\ 0 \end{pmatrix}$, $\mathbf{u}_4 = \begin{pmatrix} 4 \\ -7 \\ 8 \end{pmatrix}$, and $\mathbf{u}_5 = \begin{pmatrix} 0 \\ 0 \\ -1 \end{pmatrix}$ are linearly dependent.

(b) What is $V := \text{span}\{\mathbf{u}_1, \dots, \mathbf{u}_5\}$ for the vectors $\mathbf{u}_i$ of part (a)? Find two different orthonormal bases for V.

(c) Find a vector $\mathbf{b} \in \mathbb{R}^3$ that does not lie in $W = \text{span}\{\mathbf{u}_1, \mathbf{u}_2, \mathbf{u}_4\}$, if possible.

(d) Find an orthonormal basis for W in part (c).

17. Let $\{\mathbf{u}_1, \dots, \mathbf{u}_k\}$ be an orthogonal basis for a subspace U of $\mathbb{R}^n$ and let $\mathbf{x} \in \mathbb{R}^n$.

(a) Show that $k = \dim(U) \le n$.

(b) If $\alpha_i = \mathbf{x}^T\mathbf{u}_i \in \mathbb{R}$ for $i = 1, \dots, k$ and

$$\mathbf{a} = \begin{pmatrix} | & & | \\ \mathbf{u}_1 & \cdots & \mathbf{u}_k \\ | & & | \end{pmatrix} \begin{pmatrix} \alpha_1 \\ \vdots \\ \alpha_k \end{pmatrix},$$ show that $\mathbf{a} \in U$.

(c) Show that $\mathbf{x} - \mathbf{a} \perp U$ for a from part (b).

(d) Find the **orthogonal projection a** of $\mathbf{x} = \begin{pmatrix} -1 \\ -3 \\ 2 \\ 1 \end{pmatrix}$ onto the subspace U of $\mathbb{R}^4$ spanned by

$$\left\{ \begin{pmatrix} 1 \\ 1 \\ -1 \\ -1 \end{pmatrix}, \begin{pmatrix} 1 \\ 1 \\ 2 \\ 0 \end{pmatrix}, \begin{pmatrix} -2 \\ 0 \\ 1 \\ -3 \end{pmatrix} \right\}.$$

(e) What is the distance between $\mathbf{x}$ and the subspace U?

18. Prove the following: If U is a subspace of $\mathbb{R}^n$ that has an orthonormal basis of n vectors, then $U = \mathbb{R}^n$.

19. (a) Find a basis of $\mathbb{R}^3$ that contains the vector $\mathbf{u}_1 = \begin{pmatrix} 1 \\ 2 \\ -3 \end{pmatrix}$.

(b) Find an orthonormal basis of $\mathbb{R}^3$ that contains the vector $\mathbf{v}_1 = \frac{1}{\sqrt{5}} \begin{pmatrix} 1 \\ 0 \\ -2 \end{pmatrix}$.

20. Find matrix entries so that

$$\mathbf{U} = \begin{pmatrix} 1/\sqrt{3} & 0 & * & * \\ 1/\sqrt{3} & 1/\sqrt{2} & * & * \\ 1/\sqrt{3} & -1/\sqrt{2} & * & * \\ 0 & 0 & * & * \end{pmatrix}$$

becomes an orthogonal 4×4 matrix.

21. Show that the product of two or more orthogonal matrices is orthogonal.

22. (a) Show that every real matrix of the form $\mathbf{U} = \begin{pmatrix} \cos(x) & \sin(x) \\ -\sin(x) & \cos(x) \end{pmatrix}$ is real orthogonal (i.e., $\mathbf{U}^T\mathbf{U} = \mathbf{I}_2$).

(b) For $e^{i\theta} := \cos(\theta) + i\sin(\theta) \in \mathbb{C}$, what is the complex conjugate $\overline{e^{i\theta}}$?

(c) Show that every complex matrix of the form $\mathbf{V} = \frac{1}{\sqrt{2}} \begin{pmatrix} e^{i\theta} & -e^{i\theta} \\ e^{i\theta} & e^{i\theta} \end{pmatrix}$ with $\theta \in \mathbb{R}$ is unitary (i.e., $\mathbf{V}^*\mathbf{V} = \mathbf{I}_2$).

23. Prove the **Cauchy–Schwarz inequality**

$$|\mathbf{x} \cdot \mathbf{y}| \leq \|\mathbf{x}\| \, \|\mathbf{y}\|$$

for all $\mathbf{x}, \mathbf{y} \in \mathbb{R}^n$. (Use the hint in Section 10.1(a)).

24. Given an ONB $\{\mathbf{u}_i\}_{i=1}^k$ of a proper subspace $U \subset \mathbb{R}^n$, prove that the vector $\mathbf{a} := \mathbf{U}\mathbf{U}^T\mathbf{b} \in \mathbb{R}^n$ describes the vector in $U = \text{span}\{\mathbf{u}_i\}$ that is closest to $\mathbf{b} \in \mathbb{R}^n$. Do so in the following steps:

(a) Show that $\mathbf{a} \in U$, using Section 6.2, for example.

(b) Show that $\mathbf{b} - \mathbf{a} \perp \mathbf{u}_i$ for each basis vector $\mathbf{u}_i$ of U.

(c) Conclude that $\min_{\mathbf{x} \in U} \|\mathbf{b} - \mathbf{x}\| = \|\mathbf{b} - \mathbf{a}\|$.

25. (a) Find the closest vector to $\mathbf{b} = \begin{pmatrix} 1 \\ 0 \\ -1 \\ 0 \end{pmatrix} \in \mathbb{R}^4$ in the subspace U of $\mathbb{R}^4$ with the ONB

$$\left\{ \frac{1}{\sqrt{2}} \begin{pmatrix} 1 \\ 0 \\ 1 \\ 0 \end{pmatrix}, \frac{1}{\sqrt{5}} \begin{pmatrix} 0 \\ -1 \\ 0 \\ 2 \end{pmatrix}, \frac{1}{\sqrt{7}} \begin{pmatrix} 1 \\ 2 \\ -1 \\ 1 \end{pmatrix} \right\}.$$

What is the distance of $\mathbf{b}$ to U?

(b) Find the closest vector to $\mathbf{b} = \begin{pmatrix} -2 \\ 4 \end{pmatrix} \in \mathbb{R}^2$ in the subspace $U = \text{span}\left\{ \begin{pmatrix} 3 \\ 4 \end{pmatrix} \right\} \subset \mathbb{R}^2$.

26. Prove the **triangle inequality**

$$\|\mathbf{x} + \mathbf{y}\| \leq \|\mathbf{x}\| + \|\mathbf{y}\|$$

for all $\mathbf{x}$ and $\mathbf{y} \in \mathbb{R}^n$. (*Hint*: Use the Cauchy–Schwarz inequality of Problem 23.)

27. Let $\mathbf{X}_{nn}$ be an invertible matrix with inverse $\mathbf{Y}_{nn}$. Show that

(a) each row of $\mathbf{Y}$ is orthogonal to every column of $\mathbf{X}$ that does not occupy the same position, and

(b) each row of $\mathbf{X}$ is orthogonal to every column of $\mathbf{Y}$ that does not lie in the same position.

28. Determine the four entries a_{ij} of a 2×2 orthogonal matrix $\mathbf{A} = (a_{ij}) \neq \mathbf{I}_2$ that is symmetric. How many such matrices exist?

29. Let $\mathbf{u} = \begin{pmatrix} 1 \\ 1 \\ 1 \end{pmatrix}$ and $\mathbf{v} = \begin{pmatrix} 1 \\ 0 \\ -1 \end{pmatrix} \in \mathbb{R}^3$.

(a) Show that $\mathbf{u} \perp \mathbf{v}$ and find $\|\mathbf{u}\|$ and $\|\mathbf{v}\|$.

(b) Find an ONB for $\mathbb{R}^3$ that contains two vectors in the direction of $\mathbf{u}$ and $\mathbf{v}$. Check your answer.

30. Assume that the vectors $\mathbf{x}$, $\mathbf{y}$, and $\mathbf{z} \in \mathbb{R}^n$ are an ONB for a subspace $U \subset \mathbb{R}^n$.

(a) Show that $\mathbf{u} = \mathbf{x} + \mathbf{y} + \mathbf{z}$ and $\mathbf{v} = \mathbf{x} - \mathbf{z}$ are orthogonal. Find the lengths of $\mathbf{u}$ and $\mathbf{v}$.

(b) Find an ONB for U that contains vectors in the direction of $\mathbf{u}$ and $\mathbf{v}$. Check your answer.

(*Hint*: Look at the previous problem.)

31. Investigate the orthogonality of the vectors $\mathbf{v}_1$ and $\mathbf{v}_2 := \mathbf{v}_1 \cdot \mathbf{v}_1 \mathbf{u}_2 - \mathbf{u}_2 \cdot \mathbf{v}_1 \mathbf{v}_1$ as follows:

(a) For $\mathbf{v}_1 = \begin{pmatrix} 1 \\ -2 \\ 4 \end{pmatrix}$ and $\mathbf{u}_2 = \begin{pmatrix} 3 \\ 1 \\ -2 \end{pmatrix}$, compute $\mathbf{v}_2$ in the preceding formula.

(b) Show that the vectors $\mathbf{v}_1$ and $\mathbf{v}_2$ of part (a) are orthogonal.

(c) Show that the vectors $\mathbf{v}_1$ and $\mathbf{v}_2$ computed by the formula are orthogonal for all vectors $\mathbf{v}_1$ and $\mathbf{u}_2 \in \mathbb{R}^n$.

32. Find the distance of $\mathbf{u} = \begin{pmatrix} 2 \\ 1 \\ 0 \\ -1 \end{pmatrix}$ from the subspace

$$V = \text{span} \left\{ \begin{pmatrix} 1 \\ 1 \\ 0 \\ 1 \end{pmatrix}, \begin{pmatrix} 1 \\ 0 \\ 1 \\ 0 \end{pmatrix}, \begin{pmatrix} 0 \\ -1 \\ -1 \\ 0 \end{pmatrix} \right\}$$

of $\mathbb{R}^4$. (Be careful: Find an ONB for V first.)

33. Let $V = \text{span} \left\{ \begin{pmatrix} 0 \\ 1 \\ 0 \\ 0 \end{pmatrix}, \begin{pmatrix} 0 \\ 0 \\ 1 \\ 0 \end{pmatrix}, \begin{pmatrix} -1 \\ 1 \\ 0 \\ 1 \end{pmatrix} \right\}$.

(a) Find an ONB $\mathcal{V}$ for V.

(b) Find the matrix representation $\mathbf{P}_{\mathcal{V}}$ for the orthogonal projection of $\mathbb{R}^4$ onto V from the vectors in $\mathcal{V}$.

(c) Find the matrix representation $\mathbf{P}_{\mathcal{U}}$ for the orthogonal projection of $\mathbb{R}^4$ onto V, this time for the ONB

$$\mathcal{U} = \left\{ \begin{pmatrix} 0 \\ 1/\sqrt{2} \\ -1/\sqrt{2} \\ 0 \end{pmatrix}, \begin{pmatrix} 0 \\ 1/\sqrt{2} \\ 1/\sqrt{2} \\ 0 \end{pmatrix}, \begin{pmatrix} -1/\sqrt{2} \\ 0 \\ 0 \\ 1/\sqrt{2} \end{pmatrix} \right\}$$

of the subspace V.

(d) Does $\begin{pmatrix} 2 \\ 1 \\ 3 \\ -2 \end{pmatrix}$ lie in V?

(e) How far are the first and the fourth unit vectors $\mathbf{e}_1$ and $\mathbf{e}_4$ from V?

34. Assume that $\mathcal{U}$ and $\mathcal{V}$ are two orthonormal bases for a subspace $W \subset \mathbb{R}^n$ of dimension k. Show that $(\mathbf{P}_U)_{\mathcal{E}} = (\mathbf{P}_V)_{\mathcal{E}}$— that is, that the matrix representation of the orthogonal projection of $\mathbb{R}^n$ onto W is independent of the chosen ONB for W. (*Hint*: Use Section 7.3, namely, the formula $\mathbf{U} = \mathbf{V}\mathbf{X}_{kk}$, and show that $\mathbf{X}$ is an orthogonal matrix in this case. Finally, conclude from this that $(\mathbf{P}_U)_{\mathcal{E}} = (\mathbf{P}_V)_{\mathcal{E}}$.)

35. Assume that U is a subspace of $\mathbb{R}^n$ and that $\mathbf{a} \in \mathbb{R}^n$ is a vector that does not lie in U. Show that there is a nonzero vector $\mathbf{b} \in \mathbb{R}^n$ with $\mathbf{b} \perp U$ and $\mathbf{b} \notin U$. (*Hint*: Use orthogonal projections.) See the corollary at the end of Section 7.2 for a more elementary proof.

36. Prove or disprove the following:

(a) If $\mathbf{u} \perp \mathbf{v}$ and $\mathbf{v} \perp \mathbf{w}$, then $\mathbf{u} \perp \mathbf{w}$.

(b) If $\mathbf{u}$, $\mathbf{v}$, and $\mathbf{w}$ are linearly independent vectors in $\mathbb{R}^n$ with $\mathbf{u} \perp \mathbf{v}$ and $\mathbf{v} \perp \mathbf{w}$, then $\mathbf{u} \perp \mathbf{w}$.

37. Show that for any subspace $U \subset \mathbb{R}^n$, the set of vectors that lie in U <u>and</u> are orthogonal to U consists only of the zero vector.

38. Find the point on the line $y = 3x$ in the plane that is closest to $\mathbf{u} = (-1, 4)$. What is its distance from the line and from the origin?

39. (a) Find a matrix $\mathbf{A}_{2,3}$ with the row $\mathbf{r} = \begin{pmatrix} 1 & -1 & 3 \end{pmatrix}$ for which $\mathbf{x} = \begin{pmatrix} 2 \\ 1 \\ -1 \end{pmatrix} \in \ker(\mathbf{A})$, if possible.

(b) Find a matrix $\mathbf{A}_{2,3}$ with the row $\mathbf{r} = \begin{pmatrix} 1 & -1 & 3 \end{pmatrix}$ and a constant a for which $\mathbf{y} = \begin{pmatrix} 2 \\ 1 \\ a \end{pmatrix} \in \ker(\mathbf{A})$.

Teacher's Problem Making Exercise

T 10. (A) We have learned how to construct an integer orthogonal basis for the span of a given set of integer vectors $\mathbf{v}_1, \dots, \mathbf{v}_k \in \mathbb{R}^n$ via Gram–Schmidt.

An alternative approach for constructing vector orthogonalization problems comes from looking at a succession of kernels. This method computes a set of generally nonorthogonal integer vectors $\mathbf{v}_i$ that can be used as starting vectors for the orthogonalization process. Then the orthogonal basis that is computed via Gram–Schmidt consists of the original integer vectors $\mathbf{w}_i$ that were used in defining the matrix kernels: For an arbitrary integer vector $\mathbf{w}_1 \neq \mathbf{0} \in \mathbb{R}^n$, compute an integer nonzero vector $\mathbf{w}_2 \in \ker\begin{pmatrix} - & \mathbf{w}_1^T & - \end{pmatrix}$. Continue this process by finding a third nonzero integer vector $\mathbf{w}_3 \in \ker\begin{pmatrix} - & \mathbf{w}_1^T & - \\ - & \mathbf{w}_2^T & - \end{pmatrix}$, etc., until k mutually orthogonal vectors $\mathbf{w}_1, \dots, \mathbf{w}_k \in \mathbb{R}^n$ have been found. If we define our "starting vectors" $\mathbf{v}_i$ by setting

$$\begin{pmatrix} | & & | \\ \mathbf{v}_1 & \cdots & \mathbf{v}_k \\ | & & | \end{pmatrix} := \begin{pmatrix} | & & | \\ \mathbf{w}_1 & \cdots & \mathbf{w}_k \\ | & & | \end{pmatrix} \begin{pmatrix} d_1 & * & \cdots & * \\ 0 & \ddots & \ddots & \vdots \\ \vdots & & \ddots & * \\ 0 & \cdots & 0 & d_k \end{pmatrix}$$

for arbitrary integers $d_i \neq 0$ and arbitrary upper triangular integer entries $*$, then the $\mathbf{v}_i$ are not mutually orthogonal while $\operatorname{span}\{\mathbf{v}_i\}_{i=1}^j = \operatorname{span}\{\mathbf{w}_i\}_{i=1}^j$ for each $j = 1, 2, \dots, k$, and the vectors $\{\mathbf{w}_i\}$ are one outcome of an integer Gram–Schmidt process, minus its normalization step, performed on the $\mathbf{v}_i$. If some $d_i = 0$, then the Gram–Schmidt orthogonalization process for the vectors $\{\mathbf{v}_1, \ldots, \mathbf{v}_k\}$ will produce at least one redundant zero vector that is discarded.

EXAMPLE

(a) Given $\mathbf{w}_1 = \begin{pmatrix} 1 \\ 1 \\ 1 \\ 1 \end{pmatrix}$, we have $\mathbf{w}_2 = \begin{pmatrix} 1 \\ 1 \\ 1 \\ -3 \end{pmatrix} \in \ker\left(- \quad \mathbf{w}_1^T \quad - \right)$, for example, and $\mathbf{w}_3 = \begin{pmatrix} 0 \\ 1 \\ -1 \\ 0 \end{pmatrix} \in \ker \begin{pmatrix} - & \mathbf{w}_1^T & - \\ - & \mathbf{w}_2^T & - \end{pmatrix}$. Now set $\mathbf{v}_1 := \mathbf{w}_1$, $\mathbf{v}_2 := \mathbf{w}_2 - \mathbf{w}_1$, and $\mathbf{v}_3 := \mathbf{w}_3 - \mathbf{w}_1 + 2\mathbf{w}_2$, or

$$\begin{pmatrix} | & & | \\ \mathbf{v}_1 & \cdots & \mathbf{v}_3 \\ | & & | \end{pmatrix} := \begin{pmatrix} | & & | \\ \mathbf{w}_1 & \cdots & \mathbf{w}_3 \\ | & & | \end{pmatrix} \begin{pmatrix} 1 & -1 & -1 \\ 0 & 1 & 2 \\ 0 & 0 & 1 \end{pmatrix}.$$

Orthogonalize the $\mathbf{v}_i$ via Gram–Schmidt. What integer orthogonal vectors will be obtained? (All $d_i = 1 \neq 0$ here.)

(b) Perform the integer Gram–Schmidt process for $\mathbf{v}_1 := \mathbf{w}_1$, $\mathbf{v}_2 := \mathbf{w}_2 - \mathbf{w}_1$, and $\mathbf{v}_3 := 2\mathbf{w}_1 - 3\mathbf{w}_2$. Here, $d_3 = 0$. ◀

(B) In subsequent chapters, we need to construct integer test problems involving orthogonal matrices. In order to compute almost integer orthogonal matrices, we need to complete the null-space process of part (A) until $k = n$ (i.e., until we have found n mutually orthogonal integer vectors $\mathbf{w}_1, \ldots, \mathbf{w}_n \neq \mathbf{0} \in \mathbb{R}^n$.) Then the matrix product

$$\begin{pmatrix} | & & | \\ \mathbf{w}_1 & \cdots & \mathbf{w}_n \\ | & & | \end{pmatrix} \begin{pmatrix} \frac{1}{\|\mathbf{w}_1\|} & & 0 \\ & \ddots & \\ 0 & & \frac{1}{\|\mathbf{w}_n\|} \end{pmatrix}$$

has mutually orthonormal columns. Thus, the product is an orthogonal $n \times n$ matrix that differs from an integer matrix only by the normalizing constants $\frac{1}{\|\mathbf{w}_i\|}$ that affect each integer column $\mathbf{w}_i$ separately.

EXAMPLE The three vectors $\mathbf{w}_1 = \begin{pmatrix} 1 \\ 1 \\ 2 \end{pmatrix}$, $\mathbf{w}_2 = \begin{pmatrix} 1 \\ 1 \\ -1 \end{pmatrix}$, and $\mathbf{w}_3 = \begin{pmatrix} 1 \\ -1 \\ 0 \end{pmatrix} \in \mathbb{R}^3$ are mutually orthogonal and have lengths $\sqrt{6}$, $\sqrt{3}$, and $\sqrt{2}$, respectively. The matrix product

$$\mathbf{U} = \begin{pmatrix} 1 & 1 & 1 \\ 1 & 1 & -1 \\ 2 & -1 & 0 \end{pmatrix} \begin{pmatrix} \frac{1}{\sqrt{6}} & & \\ & \frac{1}{\sqrt{3}} & \\ & & \frac{1}{\sqrt{2}} \end{pmatrix}$$

is an orthogonal matrix, as can be readily checked. ◀

10.2 Theory

Matrix Generation, Rank 1 and Householder Matrices

We study Householder matrices as generators of the orthogonal ones, all in view of general orthogonal matrix factorizations.

Our early chapters were built on several twofold, or **dual, concepts**:

> A matrix $\mathbf{A}$ was defined by its entries a_{ij}, or as the representation of a linear transformation $\mathbf{x} \mapsto \mathbf{A}\mathbf{x}$.
> The solvability and unique solvability of a linear system $\mathbf{A}\mathbf{x} = \mathbf{b}$ were determined by row and by column properties of a REF of $\mathbf{A}$.
> The product of two matrices was viewed as a linear replacement of the columns of the first factor or as the same for the rows of the second matrix factor.

Here is another dual concept for matrices: that of an **additive** and a **multiplicative generation** of matrices. First we describe a simple additive representation of $m \times n$ matrices $\mathbf{A}$: Let $\mathbf{E}_{(ij)}$ denote the $m \times n$ matrix, all of whose entries are zero, except for a one in position (i, j) for an index pair $1 \leq i \leq m$ and $1 \leq j \leq n$. Then

$$\mathbf{A} = \left(a_{ij}\right)_{mn} = \sum_{i,j=1}^{m,n} a_{ij}\mathbf{E}_{(ij)};$$

that is, each matrix $\mathbf{A}_{mn}$ is the unique linear combination of the $m \cdot n$ **basic matrices** $\mathbf{E}_{(ij)}$ that generate $\mathbb{R}^{m,n}$. The basic matrices $\mathbf{E}_{(ij)}$ are different from, but related to, the Gaussian elimination matrices.

Recall Section 6.2 and the three kinds of Gaussian elimination matrices $\mathbf{E}_{..}$ in formulas (6.3) to (6.5): Every matrix $\mathbf{A}_{mn}$ can be written as the product

$$\mathbf{A}_{mn} = \mathbf{E}_M \cdots \mathbf{E}_1 \mathbf{R}_{mn} \tag{10.1}$$

of a finite number M of elementary $m \times m$ elimination matrices $\mathbf{E}_i$ and its reduced row echelon form matrix $\mathbf{R}_{mn}$.

By looking more closely at the multiplicative representation (10.1) of a matrix $\mathbf{A}$, one can gain a broader understanding of matrices.

What matrices are the elementary Gauss elimination matrices (6.3) to (6.5) used in (10.1)?

Each of the three types of Gaussian $m \times m$ elimination matrices can respectively be written as

$$\begin{aligned} \mathbf{E}_{ik}(c) &= \mathbf{I}_m + c\mathbf{E}_{(ik)} && \text{in (6.3)},\\ \mathbf{E}_k(c) &= \mathbf{I}_m + (c-1)\mathbf{E}_{(kk)} && \text{in (6.4)}, \end{aligned}$$

and as

$$\mathbf{E}_{jk} = \mathbf{I}_m + \mathbf{E}_{(jk)} + \mathbf{E}_{(kj)} - (\mathbf{E}_{(jj)} + \mathbf{E}_{(kk)}) \qquad \text{in (6.5)}$$

in terms of the basic matrices $\mathbf{E}_{(ik)}$. The first two elimination matrices $\mathbf{E}_{ik}(c)$ and $\mathbf{E}_k(c)$ are rank 1 modifications of the multiplicative identity matrix $\mathbf{I}_m$. The modifying matrices are the rank 1 matrices $c\mathbf{E}_{(ik)}$ and $(c-1)\mathbf{E}_{(kk)}$, respectively. We shall soon see that the third elimination matrix $\mathbf{E}_{jk}$ also differs from $\mathbf{I}_m$ by a rank 1 matrix. Thus, the Gaussian reduction of a matrix $\mathbf{A}$ to its reduced row echelon form $\mathbf{R}$ in Formula (10.1) proceeds in very small steps by rank 1 updates.

This leads us to study the most simple nonzero matrices, namely, **rank 1 matrices**.

Any matrix $\mathbf{B}_{mn}$ of rank 1 must have multiples of one nonzero vector $\mathbf{v}$ in each of its rows, since the REF of $\mathbf{B}$ has precisely one pivot and any two or more of $\mathbf{B}$'s rows are linearly dependent. Thus, if $\text{rank}(\mathbf{B}_{mn}) = 1$, then

$$\mathbf{B}_{mn} = \begin{pmatrix} - & \alpha_1 \mathbf{v} & - \\ & \vdots & \\ - & \alpha_m \mathbf{v} & - \end{pmatrix} = \begin{pmatrix} \alpha_1 \\ \vdots \\ \alpha_m \end{pmatrix} \begin{pmatrix} - & \mathbf{v} & - \end{pmatrix} = \mathbf{av}$$

for a nonzero column vector $\mathbf{a} = \begin{pmatrix} \alpha_1 \\ \vdots \\ \alpha_n \end{pmatrix} \in \mathbb{R}^m$ and a nonzero row vector $\mathbf{v} \in \mathbb{R}^n$.

DEFINITION 3 The matrix product $\mathbf{av}$ of a column vector $\mathbf{a} \in \mathbb{R}^m$ and a row vector $\mathbf{v} \in \mathbb{R}^n$ is called a **dyad**. It is an $m \times n$ matrix of rank 1 if $\mathbf{a}$ and $\mathbf{v} \neq \mathbf{0}$.

The correction terms of $\mathbf{I}_m$ in (6.3) and (6.4) are $m \times m$ dyads. For the Gaussian swap matrix $\mathbf{E}_{jk}$ in (6.5), note that the correction term

$$\mathbf{E}_{(jk)} + \mathbf{E}_{(kj)} - (\mathbf{E}_{(jj)} + \mathbf{E}_{(kk)}) = \begin{pmatrix} 0 & & & & & & & 0 \\ & \ddots & & & & & & \\ & & -1 & & & 1 & & \\ & & & 0 & & & & \\ & & & & \ddots & & & \\ & & & & & 0 & & \\ & & 1 & & & -1 & & \\ & & & & & & 0 & \\ & & & & & & & \ddots \\ & & & & & & & 0 \\ 0 & & & & & & & \end{pmatrix}_{mm}$$

is also a matrix of rank 1. Assuming, for simplicity, that $j < k$, it can be written as the dyad

$$\mathbf{E}_{(jk)} + \mathbf{E}_{(kj)} - (\mathbf{E}_{(jj)} + \mathbf{E}_{(kk)}) = \begin{pmatrix} 0 \\ \vdots \\ 0 \\ -1 \\ 0 \\ \vdots \\ 0 \\ 1 \\ 0 \\ \vdots \\ 0 \end{pmatrix}_{m,1} \begin{matrix} \\ \\ \\ \leftarrow j \\ \\ \\ \\ \leftarrow k \\ \\ \\ \\ \end{matrix} \underset{\substack{\uparrow \quad\quad\quad\quad \uparrow \\ j \quad\quad\quad\quad k}}{\begin{pmatrix} 0 & \cdots & 1 & 0 & \cdots & 0 & -1 & 0 & \cdots & 0 \end{pmatrix}_{1,m}}$$

$$= (-\mathbf{e}_j + \mathbf{e}_k)(\mathbf{e}_j - \mathbf{e}_k)^T.$$

Consequently, writing an arbitrary $m \times n$ matrix $\mathbf{A}$ as $\mathbf{E}_M \cdots \mathbf{E}_1 \mathbf{R}_{mn}$ in (10.1) via Gaussian elimination matrices involves only its RREF $\mathbf{R}$ and a certain number M of rank 1 modifications of the multiplicative identity $\mathbf{I}_m$.

Our quest in this section is to learn and work with simple orthogonal matrices instead of elimination matrices. Thus, our questions are as follows:

(a) What are the orthogonal rank 1 dyad modifications of the identity matrix $\mathbf{I}_m$, if any?
(b) How can orthogonal rank 1 modifications of the identity matrix $\mathbf{I}_m$ be used to obtain a factorization of an arbitrary matrix $\mathbf{A}_{mn}$ in terms of its REF $\mathbf{R}_{mn}$?

The following theorem answers the first question:

Theorem 10.3 For a column vector $\mathbf{u} \neq \mathbf{0} \in \mathbb{R}^m$ and a scalar $\alpha \neq 0$, the matrix $\mathbf{I}_m - \alpha \mathbf{u}\mathbf{u}^T$ is orthogonal if and only if $\alpha \mathbf{u} \cdot \mathbf{u} = \alpha \|\mathbf{u}\|^2 = 2$. ◀

Proof If $\mathbf{B} = \mathbf{I}_m - \alpha \mathbf{u}\mathbf{u}^T$ is orthogonal for a column vector $\mathbf{u} \in \mathbb{R}^m$, then $\mathbf{B}\mathbf{B}^T = \mathbf{I}_m$. Hence, by using the distributive *f-o-i-l* rule, we obtain

$$\mathbf{B}\mathbf{B}^T = (\mathbf{I} - \alpha \mathbf{u}\mathbf{u}^T)(\mathbf{I} - \alpha \mathbf{u}\mathbf{u}^T)^T = \mathbf{I} - 2\alpha \mathbf{u}\mathbf{u}^T + \alpha^2 \mathbf{u}\mathbf{u}^T\mathbf{u}\mathbf{u}^T = \mathbf{I}. \tag{10.2}$$

Thus, $2\alpha \mathbf{u}\mathbf{u}^T = \alpha^2 \mathbf{u}\mathbf{u}^T\mathbf{u}\mathbf{u}^T$ for $\alpha \neq 0$, or

$$2(\mathbf{u}\mathbf{u}^T)_{mm} = \alpha \mathbf{u}(\mathbf{u}^T\mathbf{u})\mathbf{u}^T = \alpha \mathbf{u}^T\mathbf{u}(\mathbf{u}\mathbf{u}^T)_{mm},$$

since $\mathbf{u}^T\mathbf{u} = \mathbf{u} \cdot \mathbf{u} \in \mathbb{R}$ and $\alpha \neq 0$. Hence, $2 = \alpha \mathbf{u}^T\mathbf{u} = \alpha \mathbf{u} \cdot \mathbf{u} = \alpha \|\mathbf{u}\|^2$, since the $m \times m$ matrix $\mathbf{u}\mathbf{u}^T \neq \mathbf{O}_m$ for $\mathbf{u} \neq \mathbf{0}$.

The converse implication is obvious from the expansion in (10.2). ■

If we choose $\mathbf{u}$ to be a unit vector in $\mathbb{R}^m$ (i.e., if $\mathbf{u} \cdot \mathbf{u} = \mathbf{u}^T\mathbf{u} = 1 \in \mathbb{R}$), then, by Theorem 10.3, the matrix $\mathbf{I}_m - 2\mathbf{u}\mathbf{u}^T$ is orthogonal. And $\mathbf{I}_m - 2\mathbf{u}\mathbf{u}^T$ differs additively from the identity matrix $\mathbf{I}_m$ only by the dyad $2\mathbf{u}\mathbf{u}^T \in \mathbb{R}^{m,m}$ of rank 1.

DEFINITION 4 For each unit vector $\mathbf{u} \in \mathbb{R}^m$, the $m \times m$ matrix $\mathbf{I}_m - 2\mathbf{u}\mathbf{u}^T$ is called a **Householder matrix**.

As demonstrated, Householder matrices are orthogonal. To solve the second quest of using Householder matrices in row echelon form reductions, we study how Householder matrices map m-space.

What are the intrinsic qualities of Householder matrices? For this, we embark on a complete eigenanalysis of a generic Householder matrix $\mathbf{H}_{mm} = \mathbf{I} - 2\mathbf{u}\mathbf{u}^T$ with $\mathbf{u} \cdot \mathbf{u} = 1$. As we know very little about $\mathbf{H}$ other than its defining unit vector $\mathbf{u}$, we need to think differently than using the eigenvalue methods of Chapter 9 here.

Assume that $\mathbf{H}_{mm} = \mathbf{I} - 2\mathbf{u}\mathbf{u}^T$ for a unit vector $\mathbf{u} \in \mathbb{R}^m$ with $\mathbf{u} \cdot \mathbf{u} = \|\mathbf{u}\| = \mathbf{u}^T\mathbf{u} = 1 \in \mathbb{R}$. Are there any specific vectors in $\mathbb{R}^m$ that $\mathbf{H}$ maps predictably?

As $\mathbf{H}$ is defined from the single unit vector $\mathbf{u} \in \mathbb{R}^m$, we first ask how $\mathbf{H}$ maps this specific vector $\mathbf{u}$. Note that

$$\mathbf{Hu} = (\mathbf{I} - 2\mathbf{uu}^T)\mathbf{u} = \mathbf{Iu} - 2\mathbf{u}(\mathbf{u}^T\mathbf{u}) = \mathbf{u} - 2\mathbf{u} = -\mathbf{u}, \tag{10.3}$$

since $\mathbf{u}^T\mathbf{u} = \mathbf{u} \cdot \mathbf{u} = 1$ is assumed.

Thus, the defining unit vector $\mathbf{u}$ of the Householder matrix $\mathbf{H} = \mathbf{I} - 2\mathbf{uu}^T$ is an eigenvector of $\mathbf{H}$ for the eigenvalue -1. (See Chapter 9 for the eigenvalue–eigenvector equation.)

Taking in this result and Equation (10.3) a bit more, we notice the miraculous effect of $\mathbf{u}^T\mathbf{u} = \mathbf{u} \cdot \mathbf{u}$ being a scalar equal to 1. Besides the scalar $1 \in \mathbb{R}$, the next most important scalar of the reals probably is 0.

What happens to vectors $\mathbf{v} \in \mathbb{R}^m$ with $\mathbf{u}^T\mathbf{v} = \mathbf{u} \cdot \mathbf{v} = 0$ instead under the mapping $\mathbf{H} = \mathbf{I} - 2\mathbf{uu}^T$? Where will a vector $\mathbf{v}$ perpendicular to $\mathbf{u}$ end up under the mapping by $\mathbf{H}$?

Note that

$$\mathbf{Hv} = (\mathbf{I} - 2\mathbf{uu}^T)\mathbf{v} = \mathbf{v} - 2\mathbf{u}(\mathbf{u}^T\mathbf{v}) = \mathbf{v} - \mathbf{0} = \mathbf{v} \text{ if } \mathbf{v} \perp \mathbf{u}, \text{ or if } \mathbf{u}^T\mathbf{v} = \mathbf{u} \cdot \mathbf{v} = 0.$$

Thus, every vector $\mathbf{v} \in \mathbb{R}^m$ that is orthogonal to the defining unit vector $\mathbf{u} \in \mathbb{R}^m$ of $\mathbf{H}$ is an eigenvector of $\mathbf{H} = \mathbf{I} - 2\mathbf{uu}^T$ as well. The corresponding eigenvalue is 1.

Are there any other eigenvalues or eigenvectors for $\mathbf{H}$? Can there be? The answer is, definitely not, because the subspace

$$\{\mathbf{v} \mid \mathbf{v} \perp \mathbf{u}\} = \ker\left(\, - \quad \mathbf{u}^T \quad - \,\right) \subset \mathbb{R}^m$$

has dimension $m - 1$ in $\mathbb{R}^m$, according to Chapter 5. Thus according to Section 10.1, the subspace $\{\mathbf{v} \mid \mathbf{v} \perp \mathbf{u}\}$ has an ONB consisting of $m-1$ vectors, each of which is an eigenvector of $\mathbf{H} = \mathbf{I} - 2\mathbf{uu}^T$ for the eigenvalue 1. Together with the eigenvalue -1 for the eigenvector $\mathbf{u}$ of $\mathbf{H}$, this completes the eigenanalysis of an arbitrary Householder matrix $\mathbf{H} = \mathbf{I} - 2\mathbf{uu}^T$.

Theorem 10.4 For a unit vector $\mathbf{u} \in \mathbb{R}^m$, the Householder matrix $\mathbf{H} = \mathbf{I}_m - 2\mathbf{uu}^T$ has the single eigenvalue -1 for the eigenvector $\mathbf{u}$ and the $(m-1)$-fold eigenvalue 1 for any vector $\mathbf{v} \perp \mathbf{u}$ as eigenvector. ◀

Our ultimate aim of this section is to establish that orthogonal rank 1 modifications of the identity matrix act much like the Gaussian elimination matrices of Section 6.2. More specifically, the orthogonal rank 1 modifications of the identity matrix generate every matrix $\mathbf{A}_{mn}$ multiplicatively from a row echelon form of $\mathbf{A}$. To prove this, we row reduce a given matrix $\mathbf{A}_{mn}$ orthogonally via a judicious sequence of Householder matrices $\mathbf{H}_i = \mathbf{I} - 2\mathbf{u}_i\mathbf{u}_i^T$ next.

Copying the action of one Gaussian elimination that zeros out one subpivotal nonzero entry in $\mathbf{A}$, we are interested in reducing the entries 2 through m in the first column $\mathbf{a}_1 \in \mathbb{R}^m$ of the given matrix $\mathbf{A}_{mn}$ to zero. For this, we desire to find—if possible—a Householder matrix $\mathbf{H}_1$ such that

$$\mathbf{H}_1\mathbf{a}_1 = (\mathbf{I}_m - 2\mathbf{u}_1\mathbf{u}_1^T)\mathbf{a}_1 = \begin{pmatrix} * \\ 0 \\ \vdots \\ 0 \end{pmatrix}_m = *\,\mathbf{e}_1 \tag{10.4}$$

for a "clever choice" of $\mathbf{u}_1 \in \mathbb{R}^m$ and the standard unit vector $\mathbf{e}_1 \in \mathbb{R}^m$. Then

$$\mathbf{H}_1\mathbf{A} = \begin{pmatrix} | & * & \cdots & * \\ \mathbf{H}_1\mathbf{a}_1 & & * & \\ | & * & \cdots & * \end{pmatrix} = \begin{pmatrix} * & * & \cdots & * \\ 0 & * & \cdots & * \\ \vdots & \vdots & & \vdots \\ 0 & * & \cdots & * \end{pmatrix}_{mn},$$

and we can use induction on the second to last rows and columns of the updated matrix $\mathbf{H}_1\mathbf{A}$.

The following lemma helps us find $\mathbf{u}_1$ from the first column $\mathbf{a}_1$ of $\mathbf{A}$. Besides, it sheds light on the eigenvalues of orthogonal matrices.

Lemma 3 Let $\mathbf{U}$ be an $m \times m$ orthogonal matrix. Then $\|\mathbf{Ux}\| = \|\mathbf{x}\|$ for all $\mathbf{x} \in \mathbb{R}^m$. If $\mathbf{x}$ and $\mathbf{y} \in \mathbb{R}^m$ are orthogonal vectors, then the vectors $\mathbf{Ux}$ and $\mathbf{Uy}$ are also orthogonal. Moreover, an orthogonal matrix $\mathbf{U}_{mm}$ preserves angles (i.e., $\angle(\mathbf{x}, \mathbf{y}) = \angle(\mathbf{Ux}, \mathbf{Uy})$ for all $\mathbf{x}, \mathbf{y} \in \mathbb{R}^m$). ◀

Proof We have

$$\|\mathbf{Ux}\|^2 = (\mathbf{Ux})^T\mathbf{Ux} = \mathbf{x}^T\mathbf{U}^T\mathbf{Ux} = \mathbf{x}^T\mathbf{I}_m\mathbf{x} = \mathbf{x}^T\mathbf{x} = \|\mathbf{x}\|^2$$

from Definition 1(a) and Lemma 1.

If $\mathbf{x}^T\mathbf{y} = 0$ then $(\mathbf{Ux})^T\mathbf{Uy} = \mathbf{x}^T\mathbf{U}^T\mathbf{Uy} = \mathbf{x}^T\mathbf{y} = 0$.

Finally, due to the cosine definition for the angle between two vectors, we have

$$\cos(\angle(\mathbf{x}, \mathbf{y})) = \frac{\mathbf{x} \cdot \mathbf{y}}{\|\mathbf{x}\|\ \|\mathbf{y}\|} = \frac{\mathbf{x}^T\mathbf{I}_m y}{\|\mathbf{Ux}\|\ \|\mathbf{Uy}\|} = \frac{\mathbf{x}^T\mathbf{U}^T\mathbf{Uy}}{\|\mathbf{Ux}\|\ \|\mathbf{Uy}\|} = \cos(\angle(\mathbf{Ux}, \mathbf{Uy})).$$

■

Corollary If $\mathbf{U}$ is an $m \times m$ orthogonal matrix, then all eigenvalues λ of $\mathbf{U}$ lie on the unit circle in $\mathbb{C}$. In other words, all eigenvalues $\lambda \in \mathbb{C}$ of an orthogonal matrix $\mathbf{U}$ satisfy $|\lambda| = 1$. ◀

Proof If $\mathbf{U}$ is an orthogonal matrix with $\mathbf{U}^T\mathbf{U} = \mathbf{I}_m$ and if $\mathbf{Ux} = \lambda\mathbf{x}$ for $\lambda \in \mathbb{C}$ and $\mathbf{x} \neq \mathbf{0}$, then $\|\mathbf{Ux}\| = \|\lambda\mathbf{x}\| = |\lambda|\|\mathbf{x}\|$ and $\|\mathbf{Ux}\| = \|\mathbf{x}\|$ from Lemma 3. Thus, $|\lambda| = 1$, since $\mathbf{x} \neq \mathbf{0}$. ■

Our desired Householder matrix $\mathbf{H}_1$ must map the column $\mathbf{a}_1$ to a multiple of the first unit vector $\mathbf{e}_1$. Since $\mathbf{H}_1$ is orthogonal and preserves length by Lemma 3, we must have $\|\mathbf{a}_1\| = |*|$ in (10.4). Thus, $\mathbf{H}_1$ must map the vector $\mathbf{a}_1$ of length $\|\mathbf{a}_1\|$ either to $\|\mathbf{a}_1\|\mathbf{e}_1$ or to $-\|\mathbf{a}_1\|\mathbf{e}_1 \in \mathbb{R}^m$, as depicted in Figure 10-6.

According to Theorem 10.4, $\mathbf{H}_1 = \mathbf{I}_m - 2\mathbf{u}_1\mathbf{u}_1^T$ acts like a mirror on $\mathbb{R}^m$: The mirror surface is $\mathbf{u}_1^\perp := \{\mathbf{v} \mid \mathbf{v} \perp \mathbf{u}_1\}$, of dimension $m-1$. Points on this plane remain fixed under $\mathbf{H}_1$, while any multiple $\alpha\mathbf{u}_1$ of $\mathbf{u}_1$ gets "flipped across" the mirror surface $\mathbf{u}_1^\perp$ to $-\alpha\mathbf{u}_1$ under $\mathbf{H}_1$. If we assume that $\mathbf{H}_1$ maps $\mathbf{a}_1$ to $\|\mathbf{a}_1\|\mathbf{e}_1$, then the mapping of $\mathbb{R}^m$ under $\mathbf{H}_1$ must be the **reflection** of $\mathbb{R}^m$ across the mirror surface $\mathbf{u}_1^\perp$ for the unit vector $\mathbf{u}_1 = \frac{\mathbf{H}_1\mathbf{a}_1 - \mathbf{a}_1}{\|\mathbf{H}_1\mathbf{a}_1 - \mathbf{a}_1\|} = \frac{\|\mathbf{a}_1\|\mathbf{e}_1 - \mathbf{a}_1}{\|\ \|\mathbf{a}_1\|\mathbf{e}_1 - \mathbf{a}_1\|}$. Figure 10-7 illustrates the action of $\mathbf{H}_1 = \mathbf{I}_m - 2\mathbf{u}_1\mathbf{u}_1^T$ on $\mathbf{a}_1$ with the defining vector $\mathbf{u}_1$ chosen parallel to $\mathbf{a}_1 - \|\mathbf{a}_1\|\mathbf{e}_1$:

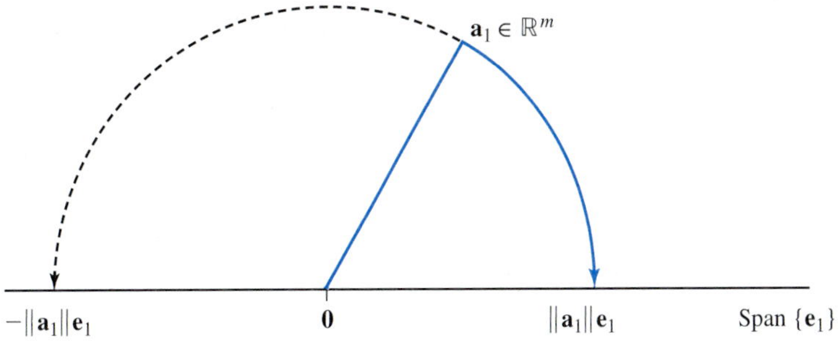

Figure 10-6

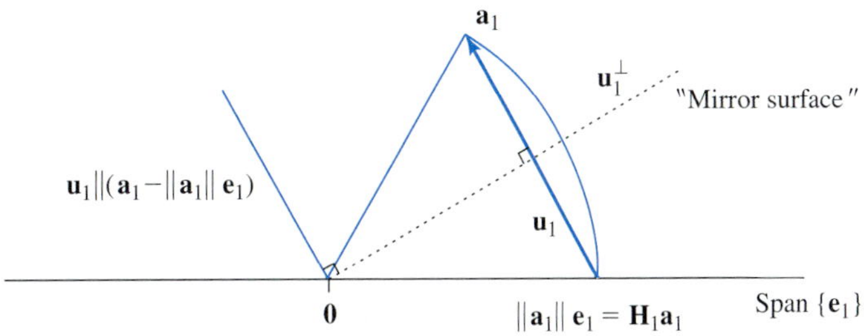

Figure 10-7

Thus, we have proved the following theorem:

Theorem 10.5 Let $\mathbf{a} \in \mathbb{R}^m$ be a nonzero vector. Set $\sigma := \pm\|\mathbf{a}\|$, $\mathbf{u} := \mathbf{a} + \sigma\mathbf{e}_1$ and $\beta := \dfrac{\|\mathbf{u}\|^2}{2}$. Then $\mathbf{H}_{mm} := \mathbf{I}_m - \beta^{-1}\mathbf{u}\mathbf{u}^T$ is a Householder transformation that satisfies $\mathbf{Ha} = -\sigma\mathbf{e}_1$, provided that $\mathbf{a} \neq -\sigma\mathbf{e}_1$. ◀

Using Theorem 10.5 for a given matrix $\mathbf{A}_{mn}$ with nonzero first column $\mathbf{a}_1$, we can thus find a Householder transformation $\mathbf{H}_1$ such that

$$\mathbf{H}_1\mathbf{A} = \begin{pmatrix} * & * & \cdots & * \\ 0 & | & & \vdots \\ \vdots & \widetilde{\mathbf{a}_2} & & \vdots \\ 0 & | & \cdots & * \end{pmatrix}.$$

Now continue with column 2 and the rows $2, \ldots, m$ of $\mathbf{H}_1\mathbf{A}_{mn}$, and find a Householder transformation $\mathbf{H}_2$ such that

$$\mathbf{H}_2\mathbf{H}_1\mathbf{A} = \mathbf{H}_2 \begin{pmatrix} * & * & \\ 0 & | & \\ \vdots & \widetilde{\mathbf{a}_2} & * \\ 0 & | & \end{pmatrix} = \begin{pmatrix} * & * & \\ 0 & * & \\ \vdots & 0 & * \\ 0 & 0 & \end{pmatrix}_{mn}$$

for $\mathbf{H}_1\mathbf{A}$'s second column $\widetilde{\mathbf{a}_2} \neq \mathbf{0} \in \mathbb{R}^{m-1}$ (omitting the updated $(1, 2)$ entry). If $\mathbf{a}_1 = \mathbf{0} \in \mathbb{R}^m$ or $\widetilde{\mathbf{a}_2} = \mathbf{0} \in \mathbb{R}^{m-1}$, we can skip to the next nonzero column of $\mathbf{A}$

in $\mathbb{R}^m$ or to the subcolumn of $\mathbf{H}_1\mathbf{A}$ in $\mathbb{R}^{m-1}$, respectively . Thus, we have shown the following theorem:

Theorem 10.6 Every real $m \times n$ matrix $\mathbf{A}$ can be reduced to a REF $\mathbf{R} \in \mathbb{R}^{m,n}$ using at most n orthogonal Householder matrices $\mathbf{H}_i = \mathbf{I}_m - \mathbf{u}_i\mathbf{u}_i^T \in \mathbb{R}^{m,m}$ for real unit vectors $\mathbf{u}_i$; that is,

$$\mathbf{H}_n \cdots \mathbf{H}_1\mathbf{A} = \mathbf{R}.$$

If $m \leq n$, then $m-1$ Householder matrices suffice to reduce $\mathbf{A}_{mn}$ to a REF $\mathbf{R}$. ◀

For the proof, note that if $m > n$ (i.e., if $\mathbf{A}$ has more rows than columns), then each of $\mathbf{A}$'s n columns must be reduced, while for $m \leq n$, the row echelon form reduction is finished once column $m-1$ has been reduced.

For completeness, we note the following:

Lemma 4 Householder matrices $\mathbf{H} = \mathbf{I} - 2\mathbf{u}\mathbf{u}^T$ for $\mathbf{u}^T\mathbf{u} = \mathbf{u}\cdot\mathbf{u} = 1$ are both symmetric and orthogonal (i.e., $\mathbf{H}^{-1} = \mathbf{H}^T = \mathbf{H}$). ◀

Proof We have

$$\mathbf{H}^T = (\mathbf{I} - 2\mathbf{u}\mathbf{u}^T)^T = \mathbf{I}^T - 2\mathbf{u}\mathbf{u}^T = \mathbf{H}$$

by Theorem 6.3 applied to $(\mathbf{u}\mathbf{u}^T)^T$. Thus, $\mathbf{H}\mathbf{H}^T = \mathbf{I} = \mathbf{H}\mathbf{H}$ by (10.2). ■

Householder matrices are important building blocks of the orthogonal ones.

10.2.P Problems

1. (a) Express $\mathbf{A} = \begin{pmatrix} 1 & 0 & 2 & 4 \\ 0 & 0 & 0 & 0 \\ -2 & 0 & -4 & -8 \end{pmatrix}$ as a dyad $\mathbf{u}\mathbf{v}^T$ for two column vectors $\mathbf{u} \in \mathbb{R}^3$ and $\mathbf{v} \in \mathbb{R}^4$.

(b) What is the rank of $\mathbf{A}$? Are the two vectors in part (a) uniquely defined?

2. (a) Show that the sum of two dyads $\mathbf{B} = \begin{pmatrix} 1 \\ 2 \\ 3 \end{pmatrix} \begin{pmatrix} 2 & 0 & 1 \end{pmatrix} + \begin{pmatrix} -2 \\ -4 \\ -6 \end{pmatrix} \begin{pmatrix} 0 & 1 & 2 \end{pmatrix}$ has rank 1.

(b) Express $\mathbf{B}$ from part (a) as a single dyad and as a sum of two dyads different from those in part (a).

3. Find a Householder matrix $\mathbf{H} = \mathbf{I} - 2\mathbf{u}\mathbf{u}^T$ by specifying a column vector $\mathbf{u}$ with $\mathbf{u}^T\mathbf{u} = \mathbf{u}\cdot\mathbf{u} = 1$ so that $\mathbf{H}$ maps $\begin{pmatrix} 1 \\ 0 \\ 2 \\ -4 \end{pmatrix}$ to $\begin{pmatrix} * \\ 0 \\ 0 \\ 0 \end{pmatrix}$. What is the value of $*$ for your $\mathbf{H}$. Check your answer.

4. Find a Householder matrix $\mathbf{H}_\mathbf{w} = \mathbf{I} - \frac{2}{\mathbf{w}\cdot\mathbf{w}}\mathbf{w}\mathbf{w}^T$, so that

$$\mathbf{H}_\mathbf{w}\begin{pmatrix} 0 \\ 12 \\ 0 \\ -5 \end{pmatrix} = \begin{pmatrix} * \\ 0 \\ 0 \\ 0 \end{pmatrix}.$$

(a) What is $\mathbf{w}$ here?

(b) Find the matrix $\mathbf{H}_\mathbf{w}$ explicitly.

(c) What value does $*$ have?

(d) What are the matrix powers $\mathbf{H}_\mathbf{w}^2$, $\mathbf{H}_\mathbf{w}^3$, and $\mathbf{H}_\mathbf{w}^{117}$?

5. If $\mathbf{H} = \mathbf{I} - 2\mathbf{u}\mathbf{u}^T$ for a unit column vector $\mathbf{u}$, show that $\mathbf{W}\mathbf{H}\mathbf{W}^{-1}$ is diagonal if $\mathbf{W}^T\mathbf{W} = \mathbf{I}$ with $\mathbf{W}\mathbf{u} = \mathbf{e}_1$.

6. Show that the matrix sum $\mathbf{A} = \mathbf{E}_{(jk)} + \mathbf{E}_{(kj)} - \mathbf{E}_{(jj)} - \mathbf{E}_{(kk)}$ composed of basic matrices has rank 1 for all $j \neq k$.

7. Prove that every orthogonal matrix $\mathbf{A}$ is invertible.

8. Let $\mathbf{V}$ be the matrix obtained from swapping some of the rows of an orthogonal matrix $\mathbf{U}$. When is $\mathbf{V}$ orthogonal?

9. Show that if the rows of a matrix $\mathbf{U}$ are mutually orthonormal, then the matrix product $\mathbf{U}\mathbf{U}^T$ is orthogonal.

10. Give examples or disprove the following:

(a) A 2×4 matrix $\mathbf{A}$ can contain mutually orthogonal columns.

(b) A 2×4 matrix $\mathbf{B}$ can contain mutually orthogonal rows.

(c) Repeat parts (a) and (b) by 4×4 matrix $\mathbf{C}$.

(d) Repeat parts (a) and (b) by 6×4 matrix $\mathbf{F}$.

(e) Repeat parts (a) by (b) mutually orthonormal rows and columns and the sizes as given in parts (a) to (c).

11. Prove that if $\mathbf{U}_{nn}$ is an orthogonal matrix, then so is $\mathbf{V}_{n+1,n+1} = \begin{pmatrix} 1 & \\ & \mathbf{U} \end{pmatrix}$.

12. Show that if $\mathbf{a}\mathbf{b}^T = \mathbf{b}\mathbf{c}^T$ for three column vectors $\mathbf{a}, \mathbf{b}, \mathbf{c} \in \mathbb{R}^n$ with $\mathbf{b} \neq \mathbf{0}$, then $\mathbf{a} = \mathbf{c} = \alpha\mathbf{b}$ for some $\alpha \in \mathbb{R}$.

13. Show that every rank 1 perturbation $\mathbf{A}_{nn} = \mathbf{I}_n - \mathbf{u}\mathbf{v}^T$ of the identity matrix $\mathbf{I}_n$ for $\mathbf{u}, \mathbf{v} \neq \mathbf{0} \in \mathbb{R}^n$ that is orthogonal is a Householder matrix $\mathbf{A} = \mathbf{H} = \mathbf{I}_n - 2\mathbf{w}\mathbf{w}^T$ for a unit vector $\mathbf{w} \in \mathbb{R}^n$ with $\mathbf{w}^T\mathbf{w} = 1$.

What is the "Householder vector" $\mathbf{w}$, depending on $\mathbf{u}$ and $\mathbf{v}$? (*Hint*: Use Problem 12.)

10.3 Applications

QR Decomposition, MATLAB, and Least Squares

In Section 10.2 we learned how to obtain a REF $\mathbf{R}$ of any matrix $\mathbf{A}_{mn}$ via at most n successive Householder eliminations $\mathbf{H}_i \in \mathbb{R}^{m,m}$ as $\mathbf{H}_n \cdots \mathbf{H}_1\mathbf{A}_{mn} = \mathbf{R}_{mn}$. Each $\mathbf{H}_i$ affects the ith and subsequent columns of $\mathbf{A}$. By transferring the Householder factors to the right-hand side of the matrix equation by inversion and using $\mathbf{H}_i^{-1} = \mathbf{H}_i$ according to Lemma 4, we can express any matrix $\mathbf{A}_{mn}$ as the product

$$\mathbf{A}_{mn} = \mathbf{H}_1 \cdots \mathbf{H}_n\mathbf{R}_{mn}$$

of at most n Householder matrices $\mathbf{H}_i$ and a REF $\mathbf{R}$ of $\mathbf{A}$. Set $\mathbf{Q} := \mathbf{H}_1 \cdots \mathbf{H}_{n-1}$. The matrix product $\mathbf{Q}$ is orthogonal. This is quickly checked. Thus, every $m \times n$ matrix $\mathbf{A}$ can be written as the product of an orthogonal matrix $\mathbf{Q} \in \mathbb{R}^{m,m}$ and a row echelon form matrix $\mathbf{R} \in \mathbb{R}^{m,n}$ i.e., as $\mathbf{A} = \mathbf{QR}$. This matrix factorization is called the **QR factorization** of $\mathbf{A}$ in analogy to the earlier LR matrix factorization $\mathbf{PA} = \mathbf{LR}$ of Section 6.2.

The QR factorization of a general $m \times n$ matrix $\mathbf{A} = \mathbf{QR}$ allows us to find an ONB of the column space or the image of $\mathbf{A}$ as follows: We have

$$\mathbf{A}_{mn} = \mathbf{QR} = \begin{pmatrix} | & \cdots & | \\ \mathbf{q}_1 & \cdots & \mathbf{q}_m \\ | & & | \end{pmatrix}_{mm} \begin{pmatrix} r_{11} & r_{12} & \cdots & r_{1n} \\ 0 & r_{22} & & \vdots \\ \vdots & \ddots & \ddots & \vdots \\ 0 & \cdots & 0 & r_{mn} \end{pmatrix}_{mn}$$

$$= \begin{pmatrix} | & | & \\ r_{11}\mathbf{q}_1 & r_{12}\mathbf{q}_1 + r_{22}\mathbf{q}_2 & \cdots \\ | & | & \end{pmatrix} = \begin{pmatrix} | & & | \\ \mathbf{a}_1 & \cdots & \mathbf{a}_n \\ | & & | \end{pmatrix} = \mathbf{A}.$$

Consequently, the columns $\mathbf{a}_i \in \mathbb{R}^m$ of $\mathbf{A}$ are given as

$$\begin{aligned} \mathbf{a}_1 &= r_{11}\mathbf{q}_1 && \text{(i.e.,} \quad \text{span}\{\mathbf{a}_1\} = \text{span}\{\mathbf{q}_1\}), \\ \mathbf{a}_2 &= r_{12}\mathbf{q}_1 + r_{22}\mathbf{q}_2 && \text{(i.e.,} \quad \text{span}\{\mathbf{a}_1, \mathbf{a}_2\} = \text{span}\{\mathbf{q}_1, \mathbf{q}_2\}), \quad \text{etc.,} \end{aligned}$$

in terms of the orthonormal columns $\mathbf{q}_i$ of $\mathbf{Q}$. Thus, the **QR** factorization $\mathbf{A} = \mathbf{QR}$ of $\mathbf{A}_{mn}$ produces an ONB for the column space or the image of $\mathbf{A}$ in the columns

$\mathbf{q}_i$ of $\mathbf{Q}$. Here, we need to take only those vectors $\mathbf{q}_i$ into the basis of im($\mathbf{A}$) that correspond to pivot rows of $\mathbf{A}$'s REF $\mathbf{R}$. These $\mathbf{q}_i$ have unit length, are mutually orthogonal, and form an ONB for im($\mathbf{A}$).

The QR decomposition of a matrix is easy to compute in MATLAB. The commands

`Q = qr(A)` or `[Q R] = qr(A)`

will produce the orthogonal factor $\mathbf{Q}$ of $\mathbf{A}$, as well as both the orthogonal $\mathbf{Q}$ and a REF $\mathbf{R}$ of $\mathbf{A}$ via orthogonal elimination, respectively.

The QR factorization of a matrix $\mathbf{A}_{mn}$ is used in MATLAB to compute "least squares solutions" that solve inconsistent linear systems of equations $\mathbf{Ax} = \mathbf{b}$. For system matrices of full rank and $m \neq n$, the MATLAB command `A\b` computes the vector $\mathbf{x} \in \mathbb{R}^n$ with minimal residual $\|\mathbf{Ax} - \mathbf{b}\|$ from the QR factorization of $\mathbf{A}_{mn}$.

Inconsistent or unsolvable linear systems $\mathbf{A}_{mn}\mathbf{x} = \mathbf{b}$ often occur with repeated measurements of n unknown quantities $x_1, \ldots, x_n$. If m, the number of measurements, is greater than the number n of unknowns, then the system $\mathbf{A}_{mn}\mathbf{x} = \mathbf{b}$ is overdetermined, and it will generally have no exact solution, since the coefficients in $\mathbf{A}$ are tainted by various measurement errors. However, there will be a **least squares solution** $\mathbf{x}_{LS}$ that has the smallest residual error

$$\|\mathbf{Ax}_{LS} - \mathbf{b}\| = \min_{\mathbf{x}\in\mathbb{R}^n} \|\mathbf{Ax} - \mathbf{b}\|.$$

The least squares solution $\mathbf{x}_{LS}$ can readily be found from the QR factorization of $\mathbf{A}$ if $\mathbf{A}_{mn}$ has **full rank** equal to $n = \min\{m, n\}$. If rank($\mathbf{A}_{mn}$) $< \min\{m, n\}$, then the n unknown quantities x_i are related. (See Example 2 of Section 5.1 for example, and also Section 12.2.)

If $\mathbf{A}_{mn} = \mathbf{Q}_{mm}\mathbf{R}_{mn}$ for an orthogonal matrix $\mathbf{Q}$ and an upper triangular $\mathbf{R}$, then

$$\|\mathbf{Ax} - \mathbf{b}\| = \|\mathbf{Q}^T\mathbf{Ax} - \mathbf{Q}^T\mathbf{b}\| = \|\mathbf{Rx} - \mathbf{Q}^T\mathbf{b}\|,$$

according to Lemma 4. Since $\mathbf{A}$ has full rank n and $m > n$, the matrix $\mathbf{R}$ contains a nonsingular $n \times n$ upper triangular block $\mathbf{R}_1$ at the top with $m-n$ zero rows below:

$$\mathbf{R}_{mn} = \left(\frac{\mathbf{R}_1}{\mathbf{O}_{m-n}}\right).$$

If we partition the vector $\mathbf{Q}^T\mathbf{b} \in \mathbb{R}^m$ conformally into $\mathbf{Q}^T\mathbf{b} = \begin{pmatrix} \mathbf{w} \\ \mathbf{z} \end{pmatrix}$ with $\mathbf{w} \in \mathbb{R}^n$ and $\mathbf{z} \in \mathbb{R}^{m-n}$, then the length of the error vector $\mathbf{Ax} - \mathbf{b}$ is

$$\|\mathbf{Ax} - \mathbf{b}\| = \|\mathbf{Rx} - \mathbf{Q}^T\mathbf{b}\| = \left\| \begin{pmatrix} \mathbf{R}_1\mathbf{x} \\ \mathbf{0} \end{pmatrix} - \begin{pmatrix} \mathbf{w} \\ \mathbf{z} \end{pmatrix} \right\|.$$

This is minimized if $\mathbf{x}$ solves $\mathbf{R}_1\mathbf{x} = \mathbf{w}$. The linear system involves n equations in the n unknowns $x_1, \ldots, x_n$. Its system matrix $\mathbf{R}_1$ is nonsingular and upper triangular. Hence, by using backsubstitution, one can find the unique vector $\mathbf{x} = \mathbf{x}_{LS}$ with $\mathbf{R}_1\mathbf{x} = \mathbf{w}$ so that $\|\mathbf{Ax}_{LS} - \mathbf{b}\| = \|\mathbf{z}\|$ is minimal. The real quantity $\|\mathbf{z}\|$ is called the **unavoidable error** of the full-rank least squares problem $\mathbf{Ax} = \mathbf{b}$, since no choice of $\mathbf{x}$ can reduce this error.

MATLAB finds the least squares solution $\mathbf{x}_{LS}$ by using the linear systems solving command `A\b` for singular system matrices $\mathbf{A}$ of full rank.

EXAMPLE 3 (Curve fitting) Assume that we are given the experimental data x_i and $f(x_i) = y_i \in \mathbb{R}$ for $i = 1, \ldots, 4$ as follows:

x_i	-1	0	1	3
y_i	2	1.5	4	6

Suppose we are not certain what functional relation f holds between the x and y data. There are many elementary choices for f. For example, we can choose a low-degree polynomial for f, such as $f_1(x) := a+bx$, $f_2(x) := a+bx+cx^2$, or $f_3(x) := a+bx+cx^2+dx^3$, or we can try $f_4(x) := a+b\sin(1.3x)+c\cos(1.3x)$.

Our task is to find the optimal parameters $a, b, \ldots$ for each of our model functions f_j for the specific problem. In other words, we need to find the parameters $a, b, \ldots$ for which

$$\left\| \begin{pmatrix} f_j(x_1) \\ \vdots \\ f_j(x_4) \end{pmatrix} - \begin{pmatrix} y_1 \\ \vdots \\ y_4 \end{pmatrix} \right\|$$

is minimal. This is a least squares problem. In order to fit the curve f_1 optimally to the given data, we need to find the values of a and $b \in \mathbb{R}$ that minimize $\left\| \begin{pmatrix} 1 & -1 \\ 1 & 0 \\ 1 & 1 \\ 1 & 3 \end{pmatrix} \begin{pmatrix} a \\ b \end{pmatrix} - \begin{pmatrix} 2 \\ 1.5 \\ 4 \\ 6 \end{pmatrix} \right\|$. Here, the system matrix $\mathbf{A} = \begin{pmatrix} 1 & -1 \\ 1 & 0 \\ 1 & 1 \\ 1 & 3 \end{pmatrix}$ comes about by writing

$$\begin{pmatrix} f_1(x_1) \\ \vdots \\ f_1(x_4) \end{pmatrix} = \begin{pmatrix} a+bx_1 \\ \vdots \\ a+bx_4 \end{pmatrix} = \begin{pmatrix} 1 & x_1 \\ \vdots & \vdots \\ 1 & x_4 \end{pmatrix} \begin{pmatrix} a \\ b \end{pmatrix} = \mathbf{A} \begin{pmatrix} a \\ b \end{pmatrix}.$$

For f_2, the analogous problem is to minimize $\left\| \begin{pmatrix} 1 & -1 & 1 \\ 1 & 0 & 0 \\ 1 & 1 & 1 \\ 1 & 3 & 9 \end{pmatrix} \begin{pmatrix} a \\ b \\ c \end{pmatrix} - \begin{pmatrix} 2 \\ 1.5 \\ 4 \\ 6 \end{pmatrix} \right\|$, for f_3, it involves $\left\| \begin{pmatrix} 1 & -1 & 1 & -1 \\ 1 & 0 & 0 & 0 \\ 1 & 1 & 1 & 1 \\ 1 & 3 & 9 & 27 \end{pmatrix} \begin{pmatrix} a \\ b \\ c \\ d \end{pmatrix} - \begin{pmatrix} 2 \\ 1.5 \\ 4 \\ 6 \end{pmatrix} \right\|$, while, for f_4, we are looking at minimizing $\left\| \begin{pmatrix} 1 & \sin(-1.3) & \cos(-1.3) \\ 1 & 0 & 0 \\ 1 & \sin(1.3) & \cos(1.3) \\ 1 & \sin(3.9) & \cos(3.9) \end{pmatrix} \begin{pmatrix} a \\ b \\ c \end{pmatrix} - \begin{pmatrix} 2 \\ 1.5 \\ 4 \\ 6 \end{pmatrix} \right\|$.

Note that in each instance the problem has been written as $\min \|\mathbf{Ax} - \mathbf{b}\|$ for a certain $4 \times k$ system matrix $\mathbf{A}$ of full rank k. For f_3, the system matrix is square and nonsingular (i.e., the optimal solution f_3 will satisfy each of the

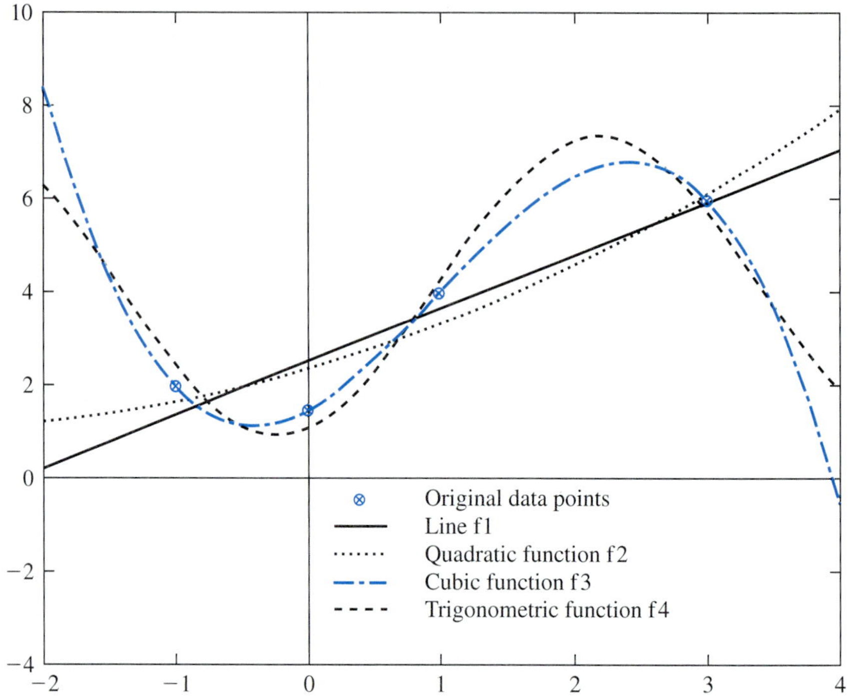

Figure 10-8

four equations $f_3(x_i) = y_i$ exactly). The optimal f_3 is the cubic interpolating polynomial for the four data points. For all other model functions f_ℓ, errors will be committed.

Using the MATLAB command `A\b` and the MATLAB plot routines (use `help plot`) for the four different system matrices **A** and the same vector **b** of y-values, we obtain the graphs in Figure 10-8 of the optimal fits.

Note the differing fits of the curves $(x, f_j(x))$ to the given data. The best line approximation is computed in MATLAB as $f_1(x) = 2.5286 + 1.1286x$, for example. When using f_1, we commit an error of size 1.2421. With f_2, the optimally fitting quadratic polynomial, the error is reduced slightly to 1.1442. For f_3 the error is zero, while with f_4 it is 0.69319. f_1 and f_2 give graphically similar approximations to the data, as do f_3 and f_4.

Further experiments will yield further data points. Where will they lie? Near the graphs of f_1 and f_2, or near those of f_3 and f_4? Their location will determine the usefulness of one model function over any other. ◀

We close by observing that the row echelon forms **U** and **R** that are computed for **A** via the two MATLAB commands `[L U P]=lu(A)` and `[Q R]=qr(A)` will usually differ. The first command generates an LR factorization of **PA** = **LU** for a permutation matrix **P** via Gaussian elimination, see Theorem 6.9 in Section 6.2, while the latter command finds a row echelon form **R** of **A** via orthogonal eliminations of **A**. The two computed row echelon forms **R** and **U** of $\mathbf{A}$ need not be the same. However, according to Chapter 2, they must be row equivalent to $\mathbf{A}$ and to each other.

The orthogonal QR matrix factorization enables us to solve otherwise inconsistent linear systems optimally.

10.3.P Problems

1. Find an orthonormal basis for the span of

$$\left\{ \begin{pmatrix} 2 \\ -1 \\ -1 \\ 2 \\ 1 \\ 1 \end{pmatrix}, \begin{pmatrix} 3 \\ 1 \\ 4 \\ -1 \\ 2 \\ -2 \end{pmatrix}, \begin{pmatrix} 3 \\ -4 \\ -7 \\ 7 \\ 1 \\ 5 \end{pmatrix}, \begin{pmatrix} 1 \\ -1 \\ -1 \\ 1 \\ 1 \\ 1 \end{pmatrix} \right\},$$

using MATLAB.

2. Assume that $\mathbf{A}_{mn} = \mathbf{Q}_{mn}\mathbf{R}_{nn}$, that $\mathbf{A}$ has n linear independent columns, and that $\mathbf{Q}$ is composed of mutually orthonormal columns.

(a) Show that $n \le m$ in this case.

(b) Show that $\text{rank}(\mathbf{R}) = n = \text{rank}(\mathbf{Q}) = \text{rank}(\mathbf{A})$.

3. Use MATLAB to find an orthonormal basis for the subspace $\mathbf{U}^{\perp}$ that is perpendicular to

$$\mathbf{U} = \text{span}\left\{ \begin{pmatrix} 1 \\ -2 \\ 3 \\ 4 \\ -1 \end{pmatrix}, \begin{pmatrix} 2 \\ 0 \\ 0 \\ 4 \\ 1 \end{pmatrix}, \begin{pmatrix} 0 \\ 0 \\ 1 \\ -1 \\ 0 \end{pmatrix} \right\}.$$ Check your answer.

4. Find the explicit QR factorization by hand for the matrices

(a) $\mathbf{A} = \begin{pmatrix} -4 & 0 \\ -3 & 4 \end{pmatrix}$,

(b) $\mathbf{B} = \begin{pmatrix} 1 & 1 \\ 1 & 1 \\ 1 & 1 \\ 1 & 1 \end{pmatrix}$, and

(c) $\mathbf{C} = \begin{pmatrix} 5 & 6 \\ 0 & 3 \end{pmatrix}$.

5. Try to find integer matrices $\mathbf{A}_{mn}$ so that the refs $\mathbf{U}$ and $\mathbf{R}$ obtained in MATLAB via the two commands `[L U P] = lu(A)` and `[Q R] = qr(A)`

(a) coincide or

(b) differ.

(Use `format long e`.)

6. (a) Show that the matrix $\mathbf{A} = \begin{pmatrix} 1 & 1 \\ 1 & 1 \\ 1 & 1 \end{pmatrix}$ does not have full rank.

(b) Use MATLAB to find the least squares solution $\mathbf{x}_{LS}$ for the problem $\mathbf{A}\mathbf{x} = \begin{pmatrix} 1 \\ 2 \\ 0 \end{pmatrix} = \mathbf{b}$ with $\mathbf{A}$ from part (a).

(c) Repeat part (b) for the right-hand sides $\mathbf{0} \in \mathbb{R}^3$ and $\mathbf{c} = \begin{pmatrix} 1 \\ -4 \\ 2 \end{pmatrix}$.

(See Section 12.3 to learn what MATLAB does in the case that the matrix is not of full rank.)

7. Use MATLAB's `A\b` command to find the least squares solution $\mathbf{x}_{LS}$ for the problem

$$\begin{pmatrix} 2 & -1 & 0 \\ 3 & -1 & 1 \\ 2 & 1 & 1 \\ 1 & 3 & 2 \\ 0 & 1 & 1 \\ 1 & 0 & -1 \end{pmatrix} \mathbf{x} = \begin{pmatrix} 2 \\ 1 \\ -3 \\ 2 \\ 1 \\ 0 \end{pmatrix}.$$

What is the length of the residual vector $\mathbf{A}\mathbf{x}_{LS} - \mathbf{b}$?

8. Consider the overdetermined linear system

$$\begin{pmatrix} 1 & 4 & 3 & 2 \\ 2 & 1 & 1 & -1 \\ -1 & 2 & -2 & 1 \\ 4 & 0 & 2 & -1 \\ 0 & 2 & -2 & 0 \end{pmatrix} \mathbf{x} = \begin{pmatrix} 9 \\ 1 \\ 1 \\ 1 \\ 0 \end{pmatrix}.$$

(a) Find the value of $\|\mathbf{A}\mathbf{x} - \mathbf{b}\|$ for three distinct vectors $\mathbf{x} \in \mathbb{R}^4$ of your choice. Try to find vectors $\mathbf{x}$ with small errors $\|\mathbf{A}\mathbf{x} - \mathbf{b}\|$ by guessing.

(b) Find the least squares solution $\mathbf{x}_{LS}$ of this linear system of equations by using MATLAB.

(c) What is the size of the smallest possible residual error $\min_{\mathbf{x} \in \mathbb{R}^4} \|\mathbf{A}\mathbf{x} - \mathbf{b}\|$ for this problem?

9. Find the optimal functions f_5 and f_6 for the data-fitting problem of Example 3, where $f_5(x) := ax^2 + b\sin^2 x + c\cos^2 x$ and $f_6(x) := ax^2 + b\sin^2(x-1) + c\cos^2(x+1)$. What are the coefficients for the optimal functions in the model classes f_5 and f_6? What are the unavoidable errors for these model functions?

10. Repeat the previous problem for $f_7(x) = a\sqrt{x+1} + bx + c$.

11. Consider the data

x_i	1	5	10	18	20
y_i	2	4	6	8	10

and use the model functions $f_1, \ldots, f_4$ from Example 3 to find the unavoidable error of these model functions for the new data. What are the respective system matrices $\mathbf{A}$ in the least squares problems?

12. What is the QR factorization of an orthogonal matrix $\mathbf{A}_{nn}$?

13. Show that if the columns $\mathbf{a}_i$ of a matrix $\mathbf{A}_{nk}$ are mutually orthogonal, then the matrix $\mathbf{R}$ in the QR factorization of $\mathbf{A}$ is $\mathbf{R} = \text{diag}(\|\mathbf{a}_1\|, \ldots, \|\mathbf{a}_k\|)_{kk}$.

14. Find the least squares solution of the linear system

$$x_1 + x_2 = 1,$$
$$x_1 - x_2 = 1,$$
$$\text{and } x_1 - x_2 = 3.$$

10.R Review Problems

1. (a) Define what is meant by calling two vectors $\mathbf{u}$ and $\mathbf{v} \in \mathbb{R}^n$ orthogonal and orthonormal.

(b) Define what is meant by calling a matrix $\mathbf{V} \in \mathbb{R}^{n,n}$ orthogonal.

(c) Show that every vector in $\mathbb{R}^n$ is orthogonal to the zero vector.

(d) Find all vectors in $\mathbb{R}^4$ that are orthogonal to $\mathbf{a} = (2, -1, 1, 1)$.

(e) Find an ONB for the vectors in part (d).

(f) Show that all vectors that are orthogonal to a given vector $\mathbf{u} \in \mathbb{R}^n$ form a subspace.

(g) Find all vectors that are orthogonal to both $\mathbf{a} = (2, -1, 1, 1)$ and $\mathbf{b} = (3, 0, -2, 1)$.

(h) Find all vectors in $\mathbb{R}^n$ that are orthogonal to each vector of a given basis of $\mathbb{R}^n$.

2. Which of the following sets are an ONB for $\mathbb{R}^2$?

(a) $\left\{\begin{pmatrix} 1 \\ -1 \end{pmatrix}, \begin{pmatrix} 1/\sqrt{2} \\ 1/-\sqrt{2} \end{pmatrix}\right\}$;

(b) $\left\{\begin{pmatrix} 1 \\ 1 \end{pmatrix}, \begin{pmatrix} 1/\sqrt{2} \\ 1/-\sqrt{2} \end{pmatrix}\right\}$;

(c) $\left\{\begin{pmatrix} 1/\sqrt{2} \\ 1/(-\sqrt{2}) \end{pmatrix}, \begin{pmatrix} 1/\sqrt{2} \\ 1/(-\sqrt{2}) \end{pmatrix}\right\}$;

(d) $\left\{\begin{pmatrix} -1/\sqrt{2} \\ 1/(-\sqrt{2}) \end{pmatrix}, \begin{pmatrix} 1/\sqrt{2} \\ 1/(-\sqrt{2}) \end{pmatrix}\right\}$;

(e) $\left\{\begin{pmatrix} \sqrt{2} \\ -\sqrt{2} \end{pmatrix}, \begin{pmatrix} \sqrt{2} \\ \sqrt{2} \end{pmatrix}\right\}$;

(f) $\left\{\begin{pmatrix} 1/\sqrt{5} \\ 2/\sqrt{5} \end{pmatrix}, \begin{pmatrix} -2/\sqrt{5} \\ 1/(-\sqrt{5}) \end{pmatrix}\right\}$;

(g) $\left\{\begin{pmatrix} \cos(x) \\ -\sin(x) \end{pmatrix}, \begin{pmatrix} \sin(x) \\ \cos(x) \end{pmatrix}\right\}$ for arbitrary $x \in \mathbb{R}$;

(h) $\left\{\begin{pmatrix} \cot(x) \\ -\tan(x) \end{pmatrix}, \begin{pmatrix} \sin(x) \\ \cos(x) \end{pmatrix}\right\}$ for arbitrary $x \in \mathbb{R}$.

3. Prove that the transpose $\mathbf{U}^T$ of a square matrix $\mathbf{U}$ is orthogonal if $\mathbf{U}$ is.

4. Find the entries of the last row for the matrix $\mathbf{U} = \begin{pmatrix} 3/5 & 0 & 4/5 \\ 0 & 1 & 0 \\ - & - & - \end{pmatrix}$ so that $\mathbf{U}$ becomes orthogonal. Is this last row unique?

5. Find an ONB $\{\mathbf{v}_i\}$ for $U = \text{span}\{\mathbf{u}_i\} \subset \mathbb{R}^4$ with $\mathbf{u}_1 = \begin{pmatrix} 1 \\ 4 \\ 0 \\ 1 \end{pmatrix}$, $\mathbf{u}_2 = \begin{pmatrix} 1 \\ 1 \\ 1 \\ 1 \end{pmatrix}$, and $\mathbf{u}_3 = \begin{pmatrix} 0 \\ 1 \\ 0 \\ 0 \end{pmatrix}$.

6. For a given set of vectors $\{\mathbf{u}_i\}_{i=1}^k \subset \mathbb{R}^n$ with rational coefficients of the form $\frac{a}{b}$, where a is an integer and $b \neq 0$ is an integer as well, show that there are mutually orthogonal vectors $\{\mathbf{v}_j\}$ with integer coefficients such that $\text{span}\{\mathbf{v}_1, \ldots, \mathbf{v}_\ell\} = \text{span}\{\mathbf{u}_1, \ldots, \mathbf{u}_\ell\}$ for each $\ell = 1, 2, \ldots, k$.

7. Find a subspace of $\mathbb{R}^n$ that does not have an ONB.

8. Is the matrix $\mathbf{A} = \begin{pmatrix} 0 & 0 & 1 \\ 1 & 0 & 0 \\ 0 & 1 & 0 \end{pmatrix}$ orthogonal? What is $\mathbf{A}^{-1}$?

9. (a) If $\mathbf{A}$ is an orthogonal matrix, show that both $\mathbf{A}^T$ and $\mathbf{A}^{-1}$ also orthogonal.

(b) If $\mathbf{A}_{nn}$ is orthogonal, which of $\mathbf{A}^2$ and $\mathbf{A}^{-2}$ is also orthogonal?

10. If $\mathbf{Q}^T\mathbf{Q} = \mathbf{I}_n$, simplify the matrix expressions $(\mathbf{I}+\mathbf{Q})^T(\mathbf{I}-\mathbf{Q})$ and $(\mathbf{I}-\mathbf{Q})^T(\mathbf{I}-\mathbf{Q})$.

Standard Questions and Tasks

1. Is a given set of vectors orthogonal? Orthonormal?

2. Find a unit vector in the direction of $\mathbf{u} \neq \mathbf{0} \in \mathbb{R}^n$.

3. Find an orthonormal basis for the span of several given vectors.

4. Explain the steps of the Gram–Schmidt orthogonalization process.

5. Find the distance of a point in $\mathbb{R}^n$ to a given line through the origin.

6. How can a matrix $\mathbf{A}_{mn}$ be written as a matrix product?

7. Explain what rank 1 matrices look like.

8. What is the definition and the eigen–structure of a Householder matrix?

Subheadings of Lecture Ten

Basic Equations

p. 318 $\mathbf{U}^T\mathbf{U} = \mathbf{I}_n$ (orthogonal matrix)

p. 319 $\mathbf{v}_2 = \mathbf{u}_2 - \dfrac{\mathbf{u}_2 \cdot \mathbf{v}_1}{\mathbf{v}_1 \cdot \mathbf{v}_1}\mathbf{v}_1$ (orthogonal projection of u_2 onto v_1)

p. 326 $\mathbf{P}_{\mathcal{V}} = \mathbf{V}\mathbf{V}^T$ (orthogonal projection)

City symmetry

CHAPTER

11

Eigenvalues of Symmetric and Normal Matrices

We study matrix representations of special linear transformations with respect to orthonormal bases.

11.1 Lecture Eleven

Matrix Representations with Respect to One Orthonormal Basis

Every real symmetric matrix can be represented diagonally with respect to an orthonormal eigenvector basis.

(a) Matrix Representation with Respect to an Orthonormal Basis

If we represent a real square matrix $\mathbf{A}_{nn}$ with respect to an orthonormal basis $\mathcal{U}$ of $\mathbb{R}^n$, we should expect to gain less insight into $\mathbf{A}$'s action on $\mathbb{R}^n$ than if we used a more general basis for representing $\mathbf{A}$.

For certain matrices, however, more can be achieved by this more stringent requirement that the basis $\mathcal{U}$ in which we represent $\mathbf{A}$ is orthonormal. Recall that a basis $\{\mathbf{u}_1, \dots, \mathbf{u}_n\}$ of $\mathbb{R}^n$ is orthonormal if it consists of n unit vectors that are mutually orthogonal—that is, if $\mathbf{u}_i \cdot \mathbf{u}_j = \delta_{ij} = \begin{cases} 0 & \text{if } i \neq j \\ 1 & \text{if } i = j \end{cases}$ for all i and j.

Our fundamental result is the following theorem:

Theorem 11.1 **(Schur Normal Form)**
For every square matrix $\mathbf{A} = \mathbf{A}_\mathcal{E} \in \mathbb{R}^{n,n}$ or $\mathbb{C}^{n,n}$, there exists an orthonormal basis $\mathcal{U}$ of $\mathbb{C}^n$ so that for the associated unitary matrix $\mathbf{U}_{nn} \in \mathbb{C}^{n,n}$ consisting of the basis vectors as columns, the matrix

$$\mathbf{A}_\mathcal{U} = \mathbf{U}^*\mathbf{A}_\mathcal{E}\mathbf{U} = \begin{pmatrix} \lambda_1 & & * \\ & \ddots & \\ 0 & & \lambda_n \end{pmatrix} = \mathbf{R} \qquad (11.1)$$

is upper triangular with $\mathbf{A}$'s eigenvalues $\lambda_i \in \mathbb{C}$ displayed on the diagonal. Specifically, if $\mathbf{A}$ is real and has only real eigenvalues λ_i, then the orthonormal basis $\mathcal{U}$ can be taken as real, too. ◀

Note, however, that according to the Fundamental Theorem of Algebra, the Schur normal form (11.1) of a matrix $\mathbf{A} \in \mathbb{R}^{n,n}$ can generally be obtained only over $\mathbb{C}^n$.

The Schur Normal Form Theorem says that

> Every real or complex square matrix $\mathbf{A}_{nn}$ can be represented by an upper triangular matrix with respect to some orthonormal basis of $\mathbb{C}^n$.
> This basis is called the **Schur basis** of $\mathbf{A}$.

In this section, we will apply Schur's theorem especially to the class of real symmetric matrices.

Before giving a proof over $\mathbb{C}^n$, we need to extend the notion of a dot product and the concept of orthogonality from $\mathbb{R}^n$ to $\mathbb{C}^n$. For example, for one given vector $\mathbf{u}_1 \neq \mathbf{0} \in \mathbb{C}^n$, a complex ONB can be obtained from $\mathbf{u}_1$ by complex orthonormal fill-in as in Section 10.2: Simply find a basis $\mathbf{v}_2, \dots, \mathbf{v}_{n-1} \in \mathbb{C}^n$ for $\ker(\; - \;\; \mathbf{u}_1^* \;\; - \;)$, and use the Gram–Schmidt process of Section 10.1 in its complex version. Here **complex conjugation** $..^*$ of complex vectors $\mathbf{x} \in \mathbb{C}^n$, or of matrices $\mathbf{A} \in \mathbb{C}^{mn}$ is defined as $\mathbf{x}^* := (\overline{\mathbf{x}})^T$ and $\mathbf{A}^* := (\overline{\mathbf{A}})^T$, respectively. The complex Gram–Schmidt algorithm for $\mathbf{v}_2, \mathbf{v}_3, \cdots, \mathbf{v}_{n-1}$ consists of the following familiar two steps:

(a) Orthogonalize

$$\mathbf{x}_2 := \mathbf{v}_2;$$

$$\mathbf{x}_3 := \mathbf{x}_2^*\mathbf{x}_2\; \mathbf{v}_3 - \mathbf{v}_3^*\mathbf{x}_2\; \mathbf{x}_2, \quad \text{etc.}$$

(b) Normalize

$$\mathbf{u}_2 := \frac{\mathbf{x}_2}{\|\mathbf{x}_2\|}, \quad \mathbf{u}_3 := \frac{\mathbf{x}_3}{\|\mathbf{x}_3\|}, \quad \text{etc.}$$

Complex Gram–Schmidt uses the **complex conjugate dot product** of two complex vectors $\mathbf{a}$ and $\mathbf{y} \in \mathbb{C}^n$, namely,

$$\mathbf{a}^*\mathbf{y} := \bar{\mathbf{a}} \cdot \mathbf{y} = (\overline{a_1}, \dots, \overline{a_n}) \begin{pmatrix} y_1 \\ \vdots \\ y_n \end{pmatrix} = \sum_{i=1}^{n} \overline{a_i}\, y_i,$$

where $\overline{a_j} = \alpha_j - \beta_j\sqrt{-1}$ denotes **complex conjugation** in $\mathbb{C}$ for $a_j = \alpha_j + \beta_j\sqrt{-1} \in \mathbb{C}$. And for complex vectors $\mathbf{x}$, we use the **complex vector norm** $\|\mathbf{x}\| := \sqrt{\mathbf{x}^*\mathbf{x}} = \sqrt{\overline{\mathbf{x}} \cdot \mathbf{x}}$. (For more details and examples, see Appendix A on complex numbers and vectors at the end of the book.)

Proof (Theorem 11.1) The given matrix $\mathbf{A} \in \mathbb{C}^{n,n}$ must have an eigenvalue $\lambda \in \mathbb{C}$ corresponding to a normalized unit eigenvector $\mathbf{u}_1 \in \mathbb{C}^n$ (i.e., with $\|\mathbf{u}_1\| = \sqrt{\mathbf{u}_1^*\mathbf{u}_1} = 1$) according to Sections 9.1 or 9.1.D. Form the $n \times n$ complex matrix

$$\mathbf{U}_{nn} = \begin{pmatrix} | & | & & | \\ \mathbf{u}_1 & \mathbf{u}_2 & \cdots & \mathbf{u}_n \\ | & | & & | \end{pmatrix},$$

where the column vectors $\{\mathbf{u}_i\}_{i=1}^n$ are an ONB for $\mathbb{C}^n$ as explained earlier. Now look at $\mathbf{U}^*\mathbf{A}\mathbf{U}$:

$$\mathbf{U}^*\mathbf{A}\mathbf{U} = \begin{pmatrix} - & \mathbf{u}_1^* & - \\ & \vdots & \\ - & \mathbf{u}_n^* & - \end{pmatrix} \begin{pmatrix} | & | & & | \\ \mathbf{A}\mathbf{u}_1 & \mathbf{A}\mathbf{u}_2 & \cdots & \mathbf{A}\mathbf{u}_n \\ | & | & & | \end{pmatrix}$$

$$= \begin{pmatrix} \mathbf{u}_1^*\mathbf{A}\mathbf{u}_1 & \\ \mathbf{u}_2^*\mathbf{A}\mathbf{u}_1 & * \\ \vdots & \\ \mathbf{u}_n^*\mathbf{A}\mathbf{u}_1 & \end{pmatrix} = \begin{pmatrix} \lambda_1 & - \; * \; - \\ 0 & \\ \vdots & \widetilde{\mathbf{A}} \in \mathbb{C}^{n-1,n-1} \\ 0 & \end{pmatrix}_{nn}.$$

Note that $\mathbf{U}^*\mathbf{A}\mathbf{U}$ is partially upper triangular, since $\mathbf{u}_2, \ldots, \mathbf{u}_n \perp \mathbf{A}\mathbf{u}_1 = \lambda_1\mathbf{u}_1$ by construction of the ONB $\{\mathbf{u}_i\}$ of $\mathbb{C}^n$. Subsequently, we need only work on $\widetilde{\mathbf{A}}_{n-1,n-1}$ as we have done on $\mathbf{A}_{nn}$: Let $\widetilde{\mathbf{V}}_{n-1,n-1}$ be unitary, so that

$$\widetilde{\mathbf{V}}^*\widetilde{\mathbf{A}}_{n-1,n-1}\widetilde{\mathbf{V}} = \begin{pmatrix} \lambda_2 & * & * \\ 0 & & \\ \vdots & & * \\ 0 & & \end{pmatrix} \in \mathbb{C}^{n-1,n-1}.$$

Then $\mathbf{W} := \mathbf{U}\begin{pmatrix} 1 & 0 & \ldots & 0 \\ 0 & & & \\ \vdots & & \widetilde{\mathbf{V}} & \\ 0 & & & \end{pmatrix}$ is $n \times n$ unitary according to Lemma 1

and

$$\mathbf{W}^*\mathbf{A}\mathbf{W} = \begin{pmatrix} \lambda_1 & * & \ldots & \ldots & * \\ 0 & \lambda_2 & * & \ldots & * \\ \vdots & 0 & & & \\ \vdots & \vdots & & \widetilde{\mathbf{A}}_{n-2,n-2} & \\ 0 & 0 & & & \end{pmatrix}.$$

This process eventually reduces $\mathbf{A}_{nn}$, by unitary similarity $\mathbf{U}^* \ldots \mathbf{U}$ in $n-1$ orthogonal reduction steps to upper triangular form due to Lemma 1. Specifically, if $\mathbf{A}$ is real and has only real eigenvalues, then the corresponding eigenvectors can be chosen in $\mathbb{R}^n$ and the corresponding Schur basis $\mathcal{U}$ of $\mathbf{A}$ will achieve $\mathbf{A}$'s Schur normal form over the reals. ■

Lemma 1 (a) If $\mathbf{B}_{mm}$ is orthogonal or unitary, then the block diagonal matrix $\mathbf{A} = \begin{pmatrix} \mathbf{I}_{n-m} & \mathbf{O} \\ \mathbf{O} & \mathbf{B} \end{pmatrix}_{nn}$ is also orthogonal or unitary.

(b) If $\mathbf{U}$ and $\mathbf{V}$ are two $n \times n$ orthogonal (or unitary) matrices, then their matrix product $\mathbf{U}\mathbf{V}$ is also orthogonal (or unitary). ◀

Proof (a) If we assume that $\mathbf{B}^*\mathbf{B} = \mathbf{I}_m$, then $\mathbf{A}^*\mathbf{A} = \begin{pmatrix} \mathbf{I}_{n-m} & \\ & \mathbf{B}^*\mathbf{B} \end{pmatrix} = \mathbf{I}_n$.

(b) If $\mathbf{U}^*\mathbf{U} = \mathbf{I}_n = \mathbf{V}^*\mathbf{V}$, then

$$(\mathbf{UV})^*\mathbf{UV} = \mathbf{V}^*\mathbf{U}^*\mathbf{UV} = \mathbf{V}^*\mathbf{I}_n\mathbf{V} = \mathbf{I}_n$$

as well by the complex generalization of Theorem 6.3. ■

(b) Symmetric Matrices

We apply Schur's theorem to real symmetric matrices.

DEFINITION 1 A matrix $\mathbf{A} = (a_{ij}) \in \mathbb{R}^{n,n}$ is **symmetric** if $\mathbf{A} = \mathbf{A}^T$, or equivalently, if $\mathbf{A}$'s entries satisfy $a_{ij} = a_{ji}$ for all indices i and j.

Pictorially speaking, symmetric matrices $\mathbf{A}$ have the same entries in positions that lie symmetrically with respect to their diagonal entries $\{a_{ii}\}$. For example, the matrix $\mathbf{A} = \begin{pmatrix} 0 & 2 & -1 \\ 2 & 4 & 0 \\ -1 & 0 & 2 \end{pmatrix}$ is symmetric, while $\mathbf{B} = \begin{pmatrix} 1 & 2 & 0 \\ 2 & 1 & 0 \end{pmatrix}$, $\mathbf{C} = \begin{pmatrix} 1 & -1 \\ 1 & 4 \end{pmatrix}$, and $\mathbf{D} = \begin{pmatrix} 1 & 3 \\ -2 & 4 \end{pmatrix}$ are not. Clearly, only square matrices can be symmetric.

Symmetry is a strong and prevalent attribute of nature. Symmetric matrices and symmetric problems occur often in applications. Moreover, symmetric real matrices have very nice mathematical properties.

Theorem 11.2 All eigenvalues of a real symmetric matrix $\mathbf{A} = \mathbf{A}^T \in \mathbb{R}^{n,n}$ are real, and the corresponding eigenvectors can be chosen to be real as well. ◀

Proof Let $\mathbf{A} = \mathbf{A}^T \in \mathbb{R}^{n,n}$ be symmetric, and assume that it has the eigenvalue λ. Then $\lambda \in \mathbb{R}$ or $\lambda \in \mathbb{C}$, according to the Fundamental Theorem of Algebra. Hence,

$$\mathbf{Ax} = \lambda\mathbf{x} \quad \text{for some (possibly complex) eigenvector } \mathbf{x} \neq \mathbf{0}.$$

If we take complex conjugates of both sides of the eigenvalue–eigenvector equation, we obtain

$$\overline{\mathbf{Ax}} = \overline{\lambda\mathbf{x}} = \overline{\lambda}\overline{\mathbf{x}}.$$

Transposing yields

$$\left(\overline{\mathbf{Ax}}\right)^T = \overline{\mathbf{x}}^T\overline{\mathbf{A}}^T = \overline{\lambda}\overline{\mathbf{x}}^T,$$

from the rule $(\mathbf{AB})^T = \mathbf{B}^T\mathbf{A}^T$ for matrix transposition. (See Theorem 6.3.)

Using the complex conjugate $\cdots^*$ notation for the conjugate transpose $\overline{\cdots}^T$ as detailed earlier, we have

$$(\mathbf{Ax})^* = \left(\overline{\mathbf{Ax}}\right)^T = \overline{\mathbf{x}}^T\overline{\mathbf{A}}^T = \mathbf{x}^*\mathbf{A}^* = \overline{\lambda}\mathbf{x}^* \text{ with } \mathbf{x} \neq \mathbf{0}.$$

Note that if $\mathbf{A} = \mathbf{A}^T \in \mathbb{R}^{n,n}$, then, clearly, $\mathbf{A}^* = \overline{\mathbf{A}}^T = \mathbf{A}^T = \mathbf{A}$.

Now look at

$$\|\mathbf{Ax}\|^2 \overset{def.}{=} \mathbf{x}^*\mathbf{A}^*\mathbf{Ax} = \overline{\lambda}\mathbf{x}^*\lambda\mathbf{x} = \overline{\lambda}\lambda\mathbf{x}^*\mathbf{x}. \tag{11.2}$$

Since $\mathbf{A}^* = \mathbf{A}^T$ and $\mathbf{A}^T = \mathbf{A}$ for a real symmetric matrix $\mathbf{A}$, we can rewrite (11.2) as

$$\begin{aligned}\|\mathbf{Ax}\|^2 &= \mathbf{x}^*\mathbf{A}^*\mathbf{Ax} = \mathbf{x}^*\mathbf{A}^T\mathbf{Ax} = \mathbf{x}^*\mathbf{A}^2\mathbf{x}\\ &= \mathbf{x}^*\lambda^2\mathbf{x} = \lambda^2\mathbf{x}^*\mathbf{x},\end{aligned} \tag{11.3}$$

because $\mathbf{A}^2\mathbf{x} = \mathbf{A}(\mathbf{Ax}) = \mathbf{A}(\lambda\mathbf{x}) = \lambda\mathbf{Ax} = \lambda\lambda\mathbf{x} = \lambda^2\mathbf{x}$ for the eigenvalue λ and corresponding eigenvector $\mathbf{x} \neq \mathbf{0}$ of $\mathbf{A}$. Now, $\mathbf{x}^*\mathbf{x} \neq 0$, and thus $\overline{\lambda}\lambda = \lambda^2$ by comparing (11.2) and (11.3). For a complex number $\lambda = a + bi \in \mathbb{C}$ with $i = \sqrt{-1} \notin \mathbb{R}$ and $a, b \in \mathbb{R}$, this yields

$$\overline{\lambda}\lambda = a^2 + b^2 = a^2 - b^2 + 2abi = \lambda^2,$$

or $b = 0$. Thus, a real symmetric matrix $\mathbf{A} = \mathbf{A}^T$ can only have real eigenvalues λ. As an immediate consequence of Section 9.1 or 9.1.D, the symmetric matrix $\mathbf{A} = \mathbf{A}^T \in \mathbb{R}^{n,n}$ must have real eigenvectors in $\ker(\mathbf{A} - \lambda\mathbf{I}) \subset \mathbb{R}^n$ as well. ■

The Schur normal form and its proof, applied to a real symmetric matrix $\mathbf{A} = \mathbf{A}^T$, gives us an ONB $\mathcal{U}$ of $\mathbb{R}^n$ for which $\mathbf{U}^T\mathbf{AU} = \begin{pmatrix} \lambda_1 & & * \\ & \ddots & \\ 0 & & \lambda_n \end{pmatrix} \in \mathbb{R}^{n,n}$ is upper triangular and real. The following lemma is most illuminating for the symmetric matrix case.

Lemma 2 If $\mathbf{A} = \mathbf{A}^T \in \mathbb{R}^{n,n}$ is symmetric, then, for any matrix $\mathbf{X} \in \mathbb{R}^{n,n}$, the matrix $\mathbf{X}^T\mathbf{AX}$ is likewise symmetric. ◀

Proof Clearly,

$$(\mathbf{X}^T\mathbf{AX})^T = \mathbf{X}^T\mathbf{A}^T\mathbf{X}^{T^T} = \mathbf{X}^T\mathbf{AX},$$

by Theorem 6.3, since $\left(\mathbf{X}^T\right)^T = \mathbf{X}$. ■

Thus, the Schur normal form for a real symmetric matrix $\mathbf{A}$, that is, the matrix

$$\mathbf{U}^T\mathbf{AU} = \begin{pmatrix} \lambda_1 & & * \\ & \ddots & \\ 0 & & \lambda_n \end{pmatrix} \in \mathbb{R}^{n,n},$$

is necessarily symmetric. Hence, it must be diagonal, and thus we have proved the following theorem:

Theorem 11.3 Every real symmetric matrix $\mathbf{A} = \mathbf{A}^T \in \mathbb{R}^{n,n}$ can be diagonalized with respect to a real orthonormal basis of eigenvectors. ◀

Real symmetric matrices can always be diagonalized over the reals, and their eigenvector bases can always be chosen to be real orthonormal. This is a much stronger result than is possible for general matrices $\mathbf{A} \in \mathbb{R}^{n,n}$ and their diagonalization. Even if a symmetric matrix $\mathbf{A}$ has a repeated eigenvalue, it is always orthogonally diagonalizable according to Theorem 11.3. This is in contrast to a general matrix with repeated eigenvalues, which may or may not be diagonalizable, even when using a general basis representation. (See also Chapter 14.)

EXAMPLE 1 The matrix

$$\mathbf{A} = \begin{pmatrix} -1 & -3 & -1 & -2 \\ -3 & 7 & 3 & 6 \\ -1 & 3 & -1 & 2 \\ -2 & 6 & 2 & 2 \end{pmatrix}$$

is real and symmetric. Its characteristic polynomial is

$$\begin{aligned} f_A(\lambda) = \det(\lambda \mathbf{I} - \mathbf{A}) &= \lambda^4 - 7\lambda^3 - 66\lambda^2 - 148\lambda - 104 \\ &= (\lambda + 2)^3(\lambda - 13), \end{aligned}$$

according to Section 9.1.D. Thus, $\mathbf{A}$ has one triple eigenvalue $\lambda = -2$ and one simple eigenvalue 13.

Alternatively, using Section 9.1, we can form the vector iteration sequence

$$\mathbf{y},\ \mathbf{Ay},\ \mathbf{A}^2\mathbf{y} := \mathbf{A}(\mathbf{Ay}),\ \mathbf{A}^3\mathbf{y} := \mathbf{A}(\mathbf{A}(\mathbf{Ay})),\ \mathbf{A}^4\mathbf{y} := \mathbf{A}(\mathbf{A}(\mathbf{Ay})))$$

in MATLAB and compute the RREF of the corresponding vector iteration matrix by using the following commands:

```
a = A*y, b = A*a, c = A*b, d = A*c, rref([y a b c d])
```

For various starting vectors, such as $\mathbf{y} = \begin{pmatrix} 2 \\ 1 \\ -1 \\ -1 \end{pmatrix}, \begin{pmatrix} 2 \\ 1 \\ -1 \\ 1 \end{pmatrix}$ or $\begin{pmatrix} 0 \\ 1 \\ -1 \\ -1 \end{pmatrix}$, the RREF of the vector iteration matrix is computed to be $\begin{pmatrix} \boxed{1} & 0 & 26 & * & * \\ 0 & \boxed{1} & 11 & * & * \\ 0 & 0 & 0 & 0 & 0 \\ 0 & 0 & 0 & 0 & 0 \end{pmatrix}$.

Thus, $\mathbf{A}^2\mathbf{y} = 11\mathbf{Ay} + 26\mathbf{y}$ for each of these starting vectors, and the vanishing polynomial is equal to $p_y(x) = x^2 - 11x - 26 = (x - 13)(x + 2)$ for each $\mathbf{y}$. Hence, $\mathbf{A}$ has at least two distinct eigenvalues, -2 and 13, according to Section 9.1.

Let us find a basis of eigenvectors for **A**'s repeated eigenvalue $\lambda = -2$ first:

$$\ker(\mathbf{A}+2\mathbf{I}):\quad \begin{array}{cccc|c|l} \boxed{1} & -3 & -1 & -2 & 0 & \\ \boxed{-3} & 9 & 3 & 6 & 0 & +3\,row_1 \\ \boxed{-1} & 3 & 1 & 2 & 0 & +\,row_1 \\ \boxed{-2} & 6 & 2 & 4 & 0 & +\,2\,row_1 \\ \hline \boxed{1} & -3 & -1 & -2 & 0 & \\ 0 & 0 & 0 & 0 & 0 & \\ 0 & 0 & 0 & 0 & 0 & \\ 0 & 0 & 0 & 0 & 0 & \\ & \uparrow & \uparrow & \uparrow & & \\ & \alpha & \beta & \gamma & & \end{array}$$

We notice immediately that $\mathbf{A}+2\mathbf{I}$ has a three–dimensional kernel. Using backsubstitution, we see that each vector $\mathbf{x} \in \ker(\mathbf{A}+2\mathbf{I})$ has the form

$$\mathbf{x} = \begin{pmatrix} 3\alpha+\beta+2\gamma \\ \alpha \\ \beta \\ \gamma \end{pmatrix} = \alpha \begin{pmatrix} 3 \\ 1 \\ 0 \\ 0 \end{pmatrix} + \beta \begin{pmatrix} 1 \\ 0 \\ 1 \\ 0 \end{pmatrix} + \gamma \begin{pmatrix} 2 \\ 0 \\ 0 \\ 1 \end{pmatrix},$$

making $U = \{\mathbf{u}_i\}_{i=1}^3 = \left\{ \begin{pmatrix} 3 \\ 1 \\ 0 \\ 0 \end{pmatrix}, \begin{pmatrix} 1 \\ 0 \\ 1 \\ 0 \end{pmatrix}, \begin{pmatrix} 2 \\ 0 \\ 0 \\ 1 \end{pmatrix} \right\}$ a basis for the eigenspace $E(-2)$ of **A**.

The three linearly independent eigenvectors $\mathbf{u}_i$ of **A** are not orthogonal. In order to diagonalize **A** orthogonally, as in Theorem 11.3, we must use the Gram–Schmidt process of Section 10.1 to obtain an ONB for the eigenspace $E(-2)$ of **A**.

At the first level of Gram–Schmidt, we orthogonalize $\mathbf{u}_2$ and $\mathbf{u}_3$ with respect to $\mathbf{u}_1$:

$$\mathbf{v}_1 := \mathbf{u}_1 = \begin{pmatrix} 3 \\ 1 \\ 0 \\ 0 \end{pmatrix};$$

$$\mathbf{v}_2 := \mathbf{v}_1 \cdot \mathbf{v}_1\ \mathbf{u}_2 - \mathbf{u}_2 \cdot \mathbf{v}_1\ \mathbf{v}_1 = 10 \begin{pmatrix} 1 \\ 0 \\ 1 \\ 0 \end{pmatrix} - 3 \begin{pmatrix} 3 \\ 1 \\ 0 \\ 0 \end{pmatrix}$$

$$= \begin{pmatrix} 10-9 \\ -3 \\ 10 \\ 0 \end{pmatrix} = \begin{pmatrix} 1 \\ -3 \\ 10 \\ 0 \end{pmatrix} \perp \mathbf{v}_1 ;$$

$$\mathbf{v}_3 := \mathbf{v}_1 \cdot \mathbf{v}_1 \ \mathbf{u}_3 - \mathbf{u}_3 \cdot \mathbf{v}_1 \ \mathbf{v}_1 = 10 \begin{pmatrix} 2 \\ 0 \\ 0 \\ 1 \end{pmatrix} - 6 \begin{pmatrix} 3 \\ 1 \\ 0 \\ 0 \end{pmatrix}$$

$$= \begin{pmatrix} 20 - 18 \\ -6 \\ 0 \\ 10 \end{pmatrix} = \begin{pmatrix} 2 \\ -6 \\ 0 \\ 10 \end{pmatrix} \simeq \begin{pmatrix} 1 \\ -3 \\ 0 \\ 5 \end{pmatrix} =: \mathbf{v}_3 \perp \mathbf{v}_1.$$

Here, we have scaled $\mathbf{v}_3$ by $\frac{1}{2}$ to obtain an equivalent vector with small integer components.

At the second level of Gram–Schmidt, we orthogonalize $\mathbf{v}_3$ with respect to $\mathbf{v}_2$:

$$\mathbf{v}_3 := \mathbf{v}_2 \cdot \mathbf{v}_2 \ \mathbf{v}_3 - \mathbf{v}_3 \cdot \mathbf{v}_2 \ \mathbf{v}_2 = 110 \begin{pmatrix} 1 \\ -3 \\ 0 \\ 5 \end{pmatrix} - 10 \begin{pmatrix} 1 \\ -3 \\ 10 \\ 0 \end{pmatrix}$$

$$= \begin{pmatrix} 110 - 10 \\ -330 + 30 \\ -100 \\ 550 \end{pmatrix} = \begin{pmatrix} 100 \\ -300 \\ -100 \\ 550 \end{pmatrix} \simeq \begin{pmatrix} 2 \\ -6 \\ -2 \\ 11 \end{pmatrix} =: \mathbf{v}_3 \perp \mathbf{v}_2, \ \mathbf{v}_1.$$

Again, we have scaled the final $\mathbf{v}_3$, this time by $\frac{1}{50}$, to obtain small relatively prime integer coordinates in $\mathbf{v}_3$. The three vectors $\mathbf{v}_1$, $\mathbf{v}_2$, and $\mathbf{v}_3$ form an orthogonal eigenvector basis for the triple eigenvalue $\lambda = -2$ of $\mathbf{A}$. Upon their normalization and after computing a unit length eigenvector $\mathbf{z}$ as $\frac{1}{\sqrt{15}} \begin{pmatrix} 1 \\ -3 \\ -1 \\ -2 \end{pmatrix}$ for the eigenvalue $\lambda = 13$ of $\mathbf{A}$, we can diagonalize $\mathbf{A}$ orthonormally with respect to its ONB of eigenvectors $\mathcal{W} = \{\mathbf{w}_i\} = \left\{ \mathbf{z}, \frac{\mathbf{v}_1}{\|\mathbf{v}_1\|}, \frac{\mathbf{v}_2}{\|\mathbf{v}_2\|}, \frac{\mathbf{v}_3}{\|\mathbf{v}_3\|} \right\}$. With this choice and ordering of the eigenvector basis for $\mathbf{A}$, the underlying linear transformation has the diagonal matrix representation

$$\mathbf{A}_{\mathcal{W}} = \begin{pmatrix} 13 & & & 0 \\ & -2 & & \\ & & -2 & \\ 0 & & & -2 \end{pmatrix}$$

for the ONB $\mathcal{W}$.

Another approach to finding an ONB $\mathcal{W}$ for the eigenspace $E(-2) = \ker(\mathbf{A} + 2\mathbf{I})$ of $\mathbf{A}$ would be to use the QR decomposition of Section 10.3: Let $\mathbf{U} = \begin{pmatrix} 3 & 1 & 2 \\ 1 & 0 & 0 \\ 0 & 1 & 0 \\ 0 & 0 & 1 \end{pmatrix}$ be made up columnwise out of our initially computed linearly independent eigenvectors $\mathbf{u}_i \in \ker(\mathbf{A}+2\mathbf{I})$. Then the MATLAB command `[Q R] = qr(U)` produces

$$\mathbf{Q} = \begin{pmatrix} -0.94868 & 0.09535 & 0.15570 & -0.25820 \\ -0.31623 & -0.28604 & -0.46710 & 0.77460 \\ 0 & 0.95346 & -0.15570 & 0.25820 \\ 0 & 0 & 0.85635 & 0.51640 \end{pmatrix}$$

and

$$\mathbf{R} = \begin{pmatrix} \boxed{-3.1623} & -0.94868 & -1.8974 \\ 0 & \boxed{1.0488} & 0.19069 \\ 0 & 0 & \boxed{1.1677} \\ 0 & 0 & 0 \end{pmatrix}$$

with $\mathbf{U} = \mathbf{QR}$. (See `help qr`.) Taking columns 1, 2, and 3 of $\mathbf{Q}$ that correspond to the pivot columns 1, 2, and 3 of $\mathbf{R}$ gives us an orthonormal basis of $\ker(\mathbf{A}+2\mathbf{I})$:

$$\mathbf{q}_1 = \begin{pmatrix} -0.94868 \\ -0.31623 \\ 0 \\ 0 \end{pmatrix}, \mathbf{q}_2 = \begin{pmatrix} 0.09535 \\ -0.28604 \\ 0.95346 \\ 0 \end{pmatrix}, \text{ and } \mathbf{q}_3 = \begin{pmatrix} 0.15570 \\ -0.46710 \\ -0.15570 \\ 0.85635 \end{pmatrix}.$$

Note that $\mathbf{q}_1$ is parallel to $\mathbf{v}_1$, but points in the opposite direction, while $\mathbf{q}_2 \| \mathbf{v}_2$ and $\mathbf{q}_3 \| \mathbf{v}_3$ both point in the same direction as our earlier computed $\mathbf{v}_i$. ◀

We have shown that every real symmetric matrix has a real orthonormal eigenvector basis.

11.1.P Problems

1. Which of the following matrices is symmetric:

(a) $\mathbf{A} = \begin{pmatrix} 1 & 2 & -1 \\ 2 & 0 & 0 \\ 1 & 0 & 0 \end{pmatrix}$;

(b) $\mathbf{B} = \begin{pmatrix} 0 & 1 & -1 \\ -1 & 0 & 0 \\ 1 & 0 & 1 \end{pmatrix}$;

(c) $\mathbf{C} = \begin{pmatrix} 2 & 3 & 0 \\ 3 & 4 & 0 \end{pmatrix}$;

(d) $\mathbf{D} = \begin{pmatrix} 1 & 0 & 1 \\ 0 & 0 & -1 \\ a & -1 & 0 \end{pmatrix}$.

2. If $\mathbf{A} = \mathbf{A}^T$ and $\mathbf{B} = \mathbf{B}^T \in \mathbb{R}^{n,n}$ are symmetric and $\mathbf{A}$ is nonsingular, which of the following matrices is symmetric:

(a) $\mathbf{ABA}$; (b) $\mathbf{AB}$;
(c) $\mathbf{AA}$; (d) $\mathbf{ABB}^T\mathbf{A}$;
(e) $\mathbf{B}^2\mathbf{AB}^2$; (f) $\mathbf{ABA}^{-1}$;
(g) $\mathbf{A}^T\mathbf{BA}^{-1}$; (h) $\mathbf{BABA}$.

3. Find all eigenvalues and the corresponding eigenvectors of $\mathbf{A} = \begin{pmatrix} 1 & 1 & 1 \\ 1 & 1 & 1 \\ 1 & 1 & 1 \end{pmatrix} = \mathbf{A}^T$.

4. Let $\mathbf{B} = \begin{pmatrix} 0 & 1 & 1 \\ 1 & 0 & 1 \\ 1 & 1 & 0 \end{pmatrix} = \mathbf{B}^T$.

(a) Find all eigenvalues and corresponding eigenvectors of $\mathbf{B}$.

(b) Why is $\mathbf{B}$ diagonalizable?

(c) What matrix $\mathbf{X}$ will achieve $\mathbf{X}^{-1}\mathbf{BX} = \mathbf{D}$ with a diagonal matrix $\mathbf{D}$? What is the matrix $\mathbf{D}$?

5. Find an orthonormal basis of eigenvectors for the symmetric matrix $\mathbf{A} = \begin{pmatrix} -1 & -1 & 1 \\ -1 & 2 & 4 \\ 1 & 4 & 2 \end{pmatrix}$.

6. Show that the matrix $\mathbf{A} = \begin{pmatrix} -15 & 5 & 5 \\ 5 & -6 & -4 \\ 5 & -4 & -6 \end{pmatrix}$ has only negative eigenvalues, namely, -20, -5, and -2.

7. Find an ONB $\mathcal{U}(a)$ of eigenvectors for the matrix $\mathbf{A}(a) = \begin{pmatrix} 1 & a \\ a & 1 \end{pmatrix}$. Note that $\mathbf{A}$ and $\mathcal{U}$ both depend on the real parameter $a \in \mathbb{R}$ here. Represent $\mathbf{A}(a)$ with respect to $\mathcal{U}(a)$ for each a.

8. Find the minimal or the characteristic polynomial $p_A(x)$ or $f_A(x)$ (without or with determinants) and all eigenvalues of the matrix $\mathbf{A} = \begin{pmatrix} -1 & 1 & 1 & 0 & 0 \\ 1 & -1 & 1 & 0 & 0 \\ 1 & 1 & -1 & 0 & 0 \\ 0 & 0 & 0 & 3 & 1 \\ 0 & 0 & 0 & 1 & 3 \end{pmatrix}$.

9. (a) Show that if $\mathbf{A}_{nn}$ is symmetric and has the eigenvalues $\lambda_i \in \mathbb{R}$ for an ONB of eigenvectors $\mathbf{x}_i \in \mathbb{R}^n$ and $i = 1, \ldots, n$, then

$$\mathbf{A} = \lambda_1 \mathbf{x}_1 \mathbf{x}_1^T + \ \ldots\ + \lambda_n \mathbf{x}_n \mathbf{x}_n^T.$$

(b) Write $\mathbf{A} = \begin{pmatrix} 0 & 2 \\ 2 & 0 \end{pmatrix}$ as a sum of two eigendyads as in part (a).

10. Diagonalize the matrices $\mathbf{A} = \begin{pmatrix} 0 & 1 \\ 1 & 0 \end{pmatrix}$ and $\mathbf{B} = \begin{pmatrix} 0 & 0 & 1 \\ 0 & 1 & 0 \\ 1 & 0 & 0 \end{pmatrix}$ by finding their eigenvalues and orthonormal bases of eigenvectors.

11. Assume that the matrix $\mathbf{W} \in \mathbb{R}^{n,n}$ satisfies the matrix equation $\mathbf{W}^T\mathbf{W} = \mathbf{I}_n$ (i.e., $\mathbf{W}$ is orthogonal).

(a) If $\mathbf{R} \in \mathbb{C}^{n,n}$ is the Schur normal form of $\mathbf{W}$, write out an equation for $\mathbf{W}$ involving $\mathbf{R}$ and an orthogonal matrix $\mathbf{U}$, and likewise for $\mathbf{W}^T$ in terms of $\mathbf{R}^*$ and the orthogonal $\mathbf{U}$.

(b) Use the equations from part (a) to expand $\mathbf{I} = \mathbf{W}^T\mathbf{W} = \ldots$. and conclude that if $\mathbf{W}^T\mathbf{W} = \mathbf{I}$ is assumed, then $\mathbf{R}\mathbf{R}^* = \mathbf{I}$ and $\mathbf{R}^*\mathbf{R} = \mathbf{I}$ as well.

(c) Show that if $\mathbf{R}^*\mathbf{R} = \mathbf{I}$ for an upper triangular matrix $\mathbf{R}$, then $\mathbf{R}$ is in fact diagonal.

(d) Conclude that all matrices $\mathbf{W}_{nn}$ with $\mathbf{W}^T\mathbf{W} = \mathbf{I}_n$ are orthogonally diagonalizable, just as symmetric matrices are.

12. State the complete theorem of Schur on matrix triangularization.

13. Compute the Schur normal forms of $\mathbf{A} = \begin{pmatrix} 1 & 1 & 1 \\ 1 & 1 & 1 \\ 1 & 1 & 1 \end{pmatrix}$ and $\mathbf{B} = \begin{pmatrix} 1 & 0 & 0 \\ 1 & 2 & 0 \\ 1 & 1 & 3 \end{pmatrix}$.

14. (a) Prove that if $\mathbf{A} = \mathbf{A}^T$, then the two dot products of $\mathbf{Ax}$ with $\mathbf{y}$ and of $\mathbf{Ay}$ with $\mathbf{x}$ are always identical.

(b) For $\mathbf{A} = \begin{pmatrix} 2 & 3 & -3 \\ 3 & 17 & 5 \\ -3 & 5 & 0 \end{pmatrix} = \mathbf{A}^T$, compute $\mathbf{Ax}$ and $\mathbf{Ay}$ for $\mathbf{x} = \begin{pmatrix} -1 \\ 3 \\ 2 \end{pmatrix}$ and $\mathbf{y} = \begin{pmatrix} 0 \\ -2 \\ 3 \end{pmatrix}$ and both of the dot products $\mathbf{Ax} \cdot \mathbf{y}$ and $\mathbf{Ay} \cdot \mathbf{x}$ of part (a).

15. (a) Show that $\mathbf{C} = \begin{pmatrix} 4 & 1 & 1 \\ 1 & 4 & 1 \\ 1 & 1 & 4 \end{pmatrix}$ has the eigenvalue 6.

(b) Is $\mathbf{x} = \begin{pmatrix} 1 \\ -1 \\ 0 \end{pmatrix}$ an eigenvector of $\mathbf{C}$?

(c) Complete the eigenanalysis of $\mathbf{C}$. Is $\mathbf{C}$ diagonalizable with respect to an ONB $\mathcal{U}$ of eigenvectors?

(d) Find $\mathcal{U}$ and $\mathbf{C}_{\mathcal{U}}$.

16. If $\mathbf{A} = \mathbf{A}^T$ is nonsingular and $\mathbf{A}^2 = \mathbf{A}$, find $\mathbf{A}^{-1}$.

17. Verify that the matrix $\mathbf{A} = \frac{1}{10}\begin{pmatrix} 6 & 8 \\ 8 & -6 \end{pmatrix}$ is orthogonal.

18. (a) Find all eigenvalues of $\mathbf{B}(x) = \begin{pmatrix} \sin(x) & \cos(x) \\ \cos(x) & -\sin(x) \end{pmatrix}$ for arbitrary $x \in \mathbb{R}$.

(b) How does $\mathbf{B}(x)$ map the standard unit vectors $\mathbf{e}_1$ and $\mathbf{e}_2$ of $\mathbb{R}^2$?

(c) Try to express $\mathbf{B}(x)$ as the matrix product of a reflection of $\mathbb{R}^2$ followed by a rotation.

(d) Use part (c) to describe the action of $\mathbf{B}$ on $\mathbb{R}^2$ geometrically.

(e) Find all eigenvectors of $\mathbf{B}(x)$.

(f) How do the eigenvectors found in part (e) correlate with part (d)?

19. Perform a complete eigenanalysis of the matrix $\mathbf{A} = \begin{pmatrix} 1 & 0 & 0 \\ 0 & 17 & 0 \\ 0 & 0 & -31 \end{pmatrix}$. Is $\mathbf{A}$ diagonalizable? Is it symmetric?

20. Show that the following two matrices $\mathbf{H} = \begin{pmatrix} 0 & 0 & 1 \\ 0 & 0 & 2 \\ 1 & 1 & 0 \end{pmatrix}$ and $\mathbf{K} = \begin{pmatrix} 0 & 0 & 1 \\ 0 & 0 & 1 \\ 1 & 1 & 0 \end{pmatrix}$ are both diagonalizable over the reals.

21. Compute an eigenvector for the eigenvalue $\lambda = 13$ of the matrix **A** in Example 1.

22. If the matrix $\mathbf{A} = \mathbf{A}^T$ is real symmetric, show that the matrices $\mathbf{A}^2, \mathbf{A}^3, \mathbf{A}^4, \ldots$ are also real symmetric.

23. Consider $\mathbf{A} = \begin{pmatrix} 1 & 2 & -1 & 3 \\ -1 & 2 & 0 & -1 \\ 0 & 1 & 2 & 2 \\ 0 & 0 & -1 & 1 \end{pmatrix}$,

$\mathbf{B} = \begin{pmatrix} 0 & 1 & -1 & 2 \\ 1 & 0 & 0 & -1 \\ -1 & 0 & -4 & 1 \\ 2 & -1 & 1 & 2 \end{pmatrix}$, and

$\mathbf{C} = \begin{pmatrix} 0 & 0 & 1 & 0 \\ 0 & 0 & 0 & 0 \\ 0 & 0 & 0 & 0 \end{pmatrix}$.

(a) Which matrix, **A**, **B**, or **C**, is diagonalizable just by the looks of it? (Why?)

(b) Which of the three matrices is obviously not diagonalizable? (Why?)

(c) Which of the matrices **A**, **B**, and **C** will require the most hand computations to determine whether it is diagonalizable? (Why?)

24. If a matrix $\mathbf{A} = \mathbf{A}^T$ is real symmetric and nonsingular, show that its inverse $\mathbf{A}^{-1}$ exists and is also symmetric.

25. Prove the converse of Theorem 11.3: The matrix $\mathbf{A} = \mathbf{U}\mathbf{D}\mathbf{U}^T$ is real symmetric if $\mathbf{U} \in \mathbb{R}^{n,n}$ is orthogonal and $\mathbf{D} \in \mathbb{R}^{n,n}$ is real diagonal.

26. Show that if $\mathbf{A} = \mathbf{A}^T$ has the real eigenvalues $\lambda_1 \leq \lambda_2 \leq \ldots \leq \lambda_n$, then

(a) $\max_{\|\mathbf{x}\|=1} \mathbf{x}^T\mathbf{A}\mathbf{x} = \lambda_n$ and

(b) $\min_{\|\mathbf{x}\|=1} \mathbf{x}^T\mathbf{A}\mathbf{x} = \lambda_1$.

27. Show that if $\mathbf{B} = \mathbf{B}^T$ has the eigenvalues $0 \leq \lambda_1 \leq \lambda_2 \leq \ldots \leq \lambda_n$ and if **A** is symmetric, then $\mathbf{x}^T\mathbf{A}\mathbf{x} \leq \mathbf{x}^T(\mathbf{A}+\mathbf{B})\mathbf{x}$ and $\mathbf{x}^T\mathbf{A}\mathbf{x} \geq \mathbf{x}^T(\mathbf{A}-\mathbf{B})\mathbf{x}$ for all vectors $\mathbf{x} \in \mathbb{R}^n$.

28. Show that if $\mathbf{A}_{nn} = \mathbf{B}\mathbf{B}^T$ for an $n \times m$ matrix **B**, then **A** is symmetric.

29. If for an $n \times n$ matrix **A** we have $\mathbf{A}\mathbf{x} = \lambda\mathbf{x}$ for $\mathbf{x} \neq \mathbf{0}$ and $\mathbf{A}^T\mathbf{y} = \mu\mathbf{y}$ for $\mathbf{y} \neq \mathbf{0}$ and $\lambda \neq \mu$, show that $\mathbf{x} \perp \mathbf{y}$.

30. True or false? A matrix $\mathbf{A} \in \mathbb{R}^{n,n}$ is diagonalizable over the reals if and only if **A** is similar to a real symmetric matrix.

31. True or false?

(a) The product of any two symmetric $n \times n$ matrices is symmetric.

(b) The product of two symmetric $n \times n$ matrices may be symmetric.

(c) The product of any two orthogonal $n \times n$ matrices is orthogonal.

(d) The product of any two diagonal $n \times n$ matrices is diagonal.

(e) The product of two orthogonal matrices may be symmetric.

32. (a) Show that for $\mathbf{X} \in \mathbb{R}^{n,k}$ the products $\mathbf{X}^T\mathbf{X}$ and $\mathbf{X}\mathbf{X}^T$ can both be formed and both are symmetric.

(b) Show that for $\mathbf{A} = \mathbf{A}^T \in \mathbb{R}^{n,n}$, $\mathbf{B} := \mathbf{X}^T\mathbf{A}\mathbf{X}$ is symmetric for all matrices $\mathbf{X} \in \mathbb{R}^{n,n}$.

33. True or false?

(a) Any diagonalizable $n \times n$ matrix is the product of two symmetric ones.

(b) The matrix $\mathbf{J} = \begin{pmatrix} 2 & 1 \\ 0 & 2 \end{pmatrix}$ is the product of two symmetric matrices. Is **J** diagonalizable?

34. Assume that $\mathbf{B}_{nn} = \mathbf{X}^{-1}\mathbf{A}\mathbf{X}$ and $\mathbf{B} = \mathbf{S}_1\mathbf{S}_2$ for two symmetric matrices $\mathbf{S}_1$ and $\mathbf{S}_2$. Prove or disprove that **A** is also the product of two symmetric matrices.

35. Assume that the matrix **A** has the eigenvalue λ with algebraic multiplicity n_λ and geometric multiplicity $k_\lambda = \dim(E(\lambda))$. Show that for any $\alpha \in \mathbb{C}$, the matrix $\mathbf{A} + \alpha\mathbf{I}$ has the eigenvalue $\lambda + \alpha$ with the same algebraic multiplicity n_λ and with the same geometric multiplicity k_λ as **A** does for λ. (*Hint*: Apply the Schur Normal Form to **A** and $\mathbf{A} + \alpha\mathbf{I}$ simultaneously.) Compare with Problem 17 of Section 9.2.P.

Teacher's Problem Making Exercise

T 11. For ease with hand computations, we must be able to prescribe the entries of an integer symmetric matrix $\mathbf{A}_{nn}$ so that it has known integer eigenvalues and, subsequently, integer eigenvectors.

If $\mathbf{W}_{nn}$ is a matrix with mutually orthogonal integer columns, as constructed in Problem T 10 for example, and if $\mathbf{D} = \text{diag}(d_i)$ is an integer diagonal matrix, then $\mathbf{A} := \mathbf{W}\mathbf{D}\mathbf{W}^T$ is integer and symmetric by Lemma 2. Now, $\mathbf{A} = \mathbf{W}\mathbf{D}\mathbf{W}^T = \mathbf{W}\ (\mathbf{N}\mathbf{D}_0\mathbf{N})\mathbf{W}^T =$

$$\begin{pmatrix} | & & | \\ \mathbf{w}_1 & \dots & \mathbf{w}_n \\ | & & | \end{pmatrix} \begin{pmatrix} \frac{1}{\|\mathbf{w}_1\|} & & \\ & \ddots & \\ & & \frac{1}{\|\mathbf{w}_n\|} \end{pmatrix} \text{diag}(d_i^\circ) \begin{pmatrix} \frac{1}{\|\mathbf{w}_1\|} & & \\ & \ddots & \\ & & \frac{1}{\|\mathbf{w}_n\|} \end{pmatrix} \begin{pmatrix} - & \mathbf{w}_1^T & - \\ & \vdots & \\ - & \mathbf{w}_n^T & - \end{pmatrix},$$

where $\mathbf{N} = \text{diag}\left(\frac{1}{\|\mathbf{w}_i\|}\right)$. The product $\mathbf{W}\mathbf{N}$ is orthogonal and the $d_i^\circ = d_i\|\mathbf{w}_i\|^2$ are integers. The eigenvalues of $\mathbf{A}$ are the eigenvalues d_i° of $\mathbf{D}_0$, since $\mathbf{A}$ is similar to $\mathbf{D}_0 = \text{diag}(d_i^\circ)$.

EXAMPLE The two integer matrices $\mathbf{W}_1 = \begin{pmatrix} 3 & -1 & 3 \\ 1 & 3 & 1 \\ -5 & 0 & 2 \end{pmatrix}$ and $\mathbf{W}_2 = \begin{pmatrix} 1 & 1 & -1 \\ -1 & 1 & 1 \\ 1 & 0 & 2 \end{pmatrix}$ each have mutually orthogonal columns of lengths $\sqrt{35}, \sqrt{10}, \sqrt{14}$ and $\sqrt{3}, \sqrt{2}, \sqrt{6}$, respectively. Choosing $\mathbf{D}_1 = \text{diag}(1, -1, 1)$ and working with $\mathbf{W}_1$ creates the symmetric integer matrix $\mathbf{A}_1 = \mathbf{W}_1\mathbf{D}_1\mathbf{W}_1^T = \begin{pmatrix} 17 & 9 & 9 \\ 9 & -7 & -3 \\ -9 & -3 & 29 \end{pmatrix}$. The squared lengths of the columns of $\mathbf{W}_1$, each multiplied by the respective d_i in $\mathbf{D}_1$, are the eigenvalues 35, -10, and 14 of $\mathbf{A}_1$. Repeating the same process for $\mathbf{W}_2$ and $\mathbf{D}_2 = \text{diag}(-1, 1, 1)$ gives us $\mathbf{A}_2 = \mathbf{W}_2\mathbf{D}_2\mathbf{W}_2^T = \begin{pmatrix} 1 & 1 & -3 \\ 1 & 1 & 3 \\ -3 & 3 & 3 \end{pmatrix} = \mathbf{A}_2^T$ with integer eigenvalues -3, 2, and 6.

Note: The smallest magnitude diagonal entries that are possible in $\mathbf{D}$ are ± 1 and 0, and any mutually orthogonal integer and nonzero vectors in the columns of $\mathbf{W}$ that have the smallest possible lengths are just permutations of the standard unit vectors $\mathbf{e}_i$. These, however, would give rise only to uninspiring sparse integer matrix examples $\mathbf{A} = \mathbf{W}\mathbf{D}\mathbf{W}^T = \mathbf{A}^T$. Every other choice of column vectors for $\mathbf{W}$ will raise the eigenvalues $d_i^\circ = d_i\|\mathbf{w}_i\|^2$ of $\mathbf{A} = \mathbf{W}\mathbf{D}\mathbf{W}^T$. For simple hand computations, this illustrates the need for selecting mutually orthogonal integer vectors of small lengths for the columns of $\mathbf{W}$. Generally speaking, integer dense and symmetric matrices will likely have sizable eigenvalues if we require them to have integer ones. ◀

11.2 Theory

Normal Matrices

Normal matrices are orthogonally diagonalizable.

Normal matrices generalize the symmetric ones in the sense that each normal matrix $\mathbf{A}_{nn} \in \mathbb{R}^{n,n}$ can be diagonalized with respect to an orthonormal eigenvector basis of $\mathbb{C}^n$.

DEFINITION 2 A real matrix $\mathbf{A}_{nn}$ is **normal** if

$$\mathbf{A}^T\mathbf{A} = \mathbf{A}\mathbf{A}^T.$$

A complex matrix $\mathbf{A}_{nn}$ is **normal** if

$$\mathbf{A}^*\mathbf{A} = \mathbf{A}\mathbf{A}^*.$$

Real symmetric matrices $\mathbf{A} = \mathbf{A}^T$ or complex **hermitian matrices** $\mathbf{A} = \mathbf{A}^*$ are obviously normal. So are real orthogonal matrices $\mathbf{A}$ with $\mathbf{A}^T\mathbf{A} = \mathbf{I}_n = \mathbf{A}\mathbf{A}^T$ and complex unitary ones with $\mathbf{A}^*\mathbf{A} = \mathbf{I}_n = \mathbf{A}\mathbf{A}^*$. Other subsets of normal matrices are the set of **skew symmetric** and the set of **skew hermitian matrices** with $\mathbf{A}^T = -\mathbf{A} \in \mathbb{R}^{n,n}$ and $\mathbf{A}^* = -\mathbf{A} \in \mathbb{C}^{n,n}$, respectively.

If $\mathbf{A}_{nn}$ is normal and has the Schur normal form

$$\mathbf{S} = \mathbf{U}^*\mathbf{A}\mathbf{U} = \begin{pmatrix} \lambda_1 & & * \\ & \ddots & \\ 0 & & \lambda_n \end{pmatrix}$$

for a unitary matrix $\mathbf{U}$ with $\mathbf{U}^*\mathbf{U} = \mathbf{I}$ according to Theorem 11.1, then $\mathbf{S}$ is normal, due to the following lemma.

Lemma 3 If $\mathbf{A}_{nn}$ is normal and $\mathbf{U}_{nn}$ is unitary, then the matrix $\mathbf{S} = \mathbf{U}^*\mathbf{A}\mathbf{U}$ is likewise normal. ◀

Proof For normality, we must show that $\mathbf{S}^*\mathbf{S} = \mathbf{S}\mathbf{S}^*$. The matrices

$$\begin{aligned}\mathbf{S}^*\mathbf{S} &= (\mathbf{U}^*\mathbf{A}\mathbf{U})^*\mathbf{U}^*\mathbf{A}\mathbf{U} = \mathbf{U}^*\mathbf{A}^*\mathbf{U}\mathbf{U}^*\mathbf{A}\mathbf{U} = \mathbf{U}^*\mathbf{A}^*\mathbf{A}\mathbf{U}\\ \mathbf{S}\mathbf{S}^* &= \mathbf{U}^*\mathbf{A}\mathbf{U}(\mathbf{U}^*\mathbf{A}\mathbf{U})^* = \mathbf{U}^*\mathbf{A}\mathbf{U}\mathbf{U}^*\mathbf{A}^*\mathbf{U} = \mathbf{U}^*\mathbf{A}\mathbf{A}^*\mathbf{U}\end{aligned}$$

clearly coincide, since $(\mathbf{U}^*)^* = \mathbf{U}$ and $\mathbf{A}^*\mathbf{A} = \mathbf{A}\mathbf{A}^*$ was assumed. ■

Next, we study Schur normal forms (i.e., upper triangular matrices $\mathbf{S} \in \mathbb{C}^{n,n}$) that are normal as matrices. Which upper triangular matrices $\mathbf{S}$ of the form (11.1) are normal? To answer this question, we form the two matrix products

$$\mathbf{S}^*\mathbf{S} = \begin{pmatrix} \overline{\lambda_1} & & & 0 \\ | & \ddots & & \\ \mathbf{a}_1^* & & \ddots & \\ | & * & & \overline{\lambda_n} \end{pmatrix} \begin{pmatrix} \lambda_1 & - & \mathbf{a}_1 & - \\ & \ddots & & * \\ & & \ddots & \\ 0 & & & \lambda_n \end{pmatrix}$$

$$= \begin{pmatrix} |\lambda_1|^2 & - & \lambda_1 \mathbf{a}_1 & - \\ | & & & \\ \lambda_1 \mathbf{a}_1^* & & * & \\ | & & & \end{pmatrix}_{nn}$$

and

$$\mathbf{S}\mathbf{S}^* = \begin{pmatrix} \lambda_1 & - & \mathbf{a}_1 & - \\ & \ddots & & * \\ & & \ddots & \\ 0 & & & \lambda_n \end{pmatrix} \begin{pmatrix} \overline{\lambda_1} & & & 0 \\ | & \ddots & & \\ \mathbf{a}_1^* & & \ddots & \\ | & * & & \overline{\lambda_n} \end{pmatrix}$$

$$= \begin{pmatrix} |\lambda_1|^2 + \|\mathbf{a}_1\|^2 & - & * & - \\ | & & & \\ * & & * & \\ | & & & \end{pmatrix}_{nn}.$$

Here $\mathbf{a}_1$ denotes the first partial row $\mathbf{a}_1 = (s_{12}, \ldots, s_{1n}) \in \mathbb{C}^{n-1}$ of $\mathbf{S}$ with $n-1$ entries obtained by leaving off the $(1, 1)$ diagonal entry λ_1. Comparing the (1,1) entries of the two matrix products $\mathbf{S}^*\mathbf{S}$ and $\mathbf{S}\mathbf{S}^*$ for a normal Schur matrix $\mathbf{S}$ with $\mathbf{S}^*\mathbf{S} = \mathbf{S}\mathbf{S}^*$, we see that $\|\mathbf{a}_1\|^2 = 0$ or $\mathbf{a}_1 = \mathbf{0} \in \mathbb{C}^{n-1}$, making

$$\mathbf{S} = \begin{pmatrix} \lambda_1 & 0 & \cdots & 0 \\ & \ddots & & * \\ & & \ddots & \\ 0 & & & \lambda_n \end{pmatrix}.$$

Subsequently, for the second partial row $\mathbf{a}_2 \in \mathbb{C}^{n-2}$ of $\mathbf{S}$, where we leave off the (2,1) entry zero and the (2,2) entry λ_2, we have

$$\mathbf{S}^*\mathbf{S} = \begin{pmatrix} \overline{\lambda_1} & & & & 0 \\ 0 & \overline{\lambda_2} & & & \\ \vdots & | & \ddots & & \\ \vdots & \mathbf{a}_2^* & & \ddots & \\ 0 & | & * & & \overline{\lambda_n} \end{pmatrix} \begin{pmatrix} \lambda_1 & 0 & \dots & \dots & 0 \\ & \lambda_2 & - & \mathbf{a}_2 & - \\ & & \ddots & & * \\ & & & \ddots & \\ 0 & & & & \lambda_n \end{pmatrix}$$

$$= \begin{pmatrix} |\lambda_1|^2 & 0 & \dots & 0 \\ 0 & |\lambda_2|^2 & * & \\ \vdots & * & & \\ 0 & & & * \end{pmatrix}$$

and

$$\mathbf{S}\mathbf{S}^* = \begin{pmatrix} \lambda_1 & 0 & \dots & \dots & 0 \\ 0 & \lambda_2 & - & \mathbf{a}_2 & - \\ & & \ddots & & * \\ & & & \ddots & \\ 0 & & & & \lambda_n \end{pmatrix} \begin{pmatrix} \overline{\lambda_1} & & & & 0 \\ 0 & \overline{\lambda_2} & & & \\ \vdots & | & \ddots & & \\ \vdots & \mathbf{a}_2^* & & \ddots & \\ 0 & | & * & & \overline{\lambda_n} \end{pmatrix}$$

$$= \begin{pmatrix} |\lambda_1|^2 & 0 & \dots & 0 \\ 0 & |\lambda_2|^2 + \|\mathbf{a}_2\|^2 & * & \\ \vdots & * & & \\ 0 & & & * \end{pmatrix}.$$

Thus, assuming that $\mathbf{S}^*\mathbf{S} = \mathbf{S}\mathbf{S}^*$, the second shorted subrow $\mathbf{a}_2 = (s_{23}, \dots, s_{2n}) \in \mathbb{C}^{n-2}$ of $\mathbf{S}$ (with its first two entries 0 and λ_2 in positions (2,1) and (2,2) omitted) is also zero, or $\mathbf{a}_2 = \mathbf{0}$. We can repeat this process on the third and each subsequent row of $\mathbf{S}$ in Schur normal form (11.1). This proves the first part of the following theorem:

Theorem 11.4 Every normal square matrix $\mathbf{A} \in \mathbb{R}^{n,n}$ (or $\mathbb{C}^{n,n}$) with $\mathbf{A}^*\mathbf{A} = \mathbf{A}\mathbf{A}^*$ can be diagonalized orthogonally (i.e., there exists an ONB $\mathcal{U}$ of $\mathbb{C}^n$ so that $\mathbf{U}^*\mathbf{A}\mathbf{U}$ is a diagonal matrix, possibly complex). Here, the unitary matrix $\mathbf{U}$ contains the basis vectors of $\mathcal{U}$ as columns.

Conversely, every orthogonally diagonalizable $n \times n$ matrix $\mathbf{A}$ is normal. ◀

Proof Only the second assertion still needs to be proved. If $\mathbf{U}^*\mathbf{A}_{nn}\mathbf{U} = \mathbf{D} = \operatorname{diag}(\lambda_i)$ for $\mathbf{U}^*\mathbf{U} = \mathbf{I}_n$ and $\lambda_i \in \mathbb{C}$, then $\mathbf{A} = \mathbf{U}\mathbf{D}\mathbf{U}^*$ and

$$\mathbf{A}^*\mathbf{A} = \mathbf{U}\mathbf{D}^*\mathbf{U}^*\mathbf{U}\mathbf{D}\mathbf{U}^* = \mathbf{U}\mathbf{D}^*\mathbf{D}\mathbf{U}^*,$$

while

$$\mathbf{AA}^* = \mathbf{UDU}^*\mathbf{UD}^*\mathbf{U}^* = \mathbf{UDD}^*\mathbf{U}^*.$$

And we are done, due to the following lemma. ■

Lemma 4 If $\mathbf{A}_{nn}$ and $\mathbf{B}_{nn}$ are two diagonal matrices, $\mathbf{A} = \text{diag}(\lambda_i)$, and $\mathbf{B} = \text{diag}(\mu_i)$, then $\mathbf{AB} = \text{diag}(\lambda_i \mu_i) = \mathbf{BA}$ (i.e., any two diagonal matrices of the same square size commute). ◀

EXAMPLE 2 (a) The matrix $\mathbf{A} = \begin{pmatrix} 0 & -1 \\ 1 & 0 \end{pmatrix}$ is skew symmetric, since $\mathbf{A}^T = \begin{pmatrix} 0 & 1 \\ -1 & 0 \end{pmatrix} = -\mathbf{A}$. We can find the eigenvalues of $\mathbf{A}$ from a vector iteration starting with $\mathbf{y} = \begin{pmatrix} 1 \\ 2 \end{pmatrix}$ and a subsequent row reduction according to Section 9.1:

A		y	Ay	$\mathbf{A}^2\mathbf{y} = \mathbf{A}(\mathbf{Ay})$	
0	−1	1	−2	−1	
1	0	2	1	−2	$-2\ row_1$
		1	−2	−1	
		0	5	0	÷5
		1	−2	−1	$+2\ row_2$
		0	1	0	
		1	0	−1	
		0	1	0	

Thus, $\mathbf{A}^2\mathbf{y} = -\mathbf{y}$ and $\mathbf{y}$ has the vanishing polynomial $p_y(x) = x^2 + 1$, which has two complex conjugate roots i and $-i$. Alternatively, using Section 9.1.D, we evaluate $\det(\lambda\mathbf{I} - \mathbf{A}) = \begin{pmatrix} \lambda & 1 \\ -1 & \lambda \end{pmatrix} = \lambda^2 + 1$ for $\mathbf{A}$. Thus, $\mathbf{A}$ has the eigenvalues i and $-i \in \mathbb{C}$ with corresponding eigenvectors $\mathbf{x} = \begin{pmatrix} i \\ 1 \end{pmatrix}$ and $\mathbf{y} = \begin{pmatrix} 1 \\ i \end{pmatrix}$. (This result should be verified by the students.) Note that these two eigenvectors of $\mathbf{A}$ are indeed orthogonal in $\mathbb{C}^2$, since, in complex spaces, we need to use the complex conjugate dot product

$$\mathbf{x}^*\mathbf{y} := \overline{\mathbf{x}} \cdot \mathbf{y} = \begin{pmatrix} -i & 1 \end{pmatrix} \begin{pmatrix} 1 \\ i \end{pmatrix} = -i + i = 0$$

of Section 11.1 to check for complex orthogonality.

(b) The matrix $\mathbf{A} = \begin{pmatrix} 0 & -1 \\ 1 & 0 \end{pmatrix}$ from part (a) is orthogonal, since

$$\mathbf{A}^T\mathbf{A} = \begin{pmatrix} 0 & 1 \\ -1 & 0 \end{pmatrix} \begin{pmatrix} 0 & -1 \\ 1 & 0 \end{pmatrix} = \begin{pmatrix} 1 & 0 \\ 0 & 1 \end{pmatrix} = \mathbf{I}_2.$$

(c) The identity matrix $\mathbf{I}_n$ is both orthogonal and hermitian.

(d) The zero matrix $\mathbf{O}_n$ is the only matrix that is both symmetric and skew symmetric; it is normal, but not orthogonal.

(e) The matrix $\mathbf{B} = \begin{pmatrix} 1 & 1 \\ 0 & 2 \end{pmatrix}$ is diagonalizable, but not orthogonally diagonalizable.

Clearly, $\mathbf{B}$ is diagonalizable according to the Corollary following Theorems 9.4 or 9.2.D, since it has two distinct eigenvalues 1 and 2. (See Theorem 9.6(b) in Section 9.2.) $\mathbf{B}$ is not orthogonally diagonalizable according to Theorem 11.4, because $\mathbf{B}$ is not a normal matrix. To see this, observe that

$$\mathbf{B}^T\mathbf{B} = \begin{pmatrix} 1 & 0 \\ 1 & 2 \end{pmatrix}\begin{pmatrix} 1 & 1 \\ 0 & 2 \end{pmatrix} = \begin{pmatrix} 1 & 1 \\ 1 & 5 \end{pmatrix},$$

while

$$\mathbf{B}\mathbf{B}^T = \begin{pmatrix} 1 & 1 \\ 0 & 2 \end{pmatrix}\begin{pmatrix} 1 & 0 \\ 1 & 2 \end{pmatrix} = \begin{pmatrix} 2 & 2 \\ 2 & 4 \end{pmatrix} \neq \mathbf{B}^T\mathbf{B}.$$

(f) The matrix $\mathbf{C} = \begin{pmatrix} 0 & 2i \\ 2i & 0 \end{pmatrix}$ is normal, but not hermitian, since $\mathbf{C}^* \neq \mathbf{C}$ and

$$\mathbf{C}^*\mathbf{C} = \begin{pmatrix} 0 & -2i \\ -2i & 0 \end{pmatrix}\begin{pmatrix} 0 & 2i \\ 2i & 0 \end{pmatrix} = \begin{pmatrix} 4 & 0 \\ 0 & 4 \end{pmatrix} = \mathbf{C}\mathbf{C}^*.$$

(g) The matrix $\mathbf{G} = \begin{pmatrix} 0 & \frac{1}{\sqrt{2}} & -\frac{1}{\sqrt{2}} \\ 0 & \frac{1}{\sqrt{2}} & \frac{1}{\sqrt{2}} \\ 1 & 0 & 0 \end{pmatrix}$ is orthogonal since $\mathbf{G}^T\mathbf{G} = \mathbf{I}_3$, but it is neither symmetric nor skew symmetric. (This should be checked.)

(h) The matrix $\mathbf{H} = \begin{pmatrix} 4 & 0 \\ 0 & 4 \end{pmatrix}$ is symmetric, but not orthogonal, while $\mathbf{K} = \begin{pmatrix} 0 & 4 \\ -4 & 0 \end{pmatrix}$ is skew symmetric but not orthogonal.

(i) The matrix $\mathbf{J} = \begin{pmatrix} 0 & 1 \\ 0 & 0 \end{pmatrix}$ is not diagonalizable for any basis $\mathcal{U}$ of $\mathbb{R}^2$ or $\mathbb{C}^2$. This is so by Theorem 9.6(b) in Section 9.2 which gives $\mathbf{J}$ the double eigenvalue $\lambda = 0$. If $\mathbf{J}$ were diagonalizable when represented with respect to a basis $\mathcal{U}$ of $\mathbb{R}^2$ or $\mathbb{C}^n$, then we would have

$$\mathbf{U}^{-1}\mathbf{J}\mathbf{U} = \mathbf{D} = \mathbf{O}_2,$$

since $\mathbf{D} = \mathbf{O}_2$ is the only diagonal matrix in $\mathbb{R}^{2,2}$ (or $\mathbb{C}^{2,2}$) with double eigenvalue zero. But then $\mathbf{J} = \mathbf{U}\mathbf{O}_2\mathbf{U}^{-1} = \mathbf{O}_2$, which contradicts $\mathbf{J} = \begin{pmatrix} 0 & 1 \\ 0 & 0 \end{pmatrix} \neq \mathbf{O}_2$. ◀

The foregoing 2×2 and 3×3 examples easily generalize to the $n \times n$ matrix case. Hence, we have the containment "map" of matrix diagonalizability and special matrix classes as shown in Figure 11-1.

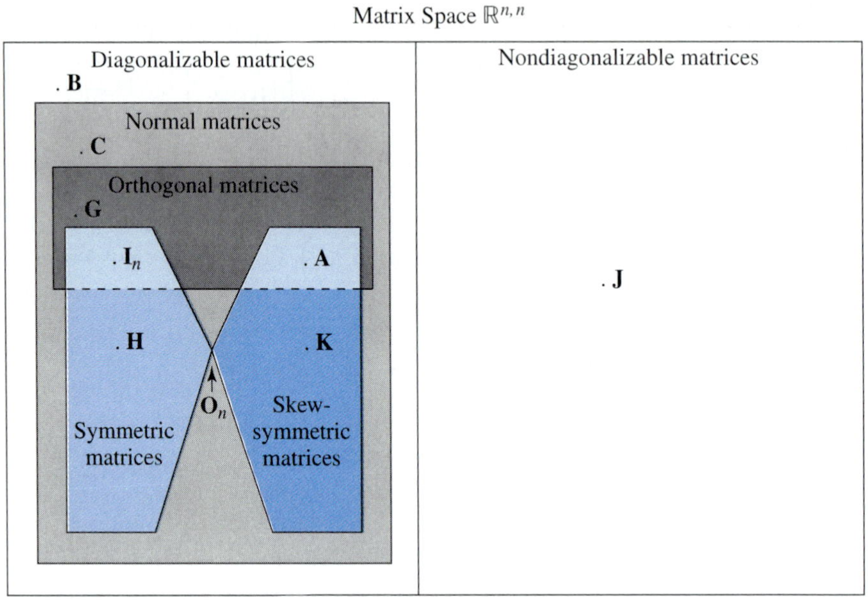

Figure 11-1

In the figure, the various matrices **A** through **O** refer to the matrices defined in Example 2. Note that the set of all $n \times n$ matrices $\mathbb{R}^{n,n}$ forms a linear space of dimension n^2, since each matrix $\mathbf{A}_{nn} = (a_{ij})$ is generated by the matrices $\mathbf{E}_{(ij)}$ with a single entry of 1 in position (i, j) and zeros elsewhere (i.e., $\mathbf{A} = \sum_{i,j=1}^{n} \mathbf{E}_{(ij)}$). (See Section 10.2.) The symmetric and the skew-symmetric real matrices form subspaces of $\mathbb{R}^{n,n}$, while all other subsets in Figure 11-1, such as the diagonalizable, the nondiagonalizable, the normal, and the orthogonal matrices, merely form subsets of $\mathbb{R}^{n,n}$. Neither of these latter sets is closed under matrix addition. (See Problem 16).

We close this section with a description of the eigenvalues of skew-symmetric matrices.

Lemma 5 If $\mathbf{A} = -\mathbf{A}^T \in \mathbb{R}^{n,n}$ is skew symmetric, then the eigenvalues of **A** lie on the imaginary axis $i\mathbb{R} = \{\alpha i \mid \alpha \in \mathbb{R}\} \subset \mathbb{C}$ with $i = \sqrt{-1}$. Moreover, if $\mathbf{A} = -\mathbf{A}^T$, then $i\mathbf{A}$ is hermitian. ◀

Proof Clearly, the matrix $\mathbf{B} = i\mathbf{A} \in \mathbb{C}^{n,n}$ is hermitian if $\mathbf{A} = -\mathbf{A}^T \in \mathbb{R}^{n,n}$ is skew symmetric, since $\mathbf{B}^* = (i\mathbf{A})^* = -i\mathbf{A}^* = -i\mathbf{A}^T = i\mathbf{A} = \mathbf{B}$. Thus, **B** has only real eigenvalues λ_j according to the complex version of Theorem 11.2 for hermitian matrices. (See Problem 14.) This gives **A** the purely imaginary eigenvalues $-i \cdot \lambda_j$, since $\mathbf{B}\mathbf{x}_j = \lambda_j \mathbf{x}_j$ implies that $i\mathbf{A}\mathbf{x}_j = \lambda_j \mathbf{x}_j$, or $\mathbf{A}\mathbf{x}_j = -i\lambda_j \mathbf{x}_j$, because $1/i = -i \in \mathbb{C}$. (For the rules of complex arithmetic, see Appendix C.) ■

Normality of matrices is the natural extension of matrix symmetry.

11.2.P Problems

1. If $\mathbf{A}$ is symmetric or skew symmetric, which of $\mathbf{A}+\mathbf{A}^T$ and $\mathbf{A}-\mathbf{A}^T$ are symmetric or skew symmetric (There are eight questions here.)

2. If $\mathbf{U}_{nn}$ is an orthogonal matrix, can the matrix sum $\mathbf{U}+\mathbf{U}^T$ have mutually orthogonal columns or rows?

3. True or false? If $\mathbf{U}_{mm}$ has mutually orthogonal columns, then the rows of $\mathbf{U}$ are also mutually orthogonal.

4. (a) Find the eigenvalues and corresponding eigenvectors for each matrix of the form $\mathbf{A}=\begin{pmatrix} a & 1 & 0 \\ 1 & a & 1 \\ 0 & -1 & a \end{pmatrix}$. Can this matrix be diagonalized for any $a \in \mathbb{R}$?

   (b) Can the matrix $\mathbf{C}=\begin{pmatrix} a & 1 & 0 \\ 1 & a & 1 \\ 0 & 1 & a \end{pmatrix}$ be diagonalized for any (or even all) $a \in \mathbb{R}$? Explain.

   (c) How about the matrix $\mathbf{G}=\begin{pmatrix} a & 1 & 0 \\ 1 & a & 1 \\ 0 & 1 & 15 \end{pmatrix}$?

   (d) Show that the $\mathbf{H}=\begin{pmatrix} a & 1 & 0 \\ -1 & a & 1 \\ 0 & -1 & a \end{pmatrix}$ can be diagonalized for all a.

5. Show that if $\mathbf{A}=-\mathbf{A}^T \in \mathbb{R}^{n,n}$ is skew symmetric, then every matrix of the form $\mathbf{B}_\alpha=\mathbf{A}+\alpha\mathbf{I}_n$, where $\alpha \in \mathbb{C}$ is arbitrary, can be diagonalized. What are the eigenvalues of $\mathbf{B}_\alpha$ in terms of those of $\mathbf{A}=-\mathbf{A}^T$? What are the eigenvectors of $\mathbf{B}_\alpha$?

6. Let $\mathbf{A}=(a_{ij})$ be a real symmetric tridiagonal matrix. Show that $\mathbf{A}$ has distinct real eigenvalues if all its codiagonal entries $a_{i,i+1}$ are nonzero. (*Hint*: Find rank $(\mathbf{A}-\lambda\mathbf{I})$ for any $\lambda \in \mathbb{R}$.)

7. Prove: If $\mathbf{A}$ has the eigenvalue λ, then the matrix $c\mathbf{A}$ has the eigenvalue $c\lambda$ for any $c \in \mathbb{C}$.

8. Show that a normal matrix $\mathbf{A}_{nn}$ has precisely one n-fold eigenvalue if and only if $\mathbf{A}=\alpha\mathbf{I}_n$ for some constant $\alpha \in \mathbb{C}$.

9. (a) Show that every matrix $\mathbf{A}_{nn} \in \mathbb{R}^{n,n}$ can be written as the sum $\mathbf{A}=\mathbf{S}+\mathbf{T}$ of a symmetric matrix $\mathbf{S}=\mathbf{S}^T$ and a skew-symmetric matrix $\mathbf{T}=-\mathbf{T}^T$.

   (b) Find the symmetric/skew-symmetric decomposition of part (a) for $\mathbf{A}=\begin{pmatrix} 4 & 2 & 4 \\ -2 & 4 & 0 \\ 4 & 0 & 6 \end{pmatrix}$.

   (c) Find all eigenvalues for $\mathbf{A}$ from part (b) and those of its symmetric and skew-symmetric parts. (*Hint*: $\mathbf{A}$ has the eigenvalue 2, and its symmetric part has the eigenvalue 8.)

10. Factor the matrix $\mathbf{A}=\begin{pmatrix} 2 & 4 \\ 1 & 1 \end{pmatrix}$ as a product of two symmetric matrices $\mathbf{A}=\mathbf{S}_1\mathbf{S}_2$ with $\mathbf{S}_i^T=\mathbf{S}_i$.

11. Give conditions for the coefficients of a real 2×2 matrix $\mathbf{A}=\begin{pmatrix} a & b \\ c & d \end{pmatrix}$ so that

    (a) $\mathbf{A}$ is symmetric;

    (b) $\mathbf{A}$ is skew symmetric;

    (c) $\mathbf{A}$ is normal;

    (d) $\mathbf{A}$ is invertible;

    (e) $\mathbf{A}$ is orthogonal; and

    (f) $\mathbf{A}$ is singular.

12. Let $\mathbf{A}=-\mathbf{A}^T \in \mathbb{R}^{n,n}$ be skew symmetric.

    (a) Show that the matrices $\mathbf{I}-\mathbf{A}$ and $\mathbf{I}+\mathbf{A}$ are both nonsingular.

    (b) Show that the matrix pair $\mathbf{I}+\mathbf{A}$ and $(\mathbf{I}-\mathbf{A})^{-1}$ commutes if $\mathbf{I}-\mathbf{A}$ is nonsingular. Show the same for $\mathbf{I}-\mathbf{A}$ and $(\mathbf{I}+\mathbf{A})^{-1}$ if $\mathbf{I}+\mathbf{A}$ is nonsingular.

    (c) Using parts (a) and (b), show that $\mathbf{U}=(\mathbf{I}-\mathbf{A})^{-1}(\mathbf{I}+\mathbf{A})$ is orthogonal, provided that $\mathbf{A}$ is skew symmetric.

13. Is the linear transformation $T\begin{pmatrix} x_1 \\ x_2 \end{pmatrix}=\begin{pmatrix} -x_2 \\ x_1 \end{pmatrix}$ normal, symmetric, skew symmetric, or none of these?

14. Rewrite the proof of Theorem 11.2 for a hermitian matrix $\mathbf{A}=\mathbf{A}^* \in \mathbb{C}^{n,n}$ and show that hermitian matrices have only real eigenvalues for an ONB of complex eigenvectors.

15. (a) Show that for any square matrix $\mathbf{A}_{nn}$, the block matrix $\mathbf{B}=\begin{pmatrix} \mathbf{A} & \mathbf{A}^* \\ \mathbf{A}^* & \mathbf{A} \end{pmatrix}$ is normal.

    (b) For which matrices $\mathbf{A}_{nn}$ is $\mathbf{C}=\begin{pmatrix} \mathbf{A} & \mathbf{A}^* \\ \mathbf{A}^* & \mathbf{A} \end{pmatrix}$ hermitian?

    (c) Construct a $2n \times 2n$ block matrix $\mathbf{H}$ from $\mathbf{A}_{nn}$ that is skew hermitian for every skew-hermitian $\mathbf{A}_{nn}$. (*Hint*: Rework part (b).)

16. (a) Show that the sets of real symmetric and of real skew-symmetric matrices in $\mathbb{R}^{n,n}$ each form a subspace of $\mathbb{R}^{n,n}$.

(b) Show that neither of the following subsets is a subspace of $\mathbb{R}^{n,n}$: the diagonalizable, the nondiagonalizable, the normal, and the orthogonal matrices.

17. Consider the set $\mathcal{T}$ of upper triangular matrices in $\mathbb{R}^{n,n}$. Is there

(a) a nondiagonalizable matrix $\mathbf{T} \in \mathcal{T}$?

(b) a diagonalizable one?

(c) a diagonalizable one with no zero entry in its upper triangle?

(d) a normal one?

(e) a symmetric one?

(f) a normal, but not symmetric one?

(g) a skew-symmetric one?

Give examples if there are such matrices, or prove that no such matrix can exist in $\mathcal{T}$.

18. Show that if $\mathbf{Q}$ is symmetric and orthogonal, then $\mathbf{Q}$ can only have the eigenvalues 1 and -1.

19. Show that for two real $n \times n$ matrices $\mathbf{S}$ and $\mathbf{T}$, the complex matrix $\mathbf{A} = \mathbf{S} + i\mathbf{T}$ is hermitian if and only is $\mathbf{S} = \mathbf{S}^T$ and $\mathbf{T} = -\mathbf{T}^T$.

20. Find matrices $\mathbf{A}, \mathbf{B}, \ldots, \mathbf{J}, \mathbf{K}$ of the lowest possible dimensions for Figure 11-1.

11.3 Applications

Polar Decomposition, Volume, ODEs, and Quadrics

Real-world problems often have normal matrix representations such as representations by symmetric, skew-symmetric, or orthogonal matrices. We study normal matrix representations because they often yield solutions that match the real-world phenomena better than general matrix representations do.

Matrix theory thrives on additive and, more prevalently, multiplicative matrix factorizations. For example, the **polar decomposition** of a matrix together with Theorem 11.2 enables us to prove the Volume Transform Theorem 8.3 of Section 8.3. We start this section with special symmetric matrices, namely, those symmetric matrices that have only positive real eigenvalues.

Lemma 6 For every nonsingular matrix $\mathbf{A} \in \mathbb{R}^{n,n}$, the matrix product $\mathbf{A}^T\mathbf{A} \in \mathbb{R}^{n,n}$ is symmetric and has positive real eigenvalues. ◀

Proof From Theorem 6.3, we have $(\mathbf{A}^T\mathbf{A})^T = \mathbf{A}^T\mathbf{A}$ (i.e., symmetry). Moreover, if $\mathbf{A}^T\mathbf{A}\mathbf{x} = \lambda\mathbf{x}$ for $\mathbf{x} \neq \mathbf{0}$, then $\lambda \neq 0$, since $\mathbf{A}^T\mathbf{A}$ is nonsingular as the product of nonsingular matrices, due to Lemma 1 of Section 7.2. Thus, $\mathbf{A}\mathbf{x} \neq \mathbf{0}$ and consequently,

$$0 < \|\mathbf{A}\mathbf{x}\|^2 = \mathbf{x}^T\mathbf{A}^T\mathbf{A}\mathbf{x} = \mathbf{x}^T\lambda\mathbf{x} = \lambda\mathbf{x}^T\mathbf{x} = \lambda\|\mathbf{x}\|^2.$$

Hence, all eigenvalues λ of $\mathbf{A}^T\mathbf{A}$ are positive real numbers. ■

Symmetric matrices, all of whose eigenvalues are positive—or at least nonnegative—play a special role in applications.

DEFINITION 3 A symmetric (or hermitian) matrix $\mathbf{A} = \mathbf{A}^T \in \mathbb{R}^{n,n}$ (or $\mathbf{A} = \mathbf{A}^* \in \mathbb{C}^{n,n}$) is **positive definite** if all eigenvalues of $\mathbf{A}$ are positive real numbers. Such a matrix is **positive semidefinite** if all eigenvalues of $\mathbf{A}$ are nonnegative real numbers (i.e., if they are greater than or equal to zero).

The converse to Lemma 6 holds as well, namely, every positive definite matrix $\mathbf{A} = \mathbf{A}^T$ can be written as $\mathbf{A} = \mathbf{X}^T\mathbf{X}$ for a nonsingular matrix $\mathbf{X}$. (See Problem 7 in Section 11.R.) Thus, positive definite matrices generalize positive real numbers, which are always squares.

According to Theorem 11.2, every real symmetric matrix has real eigenvalues, some of which may be zero or negative.

The following theorem introduces an orthogonal factorization of a nonsingular matrix $\mathbf{A}$ that differs in design from the QR matrix factorization of Section 10.3:

Theorem 11.5 **(Polar Decomposition)** Every nonsingular matrix $\mathbf{A} \in \mathbb{R}^{n,n}$ can be written as the product $\mathbf{A} = \mathbf{UP}$ of an orthogonal matrix $\mathbf{U} \in \mathbb{R}^{n,n}$ and a positive definite symmetric matrix $\mathbf{P} = \mathbf{P}^T \in \mathbb{R}^{n,n}$. ◀

Proof According to Theorems 11.2 and 11.3, applied to the symmetric matrix $\mathbf{A}^T\mathbf{A}$, we can find an ONB $\mathcal{V}$ of $\mathbb{R}^n$ so that with $\mathbf{V} \in \mathbb{R}^{n,n}$ composed columnwise of the vectors in $\mathcal{V}$, we have

$$\mathbf{V}^T(\mathbf{A}^T\mathbf{A})\mathbf{V} = \mathbf{D} = \text{diag}\ (\lambda_i)$$

for the positive and real eigenvalues λ_i of $\mathbf{A}^T\mathbf{A}$. Thus, $\mathbf{A}^T\mathbf{A} = \mathbf{VDV}^T$. We define the **matrix square root** of $\mathbf{D}$ as $\sqrt{\mathbf{D}} := \text{diag}(\sqrt{\lambda_i}) \in \mathbb{R}^{n,n}$. Then $(\sqrt{\mathbf{D}})^2 = \mathbf{D}$. Likewise, we define a **matrix square root** $\mathbf{P}$ of $\mathbf{A}^T\mathbf{A}$ by setting $\mathbf{P} := \mathbf{V}\sqrt{\mathbf{D}}\mathbf{V}^T \in \mathbb{R}^{n,n}$. Clearly, $\mathbf{P}$ is symmetric positive definite, since it is orthogonally similar to the positive diagonal matrix $\sqrt{\mathbf{D}}$. Moreover, $\mathbf{P}$ satisfies

$$\mathbf{P}^2 = \mathbf{V}\sqrt{\mathbf{D}}\mathbf{V}^T\mathbf{V}\sqrt{\mathbf{D}}\mathbf{V}^T = \mathbf{VDV}^T = \mathbf{A}^T\mathbf{A},$$

since $\mathbf{V}^T\mathbf{V} = \mathbf{I}$. Thus, calling $\mathbf{P}$ a "matrix square root" of $\mathbf{A}^T\mathbf{A}$ is justified.

Set $\mathbf{U} := \mathbf{AP}^{-1}$, where $\mathbf{P}$, the matrix square root of $\mathbf{A}^T\mathbf{A}$, is invertible and symmetric. Clearly, $\mathbf{P}^{-1} = \mathbf{V}(\sqrt{\mathbf{D}})^{-1}\mathbf{V}^T = (\mathbf{P}^{-1})^T \in \mathbb{R}^{n,n}$. Note that $\mathbf{U}^T\mathbf{U} =$

$$\mathbf{P}^{-T}\mathbf{A}^T\mathbf{AP}^{-1} = \mathbf{V}\sqrt{\mathbf{D}}^{-1}\mathbf{V}^T\mathbf{A}^T\mathbf{AV}\sqrt{\mathbf{D}}^{-1}\mathbf{V}^T = \mathbf{V}\sqrt{\mathbf{D}}^{-1}\mathbf{D}\sqrt{\mathbf{D}}^{-1}\mathbf{V}^T = \mathbf{VV}^T = \mathbf{I}_n$$

(i.e., $\mathbf{U}$ is orthogonal and $\mathbf{A} = \mathbf{UP}$). Here, we have used the notation $\mathbf{X}^{-T}$ for the **transposed inverse matrix** $(\mathbf{X}^{-1})^T$. ■

To prove the Volume Transform Theorem 8.3, we need to establish some properties for certain linear transformations of volumes.

Lemma 7 (a) If $\mathbf{W} \in \mathbb{R}^{n,n}$ is an orthogonal matrix, then

$$vol_n(\mathbf{W}(S)) = vol_n(S)$$

for any solid $S \subset \mathbb{R}^n$.

(b) If $\mathbf{P} \in \mathbb{R}^{n,n}$ is a positive definite and symmetric matrix, then

$$vol_n(\mathbf{P}(S)) = \det(\mathbf{P}) \cdot vol_n(S) = \left(\prod_i \lambda_i\right) \cdot vol_n(S)$$

for the eigenvalues λ_i of $\mathbf{P}$ and any solid $S \subset \mathbb{R}^n$.

(c) If $\mathbf{A} \in \mathbb{R}^{n,n}$ is a singular matrix, then $vol_n(\mathbf{A}(S)) = 0$ for any solid $S \subset \mathbb{R}^n$. ◀

Before proving Lemma 7, we try to understand how the volume of arbitrary solids $S \subset \mathbb{R}^n$ is measured. In calculus, irregular areas and volumes in $\mathbb{R}^2$ and $\mathbb{R}^3$ are described as areas or volumes enclosed by graphs or surfaces defined by functions. For their measurement, one uses limits of **Riemann sums**. These sums are formed from approximations of the geometric object S by rectangular inner and outer coverings. This indicates how the n-dimensional volume of solids S in $\mathbb{R}^n$ should be measured, namely, by using finer and finer inner and outer rectangular n-dimensional coverings of S that converge to S. These can be constructed out of n-dimensional orthogonal parallelepipeds in a Riemann-style multidimensional sum approximation. Building on this understanding of volume in $\mathbb{R}^n$, we now prove Lemma 7 and the Volume Transform Theorem 8.3.

Proof (Lemma 7) Following the previous remarks, it is enough to prove the Lemma for orthogonal parallelepipeds:

(a) If $S \subset \mathbb{R}^n$ is a parallelepiped with mutually orthogonal edges of lengths $x_1, \dots, x_n \in \mathbb{R}$, then $vol_n(S) = \prod_i x_i$ from elementary geometry. If $\mathbf{W}_{nn}$ is an orthogonal matrix, then by Lemma 3 of Section 10.2 we know that the edges of $\mathbf{W}(S)$ keep their mutual orthogonality and their lengths at x_i, so that $vol_n(\mathbf{W}(S)) = \prod_i x_i$ as well.

(b) To prove this part, we restrict ourselves to finding the volume of the image of the unit cube in $\mathbb{R}^n$ under the mapping by $\mathbf{P}_{nn}$, a given positive definite symmetric matrix. Clearly, $vol_n(\mathbf{P}(unit\ cube)) = \det(\mathbf{P}) = \prod_i \lambda_i$ for the eigenvalues λ_i of $\mathbf{P}$, where the first equality follows from Theorem 8.2 and the second one from Theorem 9.5(b).

(c) If $S \subset \mathbb{R}^n$ is a parallelepiped and $\mathbf{A} \in \mathbb{R}^{n,n}$ is singular, then the image $\mathbf{A}(S)$ of S will lie inside the column space of $\mathbf{A}$, which is proper (i.e., less than n-dimensional). Thus, $\mathbf{A}(S)$ has the n-dimensional volume $0 = \det(\mathbf{A})$. ■

Proof (Theorem 8.3, the Volume Transform Theorem)
We need to show that $vol_n(\mathbf{A}(S)) = |\det(\mathbf{A})| \cdot vol_n(S)$ for any solid $S \subset \mathbb{R}^n$ and any matrix $\mathbf{A} \in \mathbb{R}^{n,n}$. If $\mathbf{A}$ is singular, we are done by Lemma 7(c). For nonsingular $\mathbf{A}_{nn}$, we use the polar decomposition of Theorem 11.5 to express $\mathbf{A} = \mathbf{UP}$ with $\mathbf{U}^T\mathbf{U} = \mathbf{I}_n$ and $\mathbf{P}$ positive definite symmetric. Then

$$vol_n(\mathbf{A}(S)) = vol_n(\mathbf{UP}(S)) = vol_n(\mathbf{P}(S)) = \det(\mathbf{P}) \cdot vol_n(S),$$

according to Lemma 7(a) and (b). Now we use that $\mathbf{P}^2 = \mathbf{A}^T\mathbf{A}$, since $\mathbf{P}$ is a square root of $\mathbf{A}^T\mathbf{A}$ by construction. Thus

$$(\det(\mathbf{P}))^2 = \det(\mathbf{P}^2) = \det(\mathbf{A}^T\mathbf{A}) = \det(\mathbf{A}^T)\det(\mathbf{A}) = (\det(\mathbf{A}))^2$$

by the determinant Proposition of Section 8.1. Consequently, we have $0 < \det(\mathbf{P}) = |\det(\mathbf{A})|$, which makes $vol_n(\mathbf{A}(S)) = |\det(\mathbf{A})| \cdot vol_n(S)$, as stipulated. ■

EXAMPLE 3 What is the image of the unit cube $S \subset \mathbb{R}^3$ with its eight corners $C = \{\mathbf{c}_i\}_{i=1}^8 = \{\mathbf{0}, \mathbf{e}_1, \mathbf{e}_2, \mathbf{e}_3, \mathbf{e}_1+\mathbf{e}_2, \mathbf{e}_1+\mathbf{e}_3, \mathbf{e}_2+\mathbf{e}_3, \mathbf{e}_1+\mathbf{e}_2+\mathbf{e}_3\}$ under the linear mapping induced by $\mathbf{A} = \begin{pmatrix} 2 & 1 & 0 \\ 1 & 2 & -1 \\ 0 & 1 & 2 \end{pmatrix}$? What is the volume of its image $\mathbf{A}(S)$?

We have $\mathbf{A}(\mathbf{0}) = \mathbf{0}$, $\mathbf{A}(\mathbf{e}_1) = \begin{pmatrix} 2 \\ 1 \\ 0 \end{pmatrix}$, the first column of $\mathbf{A}$, and so forth, making $\mathbf{A}(S)$ a parallelepiped in $\mathbb{R}^3$ with the eight corners $\{\mathbf{Ac}_i\}$.

The unit cube $S \subset \mathbb{R}^3$ has three-dimensional volume equal to 1. In order to apply Theorem 8.3 for finding the volume of $\mathbf{A}(S)$, we find the determinant of $\mathbf{A}$ from its minimal or characteristic polynomial and Theorem 9.5(b) for a change. For this, we can evaluate $\det(\lambda\mathbf{I}-\mathbf{A}) = \lambda^3+a_2\lambda^2+a_1\lambda+a_0 = \lambda^3-6\lambda^2+12\lambda-8$ and find $\det(\mathbf{A})$ as its negative constant coefficient $-a_0 = 8$. Alternatively, we can form the vector iteration sequence $\mathbf{x}, \mathbf{Ax}, \mathbf{A}(\mathbf{Ax}), \mathbf{A}(\mathbf{A}(\mathbf{Ax}))$ for $\mathbf{x} \neq \mathbf{0} \in \mathbb{R}^3$ and row reduce the corresponding iteration vector matrix to find $f_A(\lambda) = p_A(\lambda)$ and $a_0 = -\det(\mathbf{A})$ for $n = 3$ from Theorem 9.5(b). Specifically, for $\mathbf{x} = \begin{pmatrix} 1 \\ 1 \\ -1 \end{pmatrix}$,

we compute $\mathbf{Ax} = \begin{pmatrix} 3 \\ 4 \\ -1 \end{pmatrix}$, $\mathbf{A}^2\mathbf{x} = \begin{pmatrix} 10 \\ 12 \\ 2 \end{pmatrix}$, and $\mathbf{A}^3\mathbf{x} = \begin{pmatrix} 32 \\ 32 \\ 16 \end{pmatrix}$.

$\mathbf{A}$			$\mathbf{x}$	$\mathbf{Ax}$	$\mathbf{A}^2\mathbf{x}$	$\mathbf{A}^3\mathbf{x}$	
2	1	0	1	3	10	32	
1	2	−1	1	4	12	32	$-\ row_1$
0	1	2	−1	−1	2	16	$+\ row_1$
			1	3	10	32	
			0	1	2	0	
			0	2	12	48	$-2\ row_2$
			1	3	10	32	
			0	1	2	0	
			0	0	8	48	$\div(8)$
			1	3	10	32	$-10\ row_3$
			0	1	2	0	$-2\ row_3$
			0	0	1	6	
			1	3	0	−28	$-3\ row_2$
			0	1	0	−12	
			0	0	1	6	
			1	0	0	8	$(= -(-1)^3\ det\ (\mathbf{A}) = det\ (\mathbf{A}))$
			0	1	0	−12	
			0	0	1	6	$(= trace(\mathbf{A}) = 6)$

Either way, the determinant of $\mathbf{A}$ is 8. Hence, the volume of the image $\mathbf{A}(S) \subset \mathbb{R}^3$ of the unit cube $S \subset \mathbb{R}^3$ under $\mathbf{A}$ is 8. Mapping objects in $\mathbb{R}^3$ by $\mathbf{A}$ enlarges their volume by the factor 8. ◀

The Schur normal form in Theorem 11.1 has many useful applications. For example, it supplies a proof of the trace condition in Theorem 9.5(a) that does not rely on determinants. We start this proof with a lemma on the trace of matrix products.

Lemma 8 If $\mathbf{A}$ and $\mathbf{B}$ are two $n \times n$ matrices, then the traces of $\mathbf{AB}$ and $\mathbf{BA}$ (i.e., the sums of the diagonal entries of the two matrix products $\mathbf{AB}$ and $\mathbf{BA}$) are equal. ◀

Proof If $\mathbf{A} = (a_{ij})$ and $\mathbf{B} = (b_{ij})$ are both $n \times n$ matrices, then the jth term in the dot product of row a_i of $\mathbf{A}$ and column $\mathbf{b}_k$ of $\mathbf{B}$ is $a_{ij}b_{jk}$. The trace of $\mathbf{AB}$ is the sum of all such dot-product terms, where $i = k$. Thus, all terms of the form $a_{ij}b_{ji}$ occur as summands in the trace of $\mathbf{AB}$. Consequently,

$$\text{trace }(\mathbf{AB}) = \sum_{i,j} a_{ij}b_{ji} = \sum_{i,j} b_{ji}a_{ij} = \text{trace}(\mathbf{BA}). \qquad \blacksquare$$

Much more can be said about the matrix products $\mathbf{AB}$ and $\mathbf{BA}$ of two $n \times n$ matrices. For example, $\mathbf{AB}$ and $\mathbf{BA}$ are similar as matrices. (See Problems 13–15 in Section 11.3.P.)

EXAMPLE 4 (Proof of Theorem 9.5(a); the Trace Condition, via Schur) We need to prove that trace $(\mathbf{A}) := \sum_i a_{ii} = \sum_i \lambda_i$ for the eigenvalues λ_i of $\mathbf{A}$. For this, assume that $\mathbf{A}$ has the Schur normal form $\mathbf{U}^*\mathbf{AU} = \begin{pmatrix} \lambda_1 & & * \\ & \ddots & \\ 0 & & \lambda_n \end{pmatrix}$ with $\mathbf{U}^*\mathbf{U} = \mathbf{I}_n$. Then $\sum_i \lambda_i = \text{trace }(\mathbf{U}^*\mathbf{AU}) = \text{trace}((\mathbf{U}^*\mathbf{A})\mathbf{U}) = \text{trace }(\mathbf{U}(\mathbf{U}^*\mathbf{A})) = \text{trace}(\mathbf{I}_n\mathbf{A}) = \text{trace}(\mathbf{A})$ by Lemma 8. ◀

Many phenomena of the physical sciences and of numerical analysis can be modeled as matrix problems. In the remainder of this section, we investigate two applications wherein a normal or diagonalizable matrix representation of the problem yields a more realistic solution (i.e., one that tracks a real-world problem more closely). Our first example deals with solving an ordinary differential equation by using difference quotient approximations.

EXAMPLE 5 Solve the differential equation $f' + 2f = \cos x$ for f with $f(0) = 0.4$ on the interval $[0, 2\pi] \subset \mathbb{R}$. This problem was solved theoretically in Example 6(c) of Section 7.3. To solve such an **initial-value problem** numerically means to compute approximate values of its solution $f(x)$ on the interval $[0, 2\pi]$ from the differential equation and the initial value $f(0) = 0.4$.

For this we consider a **partition** $\{x_j\}$ of $[0, 2\pi] \subset \mathbb{R}$ consisting of the 19 points

$$x_j = j \cdot h \quad \text{for} \quad h = \frac{2\pi}{18} \text{ and } j = 0, \ldots, 18.$$

Then we define $f_j := f(x_j)$ for each $j = 0, \ldots, 18$ as the value of a solution $f(x)$ of $f' + 2f = \cos x$, with the starting value $f(0) = 0.4$ at the point x_j of the partition.

For a numerical solution, we approximate the exact derivative $f'(x_j)$ at each x_j by a difference quotient involving several function values f_k and the step size h of the partition. In particular,

$$f'(x_j) = f_j' \approx f_{\ell j}' := \frac{f_j - f_{j-1}}{h} \text{ for all } j, \tag{11.4}$$

where we replace the exact derivative f_j' at x_j by the left-side secant slope $f_{\ell j}' := \dfrac{f_j - f_{j-1}}{h}$.

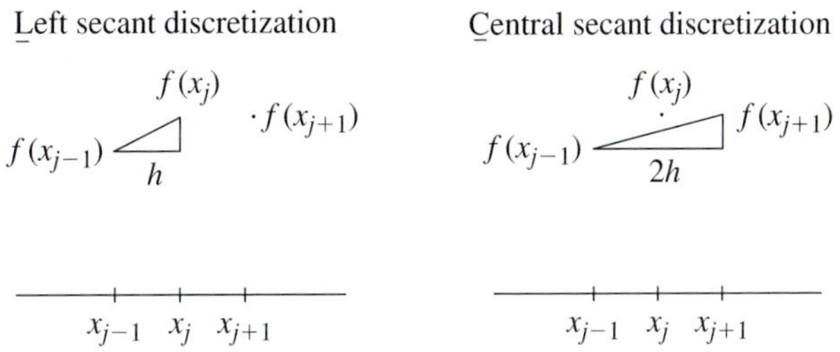

Figure 11-2

A different approximation of f_j' is obtained symmetrically from two values f_{j-1} and f_{j+1} on either side of $(x_j, f_j) \in \mathbb{R}^2$ on the graph of f by using the symmetric or central secant slope f_{cj}' instead; that is,

$$f'(x_j) = f_j' \approx f_{cj}' := \frac{f_{j+1} - f_{j-1}}{2h}. \tag{11.5}$$

(See Figure 11-2.)

At each partition point x_j, the exact solution of the differential equation $f' + 2f = \cos x$ satisfies

$$f_j' + 2f_j = \cos(x_j) \quad \text{for } j = 0, \ldots, 18.$$

Consequently, a numerical solution will approximately satisfy the following **discretized equations**:

$$f_{\ell j}' + 2f_j = \cos(x_j) \text{ for } j = 0, \ldots, 18$$
$$\text{(for the left secant slope approximation)} \tag{11.6}$$

and

$$f_{cj}' + 2f_j = \cos(x_j) \text{ for all } j = 0, \ldots, 18$$
$$\text{(for the central secant slope approximation).} \tag{11.7}$$

By plugging (11.4) into (11.6) and (11.5) into (11.7), we obtain the following two systems of linear equations relating various functional values f_k and the values $\cos(x_j)$ on the right-hand side of the differential equation:

Using left secants, we obtain

$$\frac{f_j - f_{j-1}}{h} + 2f_j = \cos x_j \text{ for } j = 1, \ldots, 18, \text{ or}$$

$$(1 + 2h) f_j - f_{j-1} = h \cos x_j \tag{11.8}$$

with $f_0 = f(x_0) = f(0) = 0.4$ by assumption.
Using central secants, we get

$$\frac{f_{j+1} - f_{j-1}}{2h} + 2f_j = \cos x_j, \text{ or}$$

$$-f_{j-1} + 4hf_j + f_{j+1} = 2h \cos x_j \text{ for } j = 1, \ldots, 18 \tag{11.9}$$

with $f_0 = f(x_0) = f(0) = 0.4$.

Formula (11.8) gives rise to the matrix equation

$$\begin{pmatrix} 1+2h & & & 0 \\ -1 & 1+2h & & \\ & \ddots & \ddots & \\ 0 & & -1 & 1+2h \end{pmatrix}_{18\times 18} \begin{pmatrix} f_1 \\ \vdots \\ f_{18} \end{pmatrix} = \begin{pmatrix} h\cos(h) + 0.4 \\ h\cos(2h) \\ \vdots \\ h\cos(18h) \end{pmatrix}, \tag{11.10}$$

while (11.9) corresponds to the system of linear equations

$$\begin{pmatrix} 4h & 1 & & 0 \\ -1 & \ddots & \ddots & \\ & \ddots & \ddots & 1 \\ 0 & & -1 & 4h \end{pmatrix}_{m\times m} \begin{pmatrix} f_1 \\ \vdots \\ f_m \end{pmatrix} = \begin{pmatrix} 2h\cos(h) + 0.4 \\ 2h\cos(2h) \\ \vdots \\ 2h\cos(mh) - f_{m+1} \end{pmatrix}. \tag{11.11}$$

Here, we have formulated the linear system (11.11) for central difference quotients in terms of $m \geq 18$, since the 18th equation,

$$-f_{17} + 4hf_{18} + f_{19} = 2h \cos x_{18},$$

in (11.9) involves the unknown term f_{19} for $x_{19} > 2\pi$. It would be wise to integrate the DE via (11.11) for an extended partition with $m > 18$, in order to reduce the effect of the uncertainty involving f_{m+1} at $x_{m+1} > 2\pi$. If we have chosen $m \gg 18$, we can seemingly set $f_{m+1} = 0$ without much harm.

Note that the system matrix in (11.10) is highly nonnormal: It has the 18–fold eigenvalue $1 + 2h$, but only a one-dimensional corresponding eigenspace. The system matrix in (11.11), however, is close to being normal as the sum of a skew-symmetric matrix and a multiple of the identity matrix. In fact, the system matrix of (11.11) is diagonalizable. (See Problems 4(d) and 5 in Section 11.2.P).

The plot shown in Figure 11-3 uses MATLAB to solve the two linear systems (11.10) and (11.11), the latter with $m = 27$ and $f_{28} := 0$, or by integrating out to

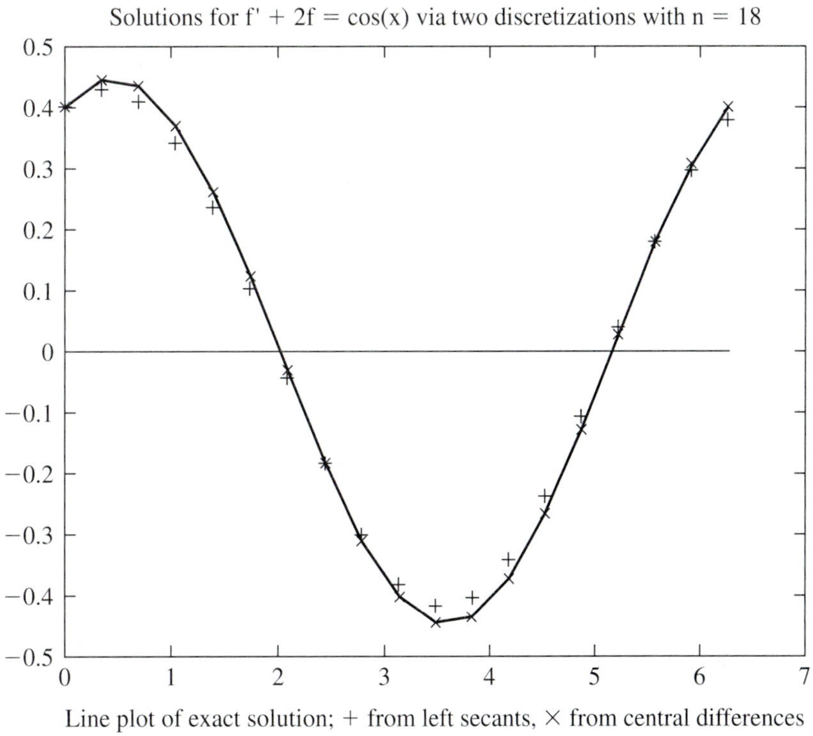

Figure 11-3

3π. The left secant solution values of (11.10) are plotted with "+" signs, while the central-difference-induced solution of (11.11) is marked by "×" signs, and the exact solution is plotted by line segments connecting the discrete points of the graph of the exact theoretical solution.

The plot shows that using central difference quotient approximations for the derivative gives us a numerical model of the DE whose solution tracks the exact solution $f(x) = 0.2\sin(x) + 0.4\cos(x)$ very accurately. (See Example 6(c) in Section 7.3.) Using one-sided difference quotient approximations for the derivative gives much worse numerical output.

The left sided difference quotients "lag" behind reality here, and the corresponding system matrix in (11.10) is not normal. Central difference quotients help make the system matrix in (11.11) become skew symmetric, except for its constant diagonal entries. This allows us to trace the true solution $f(x)$ very accurately, even for our rather coarse partition x_j. ◀

Note that each odd-order derivative intrinsically is an "odd process" induced by the asymmetric progression of the real-number line. The intrinsic "directionality" of $\mathbb{R}$ assigns opposite signs to left and right function values in all unbalanced odd-order discretizations, such as in f'_{lj}. On the other hand, even-order derivatives are symmetric operations on the functional values f_k used with symmetric differences. For example,

$$\begin{aligned} f''(x_j) \approx f''_j &:= \frac{f'_{rj} - f'_{\ell j}}{h} \\ &= \frac{\dfrac{f_{j+1} - f_j}{h} - \dfrac{f_j - f_{j+1}}{h}}{h} \\ &= \frac{f_{j+1} - 2f_j + f_{j-1}}{h^2} \end{aligned} \tag{11.12}$$

expresses the second-order difference quotient symmetrically. From Example 5, it appears to be advantageous to use symmetry and skew symmetry as much as possible in mathematical models.

Our next example deals with **quadratic forms**. These generalize the concept of a vector norm to arbitrary metrics. Quadratic forms, or **quadrics** for short, have many applications in physics, geometry, and statistics.

EXAMPLE 6 Plot the points $\mathbf{x} = \begin{pmatrix} x_1 \\ x_2 \end{pmatrix} \in \mathbb{R}^2$ that satisfy the quadratic equation $5x_1^2 + 4x_1x_2 + 2x_2^2 = 6$.

Quadratic forms Q: $\mathbb{R}^n \to \mathbb{R}$ are defined as mappings of $\mathbb{R}^n$ to $\mathbb{R}$ that involve only sums of quadratic, or second-degree, terms, such as αx_i^2 and $\beta x_j x_k$ for arbitrary $\alpha, \beta \in \mathbb{R}$ and $1 \le i, j, k \le n$. Thus, quadrics generalize the concept of the vector norm

$$\|\mathbf{x}\| = \sqrt{x_1^2 + \cdots + x_1^2}$$

in the sense that

$$\|\mathbf{x}\|^2 = \mathbf{x}^T\mathbf{x} = \mathbf{x}^T\mathbf{I}_n\mathbf{x} = \sum_{i=1}^{n} x_i^2$$

is a quadratic form.

It is easy to see that all quadratic forms $Q: \mathbb{R}^n \to \mathbb{R}$ can be expressed as $Q(\mathbf{x}) = \mathbf{x}^T\mathbf{A}\mathbf{x}$ for a matrix $\mathbf{A}_{nn} = (a_{ij})$ if each diagonal entry a_{ii} is chosen as the coefficient of x_i^2 in $Q(\mathbf{x})$, and if each pair of entries a_{ij} and a_{ji} with symmetric indices is chosen so that their sum $a_{ij} + a_{ji}$ equals the coefficient of $x_i x_j$ in the expression of Q. For example, for the chosen quadratic form $Q(\mathbf{x}) = 5x_1^2 + 4x_1x_2 + 2x_2^2$ and a generic 2×2 matrix $\mathbf{A} = \begin{pmatrix} a & b \\ c & d \end{pmatrix}$, we have

$$Q(\mathbf{x}) = 5x_1^2 + 4x_1x_2 + 2x_2^2 = \begin{pmatrix} x_1 & x_2 \end{pmatrix} \begin{pmatrix} a & b \\ c & d \end{pmatrix} \begin{pmatrix} x_1 \\ x_2 \end{pmatrix}$$

$$= \begin{pmatrix} x_1 & x_2 \end{pmatrix} \begin{pmatrix} ax_1 + bx_2 \\ cx_1 + dx_2 \end{pmatrix} = ax_1^2 + (b+c)x_1x_2 + dx_2^2.$$

Thus, $Q(\mathbf{x}) = \mathbf{x}^T\mathbf{A}\mathbf{x}$ is true precisely when $a = 5,\ d = 2$, and $b + c = 4$, and every 2×2 matrix $\mathbf{A} = \begin{pmatrix} 5 & b \\ c & 2 \end{pmatrix}$ with $b + c = 4$ represents the quadratic form $Q(\mathbf{x})$ as $\mathbf{x}^T\mathbf{A}\mathbf{x}$ for all $\mathbf{x} = \begin{pmatrix} x_1 \\ x_2 \end{pmatrix} \in \mathbb{R}^2$.

(a) If we choose $b = c = 2$ here, then $\mathbf{S} := \begin{pmatrix} 5 & 2 \\ 2 & 2 \end{pmatrix}$ represents $Q(\mathbf{x}) = \mathbf{x}^T\mathbf{S}\mathbf{x}$ symmetrically. In order to understand Q we study $\mathbf{S}$.
For this, we embark on an eigenanalysis of $\mathbf{S}$. We use vector iteration according to Section 9.1 with $\mathbf{y} = \begin{pmatrix} 1 \\ -1 \end{pmatrix}$ and $\mathbf{S}$:

$\mathbf{S}$		$\mathbf{y}$	$\mathbf{S}\mathbf{y}$	$\mathbf{S}^2\mathbf{y} = \mathbf{S}(\mathbf{S}\mathbf{y})$	
5	2	[1]	3	15	
2	2	[−1]	0	6	$+\ row_1$
		[1]	3	15	$-\ row_2$
		0	[3]	21	$\div(3)$
		[1]	0	−6	
		0	[1]	7	

Thus $\mathbf{S}^2\mathbf{y} = 7\mathbf{S}\mathbf{y} - 6\mathbf{y}$, giving $\mathbf{y}$ the vanishing polynomial $p_y(x) = f_S(x) = x^2 - 7x + 6 = (x-6)(x+1) = p_S(x)$ by the degree argument used repeatedly in Section 9.1.
Alternatively, we can find $f_S(\lambda)$ according to Section 9.1.D:

$$\det(\lambda\mathbf{I} - \mathbf{S}) = \det\begin{pmatrix} \lambda - 5 & -2 \\ -2 & \lambda - 2 \end{pmatrix} = (\lambda - 5)(\lambda - 2) - 4$$
$$= \lambda^2 - 7\lambda + 6 = (\lambda - 6)(\lambda - 1).$$

We diagonalize $\mathbf{S}$ as in Section 11.1. For the eigenvalue $\lambda_1 = 6$ of $\mathbf{S}$, we have

$\mathbf{S} - 6\mathbf{I}$:	[−1]	2	0	
	[2]	−4	0	$+2\ row_1$
	[−1]	2	0	
	0	0	0	
		↑		

Thus, $\mathbf{y}_1 = \begin{pmatrix} 2 \\ 1 \end{pmatrix}$ is an eigenvector of $\mathbf{S}$ for its eigenvalue $\lambda_1 = 6$. Similarly, $\mathbf{y}_2 = \begin{pmatrix} -1 \\ 2 \end{pmatrix}$ is an eigenvector for $\lambda_2 = 1$. Normalizing the eigenvectors $\mathbf{y}_i$ then diagonalizes $\mathbf{S}$ orthogonally, that is,

$$\frac{1}{\sqrt{5}}\begin{pmatrix} 2 & 1 \\ -1 & 2 \end{pmatrix} \mathbf{S} \frac{1}{\sqrt{5}}\begin{pmatrix} 2 & -1 \\ 1 & 2 \end{pmatrix} = \begin{pmatrix} 6 & 0 \\ 0 & 1 \end{pmatrix}.$$

The eigenanalysis of $\mathbf{S}$ tells us the following about Q: If we change the standard basis $\mathcal{E} = \{\mathbf{e}_1, \mathbf{e}_2\}$ of $\mathbb{R}^2$ in which $\mathbf{S}$ represents Q to the orthonormal basis $\mathcal{W} = \{\mathbf{w}_1, \ \mathbf{w}_2\} = \left\{\frac{1}{\sqrt{5}}\begin{pmatrix} 2 \\ 1 \end{pmatrix}, \frac{1}{\sqrt{5}}\begin{pmatrix} -1 \\ 2 \end{pmatrix}\right\}$ of $\mathbb{R}^2$, we obtain

$$Q(\mathbf{x}) = \mathbf{x}^T \mathbf{S} \mathbf{x} = \mathbf{u}^T \begin{pmatrix} 6 & 0 \\ 0 & 1 \end{pmatrix} \mathbf{u}$$

for $\mathbf{u} = \frac{1}{\sqrt{5}}\begin{pmatrix} 2 & -1 \\ 1 & 2 \end{pmatrix} \mathbf{x}$ according to Sections 7.1 and 11.1. Consequently, the original quadratic form Q, rewritten in the orthonormal eigenvector basis $\{\mathbf{w}_1, \mathbf{w}_2\}$ of $\mathbf{S}$ becomes, most simply,

$$6u_1^2 + u_2^2 = 6 \text{ in terms of } \mathcal{W}\text{–coordinate vectors } \mathbf{u}_{\mathcal{W}} = \begin{pmatrix} u_1 \\ u_2 \end{pmatrix}.$$

This last equation describes an **ellipse** in the $\mathcal{W}$–coordinate system, namely,

$$\frac{u_1^2}{1^2} + \frac{u_2^2}{\left(\sqrt{6}\right)^2} = 1,$$

with **minor** and **major axis** in the directions of $\mathbf{w}_1 = \frac{1}{\sqrt{5}}\begin{pmatrix} 2 \\ 1 \end{pmatrix}$ and $\mathbf{w}_2 = \frac{1}{\sqrt{5}}\begin{pmatrix} -1 \\ 2 \end{pmatrix}$ of lengths 1 and $\sqrt{6}$, respectively. (See Figure 11-4).

Representing Q symmetrically via $\mathbf{S}$ thus allows us to see all symmetric features and measurements of the quadratic curve $Q(\mathbf{x}) = \mathbf{u}^T \begin{pmatrix} 6 & 0 \\ 0 & 1 \end{pmatrix} \mathbf{u} = 6$ in the plane clearly.

(b) On the other hand, if we represent Q by an unsymmetric matrix $\mathbf{A}$, such as $\mathbf{A} := \begin{pmatrix} 5 & 0 \\ 4 & 2 \end{pmatrix}$, and perform the analogous eigenanalysis on $\mathbf{A}$, we obtain the following: $\mathbf{A}$ is triangular, and thus its eigenvalues 5 and 2 appear on the diagonal by Theorem 9.6(b) in Section 9.2. The corresponding eigenvectors are $\mathbf{v}_1 = \begin{pmatrix} 3 \\ 4 \end{pmatrix}$ and $\mathbf{v}_2 = \begin{pmatrix} 0 \\ 1 \end{pmatrix}$, respectively. Thus, the basis $\mathcal{V} = \{\mathbf{v}_1, \mathbf{v}_2\}$

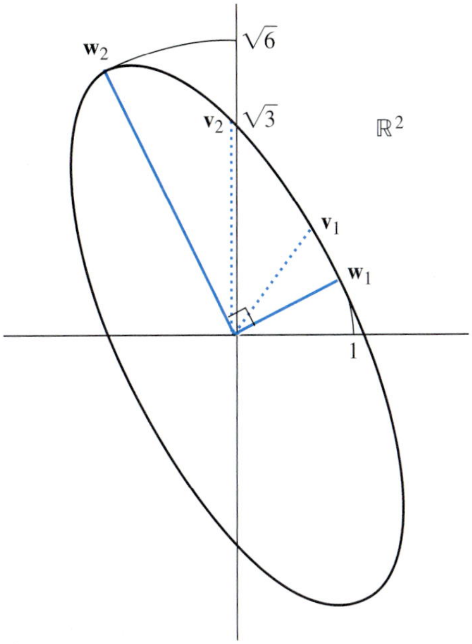

Figure 11-4

represents Q as

$$5z_1^2 + 2z_2^2 = 6,$$

or

$$\frac{z_1^2}{\left(\sqrt{\frac{6}{5}}\right)^2} + \frac{z_2^2}{\left(\sqrt{3}\right)^2} = 1 \text{ for } \mathcal{V}\text{–coordinate vectors } \mathbf{z}_{\mathcal{V}} = \begin{pmatrix} z_1 \\ z_2 \end{pmatrix}.$$

Q is described by $\mathbf{A}$ as an ellipse with two **conjugate radii** $\mathbf{v}_1, \mathbf{v}_2$ in the directions $\begin{pmatrix} 3 \\ 4 \end{pmatrix}$ and $\begin{pmatrix} 0 \\ 1 \end{pmatrix}$ of respective lengths $\sqrt{\frac{6}{5}}$ and $\sqrt{3}$. In Figure 11-4, note that the conjugate radii $\frac{\sqrt{6}}{5\sqrt{5}}\begin{pmatrix} 3 \\ 4 \end{pmatrix}$ and $\sqrt{3}\begin{pmatrix} 0 \\ 1 \end{pmatrix}$ do not give as much useful information on the shape or the geometric symmetry of the ellipse $Q(\mathbf{x}) = 6$ as the matrix $\mathbf{S}$ does in part (a) with respect to $\mathcal{W}$. The description of Q via $\mathbf{A}$ contains no obvious geometric invariants of the depicted ellipse that is given by $\mathbf{A}$'s quadratic form $\mathbf{z}^T\begin{pmatrix} 5 & 0 \\ 0 & 2 \end{pmatrix}\mathbf{z} = 6$ algebraically for $\mathcal{V}$.

(c) We were lucky in part (b) and retrieved some useful information of Q from its unsymmetric representation by $\mathbf{A} = \begin{pmatrix} 5 & 0 \\ 4 & 2 \end{pmatrix}$. We could have fared worse: The matrix $\mathbf{B} := \begin{pmatrix} 5 & 4.5 \\ -0.5 & 2 \end{pmatrix}$ also represents Q in a nonsymmetric fashion. For this $\mathbf{B}$, we have $f_B(\lambda) = p_B(\lambda) = \lambda^2 - 7\lambda + 12.15 =$

$(\lambda - 3.5)^2$, as can be checked by either one of our standard eigenvalue methods from Section 9.1 or 9.1.D. Thus, **B** has the double eigenvalue 3.5. Looking at ker(**B** – 3.5**I**), namely,

B – 3.5**I** :				
	1.5	4.5	0	
	−0.5	−1.5	0	$+1/3\ row_1$,
	1.5	4.5	0	
	0	0	0	
		↑		

we observe that **B** has only one eigenvector for its double eigenvalue 3.5. Thus **B** cannot be diagonalized. Consequently, we cannot reduce the matrix **B** with $Q(\mathbf{x}) = \mathbf{x}^T\mathbf{B}\mathbf{x}$ to any algebraic form that remotely reveals the object it describes: an ellipse of the general form

$$\frac{x^2}{a^2} + \frac{y^2}{b^2} = 1.$$

This is so for our matrix representation **B** of Q, since any similar transformation $\mathbf{X}^{-1}\mathbf{B}\mathbf{X}$ performed on **B** (a matrix that cannot be diagonalized) will always contain at least one nonzero off-diagonal term, leading to mixed terms u_1u_2 in all of its quadratic-form representations $\mathbf{u}^T\mathbf{X}^{-1}\mathbf{B}\mathbf{X}\mathbf{u}$. ◀

These examples show the power and usefulness of diagonalizable and symmetric matrix representations.

11.3.P Problems

1. (a) Find the polar decomposition of the 1×1 matrices (5), (-3), and $(4i - 3)$.

(b) Find the polar decomposition of the matrix $\mathbf{A} = \begin{pmatrix} 1 & 0 \\ 1 & 1 \end{pmatrix}$.

2. Which of the following matrices are positive definite and which positive semidefinite?

(a) $\mathbf{A} = \begin{pmatrix} 2 & 2 \\ 2 & 3 \end{pmatrix}$, $\mathbf{H} = \begin{pmatrix} 2 & 2 \\ 2 & 2 \end{pmatrix}$;

(b) $\mathbf{B} = \begin{pmatrix} 4 & 0 & 0 \\ 0 & 3 & 1 \\ 0 & 1 & 3 \end{pmatrix}$;

(c) $\mathbf{C} = \begin{pmatrix} 1 & 2 & 2 \\ 2 & 1 & 2 \\ 2 & 2 & 1 \end{pmatrix}$;

(d) $\mathbf{D} = \begin{pmatrix} 5 & -2 \\ -2 & 2 \end{pmatrix}$;

(e) $\mathbf{E} = \begin{pmatrix} 2 & 1 \\ -1 & 3 \end{pmatrix}$;

(f) $\mathbf{F} = \begin{pmatrix} 12 & 2 & 2 \\ 2 & 12 & 1 \\ 2 & 2 & 12 \end{pmatrix}$.

3. Find two matrix representations for each of the following quadratic forms, if possible:

(a) $2x_1^2 - 12x_1x_2 + 4x_2^2$;

(b) $2x^2 - 4y^2 + 16z^2$;

(c) $2x^2 - 4y^2 + 16z^2 - 6xy + 24yz$;

(d) $2x_1x_2 - 7x_2^2$.

4. Find the two major axes of symmetry of the conic $Q(\mathbf{x}) = x_1^2 - x_1x_2 + x_2^2 = 4$.

5. Solve the differential equation $f'' - 2f = e^x + e^{-x}$ with initial condition $f(-1) = -e - \frac{1}{e} = -3.0862\ldots$, using (11.12) on a partition $\{x_i\}$ with $-1 \le x_i \le 1$ and step size $h = 1/10$.

(a) Write out the discretized equations for f_j in the vein of (11.6) through (11.9) for the given second-order differential equation.

(b) Write out the matrix equations analogous to (11.10) and (11.11) for this problem.

(c) Solve these equations using MATLAB. (Make sure that the discretization goes well beyond the interval's right endpoint of 1 when using (11.11).)

(d) Study the plot routines available in MATLAB using `help plot`, and generate a picture like Figure 11-3 from your computed data.

6. Find a symmetric matrix representation for the quadratic form $Q(\mathbf{x}) = 2x_1^2 + 4x_1x_2 - x_2^2 = 6$, and decide what type of conic this form represents on $\mathbb{R}^2$. (Do an eigenanalysis on the matrix; plot the axes of symmetry of the curve, and draw some significant points on it.)

7. Repeat the previous problem for $P(\mathbf{x}) = 2x_1^2 - 4x_1x_2 - x_2^2 = 6$.

8. If $\mathbf{A}_{nn}$ is positive definite, show that $\mathbf{A}$ is symmetric and nonsingular. Moreover, show that $\mathbf{A}^{-1}$ is positive definite as well.

9. Prove or disprove the following statement: If, in the polar decomposition of a nonsingular matrix $\mathbf{A}_{nn} = \mathbf{UP}$, the two factors $\mathbf{U}$ and $\mathbf{P}$ commute, then $\mathbf{A}$ is normal. Does the converse hold?

10. Show that the matrix $\mathbf{A} = \begin{pmatrix} 2 & i \\ -i & 4 \end{pmatrix}$ is normal. Find the eigenvalues of $\mathbf{A}$ and write $\mathbf{A} = \mathbf{C}^*\mathbf{C}$ for a matrix $\mathbf{C} \in \mathbb{C}^{2,2}$, if possible.

11. Show that the quadratic form $Q(\mathbf{x}) = 6x_1^2 - 2x_1x_2 + 2x_2^2$ is positive for all $\mathbf{x} = \begin{pmatrix} x_1 \\ x_2 \end{pmatrix} \neq \mathbf{0} \in \mathbb{R}^2$. (*Hint*: Diagonalize the symmetric matrix representation $\mathbf{A}$ of Q.)

12. (a) Is the matrix $\mathbf{A} = \begin{pmatrix} 0 & 1 \\ 2 & 7 \end{pmatrix}$ similar to a symmetric matrix?

(b) Is the matrix $\mathbf{B} = \begin{pmatrix} 0 & 1 & 1 \\ 2 & -1 & 1 \\ 2 & 1 & 0 \end{pmatrix}$ similar to a real symmetric matrix?

(c) Construct a real matrix $\mathbf{C}_{3,3}$ that is not similar to a symmetric one.

13. In Example 1 of Section 6.1 the matrices $\mathbf{H} = \begin{pmatrix} 1 & 2 \\ -1 & -1 \end{pmatrix}$ and $\mathbf{K} = \begin{pmatrix} -1 & 2 \\ -1 & 1 \end{pmatrix}$ were found to represent the composition of two linear transformations $\mathbb{R}^2 \to \mathbb{R}^2$, but taken in different order.

(a) Find the eigenvalues of $\mathbf{H}$ and $\mathbf{K}$. What are their traces and determinants?

(b) Find the eigenvectors of $\mathbf{H}$ and $\mathbf{K}$.

(c) Find a nonsingular matrix $\mathbf{X}$ such that $\mathbf{HX} = \mathbf{XK}$, thereby showing that $\mathbf{H}$ and $\mathbf{K}$ are similar.

14. For any two nonsingular $n \times n$ matrices $\mathbf{A}$ and $\mathbf{B}$, show that the two matrix products $\mathbf{AB}$ and $\mathbf{BA}$ are always similar.

15. Generalize the previous problem:
For any two matrices $\mathbf{A}_{mn}$ and $\mathbf{B}_{nm}$, the nonzero eigenvalues of $(\mathbf{AB})_{mm}$ and $(\mathbf{BA})_{nn}$ are the same. [*Hint*: Evaluate the block matrix products $\begin{pmatrix} \mathbf{AB} & \mathbf{0} \\ \mathbf{B} & \mathbf{0} \end{pmatrix} \begin{pmatrix} \mathbf{I}_n & \mathbf{A} \\ \mathbf{0} & \mathbf{I}_m \end{pmatrix}$ and $\begin{pmatrix} \mathbf{I}_n & \mathbf{A} \\ \mathbf{0} & \mathbf{I}_m \end{pmatrix} \begin{pmatrix} \mathbf{0} & \mathbf{0} \\ \mathbf{B} & \mathbf{BA} \end{pmatrix}$.]

16. Find a real matrix $\mathbf{X}$ with $\mathbf{X}^T\mathbf{X} = \begin{pmatrix} 4 & 1 \\ 1 & 4 \end{pmatrix} = \mathbf{A}$, if possible.

17. Find a real matrix $\mathbf{X}$ with $\mathbf{X}^T\mathbf{X} = \begin{pmatrix} -4 & 1 \\ 1 & 4 \end{pmatrix} = \mathbf{A}$, if possible.

18. Show that if $\mathbf{A}$ is positive definite and real, then $\mathbf{x}^T\mathbf{A}\mathbf{x} > 0$ for all vectors $\mathbf{x} \neq \mathbf{0}$.

19. Show that if $\mathbf{A}$ is positive definite, then all diagonal entries a_{ii} of $\mathbf{A}$ are positive.

20. Show that if $\mathbf{A}$ is positive definite and $\mathbf{X}$ is nonsingular, then the matrix $\mathbf{X}^T\mathbf{AX}$ is also positive definite.

21. Prove or disprove the following statement: If, for a complex matrix $\mathbf{B} \in \mathbb{C}^{n,n}$, we have $\mathbf{x}^*\mathbf{Bx} = 0$ for all $\mathbf{x} \in \mathbb{C}^n$, then $\mathbf{B} = \mathbf{O}_n$.

22. Prove or disprove the following statement: If, for a complex matrix $\mathbf{B} \in \mathbb{C}^{n,n}$, we have $\mathbf{x}^*\mathbf{Bx} \in \mathbb{R}$ for all $\mathbf{x} \in \mathbb{C}^n$, then $\mathbf{B} = \mathbf{B}^*$.

23. Prove or disprove the following statement: If, for complex matrix $\mathbf{B} \in \mathbb{C}^{n,n}$, we have $\mathbf{x}^*\mathbf{Bx} > 0$ for all $\mathbf{x} \neq \mathbf{0} \in \mathbb{C}^n$, then $\mathbf{B} = \mathbf{B}^*$ and $\mathbf{B}$ is positive definite.

11.R Review Problems

1. True or false? (Give reasons or counterexamples.)
   (a) Every normal real matrix has a basis of real eigenvectors.
   (b) Every normal matrix is square.
   (c) Every square matrix is normal.
   (d) Every symmetric matrix is nonsingular.
   (e) Every diagonalizable matrix is normal.
   (f) Every orthogonally diagonalizable matrix is normal.
   (g) Every skew symmetric matrix is diagonalizable.
   (h) Every matrix is diagonalizable.
   (i) If a matrix $\mathbf{A}_{nn}$ is not diagonalizable, then $n >$ ________ .
   (j) Orthogonal matrices are orthogonally diagonalizable.
   (k) Symmetric matrices are orthogonally diagonalizable.
   (l) Symmetric matrices are diagonalizable via normal matrices.
   (m) Orthogonal matrices are diagonalizable over $\mathbb{C}$.
   (n) Some symmetric matrices can be diagonalized via symmetric matrices.
2. Prove that if $\mathbf{A} = \mathbf{A}^T \in \mathbb{R}^{n,n}$, then $\mathbf{A}$ has only real eigenvalues and a complete set of n mutually orthogonal real eigenvectors.
3. Prove that if $\mathbf{A} \in \mathbb{C}^{n,n}$ is normal, then $\mathbf{A}$ can be diagonalized via a unitary similarity transformation and vice versa.
4. Prove that if $\mathbf{A}$ is symmetric, then $\mathbf{A}^2$ is positive semidefinite.
5. Prove that if $\mathbf{A} = \mathbf{A}^T$ is symmetric and if $\mathbf{Au} = -\mathbf{u}$ while $\mathbf{Av} = 2\mathbf{v}$ for two vectors $\mathbf{u}, \mathbf{v} \neq \mathbf{0}$, then
   (a) $\mathbf{u}$ and $\mathbf{v}$ are linearly independent,
   (b) $\mathbf{u}$ and $\mathbf{v}$ are orthogonal, and
   (c) $\text{rank}(\mathbf{A}) \geq 2$.
6. Let $\mathcal{U} = \{\mathbf{u}_1, \ldots, \mathbf{u}_n\}$ be an ONB of column vectors in $\mathbb{R}^n$. Define the matrix $\mathbf{A}$ as $\mathbf{A} := \alpha_1 \mathbf{u}_1 \mathbf{u}_1^T + \cdots + \alpha_n \mathbf{u}_n \mathbf{u}_n^T$ for arbitrary, but fixed $\alpha_i \in \mathbb{R}$.
   (a) What are the eigenvalues of $\mathbf{A}$?
   (b) What are the corresponding eigenvectors?
   (c) Is $\mathbf{A}$ diagonalizable?
   (d) Is $\mathbf{A}$ normal?
   (e) Is $\mathbf{A}$ symmetric?
7. Show that every positive definite matrix $\mathbf{A} = \mathbf{A}^T$ can be written as $\mathbf{A} = \mathbf{X}^T\mathbf{X}$ for a nonsingular matrix $\mathbf{X}$, and conversely.
8. (a) Find a nonsingular 2×2 symmetric matrix $\mathbf{A}$, all of whose entries are positive, but which is not positive definite.
   (b) Repeat part (a) for a singular symmetric matrix $\mathbf{A}$, if possible.
9. True or false?
   (a) If $\mathbf{S} = \mathbf{S}^T$, then $\mathbf{S}^k$ is symmetric for any integer $k \geq 0$, and $\mathbf{S}^k$ in symmetric for any integer $k < 0$ if $\mathbf{S}$ in nonsingular.
   (b) If $\mathbf{U}$ is orthogonal, then $\mathbf{U}^k$ is orthogonal for every integer k.
   (c) If $\mathbf{S}$ is skew symmetric, then there are some integer powers of $\mathbf{S}$ that are skew symmetric and some that are symmetric. (Which are which?)

Standard Questions and Tasks

1. When is a given matrix $\mathbf{A} \in \mathbb{R}^{m,n}$ symmetric?
2. Can a real symmetric matrix $\mathbf{A}$ have nonreal eigenvalues or eigenvectors?
3. Find an ONB $\mathcal{U} \subset \mathbb{R}^n$ of eigenvectors for a symmetric matrix $\mathbf{A} = \mathbf{A}^T \in \mathbb{R}^{n,n}$.
4. Find an ONB $\mathcal{U} \subset \mathbb{C}^n$ of eigenvectors for a normal matrix $\mathbf{A} \in \mathbb{R}^{n,n}$.
5. When is a matrix $\mathbf{A}_{nn}$ normal?
6. Are normal matrices diagonalizable? How?
7. Are there non-normal matrices that can be diagonalized?
8. Can a matrix $\mathbf{A}$ be symmetric and skew symmetric at the same time?
9. Find the orthogonal and the positive definite matrix factor in the polar decomposition of a nonsingular matrix $\mathbf{A} \in \mathbb{R}^{n,n}$.

Subheadings of Lecture Eleven

Basic Equations

p. 349 $\mathbf{A}_{\mathcal{U}} = \mathbf{U}^*\mathbf{A}_{\mathcal{E}}\mathbf{U} = \mathbf{R}$ (Schur Normal Form)

p. 352 $\mathbf{A} = \mathbf{A}^T$ (symmetric matrix)

Now you see it, now you don't

CHAPTER

12

Singular Values

We study the revealing view of a linear transformation that is offered by its matrix representation with respect to two different orthonormal bases.

12.1 Lecture Twelve

Matrix Representations with Respect to Two Orthonormal Bases, the SVD, and ONBs for Image and Kernel

The singular value decomposition of a matrix shows many properties of the underlying linear transformation T. It represents T with respect to two different orthonormal bases—one for its domain and one for its range.

(a) Matrix Representation with Respect to Two Different Bases

Recall the notion of **similarity** for square matrices $\mathbf{A}_{nn}$ that map n-space to itself, namely,

$$\boxed{\mathbf{A}_{\mathcal{U}} = \mathbf{U}^{-1}\mathbf{A}_{\mathcal{E}}\mathbf{U}} \,. \tag{12.1}$$

According to Chapter 7, the matrix $\mathbf{U} = \mathbf{X}_{\mathcal{E}\leftarrow\mathcal{U}}$ is the basis change matrix from $\mathcal{U}$-coordinate vectors in $\mathbb{R}^n$ to their standard $\mathcal{E}$-coordinates. Note that the right-most factor $\mathbf{U}$ in (12.1) operates on the domain of $\mathbf{A}$ and the left factor $\mathbf{U}^{-1} = \mathbf{X}_{\mathcal{U}\leftarrow\mathcal{E}}$ changes the transformed $\mathcal{E}$-coordinate vector $\mathbf{A}_{\mathcal{E}}\mathbf{U}\mathbf{x}_{\mathcal{U}}$ in (12.1) back to $\mathcal{U}$-coordinates. While the right-hand factor $\mathbf{U}$ affects the domain $\mathbb{R}^n$ of $\mathbf{A}$, the left-hand factor $\mathbf{U}^{-1}$ in (12.1) operates on the image of $\mathbf{A}$, or on its range $\mathbb{R}^n$. (See Section 7.1 and the development of Formula (7.4) for more details). For a matrix similarity (12.1) of a square matrix $\mathbf{A}_{nn}$, there is no distinction between the domain and the range, since they are thought of as identical.

In this chapter, we develop a separate view of the role of the range and domain of a linear transformation $T : \mathbb{R}^n \to \mathbb{R}^m$ by allowing different bases for its domain $\mathbb{R}^n$ and for its range $\mathbb{R}^m$, even if $m = n$. Both of these bases, however, will be chosen as orthonormal.

The standard matrix representation $\mathbf{A}_{mn} = \mathbf{A}_{\mathcal{E}}$ represents a linear transformation $T : \mathbb{R}^n \to \mathbb{R}^m$ with respect to two (naturally different, if $m \neq n$) standard unit vector bases $\mathcal{E}$ of $\mathbb{R}^n$ and $\mathbb{R}^m$. Altering the tack from that of Chapter 9, we now study sparse, nice, and simple forms that can be achieved for matrices of the form

$$\boxed{\boldsymbol{\Sigma}_{mn} := \mathbf{U}^T \mathbf{A}_{mn} \mathbf{V}}. \tag{12.2}$$

Here, the matrix $\mathbf{V}_{nn}$ consists of n orthonormal columns for the domain $\mathbb{R}^n$ of T, while $\mathbf{U}_{mm}$ contains m orthonormal columns for $\mathbf{T}$'s range $\mathbb{R}^m$. As $\mathbf{U}^{-1} = \mathbf{U}^T$ for orthogonal matrices, formula (12.2) is formally analogous to matrix similarity (12.1) in case $m = n$ and if the two domain and range bases agree (i.e., if $\mathcal{U} = \mathcal{V}$ and $m = n$). In general, $\boldsymbol{\Sigma}_{mn}$ in formula (12.2) represents T with respect to the basis $\mathcal{V}$ of its domain $\mathbb{R}^n$ and with respect to the basis $\mathcal{U}$ of its range $\mathbb{R}^m$.

(b) The Singular Value Decomposition of a Matrix

The following form can always be achieved for $\boldsymbol{\Sigma}_{mn}$ in (12.2) for any matrix $\mathbf{A}_{mn}$.

Theorem 12.1 **(Singular Value Decomposition)** For every matrix $\mathbf{A}_{mn} \in \mathbb{R}^{m,n}$ with $m \geq n$, there exist real orthogonal matrices $\mathbf{U}_{mm}$ and $\mathbf{V}_{nn}$ such that

$$\boldsymbol{\Sigma}_{mn} = \mathbf{U}^T \mathbf{A} \mathbf{V} = \operatorname{diag}(\sigma_1, \ldots, \sigma_n)_{mn} = \begin{pmatrix} \sigma_1 & & 0 \\ & \ddots & \\ 0 & & \sigma_n \\ 0 & \ldots & 0 \\ \vdots & & \vdots \\ 0 & \ldots & 0 \end{pmatrix}_{mn} \tag{12.3}$$

is diagonal for $\sigma_i \in \mathbb{R}$ and $\sigma_1 \geq \ldots \geq \sigma_n \geq 0$. ◀

Note that $\mathbf{U}^T\mathbf{U} = \mathbf{I}_m$ and $\mathbf{V}^T\mathbf{V} = \mathbf{I}_n$, and that the apparent restriction to $m \geq n$ in Theorem 12.1 is not really necessary, other than to allow us to describe $\boldsymbol{\Sigma}_{mn}$ as having n diagonal entries σ_i. If $n > m$, the description of $\boldsymbol{\Sigma}$ involves $m = \min\{n, m\}$ diagonal entries σ_i instead.

The real nonnegative numbers σ_i and the column vectors of $\mathbf{U}$ and $\mathbf{V}$ in (12.3) carry special names:

DEFINITION 1 The nonnegative real numbers σ_i of the singular value decomposition (12.3) $\boldsymbol{\Sigma} = \mathbf{U}^T\mathbf{A}\mathbf{V} = \operatorname{diag}(\sigma_i)$, of a real matrix $\mathbf{A}_{mn}$ are called the **singular values** of $\mathbf{A}$.

The column vectors of $\mathbf{U}_{mm}$ and $\mathbf{V}_{nn}$ in the singular value decomposition (12.3) of $\mathbf{A}$ are called the **left** and **right singular vectors** of $\mathbf{A}$, respectively.

Theorem 12.1 states that with respect to two different orthonormal bases $\mathcal{V}$ and $\mathcal{U}$ for the domain and range, every linear transformation can be represented by a diagonal matrix. This is in clear contrast to the more restrictive matrix similarity (12.1) or (7.4), under which not all matrices are diagonalizable. (See Theorems 9.4 (or 9.2.D) and 11.4.) Moreover, while a diagonalization of a real matrix by similarity generally involves complex eigenvalues and eigenvectors due to the Fundamental Theorem of Algebra, the SVD of a real matrix can always be achieved entirely over $\mathbb{R}$.

To illustrate further, we recall the two matrix representation diagrams of Section 7.1 for a matrix similarity $\mathbf{A}_{\mathcal{U}} = \mathbf{U}^{-1}\mathbf{A}_{\mathcal{E}}\mathbf{U}$. In Theorem 12.1, we have established the somewhat analogous singular value equation $\mathbf{U}^T\mathbf{A}_{mn}\mathbf{V} = \mathbf{\Sigma}_{mn}$, or equivalently, $\mathbf{A}_{mn} = \mathbf{U}\mathbf{\Sigma}_{mn}\mathbf{V}^T$. According to Section 7.1, multiplying an $\mathcal{E}$-coordinate vector in $\mathbb{R}^n$ by $\mathbf{V}^T = \mathbf{V}^{-1}$ gives rise to the $\mathcal{V}$-coordinate vector of the same point in $\mathbb{R}^n$. Next, the matrix $\mathbf{\Sigma}_{mn} = \text{diag}(\sigma_i)$ in the product expression $\mathbf{U}\mathbf{\Sigma}_{mn}\mathbf{V}^T$ for $\mathbf{A}$ maps n-vectors to m-vectors. Therefore, it transforms general $\mathcal{V}$-coordinate vectors in $\mathbb{R}^n$ to $\mathcal{U}$-coordinate vectors in $\mathbb{R}^m$, where $\mathcal{U} \subset \mathbb{R}^m$ is the left singular vector basis for $\mathbf{A}$. (Look ahead to Example 4 in Section 12.3 to see how each right singular vector $\mathbf{v_i}$ of $\mathbf{A}$ is mapped by $\mathbf{\Sigma}$ to the corresponding stretched left singular vector $\sigma_i\mathbf{u_i}$ of $\mathbf{A}$.) Finally, multiplying by $\mathbf{U}$ converts the $\mathcal{U}$ vector $\mathbf{\Sigma}\mathbf{V}^T\mathbf{x}_{\mathcal{E}}$ back to its standard $\mathcal{E}$-coordinate vector in $\mathbb{R}^m$ due to the SVD equation $\mathbf{A}_{mn} = \mathbf{U}\mathbf{\Sigma}_{mn}\mathbf{V}^T$. The following is the matrix mapping diagram for Theorem 12.1, an $m \times n$ matrix $\mathbf{A} = \mathbf{A}_{'\mathcal{E}'}$, and its singular value decomposition:

$$
\begin{array}{ccc}
\mathcal{E} \textbf{ vector in } \mathbb{R}^n & & \mathcal{E} \textbf{ vector in } \mathbb{R}^m \\
x_{\mathcal{E}} & \xrightarrow{\quad \mathbf{A}_{'\mathcal{E}'} \quad} & \mathbf{A}_{'\mathcal{E}'}\mathbf{x}_{\mathcal{E}} = \mathbf{U}\mathbf{\Sigma}\mathbf{V}^T\mathbf{x}_{\mathcal{E}} \\
\mathbf{V}^T = \mathbf{V}^{-1} \Big\downarrow & & \Big\uparrow \mathbf{U} \\
\mathbf{V}^T\mathbf{x}_{\mathcal{E}} = \mathbf{x}_{\mathcal{V}} & \xrightarrow{\quad \mathbf{\Sigma} \quad} & \mathbf{\Sigma}\mathbf{x}_{\mathcal{V}} = \mathbf{\Sigma}\mathbf{V}^T\mathbf{x}_{\mathcal{E}} \\
\mathcal{V} \text{ vector in } \mathbb{R}^n & & \mathcal{U} \text{ vector in } \mathbb{R}^m
\end{array}
$$

The singular value decomposition can be derived somewhat analogously to the proof of the Schur normal form in Theorem 11.1. Before doing so, we define the norm $\|\mathbf{A}\|$ of an arbitrary matrix $\mathbf{A}_{mn}$ and deduce some simple properties.

DEFINITION 2 The **matrix norm** $\|\mathbf{A}\|$ of a real $m \times n$ matrix $\mathbf{A}_{mn}$ is

$$\|\mathbf{A}\| := \max_{\substack{\|\mathbf{x}\|=1 \\ \mathbf{x}\in\mathbb{R}^n}} \|\mathbf{A}\mathbf{x}\|.$$

For example, $\|\mathbf{I}_n\| = 1$ and $\|\mathbf{O}_{mn}\| = 0$ for all m and n.

Lemma 1 (a) If $\mathbf{U}_{mm}$ is an orthogonal matrix, then $\|\mathbf{U}\mathbf{x}\| = \|\mathbf{x}\|$ for all $\mathbf{x} \in \mathbb{R}^m$ (i.e., the length of a vector $\mathbf{x}$ does not change under an orthogonal transformation $\mathbf{U}$).

(b) The matrix norm $\|\mathbf{U}\|$ of an orthogonal matrix $\mathbf{U}_{mm}$ is equal to one.

(c) If $\mathbf{B}_{mn} = \mathbf{U}^T\mathbf{A}_{mn}\mathbf{V}$ for two orthogonal matrices $\mathbf{U}_{mm}$ and $\mathbf{V}_{nn}$, then $\|\mathbf{B}\| = \|\mathbf{A}\|$ (i.e., the matrix norm of an orthogonal product modification $\mathbf{B} = \mathbf{U}^T A\mathbf{V}$ of a matrix $\mathbf{A}$ does not change). ◀

Recall from Chapter 10 that an $m \times m$ matrix $\mathbf{U}$ is orthogonal if $\mathbf{U}^T\mathbf{U} = \mathbf{I}_m = \mathbf{U}\mathbf{U}^T$.

Proof (a) Part (a) was proved in Lemma 3 of Section 10.2.

(b) According to the definition of the matrix norm and part (a), we have

$$\|\mathbf{U}\| = \max_{\|\mathbf{x}\|=1} \|\mathbf{Ux}\| = \max_{\|\mathbf{x}\|=1} \|\mathbf{x}\| = 1$$

for orthogonal matrices $\mathbf{U}$.

(c) From part (a), we have

$$\|\mathbf{B}\| = \max_{\|\mathbf{x}\|=1} \|\mathbf{U}^T\mathbf{AVx}\| = \max_{\|\mathbf{Vx}\|=1} \|\mathbf{U}^T\mathbf{AVx}\| = \max_{\|\mathbf{y}\|=1} \|\mathbf{U}^T\mathbf{Ay}\|$$
$$= \max_{\|\mathbf{y}\|=1} \|\mathbf{Ay}\| = \|\mathbf{A}\|$$

for $\mathbf{y} := \mathbf{Vx}$. ■

Proof (Theorem 12.1, the Singular Value Decomposition) Assume without loss of generality that $\mathbf{A} \neq \mathbf{0}$. (Otherwise any two orthogonal matrices $\mathbf{U}_{mm}$ and $\mathbf{V}_{nn}$ work for $\mathbf{\Sigma} := \mathbf{O}_{mn}$.)

If $\mathbf{A} \neq \mathbf{0}$, then $\sigma := \|\mathbf{A}\| = \max_{\|\mathbf{x}\|=1} \|\mathbf{Ax}\| > 0$. Choose a unit vector $\mathbf{x} \in \mathbb{R}^n$ for which this maximum is achieved, and set $\mathbf{y} = \dfrac{\mathbf{Ax}}{\|\mathbf{Ax}\|} \in \mathbb{R}^m$. Then $\|\mathbf{x}\| = \|\mathbf{y}\| = 1$ (i.e., $\mathbf{x}$ and $\mathbf{y}$ are both unit vectors in their respective spaces), and

$$\mathbf{Ax} = \|\mathbf{Ax}\|\frac{\mathbf{Ax}}{\|\mathbf{Ax}\|} = \sigma\mathbf{y}$$

by the definition of σ and $\mathbf{y}$.

Let $\mathbf{V} \in \mathbb{R}^{n,n}$ be an orthogonal matrix with first column $\mathbf{x}$ and (i.e., $\mathbf{V} := \begin{pmatrix} \mathbf{x} & \mathbf{V}_1 \end{pmatrix}$, where $\mathbf{V}_1$ is $n \times (n-1)$), and let $\mathbf{U} \in \mathbb{R}^{m,m}$ be orthogonal with first column $\mathbf{y}$, or $\mathbf{U} := \begin{pmatrix} \mathbf{y} & \mathbf{U}_1 \end{pmatrix}$, where $\mathbf{U}_1$ is an $m \times (m-1)$ orthonormal fill-in; (See Proposition 1 in Section 10.1). Then

$$\mathbf{A}_1 := \mathbf{U}^T\mathbf{A}_{mn}\mathbf{V} = \begin{pmatrix} \mathbf{y}^T \\ \mathbf{U}_1^T \end{pmatrix} \mathbf{A} \begin{pmatrix} \mathbf{x} & \mathbf{V}_1 \end{pmatrix} = \begin{pmatrix} \mathbf{y}^T \\ \mathbf{U}_1^T \end{pmatrix} \begin{pmatrix} \mathbf{Ax} & \mathbf{AV}_1 \end{pmatrix}$$
$$= \begin{pmatrix} \mathbf{y}^T\mathbf{Ax} & \mathbf{y}^T\mathbf{AV}_1 \\ \mathbf{U}_1^T\mathbf{Ax} & \mathbf{U}_1^T\mathbf{AV}_1 \end{pmatrix} = \begin{pmatrix} \sigma & \mathbf{w}^T \\ \mathbf{0} & \mathbf{B} \end{pmatrix}_{mn},$$

where $\mathbf{w}^T \in \mathbb{R}^{n-1}$ is a row vector and $\mathbf{B}$ is $(m-1) \times (n-1)$. Here, we have used $\mathbf{Ax} = \sigma\mathbf{y}$. This makes the (1,1) entry of $\mathbf{A}_1$ equal to $\mathbf{y}^T\mathbf{Ax} = \sigma\mathbf{y}^T\mathbf{y} = \sigma$, since $\|\mathbf{y}\|^2 = \mathbf{y}^T\mathbf{y} = 1$. By the definition of $\mathbf{U}$, the vector $\mathbf{Ax} = \sigma\mathbf{y}$ is perpendicular to the column space of $\mathbf{U}_1$, or to the rows in $\mathbf{U}_1^T$. This ensures us of $n-1$ zero entries below σ in the first column of $\mathbf{A}_1 = \mathbf{U}^T\mathbf{AV}$.

On the one hand, $\|\mathbf{A}_1\|^2 = \sigma^2 = \|\mathbf{A}\|^2$ by Lemma 1(c), while the matrix norm definition makes

$$\|\mathbf{A}_1\|^2 = \max_{\|\mathbf{x}\|=1} \|\mathbf{A}_1\mathbf{x}\|^2 = \max_{\mathbf{x}\neq\mathbf{0}} \frac{\|\mathbf{A}_1\mathbf{x}\|^2}{\|\mathbf{x}\|^2} \geq \frac{\left\|\mathbf{A}_1\binom{\sigma}{\mathbf{w}}\right\|^2}{\left\|\binom{\sigma}{\mathbf{w}}\right\|^2} \geq \frac{(\sigma^2 + \mathbf{w}^T\mathbf{w})^2}{\sigma^2 + \mathbf{w}^T\mathbf{w}}$$
$$= \sigma^2 + \mathbf{w}^T\mathbf{w},$$

since $\begin{pmatrix} \sigma \\ \mathbf{w} \end{pmatrix} \neq \mathbf{0} \in \mathbb{R}^n$. The last inequality in this chain follows from evaluating

$$\mathbf{A}_1\begin{pmatrix} \sigma \\ \mathbf{w} \end{pmatrix} = \begin{pmatrix} \sigma & \mathbf{w}^T \\ \mathbf{0} & \mathbf{B} \end{pmatrix}\begin{pmatrix} \sigma \\ \mathbf{w} \end{pmatrix} = \begin{pmatrix} \sigma^2 + \mathbf{w}^T\mathbf{w} \\ \mathbf{Bw} \end{pmatrix}.$$

Hence,

$$\sigma^2 = \|\mathbf{A}_1\|^2 \geq \frac{\left\|\mathbf{A}_1 \binom{\sigma}{\mathbf{w}}\right\|^2}{\left\|\binom{\sigma}{\mathbf{w}}\right\|^2} \geq \sigma^2 + \mathbf{w}^T\mathbf{w}.$$

Consequently, $\mathbf{w}^T\mathbf{w} = 0$, or $\mathbf{w} = \mathbf{0}$, making $\mathbf{A}_1 = \begin{pmatrix} \sigma & \mathbf{0} \\ \mathbf{0} & \mathbf{B} \end{pmatrix}$ have zero entries in its first row beyond the (1,1) diagonal entry σ. Induction on $\mathbf{B}$ of smaller size completes the proof due to Lemma 1 of Section 11.1. ■

(c) Finding the Singular Value Decomposition of a Matrix

We study how to compute the singular value decomposition of modest-size matrices $\mathbf{A}_{mn}$ by elementary hand computations. The following theorem reduces this task to an eigenanalysis of the related matrix $\mathbf{A}^T\mathbf{A}$.

Theorem 12.2 If $m \geq n$, then the singular values of a matrix $\mathbf{A} \in \mathbb{R}^{m,n}$ are the nonnegative square roots $\sqrt{\lambda_i}$ of the eigenvalues λ_i of the symmetric positive semidefinite matrix $(\mathbf{A}^T\mathbf{A})_{nn}$.

If the right singular vectors for $\mathbf{A}_{mn}$ are chosen as an ONB $\mathcal{V}$ of eigenvectors of $\mathbf{A}^T\mathbf{A}$ and if $\mathbf{\Sigma}_{mn} = \text{diag}(\sqrt{\lambda_i})$, then the left singular vectors of $\mathbf{A}$ can be chosen as the columns of an orthogonal matrix $\mathbf{U}_{mm}$ with $\mathbf{AV} = \mathbf{U\Sigma}$. ◀

Proof If $\mathbf{A} = \mathbf{O}_{mn}$, there is nothing to prove. Hence, we assume that $\mathbf{A} \neq \mathbf{O}_{mn}$.

According to Lemma 6 and Definition 3 of Section 11.3, the matrix product $\mathbf{A}^T\mathbf{A} = \left(\mathbf{A}^T\right)_{nm}\mathbf{A}_{mn}$ is $n \times n$, positive semidefinite, and symmetric. Following Section 11.1, we can diagonalize $\mathbf{A}^T\mathbf{A}$ as $\mathbf{V}^T(\mathbf{A}^T\mathbf{A})\mathbf{V} = \text{diag}(\lambda_i)_{nn}$, using an orthogonal eigenvector matrix $\mathbf{V}$. Since a positive semidefinite matrix has only nonnegative real eigenvalues, we may assume that they are ordered so that $\lambda_1 \geq \lambda_2 \geq \cdots \geq \lambda_k > \lambda_{k+1} = \cdots = \lambda_n = 0$. Here, the index k is equal to the rank of $\mathbf{A}$. In particular, $k = n$ if $\mathbf{A}^T\mathbf{A}$ is positive definite.

Since $(\mathbf{AV})^T\mathbf{AV} = \mathbf{V}^T\mathbf{A}^T\mathbf{AV} = \text{diag}(\lambda_i)$ and since $\lambda_1, \ldots, \lambda_k > 0$, the first k columns of $\mathbf{AV}$ are nonzero and mutually orthogonal with respective lengths $\sqrt{\lambda_i}$, while the last $n-k$ columns of $\mathbf{AV}$ are zero. We scale each of the nonzero columns of $\mathbf{AV}$ to have unit length by matrix multiplication on the right:

$$\left(\widehat{\mathbf{AV}}\right)_{mk} := \begin{pmatrix} | & & | \\ (\mathbf{AV})_1 & \cdots & (\mathbf{AV})_k \\ | & & | \end{pmatrix} \begin{pmatrix} \frac{1}{\sqrt{\lambda_1}} & & 0 \\ & \ddots & \\ 0 & & \frac{1}{\sqrt{\lambda_k}} \end{pmatrix}$$

$$= \begin{pmatrix} | & & | \\ \mathbf{Av}_1 & \cdots & \mathbf{Av}_k \\ | & & | \end{pmatrix} \begin{pmatrix} \frac{1}{\sqrt{\lambda_1}} & & 0 \\ & \ddots & \\ 0 & & \frac{1}{\sqrt{\lambda_k}} \end{pmatrix}.$$

Now we augment $\widehat{\mathbf{AV}}$ with orthonormal fill-in $\mathbf{W}$ of size m by $m-k$ to obtain an $m \times m$ orthogonal matrix $\mathbf{U} := \left(\widehat{\mathbf{AV}}\ \mathbf{W}\right)$. For this $\mathbf{U}$ and $\mathbf{V}$, we have

$$\mathbf{U}^T\mathbf{A}\mathbf{V} = \begin{pmatrix} \begin{pmatrix} \frac{1}{\sqrt{\lambda_1}} & & 0 \\ & \ddots & \\ 0 & & \frac{1}{\sqrt{\lambda_k}} \end{pmatrix} \begin{pmatrix} - & \mathbf{v}_1^T\mathbf{A}^T & - \\ & \vdots & \\ - & \mathbf{v}_k^T\mathbf{A}^T & - \end{pmatrix} \\ \mathbf{W}^T_{m-k,m} \end{pmatrix}_{mm} \mathbf{A}_{mn}\mathbf{V}_{nn}$$

$$= \begin{pmatrix} \frac{1}{\sqrt{\lambda_1}} & & & & 0 \\ & \ddots & & & \\ & & \frac{1}{\sqrt{\lambda_k}} & & \\ & & & 0 & \\ & & & & \ddots \\ 0 & & & & 0 \\ \vdots & & & & \vdots \\ 0 & & & & 0 \end{pmatrix}_{mm} \operatorname{diag}(\lambda_i)_{mn}$$

$$= \begin{pmatrix} \sqrt{\lambda_1} & & & & 0 \\ & \ddots & & & \\ & & \sqrt{\lambda_k} & & \\ & & & 0 & \\ & & & & \ddots \\ 0 & & & & 0 \\ \vdots & & & & \vdots \\ 0 & & & & 0 \end{pmatrix} = \mathbf{\Sigma}_{mn},$$

since $\mathbf{W}^T\mathbf{A}\mathbf{V} = \mathbf{O}_{m-k,n}$ and $\mathbf{v_i}^T\mathbf{A}^T\mathbf{A}\mathbf{v_j} = \lambda_i\delta_{ij}$ for the Kronecker δ function. Thus, $\mathbf{U}^T\mathbf{A}\mathbf{V} = \mathbf{\Sigma} = \operatorname{diag}(\sqrt{\lambda_i})$ is the singular value decomposition of $\mathbf{A}$. ■

The following are the steps for finding the singular value decomposition of a real matrix $\mathbf{A}_{mn}$ by hand:

Steps for obtaining the singular value decomposition of a real matrix $\mathbf{A}_{mn}$ with $m \geq n$:

1. Perform an eigenanalysis of the matrix product $(\mathbf{A}^T\mathbf{A})_{nn}$ according to Chapters 9 and 11 and obtain its eigenvalues $\lambda_1 \geq \lambda_2 \geq \cdots \geq \lambda_k > \lambda_{k+1} = \cdots = \lambda_n = 0$. Find the associated orthonormal eigenvector basis $\mathcal{V} = \{\mathbf{v}_1, \ldots, \mathbf{v}_n\} \subset \mathbb{R}^n$ of $\mathbf{A}^T\mathbf{A}$.
2. Compute $\mathbf{A}\mathbf{v_j}$ for each index j with $\lambda_j \neq 0$, and normalize the vectors $\{\mathbf{A}\mathbf{v_j}\}_{j=1}^k$ by scaling each vector by its length $\frac{1}{\sqrt{\lambda_j}}$.
3. Extend the scaled mutually orthogonal unit vectors $\left\{\mathbf{A}\mathbf{v_j}/\sqrt{\lambda_j}\right\}_{j=1}^k$ from step 2 to an ONB $\mathcal{U}$ of $\mathbb{R}^m$ by orthonormal fill-in, as indicated in Section 10.1.

4. The singular value decomposition of $\mathbf{A}_{mn}$ is $\mathbf{A}_{mn} = \mathbf{U}_{mm}\mathbf{\Sigma}_{mn}\mathbf{V}_{nn}^T$, or equivalently, $\mathbf{\Sigma}_{mn} = \mathbf{U}^T A_{mn}\mathbf{V}$ with $\mathbf{\Sigma} = \text{diag}(\sqrt{\lambda_i})_{mn}$.

(If $n > m$, work on $\mathbf{A}^T$ instead, and use $\mathbf{AA}^T$ of smaller size $m \times m$ in step 1. The original matrix $\mathbf{A}_{mn}$ has the same singular values as $(\mathbf{A}^T)_{nm}$. The reverse-order product $\mathbf{A}^T\mathbf{A} \in \mathbb{R}^{n,n}$ is larger if $n > m$ and has an extra $n - m$ eigenvalues equal to zero.)

Note that in many cases the right singular vectors of $\mathbf{A}$ (collected in $\mathbf{V}$) can be taken as an ONB of eigenvectors of $\mathbf{A}^T\mathbf{A}$, while additionally, the left singular vectors of $\mathbf{A}$ can be taken as an orthonormal set of eigenvectors of the reverse-order matrix product $\mathbf{AA}^T$. But this is not always the case; observe that this simplistic "rule" fails for an orthogonal matrix $\mathbf{A}_{nn}$.

The following lemma is self-explanatory; simply transpose the singular value equation using Theorem 6.3.

Lemma 2 For a real matrix $\mathbf{A}_{mn}$ with $m \geq n$, the two matrices $\mathbf{A}^T$ and $\mathbf{A}$ have the same singular values. The respective left and right singular vectors of $\mathbf{A}^T$ and $\mathbf{A}$ exchange roles, or positions, in their respective singular-value matrix equations (i.e., if $\mathbf{\Sigma}_{mn} = \mathbf{U}^T\mathbf{AV}$ is the SVD of $\mathbf{A}$, then $\mathbf{\Sigma}^T = \mathbf{V}^T\mathbf{A}^T\mathbf{U}$ is the SVD of $\mathbf{A}^\mathbf{T}$). ◀

EXAMPLE 1 Find the singular-value decomposition of $\mathbf{A}_{2\times 3} = \begin{pmatrix} 0 & -2 & 1 \\ -1 & 1 & 0 \end{pmatrix}$.

We work on $\mathbf{A}^T$ instead of $\mathbf{A}$. Then $\mathbf{AA}^T = \begin{pmatrix} 5 & -2 \\ -2 & 2 \end{pmatrix}_{2\times 2}$, while $\mathbf{A}^T\mathbf{A}$ is 3×3 with one extra eigenvalue equal to zero.

Step 1: In order to find the singular values σ_i and both types of singular vectors for $\mathbf{A}^T$, we perform an eigenanalysis of $\mathbf{AA}^T$.

For $\mathbf{B}_{2\times 2} := \mathbf{AA}^T$, we have $f_B(\lambda) = \det(\lambda\mathbf{I}_2 - \mathbf{B}) = \det\begin{pmatrix} \lambda - 5 & 2 \\ 2 & \lambda - 2 \end{pmatrix} = (\lambda - 5)(\lambda - 2) - 4 = \lambda^2 - 7\lambda + 6 = (\lambda - 1)(\lambda - 6)$, according to Section 9.1.D. Or, using vector iteration as in Section 9.1 for the starting vector $\mathbf{y} = \begin{pmatrix} 1 \\ 1 \end{pmatrix}$, we have

$\mathbf{B} = \mathbf{AA}^T$		$\mathbf{y}$	$\mathbf{By}$	$\mathbf{B}^2\mathbf{y} = \mathbf{B}(\mathbf{By})$	
5	−2	[1]	3	15	
−2	2	[1]	0	−6	$-\ row_1$
		[1]	[3]	15	$+\ row_2$
		0	[−3]	−21	$\div(-3)$
		1	0	−6	
		0	1	7	

Thus, $\mathbf{B}^2\mathbf{y} = 7\mathbf{By} - 6\mathbf{y}$, or $p_y(\lambda) = \lambda^2 - 7\lambda + 6 = (\lambda - 1)(\lambda - 6)$.

For the eigenvalue $\lambda_1 = 1$ of $\mathbf{B} = \mathbf{A}\mathbf{A}^T$, the row reduction

$$\ker(\mathbf{B} - \mathbf{I}):\quad \begin{array}{cc|cl} \boxed{4} & -2 & 0 & \\ \boxed{-2} & 1 & 0 & +1/2\, row_1 \\ \hline \boxed{4} & -2 & 0 & \\ 0 & 0 & 0 & \\ & \uparrow & & \end{array}$$

yields the eigenvector $\mathbf{x_1} = \begin{pmatrix} 1 \\ 2 \end{pmatrix}$, while for the dominant eigenvalue $\lambda_2 = 6$ of $\mathbf{B} = \mathbf{A}\mathbf{A}^T$, we observe that

$$\ker(\mathbf{B} - 6\mathbf{I}):\quad \begin{array}{cc|cl} \boxed{-1} & -2 & 0 & \\ \boxed{-2} & -4 & 0 & -2\, row_1 \\ \hline \boxed{-1} & -2 & 0 & \\ 0 & 0 & 0 & \\ & \uparrow & & \end{array}$$

This yields $\mathbf{x_2} = \begin{pmatrix} -2 \\ 1 \end{pmatrix}$ as an eigenvector for $\lambda_2 = 6$. Note that both eigenvalues of $\mathbf{B} = \mathbf{A}\mathbf{A}^T$ are real and positive, since $\mathbf{A}$, and hence $\mathbf{B}$, is nonsingular.

Normalizing the mutually orthogonal eigenvectors $\{\mathbf{x}_1, \mathbf{x}_2\}$ of the symmetric matrix $\mathbf{B} = \mathbf{A}\mathbf{A}^T$ gives us the right singular vectors of $\mathbf{A}^T$.

Steps 2 and 3: The left singular vectors of $\mathbf{A}^T$ are the normalized versions of $\mathbf{A}^T\mathbf{x_1}$ and $\mathbf{A}^T\mathbf{x}_2$, that is, $\mathbf{y_1} = \frac{1}{\sqrt{5}}\begin{pmatrix} -2 \\ 0 \\ 1 \end{pmatrix}$ and $\mathbf{y_2} = \frac{1}{\sqrt{30}}\begin{pmatrix} -1 \\ 5 \\ -2 \end{pmatrix}$, augmented by one more orthonormal fill-in unit vector such as $\mathbf{y_3} = \frac{1}{\sqrt{6}}\begin{pmatrix} 1 \\ 1 \\ 2 \end{pmatrix} \in \mathbb{R}^3$.

Thus, $\mathbf{A}^T = \begin{pmatrix} | & | & | \\ \mathbf{y}_2 & \mathbf{y}_1 & \mathbf{y}_3 \\ | & | & | \end{pmatrix} \begin{pmatrix} \sqrt{6} & 0 \\ 0 & 1 \\ 0 & 0 \end{pmatrix} \begin{pmatrix} - & \frac{\mathbf{x}_2^T}{\|\mathbf{x}_2\|} & - \\ - & \frac{\mathbf{x}_1^T}{\|\mathbf{x}_1\|} & - \end{pmatrix}$, as can be readily checked. Finally, by transposing this matrix equation according to Lemma 2, we obtain

$$\mathbf{A} = \begin{pmatrix} | & | \\ \frac{\mathbf{x}_2}{\|\mathbf{x}_2\|} & \frac{\mathbf{x}_1}{\|\mathbf{x}_1\|} \\ | & | \end{pmatrix} \begin{pmatrix} \sqrt{6} & 0 & 0 \\ 0 & 1 & 0 \end{pmatrix} \begin{pmatrix} - & \mathbf{y}_2^T & - \\ - & \mathbf{y}_1^T & - \\ - & \mathbf{y}_3^T & - \end{pmatrix}$$

$$= \begin{pmatrix} \frac{-2}{\sqrt{5}} & \frac{1}{\sqrt{5}} \\ \frac{1}{\sqrt{5}} & \frac{2}{\sqrt{5}} \end{pmatrix} \begin{pmatrix} \sqrt{6} & 0 & 0 \\ 0 & 1 & 0 \end{pmatrix} \begin{pmatrix} \frac{-1}{\sqrt{30}} & \frac{5}{\sqrt{30}} & \frac{-2}{\sqrt{30}} \\ \frac{-2}{\sqrt{5}} & 0 & \frac{1}{\sqrt{5}} \\ \frac{1}{\sqrt{6}} & \frac{1}{\sqrt{6}} & \frac{1}{\sqrt{6}} \end{pmatrix}$$

as the desired singular value decomposition of $\mathbf{A}$.

Note that we use the singular vectors of $\mathbf{A}^T$ in the order $\mathbf{x}_2, \mathbf{x}_1$ and $\mathbf{y}_2, \mathbf{y}_1, \mathbf{y}_3$ here to achieve the proper order $\sigma_1 \geq \sigma_2 \geq 0$ for the singular values, since $\lambda_2 = 6$ is the dominant eigenvalue of $\mathbf{B} = \mathbf{A}\mathbf{A}^T$ with corresponding eigenvector $\mathbf{x}_2$. (In this example, $m = 2 < n = 3$, and the results of this section [stated for $m \geq n$] were reinterpreted according to Lemma 2.) ◀

(d) Applications of the SVD to Matrix-Generated Subspaces

The singular value decomposition of a matrix $\mathbf{A}$ has many useful applications. Let us emphasize one.

If $\mathbf{\Sigma}_{mn} = \mathbf{U}^T \mathbf{A}_{mn} \mathbf{V} = \operatorname{diag}(\sigma_1, \ldots, \sigma_k, 0, \ldots, 0)_{mn}$ is the singular-value decomposition of $\mathbf{A}$, and if $\sigma_i > 0$ for all $i = 1, \ldots, k$ while $\sigma_{k+1} = \ldots = \sigma_n = 0$, then $\mathbf{A}_{mn} = \mathbf{U}_{mm} \mathbf{\Sigma}_{mn} \mathbf{V}_{nn}^T$, or

$$\mathbf{A} = \begin{pmatrix} | & & | \\ \mathbf{u}_1 & \ldots & \mathbf{u}_m \\ | & & | \end{pmatrix} \begin{pmatrix} \sigma_1 & & & & & 0 \\ & \ddots & & & & \\ & & \sigma_k & & & \\ & & & 0 & & \\ & & & & \ddots & \\ 0 & & & & & 0 \end{pmatrix}_{mn} \begin{pmatrix} - & \mathbf{v}_1^T & - \\ & \vdots & \\ - & \mathbf{v}_n^T & - \end{pmatrix}. \tag{12.4}$$

By multiplying the last two factors of (12.4) first, we obtain the following matrix product representation of $\mathbf{A}$ from its singular value decomposition as

$$\mathbf{A} = \mathbf{U} \begin{pmatrix} - & \sigma_1 \mathbf{v}_1^T & - \\ & \vdots & \\ - & \sigma_k \mathbf{v}_k^T & - \\ & 0 & \\ & \vdots & \\ & 0 & \end{pmatrix},$$

where $\sigma_1, \ldots, \sigma_k > 0$. Since $\mathbf{U}_{mm}$ is an orthogonal matrix and thus nonsingular, the **kernel** of $\mathbf{A}$ coincides with the kernel of $\mathbf{U}^{-1}\mathbf{A} = \begin{pmatrix} - & \sigma_1 \mathbf{v}_1^T & - \\ & \vdots & \\ - & \sigma_k \mathbf{v}_k^T & - \\ & 0 & \\ & \vdots & \\ & 0 & \end{pmatrix}$,

(See Problem 13). This kernel is spanned by the mutually orthogonal vectors $\mathbf{v}_{k+1}, \ldots, \mathbf{v}_n \in \mathbb{R}^n$, each of which is orthogonal to the vectors $\mathbf{v}_1, \ldots, \mathbf{v}_k$, since $\mathbf{V}^T\mathbf{V} = \mathbf{I}_n$. Thus, an orthonormal basis of the **null space** or the **kernel** of $\mathbf{A}$ is given by the right singular vectors $\{\mathbf{v}_{k+1}, \ldots, \mathbf{v}_n\}$ of $\mathbf{A}$ that correspond to the singular value zero of $\mathbf{A}$, that is,

$$\ker(\mathbf{A}) = \operatorname{span}\{\mathbf{v}_{k+1}, \ldots, \mathbf{v}_n\}.$$

Likewise, the SVD describes an orthonormal basis for the **image** im(**A**) or the **column space** of $\mathbf{A}_{mn}$ in $\mathbb{R}^m$.

This follows by multiplying the first two factors in (12.4) first. Doing so, we have

$$\mathbf{A} = \begin{pmatrix} | & & | & | & & | \\ \sigma_1\mathbf{u}_1 & \dots & \sigma_k\mathbf{u}_k & \mathbf{0} & \dots & \mathbf{0} \\ | & & | & | & & | \end{pmatrix}_{mn} \mathbf{V}^T.$$

Thus, the column space $\text{im}(\mathbf{A}) = \{\mathbf{y} = \mathbf{Ax} \mid \mathbf{x} \in \mathbb{R}^n\} \subset \mathbb{R}^m$ of **A** is spanned by the orthonormal set of left singular vectors $\{\mathbf{u}_1, \dots, \mathbf{u}_k\}$, provided that **A** has precisely k nonzero singular values $\sigma_1, \dots, \sigma_k$. This follows from our interpretation of matrix products as column alterations of the first matrix factor in Section 6.2. In other words,

$$\text{im}(\mathbf{A}) = \text{span}\{\mathbf{u}_1, \dots, \mathbf{u}_k\}.$$

Thus, the SVD gives us tools for finding orthonormal bases for two important subspaces associated with a linear transformation: the kernel and the image.

EXAMPLE 2 Find an orthonormal basis for the kernel and one for the image of the matrix $\mathbf{A} = \begin{pmatrix} 0 & -2 & 1 \\ -1 & 1 & 0 \end{pmatrix}$. The SVD of **A** in Example 1 comes in handy. An ONB for the image of **A** is given by the set of left singular vectors

$$\left\{\frac{\mathbf{x}_1}{\|\mathbf{x}_1\|}, \frac{\mathbf{x}_2}{\|\mathbf{x}_2\|}\right\} = \left\{1/\sqrt{5}\begin{pmatrix} 1 \\ 2 \end{pmatrix}, 1/\sqrt{5}\begin{pmatrix} -2 \\ 1 \end{pmatrix}\right\}$$

that correspond to the nonzero singular values 1 and $\sqrt{6}$ of **A**. These were derived in Example 1. Similarly, an ONB for the kernel of **A** is given by the single normalized right singular vector $\mathbf{y}_3 = 1/\sqrt{6}\begin{pmatrix} 1 \\ 1 \\ 2 \end{pmatrix}$ of **A**. ◀

The singular value decomposition of a matrix involves many fundamental matrix techniques and gives many insights into linear transformations.

12.1.P Problems

1. Find the singular-value decompositions of the following matrices:

(a) $\mathbf{A} = \begin{pmatrix} 1 & -1 \\ 2 & -2 \end{pmatrix}$;

(b) $\mathbf{B} = \begin{pmatrix} -2 & 2 \\ -1 & -2 \end{pmatrix}$;

(c) $\mathbf{C} = \begin{pmatrix} 3 & 0 & 0 \\ 0 & 1 & 2 \\ 0 & 2 & 1 \\ 0 & 0 & 0 \end{pmatrix}$;

(d) $\mathbf{F} = \begin{pmatrix} 2 & 0 \\ 0 & 1 \\ 0 & 1 \end{pmatrix}$;

(e) $\mathbf{G} = \begin{pmatrix} \sqrt{5} & 0 \\ 1 & \sqrt{5} \\ 0 & 0 \end{pmatrix}$;

(f) $\mathbf{H} = \begin{pmatrix} 0 & 1 \\ 1 & 1 \\ -1 & 0 \end{pmatrix}$;

(g) $\mathbf{K} = \begin{pmatrix} 1 & 2 & 2 \\ 2 & 1 & -2 \end{pmatrix}$.

2. What are the ranks of the matrices in Problem 1?

3. Let $\mathbf{A}$ be a nonsingular matrix with SVD $\mathbf{A} = \mathbf{U}\boldsymbol{\Sigma}\mathbf{V}^T$. Find the SVD of $\mathbf{A}^{-1}$.

4. Show that if $\mathbf{W}_{mm}$ is orthogonal, then $\mathbf{WA}$ and $\mathbf{A}$ have the same SVD for each matrix $\mathbf{A}_{mn}$.

5. Find an explicit singular-value decomposition $\mathbf{U}^T\mathbf{A}\mathbf{V} = \boldsymbol{\Sigma}$ for $\mathbf{A} = \mathbf{O}_{mn}$ and $\mathbf{A} = \mathbf{I}_n$.

6. Find the singular values of a unitary matrix $\mathbf{U}_{nn}$ with $\mathbf{U}^*\mathbf{U} = \mathbf{U}\mathbf{U}^* = \mathbf{I}_n$.

7. Compute the explicit SVD of a row and of a column vector. Be careful with the zero vector.

8. If $\mathbf{A}_{nn} = \mathbf{U}\boldsymbol{\Sigma}\mathbf{V}^T$ is the singular value decomposition of a nonsingular matrix $\mathbf{A}$, show that $\mathbf{A} = \mathbf{QP}$ is a polar decomposition of $\mathbf{A}$ for $\mathbf{Q} = \mathbf{U}\mathbf{V}^T$ and $\mathbf{P} = \mathbf{V}\boldsymbol{\Sigma}\mathbf{V}^T$.

9. How are the singular value decompositions of a matrix $\mathbf{A}$ and its transpose $\mathbf{A}^T$ related?

10. True or false?

(a) The singular vectors of a matrix $\mathbf{A}_{mn}$ all have length one.

(b) The left singular vectors of a matrix $\mathbf{A}$ are mutually orthogonal.

(c) The right singular vectors of any matrix $\mathbf{A}_{mn}$ are a basis of $\mathbb{R}^m$.

(d) Every left singular vector of a square matrix $\mathbf{A}_{nn}$ is orthogonal to every right singular vector of $\mathbf{A}$.

(e) At least one left singular vector of a matrix $\mathbf{A}_{nn}$ is orthogonal to a right singular vector of $\mathbf{A}$.

11. For a square matrix $\mathbf{A}$, show that $\prod$ *(singular values of* $\mathbf{A}$*)* $= \left|\prod(\textit{eigenvalues of } \mathbf{A})\right|$.

12. Consider the linear transformation $T : \mathbb{R}^4 \to \mathbb{R}^4$ defined by $T\begin{pmatrix} x_1 \\ x_2 \\ x_3 \\ x_4 \end{pmatrix} = \begin{pmatrix} 0 \\ 3x_1 \\ 2x_2 \\ -3x_4 \end{pmatrix}$.

(a) Find the eigenvalues of T.

(b) Find the matrix representation $\mathbf{A} = \mathbf{A}_{\mathcal{E}}$ of T. Compute $\mathbf{A}\mathbf{A}^T$.

(c) Find the eigenvalues of $\mathbf{A}\mathbf{A}^T$.

(d) Find the eigenvalues of $\mathbf{A}^T\mathbf{A}$. Is $\mathbf{A}$ normal? Is it diagonalizable?

(e) Find the singular value decomposition of $\mathbf{A}$.

13. If $\mathbf{X}_{mm}$ is nonsingular, show that for every matrix $\mathbf{A}_{mn}$, we have $\ker(\mathbf{A}) = \ker(\mathbf{XA})$.

14. Derive Lemma 2.

15. Let $\mathbf{A} = \dfrac{1}{\sqrt{30}}\mathbf{U}_0\boldsymbol{\Sigma}\mathbf{V}_0^T =$

$$\frac{1}{\sqrt{30}}\begin{pmatrix} \sqrt{2} & -1 & -\sqrt{3} \\ \sqrt{2} & 2 & 0 \\ \sqrt{2} & -1 & \sqrt{3} \end{pmatrix}\begin{pmatrix} 3 & 0 \\ 0 & 2 \\ 0 & 0 \end{pmatrix}\begin{pmatrix} 1 & -2 \\ 2 & 1 \end{pmatrix}.$$

(a) Find the singular values of $\mathbf{A}$.

(b) What are the singular vectors of $\mathbf{A}$?

(c) Write out the SVD for $\mathbf{A}$.

16. Extend Problem 8 to find the polar decomposition $\mathbf{A} = \mathbf{QP}$ for the singular matrix $\mathbf{A} = \begin{pmatrix} 1 & -1 \\ 2 & -2 \end{pmatrix}$.

17. True or false?

(a) If $\mathbf{A}_{nn}$ and $\mathbf{B}_{nn}$ are similar, then $\mathbf{A}$ and $\mathbf{B}$ have the same eigenvalues.

(b) If $\mathbf{A}_{nn}$ and $\mathbf{B}_{nn}$ are similar, then $\mathbf{A}^2$ and $\mathbf{B}^2$ have the same eigenvalues.

(c) If $\mathbf{A}_{nn}$ and $\mathbf{B}_{nn}$ are similar, then $\mathbf{A}$ and $\mathbf{B}$ have the same singular values.

(d) If $\mathbf{A}_{nn}$ has the singular value σ, then $\mathbf{A}^2$ has the singular value σ^2.

(e) If one singular value of $\mathbf{A}$ is zero, then $\mathbf{A}$ is singular.

(f) If all singular values of $\mathbf{A}$ are zero, then $\mathbf{A}$ is invertible.

(g) There is only one matrix $\mathbf{A}_{mn}$, all of whose singular values are zero.

(h) If all singular values of a square matrix $\mathbf{A}_{nn}$ are 1, then $\mathbf{A}$ is orthogonal.

18. Find the singular values of the following matrices:

(a) $\mathbf{A} = \begin{pmatrix} -2 & 0 \\ 0 & 1 \end{pmatrix}$.

(b) $\mathbf{B} = \begin{pmatrix} -2 & 1 \\ 0 & 1 \end{pmatrix}$.

(c) $\mathbf{C} = \begin{pmatrix} 1 & 1 \\ -1 & 1 \end{pmatrix}$.

19. Let $\mathbf{A} = \begin{pmatrix} 4 & 2 \\ 2 & 1 \end{pmatrix}$.

(a) Find the singular values of $\mathbf{A}$ and $\mathbf{A}^2$.

(b) Find a unit vector $\mathbf{u} \in \mathbb{R}^2$ with $\|\mathbf{Au}\| = \sigma_1$, the largest singular value of $\mathbf{A}$.

20. Assume that $\mathbf{A}_{mn}$ has the SVD $\mathbf{U}^T\mathbf{A}\mathbf{V} = \boldsymbol{\Sigma}$.

(a) Find the singular value decomposition of $4\mathbf{A}$.

(b) What is the SVD of $-2\mathbf{A}$?

(c) If $m = n$ and $\mathbf{A}$ is invertible, what is the SVD of $\mathbf{A}^{-1}$?

21. True or false?

(a) If $\mathbf{B}_{nn} = \mathbf{X}^{-1}\mathbf{A}_{nn}\mathbf{X}$, then $\mathbf{A}$ and $\mathbf{B}$ have the same singular values.

(b) If $\mathbf{B}_{nn} = \mathbf{U}^{-1}\mathbf{A}_{nn}\mathbf{U}$ for an orthogonal matrix $\mathbf{U}$, then $\mathbf{A}$ and $\mathbf{B}$ have the same SVD.

(c) If $\mathbf{A}_{nn}$ and $\mathbf{B}_{nn}$ have the same SVD, then $\mathbf{B} = \mathbf{W}^T\mathbf{A}\mathbf{Z}$ for two orthogonal matrices $\mathbf{W}$ and $\mathbf{Z} \in \mathbb{R}^{n,n}$.

(d) The singular values of $\mathbf{A}^3$ are the cubes of the singular values of $\mathbf{A}$, provided that $\mathbf{A}$ is square.

22. Find the SVD of a dyad $\mathbf{A}_{mn} = \mathbf{u}\mathbf{v}^T$ for two column vectors $\mathbf{u} \in \mathbb{R}^m$ and $\mathbf{v} \in \mathbb{R}^n$.

23. What are the singular values of a real diagonal matrix $\mathbf{D} = \text{diag}(d_{ii})$? What are the singular vectors?

24. What are the singular values of a real counterdiagonal matrix $\mathbf{C} = \begin{pmatrix} 0 & & a_1 \\ & \iddots & \\ a_n & & 0 \end{pmatrix}$? What are the singular vectors?

25. Show that if $\mathbf{A}_{nn}$ has the n-fold singular value 1, then $\mathbf{A}$ is orthogonal.

26. Show that the extreme singular values σ_1 and σ_n of a matrix $\mathbf{A}_{mn}$ with $m \geq n$ provide the following bounds for the length of $\mathbf{Ax} \in \mathbb{R}^m$ for any vector $\mathbf{x} \in \mathbb{R}^n$:

$$\sigma_n\|\mathbf{x}\| \leq \|\mathbf{Ax}\| \leq \sigma_1\|\mathbf{x}\|.$$

27. True or false?

(a) A matrix $\mathbf{A}_{nn}$ is singular if and only if zero is a singular value of $\mathbf{A}$.

(b) If $\mathbf{A}$ is symmetric with the eigenvalue λ, then $|\lambda|$ is a singular value of $\mathbf{A}$.

(c) If $\mathbf{A}$ is skewsymmetric with the eigenvalue λ, then $|\lambda|$ is a singular value of $\mathbf{A}$.

(d) If $\mathbf{A} = \mathbf{A}^T \in \mathbb{R}^{n,n}$ and $\lambda > 0$, then λ or $-\lambda$ is an eigenvalue of $\mathbf{A}$ if and only if λ is a singular value of $\mathbf{A}$.

28. Is it true for a square matrix $\mathbf{A}_{nn}$ that the singular values of $\mathbf{A}^2$ are the squares of the singular values of $\mathbf{A}$? (*Hint*: Look at $\mathbf{A} = \begin{pmatrix} 0 & 1 \\ 0 & 0 \end{pmatrix}$.)

29. Prove that a square matrix is invertible if and only if zero is not a singular value of $\mathbf{A}$.

30. Let $\mathbf{A} = \begin{pmatrix} 2 & 0 \\ 0 & -3 \end{pmatrix}$.

(a) Find the SVD of $\mathbf{A}$.

(b) Find the polar decomposition of $\mathbf{A}$ from $\mathbf{A}$'s SVD. (*Hint*: See Problem 8.)

31. Prove or disprove the following by counterexample: If σ is a singular value of $\mathbf{A}^2$, then $\sqrt{\sigma}$ is a singular value of $\mathbf{A}$.

32. Let $\mathbf{A}_{mn}$ have the singular values $\sigma_1 \geq \ldots \geq \sigma_n$, and assume that $m \geq n$. Then, for any unit vector $\mathbf{x} \in \mathbb{R}^n$, show that $\sigma_n \leq \|\mathbf{Ax}\| \leq \sigma_1$.

33. Show that if $\sigma_1 \geq \ldots \geq \sigma_n$ are the singular values of $\mathbf{A}_{mn}$ for $m \geq n$, then $\min_{\|\mathbf{x}\|=1} \|\mathbf{Ax}\| = \sigma_n$.

34. Prove that all singular values of an orthogonal matrix $\mathbf{Q}$ are equal to 1.

35. Show that for a left and right singular vector pair $\mathbf{u}$ and $\mathbf{v}$ for the same singular value σ of a given matrix $\mathbf{A}_{mn}$, we have $\mathbf{A}^T\mathbf{u} = \sigma\mathbf{v}$.

36. Let $\mathbf{A} = \begin{pmatrix} 2 & 0 \\ 1 & 1 \\ -1 & 2 \\ 0 & 1 \end{pmatrix}$.

(a) Compute the left singular vectors of $\mathbf{A}$. How many are there?

(b) Verify that the two left singular vectors from part (a) that correspond to the nonzero singular values of $\mathbf{A}$ span the image of $\mathbf{A}$ orthogonally.

Teacher's Problem Making Exercise

T 12. How can the entries of an integer matrix $\mathbf{A}_{mn}$ be prescribed so that it has known "almost integer" singular values and subsequently "nearly integer" singular vectors (before normalization) for easy hand computations?

If $\mathbf{V}_{nn}$ is a matrix with mutually orthogonal integer columns $\mathbf{v}_i$ in $\mathbb{R}^n$ (as constructed in Problem T 10 for example), if $\mathbf{U}_{mm}$ is likewise for $\mathbb{R}^m$, and if $\mathbf{\Sigma} = \text{diag}(\sigma_i)_{mn}$ is an integer diagonal matrix, then $\mathbf{A} := \mathbf{U\Sigma V}^T$ is an integer $m \times n$ matrix. Note that $\mathbf{A} = \mathbf{U\Sigma V}^T$ is not the SVD of $\mathbf{A}$, since neither $\mathbf{U}$ nor $\mathbf{V}$ is assumed to be an orthogonal matrix.

Normalizing the columns in $\mathbf{U}$ and $\mathbf{V}$ leads to $\mathbf{A} = \mathbf{U\Sigma V}^T = \mathbf{U}(\mathbf{N}_U\mathbf{\Sigma}_0\mathbf{N}_V^T)\mathbf{V}^T :=$

$$\begin{pmatrix} | & & | \\ \mathbf{u}_1 & \dots & \mathbf{u}_m \\ | & & | \end{pmatrix} \begin{pmatrix} \frac{1}{\|\mathbf{u}_1\|} & & \\ & \ddots & \\ & & \frac{1}{\|\mathbf{u}_m\|} \end{pmatrix} \text{diag}(\sigma_i^o) \begin{pmatrix} \frac{1}{\|\mathbf{v}_1\|} & & \\ & \ddots & \\ & & \frac{1}{\|\mathbf{v}_n\|} \end{pmatrix} \begin{pmatrix} - & \mathbf{v}_1^T & - \\ & \vdots & \\ - & \mathbf{v}_n^T & - \end{pmatrix},$$

where $\mathbf{N}_U := \text{diag}\left(\frac{1}{\|\mathbf{u}_i\|}\right)$ makes $\mathbf{UN}_U$ orthogonal, and likewise for $\mathbf{N}_V = \mathbf{N}_V^T$ and $\mathbf{VN}_V$, (See Problem T 10 (B) for details.) Furthermore, $\sigma_i^o := \sigma_i\|\mathbf{u}_i\|\|\mathbf{v}_i\|$. The singular values of $\mathbf{A}$ are the diagonal entries σ_i^o of $\mathbf{\Sigma}_0 = \text{diag}(\sigma_i^o)$, since $\mathbf{A}$ and $\mathbf{\Sigma}_0$ are related by the singular-value equation $\mathbf{A} = (\mathbf{UN}_U)\mathbf{\Sigma}_0(\mathbf{N}_V^T\mathbf{V}^T)$ for the two orthogonal matrices $\mathbf{UN}_U \in \mathbb{R}^{m,m}$ and $\mathbf{VN}_V \in \mathbb{R}^{n,n}$. Note that the singular values σ_i^0 of the integer matrix $\mathbf{A}$ will generally involve real square roots, due to their defining formula $\sigma_i^o = \sigma_i\|\mathbf{u}_i\|\|\mathbf{v}_i\|$. When performing an eigenanalysis on $\mathbf{A}^T\mathbf{A}$, we will, however, compute the squares of the singular values as the eigenvalues of the integer symmetric matrix $\mathbf{A}^T\mathbf{A} = \mathbf{VN}_V\mathbf{\Sigma}_0^T\mathbf{\Sigma}_0(\mathbf{N}_V\mathbf{V})^T$. This matrix has the same integer eigenvalues $\sigma_0^2 = \sigma_i^2\|\mathbf{u}_i\|^2\|\mathbf{v}_i\|^2$ as $\mathbf{\Sigma}_0^T\mathbf{\Sigma}_0 = (\mathbf{\Sigma}_0)^2$ does.

EXAMPLE Let $\mathbf{V} = \begin{pmatrix} 1 & 1 & -1 \\ -1 & 1 & 1 \\ 1 & 0 & 2 \end{pmatrix}$, $\mathbf{\Sigma} = \begin{pmatrix} 2 & 0 & 0 \\ 0 & 1 & 0 \end{pmatrix}$, and $\mathbf{U} = \begin{pmatrix} 1 & -1 \\ 1 & 1 \end{pmatrix}$, where $\mathbf{U}$ and V each have mutually orthogonal columns. Then $\mathbf{A} := \mathbf{U\Sigma V}^T = \begin{pmatrix} 1 & -3 & 2 \\ 3 & -1 & 2 \end{pmatrix}_{2\times 3}$ has the singular values $\sigma_1^o = \sigma_1\|\mathbf{u}_1\|\|\mathbf{v}_1\| = 2\sqrt{6}$ and $\sigma_2^o = \sigma_2\|\mathbf{u}_2\|\|\mathbf{v}_2\| = 2$. These are the square roots of the nonzero integer eigenvalues of $\mathbf{A}^T\mathbf{A} = \begin{pmatrix} 10 & -6 & 8 \\ -6 & 10 & -8 \\ 8 & -8 & 8 \end{pmatrix}_{3\times 3}$.

Note that in this example the eigenvalue computations by hand become a little simpler for $\mathbf{A}^T\mathbf{A}$ if we compute those of $\frac{1}{2}\mathbf{AA}^T$ instead, since $m = 2 < n = 3$. (See Lemma 2.) The 2×2 matrix $\frac{1}{2}\mathbf{AA}^T = \begin{pmatrix} 7 & 5 \\ 5 & 7 \end{pmatrix}$ has the eigenvalues 2 and 12, giving $\mathbf{A}^T\mathbf{A}$ the eigenvalues 0, $2 \cdot 2$, and $2 \cdot 12$. Hence, $\mathbf{A}$ has the singular values $\sqrt{4} = 2$ and $\sqrt{24} = 2\sqrt{6}$. Finally, the left and right singular vectors of $\mathbf{A}$ are the respective mutually orthogonal and normalized columns of $\mathbf{U}$ and $\mathbf{V}$. ◀

12.2 Theory

Matrix Approximation, Least Squares

The singular value decomposition allows us to measure distances between sets of matrices and vectors, it lets us approximate matrices, and it finds approximate solutions to unsolvable linear systems of equations.

Intrinsically, the singular-value decomposition $\mathbf{A}_{mn} = \mathbf{U}\boldsymbol{\Sigma}\mathbf{V}^T$ is a matrix factorization of $\mathbf{A}$. But at the same time, it allows us to represent every matrix $\mathbf{A}$ additively as a sum of meaningful dyads.

Lemma 3 If $\mathbf{U}^T\mathbf{A}\mathbf{V} = \text{diag}\,(\sigma_i)$ is the SVD of a real matrix $\mathbf{A}_{mn}$ for $m \geq n$, then $\mathbf{A}$ is the sum of n dyads; that is,

$$\mathbf{A} = \sum_{i=1}^{n} \sigma_i \mathbf{u}_i \mathbf{v}_i^T, \text{ where } \mathbf{U}_{mm} = \begin{pmatrix} | & & | \\ \mathbf{u}_1 & \cdots & \mathbf{u}_m \\ | & & | \end{pmatrix} \text{ and } \mathbf{V}_{nn} = \begin{pmatrix} | & & | \\ \mathbf{v}_1 & \cdots & \mathbf{v}_n \\ | & & | \end{pmatrix}.$$

◀

Proof Every matrix product can be written as the sum of dyads, generated by the columns of the first factor and the like-numbered rows of the second one. Specifically, for the SVD $\mathbf{A} = \mathbf{U}\boldsymbol{\Sigma}\mathbf{V}^T$, we have

$$\mathbf{A}_{mn} = \begin{pmatrix} | & & | \\ \mathbf{u}_1 & \cdots & \mathbf{u}_m \\ | & & | \end{pmatrix} \begin{pmatrix} \sigma_1 & & 0 \\ & \ddots & \\ 0 & & \sigma_n \\ 0 & \cdots & 0 \\ \vdots & & \vdots \\ 0 & \cdots & 0 \end{pmatrix}_{mn} \begin{pmatrix} - & \mathbf{v}_1^T & - \\ & \vdots & \\ - & \mathbf{v}_n^T & - \end{pmatrix}$$

$$= \begin{pmatrix} | & & | & | & & | \\ \sigma_1\mathbf{u}_1 & \cdots & \sigma_n\mathbf{u}_n & \mathbf{0} & \cdots & \mathbf{0} \\ | & & | & | & & | \end{pmatrix} \begin{pmatrix} - & \mathbf{v}_1^T & - \\ & \vdots & \\ - & \mathbf{v}_n^T & - \end{pmatrix}$$

$$= \left[\begin{pmatrix} | & | & & | \\ \sigma_1\mathbf{u}_1 & \mathbf{0} & \cdots & \mathbf{0} \\ | & | & & | \end{pmatrix} + \begin{pmatrix} | & | & | & & | \\ \mathbf{0} & \sigma_2\mathbf{u}_2 & \mathbf{0} & \cdots & \mathbf{0} \\ | & | & | & & | \end{pmatrix} + \cdots \right] \cdot$$

$$\cdot \left[\begin{pmatrix} - & \mathbf{v}_1^T & - \\ - & \mathbf{0} & - \\ & \vdots & \\ - & \mathbf{0} & - \end{pmatrix} + \begin{pmatrix} - & \mathbf{0} & - \\ - & \mathbf{v}_2^T & - \\ - & \mathbf{0} & - \\ & \vdots & \\ - & \mathbf{0} & - \end{pmatrix} + \cdots \right]$$

$$= \sigma_1 \begin{pmatrix} | \\ \mathbf{u}_1 \\ | \end{pmatrix} \begin{pmatrix} - & \mathbf{v}_1^T & - \end{pmatrix} + \cdots + \sigma_n \begin{pmatrix} | \\ \mathbf{u}_n \\ | \end{pmatrix} \begin{pmatrix} - & \mathbf{v}_n^T & - \end{pmatrix},$$

$$\text{since} \begin{pmatrix} | & & | & | & | & & | \\ \mathbf{0} & \dots & \mathbf{0} & \sigma_i \mathbf{u}_i & \mathbf{0} & \dots & \mathbf{0} \\ | & & | & | & | & & | \end{pmatrix}_{mn} \begin{pmatrix} - & \mathbf{0} & - \\ & \vdots & \\ - & \mathbf{0} & - \\ - & \mathbf{v}_j^T & - \\ - & \mathbf{0} & - \\ & \vdots & \\ - & \mathbf{0} & - \end{pmatrix}_{nn} = \mathbf{O}_{mn} \text{ whenever } i \neq j. \qquad \blacksquare$$

This additive matrix expansion by dyads is useful for approximating a given matrix $\mathbf{A}$ by a low-rank matrix, for evaluating the matrix norm $\|\mathbf{A}\|$, and for defining the numerical rank of an arbitrary matrix reliably.

Theorem 12.3 **(Rank k Matrix Approximation)** Let $\mathbf{U}^T \mathbf{A} \mathbf{V} = \text{diag}(\sigma_i)$ be the SVD of a real matrix $\mathbf{A}_{mn}$ with $\sigma_1 \geq \sigma_2 \geq \ldots \geq \sigma_r > \sigma_{r+1} = \ldots = \sigma_n = 0$ and $m \geq n \geq r$.

For any $k < r = \text{rank}(\mathbf{A})$, set $\mathbf{A}_k := \sum_{i=1}^{k} \sigma_i \mathbf{u}_i \mathbf{v}_i^T$ with $\mathbf{U}_{mm} = \begin{pmatrix} | & & | \\ \mathbf{u}_1 & \dots & \mathbf{u}_m \\ | & & | \end{pmatrix}$ and $\mathbf{V} = \begin{pmatrix} | & & | \\ \mathbf{v}_1 & \dots & \mathbf{v}_n \\ | & & | \end{pmatrix}$. Then

$$\min_{\text{rank } (\mathbf{B}_{mn}) \leq k} \|\mathbf{A} - \mathbf{B}\| = \|\mathbf{A} - \mathbf{A}_k\| = \sigma_{k+1}.$$

◀

In Theorem 12.3, observe that $\sigma_{k+1} > 0$ by assumption. Specifically for $k = 0$, we note that $\{\mathbf{B}_{mn} \mid \text{rank}(\mathbf{B}) = 0\} = \{\mathbf{O}_{mn}\}$, and thus we have the following corollary:

Corollary 1 For every $m \times n$ matrix $\mathbf{A}$, the **matrix norm** satisfies

$$\|\mathbf{A}\| = \sigma_1;$$

that is, the matrix norm $\|\mathbf{A}\|$ of $\mathbf{A}_{mn}$ is equal to its maximal singular value σ_1.

◀

For example, the matrix $\mathbf{A}_{2\times 3}$ of Examples 1 and 2 has norm $\|\mathbf{A}\| = \sqrt{6} = \sigma_1$.

Proof (Theorem 12.3) From the definition of $\mathbf{A}_k$, we have

$$\begin{aligned} \mathbf{U}^T \mathbf{A}_k \mathbf{V} &= \mathbf{U}^T \left(\sum_{i=1}^{k} \sigma_i \mathbf{u}_i \mathbf{v}_i^T \right) \mathbf{V} \\ &= \sum_{i=1}^{k} \sigma_i \left(\mathbf{U}^T \mathbf{u}_i \mathbf{v}_i^T \mathbf{V} \right) \\ &= \sum_{i=1}^{k} \sigma_i \mathbf{e}_i \mathbf{e}_i^T \\ &= \text{diag } (\sigma_1, \ldots, \sigma_k, 0, \ldots, 0)_{mn}, \end{aligned}$$

since $\mathbf{U}^T\mathbf{u}_i = \mathbf{e}_i \in \mathbb{R}^m$ and $\mathbf{v}_j^T\mathbf{V} = \mathbf{e}_j \in \mathbb{R}^n$. Thus, $\text{rank}(\mathbf{A}_k) = k$ by Lemma 1 of Section 7.2, because the assumption $k < r$ ensures that the diagonal entries $\sigma_1, \dots, \sigma_k$ are nonzero. Moreover,

$$\begin{aligned}\|\mathbf{A} - \mathbf{A}_k\| &= \|\mathbf{U}^T(\mathbf{A} - \mathbf{A}_k)\mathbf{V}\| = \|\mathbf{U}^T\mathbf{A}\mathbf{V} - \mathbf{U}^T\mathbf{A}_k\mathbf{V}\| \\ &= \|\text{diag}(\sigma_1, \dots, \sigma_n) - \text{diag}(\sigma_1, \dots, \sigma_k, 0, \dots, 0)\| \\ &= \|\text{diag}(0, \dots, 0, \sigma_{k+1}, \dots, \sigma_n)\| = \sigma_{k+1},\end{aligned}$$

since the matrix norm is invariant under orthogonal transformations according to Lemma 1(c) and since the largest diagonal entry of a diagonal matrix defines its norm by Lemma 4, to be presented shortly.

Next, let us assume that a matrix $\mathbf{B} \in \mathbb{R}^{m,n}$ has rank less than or equal to k and satisfies $\|\mathbf{A} - \mathbf{B}\| < \sigma_{k+1}$. By the dimension Theorem 5.2, we conclude that $\dim \ker(\mathbf{B}) \geq n - k$, and for every vector $\mathbf{0} \neq \mathbf{x} \in \ker(\mathbf{B})$, we have

$$\|\mathbf{A}\mathbf{x}\| = \|(\mathbf{A} - \mathbf{B})\mathbf{x}\| \leq \|\mathbf{A} - \mathbf{B}\| \, \|\mathbf{x}\| < \sigma_{k+1}\|\mathbf{x}\|,$$

since $\mathbf{B}\mathbf{x} = \mathbf{0}$ and $\|\mathbf{A} - \mathbf{B}\| < \sigma_{k+1}$. Here, the middle inequality holds because, by definition,

$$\|\mathbf{A} - \mathbf{B}\| = \max_{\mathbf{0} \neq \mathbf{u} \in \mathbb{R}^n} \frac{\|(\mathbf{A} - \mathbf{B})\mathbf{u}\|}{\|\mathbf{u}\|} \geq \frac{\|(\mathbf{A} - \mathbf{B})\mathbf{x}\|}{\|\mathbf{x}\|}$$

for each nonzero vector $\mathbf{x}$. On the other hand, if $\mathbf{v} \in \text{span}\{\mathbf{v}_1, \dots, \mathbf{v}_{k+1}\} \subset \mathbb{R}^n$, then $\mathbf{v} = \mathbf{V}_{nn}\begin{pmatrix} y_1 \\ \vdots \\ y_{k+1} \\ 0 \\ \vdots \\ 0 \end{pmatrix}$ for the right singular vector matrix $\mathbf{V}_{nn}$. Here, $\|\mathbf{v}\| = \|\mathbf{y}\|$ for $\mathbf{y} := \begin{pmatrix} y_1 \\ \vdots \\ y_{k+1} \\ 0 \\ \vdots \\ 0 \end{pmatrix}$, according to Lemma 1(a). Since $\mathbf{A}\mathbf{V} = \mathbf{U}\mathbf{\Sigma}$ in the SVD of $\mathbf{A}$, we have

$$\begin{aligned}\|\mathbf{A}\mathbf{v}\| = \|\mathbf{A}\mathbf{V}\mathbf{y}\| = \|\mathbf{U}\mathbf{\Sigma}\mathbf{y}\| = \|\mathbf{\Sigma}\mathbf{y}\| &= \left\| \begin{pmatrix} \sigma_1 y_1 \\ \vdots \\ \sigma_{k+1} y_{k+1} \\ 0 \\ \vdots \\ 0 \end{pmatrix} \right\| \\ &= \sqrt{\sum_{i=1}^{k+1} |\sigma_i y_i|^2} \geq \sigma_{k+1}\sqrt{\sum_{i=1}^{k+1} |y_i|^2} = \sigma_{k+1}\|\mathbf{y}\| = \sigma_{k+1}\|\mathbf{v}\|.\end{aligned}$$

Thus, on the subspace ker(**B**) of $\mathbb{R}^n$ of dimension at least $n-k$, we have

$$\|\mathbf{Ax}\| < \sigma_{k+1}\|\mathbf{x}\|,$$

while on the subspace span$\{\mathbf{v}_1, \ldots, \mathbf{v}_{k+1}\} \subset \mathbb{R}^n$ of dimension $k+1$, the opposite inequality, viz.,

$$\|\mathbf{Av}\| \ \geq \sigma_{k+1}\|\mathbf{v}\|,$$

must hold. But these two subspaces intersect in a subspace of positive dimension according to the Corollary of Section 5.1, a contradiction.

Thus, no matrix **B** of rank less than or equal to k can approximate **A** better than the dyadic sum $\mathbf{A}_k$ that is derived from **A**'s SVD. ■

Lemma 4 If $\mathbf{D}_{mn} = \text{diag}(d_1, \ldots, d_n)$ is a diagonal matrix for $m \geq n$, then

$$\|\mathbf{D}\| = \max_{i=1}^{n} |d_i|.$$

◀

Proof According to the definition of the matrix norm, we have

$$\|\mathbf{D}\|^2 = \max_{\|\mathbf{x}\|=1} \|\mathbf{D}x\|^2 = \max_{\|\mathbf{x}\|=1} (d_1x_1)^2 + \cdots + (d_nx_n)^2$$

$$\leq \max_{\|\mathbf{x}\|=1} \sum_i \left(\max_j |d_j|^2 x_i^2\right) = \max_{\|\mathbf{x}\|=1} \left(\max_j |d_j|^2 \sum_i x_i^2\right) = \max_j |d_j|^2.$$

Note that the maximum for $\|\mathbf{D}\|^2$ is achieved for the standard unit vector $\mathbf{x} = \mathbf{e}_{j_0} \in \mathbb{R}^n$, provided that $|d_{j_0}| = \max_j |d_j|$. ■

EXAMPLE 3 The closest rank 1 approximation of $\mathbf{A} = \begin{pmatrix} 0 & -2 & 1 \\ -1 & 1 & 0 \end{pmatrix}$ from Example 1 is given by $\mathbf{A}_1 := \sigma_1 \mathbf{u}_1 \mathbf{v}_1^T = \sqrt{6}\dfrac{\mathbf{x}_2}{\|\mathbf{x}_2\|}\mathbf{y}_2^T$ for the dominant singular vectors $\mathbf{u}_1 := \dfrac{\mathbf{x}_2}{\|\mathbf{x}_2\|} = \dfrac{1}{\sqrt{5}}\begin{pmatrix} -2 \\ 1 \end{pmatrix}$ and $\mathbf{v}_1 := \mathbf{y}_2 = \dfrac{1}{\sqrt{30}}\begin{pmatrix} -1 \\ 5 \\ -2 \end{pmatrix}$ of **A**. Thus,

$$\mathbf{A}_1 = \frac{1}{5}\begin{pmatrix} -2 \\ 1 \end{pmatrix}\begin{pmatrix} -1 & 5 & -2 \end{pmatrix} = \frac{1}{5}\begin{pmatrix} 2 & -10 & 4 \\ -1 & 5 & -2 \end{pmatrix} = \begin{pmatrix} 0.4 & -2 & 0.8 \\ -0.2 & 1 & -0.4 \end{pmatrix}.$$

Observe that $\mathbf{A} - \mathbf{A}_1 = \begin{pmatrix} -0.4 & 0 & 0.2 \\ -0.8 & 0 & -0.4 \end{pmatrix}$ has the singular value decomposition $\mathbf{A} - \mathbf{A}_1 = \mathbf{x}_1\mathbf{y}_1^T$ for the two singular vectors $\mathbf{x}_1 = \frac{1}{\sqrt{5}}\begin{pmatrix} 1 \\ 2 \end{pmatrix} \in \mathbb{R}^2$ and $\mathbf{y}_1^T = \frac{1}{\sqrt{5}}\begin{pmatrix} -2 & 0 & 1 \end{pmatrix} \in \mathbb{R}^3$ of **A**. Thus, the matrix difference $\mathbf{A} - \mathbf{A}_1$ has rank 1, and its maximal singular value is equal to 1, (i.e., the matrix norm of $\mathbf{A} - \mathbf{A}_1$ is equal to 1 in accordance with Lemma 3 and Theorem 12.3). ◀

The **rank** of a matrix **A** was defined in Chapters 2 and 3 as the number of nonzero pivots of a row echelon form reduction of **A**. The decision when to call a small-magnitude matrix entry nonzero or zero is very difficult to make in practice. The SVD offers us a more reliable numerical criterion for assigning matrix rank.

DEFINITION 3 If $\mathbf{U}^T\mathbf{A}\mathbf{V} = \text{diag}(\sigma_i)$ is the SVD of a matrix $\mathbf{A}_{mn}$ for $m \geq n$ and $\sigma_1 \geq \sigma_2 \geq \cdots \geq \sigma_r > \sigma_{r+1} \geq \cdots \geq 0$ with r minimal so that $\sigma_r > 10^{12}\sigma_{r+1}$, then we say that $\mathbf{A}$ has the **numerical rank** r, provided that the SVD of $\mathbf{A}$ has been computed in double precision with around 16 valid digits.

A quantifiably large and sudden drop in the relative size of $\mathbf{A}$'s singular values σ_i indicates that the subsequent singular values of $\mathbf{A}$ are most likely nonzero only because of **rounding errors** in the computations. Indeed, if we replace $\mathbf{A}$ by its rank r dyadic approximation $\mathbf{A}_r := \sum_{i=1}^{r} \sigma_i \mathbf{u}_i \mathbf{v}_i^T$, with r set as in Definition 3, then, from Theorem 12.3, we have made a negligibly small error of at most $\|\mathbf{A} - \mathbf{A}_r\| = \sigma_{r-1} < 10^{-12}\sigma_r$ when using $\mathbf{A}_r$ of rank r in place of $\mathbf{A}$.

Next, we look at approximate solutions for **unsolvable linear systems**. Recall from Theorem 3.1 that a system of linear equations $\mathbf{Ax} = \mathbf{b}$ is unsolvable if and only if $\text{rank}(\mathbf{A} \mid \mathbf{b}) > \text{rank}(\mathbf{A})$. If a linear system is unsolvable, then $\mathbf{b} \notin \text{im}(\mathbf{A})$. In this case, one might solve the linear system as best as one can; that is, one may want to try and find a vector $\mathbf{x}_0 \in \mathbb{R}^n$ that minimizes the error

$$\|\mathbf{A}\mathbf{x}_0 - \mathbf{b}\| = \min_{\mathbf{x} \in \mathbb{R}^n} \|\mathbf{A}\mathbf{x} - \mathbf{b}\|.$$

The SVD gives an explicit solution to such **least squares problems**.

Theorem 12.4 **(Least Squares Solution)** Let $\mathbf{A}_{mn} = \sum_{i=1}^{r} \sigma_i \mathbf{u}_i \mathbf{v}_i^T = \mathbf{U}\Sigma\mathbf{V}^T$ be the SVD of a matrix $\mathbf{A}$ of rank r, where $\mathbf{U}_{mm} = \begin{pmatrix} | & & | \\ \mathbf{u}_1 & \dots & \mathbf{u}_m \\ | & & | \end{pmatrix}$ and $\mathbf{V}_{nn} = \begin{pmatrix} | & & | \\ \mathbf{v}_1 & \dots & \mathbf{v}_n \\ | & & | \end{pmatrix}$ are expressed column-wise.

Then, for every $\mathbf{b} \in \mathbb{R}^m$, the least squares problem $\min_{\mathbf{x} \in \mathbb{R}^n} \|\mathbf{Ax} - \mathbf{b}\|$ has

$$\mathbf{x}_{\text{LS}} := \sum_{i=,1}^{r} \left(\frac{\mathbf{u}_i^T\mathbf{b}}{\sigma_i} \right) \mathbf{v}_i \in \mathbb{R}^n$$

as its **least squares solution** closest to the origin $\mathbf{0} \in \mathbb{R}^n$. $\mathbf{x}_{\text{LS}}$ solves the linear system $\mathbf{Ax} = \mathbf{b}$, except for the **unavoidable error**

$$\sqrt{\sum_{i=r+1}^{m} (\mathbf{u}_i^T\mathbf{b})^2} = \|\mathbf{A}\mathbf{x}_{\text{LS}} - \mathbf{b}\|.$$ ◀

Proof Assume that $\mathbf{U}^T\mathbf{A}\mathbf{V} = \text{diag}(\sigma_1, \ldots, \sigma_r, 0, \ldots, 0) = \boldsymbol{\Sigma}_{mn}$ is the SVD of $\mathbf{A}$ with $\sigma_i \neq 0$ for $i = 1, \ldots, r$ and $r = \text{rank}(\mathbf{A})$. Since the length of a vector remains the same under an orthogonal mapping, we have

$$\begin{aligned}\|\mathbf{Ax} - \mathbf{b}\|^2 &= \|\mathbf{U}^T\mathbf{Ax} - \mathbf{U}^T\mathbf{b}\|^2 = \|\mathbf{U}^T\mathbf{AV}(\mathbf{V}^T\mathbf{x}) - \mathbf{U}^T\mathbf{b}\|^2 \\ &= \|\boldsymbol{\Sigma}\mathbf{V}^T\mathbf{x} - \mathbf{U}^T\mathbf{b}\|^2 = \sum_{i=1}^{r}(\sigma_i\alpha_i - \mathbf{u}_i^T\mathbf{b})^2 + \sum_{i=r+1}^{m}(\mathbf{u}_i^T\mathbf{b})^2, \qquad (12.5)\end{aligned}$$

where the $\mathbf{u}_i$ denote the columns of $\mathbf{U}$ and $\begin{pmatrix}\alpha_1\\ \vdots \\ \alpha_n\end{pmatrix} := \mathbf{V}^T\mathbf{x}$. The second summand, $\sum_{i=r+1}^{m}(\mathbf{u}_i^T\mathbf{b})^2$, in (12.5) represents the "unavoidable error" of the least squares problem. Since any solution vector $\mathbf{x} = \mathbf{V}\begin{pmatrix}\alpha_1\\ \vdots \\ \alpha_n\end{pmatrix}$ can affect only the values α_i and not the unavoidable error, we note from (12.5) that

$$\begin{aligned}\mathbf{x} = \mathbf{x}_{\text{LS}} &\text{ if and only if } \sigma_i\alpha_i = \mathbf{u}_i^T\mathbf{b} \quad \text{or} \\ &\text{if and only if } \alpha_i = (\mathbf{V}^T\mathbf{x})_i = \mathbf{v}_i^T\mathbf{x} = \frac{\mathbf{u}_i^T\mathbf{b}}{\sigma_i} \quad \text{for } i = 1, \ldots, r. \qquad (12.6)\end{aligned}$$

This is so because the first sum in (12.5) is zero for this choice of $\mathbf{x}_{\text{LS}}$.
Writing the solution vector $\mathbf{x}$ in terms of the orthonormal basis $\{\mathbf{v}_j\}$ for the domain $\mathbb{R}^n$ gives $\mathbf{x} = \mathbf{x}_{\text{LS}}$ the form $\mathbf{x}_{\text{LS}} = \mathbf{V}\begin{pmatrix}\alpha_1\\ \vdots \\ \alpha_n\end{pmatrix}$ with $\alpha_i = \dfrac{\mathbf{u}_i^T\mathbf{b}}{\sigma_i}$ for $i \leq r$ and $\alpha_i = 0$ for $i > r$. This describes the least squares solution

$$\mathbf{x}_{\text{LS}} = \sum_{i=1}^{r}\left(\frac{\mathbf{u}_i^T\,\mathbf{b}}{\sigma_i}\right)\mathbf{v}_i$$

that is closest to the origin in $\mathbb{R}^n$.

Note that if $r < n$, then there are many choices for the coordinates of $\mathbf{x}_{\text{LS}}$ in terms of the orthonormal basis $\{\mathbf{v}_i\}$ of $\mathbb{R}^n$, since, for the indices $i = r+1, \ldots, n$ the $\mathcal{V}$–coordinates α_i of $\mathbf{x}_{\text{LS}}$ can be arbitrarily assigned without changing the value of $\|\mathbf{Ax}_{\text{LS}} - \mathbf{b}\|$.

This is clearly not the case if $r = n$ (i.e., in the full rank case), when the solution to the least squares problem is unique. As the **rank-deficient** least squares problem leaves a certain freedom of choice, the customary option involves choosing the vector $\mathbf{x}_{\text{LS}}$ that is closest to the origin—that is, setting all $\mathcal{V}$-coordinates α_i of $\mathbf{x}_{\text{LS}}$ equal to zero for $i > r = \text{rank}(\mathbf{A})$. ■

Note that for this choice of $\mathbf{x}_{LS}$, both $\|\mathbf{A}\mathbf{x}_{LS} - \mathbf{b}\|$ and $\|\mathbf{x}_{LS}\|$ are minimized. If $\mathbf{A} \in \mathbb{R}^{m,n}$ is rank deficient, then there are many vectors $\mathbf{x}$ with minimal residual norm $\|\mathbf{A}\mathbf{x} - \mathbf{b}\|$, and it follows that $\|\mathbf{x}\| \geq \|\mathbf{x}_{LS}\|$ in general.

An example of the least squares solution of an unsolvable rank-deficient linear system is included in Section 12.3.

The singular value decomposition is a very powerful tool of matrix analysis.

12.2.P Problems

1. (a) Find the values of the matrix norm of each matrix in Problem 1 of Section 12.1.P.

(b) What is the best rank-1 matrix approximation of the matrices **A**, **B**, **H**, and **K** in Problem 1 of Section 12.1.P.

(c) What is the best rank-2 matrix approximation of the matrices **A**, **C**, and **F** in Problem 1 of Section 12.1.P.

2. (a) Find the singular-value decomposition $\mathbf{U}^T\mathbf{A}\mathbf{V} = \mathbf{\Sigma}$ for

$$\mathbf{A} = \begin{pmatrix} 1 & 1 \\ 2 & 1 \\ -1 & -2 \end{pmatrix}.$$

(b) How close is **A** to a rank 1 matrix?

(c) Which matrix of rank 1 is closest to **A**?

3. Assume that $\mathbf{A} =$

$$\mathbf{U} \begin{pmatrix} 22 &&&&& \\ & 13 &&&& \\ && 5 &&& \\ &&& 0.44 && \\ &&&& 0.0002 & \\ &&&&& 0 \end{pmatrix}_{6,6} \mathbf{V}^T$$

for two orthogonal matrices **U** and **V**.

(a) What is the size of **A**? What are the sizes of **U** and **V**?

(b) What is the matrix norm of **A**?

(c) How close is **A** to a rank-2 matrix? To a rank-4 matrix?

(d) Is **A** nonsingular or singular?

(e) What is **A**'s rank?

4. Solve the following unsolvable linear systems in the least squares sense without using MATLAB:

(a) $\begin{pmatrix} 2 & 2 & 2 \\ 1 & 1 & 1 \end{pmatrix} \mathbf{x} = \begin{pmatrix} 0 \\ 1 \end{pmatrix}$;

(b) $\begin{pmatrix} 1 & 4 \\ 2 & 8 \end{pmatrix} \mathbf{x} = \mathbf{e}_2$;

(c) $\begin{pmatrix} 1 & -1 \\ 2 & -2 \\ -3 & 3 \end{pmatrix} \mathbf{x} = \begin{pmatrix} 1 \\ -2 \\ 1 \end{pmatrix}$.

5. Using Theorem 12.4, solve the least squares problem $\min_{\mathbf{x} \in \mathbb{R}^2} \|\mathbf{A}\mathbf{x} - \mathbf{b}\|$ for $\mathbf{A} = \begin{pmatrix} 1 & 3 \\ 3 & 5 \\ 5 & 7 \end{pmatrix}$, $\mathbf{b} = \begin{pmatrix} 1 \\ 2 \\ 3 \end{pmatrix}$ from the SVD of **A**.

6. Let $n \geq 5$ and $\mathbf{A}_{mn}$ be given.

(a) Write out the SVD of **A** in terms of its (unknown) singular values and singular vectors. How many of **A**'s singular values are determined by the additional information that rank $\mathbf{A} = n - 4$? What are those singular values?

(b) How many rank 1 matrices suffice to describe **A** of rank $n - 4$ as a matrix sum? Using part (a), write out an explicit formula for this sum.

(c) How close is **A** to a matrix of rank $n - 5$ in terms of the matrix norm and the singular values of **A**?

7. Express the singular values of a symmetric matrix $\mathbf{A} = \mathbf{A}^T$ in terms of the eigenvalues of **A**. What are the singular vectors of a symmetric matrix $\mathbf{A} = \mathbf{A}^T$?

8. Use the previous problem to show that: if $\mathbf{A} = \mathbf{A}^T$ is positive semidefinite, then the eigenvalues of **A** are its singular values.

9. Find all singular values and singular vectors of the matrices $\mathbf{A} = \begin{pmatrix} 0 & & 1 \\ & \cdot^{\cdot^{\cdot}} & \\ 1 & & 0 \end{pmatrix}_{nn}$, $\mathbf{B} = \begin{pmatrix} 1 & \dots & 1 \\ \vdots & & \vdots \\ 1 & \dots & 1 \end{pmatrix}_{nn}$,

and $\mathbf{C} = \begin{pmatrix} 0 & 1 & \dots & \dots & 1 \\ 1 & 0 & 1 & & \vdots \\ \vdots & \ddots & \ddots & \ddots & \vdots \\ \vdots & & \ddots & 0 & 1 \\ 1 & \dots & \dots & 1 & 0 \end{pmatrix}_{nn}$ for arbitrary dimensions n. (*Hint*: Use a previous problem.)

10. Find the singular values of

$$\mathbf{A} = \begin{pmatrix} 0 & \sqrt{2} & -2 \\ 0 & \sqrt{2} & 2 \\ 0 & 0 & 0 \\ 0 & 0 & \sqrt{8} \end{pmatrix}$$

from the square roots of the eigenvalues of $\mathbf{A}^T\mathbf{A}$.

11. Consider a **Jordan block** $\mathbf{J} = \begin{pmatrix} 0 & 1 & & & 0 \\ 0 & 0 & 1 & & \\ & & \ddots & \ddots & \\ & & & \ddots & 1 \\ 0 & & & & 0 \end{pmatrix}_{nn}$ for the eigenvalue 0.

(a) Find the singular values of $\mathbf{J}$.

(b) What is the matrix norm $\|\mathbf{J}\|$?

(c) Find the powers $\mathbf{J}^k$ for $k = 1, 2, \ldots, n, n+1, \ldots$.

(d) What are the singular values of $\mathbf{J}^k$ for $k = 1, 2, \ldots, n, n+1, \ldots$.

(e) What are the matrix norms $\|\mathbf{J}^k\|$ for $k = 1, 2, \ldots, n, n+1, \ldots$.

12. Find the best rank 1 approximation $\mathbf{A}_1$ of the matrix $\mathbf{A} = \begin{pmatrix} 1 & 0 \\ 0 & 2 \\ 1 & 0 \end{pmatrix}$.

13. Express $\mathbf{A} = \begin{pmatrix} 0 & 2 & -1 \\ 0 & 0 & 1 \\ 0 & 1 & 1 \end{pmatrix}$ as two different sums of two dyads.

14. What is the distance between the two matrices $\mathbf{A} = \begin{pmatrix} 2 & -1 \\ 1 & -1 \end{pmatrix}$ and $\mathbf{I}_2$?

15. Find several best rank 1 approximations of the matrix $\mathbf{A} = \begin{pmatrix} 0 & 1 & 0 \\ 0 & 0 & 1 \\ 0 & 0 & 0 \end{pmatrix}$, if possible.

16. How many matrices of rank 1 approximate $\mathbf{B} = \begin{pmatrix} 0 & 0 & 0 \\ 1 & 0 & 0 \\ 0 & 2 & 0 \end{pmatrix}$ best?

17. Derive a dyadic expansion formula involving the eigenvalues and eigenvectors of a square diagonalizable matrix $\mathbf{A}_{nn}$. (*Hint*: Refer to the proof of Lemma 3.)

12.3 Applications

Geometry, Data Compression, Least Squares, and MATLAB

We give several applications of the singular value decomposition.

The eigenanalysis and singular value decomposition of a matrix $\mathbf{A}$ each give valuable geometric information about the linear transformation $\mathbf{x} \mapsto \mathbf{A}\mathbf{x}$. Recall from Theorem 8.3 and Section 11.3 that the volume change of solids $S \subset \mathbb{R}^n$ under a linear mapping $\mathbf{x} \mapsto \mathbf{A}\mathbf{x}$ is governed by the factor $|\det(\mathbf{A}))| = \left|\prod_i \lambda_i\right|$ (i.e., by the absolute value of the product of the eigenvalues λ_i of $\mathbf{A}_{nn}$). A similar result holds for the singular values of $\mathbf{A}_{nn}$.

Corollary 2 (a) For a square matrix $\mathbf{A}_{nn}$, the volume change factor for solids under the linear transformation $S \subset \mathbb{R}^n \mapsto \mathbf{A}(S) \subset \mathbb{R}^n$ is given by

$$|\det(\mathbf{A})| = \prod_i \sigma_i,$$

i.e., by the product of the singular values of $\mathbf{A}$.

(b) The absolute value of the product of all eigenvalues of a square matrix $\mathbf{A}_{nn}$ is equal to the product of all singular values of the matrix. ◀

Proof Look at the SVD of the square matrix $\mathbf{A}_{nn} = \mathbf{U}\mathbf{\Sigma}_{nn}\mathbf{V}^T$. Here, $\mathbf{U}_{nn}$ and $\mathbf{V}_{nn}$ are both orthogonal matrices with $|\det(\mathbf{U})| = 1 = |\det(\mathbf{V})|$, since

$$1 = \det(\mathbf{I}_n) = \det(\mathbf{U}^T\mathbf{U}) = \det(\mathbf{U}^T)\det(\mathbf{U}) = (\det(\mathbf{U}))^2.$$

Thus, for the eigenvalues λ_i and the singular values σ_i of $\mathbf{A}$, we have

$$\left|\prod_i \lambda_i\right| = |\det(\mathbf{A})| = |\det(\mathbf{U}\mathbf{\Sigma}\mathbf{V}^T)| = |\det(\mathbf{U})|\,|\det(\mathbf{\Sigma})|\,|\det(\mathbf{V}^T)|$$
$$= \det(\mathbf{\Sigma}) = \prod_i \sigma_i. \qquad ■$$

The singular values and singular vectors of a real 2×2 matrix $\mathbf{A}$ allow us to describe the transformation of the plane under the mapping by $\mathbf{A}$ visually in a similar, yet different, manner, as Example 12 of Section 1.3 did for mapping the standard unit vectors under $\mathbf{A}$ and as Example 4 of Section 9.2 did for the eigenvectors of $\mathbf{A}$.

EXAMPLE 4 If $\mathbf{U}\mathbf{\Sigma}_{mn}\mathbf{V}^T$ is the SVD of $\mathbf{A}_{mn}$, then $\mathbf{AV} = \mathbf{U\Sigma}$. Thus, $\mathbf{A}$ maps the ONB $\{\mathbf{v}_1, \dots, \mathbf{v}_n\} \subset \mathbb{R}^n$ that is contained in the columns of $\mathbf{V}$ to the orthogonal set of vectors $\{\sigma_1\mathbf{u}_1, \dots, \sigma_n\mathbf{u}_n\} \subset \mathbb{R}^m$. Here, $\sigma_1 = \max_{\substack{\|\mathbf{x}\|=1\\ \mathbf{x}\in\mathbb{R}^n}} \|\mathbf{Ax}\|$, from the proof of the SVD Theorem 12.1. Thus, $\mathbf{v}_1$ is the vector in $\mathbb{R}^n$ whose image is maximally stretched under the map $\mathbf{A}$, while $\mathbf{v}_n$ is minimally stretched under $\mathbf{x} \mapsto \mathbf{Ax}$. Specifically, for $m = n = 2$, a matrix $\mathbf{A}_{2\times2}$ maps the unit circle $\left\{\begin{pmatrix} x_1 \\ x_2 \end{pmatrix} \,\middle|\, x_1^2 + x_2^2 = 1\right\}$ of $\mathbb{R}^2$ to the planar ellipse with major and minor semiaxes $\sigma_1\mathbf{u}_1$ and $\sigma_2\mathbf{u}_2$. These observations give us a new tool to view a linear map $\mathbf{x} \mapsto \mathbf{Ax}$. We use the matrices from Example 12 of Section 1.3 and Example 4 in Section 9.2.

(a) The matrix $\mathbf{A} = \begin{pmatrix} 2 & -1 \\ -2 & -2 \end{pmatrix}$ has the SVD

$$\mathbf{U\Sigma V}^T = \begin{pmatrix} 0.44721 & -0.89443 \\ -0.89443 & -0.44721 \end{pmatrix}\begin{pmatrix} 3 & 0 \\ 0 & 2 \end{pmatrix}\begin{pmatrix} 0.89443 & 0.44721 \\ -0.44721 & 0.89443 \end{pmatrix}.$$

Thus, $\mathbf{A}$ has the singular values 3 and 2. Since $\mathbf{AV} = \mathbf{U\Sigma}$, the linear transformation $\mathbf{x} \mapsto \mathbf{Ax}$ induced by $\mathbf{A}$ maps the unit column vectors $\mathbf{v}_i$ of $\mathbf{V} = \begin{pmatrix} 0.89443 & -0.44721 \\ 0.44721 & 0.89443 \end{pmatrix}$ to the mutually orthogonal vectors $\mathbf{Av}_1 = \sigma_1\mathbf{u}_1 = 3\begin{pmatrix} 0.44721 \\ -0.89443 \end{pmatrix}$ and $\mathbf{Av}_2 = \sigma_2\mathbf{u}_2 = 2\begin{pmatrix} -0.89443 \\ -0.44721 \end{pmatrix}$. Hence, $\mathbf{A}$ maps the unit circle in $\mathbb{R}^2$ to an ellipse with major semiaxis of length 3 and 2 in the directions of $\mathbf{u}_1$ and $\mathbf{u}_2$ as in Figure 12-1.

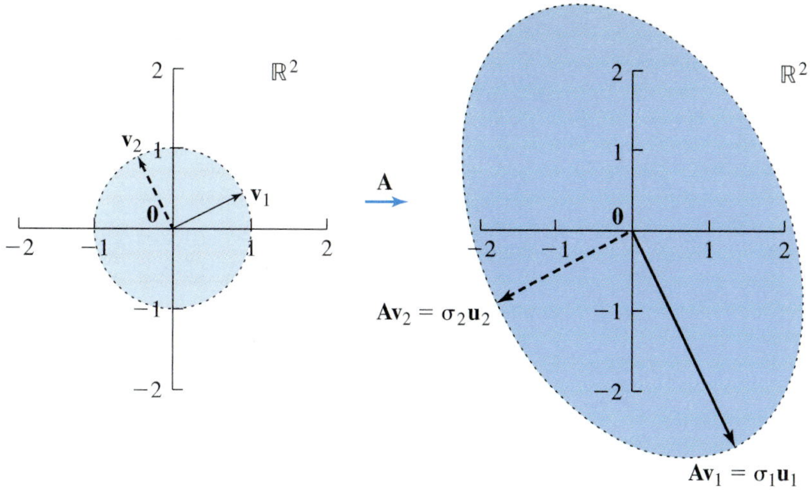

Figure 12-1

(b) The matrix $\mathbf{B} = \begin{pmatrix} 1 & 1 \\ 0 & 1.5 \end{pmatrix}$ has the SVD

$$\mathbf{U}\boldsymbol{\Sigma}\mathbf{V}^T = \begin{pmatrix} 0.67711 & 0.73588 \\ 0.73588 & -0.67711 \end{pmatrix} \begin{pmatrix} 1.9053 & 0 \\ 0 & 0.78727 \end{pmatrix} \cdot \begin{pmatrix} 0.35538 & 0.93472 \\ 0.93472 & -0.35538 \end{pmatrix}.$$

The right singular vectors in the columns of $\mathbf{V}$ map to $1.9053\mathbf{u}_1$ and $0.78727\mathbf{u}_2$ for the respective singular values $\sqrt{\frac{17}{8} \pm \sqrt{\frac{17^2}{8^2} - \frac{9}{4}}}$, as computed from the eigenvalues $\frac{17}{8} \pm \sqrt{\frac{17^2}{8^2} - \frac{9}{4}}$ of $\mathbf{B}^T\mathbf{B}$ using Theorem 12.2 (See Figure 12-2).

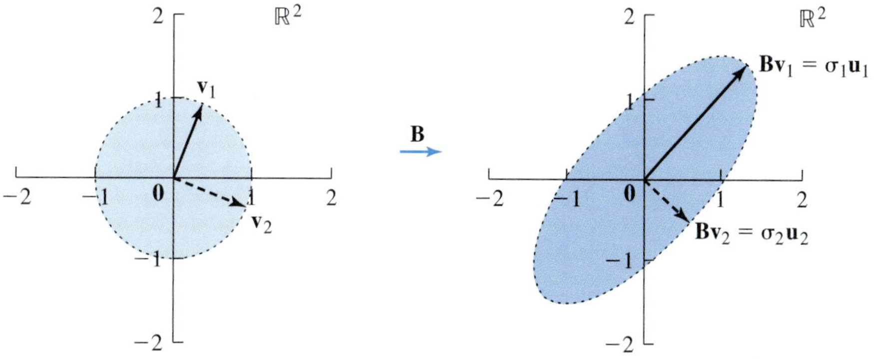

Figure 12-2

(c) For the dilation-with-rotation matrix $\mathbf{C} = \begin{pmatrix} a & -b \\ b & a \end{pmatrix}$, the SVD is given by

$$\mathbf{U}\mathbf{\Sigma}\mathbf{V}^T = \left(\frac{1}{\sqrt{a^2+b^2}}\mathbf{C}\right)\begin{pmatrix} \sqrt{a^2+b^2} & 0 \\ 0 & \sqrt{a^2+b^2} \end{pmatrix}\mathbf{I}_2.$$

Thus, both singular values of $\mathbf{C}$ are equal to the stretch factor $d := \sqrt{a^2+b^2}$ and the right singular vectors $\mathbf{e}_1$ and $\mathbf{e}_2$ of $\mathbf{C}$ and the unit circle of the plane map as shown in Figure 12-3.

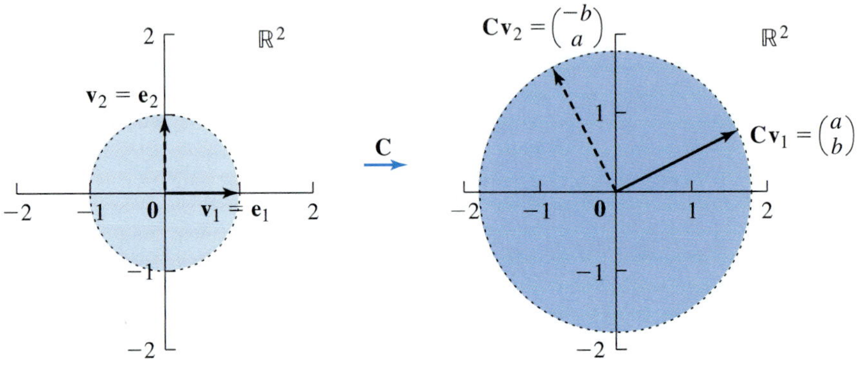

Figure 12-3

Note that the image of the unit circle in $\mathbb{R}^2$ under $\mathbf{C}$ is the dilated unit circle with radius $d = \sqrt{a^2+b^2}$.

(d) For $\mathbf{D} = \begin{pmatrix} 2 & 0 \\ 0 & 0 \end{pmatrix}$, the SVD is most trivially $\mathbf{I}_2\mathbf{D}\mathbf{I}_2$, and Figure 1-9 in Section 1.3 describes the geometry completely.

(e) The SVD of $\mathbf{E} = \begin{pmatrix} 0 & 2 \\ 0 & 0 \end{pmatrix}$ is $\mathbf{I}_2 \begin{pmatrix} 2 & 0 \\ 0 & 0 \end{pmatrix}\begin{pmatrix} 0 & 1 \\ 1 & 0 \end{pmatrix}$. The image of the unit circle in the plane under the mapping by $\mathbf{E}$ is the interval $[-2, 2]$ on the real axis in $\mathbb{R}^2$, and Figure 1-10 illustrates the geometry.

Note the difference between the figures in this example and those in Section 9.2. The SVD of a matrix $\mathbf{A}_{2\times 2}$ describes the maximally and minimally stretched vectors under the map $\mathbf{x} \mapsto \mathbf{A}\mathbf{x}$, while the eigenanalysis of a diagonalizable matrix $\mathbf{A}$ finds the vectors of $\mathbb{R}^2$ whose directions stay fixed under the mapping by $\mathbf{A}$. ◀

MATLAB uses the command `[U,S,V] = svd(A)` to compute the diagonal matrix $\mathbf{S}$ of singular values and the left and right singular vector matrices $\mathbf{U}$ and $\mathbf{V}$ for which $\mathbf{A} = \mathbf{U}\mathbf{S}\mathbf{V}^*$. A simple call of `svd(A)` returns the list of singular values of $\mathbf{A}$ only.

The next application of the SVD involves data compression.

EXAMPLE 5 (**Data transfer**) Assume that the "word" ABC is given by the 16×32 "pixel matrix" shown in Figure 12-4.

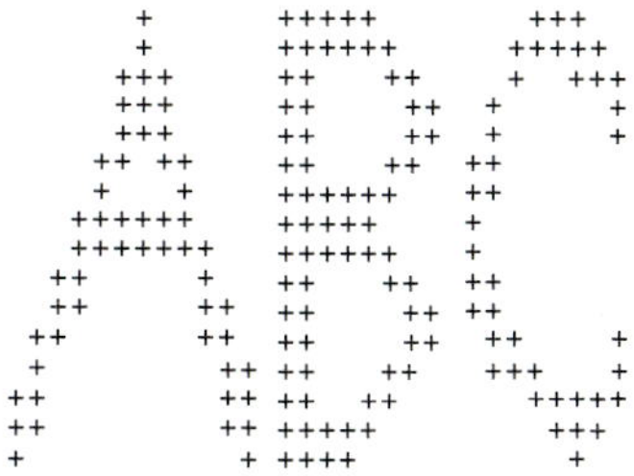

Figure 12-4

Here, the pixel entries are stored as either 1 or 0 in memory in a 16×32 integer matrix. In the figure, we have displayed the word file for ABC by invoking `format +` in MATLAB.

In order to transmit the pixel file for ABC economically, we try to reduce the number $16 \cdot 32 = 512$ of its data entries in the pixel matrix before sending the file. We need to do this in such a way that we can still retrieve the word ABC legibly at the receiving station. For this, we can compute the SVD of the pixel matrix and send out a limited rank singular value–singular vector description of the matrix. At the receiving end, we reconstruct the word ABC more or less legibly as an approximate sum of dyads from the transmitted partial data by using Theorem 12.3.

This approach is based on Lemma 3 and Theorem 12.3 from Section 12.2. For this, denote the real 16×32 matrix that represents the word ABC in zeros and ones by $\mathbf{A}$. The SVD of $\mathbf{A}_{16\times 32} = \mathbf{U}_{16\times 16}\mathbf{\Sigma}_{16\times 32}\mathbf{V}^T_{32\times 32}$ allows us to express $\mathbf{A} = \sum_{i=1}^{16} \sigma_i \mathbf{u}_i \mathbf{v}_i^T$ as the sum of 16 dyads, according to Lemma 3. To achieve the envisioned data compression, we approximate $\mathbf{A}$ by

$$\mathbf{A}_k := \sum_{i=1}^{k} \sigma_i \mathbf{u}_i \mathbf{v}_i^T$$

for $k \ll 16$. The dyadic expression for $\mathbf{A}_k$ uses the singular values $\sigma_1, \ldots, \sigma_k$ of $\mathbf{A}$ and the singular vectors $\mathbf{u}_i$ and $\mathbf{v}_i$ of $\mathbf{A}$ for $i < k$. In our case, the singular values of $\mathbf{A}$ are the 16 ordered numbers 9.0750, 5.0255, 4.6128, 3.4966, 2.6576, 2.3940, 2.0145, 1.8786, 1.6255, 1.3830, 1.0475, 0.8270, 0.7378, 0.6601, 0.4326, and 0.0000.

In view of Theorem 12.3, any short truncation of the foregoing singular vector dyadic sum expansion of $\mathbf{A}$ will create relatively large numerical errors in the approximation $\mathbf{A}_k$. However, for $k = 4$ and 6, we can already reconstruct the word ABC quite legibly from its shortened dyadic singular vector representation, as shown in Figure 12-5.

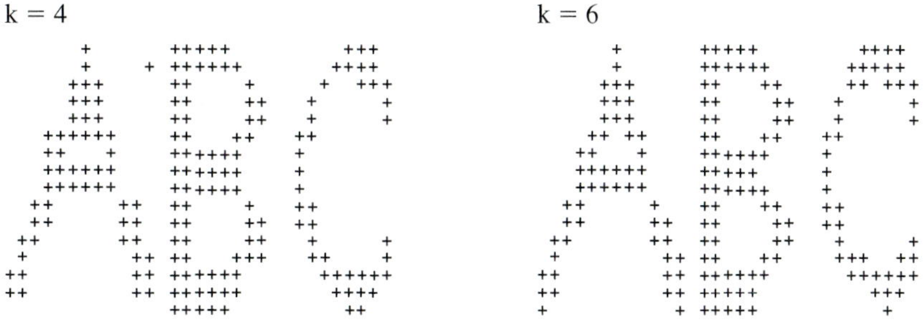

Figure 12-5

For $k = 4$, we need to transmit four singular vectors of length 16, four of length 32, and the four leading singular values of $\mathbf{A}$, requiring $4(16+32)+4 = 156$ data entries. This amounts to a data reduction rate of $\frac{512-156}{512} \approx 69.5\%$ and seems very efficient. However, in modern commercially used data transfer schemes, other compression algorithms can achieve much better reduction rates and better readability results. ◀

The final example treats systems of linear equations $\mathbf{Ax} = \mathbf{b}$ that cannot be solved in the sense of Chapter 3. It computes the least squares solution $\mathbf{x}_{\text{LS}}$ from the SVD of the system matrix $\mathbf{A}$.

EXAMPLE 6 (**Unsolvable system of linear equations**) Solve the linear system $\mathbf{Ax} = \mathbf{b}$ with

$$\mathbf{A} = \begin{pmatrix} 1 & 2 & 3 \\ 2 & 3 & 5 \\ -1 & 1 & 0 \\ 4 & 1 & 5 \end{pmatrix} \quad \text{and} \quad \mathbf{b} = \begin{pmatrix} 1 \\ 3 \\ -1 \\ 1 \end{pmatrix}.$$

The REF of the augmented matrix $(\mathbf{A} \mid \mathbf{b})$ computes as follows:

A			**b**	
[1]	2	3	1	
[2]	3	5	3	$-2\ row_1$
[−1]	1	0	−1	$+\ row_1$
[4]	1	5	1	$-4\ row_1$
[1]	2	3	1	
0	[−1]	−1	1	
0	[3]	3	0	$+3\ row_2$
0	[−7]	−7	−3	$-7\ row_2$
[1]	2	3	1	
0	[−1]	−1	1	
0	0	0	[3]	← *pivot in* $\widehat{\mathbf{b}}$: *system is unsolvable by Theorem 3.1*
0	0	0	−10	

Since the linear system $\mathbf{Ax} = \mathbf{b}$ is unsolvable, we try to ignore the trailing inconsistent rows in $(\mathbf{R} \mid \widehat{\mathbf{b}})$ at first. If we solve only the top two equations of the REF $(\mathbf{R} \mid \widehat{\mathbf{b}})$, how large is the error?

The top 2×3 row-reduced system $\begin{pmatrix} 1 & 2 & 3 \\ 0 & -1 & -1 \end{pmatrix} \mathbf{x} = \begin{pmatrix} 1 \\ 1 \end{pmatrix}$ has the particular solution $\mathbf{x} = \begin{pmatrix} 3 \\ -1 \\ 0 \end{pmatrix}$. The associated error vector $\mathbf{Ax} - \mathbf{b}$ is $\begin{pmatrix} 1 \\ 3 \\ -4 \\ 3 \end{pmatrix}$

$$- \begin{pmatrix} 1 \\ 3 \\ -1 \\ 1 \end{pmatrix} = \begin{pmatrix} 0 \\ 0 \\ -3 \\ 2 \end{pmatrix}, \text{ of length } \sqrt{9+4} = \sqrt{13} \doteq 3.6056.$$

If, however, we use the SVD induced least squares solution and solve the problem $\min_{\mathbf{x}\in\mathbb{R}^n} \|\mathbf{Ax}-\mathbf{b}\|$ instead, we can utilize MATLAB and invoke `A\b`. This command uses the SVD of $\mathbf{A}$ and finds the least squares solution $\mathbf{x}_{LS}$ for nonsquare system matrices $\mathbf{A}_{mn}$ that do not have full rank, such as our matrix $\mathbf{A}_{4,3}$ of rank 2. (See Section 10.3 on how to solve full rank least squares problems by using the QR matrix factorization.) Here, we obtain $\mathbf{x}_{LS} = \begin{pmatrix} 0 \\ 0.244 \\ 0.282 \end{pmatrix}$, with the associated unavoidable error $\|\mathbf{Ax}_{LS}-\mathbf{b}\| \doteq 1.6802$, which more than halves our previous naive error. ◀

Finally, we interpret Example 6 in terms of subspaces: The subspace $\text{im}(\mathbf{A}) = \{\mathbf{Ax} \mid \mathbf{x} \in \mathbb{R}^3\}$ is the column space of the 4×3 matrix $\mathbf{A}$. The REF of $(\mathbf{A} \mid \mathbf{b})$ shows that $\mathbf{b} \notin \text{im}(\mathbf{A})$. By using the SVD and calling `A\b` in MATLAB, we find the least squares solution $\mathbf{x}_{LS}$ and the vector $\mathbf{Ax}_{LS} \in \text{im}(\mathbf{A})$ that is closest to our given right-hand side $\mathbf{b} = \begin{pmatrix} 1 \\ 3 \\ -1 \\ 1 \end{pmatrix}$, namely, $\mathbf{Ax}_{LS} = \begin{pmatrix} 1.3349 \\ 2.1435 \\ 0.2440 \\ 1.6555 \end{pmatrix}$. That $\mathbf{Ax}_{LS}$ is closest to $\mathbf{b}$ means, geometrically, that the connecting vector $\mathbf{Ax}_{LS}-\mathbf{b}$ must be perpendicular to the plane spanned by the columns of $\mathbf{A}$ (i.e., it is perpendicular to the column space of $\mathbf{A}$). Thus, $\mathbf{Ax}_{LS}-\mathbf{b} \in \ker\mathbf{A}^T$, according to Section 4.1. This can readily be verified (to within small rounding errors) in MATLAB by using the convoluted command

```
A'*(A*(A\b)-b)
```

The output contains the dot products of the columns of $\mathbf{A}$ (in the rows of the transpose `A'`) with the projection vector $\mathbf{Ax}_{LS}-\mathbf{b}$ of $\mathbf{b}$ onto the column space of $\mathbf{A}$ (in the MATLAB expression `A*(A\b)-b`). The command produces a vector with three components, each of near machine constant order magnitude $\approx 10^{-15}$ for the data of Example 6, indicating that $\mathbf{Ax}_{LS}-\mathbf{b}$ is numerically orthogonal to each of the columns of $\mathbf{A}$.

A short note of caution is in order regarding MATLAB's inability to solve least squares problems for linear systems of equations $\mathbf{Ax} = \mathbf{b}$ with a singular <u>and</u> square system matrix $\mathbf{A}_{nn}$ via the command `A\b`.

In the square matrix case, MATLAB's `A\b` command uses neither the SVD nor the QR decomposition. (See Section 10.3.) Instead, it only uses row reduction. Thus, MATLAB will give no output, other than the reply "matrix is singular to working precision," which is correct. To circumvent this lack of output in the singular square case, one must resort to a trick: Simply augment the singular or near-singular square matrix $\mathbf{A}_{nn}$ by a row of zeros via the command `A = [A;zeros(1,n)]`, and add one extra right-hand-side entry with value zero to $\mathbf{b}$ by setting `b = [b;0]` . Then the command `A\b` works on the nonsquare system matrix $\mathbf{A}_{n+1,n}$ and produces the desired least squares solution $\mathbf{x}_{LS} \in \mathbb{R}^n$ of the original problem, using the SVD, since the matrix $\mathbf{A}$ is now rank deficient and nonsquare. It goes to show how silly MATLAB (and other software) can be.

To close this section, we note that the MATLAB command

```
norm(u)
```

computes the length, or norm $\|\mathbf{u}\| = \sqrt{\sum u_i^2}$ (or $\|\mathbf{u}\| = \sqrt{\sum |u_i|^2}$), of a given vector $\mathbf{u} \in \mathbb{R}^n$ (or in $\mathbb{C}^n$), while the command

```
norm(A)
```

computes the norm of a matrix $\mathbf{A}_{mn} \in \mathbb{R}^{m,n}$ (or $\mathbf{A}_{mn} \in \mathbb{C}^{m,n}$) as the maximal singular value of $\mathbf{A}$, even for square matrices $\mathbf{A}$.

12.3.P Problems

1. (a) Use MATLAB to find the SVD of the matrix

$$\mathbf{A} = \begin{pmatrix} 1 & -2 & 7 & 13 \\ -2 & -4 & 0 & 2 \\ 3 & 12 & -6 & -9 \\ 0 & 0 & 5 & 0 \end{pmatrix}.$$

(b) What is the rank of $\mathbf{A}$?

(c) Construct the best rank 2 and rank 3 approximations $\mathbf{A}_2$, $\mathbf{A}_3$ of $\mathbf{A}$ in MATLAB, and print out the difference matrices $\mathbf{A} - \mathbf{A}_i$ for $i = 2$ and 3.

(d) Find an ONB for the column space of $\mathbf{A}$ and of $\mathbf{A}_2$.

(e) Find an ONB for the null spaces of $\mathbf{A}$ and $\mathbf{A}_1$.

2. Solve the unsolvable linear system of equations

$$\mathbf{Ax} = \begin{pmatrix} 1 & 3 & -2 & 5 \\ 2 & 3 & 0 & -2 \\ 3 & 6 & -2 & 3 \end{pmatrix} \mathbf{x} = \begin{pmatrix} 7 \\ 1 \\ -3 \end{pmatrix} = \mathbf{b} \text{ in}$$

the least squares sense via MATLAB and the SVD, and "incorrectly" by taking $\mathbf{x}$ as the solution to the pivot row part of the REF of $(\mathbf{A} \mid \mathbf{b})$. Compare your results, and compute $\|\mathbf{A}\|$.

3. Review the following MATLAB commands:

`function, svd, size, zeros, ones, diag, disp, format, pause, fprintf, fliplr, flipud.`

(a) Here is the "pixel file" made up of zeros and ones for the letter A:

```
0 0 0 0 0 0 1 0 0 0 0 0
0 0 0 0 0 0 1 0 0 0 0 0
0 0 0 0 0 1 1 1 0 0 0 0
0 0 0 0 0 1 1 1 0 0 0 0
0 0 0 0 0 1 1 1 0 0 0 0
0 0 0 0 1 1 0 1 1 0 0 0
0 0 0 0 1 0 0 0 1 0 0 0
0 0 0 1 1 1 1 1 1 0 0 0
0 0 0 1 1 1 1 1 1 1 0 0
0 0 1 1 0 0 0 0 0 1 0 0
0 0 1 1 0 0 0 0 0 1 1 0
0 1 1 0 0 0 0 0 0 1 1 0
0 1 0 0 0 0 0 0 0 0 1 1
1 1 0 0 0 0 0 0 0 0 1 1
1 1 0 0 0 0 0 0 0 0 1 1
1 0 0 0 0 0 0 0 0 0 0 1
```

(Stand back and unfocus your eyes; the letter A should appear.)

Create the analogous file for a 16×16 zero–one pixel matrix for the letter X by entering the following MATLAB commands:

```
X = diag(ones(1,15),1);
X = X + X' + eye(16);
X = X + fliplr(X)
```

Note that the four center positions of X are occupied by twos instead of ones. Remedy this by entering `X(8:9,8:9) = ones(2)`. You can save X by entering `save X X -ascii`, and X will always be there by entering `load X` at the next computer session. Create and load the pixel file for the letter *A* as well. Display your "letters" in `format +`.

(b) Use the letters A and X to make up "words" such as `word = [A X flipud(A); X A A]` and `word2 = [zeros(1,40); A X fliplr(A); zeros(1,40)]` to give the "word" a buffer of zeros, or a visual frame. Enter `word3 = ones(size(word)) - word` to invert the word graphically. Display the various words in `format +` and watch how your "writing" looks on the screen.

(c) Use the text editor in MATLAB to create a MATLAB SVD based "writer" program SVDW.m for your words created in part (b). You need to create a function m-file SVDW.m, that is a MATLAB function ready to work on any given "word" by entering `SVDW(word)` inside MATLAB.

Here the "word" is an externally defined 0–1 rectangular pixel matrix. The task of the SVDW.m routine is to print the original word once, then to compute the singular value decomposition of the pixel matrix for "word", and, finally to re-create the partially reconstructed "word", one additional singular vector dyad at a time, with displays of the intermediate re-created "word", until the word is readable. Write your writer SVDW.m using a "for loop", and round all entries of a partially reconstructed word that are below 0.5 in absolute value to 0, with those above 0.5 in absolute value rounded to 1 for displaying in `format +`.

This MATLAB function SVDW.m should only take about a dozen lines of MATLAB code. Try runs first with the word in black and then with the surround of the word in black, as detailed with `word3`, given in part (b).

What do you find? What font, standard or inverted, gives better images? Which font takes fewer SVD steps until recognition? Experiment with your word; try nonsense random words such as `B = floor(2*rand(16))`, for example.

4. Repeat Problems 4 and 5 from Section 12.2.P, using MATLAB.

5. Using MATLAB, find the least squares solution $\mathbf{x}$ of $\mathbf{Ax} = \mathbf{b}$ for $\mathbf{A} = \begin{pmatrix} 1 & 0 \\ 0 & 2 \\ 1 & 0 \end{pmatrix}$ and $\mathbf{b} = \begin{pmatrix} -1 \\ 2 \\ 9 \end{pmatrix}$.

6. Using MATLAB, find the SVD for the following 2×2 matrices, and plot the image of the unit circle under the induced linear transformations for the following matrices:

(a) $\mathbf{A} = \begin{pmatrix} 1 & -3 \\ 1 & 4 \end{pmatrix}$;

(b) $\mathbf{B} = \begin{pmatrix} 0 & 3 \\ 2 & 4 \end{pmatrix}$;

(c) $\mathbf{C} = \begin{pmatrix} 7 & 0 \\ 12 & 4 \end{pmatrix}$;

(d) $\mathbf{D} = \begin{pmatrix} 7 & 0 \\ 1 & 4 \end{pmatrix}$.

7. (a) Use MATLAB to find the dyadic eigenvalue and eigenvector expansion and the dyadic singular value decomposition expansion of the following matrices (see Problem 17 in Section 12.2.P):

$$\mathbf{A} = \begin{pmatrix} 5 & 3 & -1 \\ 0 & 2 & 1 \\ 1 & 1 & 4 \end{pmatrix};$$

$$\mathbf{B} = \begin{pmatrix} 5 & 1 \\ 8 & -2 \end{pmatrix}.$$

(b) Approximate both $\mathbf{A}$ and $\mathbf{B}$ by the rank 1 dyads $\mathbf{As}$, $\mathbf{Bs}$, $\mathbf{Ae}$, and $\mathbf{Be}$ that are formed from the respective dominant singular value and singular vectors and from the dominant eigenvalue and eigenvector.

(c) Determine the quality of these two types of approximations; that is, find the norm of the differences $\|\mathbf{A} - \mathbf{Ae}\|$, $\|\mathbf{A} - \mathbf{As}\|$, $\|\mathbf{B} - \mathbf{Be}\|$, and $\|\mathbf{B} - \mathbf{Bs}\|$, always using MATLAB.

8. Compute the singular values of the matrix $\mathbf{A}_{nn} = \begin{pmatrix} 1 & -1 & \cdots & \cdots & -1 \\ 0 & 1 & -1 & \cdots & -1 \\ \vdots & & \ddots & \ddots & \vdots \\ \vdots & & & \ddots & -1 \\ 0 & \cdots & \cdots & 0 & 1 \end{pmatrix}$ for increasing dimensions n. Use the MATLAB commands `A=eye(n); for i=1:n-1, A=A-diag(ones(1,n-i),i); end, svd(A)` and `format long e`.

What happens as n approaches 60? How close is $\mathbf{A}_{nn}$ to a singular matrix? What is $\det(\mathbf{A})$ depending on n? Explain.

12.R Review Problems

1. If T maps $\mathbb{R}^n$ to $\mathbb{R}^m$ linearly, how can one find two bases $\mathcal{U}$ of $\mathbb{R}^m$ and $\mathcal{V}$ of $\mathbb{R}^n$ so that T is represented diagonally with respect to these two different bases.

2. (a) Solve the least squares problem $\mathbf{Ax} = \mathbf{b}$:
$$\begin{pmatrix} 1 \\ 3 \\ -2 \end{pmatrix} \mathbf{x} = \begin{pmatrix} -2 \\ 2 \\ -1 \end{pmatrix}.$$
(b) Show that $\mathbf{b} - \mathbf{A}\mathbf{x}_{LS} \perp \begin{pmatrix} 1 \\ 3 \\ -2 \end{pmatrix}$ for the least squares solution $\mathbf{x}_{\text{LS}}$.
(c) What is the distance d from the point $\begin{pmatrix} -2 \\ 2 \\ -1 \end{pmatrix}$ to the line through $\begin{pmatrix} 1 \\ 3 \\ -2 \end{pmatrix}$ and the origin?

3. Let $\mathbf{A}_{5,5} = -2 \begin{pmatrix} 1 \\ 1 \\ -4 \\ 2 \\ 0 \end{pmatrix} \begin{pmatrix} 0 & 1 & 0 & 3 & 2 \end{pmatrix} +$
$3 \begin{pmatrix} 0 \\ -1 \\ 4 \\ 2 \\ 3 \end{pmatrix} \begin{pmatrix} 2 & -1 & 0 & 3 & -2 \end{pmatrix} + 6\mathbf{e}_2\mathbf{e}_4^T.$
(a) What is the rank of $\mathbf{A}$?
(b) Find the SVD of $\mathbf{A}$.
(c) Find a basis for the column space of $\mathbf{A}$.
(d) Find the SVD of $\mathbf{A}^T$.
(e) Find a basis for the row space of $\mathbf{A}$.
(f) Find a basis for the null space of $\mathbf{A}$.
(g) $\|\mathbf{A}\| =$ ________ .
(h) What is the nearest rank 2 matrix to $\mathbf{A}$?
(i) Solve $\mathbf{Ax} = \mathbf{e}_4$ and $\mathbf{Ay} = 3\mathbf{e}_1 - \mathbf{e}_2 + 5\mathbf{e}_5$ in the least squares sense.

4. (a) Is the SVD of the identity matrix unique?
(b) Is the SVD of any matrix unique?

5. (a) Find the singular values of $\mathbf{B} = \mathbf{UAV}$ in terms of the singular values of $\mathbf{A}$, where $\mathbf{U}$ and $\mathbf{V}$ are two arbitrary orthogonal matrices.
(b) How are the singular values of a matrix $\mathbf{A}$ and those of a similar matrix $\mathbf{B} = \mathbf{X}^{-1}\mathbf{AX}$ related? (*Hint*: Work with the example matrix $\mathbf{A} = \begin{pmatrix} 4 & 0 \\ 0 & 3 \end{pmatrix}$ and $\mathbf{B} = \mathbf{X}^{-1}\mathbf{AX}$ for $\mathbf{X} = \begin{pmatrix} 1 & \alpha \\ 0 & 1 \end{pmatrix}$ and $\alpha \in \mathbb{R}$.)

6. Explain which items of the SVD of a matrix $\mathbf{A} = \mathbf{U\Sigma V}^T$ are unique and which are not.

7. If $\mathbf{A} = \mathbf{A}^T$ is positive definite, show that the SVD of $\mathbf{A}$ has the form $\mathbf{UDU}^T$ for a diagonal matrix $\mathbf{D}$ and an orthogonal $\mathbf{U}$.

8. Find the singular value decomposition of a skew-symmetric matrix $\mathbf{S} = -\mathbf{S}^T$. In particular, show that for such a matrix, the nonzero singular values always occur in pairs, while the singular value zero of $\mathbf{S}$ may appear an odd number of times. Prove this result theoretically, and then verify it for random skew-symmetric matrices by using MATLAB.

Standard Questions and Tasks

1. Find the singular values and singular vectors of a matrix $\mathbf{A}_{mn}$.

2. Find an ONB for the kernel and an ONB for the column space, or image, of a matrix $\mathbf{A}_{mn}$.

3. How close is a given matrix $\mathbf{A}_{mn}$ to a rank 2 matrix?

4. What is the best rank 5 matrix approximation of a given matrix $\mathbf{A}_{mn}$?

5. What is the least squares solution of an unsolvable linear system $\mathbf{Ax} = \mathbf{b}$?

6. What is the least squares solution of a solvable linear system $\mathbf{Ax} = \mathbf{b}$?

7. How can one compute the least squares solution of an unsolvable linear system $\mathbf{Ax} = \mathbf{b}$ in theory and by using MATLAB?

Subheadings of Lecture Twelve

Basic Equations

p. 386 $\mathbf{\Sigma} = \mathbf{U}^T\mathbf{AV}$ (singular value decomposition)

Figure it out

CHAPTER

13

Basic Numerical Linear Algebra Techniques

13.1 Lecture Thirteen

Computer Arithmetic, Stability, and the QR Algorithm

Linear algebra immerses us into vector spaces and their geometry. It studies linear transformations of such spaces and much more. Looking back over the previous chapters, we see plainly that the topics of Chapters 3, 9, 11, and 12 have the farthest reach to applications: They detail how to solve systems of linear equations and how to find eigenvalues, eigenvectors, and singular values of matrices. The remaining chapters are either technical, such as Chapters 2 and 8, or conceptual, such as Chapters 1, 4, 5, 6, 7, and 10.

Given a certain familiarity with the concepts and nature of linear spaces and the properties of linear transformations, we now assess our simple hand computational methods in light of numerical analysis. While the concepts that were developed in the exploratory lectures 1, 4, 5, 6, 7, and 10 help us understand linear algebra well, our hand computation schemes, unfortunately, do not measure up to the current state of the art in computing.

What makes our Gaussian row reduction scheme of Chapter 2 and our method for finding eigenvalues and singular values by hand via polynomials in Chapters 9, 11, and 12 of such limited use? It is the nature of the computer. If computers were infinite-precision machines, then our basic pencil-and-paper algorithms would flourish. But modern computers work with floating point precision; that is, every number is represented on a computer in the form

$$\pm 0.\underbrace{xxx\cdots x}_{m} \cdot 10^{\pm \overbrace{yyy}^{n}},$$

with a **mantissa** $\pm 0.xxx\cdots x$ that is less than 1 in magnitude and that has a fixed word length m ($m \approx 7$ in single precision; $m \approx 16$ in double precision). This number is multiplied by a power of 10 with a limited range of exponents $\pm yyy$. Here, the symbols x and y represent any of the integers from 0 to 9. If we add two such floating-point numbers such as $a = 0.9 \cdot 10^{-300}$ and $b = 0.1 \cdot 10^{+300}$ on a computer, then their sum is computed incorrectly as $a+b \overset{\text{comp.}}{=} b$, since the mantissa length m of computers is generally well below 600. Due to such **rounding errors**, all our real-number field axioms fall by the wayside in **floating-point arithmetic**. Consequently, in computer computations, we **never know whether a computed quantity really is zero** or only nearly so. Thus, pivot decisions and the rank of a matrix are difficult, if not impossible, to obtain correctly by computer computation. That makes our hand scheme for Gaussian elimination potentially unstable. Worse yet, if we accept a real entry of relative small magnitude as a pivot for further row reductions, we will have to eliminate entries of larger magnitude below this small pivot with relatively huge multiples of the pivot row, possibly overpowering and wiping out any comparatively small entries of the affected row, much as the preceding floating-point addition of a and b wiped out a. Indeed, if our basic hand schemes were installed verbatim on a computer that uses floating-point arithmetic, the computations would often lead to nonsense for general real matrix problems.

Gaussian elimination can be stabilized with more judicious pivot searches, such as a column pivot search: When looking for a pivot in a subcolumn of **A**, swap rows until the entry of largest magnitude appears in the pivot position. This technique is called **partial pivoting** and is standardly employed in all software that solves linear equations. The technique is numerically stable. However, it makes pivots generally unequal to ± 1, and thus it renders hand computations prohibitively difficult.

Turning to matrix eigenvalues, note that finding eigenvalues as the roots of polynomials in two ways, as we have done in Chapter 9, creates numerical havoc for all but the smallest dimensions: The polynomial $p(x) = \prod_{j=1}^{20}(x-j) = x^{20} + a_{19}x^{19} + \cdots + a_1x + a_0$, for example, has the integers 1, 2, ..., 20 as its roots for certain real coefficients a_i. Changing, or perturbing, the 13th coefficient a_{13} of this polynomial only slightly in its last double-precision floating-point digit to become $(1+10^{-15})a_{13}$ will change some of the roots of the perturbed polynomial by more than 1% from their original location. The reason for this is that polynomial roots are, in general, very sensitive to changes of the polynomial coefficients. Yet matrix eigenvalues must and can be found reliably for dimensions well into the hundreds and thousands. How is this done numerically if one cannot find polynomial roots reliably?

The answer is to think "basis." Slowly and stably, we can turn the standard basis $\mathcal{E}$ of $\mathbb{R}^n$ that is used to represent a given linear transformation T as $\mathbf{A}_{\mathcal{E}}$ into an eigenvector basis $\mathcal{U}$ for **A**. This involves an almost miraculous iteration that was discovered by H. Rutishauser in the 1950s: For $\mathbf{A}_{nn}$, form the double sequence of orthogonal matrices $\mathbf{Q}_i$ and upper triangular matrices $\mathbf{R}_i$ as follows:

$$\begin{aligned}
\mathbf{A} =: \mathbf{A}_1 &=: \mathbf{Q}_1\mathbf{R}_1; && \text{original QR factorization (see Section 10.2);}\\
\mathbf{A}_2 := \mathbf{R}_1\mathbf{Q}_1 &=: \mathbf{Q}_2\mathbf{R}_2; && \text{reverse-order multiply and again factor as } \mathbf{QR};\\
\mathbf{A}_3 := \mathbf{R}_2\mathbf{Q}_2 &=: \mathbf{Q}_3\mathbf{R}_3; && \text{repeat, etc.}
\end{aligned}$$

The miracle of the factorization–reverse-order-multiply **QR algorithm** is that, for **A** in **Hessenberg form**—that is, for

$$\mathbf{A} = \begin{pmatrix} * & * & \dots & \dots & * \\ * & * & \dots & \dots & * \\ 0 & * & \ddots & & \vdots \\ \vdots & & \ddots & \ddots & \vdots \\ 0 & \dots & 0 & * & * \end{pmatrix}$$

in the form of an upper triangular matrix plus one possibly nonzero lower codiagonal band—the sequence of matrices $\mathbf{A}_i$ will eventually become upper triangular (i.e., the subdiagonal entries of the $\mathbf{A}_i$ converge to zero), exposing the eigenvalues of **A** on the diagonal. Thus, approximations of **A**'s eigenvalues will appear on the diagonal, while the product of the orthogonal matrices $\mathbf{Q}_{\dots}$ can be used to describe an eigenvector basis for **A**.

Here is a quick indication why this is so: Clearly,

$$\begin{aligned} \mathbf{A}_{i+1} &= \mathbf{R}_i\mathbf{Q}_i = (\mathbf{Q}_i^{-1}\mathbf{Q}_i)\mathbf{R}_i\mathbf{Q}_i = \mathbf{Q}_i^{-1}(\mathbf{Q}_i\mathbf{R}_i)\mathbf{Q}_i \\ &= \mathbf{Q}_i^{-1}\mathbf{A}_i\mathbf{Q}_i, \end{aligned}$$

so that every iterate $\mathbf{A}_i$ is similar to the original matrix $\mathbf{A} = \mathbf{A}_\varepsilon$. Further,

$$\begin{aligned} \mathbf{A}_{i+1} &= \mathbf{Q}_i^{-1}\mathbf{A}_i\mathbf{Q}_i = \mathbf{Q}_i^{-1}(\mathbf{Q}_{i-1}^{-1}\mathbf{A}_{i-1}\mathbf{Q}_{i-1})\mathbf{Q}_i \\ &= \cdots = \left(\prod_{j=1}^{i}\mathbf{Q}_j\right)^{-1}\mathbf{A}\left(\prod_{j=1}^{i}\mathbf{Q}_j\right). \end{aligned}$$

Hence, if the Hessenberg matrices $\{\mathbf{A}_i\}$ converge to upper triangular form, then the infinite product $\prod_{j=1}^{\infty}\mathbf{Q}_j$ contains a tridiagonalizing basis (i.e., a Schur basis) for **A** in its columns. In fact, the QR algorithm computes a Schur Normal form of **A**. The Schur basis can then be used to find the eigenvectors of **A** easily.

These are only some of the basic ideas behind the QR matrix eigenvalue algorithm.

The best numerical versions of the **QR algorithm** use elaborate shift techniques to speed up convergence. For example, the QR algorithm is the "shining knight" behind MATLAB's command `eig`. It is likewise used in spirit and by performing what amounts to a double implicit implementation of QR to compute the singular values of matrices very accurately, rather than using Theorem 12.2 directly. To illustrate, matrix eigenvalue algorithms have become so reliable that the roots of polynomials p nowadays are best computed using a companion or tridiagonal matrix representation of p and finding the related eigenvalues. This is truly evidence of the "circular nature" of linear algebra.

Other widely used numerical techniques for solving linear algebra and matrix problems are iterative methods for solving huge and structured systems of linear equations when Gaussian elimination would be too costly, as well as optimized vector iterations to find eigenvalues and singular values of very large matrices that could not be computed directly via QR due to their size.

To compute data from and for matrices has become a huge field of development over the last 50 years since the advent of computers. On the one hand, the field is nourished by a deep understanding of the structure of linear spaces and linear transformations. On the other hand, numerical matrix analysis obeys quite different rules and demands and uses intuitions of its own.

13.1.P Problems

1. Perform a few steps of the unreduced and unshifted basic QR algorithm:

$$\mathbf{A}_1 =: \mathbf{Q}_1\mathbf{R}_1 \to \mathbf{A}_2 := \mathbf{R}_1\mathbf{Q}_1 =: \mathbf{Q}_2\mathbf{R}_2 \to \\ \to \mathbf{A}_3 := \mathbf{R}_2\mathbf{Q}_2 =: \mathbf{Q}_3\mathbf{R}_3 \to \ \ldots$$

etc., for the dense matrices $\mathbf{A} = \begin{pmatrix} 2 & 1 & 0 \\ -1 & 1 & 2 \\ 1 & 0 & 1 \end{pmatrix}$ and

$$\mathbf{B} = \begin{pmatrix} 14 & 2 & -1 \\ 2 & 0 & 1 \\ -3 & 1 & 0 \end{pmatrix}.$$

For this, use the MATLAB command line

```
[Q R] = qr(A); A = R*Q,
```

repeatedly, and view the iterates. For comparison, evaluate `eig(A)` and `eig(B)`. What do you observe? (In MATLAB the up ↑ key always enters the previous command on the command line, so ↑ followed by "enter" will keep supplying you with new QR iterates as long as you wish. Set up your `format` in such a way that you are able to view subtle differences.)

2. Perform a few steps of the basic QR algorithm

$$\mathbf{A}_1 =: \mathbf{Q}_1\mathbf{R}_1 \to \mathbf{A}_2 := \mathbf{R}_1\mathbf{Q}_1 =: \mathbf{Q}_2\mathbf{R}_2 \to \\ \to \mathbf{A}_3 := \mathbf{R}_2\mathbf{Q}_2 =: \mathbf{Q}_3\mathbf{R}_3 \to \ldots$$

etc., for the 3×3 upper Hessenberg matrices

$$\mathbf{A} = \begin{pmatrix} -2 & 1 & 3 \\ -2 & 11 & 2 \\ 0 & -10 & 3 \end{pmatrix} \text{ and } \mathbf{B} = \begin{pmatrix} 14 & 2 & -1 \\ 7 & -2 & 1 \\ 0 & 1 & 0 \end{pmatrix}$$

and for two dense matrices $\mathbf{W} := \mathbf{X}^T\mathbf{A}\mathbf{X}$ and $\mathbf{Z} := \mathbf{X}^T\mathbf{B}\mathbf{X}$ that are similar to $\mathbf{A}$ and $\mathbf{B}$, respectively, for a random orthogonal matrix $\mathbf{X}$ with $\mathbf{X}^T\mathbf{X} = \mathbf{I}$.

(a) Create a random orthogonal 3×3 matrix $\mathbf{X}$ via the MATLAB command sequence `Y=randn(3); [X R] = qr(Y);` and form $\mathbf{W}$ as `W = X'*A*X;`

(b) Compare the speed of convergence of our basic QR algorithm for dense and Hessenberg matrices by using the MATLAB command lines

```
[Q R] = qr(A); [Qd Rd] = qr(W);
A = R*Q; W = Rd*Qd; disp([A W]),
```

to view the iterates for each pair $(\mathbf{A}, \mathbf{W})$ and $(\mathbf{B}, \mathbf{Z})$ of similar matrices.

(c) Evaluate `eig(A)` and `eig(B)`, and likewise, find the eigenvalues of $\mathbf{W}$ and $\mathbf{Z}$. What do you observe?

3. Consider the 6×6 matrix

$$\mathbf{A} = \begin{pmatrix} 1 & 0 & 0 & 0 & 0 & 1 \\ -1 & 1 & 0 & 0 & 0 & 1 \\ -1 & -1 & 1 & 0 & 0 & 1 \\ -1 & -1 & -1 & 1 & 0 & 1 \\ -1 & -1 & -1 & -1 & 1 & 1 \\ -1 & -1 & -1 & -1 & -1 & 1 \end{pmatrix}.$$

(a) Generate this matrix by the sequence of commands `n = 6; A = eye(n); A(:,n) = ones(n,1); for i=1:n-1, A(i+1:n,i) = -ones(n-i,1); end` in MATLAB.

(b) Find a Hessenberg matrix $\mathbf{B}$ that is orthogonally similar to $\mathbf{A}$ via the MATLAB command `B = hess(A);`.

(c) Watch the basic QR algorithm without shift performed on $\mathbf{A}$ and $\mathbf{B}$ via the repeated command lines

```
[Q R] = qr(A); [Qb Rb] = qr(B);
A = R*Q; B = Rb*Qb; disp([A B]).
```

(Iterate at least 100 times.)

(d) From the convergence behavior in part (c), conclude what type of eigenvalues the matrix $\mathbf{A}$ is likely to have.

(e) What are the singular values of $\mathbf{A}$?

(*Hint*: Use the MATLAB function `svd(A)`.)

4. Adapt the singular-value decomposition of a matrix $\mathbf{A} = \mathbf{U}\Sigma\mathbf{V}^T$ from Chapter 12 to solve a linear system $\mathbf{A}\mathbf{x} = \mathbf{b}$.

(a) Express the solution $\mathbf{x}$ to $\mathbf{A}\mathbf{x} = \mathbf{b}$ in terms of the SVD of $\mathbf{A}$ and the right-hand side $\mathbf{b}$ if $\mathbf{A}$ is nonsingular.

(b) Repeat Problem 1 of Section 6.3.P for $3 \leq n \leq 30$ and the **Hilbert matrix** $\mathbf{H}_n = \left(\frac{1}{i+j}\right)$ as system matrix for the right-hand side $\mathbf{b} = \mathbf{e}_1 - \mathbf{e}_n \in \mathbb{R}^n$, as before. This time use both the MATLAB Gaussian-elimination-based solver `A\b` and your own SVD-based MATLAB solver from part (a). Compare your results.

CHAPTER

14

Nondiagonalizable Matrices, the Jordan Normal Form*

[*On the Web only at http://www.auburn.edu/~uhligfd/TLA/download/c14.pdf.]

Tree city

Epilogue

From the tasks laid out in Section 1.2, we have set out to describe the landscape, the scenery, and hopefully some of the "culture" of linear spaces and matrices. We have done so with a keen eye on fundamentals, combined with a hands-on approach to solving problems of both a theoretical and a practical nature.

Our process of teaching was, and had to be, haphazard: Many concepts were first mentioned in context, but could be defined rigorously only much later. Generally, classes fare well with such an approach if the teacher picks up questions from the class readily, builds on them, and thus fosters the students' curiosity while loosely keeping to his or her (always changing and adjusting) lesson plans. Every class will acquire this material differently. Every teacher will have to interpret the "city of matrices" differently on every tour that he or she guides through this magical place.

Matrix theory has ultimately enabled us to recognize the individual properties of each linear transformation clearly and distinctly. This is an admirable feat of "glasnost" at a time when humanity as a whole struggles to do the same for each individual human being.

We shall not cease from exploration
And the end of all our exploring
Will be to arrive where we started
and know the place for the first time.

T. S. Eliot,
Excerpt from
"Little Gidding" in
FOUR QUARTETS*

Complex wall

Appendix A

Complex Numbers and Vectors

Our number systems may have evolved in the following way:

Starting with the **positive integers** $\mathbb{N} = \{1, 2, 3, \dots\}$ that are naturally used for counting, we can freely add and multiply numbers m and n in $\mathbb{N}$, and the result will still be a "counting number" in $\mathbb{N}$. But the difference $n - m \notin \mathbb{N}$ if $m \geq n$, and the quotient $\frac{n}{m} \notin \mathbb{N}$, unless $n = k \cdot m$ for some positive integer k.

Thus, we must introduce the set of all **integers** $\mathbb{Z} = \{\dots, -2, -1, 0, 1, 2, \dots\}$ to close our number system under subtraction, and the **rational numbers** $\mathbb{Q} = \left\{ \frac{p}{q} \middle| p, q \in \mathbb{Z}; q \neq 0 \right\}$ to allow for division. Clearly,

$$\mathbb{N} \subset \mathbb{Z} \subset \mathbb{Q}.$$

However, the rational numbers $\mathbb{Q}$ are still not plentiful enough. There are sequences of rational numbers $x_i = \frac{p_i}{q_i}$ that converge, but for which $\lim_{i \to \infty} x_i \notin \mathbb{Q}$. The Greeks already knew that $\sqrt{2} \notin \mathbb{Q}$. For if it were and $\sqrt{2} = \frac{p}{q} \in \mathbb{Q}$ for two integers p and q without a common prime factor, then $2q^2 = p^2$, making p^2 even, or $p = 2k$ for some $k \in \mathbb{Z}$. Thus, $p^2 = (2k)^2 = 4k^2 = 2q^2$. Consequently, $2k^2 = q^2$, and q is even as well. This contradicts the assumption that the two integers p and q with $\sqrt{2} = \frac{p}{q}$ are relatively prime. Hence, $\sqrt{2} \notin \mathbb{Q}$.

But $\sqrt{2} \notin \mathbb{Q}$ is the limit of rational numbers. For example, starting with $x_1 = 1$ and setting $x_i = 1 + \frac{1}{x_{i-1} + 1}$ for $i = 2, 3$, etc., inductively, will create a convergent sequence of rational numbers $x_i \in \mathbb{Q}$ with $\lim_{i \to \infty} x_i = \sqrt{2}$. This can be readily checked by setting $x^* := \lim_{i \to \infty} x_i$ and finding the value of x^* as $\sqrt{2}$ from the iteration formula. The lack of closure of $\mathbb{Q}$ under limits led mankind to introduce the **real numbers** $\mathbb{R}$ that contain all limits of convergent rational sequences.

Yet not all is well with $\mathbb{R}$. It became apparent in the late middle ages that not all real (and not even all rational or integer) coefficient polynomials have their roots in $\mathbb{R}$. There are no real roots for a polynomial such as $f(x) = x^2 + 1$. Its roots $\pm i := \pm\sqrt{-1} \notin \mathbb{R}$, since for all real values of x, we have $f(x) = x^2 + 1 \geq 1 > 0$.

This dilemma led to the introduction of the **complex numbers** $\mathbb{C}$ more than 200 years ago:

$$\mathbb{C} = \{a + bi \mid a, b \in \mathbb{R}; i \notin \mathbb{R} \text{ with } i^2 = -1\}.$$

Here $i := \sqrt{-1} \notin \mathbb{R}$ is the **imaginary unit** of $\mathbb{C}$. Each complex number $z = a+bi$ has a **real part** $a \in \mathbb{R}$ and an **imaginary part** $b \in \mathbb{R}$. We have the following string of inclusions:

$\mathbb{N}$	$\subset$	$\mathbb{Z}$	$\subset$	$\mathbb{Q}$	$\subset$	$\mathbb{R}$	$\subset$	$\mathbb{C}$
counting numbers	$\subset$	integers	$\subset$	rationals	$\subset$	reals	$\subset$	complex numbers
natural numbers		closed under subtraction		closed under division		closed under limits		closed under polynomial roots

Complex numbers occur most naturally as roots of polynomials. For example, the stable version of the **quadratic formula** describes the two roots x_1 and x_2 of a quadratic polynomials $p_2(x) = ax^2 + bx + c$ in terms of its coefficients as

$$x_1 = -\frac{b + \text{sign}(b)\sqrt{b^2 - 4ac}}{2a} \quad \text{and} \quad x_2 = \frac{c}{ax_1}.$$

Here the sign of a real number x is defined as $\text{sign}(x) = \begin{cases} 1 & \text{if } x \geq 0 \\ -1 & \text{if } x < 0 \end{cases}$.

The formula for the roots of a quadratic polynomial is often given in its unstable but easier to use form as

$$x_{1,2} = \frac{-b \pm \sqrt{b^2 - 4ac}}{2a}.$$

The numerical instability occurs when $|4ac| \ll b^2$. If, on the other hand, $|4ac| \approx b^2$ for the coefficients of the quadratic polynomial p_2, then the simpler and more common $x_{1,2}$ formula can be used without harm.

EXAMPLE 1 For the polynomial $p(x) = x^2 + 20,000x + 1$, we have $a = 1, b = 20,000$ and $c = 1$. A simple calculator that carries eight digits computes the roots of p using the unstable $x_{1,2}$ quadratic formula as $x_1 = -10,000 - \sqrt{99,999,999} \doteq -19,999.999$ and $x_2 = -10,000 + \sqrt{99,999,999} \doteq -10,000 + 9,999.9999 \doteq -0.0010000$, where the approximately equal symbol ($\doteq$) points to the rounding errors of the calculations. Switching to the stable version of the quadratic formula, we see that the same calculator computes p's roots as $\tilde{x}_1 = x_1 \doteq -19,999.999$ and $\tilde{x}_2 = \dfrac{1}{-19,999.999} \doteq -0.00005000$. Now the second computed root $\tilde{x}_2$ is correct in all printed digits. The relative error of the root x_2, computed via the unstable quadratic formula, is nearly 1,000%. ◀

If a, b, and c are real numbers with $b^2 - 4ac < 0$, then the two roots x_i of $p_2 = ax^2 + bx + c$ are necessarily complex numbers. For example, the real quadratic polynomial $p(x) = 2x^2 - 3x + 2$ has the two complex roots $x_{1,2} = \frac{3 \pm \sqrt{9-16}}{4} = \frac{3}{4} \pm \frac{\sqrt{7}}{4} i$ for $i = \sqrt{-1} \in \mathbb{C}$, while $q(x) = x^2 + x + 1$ has the roots $x_{1,2} = -\frac{1}{2} \pm \frac{\sqrt{3}}{2} i \in \mathbb{C}$. But $r(x) = x^2 + x - 1$ has two real roots, since $b^2 - 4ac = 1 + 4 = 5 > 0$. They are $x_{1,2} = -\frac{1}{2} \pm \frac{\sqrt{5}}{2}$.

The complex numbers $\mathbb{C}$ are often visualized similarly to how $\mathbb{R}^2$ was in Figure I-1 of the Introduction:

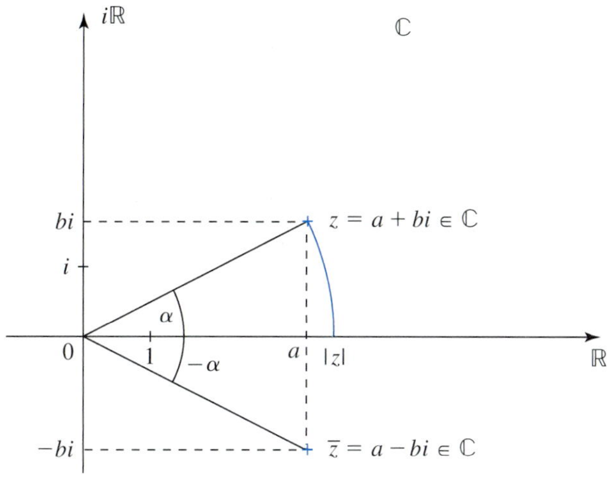

Figure A-1

The difference between the two Figures I-1 and A-1 is both deep and subtle: In the **vector space** $\mathbb{R}^2$, we only perform the two basic linear operations of vector addition and scalar multiplication. In the **field** $\mathbb{C}$, we also multiply (and divide) complex numbers $a + bi$ and $c + di$, namely,

$$(a + bi)(c + di) := (ac - bd) + (bc + ad)i,$$

by distributing multiplication and addition and by using the complex number identity $i^2 = -1$.

NOTE: There is no such multiplication of two vectors in $\mathbb{R}^2$ (or in $\mathbb{R}^n$).

EXAMPLE 2 (a) $(2 + 3i)(1 - i) = (2 + 3) + (3 - 2)i = 5 + i \in \mathbb{C}$.
(b) $i \cdot i = i^2 = -1 \in \mathbb{C}$.
(c) $2i(3 - i) = 6i - 2i^2 = 2 + 6i \in \mathbb{C}$. ◀

Complex numbers can be multiplied geometrically by using planar real rotations as follows:

A complex number $z = a + bi$ has the **magnitude** $|z| = |a + bi| := \sqrt{a^2 + b^2} \geq 0 \in \mathbb{R}$.

The product of two complex numbers has the product of their magnitudes as its magnitude. (See Problem 1.)
If the point $z = a + bi \in \mathbb{C}$ sustains an **angle** α (modulo 2π, or 360 degrees) with the positive real axis $\mathbb{R}^+ \subset \mathbb{C}$, and if $w = c + di \in \mathbb{C}$ likewise sustains an angle β (modulo 2π), then their complex number product $zw = wz \in \mathbb{C}$ sustains the angle $\alpha + \beta$ with the positive real axis $\mathbb{R}^+ \subset \mathbb{C}$. (See Problem 2.)

EXAMPLE 3
(a) $|2 + 3i| = \sqrt{4+9} = \sqrt{13}$; $|i| = 1$; $|2 - i| = \sqrt{4+1} = \sqrt{5}$.
(b) The imaginary unit $i \in \mathbb{C}$ sustains a counterclockwise angle of $\frac{\pi}{2}$, or 90° with the positive real axis from Figure A-1, while $1+i$ sustains a counterclockwise angle of $\frac{\pi}{4}$ or 45° with $\mathbb{R}^+$. And their product $i(1 + i) = -1 + i$ lies on the angular bisector of the second quadrant in $\mathbb{C}$ and thus it sustains a counterclockwise angle of $135° = 90° + 45°$ degrees with $\mathbb{R}^+$, or a clockwise angle of $360° - 135° = 225°$ with $\mathbb{R}^+$. ◀

Complex arithmetic becomes easier to handle if we introduce **complex conjugation**: If $z = a + bi \in \mathbb{C}$ is an arbitrary complex number, then its **complex conjugate** $\overline{z}$ is defined as $\overline{z} = a - bi \in \mathbb{C}$. With this notation, the **magnitude** $|z|$, or the absolute value of a complex number z is written as

$$|z| = \sqrt{\overline{z}z} = \sqrt{z\overline{z}} = \sqrt{a^2 + b^2} \geq 0 \in \mathbb{R},$$

since $\overline{z}z = (a - bi)(a + bi) = a^2 - b^2i^2 = a^2 + b^2$. Moreover, the division of two complex numbers z and $w \neq 0$ is given by the formula

$$\frac{z}{w} = \frac{z\overline{w}}{w\overline{w}} = \frac{z\overline{w}}{|w|^2} \quad \text{for } w \neq 0$$

(i.e., as the product of the numerator z and the conjugated denominator $\overline{w}$, divided by the magnitude of the denominator squared).

EXAMPLE 4
(a) If $z = 3 + i$, then $\overline{z} = 3 - i$ and $z\overline{z} = (3 + i)(3 - i) = 9 + 1 = 10 = |z|^2$.
(b) If, moreover, $w = 1 - i$, then

$$\frac{z}{w} = \frac{3+i}{1-i} = \frac{z\overline{w}}{|w|^2} = \frac{(3+i)(1+i)}{(1-i)(1+i)} = \frac{2+4i}{1+1} = 1 + 2i,$$

$$\frac{w}{z} = \frac{1}{\left(\frac{z}{w}\right)} = \frac{\overline{\left(\frac{z}{w}\right)}}{\left|\frac{z}{w}\right|^2} = \frac{1-2i}{5} = \frac{1}{5} - \frac{2}{5}i,$$

where we have used the formula $\frac{1}{v} = \frac{\overline{v}}{|v|^2}$ for complex number inverses (see Problem 8) and our knowledge of $\frac{z}{w} \in \mathbb{C}$. ◀

As an alternative to complex multiplication defined earlier as $(a+bi)(c+di) = (ac - bd) + (bc + ad)i$ and in Problems 1 and 2, we can visualize complex multiplication as a special 2×2 real matrix product:

Let $z = a + bi \in \mathbb{C}, |z| = \sqrt{a^2+b^2} \in \mathbb{R}$ and $\tan\alpha = \dfrac{b}{a}$ be as indicated in Figure A-1. Then we represent z by the matrix $\mathbf{Z}_{2,2} = |z| \begin{pmatrix} \cos\alpha & -\sin\alpha \\ \sin\alpha & \cos\alpha \end{pmatrix} \in \mathbb{R}^{2,2}$, where $|z| \geq 0 \in \mathbb{R}$ is its magnitude and the real 2×2 matrix $\begin{pmatrix} \cos\alpha & -\sin\alpha \\ \sin\alpha & \cos\alpha \end{pmatrix}$ parametrizes the counterclockwise rotation of the positive real axis $\mathbb{R}^+ \subset \mathbb{C}$ by the angle α.

If $\mathbf{W} = |w| \begin{pmatrix} \cos\beta & -\sin\beta \\ \sin\beta & \cos\beta \end{pmatrix}$ for a second complex number $w = c + di$ with $|w| = \sqrt{c^2+d^2} \in \mathbb{R}$ and $\tan\beta = \dfrac{d}{c}$, then the matrix product $\mathbf{ZW}$ evaluates as follows:

$$\begin{aligned}
\mathbf{ZW} &= |z||w| \begin{pmatrix} \cos\alpha & -\sin\alpha \\ \sin\alpha & \cos\alpha \end{pmatrix} \begin{pmatrix} \cos\beta & -\sin\beta \\ \sin\beta & \cos\beta \end{pmatrix} \\
&= |z||w| \begin{pmatrix} \cos\alpha\cos\beta - \sin\alpha\sin\beta & -\sin\alpha\cos\beta - \cos\alpha\sin\beta \\ \sin\alpha\cos\beta + \cos\alpha\sin\beta & \cos\alpha\cos\beta - \sin\alpha\sin\beta \end{pmatrix} \\
&= |z||w| \begin{pmatrix} \cos(\alpha+\beta) & -\sin(\alpha+\beta) \\ \sin(\alpha+\beta) & \cos(\alpha+\beta) \end{pmatrix}
\end{aligned}$$

by using the addition theorems for sine and cosine. Thus, the matrix product $\mathbf{ZW}$ represents or parametrizes the complex number product $zw = wz \in \mathbb{C}$. The product $zw \in \mathbb{C}$ has the magnitude $|zw| = |z||w|$ and represents a point in the complex plane that sustains the angle $\alpha + \beta$ with the positive real axis. (Special care must be taken when defining $\mathbf{Z}$ or $\mathbf{W}$ if the real parts a or c are zero in $\mathbf{Z}$ or $\mathbf{W}$; see Problem 10.)

Having learned to compute in $\mathbb{C}$, we now study **complex vectors** $\mathbf{x}, \mathbf{y} \in \mathbb{C}^n$. Vectors in $\mathbb{C}^n$ have n complex components. Adding two such vectors is done component–wise over $\mathbb{C}$, as is scalar multiplication, allowing for complex scalars as well now. If we introduce **complex vector conjugation** $\mathbf{x}^* = \overline{\mathbf{x}}^T = (\overline{x_1}, \ldots, \overline{x_n})$ as the componentwise conjugation and transposition for a vector $\mathbf{x} = \begin{pmatrix} x_1 \\ \vdots \\ x_n \end{pmatrix} \in \mathbb{C}^n$, then the **length** $\|\mathbf{x}\| \in \mathbb{R}$ of a vector $\mathbf{x} \in \mathbb{C}^n$ is determined analogously to the real case as

$$\|\mathbf{x}\| := \sqrt{\overline{\mathbf{x}} \cdot \mathbf{x}} = \sqrt{\mathbf{x}^*\mathbf{x}} \in \mathbb{R}.$$

Note that in order to evaluate the length $\|\mathbf{x}\|$ of a vector $\mathbf{x} \in \mathbb{C}^n$ with n components $x_j = a_j + b_j i \in \mathbb{C}$ and $a_j,\ b_j \in \mathbb{R}$, we have to compute twice as many squares as in the real vector case, namely, $\|\mathbf{x}\| = \sqrt{\sum_j \left(a_j^2 + b_j^2\right)}$. Moreover, $\|\alpha\mathbf{x}\| = |\alpha|\ \|\mathbf{x}\|$ for all scalars $\alpha \in \mathbb{C}$ and all vectors $\mathbf{x} \in \mathbb{C}^n$, just as in the real case.

EXAMPLE 5 $\left\| \begin{pmatrix} 1+i \\ 1-3i \end{pmatrix} \right\| = \sqrt{1+1+1+9} = \sqrt{12}$, while $\left\| i \begin{pmatrix} 1+2i \\ 2-i \end{pmatrix} \right\| = |i| \left\| \begin{pmatrix} 1+2i \\ 2-i \end{pmatrix} \right\|$ $= 1 \cdot \sqrt{1+4+4+1} = \sqrt{10}$. ◀

Complex vectors and quantities do not occur with linear equations for real data. They will, however, appear with eigenvalue problems, even for real matrices. The reason for this is the Fundamental Theorem of Algebra. (See Theorem 9.1 or Theorem 9.1.D in Section 9.1 or Section 9.1.D.) Real polynomials generally have their roots over $\mathbb{C}$. Thus, the characteristic and vanishing polynomials of a real $n \times n$ real matrix $\mathbf{A}$ will reveal their roots (i.e., the eigenvalues of $\mathbf{A}$), only over $\mathbb{C}$.

Moreover, if $\mathbf{A}\mathbf{x} = \lambda\mathbf{x} = (a + bi)\mathbf{x}$ for a real matrix $\mathbf{A}$ and a complex eigenvalue $\lambda = a + bi \notin \mathbb{R}$ (i.e., if $b \neq 0$), then the eigenvector $\mathbf{x} \neq \mathbf{0}$ must necessarily lie outside of $\mathbb{R}^n$ and in $\mathbb{C}^n$. Otherwise, the left-hand-side $\mathbf{A}\mathbf{x}$ of the eigenvalue–eigenvector equation would involve only real quantities, while its right-hand-side $\lambda\mathbf{x} = (a + bi)\mathbf{x}$ would not. On the other hand, symmetric real matrices $\mathbf{A} = \mathbf{A}^T$ have only real eigenvalues (see Chapter 11), and thus their eigenvectors can always be computed in $\mathbb{R}^n$.

Complex numbers behave like a **field** under the arithmetic operations of number addition and number multiplication just as $\mathbb{Q}$ and $\mathbb{R}$ do. There are further extensions of the natural counting set $\mathbb{N}$ beyond $\mathbb{C}$. The quaternions are a skew field that contains $\mathbb{C}$, where number multiplication, however, is no longer commutative (i.e., generally $ab \neq ba$ for quaternions a and b). Beyond the quaternions, there are the hamiltonians, but these form only a division algebra whose multiplication is no longer associative (i.e., $a(bc) \neq (ab)c$ in general for hamiltonian numbers a, b, and c). Hence, in a practical way, the complex numbers $\mathbb{C}$ are the largest useful set of numbers that we need for our studies.

A.A.P Problems

1. Show that $|(a+bi)(c+di)| = |a+bi|\ |c+di|$ for all $a, b, c, d \in \mathbb{R}$ and $i \in \mathbb{C}$ with $i^2 = -1$.

2. Show directly that if $z = a + bi \in \mathbb{C}$ sustains an angle α with the positive real axis $\mathbb{R}^+ \subset \mathbb{C}$, and if $w = c + di \in \mathbb{C}$ sustains an angle β with the positive real axis $\mathbb{R}^+ \subset \mathbb{C}$, then their complex number product $zw = wz$ sustains the angle of $\alpha + \beta$ with the positive real axis $\mathbb{R}^+ \subset \mathbb{C}$.

3. Find the four complex numbers zw, $\dfrac{z}{w}$, $\dfrac{w}{z}$, and $z - 2w$ for $z = 2 - i$ and $w = 3 + 2i \in \mathbb{C}$.

4. Do the same as in Problem 3, but for $z = -2i$ and $w = 1 - i$.

5. Find the length of the following complex vectors:

   (a) $\mathbf{w} = \begin{pmatrix} 1+i \\ 1-i \end{pmatrix} \in \mathbb{C}^2$.

   (b) $\mathbf{z} = \begin{pmatrix} i \\ -3i \end{pmatrix} \in \mathbb{C}^2$.

   (c) $\mathbf{w} + \mathbf{z}$.

   (d) $\mathbf{u} := (1+3i)\mathbf{w} - (2-i)\mathbf{z} \in \mathbb{C}^2$.

   (e) $\mathbf{u} - \mathbf{w}$.

   (f) $\mathbf{x} = \begin{pmatrix} \mathbf{w} \\ -\mathbf{z} \end{pmatrix} \in \mathbb{C}^4$.

   (g) $\mathbf{y} = \begin{pmatrix} (2+i)\mathbf{w} \\ 1+i \\ 2\mathbf{z} \end{pmatrix} \in \mathbb{C}^5$.

6. Show that complex multiplication is commutative (i.e., that $zw = wz$ for any two complex numbers z and w).

7. Show that the angle matrices $\mathbf{A}(\alpha) = \begin{pmatrix} \cos\alpha & -\sin\alpha \\ \sin\alpha & \cos\alpha \end{pmatrix} \in \mathbb{R}^{2,2}$ satisfy $\mathbf{A}(\alpha)\mathbf{A}(\beta) =$

$\mathbf{A}(\alpha+\beta) = \mathbf{A}(\beta)\mathbf{A}(\alpha)$ for all $\alpha, \beta \in \mathbb{R}$. (*Hint*: Use the addition theorems for sine and cosine.)

8. (a) Show that for $z = a + bi \neq 0 \in \mathbb{C}$, the complex number inverse z^{-1} is $\frac{1}{z} = \frac{\overline{z}}{|z|^2}$.

(b) For $z = 1 + i$, $w = 2 - 3i$, and $i \in \mathbb{C}$ draw the three respective complex inverses $\frac{1}{z}, \frac{1}{w}, \frac{1}{i}$ and the three complex conjugates $\overline{z}, \overline{w}, \overline{i}$ in the same copy of the complex plane.

9. (a) Show that in the 2×2 real matrix representation of complex numbers, the inverse $\frac{1}{z}$ of $z = a + bi \neq 0$ with $\tan\alpha = \frac{b}{a}$ is represented by $\mathbf{Z}^{-1} := \frac{1}{|z|}\begin{pmatrix} \cos(-\alpha) & -\sin(-\alpha) \\ \sin(-\alpha) & \cos(-\alpha) \end{pmatrix} = \frac{1}{|z|}\begin{pmatrix} \cos\alpha & \sin\alpha \\ -\sin\alpha & \cos\alpha \end{pmatrix} = \frac{1}{|z|^2}\mathbf{Z}^T$.

(b) Show that for $z = a + bi \neq 0$ with $\tan\alpha = \frac{b}{a}$, we have $\frac{\overline{\mathbf{Z}}}{|z|^2} = \frac{1}{|\overline{z}|}\begin{pmatrix} \cos(-\alpha) & -\sin(-\alpha) \\ \sin(-\alpha) & \cos(-\alpha) \end{pmatrix} = \frac{1}{|\overline{z}|}\begin{pmatrix} \cos\alpha & \sin\alpha \\ -\sin\alpha & \cos\alpha \end{pmatrix} = \mathbf{Z}^{-1}$ from part (a), where $\overline{\mathbf{Z}}$ is the matrix representation of the complex conjugate $\overline{z} = a - bi$ of z.

(c) Show that the two 2×2 real matrix representations of the inverse of a complex number z in parts (a) and (b) are identical.

10. Find a 2×2 real matrix representation of a complex number $z = bi$ on the imaginary axis of $\mathbb{C}$ for which the angular tangent cannot be naively defined.

11. Prove the following properties for complex numbers $z = a + bi$ and $w = c + di$ with a, b, c, and $d \in \mathbb{R}$:

(a) $\overline{zw} = \overline{z}\,\overline{w}$.

(b) $\overline{z + w} = \overline{z} + \overline{w}$.

(c) $\overline{z^m} = (\overline{z})^m$ for all integers m.

12. Solve the following equations:

(a) $3x^2 = 27$.

(b) $x^6 - 1 = 0$.

(c) $w^2 + w + 4 = 0$.

(d) $1 + \frac{2}{z} + \frac{2}{z^2} = 0$.

13. Find the magnitudes and the angles sustained with the positive real axis for the following complex numbers
(a) $z = 1 + 3i$. (b) $w = -2 - 7i$.
(c) $u = 2 - 2i$. (d) $v = -4i + 12$.
(e) $r = -23.73$. (f) $s = 14.3i$.

14. (a) Write out the 2×2 rotation matrices $\mathbf{A}$ and $\mathbf{B}$ for the rotations of $\mathbb{C}$ by 30° degrees clockwise and counterclockwise.

(b) Evaluate the two matrix products $\mathbf{AB}$ and $\mathbf{BA}$ for $\mathbf{A}$ and $\mathbf{B}$ from part (a).

(c) Which powers of $\mathbf{A}$ or $\mathbf{B}$ are equal to $-\mathbf{I}_2$, which to $\mathbf{I}_2$?

15. Consider the sequence $\{x_i\} \subset \mathbb{R}$ for $x_1 = 1$, $x_i := 1 + \frac{1}{x_{i-1} + 1}$ and $i = 2, 3, \ldots$.

(a) Write out the first seven numbers $x_1, \ldots, x_7$ of this sequence.

(b) Find the theoretical limit x^* of the sequence $\{x_i\}$.

16. Compute the roots of the following quadratic polynomials by both the stable and unstable version of the quadratic formula and compare (use a basic calculator, if possible):

(a) $p(x) = x^2 - 200,000x + 10$.

(b) $q(x) = 10x^2 + 10,000x - 1$.

(c) $r(x) = 17x^2 - 35x + 0.0001$.

(d) $s(x) = x^2 + 9,000,000x - 4$.

Precarious roots

Appendix B

Finding Integer Roots of Integer Polynomials

The roots of first- and second-degree polynomials are easy to find. If $p(x) = x + \alpha$ has degree 1, then $p(-\alpha) = 0$ (i.e., $x_0 = -\alpha$ is the root of p). If $p(x) = x^2 + \alpha x + \beta$ has degree 2, then $p(x_1) = p(x_2) = 0$ for

$$x_1 = -\left(\frac{\alpha}{2} + \text{sign}(\alpha)\sqrt{\frac{\alpha^2}{4} - \beta}\right) \quad \text{and} \quad x_2 = \frac{\beta}{x_1},$$

according to the stable version of the **monic quadratic formula**. More simply, $x_{1,2} = -\frac{\alpha}{2} \pm \sqrt{\frac{\alpha^2}{4} - \beta}$. (See Appendix A for both quadratic formulas and the definition of the sign function.)

Let

$$p(x) = x^n + a_{n-1}x^{n-1} + \cdots + a_1 x + a_0 \qquad (*)$$

be a **monic polynomial** of arbitrary **degree** n with integer coefficients a_i (i.e., assume that $a_n = 1$ and $a_i \in \mathbb{Z}$ for p). Monic integer polynomials occur naturally as the vanishing or characteristic polynomials of integer matrices. If p has only integer roots $x_1, \ldots, x_n$, then each root is a factor of its constant term $a_0 \in \mathbb{Z}$, since

$$p(x) = x^n + \cdots + a_1 x + a_0 = \prod_{i=1}^{n}(x - x_i) = x^n + \cdots + (-1)^n \prod_{i=1}^{n} x_i .$$

To find the roots of an integer coefficient polynomial p given in the form (*), we recommend evaluating $p(x)$ for (plus and minus) each integer factor of $a_0 \in \mathbb{Z}$. Even if p has some noninteger roots, we still recommend a "guessing approach": Evaluate $p(x)$ for several small integer values $x = \ldots, -3, -2, -1, 0, 1, 2, 3, \ldots$ that divide a_0 until you find an $x \in \mathbb{Z}$ with $p(x) = 0$. Note that integer polynomials need not have integer roots at all. (See Problem 4.)

If the corresponding matrix eigenvalue problem is properly posed for hand arithmetic according to Problem T 9, for example, then the guessing approach will yield one or several roots of p quickly. If it does not, check to make sure that the vanishing, or characteristic, polynomial p of A_{nn} has been formed correctly. Specifically look at Theorem 9.5 here. If you can find no error and no integer root of p by guessing, check with the instructor, or simply **give up** on hand computations: There is no reasonable way for you to find the roots of higher degree integer polynomials by hand that do not have at least a few integer roots.

Once, however, an integer root x_0 of p has been found, we use **long division** to write

$$p(x) = (x - x_0)q(x)$$

for a polynomial q of lower degree than the given p. To find the other roots of the original polynomial p, continue with q in place of p (i.e., start by factoring q's constant term over the integers and by guessing an integer root of q until you are successful). Then lower q's degree by a second long division and so forth, until a quadratic polynomial with easily computable roots is reached.

EXAMPLE 1 (a) Find $567 \div 38$ by **long division**:

$$\begin{array}{r|l} & 14.92\ldots \\ \hline 38 & 567 \\ & \underline{38} \\ & 187 \\ & \underline{152} \\ & \ \ 350 \\ & \ \ \underline{342} \\ & \ \ \ \ \ \ 8 \end{array}$$

(b) For $p(x) = x^3 - 3x^2 + 5x - 3$, observe that $p(1) = 1 - 3 + 5 - 3 = 0$, while $p(-1) = -12$, $p(3) = 12$, and $p(-3) = -72$ are all nonzero. Thus, $x_0 = 1$ is the only integer root of p. We divide $p(x)$ by $(x - 1)$ as follows, using **polynomial long division**:

$$\begin{array}{rrrrrrrr} & & & x^2 & - & 2x & + & 3 \\ \cline{2-8} p(x) \div (x - x_0):\ x - 1 & x^3 & - & 3x^2 & + & 5x & - & 3 \\ & \_x^3 & \mp & x^2 & & & & \\ \cline{2-4} & & & -2x^2 & + & 5x & & \\ & & & \mp 2x^2 & \pm & 2x & & \\ \cline{4-6} & & & & & 3x & - & 3 \\ & & & & & \_3x & \mp & 3 \\ \cline{6-8} & & & & & & & 0 \ \text{remainder} \end{array}$$

Therefore, $p(x) = x^3 - 3x^2 + 5x - 3 = (x - 1)(x^2 - 2x + 3) = (x - 1)q(x)$. Note that if x_0 is a root of p, then the long division of p by $x - x_0$ must have zero **remainder**. This serves as a check of the arithmetic. The roots of the factor polynomial $q(x)$ of smaller degree with $p(x) = (x - 1)q(x)$ are also roots of p. The quadratic formula computes them as $1 \pm \sqrt{1-3} = 1 \pm \sqrt{2}i \in \mathbb{C}$ for our example.

Such complex conjugate roots readily occur as eigenvalues of integer matrices. For example, both the integer companion matrix $A = \begin{pmatrix} 0 & 0 & 3 \\ 1 & 0 & -5 \\ 0 & 1 & 3 \end{pmatrix}$ of p and the integer matrix $B = \begin{pmatrix} 11 & -3 & 3 \\ 35 & -6 & 12 \\ -12 & 5 & -2 \end{pmatrix}$ have 1 and $1 \pm \sqrt{2}i$ as their eigenvalues. ◀

MATLAB has a built-in polynomial root finder called `roots`. If $p(x) = a_n x^n + \cdots + a_k x^k + \cdots + a_1 x + a_0$ has degree n, form the vector $\mathbf{a} \in \mathbb{R}^{n+1}$ (or $\mathbb{C}^{n+1}$) that contains the $n+1$ coefficients $a_n, a_{n-1}, \ldots, a_1$, and a_0 in descending order (i.e., place a_n first, followed by a_{n-1}, etc., and a_0 last). (Note that this order is the reverse of the order we used in Section 1.3 for representing degree n polynomials by vectors in $\mathbb{R}^{n+1}$.) Then the MATLAB command `roots(a)` computes the roots of p as the eigenvalues of the $n \times n$ companion matrix for p. (See Section 9.3 and Lecture 13.)

A.B.P Problems

1. (a) Find the roots of $p(x) = x^3 + 6x^2 - x - 30$ by hand computation.

(b) Show that $q(x) = x^3 + 6x^2 - x + 30$ has no integer roots.

(c) Does $q(x) = x^3 + 6x^2 - x + 30$ have any real roots?

2. (a) Try to find the roots of $p(x) = x^4 - 4x^3 - 7x^2 + 34x - 24$ by hand.

(b) Try to use MATLAB to find the roots of p.

3. (a) Find the roots of $p(x) = x^5 + 7x^4 - x - 7$ by hand and using MATLAB.

(b) Does $q(x) = x^5 - 7x^4 - x - 7$ have any integer roots?

(c) Does $q(x) = x^5 - 7x^4 - x - 7$ have any real roots?

4. (a) Find the coefficients of a polynomial p with four distinct integer roots.

(b) Find an integer polynomial q of degree 4 without any integer roots.

(c) Verify by hand that the polynomial in (b) is correctly chosen.

(d) What roots does MATLAB compute for the polynomials chosen in parts (a) and (b)?

5. Find all roots of $x^4 - 1$.

6. Find all roots of $x^4 + 1$ over $\mathbb{C}$. (*Hint*: Think angles: When is $z^2 = -1$ for z in the complex plane $\mathbb{C}$?)

7. Assume that $x_1, \ldots, x_n$ are the roots of an nth degree polynomial $p(x) = x^n + a_{n-1}x^{n-1} + \cdots + a_1 x + a_0$. Show that

(a) $\sum_i x_i = -a_{n-1}$.

(b) $\prod_i x_i = (-1)^n a_0$.

(c) Evaluate the sum and product of the two roots x_1, x_2 in the monic quadratic formula for $p(x) = x^2 + ax + b$. That is, show that the two roots x_1 and x_2 in the quadratic formula satisfy $x_1 + x_2 = -a$ and $x_1 x_2 = b$.

8. Find all eigenvalues of $\mathbf{A} = \begin{pmatrix} 0 & 0 & 3 \\ 1 & 0 & -5 \\ 0 & 1 & 3 \end{pmatrix}$ by hand according to Chapter 9 and, alternatively, by using MATLAB.

9. Find the eigenvalues of both $\mathbf{A} = \begin{pmatrix} 10 & 7 \\ 7 & 5 \end{pmatrix}$ and $\mathbf{B} = \begin{pmatrix} 7 & 10 \\ 5 & 7 \end{pmatrix}$ using the methods of Chapter 9 and polynomial root finding.

10. Investigate the graph of $f(x) = x^5 - 5x^3 + 5x$ as follows:

(a) Complete the following table of values for f and plot the function from this table:

x	-2	-1	0	1	2
$f(x)$					

(b) Determine the behavior of f as $x \to \pm\infty$.

(c) Determine the roots of f by hand. (Factor off one root $x = 0$, set $z = x^2$, and solve for z, then for x.) Verify that your computed roots are correct via MATLAB.

(d) Plot f for $-2.2 \le x \le 2.2$ in MATLAB, using the commands `a=2.2; x=[-a:.01:a]; y=x.^5-5*x.^3+5*x; plot(x,y), hold on, plot([-a;a],[0;0]), x=axis; plot([0;0],[x(3);x(4)]), plot([-2:2],[-2:2],'o'), hold off.`

11. Let $p(x) = a_n x^n + a_{n-1}x^{n-1} + \cdots + a_1 x + a_0$ be a polynomial of degree n (i.e., a polynomial with $a_n \neq 0$), and assume that $a_0 \neq 0$ as well.

(a) Show that $x_0 = 0$ cannot be a root of p.

(b) Show that if p has the root $x_1 \neq 0$, then the "reverse-order coefficient polynomial" $q(x) := a_0 x^n + a_1 x^{n-1} + \cdots + a_{n-1}x + a_n$ has the root $\dfrac{1}{x_1}$ and vice versa.

Spatial abstractions

Appendix C

Abstract Vector Spaces

Linear spaces occur in many contexts of mathematics and science. Oftentimes, vector spaces are finite dimensional, such as $\mathbb{R}^n$ or $\mathbb{C}^n$. More complicated examples and applications occur with finite and infinite dimensional linear spaces of functions, sequences, and the like. The closure of a vector space when taking linear combinations of vectors, is the glue for this fundamental mathematical concept.

For a given field $\mathbb{F}$ of scalars, such as $\mathbb{R}$ or $\mathbb{C}$, a vector space V satisfies the following 10 axioms:

DEFINITION

An object $(V, \mathbb{F}, +, \cdot)$, consisting of **vectors** $\mathbf{v} \in V$ and **scalars** $\alpha \in \mathbb{F}$, that allows the two operations of **vector addition** $(+)$ and **scalar multiplication** $(\cdot)$ is a **vector space** if the following axioms are satisfied:

(C. I) The vector sum $\mathbf{u} + \mathbf{v}$ belongs to V for all vectors $\mathbf{u}$ and $\mathbf{v} \in V$.

(C. II) The scalar multiple $\alpha\mathbf{u}$ belongs to V for all $\mathbf{u} \in V$ and all $\alpha \in \mathbb{F}$.

(I. III) $\mathbf{u} + \mathbf{v} = \mathbf{v} + \mathbf{u}$ for all vectors $\mathbf{u}$ and $\mathbf{v} \in V$.

(I. IV) $(\mathbf{u} + \mathbf{v}) + \mathbf{w} = \mathbf{u} + (\mathbf{v} + \mathbf{w})$ for all vectors $\mathbf{u}$, $\mathbf{v}$, and $\mathbf{w} \in V$.

(E. V) There exists a vector $\mathbf{0} \in V$, called the **zero vector**, with $\mathbf{u} + \mathbf{0} = \mathbf{u}$, for every vector $\mathbf{u} \in V$.

(E. VI) For each vector $\mathbf{u} \in V$, there is a vector $-\mathbf{u} \in V$ with $\mathbf{u} + (-\mathbf{u}) = \mathbf{0} \in V$, the zero vector.

(D. VII) $\alpha(\mathbf{u} + \mathbf{v}) = \alpha\mathbf{u} + \alpha\mathbf{v}$ for all vectors $\mathbf{u}, \mathbf{v} \in V$ and all scalars $\alpha \in \mathbb{F}$.

(D. VIII) $(\alpha + \beta)\mathbf{u} = \alpha\mathbf{u} + \beta\mathbf{u}$ for all vectors $\mathbf{u} \in V$ and all scalars $\alpha, \beta \in \mathbb{F}$.

(M. IX) $\alpha(\beta\mathbf{u}) = (\alpha\beta)\mathbf{u}$ for all vectors $\mathbf{u} \in V$ and all scalars $\alpha, \beta \in \mathbb{F}$.

(M. X) $1\mathbf{u} = \mathbf{u}$ for all vectors $\mathbf{u} \in V$, and the unit element $1 \in \mathbb{F}$.

These axioms for a vector space are grouped into

- two closure axioms (C. I), (C. II) under the linear operations;
- two vector identity axioms (I. III), (I. IV) that ensure the commutativity and associativity of vector addition;

- two existence axioms (E. V), (E. VI) for the zero and the negative vector;
- two distributive axioms (D. VII), (D. VIII) of addition and multiplication; and
- two scalar multiplication axioms (M. IX), (M. X).

EXAMPLE 1 (a) Real n-space $\mathbb{R}^n$, or $(\mathbb{R}^n, \mathbb{R}, +, \cdot)$, with its field of scalars $\mathbb{R}$ and the two vector space operations of vector addition and scalar multiplication, is a vector space over the reals, since it satisfies all 10 axioms. (Check!)

(b) $W := \left(\{(1, y) \in \mathbb{R}^2 \mid y \text{ real}\}, \mathbb{F} = \mathbb{R}, +, \cdot\right)$, consisting of all points in $\mathbb{R}^2$ that lie on the line parallel to the y-axis through $(1, 0)$, is not a vector space with the standard operations $+$ and $\cdot$ of $\mathbb{R}^2$. W does not satisfy axioms (C. I) or (C. II): For $(1, 1) \in W$, note that $(1, 1) + (1, 1) = (2, 2) \notin W$, and likewise, $3(1, 1) = (3, 3) \notin W$, since all elements of W must have their first component equal to 1. ◀

The following properties hold true in all vector spaces:

Proposition: Let $(V, \mathbb{F}, +, \cdot)$ be a vector space. Then, for any $\mathbf{u} \in V$,

(a) $0\mathbf{u} = \mathbf{0}$ (i.e., the scalar zero in $\mathbb{F}$, multiplied by any vector $\mathbf{u} \in V$, is the zero vector).

(b) If $\mathbf{u} + \mathbf{v} = \mathbf{0}$, then $\mathbf{u} = -\mathbf{v}$.

(c) $(-1)\mathbf{u} = -\mathbf{u}$ for every $\mathbf{u} \in V$. ◀

Proof We deduce these properties solely from the 10 vector space axioms.

(a) Using (M. X) and (D. VIII), we observe that

$$\mathbf{u} = 1\mathbf{u} = (1 + 0)\mathbf{u} = 1\mathbf{u} + 0\mathbf{u} = \mathbf{u} + 0\mathbf{u}.$$

Thus,

$$\mathbf{0} = -\mathbf{u} + \mathbf{u} = -\mathbf{u} + \mathbf{u} + 0\mathbf{u} = (-\mathbf{u} + \mathbf{u}) + 0\mathbf{u} = \mathbf{0} + 0\mathbf{u} = 0\mathbf{u},$$

according to the axioms (C. I), (C. II), (E. V), (E. VI), and (I. IV).

(b) If $\mathbf{u} + \mathbf{v} = \mathbf{0}$, then

$$-\mathbf{u} = -\mathbf{u} + \mathbf{0} = -\mathbf{u} + (\mathbf{u} + \mathbf{v}) = (-\mathbf{u} + \mathbf{u}) + \mathbf{v} = \mathbf{0} + \mathbf{v} = \mathbf{v},$$

using the axioms (C. I), (I, IV), (E. V), and (E. VI).

(c) By part (a), (D. VIII), (M, IX), and (M. X), we have

$$\mathbf{0} = 0\mathbf{u} = (1 + (-1))\mathbf{u} = 1\mathbf{u} + (-1)\mathbf{u} = \mathbf{u} + (-1)\mathbf{u}.$$

Hence, $(-1)\mathbf{u} = -\mathbf{u}$ by part (b). ■

Finite-dimensional vector spaces are studied in linear algebra and matrix theory. Infinite-dimensional ones are the subject of functional analysis and operator theory. The most common fields $\mathbb{F}$ used for vector spaces are either the reals $\mathbb{R}$ or the complex numbers $\mathbb{C}$. Vector spaces over general fields (and rings) lead us into modern algebra. Axiomatic definitions like the one for groups in Section 6.2 or the foregoing one for vector spaces afford mathematicians the benefit of rigor. Unfortunately, an axiomatic foundation of a subject is often too cumbersome and distracting for beginners to be of much use. We conclude with a few examples, mainly of vector spaces that do not look like the familiar $\mathbb{R}^n$.

EXAMPLE 2

(a) The set $\mathcal{F}$ of all functions f defined on $\mathbb{R}$, or on an interval of $\mathbb{R}$, becomes a vector space over the field of reals if we define the sum and scalar product of functions as $(f+g)(x) := f(x)+g(x)$ and $(\alpha f)(x) := \alpha f(x)$ for all x and $\alpha \in \mathbb{R}$.

(b) The set $\mathcal{S}$ of all real sequences $\{a_i\}$ becomes a vector space over the field of reals if we define the sum and scalar product of sequences as $\{a_i\}+\{b_i\} := \{a_i+b_i\}$ and $\alpha\{a_i\} := \{\alpha a_i\}$ for all $\alpha \in \mathbb{R}$. ◀

For a given vector space $(V, \mathbb{F}, +, \cdot)$, a subset $U \subset V$ is a vector **subspace** if

(a) $\mathbf{0} \in U$,
(b) for all $\mathbf{u}, \mathbf{v} \in U$, the vector sum $\mathbf{u}+\mathbf{v} \in U$, and
(c) for all $\mathbf{u} \in U$ and $\alpha \in \mathbb{F}$, the scalar product $\alpha\mathbf{u} \in U$.

This is the same definition as for $\mathbb{R}^n$. (Compare it with Definition 1 of Section 4.1.)

Special subspaces of the space of all one-variable real-valued functions $\mathcal{F}$, such as the differentiable or integrable functions, the polynomials, and the solutions of specific differential equations are studied in Sections 1.3, 7.3, 11.3, and 14.3, for example.

To ensure that a given subset of a vector space is a subspace, we need not check all 10 vector space axioms, but rather, we need to verify only that the subset is nonempty and closed in the sense of the aforementioned conditions (b) and (c).

EXAMPLE 3

(a) The set $\mathcal{S}_C :=$ {all convergent real sequences } $\subset \mathcal{S}$ forms a subspace of all real sequences $\mathcal{S}$, since $\lim(a_i+b_i) = \lim(a_i)+\lim(b_i)$ and $\lim(ca_i) = c\lim(a_i)$ hold for convergent sequences, as is seen in calculus.

(b) The set $\mathcal{S}_0 :=$ {all real sequences that converge to zero } $\subset \mathcal{S}$ forms a subspace of $\mathcal{S}_C$ and of $\mathcal{S}$, since all sums and scalar multiples of sequences that converge to zero also converge to zero.

(c) The set $\mathcal{F}_{0,0} :=$ {all real functions with $f(0)=0$} $\subset \mathcal{F}$ forms a subspace of the real-valued functions $\mathcal{F}$, since all sums and scalar multiples of functions $f \in \mathcal{F}_0$ also vanish at zero.

(d) The set $\mathcal{F}_{1,0} :=$ {all real functions with $f(1)=0$} $\subset \mathcal{F}$ likewise forms a subspace of $\mathcal{F}$.

(e) The set $\mathcal{F}_{1,1} := \{\text{all real functions with } f(1) = 1\} \subset \mathcal{F}$ is not closed under addition or under scalar multiplication. For example, if $f \in \mathcal{F}_{1,1}$, then $(2f)(1) = 2 \neq 1$, and thus $2f \notin \mathcal{F}_{1,1}$. Hence, the set $\mathcal{F}_{1,1}$ does not form a subspace of $\mathcal{F}$. ◀

A.C.P Problems

1. Decide which of the objects $(V, \mathbb{F}, +, \cdot)$ that follow are vector spaces. Here, the first entry V denotes the set of vectors, the second the field of scalars $\mathbb{F}$, and $+ : V \times V \to V$ and $\cdot : \mathbb{F} \times V \to V$ are ordinary vector addition and scalar multiplication:
   (a) $V_1 = \{\mathbb{R}^n, \mathbb{C}, +, \cdot\}$
   (b) $V_2 = \{\mathbb{C}^n, \mathbb{R}, +, \cdot\}$
   (c) $V_3 = \{\mathbb{C}^n, \mathbb{C}, +, \cdot\}$
   (d) $V_4 = \{\{0\}, \mathbb{C}, +, \cdot\}$
2. Let $\mathbb{Q} = \left\{ \frac{p}{q} \,\middle|\, p, q \text{ integer} \right\}$ denote the rational numbers and $\mathbb{Q}^n$ the set of n-vectors with rational components. Which of the following are vector spaces?
   (a) $W_1 = \{\mathbb{Q}^n, \mathbb{Q}, +, \cdot\}$
   (b) $W_2 = \{\mathbb{Q}^n, \mathbb{C}, +, \cdot\}$
   (c) $W_3 = \{\mathbb{Q}^n, \mathbb{R}, +, \cdot\}$
   (d) $W_4 = \{\mathbb{R}^n, \mathbb{Q}, +, \cdot\}$
   (e) $W_5 = \{\{0\}, \mathbb{Q}, +, \cdot\}$
3. Decide whether the following subsets of the real function space $\mathcal{F}$ are subspaces:
   (a) $\mathcal{F}_{0,1} = \{\text{all real functions with } f(0) = 1\}$
   (b) $\mathcal{F}_{2,0} = \{\text{all real functions with } f(2) = 0\}$
   (c) $\mathcal{F}_{even} = \{\text{all even real functions with } f(-x) = f(x) \text{ for all } x \in \mathbb{R}\}$
   (d) $\mathcal{F}_{odd} = \{\text{all odd real functions with } f(-x) = -f(x) \text{ for all } x \in \mathbb{R}\}$
4. Consider the differential equation $f''(x) + \sin(x) f(x) = h(x)$.
   (a) Show that if $h(x) = 0$ for all x, then all solutions of the differential equation form a subspace of the twice-differentiable real functions.
   (b) Investigate whether the solutions of the differential equation for $h \neq 0$ also form a subspace.
5. Show that the convergent real power series for one fixed radius of convergence $r > 0$ forms a vector space over $\mathbb{R}$ with the standard addition and scalar multiplication.
6. Define $T_{ls} : \mathcal{S} \to \mathcal{S}$ as $T_{ls}((a_1, a_2, \ldots)) := (a_2, a_3, \ldots)$ (i.e., as the left-shift operator of the space of real sequences $\{a_i\}$). Likewise define $T_{rs} : \mathcal{S} \to \mathcal{S}$ as $T_{rs}((a_1, a_2, \ldots)) := (0, a_1, a_2, \ldots)$ (i.e., as the right-shift operator of the real sequences).
   (a) Show that both shift operators are linear functions.
   (b) Show that T_{rs} has no eigenvalue.
   (c) Show that T_{ls} has every scalar as an eigenvalue. What are the corresponding "eigensequences" $\{a_i\}$?
   (d) Show that $T_{ls} \circ T_{rs} = id_{\mathcal{S}}$. What is $T_{rs} \circ T_{ls}$?
   (e) Find all real sequences $\{a_i\}$ with $T_{rs} \circ T_{ls}\{a_i\} = (0, 0, \ldots)$.
   (f) Find all real sequences $\{a_i\}$ with $T_{ls} \circ T_{rs}\{a_i\} = (0, 0, \ldots)$.
7. Decide whether the following sets are subspaces of the space of real functions $\mathcal{F}$:
   (a) All polynomials of the form $p(x) = ax^2 - b$ for arbitrary constants $a, b \in \mathbb{R}$.
   (b) All polynomials of the form $p(x) = ax^2 - b$ for arbitrary rational constants $a, b \in \mathbb{Q}$.
   (c) All polynomials of the form $p(x) = x^2 + b$.
   (d) All polynomials of the form $p(x) = (ax - c)x^2 + b$.
   (e) All polynomials p with $p(14) = -2$.
   (f) All polynomials p with $p(14) = 0$.

8. Using the vector space axioms, prove that $\mathbf{u} + \mathbf{x} = \mathbf{u} + \mathbf{y}$ implies $\mathbf{x} = \mathbf{y}$ in a vector space.

9. (a) Show that the set of diagonal matrices $\mathbf{D} \subset \mathbb{R}^{n,n}$ forms a vector space under ordinary matrix addition and scalar matrix multiplication.

(b) Do the matrices $\mathbf{A} \in \mathcal{D}$ with (1,1) entry equal to 3 form a vector space?

(c) Do the matrices $\mathbf{A} \in \mathcal{D}$ with (n, n) entry equal to 0 form a vector space?

Appendix D

Inner Product Spaces*

[*On the Web only at `http://www.auburn.edu/~uhligfd/TLA/download/AIPS.pdf`.]

All clear now

Solutions to Selected Problems

The best use of this section is <u>not to use it at all</u>.
Custom, however, dictates that we include solutions to every other problem.
The worst use is to look at a solution before reading the actual problem, before thinking about the problem, and without trying to work it out by oneself.

Learning takes self-discipline. Self-discipline takes practice.

Chapter 1

1.1.P p. 32

0. (*This problem and its proof are included to lay a solid foundation for the study of linear algebra.*)

Proof: (Parallelogram law and vector addition)

Look at the two vectors from the origin $\mathbf{0}$ in $\mathbb{R}^2$ to $\mathbf{a}$ and $\mathbf{b} \in \mathbb{R}^2$ and at the supporting triangle $\mathbf{0}A\mathbf{b}$ of $\mathbf{b}$:

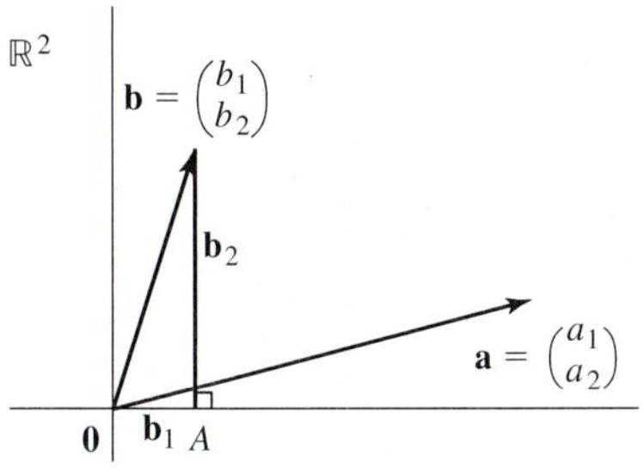

Next, draw a line though a parallel to the first coordinate axis of $\mathbb{R}^2$. On this line, measure out a segment of length $|\mathbf{0}A| = \mathbf{b}_1$ from $\mathbf{a}$ to B.

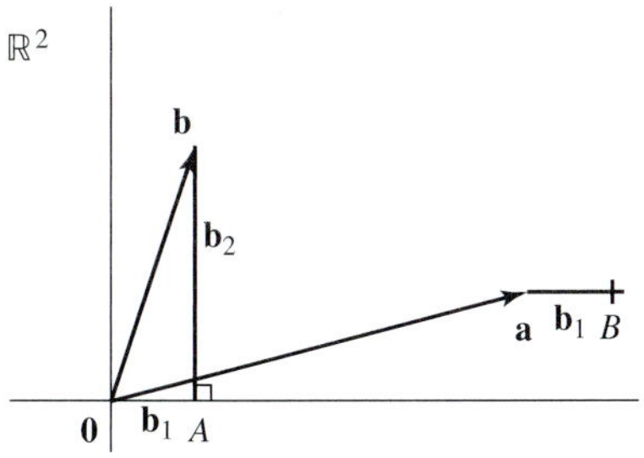

Form a right angle at B with one leg $\mathbf{aB}$, and mark the point C on the other leg so that $|BC| = |A\mathbf{b}| = \mathbf{b}_2$.

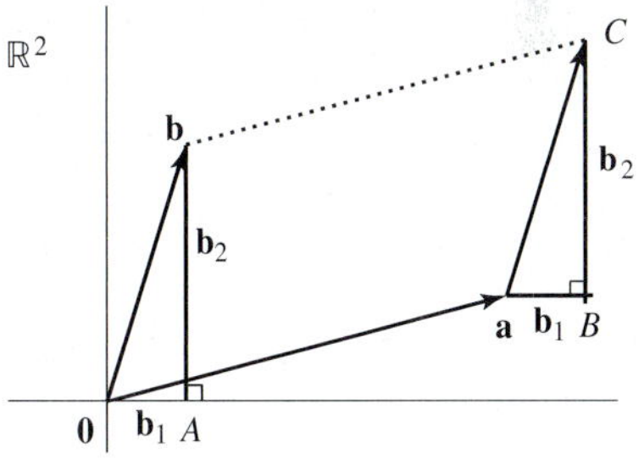

By construction, C has the coordinates $a_1 + b_1$ and $a_2 + b_2$. Thus, the vector from $\mathbf{0}$ to C is $\begin{pmatrix} a_1 + b_1 \\ a_2 + b_2 \end{pmatrix}$, the algebraic sum of the two given vectors $\begin{pmatrix} a_1 \\ a_2 \end{pmatrix}$ and $\begin{pmatrix} b_1 \\ b_2 \end{pmatrix} \in \mathbb{R}^2$. Where does C lie geometrically?

From elementary geometry, recall the properties of congruent triangles and of parallel lines cut by a traversal:

By construction, we have $|\mathbf{0}A| = |\mathbf{a}B|$, $\angle(\mathbf{0}A\mathbf{b}) = \angle(\mathbf{a}BC) = 90^o$, and $|A\mathbf{b}| = \mathbf{b}_2 = |BC|$. Thus, the two triangles $\mathbf{0}A\mathbf{b}$ and $\mathbf{a}BC$ are congruent. Hence, $\angle(A\mathbf{0}\mathbf{b}) = \angle(B\mathbf{a}C)$. Since $\mathbf{a}B \| x_1$–axis $\| \mathbf{0}A$ we have $\mathbf{0b} \| \mathbf{a}C$ and $|\mathbf{0b}| = |\mathbf{a}C|$. Therefore, the planar figure with the four corners $\mathbf{0}$, $\mathbf{a}$, C, and $\mathbf{b}$ is a parallelogram, and $\overline{\mathbf{0}C}$ is its diagonal. ∎

1. True: (a), (c), (f), (g), (h) and (j). The rest are false. (You may agree or disagree on part (j).)

3. (a) T_3, T_4 are nonlinear components of T.

(b) S_2 and S_4 are nonlinear.

(c) T_1, T_2, S_1, S_3, and S_5 are linear.

5. (a) $\mathbf{A}_{\mathcal{E}} = \begin{pmatrix} 0 & 2 & -3 \\ 1 & 3 & 7 \\ 4 & -1 & 0 \end{pmatrix}$.

(b) **Theorem 1.3:** If f and $g : \mathbb{R}^n \to \mathbb{R}^m$ are two linear functions with respective standard matrix representations $\mathbf{A}_{\mathcal{E}}$ and $\mathbf{B}_{\mathcal{E}} \in \mathbb{R}^{m,n}$, then their sum $h(\mathbf{x}) := f(\mathbf{x}) + g(\mathbf{x}) : \mathbb{R}^n \to \mathbb{R}^m$ is a linear map with the $m \times n$ standard matrix representation $\mathbf{C}_{\mathcal{E}} := \mathbf{A}_{\mathcal{E}} + \mathbf{B}_{\mathcal{E}}$.

7. (a) $\begin{pmatrix} 11 \\ -1 \\ 6 \end{pmatrix}$, (c) $\begin{pmatrix} 2 & 1 & 1 \\ 4 & 2 & 2 \\ 0 & 0 & 0 \\ -2 & -1 & 1 \end{pmatrix}$,

(f) $(8, 9, 0, -1)$. The other products cannot be formed.

9. $$\mathbf{A}(\alpha\mathbf{x} + \beta\mathbf{y}) = \begin{pmatrix} 1 & -2 & 6 \\ 2 & -1 & 3 \\ 1 & 2 & -1 \end{pmatrix} \begin{pmatrix} \alpha x_1 + \beta y_1 \\ \alpha x_2 + \beta y_2 \\ \alpha x_3 + \beta y_3 \end{pmatrix} =$$
$$\begin{pmatrix} \alpha x_1 - 2\alpha x_2 + 6\alpha x_3 + \beta y_1 - 2\beta y_2 + 6\beta y_3 \\ 2\alpha x_1 - \alpha x_2 + 3\alpha x_3 + 2\beta y_1 - \beta y_2 + 3\beta y_3 \\ \alpha x_1 + 2\alpha x_2 - \alpha x_3 + \beta y_1 + 2\beta y_2 - \beta y_3 \end{pmatrix} =$$
$$\alpha \begin{pmatrix} 1 & -2 & 6 \\ 2 & -1 & 3 \\ 1 & 2 & -1 \end{pmatrix} \mathbf{x} + \beta \begin{pmatrix} 1 & -2 & 6 \\ 2 & -1 & 3 \\ 1 & 2 & -1 \end{pmatrix} \mathbf{y}.$$

11. $$\mathbf{A}\begin{pmatrix} 1 \\ 1 \\ a \end{pmatrix} = \begin{pmatrix} 1+a \\ a^2 - 1 \\ 2a \end{pmatrix}; \mathbf{A}\begin{pmatrix} a \\ 0 \\ -a \end{pmatrix} = \begin{pmatrix} a \\ -a^2 \\ a^2 - a \end{pmatrix}.$$

13. (a) Each component function is linear.

(b) $T(\mathbf{e}_1) = (1, 0, 0, 0)^T$, $T(\mathbf{e}_2) = (2, 3, 0, 0)^T$, $T(\mathbf{e}_3) = (-1, 0, 0, 0)^T$, $T(\mathbf{e}_4) = (0, -1, 0, 0)^T$, $T(\mathbf{e}_5) = (0, 0, 0, 0)^T$, $T(\mathbf{e}_6) = (0, 0, 0, -1)^T$.

(c), (d) $\mathbf{A} = \begin{pmatrix} 1 & 2 & -1 & 0 & 0 & 0 \\ 0 & 3 & 0 & -1 & 0 & 0 \\ 0 & 0 & 0 & 0 & 0 & 0 \\ 0 & 0 & 0 & 0 & 0 & -1 \end{pmatrix}_{4\times 6}$.

15. $\mathbf{A}_{\mathcal{E}} = \begin{pmatrix} 0 & -20 & 6 & -5 \\ 1 & 3 & 0 & 1 \\ 2 & 0 & 6 & 0 \end{pmatrix}_{3\times 4}$.

17. (a) $\mathbf{A} = \begin{pmatrix} -1 & 11 & 3 \\ -1 & 0 & -4 \\ -1 & 2 & 17 \end{pmatrix}$; $\mathbf{A}\begin{pmatrix} 1 \\ -1 \\ 1 \end{pmatrix} = \begin{pmatrix} -9 \\ -5 \\ 14 \end{pmatrix}$.

(b) $\mathbf{A} = \begin{pmatrix} \frac{1}{3} & 1 & 26 \\ 0 & -1 & 2 \\ 0 & -1 & -4 \end{pmatrix}$; $\mathbf{A}\begin{pmatrix} 1 \\ -1 \\ 1 \end{pmatrix} = \begin{pmatrix} 75/3 \\ 3 \\ -3 \end{pmatrix}$.

(c) $\mathbf{A} = \begin{pmatrix} 4 & 1 & 10 \\ 0 & -\frac{1}{3} & -\frac{19}{3} \\ 1 & \frac{2}{3} & -\frac{1}{3} \end{pmatrix}$; $\mathbf{A}\begin{pmatrix} 1 \\ -1 \\ 1 \end{pmatrix} = \begin{pmatrix} 13 \\ -6 \\ 0 \end{pmatrix}$.

19. (a) $T(\mathbf{e}_1) = \begin{pmatrix} -2/3 \\ 2/3 \end{pmatrix}$; $T(\mathbf{e}_2) = \begin{pmatrix} -1/3 \\ 7/3 \end{pmatrix}$.

(b) $\mathbf{A} = \frac{1}{3}\begin{pmatrix} -2 & -1 \\ 2 & 7 \end{pmatrix}$.

(c) $T(1, -1) = \begin{pmatrix} -1/3 \\ -5/3 \end{pmatrix}$; $T(3, 2) = \begin{pmatrix} -8/3 \\ 20/3 \end{pmatrix}$.

(d) $\mathbf{x} = \mathbf{0} \in \mathbb{R}^2$.

21. (b) $f'(x) = 6x - 7 \neq f(1) = 10$.

(c) $g(x) = g(1)x$.

23. $\mathbf{A}_{\mathcal{E}} = \begin{pmatrix} 1 & 0 & 1 \\ 0 & 2 & 0 \\ 1 & 0 & 1 \end{pmatrix}$.

25. (a) Yes. (b) Linear component functions.

(c) $T(3, 19, -4) = (0, 19, 0)$; $T(19, 0, -14) = \mathbf{0} \in \mathbb{R}^3$; $T(\mathbf{e}_3) = \mathbf{0}$; $T(3\mathbf{e}_2 - 4\mathbf{e}_1) = (0, 3, 0)$; $T(\mathbf{0}) = \mathbf{0}$.

(d) $\mathbf{A}_{\mathcal{E}} = \begin{pmatrix} 0 & 0 & 0 \\ 0 & 1 & 0 \\ 0 & 0 & 0 \end{pmatrix}$.

27. (a) Yes. (b) $f(\mathbf{e}_2) = \begin{pmatrix} 1 \\ -1 \end{pmatrix}$, $g(\mathbf{e}_1) = \begin{pmatrix} 1 \\ -1 \end{pmatrix}$,

$f(\mathbf{e}_1 + \mathbf{e}_2) = \begin{pmatrix} 3 \\ -1 \end{pmatrix}$, $g(2\mathbf{e}_2 - 3\mathbf{e}_3) = \begin{pmatrix} -6 \\ 8 \end{pmatrix}$.

(c) $-2f(\mathbf{e}_3) + 3g(\mathbf{e}_3) = \begin{pmatrix} 8 \\ -10 \end{pmatrix}$.

(d) $\mathbf{C}_{\mathcal{E}} = \begin{pmatrix} -1 & -2 & 8 \\ -3 & 5 & -10 \end{pmatrix}$.

(e) $g(\mathbf{x}) = \frac{5}{3}\mathbf{e}_1$.

29. Only $h(x)$ and $m(x)$ are linear functions.

31. (b) $T(\mathbf{e}_2) = \mathbf{e}_1$. (c) $\mathbf{A}_{\mathcal{E}} = \begin{pmatrix} 0 & 1 \\ -1 & 0 \end{pmatrix}$.

33. (a) number of square feet = 9· number of square yards; number of cubic feet = 27· number of cubic yards.
(b) All three conversion formulas are linear functions.
(c) $T((1,-1)) = (27,-16)$; $T((3,2)) = (-94,62)$.
(d) $T(\mathbf{0}) = \mathbf{0}$.

35. The functions in (d) and (f) cannot be linear; the rest may be linear.

37. (a), (b) $\mathbf{A}_{\mathcal{E}} = \begin{pmatrix} | & | \\ T(\mathbf{e}_1) & T(\mathbf{e}_2) \\ | & | \end{pmatrix} = \begin{pmatrix} -8 & -35 \\ 6 & 22 \end{pmatrix}$.

39. (a) $\mathbf{A}_{\mathcal{E}} = \begin{pmatrix} 4 & -2 \\ -1 & 1 \end{pmatrix}$, $\mathbf{B}_{\mathcal{E}} = \begin{pmatrix} 3 & 9 \\ -1 & -3 \end{pmatrix}$.
(b) $\begin{pmatrix} 18 & -10 \\ -5 & 3 \end{pmatrix}$. (c) $\begin{pmatrix} 0 & 0 \\ 0 & 0 \end{pmatrix} = \mathbf{O}_2$.

1.3.P p. 49

3. (a)

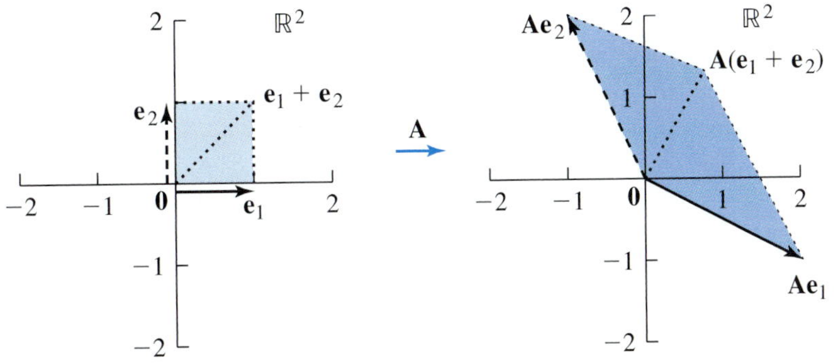

(b)

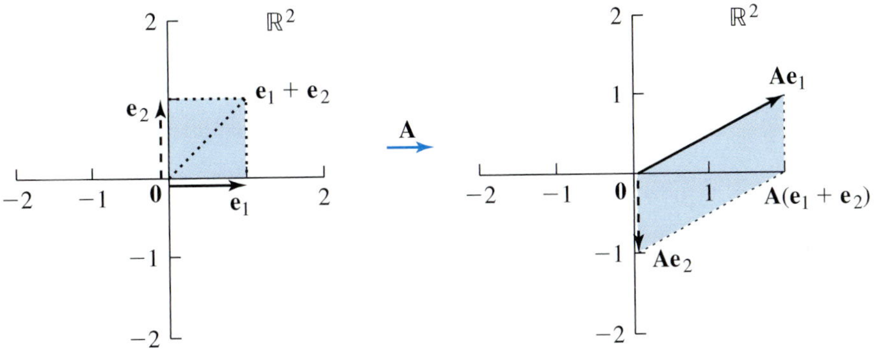

(c)

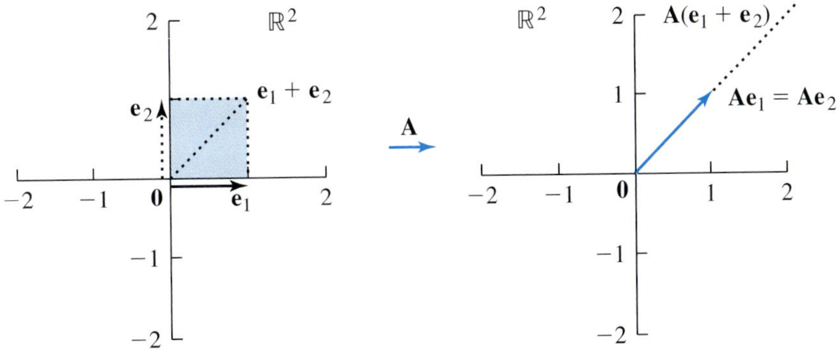

(d) The picture is the same as in part (c), except that the arrow in the range space points in the opposite direction, into the third quadrant rather than the first.

(e)

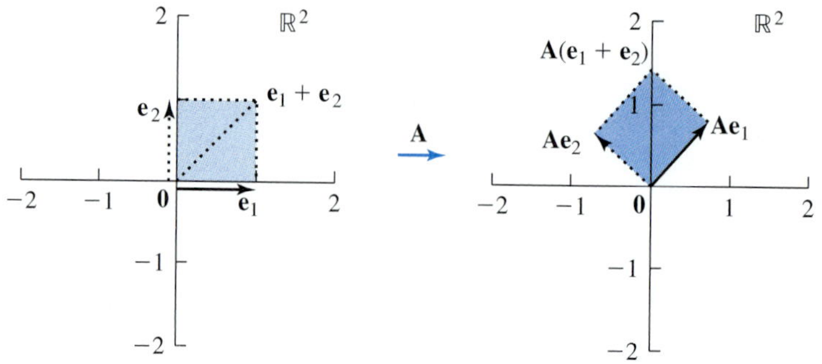

(f) $\mathbf{x} \mapsto \mathbf{Ax}$ describes a reflection of the plane across the angular bisector of quadrants one and three.

5. $\mathbf{A}_{\mathcal{E}} = \frac{1}{\sqrt{2}} \begin{pmatrix} \sqrt{2} & -\sqrt{2} \\ \sqrt{2} & \sqrt{2} \end{pmatrix}.$

7. T is linear.

9. (a) $(20, 22, -22)^T$, $(0, 0, 0)^T$, and $(4, 0, -20)^T$.

(b) Yes. (c) $\mathbf{A}_{3,n+1} =$

$$\begin{pmatrix} 1 & 1 & 1 & 1 & \dots & 1 \\ 1 & 0 & 0 & 0 & & 0 \\ 1 & -2 & 4 & -8 & \dots & (-1)^n 2^n \end{pmatrix}.$$

11. 1796.6666....

13. (a) $\mathbf{A}$ is 4×4; $\mathbf{x}$ is a row vector.

(b) Only $\mathbf{Ax}^T$, $\mathbf{xA}$, $\mathbf{xx}^T$, and $\mathbf{x}^T\mathbf{x}$ can be formed.

(c) Only $\mathbf{x} + \mathbf{x}$ and $\mathbf{A} + 2\mathbf{A}$ can be formed.

15. (a) (2); (b) (7); (c) (1); (d) (4); (e) (5); (f) (3); and (g) (6).

17. The transformations in (a), (b), and (c) are linear. The one in (d) is not linear, since $p(1) \neq 0$ is possible for p in $\mathcal{P}_4$.

1.R p. 51

1. The columns of $\mathbf{A}_{\mathcal{E}}$ lie in $\mathbb{R}^k$. They are the images $\mathbf{Ae_i} \in \mathbb{R}^k$ of the standard unit vectors $\mathbf{e_i} \in \mathbb{R}^n$. $\mathbf{A}_{\mathcal{E}}$ has k rows, n columns, and $k \cdot n$ entries.

3. (d) $\mathbf{M} = \pi \mathbf{I}_n$; $\mathbf{K} = -\mathbf{I}_n$;

$$\mathbf{J} = \begin{pmatrix} 0 & & & 1 \\ & & 0 & \\ & & \cdot^{\cdot^{\cdot}} & \\ & 0 & & \\ -1 & & & 0 \end{pmatrix}_{nn}.$$

5. (a) $F(0, 0) = (0^2, 0 \cdot 0) = (0, 0)$.

(b) The component functions are not linear.

7. T is nonlinear for $n \geq 1$.

9. $T(\mathbf{0}) = T(\mathbf{x} - \mathbf{x}) = T(\mathbf{x}) - T(\mathbf{x}) = \mathbf{0}$.

11. $\mathbf{A} = \begin{pmatrix} 7/4 & -5/2 & -1/4 \\ 2 & -3 & 2 \end{pmatrix}.$

13. f cannot be linear. Evaluate $f((2, -1) - (1, -1)) = f(1, 0)$ in two ways.

Chapter 2

2.1.P p. 68

1. A REF of $\mathbf{A}$ consists of a reordering of the rows of $\mathbf{A}$: For the first row take row_4, followed by row_1, row_2, row_5, and $row_3 - 3\, row_5$, for example.

3. As in problem 1, to start out, swap rows 5 and 4 and replace row 5 by $row_5 - 3\, row_4$. Then continue thus:

$$\begin{array}{cccccc|l} \boxed{1} & 0 & 0 & 0 & \boxed{4} & -2 & -4\, row_5 \\ 0 & \boxed{7} & 3 & -3 & \boxed{5} & 3 & -5\, row_5 \\ 0 & 0 & \boxed{1} & 4 & 0 & 0 & \\ 0 & 0 & 0 & \boxed{1} & 0 & 2 & \\ 0 & 0 & 0 & 0 & \boxed{1} & 4 & \end{array}$$

$$\begin{array}{cccccc|l}
\boxed{1} & 0 & 0 & 0 & 0 & -18 & \\
0 & \boxed{7} & 3 & \boxed{-3} & 0 & -17 & +3\ row_4 \\
0 & 0 & \boxed{1} & \boxed{4} & 0 & 0 & -4\ row_4 \\
0 & 0 & 0 & \boxed{1} & 0 & 2 & \\
0 & 0 & 0 & 0 & \boxed{1} & 4 & \\
\hline
\boxed{1} & 0 & 0 & 0 & 0 & -18 & \\
0 & \boxed{7} & \boxed{3} & 0 & 0 & -11 & -3\ row_3 \\
0 & 0 & \boxed{1} & 0 & 0 & -8 & \\
0 & 0 & 0 & \boxed{1} & 0 & 2 & \\
0 & 0 & 0 & 0 & \boxed{1} & 4 & \\
\hline
\boxed{1} & 0 & 0 & 0 & 0 & -18 & \\
0 & \boxed{7} & 0 & 0 & 0 & 13 & \div(7) \\
0 & 0 & \boxed{1} & 0 & 0 & -8 & \\
0 & 0 & 0 & \boxed{1} & 0 & 2 & \\
0 & 0 & 0 & 0 & \boxed{1} & 4 &
\end{array}$$

This will give the RREF of **B**.

5. $\mathbf{R} = \begin{pmatrix} -1 & -3 & 0 & 1 & 1 & -1 \\ 0 & -1 & -2 & 3 & 2 & -2 \\ 0 & 0 & 5 & -2 & -4 & 5 \end{pmatrix}$ and $-\mathbf{R}$ or $2\mathbf{R}$, for example, are REFs of **A**.

7. (a) Take $\mathbf{A} = \begin{pmatrix} 1 & 0 & 0 & 0 \\ 0 & 1 & 0 & 0 \\ 1 & 1 & 0 & 0 \end{pmatrix}$ and $\mathbf{B} = 2\mathbf{A}$, for example.

(b) **A** as in part (a); $\mathbf{B} = -\mathbf{A}$.

(c) and (e) are impossible to achieve; the maximal possible rank is three.

(d) $\mathbf{A} = \begin{pmatrix} 1 & 0 & 0 & 0 \\ 0 & 0 & -1 & 0 \\ 0 & 0 & 3 & 0 \end{pmatrix}$ and

$$\mathbf{B} = \begin{pmatrix} -1 & 0 & 0 & 0 \\ 0 & 0 & -1 & 0 \\ 0 & 0 & 152 & 0 \end{pmatrix}.$$

9. $\begin{pmatrix} 1 & 0 & 6 \\ 0 & 1 & 4 \end{pmatrix}$ is the RREF and has rank 2, while the RREF of the second matrix is $\begin{pmatrix} 1 & 0 & 0 & 1 & 1 \\ 0 & 1 & 0 & 0 & -2 \\ 0 & 0 & 1 & -1 & 2 \end{pmatrix}$ with rank 3.

11. (a) Row echelon forms are not unique.

(b)

$$\begin{aligned} x_1 + x_2 + 2x_3 &= 3, \\ 2x_2 - 2x_3 &= 2, \\ x_1 + 3x_2 &= 4, \end{aligned}$$

and

$$\begin{aligned} x_1 + x_2 + 2x_3 &= 3, \\ 2x_2 - 2x_3 &= 2, \\ x_1 - x_2 + 4x_3 &= 4. \end{aligned}$$

13. One REF is $\begin{pmatrix} 1 & 1 & 1 \\ 0 & 0 & -1 \\ 0 & 0 & 0 \end{pmatrix}$ of rank 2.

15. $\text{rank}(\mathbf{A}) = n - 4 \le \min\{m, n\} \le m$ makes $n - m \le 4$.

17. (a) 7. (b) 0. (c) 0. (d) $\min\{m, n\}$.

19. (a) $1 \le \text{rank}(\mathbf{A}) \le 3$. (b) $\mathbf{A} = \mathbf{0}_{mn}$. (c) Two.
(d) All entries of the first $k - 1$ columns are zero.
(e) $\text{rank}(\mathbf{A}) = m$.

21. (a), (b) $\text{rank}(\mathbf{A}) = 4 = \text{rank}(\mathbf{B})$.

23. RREF: $\begin{pmatrix} 1 & 0 & 0 & 0 & 3 \\ 0 & 1 & -2 & 0 & 1 \\ 0 & 0 & 0 & 1 & -2 \\ 0 & 0 & 0 & 0 & 0 \end{pmatrix}$.

25.

$$\begin{array}{cc|l}
\boxed{1} & -1 & \\
\boxed{2} & 1 & -2\ row_1 \\
\hline
\boxed{1} & -1 & \\
0 & \boxed{3} & \div 3 \\
\hline
\boxed{1} & \boxed{-1} & +\ row_2 \\
0 & \boxed{1} & \\
\hline
1 & 0 & \\
0 & 1 &
\end{array}$$

27. If $a \ne -10$, we have the RREF

$$\begin{pmatrix} 1 & 0 & 0 & -6 + 7\frac{b+10}{a+10} \\ 0 & 1 & 0 & 4 - 3\frac{b+10}{a+10} \\ 0 & 0 & 1 & \frac{b+10}{a+10} \end{pmatrix}.$$

If $a = -10$ and $b = -10$, then we have the RREF $\begin{pmatrix} 1 & 0 & -7 & -6 \\ 0 & 1 & 3 & 4 \\ 0 & 0 & 0 & 0 \end{pmatrix}$, while for $a = -10$ and $b \ne -10$, we obtain $\begin{pmatrix} 1 & 0 & -7 & 0 \\ 0 & 1 & 3 & 0 \\ 0 & 0 & 0 & 1 \end{pmatrix}$.

29. The two matrices have different rank: *A* REF of **A** is $\begin{pmatrix} 1 & -2 & 3 \\ 0 & 1 & 4 \\ 0 & 0 & 0 \end{pmatrix}$, while **B** is row equivalent to $\mathbf{I}_3$.

31. If $\alpha = 0$, then $\gamma = 0$. Then $\alpha + \beta + \gamma = \beta = \delta$ is possible only if $\beta = \delta = 0$ (i.e., if $\mathbf{R} = \mathbf{O}_2$).

2.2.P p. 75

3. (a) **B**; (b) RREF(2**B** − 4**A**) =
$$\begin{pmatrix} 1 & 0 & 0 & -17/11 & -13/11 \\ 0 & 1 & 0 & 133/11 & 92/11 \\ 0 & 0 & 1 & -91/11 & -56/11 \end{pmatrix}.$$

5. $\text{rank}(\mathbf{A}) = 2 = \text{rank}(\mathbf{A}^T)$.

9. Flipping apparently does not change matrix rank.

2.R p. 76

1. Just kidding, if the matrix **A** is real or complex. But what about matrices **A** made up of polynomial entries, such as $\mathbf{A} = \begin{pmatrix} x+1 & x^2 \\ x^2-1 & x^3-1 \end{pmatrix}$ or
$\mathbf{B} = \begin{pmatrix} x+1 & x^2 \\ x^2+1 & x^4-1 \end{pmatrix}$?

3. Not true.

5. Matrix rank is mathematically well defined, but numerically, its evaluation via pivots is not always well computable.

7. **A**, **B**, and **E** are row equivalent (to $\mathbf{I}_3$). **G** and **H**, and **C** and **F** are row equivalent pairs, respectively.

9.

$\boxed{1}$	2	3	
$\boxed{-4}$	−2	0	$+4\ row_1$
$\boxed{1}$	2	3	
0	$\boxed{6}$	12	$\div(-6)$
$\boxed{1}$	$\boxed{2}$	3	$+2\ row_2$
0	$\boxed{-1}$	−2	
1	0	−1	
0	−1	−2	

(a) Thus, $a = -1$ and $b = -2$ for row equivalence.
(b) Take any $a \neq -1$ or $b \neq -2$ for inequivalence.

11. If $\cos(\theta) = 0$, then $\sin(\theta) = \pm 1$ and
$\mathbf{A} = \begin{pmatrix} 0 & \pm 1 \\ \mp 1 & 0 \end{pmatrix}$ is row equivalent to $\mathbf{I}_2$.
If $\cos(\theta) \neq 0$, then

$\boxed{c}$	s	
$\boxed{-s}$	c	$+\frac{s}{c}\ row_1$
$\boxed{c}$	s	
0	$c + \frac{s^2}{c}$	$\cdot c$
$\boxed{c}$	$\boxed{s}$	$-s\ row_2$
0	$\boxed{1}$	
c	0	$\div c$
0	1	
1	0	
0	1	

13.

$\boxed{-1}$	1	1	
$\boxed{2}$	1	0	$+2\ row_1$
0	1	1	
$\boxed{1}$	0	−1	$+\ row_1$
$\boxed{-1}$	1	1	
0	3	2	*swap rows 2 and 4*
0	1	1	
0	1	0	
$\boxed{-1}$	1	1	
0	$\boxed{1}$	0	
0	$\boxed{1}$	1	$-\ row_2$
0	$\boxed{3}$	2	$-3\ row_2$
$\boxed{-1}$	1	1	
0	$\boxed{1}$	0	
0	0	$\boxed{1}$	
0	0	$\boxed{2}$	$-2\ row_3$
−1	1	1	
0	1	0	
0	0	1	
0	0	0	

This further row reduces to $\begin{pmatrix} 1 & 0 & 0 \\ 0 & 1 & 0 \\ 0 & 0 & 1 \\ 0 & 0 & 0 \end{pmatrix}$.

15. (a) and (b) : Both matrices row reduce to $\mathbf{I}_4$.

Chapter 3

3.1.P p. 94

1. $\mathbf{x} = \frac{1}{2}\begin{pmatrix} 1 \\ -3 \end{pmatrix}$. $\mathbf{x}_{hom} = \begin{pmatrix} 0 \\ 0 \end{pmatrix}$.

3. $\mathbf{x} = \frac{1}{39}\begin{pmatrix} 47 \\ 35 \\ 4 \end{pmatrix}$.

5. (a) $\mathbf{B}_{4,4}$.

(b) $\text{rank}(\mathbf{B}) = 2$ if and only if $a = 0$ and $b = 3.5$; $\text{rank}(\mathbf{B}) = 3$ otherwise.

(c) $\mathbf{Ax} = \mathbf{r}$ is solvable precisely when $\text{rank}(\mathbf{B}) = 2$ (i.e., when $a = 0$ and $b = 3.5$).

(d) $\mathbf{x} = \frac{1}{2}\begin{pmatrix} 5 \\ -3 \\ 0 \end{pmatrix} + \alpha \begin{pmatrix} -3 \\ 2 \\ 1 \end{pmatrix}$.

7. A REF of $\begin{array}{ccc|c} -1 & 3 & 4 & 2 \\ 1 & 3 & 2 & 4 \\ 4 & 4 & 0 & 4 \end{array}$ is $\begin{array}{ccc|c} \boxed{-1} & 3 & 4 & 2 \\ 0 & \boxed{1} & 1 & 1 \\ 0 & 0 & 0 & \boxed{4} \end{array}$.

Thus, the system is unsolvable.

9. A REF of the augmented matrix $(\mathbf{A} \mid \mathbf{b})$ is

$$\begin{array}{ccccc|c} \boxed{-1} & -1 & 2 & -3 & 1 & -1 \\ 0 & 0 & \boxed{1} & -2 & 1 & 0 \\ 0 & 0 & 0 & \boxed{1} & 0 & 1 \\ 0 & 0 & 0 & 0 & 0 & 0 \end{array}.$$

$$\mathbf{x}_{gen} = \begin{pmatrix} 2 \\ 0 \\ 2 \\ 1 \\ 0 \end{pmatrix} + \alpha \begin{pmatrix} -1 \\ 0 \\ -1 \\ 0 \\ 1 \end{pmatrix} + \beta \begin{pmatrix} -1 \\ 1 \\ 0 \\ 0 \\ 0 \end{pmatrix}.$$

11. For $k = 0$; the system has the solutions

$\mathbf{x} = \begin{pmatrix} 1 \\ -1 \\ 0 \end{pmatrix} + \alpha \begin{pmatrix} -1 \\ 0 \\ 1 \end{pmatrix}$; for $k \neq 0$,

$\mathbf{x} = \begin{pmatrix} 3/2 \\ -1 + k/2 \\ -1/2 \end{pmatrix}$ is the unique solution.

13. A REF is $\begin{array}{ccccc|c} \boxed{1} & 0 & 1 & 0 & 1 & 3 \\ 0 & \boxed{1} & 0 & 1 & 1 & 4 \\ 0 & 0 & \boxed{-1} & -1 & -1 & -5 \\ 0 & 0 & 0 & \boxed{1} & -1 & -4 \\ 0 & 0 & 0 & 0 & \boxed{-1} & -4 \end{array}$. Hence,

the solution $\mathbf{x} = \begin{pmatrix} -2 \\ 0 \\ 1 \\ 0 \\ 4 \end{pmatrix}$ is unique.

15. (a) Each component function is linear.

(b) $T(\mathbf{e}_1) = \mathbf{e}_1$, $T(\mathbf{e}_3) = -\mathbf{e}_1$, and $T(\mathbf{e}_5) = 2\mathbf{e}_1 + \mathbf{e}_2$.

(c) $\mathbf{A}_{\mathcal{E}} = \begin{pmatrix} 1 & 0 & -1 & 0 & 2 & 0 \\ 0 & 0 & 0 & -1 & 1 & 0 \\ 0 & 0 & 0 & 0 & 0 & 0 \\ 0 & 0 & 0 & 0 & 0 & -1 \end{pmatrix}$;

$\text{rank}(\mathbf{A}) = 3$.

(d) $T(-\mathbf{e}_1) = T(\mathbf{e}_3)$ or $T(\mathbf{0}) = T(\mathbf{e}_2)$, for example.

17. (a) $\mathbf{x}_{gen} = \begin{pmatrix} -4 \\ -4 \\ 3 \\ 0 \end{pmatrix} + \alpha \begin{pmatrix} 3 \\ -1 \\ -2 \\ 1 \end{pmatrix}$.

(b) $\ker(\mathbf{A})$ consists of all multiples of $\begin{pmatrix} 3 \\ -1 \\ -2 \\ 1 \end{pmatrix}$;

$\text{rank}(\mathbf{A}) = 3$.

(c) Never uniquely solvable, since a REF of $\mathbf{A}$ is

$\begin{array}{cccc} \boxed{1} & 1 & 3 & 4 \\ 0 & \boxed{1} & 1 & 3 \\ 0 & 0 & \boxed{1} & 2 \\ 0 & 0 & 0 & 0 \end{array}$ with one free variable.

19. Always solvable; $\mathbf{x} = \mathbf{0}$ will do.

21. $\mathbf{x} = \begin{pmatrix} 2 \\ 1 \\ -1 \\ 1 \end{pmatrix}$.

23. (b) $\mathbf{A}_{\mathcal{E}} = \begin{pmatrix} 2 & -2 & 0 & 4 \\ -4 & 4 & 3 & 0 \\ 0 & 0 & 1 & 0 \end{pmatrix}$.

(c) RREF $\mathbf{R} = \begin{pmatrix} \boxed{1} & -1 & 0 & 0 \\ 0 & 0 & \boxed{1} & 0 \\ 0 & 0 & 0 & \boxed{1} \end{pmatrix}$;

$\mathbf{x}_{hom} = \alpha \begin{pmatrix} 1 \\ 1 \\ 0 \\ 0 \end{pmatrix}$. (d) $T\begin{pmatrix} 1 \\ 1 \\ 0 \\ 0 \end{pmatrix} = \begin{pmatrix} 0 \\ 0 \\ 0 \end{pmatrix}$.

25. (b), (c) One REF

$(\mathbf{R} \mid \tilde{\mathbf{b}})$ is $\begin{array}{ccccc|c} \boxed{1} & 0 & 2 & 1 & -1 & 4 \\ 0 & \boxed{1} & 3 & 4 & 1 & 7 \\ 0 & 0 & \boxed{1} & 1 & -1 & 0 \end{array}$. Its fourth and fifth columns are free. Solvable, but not uniquely solvable.

(d), (e) $\mathbf{x}_{gen} = \begin{pmatrix} 4 \\ 7 \\ 0 \\ 0 \\ 0 \end{pmatrix} + \alpha \begin{pmatrix} 1 \\ -1 \\ -1 \\ 1 \\ 0 \end{pmatrix} + \beta \begin{pmatrix} -1 \\ -4 \\ 1 \\ 0 \\ 1 \end{pmatrix}$

$= \mathbf{x}_{part} + \mathbf{x}_{hom}$. (g) Yes.

27. If $\mathbf{Ax} = \mathbf{0}$ and $\mathbf{Ay} = \mathbf{0}$, then, by linearity, $\mathbf{A}(\alpha\mathbf{x} + \beta\mathbf{y}) = \mathbf{A}(\alpha\mathbf{x}) + \mathbf{A}(\beta\mathbf{y}) = \alpha\mathbf{Ax} + \beta\mathbf{Ay} = \mathbf{0} + \mathbf{0} = \mathbf{0}$ for all $\alpha, \beta \in \mathbb{R}$.

29. True; any REF of **A** must have free variables. Thus, if $\mathbf{Ax} = \mathbf{b}$ is solvable, then there are infinitely many solutions.

31. $\mathbf{x}_{part} = \begin{pmatrix} -1 \\ 0 \\ 1 \\ 3 \\ 0 \end{pmatrix}$; $\mathbf{x}_{hom} = \alpha \begin{pmatrix} 1 \\ 0 \\ 1 \\ -2 \\ 1 \end{pmatrix}$.

$$\ker(\mathbf{A}) = \{\mathbf{x}_{hom}\} = \left\{ \alpha \begin{pmatrix} 1 \\ 0 \\ 1 \\ -2 \\ 1 \end{pmatrix} \middle| \alpha \in \mathbb{R} \right\}.$$

33. $\begin{pmatrix} x \\ \mathbf{y} \\ z \end{pmatrix} = \frac{1}{9} \begin{pmatrix} 50 \\ 30 \\ 1 \end{pmatrix}$.

35. (a) False, the system may be unsolvable;
(b) True. (c) ..., provided that rank $(\mathbf{A}) =$ rank$(\mathbf{A} \mid \mathbf{b})$.

37. (a) The system is solvable for all right-hand sides, since the REF $\begin{matrix} \boxed{1} & 2 & 3 & 4 & 0 & -1 \\ 0 & \boxed{1} & -3 & 2 & -3 & -1 \\ 0 & 0 & \boxed{-1} & -2 & 2 & -2 \\ 0 & 0 & 0 & 0 & \boxed{2} & 12 \end{matrix}$ of its system matrix **A** has a pivot in every row.

(b) It is not uniquely solvable for any right-hand side (two free variables).

(c) For the matrix **B**, there are many right-hand sides, such as $\mathbf{c} := \mathbf{e}_1$, for which the linear system $\mathbf{Bx} = \mathbf{c}$ cannot be solved. Yet when it can be solved, the solution is unique because the REF $\begin{matrix} \boxed{1} & -2 & 3 & 0 \\ 0 & \boxed{1} & -1 & -2 \\ 0 & 0 & \boxed{-1} & 4 \\ 0 & 0 & 0 & \boxed{2} \\ 0 & 0 & 0 & 0 \\ 0 & 0 & 0 & 0 \end{matrix}$ of **B** has a pivot in every column.

39. (a) If $m \le n$ in $\mathbf{A}_{mn}$ [i.e., in cases (1), (3), (4), (5), and (8)], the linear system may be solvable for all right-hand sides. In this case, the RREF of **A** may have a pivot in every row.

(b) Not every such linear system may be solvable.

(c) Only if $m = n$ in $\mathbf{A}_{mn}$ [i.e., in cases (3) and (8)], may the linear system be uniquely solvable for all right-hand sides. For this case, every row and column of a REF of **A** must be able to contain a pivot.

(d) If $m \ge n$ in $\mathbf{A}_{mn}$ [i.e., in cases (2), (3), (6), (7), and (8)], the linear system may be uniquely solvable for some right-hand sides. Here the RREF of **A** must not contain a free column.

T 3. (a) $\mathbf{b} = \begin{pmatrix} -10 \\ 1 \\ 11 \\ -3 \end{pmatrix}$;

(b) Taking $\mathbf{c} = \mathbf{b} + \mathbf{e}_1$, for example, gives the augmented matrix $(\mathbf{A} \mid \mathbf{c})$ rank 4, while rank$(\mathbf{A}) = 3 =$ rank $(\mathbf{A} \mid \mathbf{b})$. Hence, by Theorem 3.1, the linear system $\mathbf{Ay} = \mathbf{c}$ is not solvable.

3.2.P p. 107

1.

$$3c + 1b = 2;$$
$$2c + 2b = 3.$$

A bread costs as much as 5 candles, or $1\frac{1}{4}$ silverlings.

3. (a) $(2, -1, 0)^T, (1, 0, 2)^T$ and $(-10, 0, -20)^T$, for example.
(b) All multiples of the vector $(-2.5, -2, 1)^T$ form ker $\begin{pmatrix} 2 & -2 & 1 \\ 0 & 1 & 2 \end{pmatrix}$.
(c) All linear combinations of the vectors $(1, 1, 0)^T$ and $(-1, 0, 2)^T$ belong to ker $\begin{pmatrix} 2 & -2 & 1 \end{pmatrix}$.
(d) $\mathbf{0} \in \mathbb{R}^3$.

5. The planes intersect at $\mathbf{x} = \begin{pmatrix} -1.5 \\ 0.5 \\ -1.5 \end{pmatrix}$.

7. (a) $2NaOH + CO_2 \rightarrow (Na)_2CO_3 + H_2O$;
(b) $6NaOH + (Fe)_2(SO_4)_3 \rightarrow 3(Na)_2SO_4 + 2Fe(OH)_3$;
(c) $C_3H_8 + 5O_2 \rightarrow 3CO_2 + 4H_2O$.

9. (a) $x_1 = 300$; $\mathbf{y} = \begin{pmatrix} y_1 \\ \vdots \\ y_7 \end{pmatrix} = \begin{pmatrix} 200 \\ 100 \\ -200 \\ 0 \\ 300 \\ 600 \\ 0 \end{pmatrix} +$

$$\alpha \begin{pmatrix} 200 \\ 0 \\ -200 \\ 0 \\ 400 \\ 500 \\ 100 \end{pmatrix} + \beta \begin{pmatrix} 200 \\ 500 \\ 200 \\ 100 \\ 0 \\ 0 \\ 0 \end{pmatrix}.$$

(b) $x_1 = 300$; $\mathbf{y} = \begin{pmatrix} y_1 \\ y_3 \\ \vdots \\ y_7 \end{pmatrix} = \begin{pmatrix} -300 \\ 0 \\ -100 \\ 300 \\ 600 \\ 0 \end{pmatrix} +$

$$\alpha \begin{pmatrix} 1 \\ 0 \\ 1 \\ 1 \\ -1 \\ 1 \end{pmatrix} + \beta \begin{pmatrix} 0 \\ 1 \\ 0 \\ 0 \\ 0 \\ 0 \end{pmatrix}.$$

11. Neither linear system can be solved in MATLAB via the command B\b or B\c. (See the end of Section 12.3 for a fix to this problem.) The kernel of **B** is computed as two-dimensional in MATLAB:

```
>> null(B)
ans =
   0.1198  -0.5345
  -0.5581   0.6233
   0.7568   0.3567
  -0.3185  -0.4456
```

The RREFs of the two augmented matrices are computed in MATLAB as

```
   1   0   -1   -2    2
   0   1    2    3   -2
   0   0    0    0    0
   0   0    0    0    0
```

for $(\mathbf{B} \mid \mathbf{b})$ and

```
   1   0   -1   -2   0
   0   1    2    3   0
   0   0    0    0   1
   0   0    0    0   0
```

for $(\mathbf{B} \mid \mathbf{c})$. Thus, according to Theorem 3.1, the linear system $\mathbf{Bx} = \mathbf{b}$ is solvable, while $\mathbf{By} = \mathbf{c}$ is not solvable.

13. The sequence of MATLAB commands A=[2;4],b=[6;-4],A\b gives the "solution" x as -0.2.

3.R p. 108

3. Solvable only for $k = 2$, in which case the solution

$$\mathbf{x} = \begin{pmatrix} 1 \\ 2 \\ 0 \end{pmatrix} + \alpha \begin{pmatrix} -3 \\ -5 \\ 1 \end{pmatrix} \text{ is not unique.}$$

5. (a) $\beta = 6$. (b) $\mathbf{x} = \begin{pmatrix} 0 \\ 0 \\ 2/3 \\ 0 \\ 0 \end{pmatrix} + \begin{pmatrix} y_1 \\ y_2 \\ 0 \\ y_4 \\ y_5 \end{pmatrix}$ with

$$\mathbf{A} \begin{pmatrix} y_1 \\ y_2 \\ 0 \\ y_4 \\ y_5 \end{pmatrix} = \mathbf{0}.$$

(c) For $\beta = 0$ and all entries $*$ set to 0 in **A**, one cannot solve $\mathbf{Ax} = \mathbf{b}(0)$.

7. The general solution of the linear system in terms of x, y^3, and $\ln(z)$ is $\begin{pmatrix} x \\ y^3 \\ \ln(z) \end{pmatrix} = \begin{pmatrix} -\frac{1}{4} \\ 0 \\ -\frac{1}{2} \end{pmatrix} + \alpha \begin{pmatrix} -3 \\ 2 \\ 0 \end{pmatrix}$, making $x = -1/4 - 3\alpha$, $y = \sqrt[3]{2\alpha}$, and $z = e^{-1/2}$ the solution in terms of the variables x, y, and z.

9. (a) True. (b) True.

11. Every such system can be solved; none, however, uniquely, because there are at least two free variables in every REF of **A**.

13. Unique solvability if every REF column has a pivot (i.e., if $n = m$ in this case).

15. An unsolvable linear system.

17. (a) The REF of the augmented matrix $(\mathbf{A} \mid \mathbf{b})$ has no inconsistent rows.

(b) $\mathbf{x}_{gen} = \begin{pmatrix} 10 \\ 0 \\ 3 \\ 1 \end{pmatrix} + \alpha \begin{pmatrix} 4 \\ 1 \\ 0 \\ 0 \end{pmatrix}.$

Chapter 4

4.1.P p. 128

1. All except (b) are subspaces. The respective subspaces can be represented as follows:

(a) $\ker\begin{pmatrix} 1 & 1 \\ 2 & 2 \end{pmatrix} = \operatorname{im}\begin{pmatrix} 1 & -2 & 0 \\ -1 & 2 & 0 \end{pmatrix}.$

(c) $\ker\begin{pmatrix} 0 & 1 & 1 & 0 \\ 1 & 0 & 0 & -2 \\ -1 & 2 & 2 & 2 \end{pmatrix} = \operatorname{im}\begin{pmatrix} 2 & 0 & 2 \\ 0 & 1 & -1 \\ 0 & -1 & 1 \\ 1 & 0 & 1 \end{pmatrix}.$

(d) $\ker\begin{pmatrix} 1 & -1 & 1 \\ 0 & 1 & 0 \\ 0 & 0 & 1 \end{pmatrix} = \operatorname{im}\begin{pmatrix} 0 & 0 & 0 \\ 0 & 0 & 0 \\ 0 & 0 & 0 \end{pmatrix}.$

(e) $\ker\begin{pmatrix} 2 & 1 & -3 \end{pmatrix} = \operatorname{im}\begin{pmatrix} 1 & 3 & -2 \\ -2 & 0 & -2 \\ 0 & 2 & -2 \end{pmatrix}$.

(f) $\ker\begin{pmatrix} 1 & 0 & 1 \\ 2 & 0 & 2 \end{pmatrix} = \operatorname{im}\begin{pmatrix} 1 & 0 \\ 0 & -1 \\ -1 & 0 \end{pmatrix}$.

(g) $\ker\begin{pmatrix} 0 & 0 & 1 & 0 \end{pmatrix} = \operatorname{im}\begin{pmatrix} 1 & 0 & -11 \\ 2 & 1 & 0 \\ 0 & 0 & 0 \\ 1 & 0 & 1 \end{pmatrix}$.

(Note that none of these generating matrices are unique, except for the "image" in part (d).)

3. $\alpha\begin{pmatrix} -2 \\ 1 \\ 1 \\ 0 \end{pmatrix} + \beta\begin{pmatrix} 1 \\ -1 \\ 0 \\ 1 \end{pmatrix} = \ker\begin{pmatrix} 1 & 2 & 0 & 1 \\ 0 & 1 & -1 & 1 \end{pmatrix}$ is a subspace.

5. If x and y are positive real numbers and if α and β are positive real, then $\alpha x + \beta y$ is positive real.

7. (a) $\mathbf{A}_{5,6} = \begin{pmatrix} \mathbf{0}_{5,1} & \mathbf{I}_5 \end{pmatrix}$;

(b) $\mathbf{B}_{3,4} = \begin{pmatrix} 1 & 0 & 0 & 0 \\ 0 & 4 & 0 & 0 \\ 0 & 0 & 1 & 1 \end{pmatrix}$; and

(c) $\mathbf{C}_{5,5} = \begin{pmatrix} 1 & 0 & 0 & 0 & -1 \\ 0 & -1 & 0 & 2 & 0 \\ 0 & 0 & 12 & 0 & 0 \\ 0 & 0 & 0 & 3 & 0 \\ 0 & 0 & 2 & 4 & 0 \end{pmatrix}$, for example.

9. In (a) and (b), take the appropriate zero matrix. Part (c) is impossible, since $\ker(\mathbf{C}) \subset \mathbb{R}^4$.

11. (a) $U = \ker\begin{pmatrix} -2 & 1 & 0 & 0 \\ 3 & 0 & 1 & 0 \\ 1 & 0 & 0 & 1 \end{pmatrix} = \ker\begin{pmatrix} 0 & 1 & 0 & 2 \\ 3 & 0 & 1 & 0 \\ 1 & 0 & 0 & 1 \\ 14 & 1 & 0 & 16 \end{pmatrix}$, for example.

(b) $\mathbf{A}_{k4}$ for $k \geq 3$.

13. (a) $\ker(\mathbf{A}) = \operatorname{span}\{\mathbf{0}\} \subset \mathbb{R}$;

(b) $\operatorname{span}\left\{\begin{pmatrix} -4 \\ 1 \\ 0 \\ 0 \end{pmatrix}, \begin{pmatrix} 5 \\ 0 \\ 1 \\ 0 \end{pmatrix}, \mathbf{e}_4\right\}$;

(c) $\operatorname{span}\{\mathbf{0}\} \subset \mathbb{R}^3$.

(d) $\operatorname{span}\left\{\mathbf{e}_1, \begin{pmatrix} -7 \\ 1 \\ 13 \\ -6 \\ 1 \end{pmatrix}\right\}$.

(e) $\operatorname{span}\left\{\begin{pmatrix} 3 \\ 0 \\ 1 \\ 0 \\ -2 \\ 1 \end{pmatrix}, \begin{pmatrix} -4 \\ 0 \\ -1 \\ 2 \\ 0 \\ 0 \end{pmatrix}, \begin{pmatrix} 1 \\ 1 \\ 0 \\ 0 \\ 0 \\ 0 \end{pmatrix}\right\}$.

15. (a) $U = \ker\begin{pmatrix} -1 & 0 & 1 \end{pmatrix}$;

(b) $V = \ker\begin{pmatrix} 0 & -2 & 0 & 1 \\ 1 & 1 & -1 & 0 \end{pmatrix}$;

(c) $W = \ker(\mathbf{O}_3)$.

17. Finding real coefficients y_i so that an arbitrary second-degree polynomial $p(x) = a_2x^2 + a_1x + a_0 = y_1u_1(x) + y_2u_2(x) + y_3u_3(x)$ is equivalent to solving the linear system $\begin{pmatrix} 0 & 0 & -6 \\ -1 & 2 & 3 \\ 1 & 1 & 3 \end{pmatrix}\begin{pmatrix} y_1 \\ y_2 \\ y_3 \end{pmatrix} = \begin{pmatrix} a_1 \\ a_2 \\ a_3 \end{pmatrix}$. The system matrix has rank 3, making the linear system solvable for all possible right-hand sides.

19. (a) $\mathbf{v} := \begin{pmatrix} 1 \\ 0 \\ -1 \end{pmatrix}$, for example.

(b) $\mathbf{w} = -\mathbf{v}$, for example.

(c) $\mathbf{z} = \begin{pmatrix} 1 \\ -1/2 \\ 0 \end{pmatrix}$. (d) $\mathbf{x} = \begin{pmatrix} 2 \\ 4 \\ -2 \end{pmatrix}$.

21. (a)

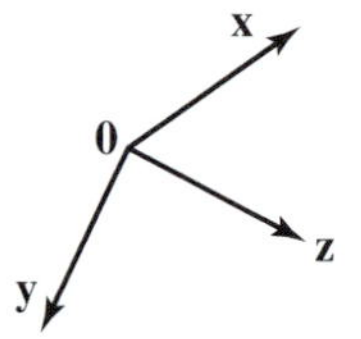

Each pair of vectors spans all of $\mathbb{R}^2$.

(b)

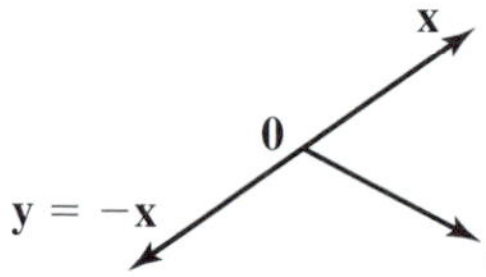

Clearly, $\mathbf{z} \notin \operatorname{span}\{\mathbf{x}, \mathbf{y}\} = \operatorname{span}\{\mathbf{x}\}$.

(c) No: span$\{\mathbf{x}, \mathbf{y}\} = \mathbb{R}^2$, unless $\mathbf{x} = \alpha\mathbf{y}$ or $\mathbf{y} = \beta\mathbf{x}$ for some $\alpha, \beta \in \mathbb{R}$.

23. First, row reduce the system matrix:

$$\begin{array}{cccc|c|l}
\boxed{1} & 4 & 2 & 0 & 0 & \\
0 & 1 & 5 & -1 & 0 & \\
\boxed{2} & 7 & -1 & 1 & 0 & -2\,row_1 \\
\hline
\boxed{1} & 4 & 2 & 0 & 0 & \\
0 & \boxed{1} & 5 & -1 & 0 & \\
0 & \boxed{-1} & -5 & 1 & 0 & +\,row_2 \\
\hline
\boxed{1} & 4 & 2 & 0 & 0 & \\
0 & \boxed{1} & 5 & -1 & 0 & \\
0 & 0 & 0 & 0 & 0 &
\end{array}$$

Use the two pivot rows, insert $\mathbf{e}_3$ and $\mathbf{e}_4$ as rows, and append them as columns:

$$\begin{array}{cccc|cc|l}
1 & 4 & 2 & 0 & 0 & 0 & \\
0 & 1 & 5 & \boxed{-1} & 0 & 0 & +\,row_4 \\
0 & 0 & 1 & 0 & 1 & 0 & \\
0 & 0 & 0 & \boxed{1} & 0 & 1 & \\
\hline
1 & 4 & \boxed{2} & 0 & 0 & 0 & -2\,row_3 \\
0 & 1 & \boxed{5} & 0 & 0 & 1 & -5\,row_3 \\
0 & 0 & \boxed{1} & 0 & 1 & 0 & \\
0 & 0 & 0 & \boxed{1} & 0 & 1 & \\
\hline
1 & \boxed{4} & 0 & 0 & -2 & 0 & -4\,row_2 \\
0 & \boxed{1} & 0 & 0 & -5 & 1 & \\
0 & 0 & \boxed{1} & 0 & 1 & 0 & \\
0 & 0 & 0 & \boxed{1} & 0 & 1 & \\
\hline
\boxed{1} & 0 & 0 & 0 & 18 & -4 & \\
0 & \boxed{1} & 0 & 0 & -5 & 1 & \\
0 & 0 & \boxed{1} & 0 & 1 & 0 & \\
0 & 0 & 0 & \boxed{1} & 0 & 1 &
\end{array}$$

Thus, $\mathbf{x}_{hom} = \alpha \begin{pmatrix} 18 \\ -5 \\ 1 \\ 0 \end{pmatrix} + \beta \begin{pmatrix} -4 \\ 1 \\ 0 \\ 1 \end{pmatrix}$.

25. (a) $\mathbf{w} = \mathbf{e}_1$ or $\mathbf{w} = \mathbf{e}_2$, for example.

(b) $\mathbf{z} = \begin{pmatrix} -1 \\ -2 \\ 3 \end{pmatrix}$ or $\mathbf{z} = 2\mathbf{e}_1$, for example.

27. (a) $\{\mathbf{0}\}$ or $\mathbb{R}$.

(b) span$\{\mathbf{0}\}$; span$\{\mathbf{x}\}$ for any $\mathbf{x} \neq \mathbf{0} \in \mathbb{R}^2$; or span$\{\mathbf{e}_1, \mathbf{e}_2\}$ for the unit vectors $\mathbf{e}_i \in \mathbb{R}^2$. $\{\mathbf{0}\} = \ker(\mathbf{I}_2)$; $\mathbb{R}^2 = \ker(\mathbf{O}_2)$.

(c) span$\{\mathbf{0}\}$; span$\{\mathbf{x}\}$ for any $\mathbf{x} \neq \mathbf{0} \in \mathbb{R}^3$; span$\{\mathbf{x}, \mathbf{y}\}$ for any two vectors $\mathbf{x}, \mathbf{y} \in \mathbb{R}^3$ with $\mathbf{x} \neq \alpha\mathbf{y}$ and $\mathbf{y} \neq \beta\mathbf{x}$; or span$\{\mathbf{e}_1, \mathbf{e}_2, \mathbf{e}_3\} = \mathbb{R}^3$.

29. $\Leftarrow$: True, since identical pivot rows determine the same kernel.

$\Rightarrow$: If $\ker(\mathbf{A}) = \ker(\mathbf{B})$, then, for the two RREFs, we have $\ker(\mathbf{R}_A) = \ker(\mathbf{R}_B)$. Hence, the pivot rows coincide in $\mathbf{R}_A$ and $\mathbf{R}_B$.

31. (a) $\mathbf{B} = \begin{pmatrix} -2 & -4 & 0 & 0 & 0 \\ -7 & -14 & 0 & 0 & 0 \\ 3 & 6 & 0 & 0 & 0 \end{pmatrix}$, for example, by putting generating vectors for $\ker(\mathbf{A})$ into the columns of $\mathbf{B}$.

(b) $\mathbf{AB} = \mathbf{O}_{4,5}$.

33. (a), (b) The two linear systems $\mathbf{Ax} = \mathbf{b}$ and $\mathbf{Ry} = \tilde{\mathbf{b}}$ for $\mathbf{A}$ and a REF $\mathbf{R}$ of $\mathbf{A}$ always have the same solutions. So if $\mathbf{Ax} = \mathbf{0}$, then $\mathbf{Rx} = \mathbf{0}$ as well, and conversely, since a row reduction update of the zero vector never alters it.

35. (a) False: $\begin{pmatrix} 1 & -1 \\ 1 & -1 \end{pmatrix}\begin{pmatrix} 1 \\ 1 \end{pmatrix} = \begin{pmatrix} 0 \\ 0 \end{pmatrix}$.

(b) True: If $\mathbf{a}_i^T\mathbf{a}_i = 0 \in \mathbb{R}$, then $\mathbf{a}_i = \mathbf{0} \in \mathbb{R}^n$.

37. No. The only vector with $\mathbf{Ax} = \mathbf{x}$ and with $\mathbf{Ay} = -\mathbf{y}$ for the given matrix $\mathbf{A}$ is $\mathbf{x} = \mathbf{y} = \mathbf{0} \in \mathbb{R}^3$.

39. (a) Use Scheme (4.1) to express the direction vector $\begin{pmatrix} 1 \\ -2 \\ 2 \end{pmatrix}$ of ℓ as a matrix kernel:

$$\begin{array}{c||ccc|l}
\boxed{1} & 1 & 0 & 0 & \\
\boxed{-2} & 0 & 1 & 0 & +2\,row_1 \\
\boxed{2} & 0 & 0 & 1 & -2\,row_1 \\
\hline
\boxed{1} & 1 & 0 & 0 & \\
\cline{2-4}
0 & 2 & 1 & 0 & \\
0 & -2 & 0 & 1 &
\end{array}$$

The row vectors of the lower right corner matrix $\mathbf{A}$ are normal to the two desired planes: $\mathbf{a}_1 = (2, 1, 0)$; $\mathbf{a}_2 = (-2, 0, 1)$. The normal equation of a plane ℓ is given by $\mathbf{a} \cdot \mathbf{x} = \mathbf{a} \cdot \mathbf{x}_0$ for $\mathbf{a} \perp \ell$, $\mathbf{x}_0 \in \ell$, and $\mathbf{x} \in \mathbb{R}^n$. Here, $\mathbf{a}_1 \cdot \mathbf{x} = \mathbf{a}_1 \cdot (0, 2, -1)$ makes $2x + y = 2$, while $\mathbf{a}_2 \cdot \mathbf{x} = \mathbf{a}_2 \cdot (0, 2, -1)$ makes $-2x + z = -1$. These two normal plane equations have the intersection ℓ.

(b) Here, the three normal plane equations are x+y=1, $-2x+z=-3$, and $-x+w=-1$ for $\begin{pmatrix} x \\ y \\ z \\ w \end{pmatrix} \in \mathbb{R}^4$.

4.2.P p. 135

1. If $\mathbf{w}$ and $\mathbf{z}$ lie in the join of two subspaces $U \subset \mathbb{R}^n$ and $V \subset \mathbb{R}^n$, then $\mathbf{w} = \mathbf{w_u} + \mathbf{w_v}$ for two vectors $\mathbf{w_u} \in U$ and $\mathbf{w_v} \in \mathbf{V}$ by definition; and likewise, $\mathbf{z} = \mathbf{z_u} + \mathbf{z_v}$ for two vectors $\mathbf{z_u} \in U$ and $\mathbf{z_v} \in V$. Consequently, $\alpha\mathbf{w} + \beta\mathbf{z} = \alpha\mathbf{w_u} + \beta\mathbf{z_u} + \alpha\mathbf{w_v} + \beta\mathbf{z_v} \in U + V$, since $\alpha\mathbf{w_u} + \beta\mathbf{z_u} \in U$ and $\alpha\mathbf{w_v} + \beta\mathbf{z_v} \in V$ by the closure of subspaces.

If $\mathbf{w}$ and $\mathbf{z}$ lie in the intersection of two subspaces $U \subset \mathbb{R}^n$ and $V \subset \mathbb{R}^n$, then both $\mathbf{w}$ and $\mathbf{z}$ lie in U and in V, which are subspaces. Thus, $\alpha\mathbf{w} + \beta\mathbf{z} \in U$ and $\alpha\mathbf{w} + \beta\mathbf{z} \in V$ by the linear combination closure of both U and V. Consequently, $\alpha\mathbf{w} + \beta\mathbf{z} \in U \cap \mathbf{V}$.

3. $U + V = \text{span}\left\{ \begin{pmatrix} 2 \\ -1 \\ 0 \end{pmatrix}, \begin{pmatrix} 1 \\ 0 \\ 1 \end{pmatrix}, \begin{pmatrix} 1 \\ 0 \\ -1 \end{pmatrix}, \begin{pmatrix} 0 \\ 2 \\ 1 \end{pmatrix} \right\} =$ span $\{\mathbf{e}_1, \mathbf{e}_2, \mathbf{e}_3\}$.

$U \cap V = \ker\begin{pmatrix} 1 & 2 & -1 \\ 2 & -1 & 2 \end{pmatrix}$.

5. (a) $U + V = \text{span}\left\{ \begin{pmatrix} -1 \\ 1 \\ 1 \\ 1 \end{pmatrix}, \begin{pmatrix} 1 \\ 0 \\ 0 \\ -2 \end{pmatrix}, \begin{pmatrix} 0 \\ 1 \\ -1 \\ 0 \end{pmatrix}, \begin{pmatrix} 2 \\ 1 \\ 0 \\ 1 \end{pmatrix} \right\} = \mathbb{R}^4$;

$U \cap V = \ker\begin{pmatrix} 0 & 1 & -1 & 0 \\ 2 & 1 & 0 & 1 \\ 0 & -1 & -1 & 1 \\ 1 & -2 & -2 & 0 \end{pmatrix} = \mathbf{0} \in \mathbb{R}^4$.

$\dim U = 2$, $\dim V = 2$, $\dim(U+V) = 4$, and $\dim(U \cap V) = 0$.

(b) $W = U = W + U = W \cap U$. All dimensions are equal to 2.

7. $U \cap V = \ker\begin{pmatrix} 1 & 0 & -1 & 0 \\ 0 & 2 & 0 & 1 \\ 1 & 1 & -3 & 0 \\ -2 & 0 & 3 & 1 \end{pmatrix} = \text{span}\ \{\mathbf{0} \in \mathbb{R}^4\}$.

$U + V = \mathbb{R}^4$.

$\dim U = 2$, $\dim V = 2$, $\dim(U+V) = 4$, and $\dim(U \cap V) = 0$.

9. (a) $U = \ker\begin{pmatrix} -2 & 1 & -4 & 0 \\ -1 & 0 & -1 & 1 \end{pmatrix}$;

$V = \ker\begin{pmatrix} 0 & -3 & 4 & 7 \\ 1 & 5 & -5 & -13 \end{pmatrix}$.

(b) $U \cap V = \text{span}\left\{ \begin{pmatrix} 1 \\ 6 \\ 1 \\ 2 \end{pmatrix} \right\}$.

11. $\ker(\mathbf{A}) \cap \text{im}(\mathbf{A}) = \text{span}\{\mathbf{e}_1\}$; $\ker(\mathbf{A}) + \text{im}(\mathbf{A}) = \text{span}\{\mathbf{e}_1, \mathbf{e}_2\}$.
$\ker(\mathbf{B}) \cap \text{im}(\mathbf{B}) = \text{span}\{\mathbf{0}\}$; $\ker(\mathbf{B}) + \text{im}(\mathbf{B}) = \mathbb{R}^3$.
$\ker(\mathbf{C}) \cap \text{im}(\mathbf{C}) = \text{span}\{\mathbf{0}\}$; $\ker(\mathbf{C}) + \text{im}(\mathbf{C}) = \mathbb{R}^2$.

4.R p. 136

3. Use Theorem 4.2: $\ker(\mathbf{A}_1) = \ker(\mathbf{A}_2) = \ker(\mathbf{A}_3) = \mathbf{a}^\perp$.

5. Use Scheme (4.1):

$$\begin{array}{cc||cccc|l}
\mathbf{u}_1 & \mathbf{u}_2 & x_1 & x_2 & x_3 & x_4 & \\
\hline
\boxed{1} & 2 & 1 & 0 & 0 & 0 & \\
0 & 0 & 0 & 1 & 0 & 0 & \\
\boxed{-1} & 0 & 0 & 0 & 1 & 0 & +\,row_1 \\
0 & 0 & 0 & 0 & 0 & 1 & \\
\hline
\boxed{1} & 2 & 1 & 0 & 0 & 0 & \\
0 & 0 & 0 & 1 & 0 & 0 & \textit{swap down} \\
0 & 2 & 2 & 0 & 1 & 0 & \textit{swap up} \\
0 & 0 & 0 & 0 & 0 & 1 & \\
\hline
\boxed{1} & 2 & 1 & 0 & 0 & 0 & \\
0 & \boxed{2} & 2 & 0 & 1 & 0 & \\
\hline
0 & 0 & 0 & 1 & 0 & 0 & \\
0 & 0 & 0 & 0 & 0 & 1 & = \mathbf{A}
\end{array}$$

Thus, $U = \ker(\mathbf{A}) = \ker\begin{pmatrix} 0 & 1 & 0 & 0 \\ 0 & 0 & 0 & 1 \end{pmatrix}$.

7. $-2\mathbf{e}_3 = 2(\mathbf{u}_1 + \mathbf{u}_2 - \mathbf{u}_3)$.

9. $U + V = V$.

11. (a) $\mathbf{AB} = \mathbf{O}_{m\ell}$. (b) $\ker(\mathbf{A}) \subset \ker(\mathbf{BA})$.

13. (a) X is a subspace, since $X = \mathbb{R}^2$. (b) $Y = \mathbb{R}^3$.
(c) Z is not a subspace, because $\mathbf{e}_3 \in Z$ for $a = b = c = 0$, but $2\mathbf{e}_3 \notin Z$, since the corresponding linear system with the augmented matrix $\begin{array}{ccc|c} 1 & 2 & 3 & 0 \\ 0 & 1 & 1 & 0 \\ -1 & 0 & -1 & 1 \end{array}$ is inconsistent.

15. If $\mathbf{A}$ has entries with $a_{11} = 3a_{22}$ and $\mathbf{B}$ with $b_{11} = 3b_{22}$, then $\mathbf{C} := \alpha\mathbf{A} + \beta\mathbf{B}$ has entries c_{ij} with $c_{11} = \alpha a_{11} + \beta b_{11} = 3\alpha a_{22} + 3\beta b_{22} = 3c_{22}$. That is, X is a matrix subspace.

Chapter 5

5.1.P p. 154

1. (a) No. (b) Yes. (c) No.

3. (a) No. (b) No. (c) Yes. (d) Yes. (e) No. (f) $0 \le \text{rank}(\mathbf{A}) \le 5$.

5. (a)

$$\begin{array}{ccc|l}
\boxed{1} & 0 & -1 & \\
\textcircled{2} & -1 & -4 & -2\,row_1\\
\textcircled{3} & 3 & 3 & -3\,row_1\\ \hline
\boxed{1} & 0 & -1 & \\
0 & \boxed{-1} & -2 & \\
0 & \textcircled{3} & 6 & +3\,row_2 \\ \hline
\boxed{1} & 0 & -1 & \\
0 & \boxed{-1} & -2 & \cdot(-1)\\
0 & 0 & 0 & \\ \hline
\boxed{1} & 0 & -1 & \\
0 & \boxed{1} & 2 & \\
0 & 0 & 0 &
\end{array}$$

The $\mathbf{u}_i$ are linearly dependent.

(b)

$$\begin{array}{ccc|l}
\boxed{1} & 0 & 2 & \\
\textcircled{2} & -1 & 7 & -2\,row_1\\
\textcircled{3} & 3 & -3 & -3\,row_1\\
\textcircled{4} & 2 & 2 & -4\,row_1\\ \hline
\boxed{1} & 0 & 2 & \\
0 & \boxed{-1} & 3 & \\
0 & \textcircled{3} & -9 & +3\,row_2\\
0 & \textcircled{2} & -6 & +2\,row_2\\ \hline
\boxed{1} & 0 & 2 & \\
0 & \boxed{-1} & 3 & \cdot(-1)\\
0 & 0 & 0 & \\
0 & 0 & 0 &
\end{array}$$

7. (a) $\mathbf{u}_1, \mathbf{u}_2$, and $\mathbf{u}_4$ are a basis for U.
(b) Only $\mathbf{e}_4 \in U$.

9. (a) No conclusion is possible. Only (d) is universally true, and (f) is true if $\mathbf{A}$ is $k \times 2$ of rank 1. The other statements are false in $\mathbb{R}^m$.

11. (a) $\mathcal{E}$, the standard basis for example.
(b) and (c): Yes. (d) and (e): Linearly dependent.

13. $\mathbf{e}_1 - \mathbf{e}_2, \mathbf{e}_1 - \mathbf{e}_3, \mathbf{e}_1 - \mathbf{e}_4$, for example.

15. (a) two. (b) Yes: $\mathbf{b} = \begin{pmatrix} 1 \\ 3 \\ 1 \end{pmatrix}$.

17. (b) $** = -14$.

(c) From (b), we use $** = -14$. If the $*$ entries of column 2 are 3, 3, and -6 and the $*$ entries in column 4 are $-1.5, -1.5, -1.5$ from the top on down, then the rank is 1.

If the $*$ entries in column 4 are all -1.5 as before, and the (1.2) entry is, for example, 0, then the rank is 2.

If all $*$ entries in $\mathbf{A}_{**}$ are zero with $** = -14$, then the rank is 3.

19. Linearly independent; dimension four.

21. $\mathbf{A}$ has n columns. In a REF of $\mathbf{A}$, each column will either have a pivot or be free. The pivot columns contribute to $\mathbf{A}$'s rank, while the number of free columns defines the dimension of $\mathbf{A}$'s kernel.

23. The vectors $\mathbf{x}$, $\mathbf{x}_k$, and $\mathbf{x}_\ell$ are linearly independent precisely if $\alpha\mathbf{x} + \beta\mathbf{x}_k + \gamma\mathbf{x}_\ell = \mathbf{0}$ implies that $\alpha = \beta = \gamma = 0$.

If $k \ne \ell$, then $\alpha\mathbf{x} + \beta\mathbf{x}_k + \gamma\mathbf{x}_\ell = \alpha\mathbf{x}_1 + \ldots + (\alpha + \beta)\mathbf{x}_k + \ldots + (\alpha + \gamma)\mathbf{x}_\ell + \ldots + \alpha\mathbf{x}_n = \mathbf{0}$ is equivalent to $\alpha = 0$, $\alpha + \beta = 0$, and $\alpha + \gamma = 0$, since the vectors $\mathbf{x}_i$ are linearly independent. Thus, in turn, $\alpha = 0 = \beta = \gamma$.

If $k = \ell$, then $\mathbf{x}_k - \mathbf{x}_\ell = \mathbf{0}$ expresses a nontrivial linear dependency among the three given vectors $\mathbf{x}$, $\mathbf{x}_k$, and $\mathbf{x}_\ell$.

25. If the points A and B lie opposite each other on the base circle of the cone, then the four points O, A, B, and C lie on a plane, and any three vectors in that plane must be linearly dependent.

27. $P_0 = \ker(a\,b\,c)$ and $\dim(P_0) = 2$. Hence, any three vectors in P_0 must be linearly dependent.

29. For 7(a) and $\begin{pmatrix} - & \mathbf{u}_1^T & - \\ - & \mathbf{u}_2^T & - \\ - & \mathbf{u}_3^T & - \\ - & \mathbf{u}_4^T & - \end{pmatrix}_{4,4}$,

$$\begin{array}{cccc|l}
\boxed{1} & 2 & -1 & 0 & \\
\boxed{2} & 5 & 0 & 1 & -2\ row_1 \\
\boxed{-1} & -3 & -1 & -1 & +\ row_1 \\
0 & 0 & 0 & 1 & \\
\hline
\boxed{1} & 2 & -1 & 0 & \\
0 & \boxed{1} & 2 & 1 & \\
0 & \boxed{-1} & -2 & -1 & +\ row_2 \\
0 & 0 & 0 & 1 & \\
\hline
\boxed{1} & 2 & -1 & 0 & \\
0 & \boxed{1} & 2 & 1 & \\
0 & 0 & 0 & 0 & \\
0 & 0 & 0 & 1 &
\end{array}$$

Thus, a basis U is the set of vectors

$$\left\{ \begin{pmatrix} 1 \\ 2 \\ -1 \\ 0 \end{pmatrix}, \begin{pmatrix} 0 \\ 1 \\ 2 \\ 1 \end{pmatrix}, \begin{pmatrix} 0 \\ 0 \\ 0 \\ 1 \end{pmatrix} \right\}.$$

For 11(a) and $\begin{pmatrix} - & \mathbf{u}_1^T & - \\ - & \mathbf{u}_2^T & - \\ - & \mathbf{u}_3^T & - \\ - & \mathbf{u}_4^T & - \end{pmatrix}_{4,3}$,

$$\begin{array}{ccc|l}
\boxed{-1} & 3 & 2 & \\
\boxed{1} & 1 & -1 & +\ row_1 \\
\boxed{-1} & -1 & -2 & -\ row_1 \\
\boxed{3} & -1 & -1 & +3\ row_1 \\
\hline
\boxed{-1} & 3 & 2 & \\
0 & 4 & 1 & \\
0 & -4 & -4 & \div(-4);\ swap\ with\ row_2 \\
0 & 8 & 5 & \\
\hline
\boxed{-1} & 3 & 2 & \\
0 & \boxed{1} & 1 & \\
0 & \boxed{4} & 1 & -4\ row_2 \\
0 & \boxed{8} & 5 & -8\ row_2 \\
\hline
\boxed{-1} & 3 & 2 & \\
0 & \boxed{1} & 1 & \\
0 & 0 & \boxed{-3} & \\
0 & 0 & \boxed{-3} & -\ row_3
\end{array}$$

A basis for U are the vectors $\begin{pmatrix} -1 \\ 3 \\ 2 \end{pmatrix}, \begin{pmatrix} 0 \\ 1 \\ 1 \end{pmatrix}, \begin{pmatrix} 0 \\ 0 \\ 1 \end{pmatrix}$.

33. (a) Assume that the two subspaces have the bases $\{\mathbf{u}_1, \ldots, \mathbf{u}_7\}$ and $\{\mathbf{v}_1, \ldots, \mathbf{v}_5\}$, respectively. Then the set $\{\mathbf{u}_1, \ldots, \mathbf{u}_7, \mathbf{v}_1, \ldots, \mathbf{v}_5\} \subset \mathbb{R}^{10}$ must be linearly dependent (i.e., there is a nontrivial linear combination of the $\mathbf{u}_i$ and $\mathbf{v}_j$ that equals the zero vector). Since each of the 12 vectors $\mathbf{u}_i, \mathbf{v}_j$ is nonzero, this nontrivial linear combination involves at least two nonzero coefficients. Because each set $\{\mathbf{u}_i\}$ and $\{\mathbf{v}_i\}$ consists of linearly independent vectors, the nonzero coefficients must appear both with some $\mathbf{u}_i$ and some $\mathbf{v}_j$ (i.e., a certain nontrivial linear combination of the $\mathbf{u}_i$ equals a certain other nontrivial linear combination of the $\mathbf{v}_j$). Thus, $U \cap V \neq \{\mathbf{0}\}$.

If $V \subset U$, then $U \cap V = V$ is five dimensional. Furthermore, at most three of the $\mathbf{v}_j$ can lie outside of U, for the space allows only 10 linearly independent vectors overall. Thus, $2 \le \dim(U \cap V) \le 5$.

(b) $U = \text{span}\{\mathbf{e}_1, \mathbf{e}_2, \mathbf{e}_4\}$ and $V = \text{span}\{\mathbf{e}_3, \mathbf{e}_2\}$, for example. $U \cap V = \text{span}\{\mathbf{e}_2\}$.

(c) $U = \text{span}\{\mathbf{e}_1, \mathbf{e}_4\}$ and $V = \text{span}\{\mathbf{e}_3, \mathbf{e}_2\}$, for example.

(d) $U = \mathbb{R}^4$ and $V = \text{span}\{\mathbf{e}_3, \mathbf{e}_1\}$, for example. $U \cap V = V$

35. (a) Since the $\mathbf{u}_i$ are assumed to be linearly dependent, we have $\sum \alpha_i \mathbf{u}_i = \mathbf{0}$ for some nonzero coefficients α_i. Thus, $\mathbf{0} = \mathbf{A}(\sum \alpha_i \mathbf{u}_i) = \sum \alpha_i \mathbf{A}\mathbf{u}_i$ by linearity, signifying that the images $\mathbf{A}\mathbf{u}_i$ are also linearly dependent.

(b) Assume that $\sum \alpha_i \mathbf{v}_i = \mathbf{0}$. Then $\mathbf{0} = \mathbf{B}(\sum \alpha_i \mathbf{v}_i) = \sum \alpha_i \mathbf{B}\mathbf{v}_i$. Since the vectors $\mathbf{B}\mathbf{v}_i$ are linearly independent, we must have $\alpha_i = 0$ for all i (i.e., the vectors v_i are linearly independent by Definition 2).

37. Taking the first, second, and last given vector forms a basis for U, which has dimension 3.

39. $\begin{pmatrix} 1 \\ 1 \end{pmatrix}$. $\mathbf{V} = \ker(\mathbf{A} - 2\mathbf{I})$.

41. (a) $\mathbf{A}_{3,6}$ is in REF with three pivots (i.e., $\text{rank}(\mathbf{A}) = 3$).
(b) The three pivot columns, numbered 1, 3, and 5.
(c) The standard basis $\mathcal{E}$. (d) The RREF of $\mathbf{A}$ is
$\begin{pmatrix} \boxed{1} & 0 & 0 & b & 0 & 4(c-a) \\ 0 & 0 & \boxed{1} & 0 & 0 & d+4 \\ 0 & 0 & 0 & 0 & \boxed{-1} & 4 \end{pmatrix}$. It depends on the values of a, b, c, and d. Hence, every row space basis for $\mathbf{A}$ depends on the values of the constants.

43. The dimensions of U and V can be 3 and 7, 4 and 6, or 5 and 5.

5.2.P p. 162

1.

$\boxed{1}$	2	3	0	
$\boxed{-1}$	1	0	3	$+\ row_1$
0	1	1	1	
$\boxed{2}$	1	3	–3	$-2\ row_1$
$\boxed{1}$	2	3	0	
0	$\boxed{3}$	3	3	
0	$\boxed{1}$	1	1	$-1/3\ row_2$
0	$\boxed{-3}$	–3	–3	$+\ row_2$

Thus, $V \subset U$. Now reverse the entries and repeat.

3. A basis for U is $\{\mathbf{y}_1, \mathbf{y}_2, \mathbf{y}_3\}$. For V, a basis is $\{\mathbf{z}_1, \mathbf{z}_2\}$, and for $U \cap V$, a basis is $\{\mathbf{z}_2\}$, while for $U + V$ it is $\{\mathbf{x}_1, \ldots, \mathbf{x}_4\}$. Explanation for $U \cap V$: If $\sum \alpha_i \mathbf{u}_i = \sum \beta_j \mathbf{z}_j$, then, in terms of the original basis $\mathbf{x}_i$, we have

$$(\alpha_1 + \alpha_2 - \alpha_3 - \beta_2)\mathbf{x}_1 + (\alpha_1 - 2\beta_2)\mathbf{x}_2 + \\ +(\alpha_2 + \alpha_3 - \beta_1)\mathbf{x}_3 - \beta_1 \mathbf{x}_4 = \mathbf{0}$$

(i.e., $\beta_1 = 0$). Moreover, $\mathbf{z}_2 = \mathbf{x}_1 + 2\mathbf{x}_2 = 2\mathbf{y}_1 - \frac{\mathbf{y}_2}{2} + \frac{\mathbf{y}_3}{2} \neq \mathbf{0} \in U \cap V$.

5. The first two rows of the RREFs of **A** and **B** are nonzero and the same.

7. The MATLAB commands `format rat, A=[3 0 1 2;0 3 -1 1;1 1 0 1;3 -3 2 1]; rref(A)` produce

```
ans =
  1   0  1/3  2/3
  0   1  -1/3 1/3
  0   0   0    0
  0   0   0    0
```

9. The MATLAB commands `A=[1 2 0 3;2 -1 4 0;4 3 4 6]; null(A,'r'); 5*ans` compute an integer basis for the kernel as follows:

```
ans =
    -8   -3
     4   -6
     5    0
     0    5
```

5.R p. 162

3. (a), (c): Yes; (b), (d): No.

(e) Yes, yes, yes, yes.

(f) No, yes, yes, no.

5. Dependent.

7. $\left\{ \begin{pmatrix} 1 \\ 0 \\ 1 \\ 1 \end{pmatrix}, \begin{pmatrix} -2 \\ 1 \\ -1 \\ 3 \end{pmatrix}, \begin{pmatrix} 1 \\ 2 \\ 3 \\ 1 \end{pmatrix} \right\}$.

9. Any three vectors in $\mathbb{R}^2$ are linearly dependent.

11. $-2\mathbf{e}_3 = -2\mathbf{u}_3 + 2\mathbf{u}_2 - 2\mathbf{u}_1$.

13. (a) If $\text{im}(\mathbf{A}) = \ker(\mathbf{A})$, then their dimensions are equal. The dimensions add up to n by the Dimension Theorem 5.2; hence, n must be even.

(b) $\mathbf{A} = \begin{pmatrix} 1 & -1 \\ 1 & -1 \end{pmatrix}$, for example.

(c) $\mathbf{B} = \begin{pmatrix} \mathbf{A} & \mathbf{O}_2 \\ \mathbf{O}_2 & \mathbf{A} \end{pmatrix}_{4,4}$, $\mathbf{C} = \begin{pmatrix} \mathbf{B} & \mathbf{O}_{4,2} \\ \mathbf{O}_{2,4} & \mathbf{A} \end{pmatrix}_{6,6}$, for example.

15. (a) $\mathbf{w}$ belongs to $\text{span}\{\mathbf{u}_i\}$.

(b) $\{\mathbf{u}_1, \ldots, \mathbf{u}_4\}$ is linearly independent; $\{\mathbf{u}_1, \ldots, \mathbf{u}_4, \mathbf{w}\}$ is linearly dependent.

(c) $\mathbf{w}_{\mathcal{U}} = \begin{pmatrix} 2 \\ -1 \\ 0 \\ 3 \end{pmatrix}$.

17. (a) $\mathbf{u}_4 = \mathbf{u}_3 - \mathbf{u}_2$. (b) $\mathbf{u}_2 = 2\mathbf{u}_1$. $\mathbf{u}_1$ and $\mathbf{u}_3$ are linearly independent.

(d) $\dim(U) = 2$. (c) $\mathbf{u}_3$ and $\mathbf{u}_4$ are linearly independent in the two-dimensional subspace U and hence are a basis as well.

19. (a) $\mathbf{x}_{gen} = \begin{pmatrix} 14 \\ 4 \\ 0 \\ 0 \end{pmatrix} + \alpha \begin{pmatrix} 10 \\ 5 \\ -2 \\ 1 \end{pmatrix}$.

(b) $\text{rank}(\mathbf{A} \mid \mathbf{b}) = 3$; $\dim(\ker(\mathbf{A})) = 1$.

(c) The first three columns of **A** are a basis for its image. The vector $\begin{pmatrix} 10 \\ 5 \\ -2 \\ 1 \end{pmatrix}$ is a basis for $\ker(\mathbf{A})$.

21. (a) $\cos^2(x) = \sin^2(x) - 1 = 0$ for all $x \in \mathbb{R}$.

(b) Assume that $a\sin(x) + b\cos(x) = c \cdot 1 = c$ for all $x \in \mathbb{R}$ and some coefficients a, b, and $c \in \mathbb{R}$. Then, for $x = 0$, we conclude that $b = c$, and for $x = \frac{\pi}{2}$, it follows that $a = c$. Thus, either $a = b = c = 0$ or $\sin(x) + \cos(x) = 1$ for all $x \in \mathbb{R}$. But for $x_0 = \pi$, we have $\sin(x_0) + \cos(x_0) = 0 - 1 \neq 1$, forcing $a = b = c = 0$. Hence, the three functions are linearly independent by Definition 2.

23. (a) The plane p with the equation $x_1 + x_2 + x_3 = 1$ contains all three standard basis vectors e_i.

(b) If $\mathbf{0} \in p$ for a plane $p \subset \mathbb{R}^3$, then p is a proper subspace of $\mathbb{R}^3$ and the points $\mathbf{x} \in p$ satisfy a linear homogeneous equation of the form $\alpha x_1 + \beta x_2 + \gamma x_3 = 0$ with at least one of the coefficients nonzero. (Otherwise $p = \mathbb{R}^3$ would not be a plane.) This equation corresponds to a linear system with a rank-1 system matrix; hence, the system has a two-dimensional kernel equal to p, and any three vectors in this two-dimensional subspace p are necessarily linearly dependent.

(c) The three standard vectors $\mathbf{e}_i$ of $\mathbb{R}^3$ are linearly independent and lie in a plane in $\mathbb{R}^3$ that is not a subspace of $\mathbb{R}^n$. So the two parts (a) and (b) do not contradict each other. The crux is being or not being a subspace here.

Chapter 6

6.1.P p. 177

1. (a) $\mathbf{AB}$ is m by k.

(b) $\mathbf{AB} = \begin{pmatrix} -17 & 13 & 29 & -3 \\ -11 & -1 & 14 & -3 \\ 16 & -26 & 16 & -17 \\ -2 & 10 & 1 & 4 \end{pmatrix}$ and

$\mathbf{BA} = \begin{pmatrix} 4 & 9 & -3 & -7 \\ 4 & 5 & -17 & 4 \\ -12 & -27 & -9 & 36 \\ 11 & -8 & -8 & 2 \end{pmatrix}$.

(c) and (d) $\mathbf{A}_{mn}\mathbf{I}_n = \mathbf{A} = \mathbf{I}_m\mathbf{A}_{mn}$; $\mathbf{A}_{mn}\mathbf{O}_n = \mathbf{O}_{mn} = \mathbf{O}_m\mathbf{A}_{mn}$.

3. (a) $\mathbf{A}^{-1} = \begin{pmatrix} -1/2 & -1/2 & 1/2 \\ 1/4 & -1/4 & -1/4 \\ 1/6 & 1/2 & 1/6 \end{pmatrix}$.

(b) $\mathbf{A}^{-1}\begin{pmatrix} 12 \\ -12 \\ 8 \end{pmatrix} = \begin{pmatrix} 4 \\ 4 \\ -8/3 \end{pmatrix}$:

$\boxed{1}$	4	3	1	0	0	
$\boxed{-1}$	-2	0	0	1	0	$+\ row_1$
$\boxed{2}$	2	3	0	0	1	$-2\ row_1$
$\boxed{1}$	4	3	1	0	0	
0	$\boxed{2}$	3	1	1	0	
0	$\boxed{-6}$	-3	-2	0	1	$+3\ row_2$
$\boxed{1}$	4	3	1	0	0	
0	$\boxed{2}$	3	1	1	0	$\div 2$
0	0	$\boxed{6}$	1	3	1	$\div 6$
$\boxed{1}$	4	$\boxed{3}$	1	0	0	$-3\ row_3$
0	$\boxed{1}$	$\boxed{\frac{3}{2}}$	$\frac{1}{2}$	$\frac{1}{2}$	0	$-\frac{3}{2}\ row_3$
0	0	$\boxed{1}$	$\frac{1}{6}$	$\frac{1}{2}$	$\frac{1}{6}$	
$\boxed{1}$	$\boxed{4}$	0	$\frac{1}{2}$	$-\frac{3}{2}$	$-\frac{1}{2}$	$-4\ row_2$
0	$\boxed{1}$	0	$\frac{1}{4}$	$-\frac{1}{4}$	$-\frac{1}{4}$	
0	0	$\boxed{1}$	$\frac{1}{6}$	$\frac{1}{2}$	$\frac{1}{6}$	
	$\mathbf{I}_3$		$-\frac{1}{2}$	$-\frac{1}{2}$	$\frac{1}{2}$	
			$\frac{1}{4}$	$-\frac{1}{4}$	$-\frac{1}{4}$	
			$\frac{1}{6}$	$\frac{1}{2}$	$\frac{1}{6}$	

5. (a) $(\mathbf{I}+\mathbf{E})(\mathbf{I}-\mathbf{E}) = \mathbf{O}_3$, since $\mathbf{E}^2 = \mathbf{I}_3$.

(b) $(\mathbf{I}+\mathbf{E})(\mathbf{I}-\mathbf{E}) = \mathbf{O}_n$ if $\mathbf{E}^2 = \mathbf{I}_n$.

(c) $\mathbf{E}^k = \mathbf{I}_3$ if k is even; $\mathbf{E}^k = \mathbf{E}$ if k is odd; $\mathbf{E}^0 = \mathbf{I}_3$.

7. $\mathbf{A}^{-1} = \begin{pmatrix} -1 & -1 & 2 \\ -1 & 0 & 1 \\ 2 & 1 & -2 \end{pmatrix}$:

$\boxed{1}$	0	1	1	0	0	
0	2	1	0	1	0	
$\boxed{1}$	1	1	0	0	1	$-\ row_1$
$\boxed{1}$	0	1	1	0	0	
0	2	1	0	1	0	*swap* row_2
0	1	0	-1	0	1	*and* row_3
$\boxed{1}$	0	1	1	0	0	
0	$\boxed{1}$	0	-1	0	1	
0	$\boxed{2}$	1	0	1	0	$-2\ row_2$
$\boxed{1}$	0	$\boxed{1}$	1	0	0	$-\ row_3$
0	$\boxed{1}$	0	-1	0	1	
0	0	$\boxed{1}$	2	1	-2	
	I_3		-1	-1	2	
			-1	0	1	
			2	1	-2	

9. Column 3 of $\mathbf{A}^{-1}$ is $\mathbf{x} = \frac{1}{17}\begin{pmatrix} 14 \\ -3 \\ 2 \end{pmatrix}$. (Solve $\mathbf{Ax} = \mathbf{e}_3$ by Gaussian elimination.)

11. If $\mathbf{XA} = \mathbf{O}_{kn}$ and $\mathbf{A}_{nn}\mathbf{A}_{nn}^{-1} = \mathbf{I}_n$, then we have $\mathbf{X} = \mathbf{XAA}^{-1} = \mathbf{OA}^{-1} = \mathbf{O}_{kn}$.

13. $\mathbf{C}^T\mathbf{B}(\mathbf{A}^{-1})^T$.

15. $\mathbf{A}$ is nonsingular for all $d \neq 1$ and $d \neq -3$:

$$
\begin{array}{cccc|l}
\boxed{1} & -1 & 2 & d & \\
0 & 4 & -2 & 0 & \\
2 & 0 & -d & 1 & -2\ row_1 \\
1 & 1 & 1 & 1 & -\ row_1 \\
\hline
\boxed{1} & -1 & 2 & d & \\
0 & \boxed{4} & -2 & 0 & \div 2 \\
0 & 2 & -4-d & 1-2d & \\
0 & 2 & -1 & 1-d & \\
\hline
\boxed{1} & -1 & 2 & d & \\
0 & \boxed{2} & -1 & 0 & \\
0 & 2 & -4-d & 1-2d & -\ row_2 \\
0 & 2 & -1 & 1-d & -\ row_2 \\
\hline
\boxed{1} & -1 & 2 & d & \\
0 & \boxed{2} & -1 & 0 & \\
0 & 0 & -3-d & 1-2d & \\
0 & 0 & 0 & 1-d &
\end{array}
$$

This REF has four pivots, and $\mathbf{A}$ is nonsingular precisely if $d \neq 3$ and $d \neq 1$.

17. (a) $q = r$ and $p = s$, or $\mathbf{A}_{pq}$ and $\mathbf{B}_{qp}$.

(b) For $\mathbf{A} = \begin{pmatrix} 0 & 1 \\ 0 & 0 \end{pmatrix}$, $\mathbf{B} = \begin{pmatrix} 0 & 0 \\ 1 & 0 \end{pmatrix}$, $\mathbf{C} = \begin{pmatrix} 1 & 0 \\ 0 & 0 \end{pmatrix}$, and $\mathbf{D} = \begin{pmatrix} 2 & 0 \\ 0 & 0 \end{pmatrix}$, we have

$$\mathbf{AB} = \begin{pmatrix} 1 & 0 \\ 0 & 0 \end{pmatrix} \neq \mathbf{BA} = \begin{pmatrix} 0 & 0 \\ 0 & 1 \end{pmatrix},$$ while

$$\mathbf{CD} = \begin{pmatrix} 2 & 0 \\ 0 & 0 \end{pmatrix} = \mathbf{DC}.$$

19. $\mathbf{A}^3 =$

$$\begin{pmatrix} 0 & -a^3 - ab^2 - ac^2 & -a^2b - bc^2 - b^3 \\ a^3 + ab^2 + ac^2 & 0 & -a^2c - b^2c - c^3 \\ a^2b + bc^2 + b^3 & a^2c + b^2c + c^3 & 0 \end{pmatrix}.$$

21. Since $S\mathbf{u}_i = \mathbf{v}_i$ and $T\mathbf{u}_j = \mathbf{v}_j$, we respectively have

$$\mathbf{A} = \begin{pmatrix} | & | & | \\ \mathbf{v}_2 & \mathbf{v}_3 & \mathbf{v}_4 \\ | & | & | \end{pmatrix} \begin{pmatrix} | & | & | \\ \mathbf{u}_2 & \mathbf{u}_3 & \mathbf{u}_4 \\ | & | & | \end{pmatrix}^{-1}$$

and $$\mathbf{B} = \begin{pmatrix} | & | & | \\ \mathbf{v}_1 & \mathbf{v}_3 & \mathbf{v}_4 \\ | & | & | \end{pmatrix} \begin{pmatrix} | & | & | \\ \mathbf{u}_1 & \mathbf{u}_3 & \mathbf{u}_4 \\ | & | & | \end{pmatrix}^{-1}.$$

The two matrix inverses exist, since the respective sets of vectors $\{\mathbf{u}_i\}$ and $\{\mathbf{u}_j\}$ are linearly independent.

23. (a) $\begin{pmatrix} a & d & g \\ b & e & h \\ c & f & i \end{pmatrix}^{-1} \begin{pmatrix} a \\ b \\ c \end{pmatrix} = \begin{pmatrix} 1 \\ 0 \\ 0 \end{pmatrix} = \mathbf{e}_1$.

(b) $2\mathbf{e}_2$.

25. (a) $\mathbf{B}^{-1} = \frac{1}{3}\begin{pmatrix} -1 & -2 & 1 \\ 1 & -1 & -1 \\ 0 & 2 & 1 \end{pmatrix}$:

$$
\begin{array}{ccc|ccc|l}
\boxed{1} & 4 & 3 & 1 & 0 & 0 & \\
-1 & -1 & 0 & 0 & 1 & 0 & +\ row_1 \\
2 & 2 & 3 & 0 & 0 & 1 & -2\ row_1 \\
\hline
\boxed{1} & 4 & 3 & 1 & 0 & 0 & \\
0 & \boxed{3} & 3 & 1 & 1 & 0 & \\
0 & -6 & -3 & -2 & 0 & 1 & +2\ row_2 \\
\hline
\boxed{1} & 4 & 3 & 1 & 0 & 0 & \\
0 & \boxed{3} & 3 & 1 & 1 & 0 & \div 3 \\
0 & 0 & \boxed{3} & 0 & 2 & 1 & \div 3 \\
\hline
\boxed{1} & 4 & 3 & 1 & 0 & 0 & -3\ row_3 \\
0 & \boxed{1} & 1 & \frac{1}{3} & \frac{1}{3} & 0 & -\ row_3 \\
0 & 0 & \boxed{1} & 0 & \frac{2}{3} & \frac{1}{3} & \\
\hline
\boxed{1} & 4 & 0 & 1 & -2 & -1 & -4\ row_2 \\
0 & \boxed{1} & 0 & \frac{1}{3} & -\frac{1}{3} & -\frac{1}{3} & \\
0 & 0 & \boxed{1} & 0 & \frac{2}{3} & \frac{1}{3} & \\
\hline
 & & & -\frac{1}{3} & -\frac{2}{3} & \frac{1}{3} & \\
 & \mathbf{I}_3 & & \frac{1}{3} & -\frac{1}{3} & -\frac{1}{3} & \\
 & & & 0 & \frac{2}{3} & \frac{1}{3} &
\end{array}
$$

(b) $\mathbf{x} = \mathbf{B}^{-1}\mathbf{b} = \frac{1}{3}\begin{pmatrix} 20 \\ 16 \\ -16 \end{pmatrix}$.

27. $\mathbf{A}_{\mathcal{E}} = \frac{1}{3}\begin{pmatrix} -2 & -1 \\ 2 & 7 \end{pmatrix}$.

29. (a) $\mathbf{O}_{5,3}$; (b) 10 values.

(c) 15 equations: $x_{i1} + 2x_{i2} = 0$; $2x_{i2} = 0$; and $-x_{i1} + x_{i2} = 0$ for each $i = 1, \dots, 5$.

(d) $\mathbf{X} = \mathbf{O}_{5,2}$.

(e) Take $\mathbf{Y}_{3,4} = \begin{pmatrix} -1 & 0 & 1 & 2 \\ 3/2 & 0 & -3/2 & -3 \\ -1 & 0 & 1 & 2 \end{pmatrix}$, for example. (The columns of $\mathbf{Y}$ must lie in the kernel of $\mathbf{A}$.)

31. $\begin{pmatrix} | & | \\ \mathbf{u} & \mathbf{v} \\ | & | \end{pmatrix} = \begin{pmatrix} | & | \\ \mathbf{x} & \mathbf{y} \\ | & | \end{pmatrix} \begin{pmatrix} a & c \\ b & d \end{pmatrix}$. If $ad - bc = 0$, then $\begin{pmatrix} a & c \\ b & d \end{pmatrix}$ is singular (i.e., there is a nonzero vector in its kernel). Hence, the matrix $\begin{pmatrix} | & | \\ \mathbf{u} & \mathbf{v} \\ | & | \end{pmatrix}$ also has a nontrivial kernel, and $\mathbf{u}$ and $\mathbf{v}$ are linearly dependent. If $ad - bc \neq 0$, then $\begin{pmatrix} a & c \\ b & d \end{pmatrix}$ is nonsingular, as is $\begin{pmatrix} | & | \\ \mathbf{u} & \mathbf{v} \\ | & | \end{pmatrix}$, making the vectors $\mathbf{u}$ and $\mathbf{v}$ linearly independent, all by Chapter 5. Since $\dim(U) = 2$, $\{\mathbf{u}, \mathbf{v}\}$ must be a basis for U in this case.

33. $$\mathbf{I}_n\mathbf{B}_{nk} = \begin{pmatrix} 1 & & 0 \\ & \ddots & \\ 0 & & 1 \end{pmatrix} \begin{pmatrix} b_{11} & \dots & b_{1k} \\ \vdots & & \vdots \\ b_{n1} & \dots & b_{nk} \end{pmatrix} = \begin{pmatrix} b_{11} & \dots & b_{1k} \\ \vdots & & \vdots \\ b_{n1} & \dots & b_{nk} \end{pmatrix}.$$

35. (a) $\mathbf{AB} = \begin{pmatrix} 1 & 1 \\ 3 & 0 \end{pmatrix}$, $\mathbf{BA} = \begin{pmatrix} -1 & 1 \\ 1 & 2 \end{pmatrix}$,

$\mathbf{AE} = \begin{pmatrix} 2 & 1 \\ 3 & 0 \end{pmatrix}$, $\mathbf{EA} = \begin{pmatrix} 0 & 3 \\ 1 & 2 \end{pmatrix}$,

$\mathbf{BC} = \begin{pmatrix} 1 & 0 \\ 0 & 0 \end{pmatrix}$, $\mathbf{CB} = \begin{pmatrix} 0 & 0 \\ -1 & 1 \end{pmatrix}$,

$\mathbf{BE} = \begin{pmatrix} 1 & -1 \\ 0 & 1 \end{pmatrix}$, $\mathbf{EB} = \begin{pmatrix} 1 & 0 \\ -1 & 1 \end{pmatrix}$,

$\mathbf{AC} = \begin{pmatrix} 2 & 0 \\ 3 & 0 \end{pmatrix}$, $\mathbf{CA} = \begin{pmatrix} 0 & 0 \\ 1 & 2 \end{pmatrix}$,

$\mathbf{CE} = \begin{pmatrix} 0 & 0 \\ 0 & 1 \end{pmatrix}$, and $\mathbf{EC} = \begin{pmatrix} 1 & 0 \\ 0 & 0 \end{pmatrix}$.

None of the matrices **A**, **B**, **C**, or **E** commute with each other.

(b) $\mathbf{A}(\mathbf{BE}) = \begin{pmatrix} 1 & 1 \\ 0 & 3 \end{pmatrix} = (\mathbf{AB})\mathbf{E}$ and $\mathbf{C}(\mathbf{AB}) = \begin{pmatrix} 0 & 0 \\ 1 & 1 \end{pmatrix} = (\mathbf{CA})\mathbf{B}$.

All triple matrix products are associative.

37. (b) If $\mathbf{AX} = \begin{pmatrix} x_{21} & x_{22} \\ 2x_{21} & 2x_{22} \end{pmatrix} = \begin{pmatrix} 1 & 0 \\ 0 & 1 \end{pmatrix} = \mathbf{I}_2$, then $x_{21} = 1$ and $2x_{21} = 0$, a contradiction. Thus, **A** is not invertible.

(c) No such **Y** can exist, since **A** is not invertible according to part (b).

39. (a) and (b) $\mathbf{A} = \begin{pmatrix} 1 & 0 \\ 0 & 2 \end{pmatrix}$; $\mathbf{A}^{-1} = \begin{pmatrix} 1 & 0 \\ 0 & \frac{1}{2} \end{pmatrix}$ and $\mathbf{B} = \begin{pmatrix} 2 & 0 \\ 0 & 1 \end{pmatrix}$; $\mathbf{B}^{-1} = \begin{pmatrix} \frac{1}{2} & 0 \\ 0 & 1 \end{pmatrix}$. For example; $(\mathbf{A}+\mathbf{B})^{-1} = \frac{1}{3}\mathbf{I}_2 \neq \mathbf{A}^{-1} + \mathbf{B}^{-1} = \frac{3}{2}\mathbf{I}_2$.

(c) $(\mathbf{A}+\mathbf{B})^{-1} = \left(\begin{pmatrix} 1 & 1 \\ 0 & 1 \end{pmatrix} + \begin{pmatrix} 0 & -1 \\ 1 & 0 \end{pmatrix}\right)^{-1} = \begin{pmatrix} 1 & 0 \\ 1 & 1 \end{pmatrix}^{-1} = \begin{pmatrix} 1 & 0 \\ -1 & 1 \end{pmatrix}$ with $\mathbf{A}^{-1} = \begin{pmatrix} 1 & 1 \\ 0 & 1 \end{pmatrix}^{-1} = \begin{pmatrix} 1 & -1 \\ 0 & 1 \end{pmatrix}$ and $\mathbf{B}^{-1} = \begin{pmatrix} 0 & -1 \\ 1 & 0 \end{pmatrix}^{-1} = \begin{pmatrix} 0 & 1 \\ -1 & 0 \end{pmatrix}$ give one such example.

(d) No such real numbers exist: Assume that $\frac{1}{x+y} = \frac{1}{x} + \frac{1}{y} = \frac{y+x}{xy}$. Then $(x+y)^2 = xy$ or $0 = x^2 + xy + y^2 = x^2\left(1 + \frac{y}{x} + \frac{y^2}{x^2}\right) = 0$. Here, $x \neq 0$. With $z := \frac{y}{x}$, we must solve $z^2 + z + 1 = 0$, making $z_{1,2} = -\frac{1}{2} \pm \sqrt{\frac{1}{4} - 1} \notin \mathbb{R}$.

41. (a) Every polynomial $r(x) = ax^2 + bx + c \in \mathcal{P}_2$ can be obtained by differentiation from $\mathcal{P}_3$: $D\left(\frac{a}{3}x^3 + \frac{b}{2}x^2 + cx\right) = r(x)$. But $D(1) = D(0) = 0$.

(b) p and $q \in S$ have the form $p(x) = ax^3 + bx^2 + cx$ and $q(x) = ex^3 + fx^2 + gx$ for real constants $a, \dots, g$. If $D(p) = p' = D(q) = q'$, then $3ax^2 + 2bx + c = 3ex^2 + 2fx + g$. Thus, $a = e$, $b = f$, and $c = g$; that is, $p = q$, or D is one-to-one on S. Finally, there is no polynomial $s \in S$ with $D(s) = x^3$ (i.e., D is not onto $\mathcal{P}_3$).

T 6. We have $(\mathbf{AB})^{-1} = \begin{pmatrix} 1 & 7 & 12 \\ -2 & -15 & -26 \\ 1 & 7 & 13 \end{pmatrix}$ and $(\mathbf{BA})^{-1} = \begin{pmatrix} -1 & 7 & 12 \\ 0 & -1 & -2 \\ 1 & 0 & 1 \end{pmatrix}$, which are both dense and integer matrices.

6.2.P p. 193

1. $\mathbf{E}(c_{jk})\mathbf{E}(-c_{jk})$

$$= \begin{pmatrix} 1 & & & \\ & \ddots & & \\ & c_{jk} & \ddots & \\ & & & 1 \end{pmatrix} \begin{pmatrix} 1 & & & \\ & \ddots & & \\ & -c_{jk} & \ddots & \\ & & & 1 \end{pmatrix}$$

$$= \begin{pmatrix} 1 & & & \\ & \ddots & & \\ & c_{jk} - c_{jk} & \ddots & \\ & & & 1 \end{pmatrix} = \mathbf{I} \text{ if } j \neq k.$$

$\mathbf{E}_k(c)\mathbf{E}_k(1/c)$

$$= \begin{pmatrix} 1 & & & & \\ & \ddots & & & \\ & & c & & \\ & & & \ddots & \\ & & & & 1 \end{pmatrix} \cdot \begin{pmatrix} 1 & & & & \\ & \ddots & & & \\ & & 1/c & & \\ & & & \ddots & \\ & & & & 1 \end{pmatrix}$$

$= \mathbf{I}$ if $c \neq 0$.

$\mathbf{E}_{jk}\mathbf{E}_{jk}$

$$= \begin{pmatrix} 1 & & & & & & \\ & \ddots & & & & & \\ & & 0 & \dots & \dots & 1 & \\ & & & 1 & & & \\ & & & & \ddots & & \\ & & 1 & & & 0 & \\ & & & & & & 1 \\ & & & & & & & \ddots \end{pmatrix} \cdot$$

$$\cdot \begin{pmatrix} 1 & & & & & & \\ & \ddots & & & & & \\ & & 0 & \dots & \dots & 1 & \\ & & & 1 & & & \\ & & & & \ddots & & \\ & & 1 & & & 0 & \\ & & & & & & 1 \\ & & & & & & & \ddots \end{pmatrix} = \mathbf{I}.$$

3. (a) $\begin{pmatrix} 1 & & & \\ & 1 & & \\ & 0 & 1 & \\ & 3 & 0 & 1 \end{pmatrix} \begin{pmatrix} 1 & & & \\ & 1 & & \\ & 2 & 1 & \\ & & & 1 \end{pmatrix}$ maps $\begin{pmatrix} 4 \\ -1 \\ 2 \\ 3 \end{pmatrix}$ to $\begin{pmatrix} 4 \\ -1 \\ 0 \\ 0 \end{pmatrix}$.

(b) $\begin{pmatrix} 1 & & & \\ & 1 & & \\ & 0 & 1 & \\ & 3 & 0 & 1 \end{pmatrix} \begin{pmatrix} 1 & & & \\ & 1 & & \\ & 0 & 1 & \\ & -1/2 & 0 & 1 \\ & & & & 1 \end{pmatrix}$ also maps $\begin{pmatrix} 4 \\ -1 \\ 2 \\ 3 \end{pmatrix}$ to $\begin{pmatrix} 4 \\ -1 \\ 0 \\ 0 \end{pmatrix}$. (Why are these two matrix products different?)

5. $(\mathbf{A} + \alpha\mathbf{I})(\mathbf{A} - \beta\mathbf{I}) = \mathbf{A}^2 + (\alpha - \beta)\mathbf{A} - \alpha\beta\mathbf{I} = (\mathbf{A} - \beta\mathbf{I})(\mathbf{A} + \alpha\mathbf{I})$.

7. $\mathbf{A}(\mathbf{e}_1) = \begin{pmatrix} 4 \\ 1 \\ 2 \end{pmatrix}$, $\mathbf{A}(\mathbf{e}_2) = \begin{pmatrix} -4 \\ 1 \\ 0 \end{pmatrix}$, and $\mathbf{A}(\mathbf{e}_3)$ $= \begin{pmatrix} 1 \\ 2 \\ 0 \end{pmatrix}$; thus, $\mathbf{A} = \begin{pmatrix} 4 & -4 & 1 \\ 1 & 1 & 2 \\ 2 & 0 & 0 \end{pmatrix}$.

$$\mathbf{A}^{-1} = \begin{pmatrix} 0 & 0 & 1/2 \\ -2/9 & 1/9 & 7/18 \\ 1/9 & 4/9 & -4/9 \end{pmatrix}:$$

4	−4	1	1	0	0	*swap rows 1, 30*
1	1	2	0	1	0	
2	0	0	0	0	1	
$\boxed{2}$	0	0	0	0	1	
$\boxed{1}$	1	2	0	1	0	$-\frac{1}{2}\ row_1$
$\boxed{4}$	−4	1	1	0	0	$-2\ row_1$
$\boxed{2}$	0	0	0	0	1	$\div 2$
0	$\boxed{1}$	2	0	1	$-\frac{1}{2}$	
0	$\boxed{-4}$	1	1	0	−2	$+4\ row_2$
$\boxed{2}$	0	0	0	0	1	$\div 2$
0	$\boxed{1}$	2	0	1	$-\frac{1}{2}$	
0	0	$\boxed{9}$	1	4	−4	$\div 9$
$\boxed{1}$	0	0	0	0	$\frac{1}{2}$	
0	$\boxed{1}$	$\boxed{2}$	0	1	$-\frac{1}{2}$	$-2\ row_3$
0	0	$\boxed{1}$	$\frac{1}{9}$	$\frac{4}{9}$	$-\frac{4}{9}$	
			0	0	$\frac{1}{2}$	
	$\mathbf{I}_3$		$-\frac{2}{9}$	$\frac{1}{9}$	$\frac{7}{18}$	
			$\frac{1}{9}$	$\frac{4}{9}$	$-\frac{4}{9}$	

9. $$\begin{pmatrix} d_1 & & \\ & \ddots & \\ & & d_m \end{pmatrix} \begin{pmatrix} - & \mathbf{b}_1^T & - \\ & \vdots & \\ - & \mathbf{b}_m^T & - \end{pmatrix} = \begin{pmatrix} - & d_1\mathbf{b}_1^T & - \\ & \vdots & \\ - & d_m\mathbf{b}_m^T & - \end{pmatrix};\ \begin{pmatrix} | & & | \\ \mathbf{a}_1 & \dots & \mathbf{a}_m \\ | & & | \end{pmatrix}$$

$$\begin{pmatrix} d_1 & & \\ & \ddots & \\ & & d_m \end{pmatrix} = \begin{pmatrix} | & & | \\ d_1\mathbf{a}_1 & \dots & d_m\mathbf{a}_m \\ | & & | \end{pmatrix}.$$

11. $$\begin{pmatrix} \mathbf{A}_{11} & \mathbf{A}_{12} \\ \mathbf{0} & \mathbf{A}_{22} \end{pmatrix}^{-1} = \begin{pmatrix} \mathbf{A}_{11}^{-1} & -\mathbf{A}_{11}^{-1}\mathbf{A}_{12}\mathbf{A}_{22}^{-1} \\ \mathbf{0} & \mathbf{A}_{22}^{-1} \end{pmatrix}.$$

$(\lambda - 1)$, giving $\mathbf{B}$ the eigenvalues 1 and 0 (double). As rank($\mathbf{B}$) = 2, there are not sufficiently many (only one) corresponding linearly independent eigenvectors for $\lambda = 0$. $\mathbf{B}$ is not diagonalizable.

17. (a) and (b): $\mathbf{x} = \begin{pmatrix} 2 \\ 3 \\ 1 \end{pmatrix}$ is an eigenvector for $\mathbf{A}$.

(c) and (d): $\mathbf{Ay} = \begin{pmatrix} 9 \\ -5 \\ -5 \end{pmatrix} \neq \lambda \begin{pmatrix} 1 \\ -1 \\ 1 \end{pmatrix}$ for any $\lambda \neq 0$, so $\mathbf{y}$ is not an eigenvector of $\mathbf{A}$, and neither is $\mathbf{z} = -\mathbf{y}$.

19. (a) $\mathbf{A}_{\mathcal{U}} = \mathbf{U}^{-1}\mathbf{A}_{\mathcal{E}}\mathbf{U} = \begin{pmatrix} 2 & 0 \\ 0 & 4 \end{pmatrix}$ with $\mathbf{U} = \begin{pmatrix} 1 & 2 \\ -1 & 2 \end{pmatrix}$ and $\mathbf{U}^{-1} = \frac{1}{4}\begin{pmatrix} 2 & -2 \\ 1 & 1 \end{pmatrix}$.

(b) The eigenvalues of $\mathbf{A}$ are 2 and 4, with eigenvectors $\begin{pmatrix} 1 \\ -1 \end{pmatrix}$ and $\begin{pmatrix} 2 \\ 2 \end{pmatrix}$, respectively.

21. For $\mathbf{y} = \mathbf{e}_1$, we have

$\mathbf{y}$	$\mathbf{Ay}$	$\mathbf{A}^2\mathbf{y}$	$\mathbf{A}^3\mathbf{y}$	
1	0	1	2	+ *row*$_3$
0	−1	−2	−6	−2 *row*$_3$
0	0	−1	−6	
1	0	0	−4	
0	−1	0	6	· (−1)
0	0	1	6	
	I_3		−4	
			−6	
			6	

Hence, $\mathbf{A}^3\mathbf{y} = 6\mathbf{A}^2\mathbf{y} - 6\mathbf{Ay} - 4\mathbf{y}$ and $p_y(\lambda) = \lambda^3 - 6\lambda^2 + 6\lambda + 4$. By trial and error, note that $p_y(2) = 0$, so that $\lambda_1 = 2$ is one eigenvalue of $\mathbf{A}$. Now use long division:

$$\begin{array}{llll} (\ \lambda^3 & -6\lambda^2 & +6\lambda & +4\ ) \div (\lambda - 2) = \\ \lambda^3 & \mp 2\lambda^2 & & \\ \hline & -4\lambda^2 & +6\lambda & \\ & \mp 4\lambda^2 & \pm 8\lambda & \\ \hline & & -2\lambda & +4 \\ & & \mp 2\lambda & \pm 4 \\ \hline & & & 0 \end{array}$$

$= \lambda^2 - 4\lambda - 2$. The factor polynomial $\lambda^2 - 4\lambda - 2 = 0$ has the roots $\lambda_{2,3} = 2 \pm \sqrt{6}$. Hence, the eigenvalues of $\mathbf{A}$ are 2, $2+\sqrt{6}$, $2-\sqrt{6}$, with corresponding eigenvectors $\begin{pmatrix} 1 \\ -2 \\ 1 \end{pmatrix}$, $\begin{pmatrix} -5+2\sqrt{6} \\ -2+\sqrt{6} \\ 1 \end{pmatrix}$, and $\begin{pmatrix} 1 \\ -2+\sqrt{6} \\ -5+2\sqrt{6} \end{pmatrix}$.

$$\mathbf{A}_{\mathcal{U}} = \begin{pmatrix} 2 & 0 & 0 \\ 0 & 2+\sqrt{6} & 0 \\ 0 & 0 & 2-\sqrt{6} \end{pmatrix}.$$

23. If $\mathbf{A}_{nn}$ is singular, then $\ker(\mathbf{A}) = \ker(\mathbf{A} - 0\mathbf{I}_n) = E(0) \neq \{\mathbf{0}\}$, so $\lambda = 0$ is an eigenvalue of $\mathbf{A}$ and conversely.

25. (a) Work with $\mathbf{B} := \frac{\mathbf{A}}{3} = \begin{pmatrix} 1 & 1 & 0 \\ 0 & 3 & -1 \\ 0 & 0 & -2 \end{pmatrix}$ for smaller numbers to compute with. The eigenvalues of $\mathbf{B}$ are one-third the eigenvalues of $\mathbf{A}$; the eigenvectors are identical. Take $\mathbf{y} = \mathbf{e}_3$, for example, and vector iterate:

$\mathbf{y}$	$\mathbf{By}$	$\mathbf{B}^2\mathbf{y}$	$\mathbf{B}^3\mathbf{y}$	
0	0	−1	−2	*swap all rows*
0	−1	−1	−7	
1	−2	4	−8	
1	−2	4	−8	+4 *row*$_3$
0	−1	−1	−7	− *row*$_3$
0	0	−1	−2	
1	−2	0	−16	−2 *row*$_2$
0	−1	0	−5	
0	0	−1	−2	· (−1)
1	0	0	−6	
0	−1	0	−5	· (−1)
0	0	1	2	
	$\mathbf{I}_3$		−6	
			5	
			2	

Thus, $\mathbf{A}^3\mathbf{y} = 2\mathbf{A}^2\mathbf{y} + 5\mathbf{Ay} - 6\mathbf{y}$, or $p_y(\lambda) = \lambda^3 - 2\lambda^2 - 5\lambda + 6$. Now, $p_y(1) = 0 = p_y(-2) = p_y(3)$, by using a little guesswork on the integer factors of the constant term 6. Hence, the eigenvalues of $\mathbf{B}$ are 1, 3, and –2, making those of $\mathbf{A}$ equal to 3, 9, and –6. The corresponding eigenvectors are $\mathbf{e}_1$, $\mathbf{e}_1 + 2\mathbf{e}_2$, and $\begin{pmatrix} -1 \\ 3 \\ 15 \end{pmatrix}$; $\mathbf{A}$ is diagonalizable.

(b) eigenvalues –1 (double) and 2; corresponding eigenvector $\mathbf{e}_1$ (only one) for $\lambda = -1$, and $-\mathbf{e}_1 + 3\mathbf{e}_2$

(b) Take $i = 3$, for example. Then $x_3 = 1$, $\frac{1}{x_3}\mathbf{x}\mathbf{a}_3$
$= \begin{pmatrix} -2 \\ 5 \\ 1 \end{pmatrix} \begin{pmatrix} -11 & -5 & 6 \end{pmatrix}$, and $\mathbf{B}_0 = \begin{pmatrix} 1 & 0 & -22 \\ -3 & -2 & 4 \\ 0 & 0 & 0 \end{pmatrix}$.

(c) The matrix $\mathbf{B}_0$ has a row of zeros in position i. If $\mathbf{B}_0\mathbf{w} = \mu\mathbf{w}$, then $w_i = 0$. If $\mathbf{A}\mathbf{x} = \lambda\mathbf{x}$, then $\mathbf{B}_0\mathbf{x} = \left(\mathbf{A} - \frac{1}{x_i}\mathbf{x}\mathbf{a}_i\right)\mathbf{x} = \mathbf{A}\mathbf{x} - \frac{1}{x_i}\mathbf{x}\mathbf{a}_i\mathbf{x} = \lambda\mathbf{x} - \lambda\mathbf{x} = \mathbf{0}$, since $\mathbf{a}_i\mathbf{x} = \lambda x_i$.

(d) $\mathbf{B}_1 = \begin{pmatrix} 1 & 0 \\ -3 & -2 \end{pmatrix}$.

$\mathbf{B}_1$		$\mathbf{y}$	$\mathbf{B}_1\mathbf{y}$	$\mathbf{B}_1^2\mathbf{y}$	
1	0	$\boxed{1}$	1	1	
–3	–2	$\boxed{1}$	–5	7	$-$ *row*$_1$
		$\boxed{1}$	1	1	
		0	$\boxed{-6}$	–6	$\div(-6)$
		$\boxed{1}$	$\boxed{1}$	1	$-$ *row*$_2$
		0	$\boxed{1}$	–1	
		$\boxed{1}$	0	2	
		0	$\boxed{1}$	–1	

Thus, $\mathbf{B}_1^2\mathbf{y} = -\mathbf{B}_1\mathbf{y} + 2\mathbf{y}$ or $p_y(x) = p_{B_1}(x) = x^2 + x - 2 = (x-1)(x+2)$. $\begin{pmatrix} 1 \\ -1 \end{pmatrix}$ and $\begin{pmatrix} 0 \\ 1 \end{pmatrix}$ are eigenvectors for the eigenvalues 1 and –2 of $\mathbf{B}_1$.

(e) If $\mathbf{A}$ is nonsingular and diagonalizable, assume that it also has the eigenvalues $\lambda_2, \ldots, \lambda_n \in \mathbb{R}$ for the eigenvectors $\mathbf{y}_2, \ldots, \mathbf{y}_n \in \mathbb{R}^n$. Then $\mathbf{B}_0\left(\mathbf{y}_i - \frac{\mathbf{a}_i\mathbf{y}_i}{\lambda_i x_i}\mathbf{x}\right) = \mathbf{B}_0(\mathbf{y}_i)$, since $\mathbf{B}_0\mathbf{x} = \mathbf{0}$, and $\mathbf{B}_0\mathbf{y}_i = \left(\mathbf{A} - \frac{1}{x_i}\mathbf{x}\mathbf{a}_i\right)\mathbf{y}_i = \lambda_i\left(\mathbf{y}_i - \frac{\mathbf{a}_i\mathbf{y}_i}{\lambda_i x_i}\mathbf{x}\right) = \mathbf{B}_0\left(\mathbf{y}_i - \frac{\mathbf{a}_i\mathbf{y}_i}{\lambda_i x_i}\mathbf{x}\right)$. So $\mathbf{B}_0$ also has the eigenvalues λ_i of $\mathbf{A}$, provided that the vectors $\mathbf{y}_i - \frac{\mathbf{a}_i\mathbf{y}_i}{\lambda_i x_i}\mathbf{x}$ are nonzero. But this is true, because the $\mathbf{y}_i$ and $\mathbf{x}$ are linearly independent; they form a diagonalizing basis for $\mathbf{A}$.

(f) $\mathbf{A}$ is diagonalizable.

11. (b) and (c): $\mathbf{A}^2 - 2\mathbf{A} = 3\mathbf{I}$ or $(\mathbf{A} - 2\mathbf{I})\mathbf{A} = 3\mathbf{I}$; thus, $\mathbf{A}^{-1} = \frac{1}{3}(\mathbf{A} - 2\mathbf{I}) = \frac{1}{3}\begin{pmatrix} -1 & 2 \\ 2 & -1 \end{pmatrix}$.

(d) eigenvalues: 3, –1. They are the roots of the vanishing polynomial $\lambda^2 - 2\lambda - 3 = (\lambda - 3)(\lambda + 1)$ given in part (a) for $\mathbf{A}$.

13. $\mathbf{A}_\mathcal{E} = \begin{pmatrix} 3 & 1 & 0 \\ 0 & 1 & 0 \\ 4 & 2 & 1 \end{pmatrix}$. For $\mathbf{y} = \begin{pmatrix} 1 \\ -1 \\ 0 \end{pmatrix}$, we have

$\mathbf{y}$	$\mathbf{A}\mathbf{y}$	$\mathbf{A}^2\mathbf{y}$	$\mathbf{A}^3\mathbf{y}$	
$\boxed{1}$	2	5	14	
$\boxed{-1}$	–1	–1	–1	$+$ *row*$_1$
0	2	8	26	
$\boxed{1}$	2	5	14	
0	$\boxed{1}$	4	13	
0	$\boxed{2}$	8	26	-2 *row*$_2$
$\boxed{1}$	$\boxed{2}$	5	14	-2 *row*$_2$
0	$\boxed{1}$	4	13	
0	0	0	0	
1	0	–3	*	
0	1	4	*	
0	0	0	0	

Thus, $\mathbf{A}^3\mathbf{y} = 4\mathbf{A}\mathbf{y} - 3\mathbf{y}$, or $(\mathbf{A}^2 - 4\mathbf{A} + 3\mathbf{I}_3)\mathbf{y} = \mathbf{0}$, and $p_y(\lambda) = \lambda^2 - 4\lambda + 3 = (\lambda - 1)(\lambda - 3)$. Next, we find the kernel of $\mathbf{A} - \mathbf{I}$:

$\boxed{2}$	1	0	0	
0	0	0	0	
$\boxed{4}$	2	0	0	-2 *row*$_1$
$\boxed{2}$	1	0	0	
0	0	0	0	
0	0	0	0	

There are two eigenvectors for $\lambda = 1$: $\begin{pmatrix} 0 \\ 0 \\ 1 \end{pmatrix}$ and $\begin{pmatrix} -1 \\ 2 \\ 0 \end{pmatrix}$. For $\lambda = 3$, we find the eigenvector $\begin{pmatrix} 1 \\ 0 \\ 2 \end{pmatrix}$.

$\mathbf{A}$ is diagonalizable, and $\mathbf{A}_\mathcal{U} = \mathbf{U}^{-1}\mathbf{A}_\mathcal{E}\mathbf{U}$
$= \begin{pmatrix} 1 & 0 & 0 \\ 0 & 1 & 0 \\ 0 & 0 & 3 \end{pmatrix}$ for $\mathbf{U} = \begin{pmatrix} 0 & -1 & 1 \\ 0 & 2 & 0 \\ 1 & 0 & 2 \end{pmatrix}$.

15. For $\mathbf{y} = \begin{pmatrix} 1 \\ -1 \\ 0 \end{pmatrix}$, we have

$$\begin{pmatrix} | & | & | & | \\ \mathbf{y} & \mathbf{A}\mathbf{y} & \mathbf{A}^2\mathbf{y} & \mathbf{A}^3\mathbf{y} \\ | & | & | & | \end{pmatrix} = \begin{pmatrix} 1 & 8 & 6 & 6 \\ -1 & 6 & 4 & 4 \\ 0 & 3 & 2 & 2 \end{pmatrix}$$

with the RREF $\begin{pmatrix} 1 & 0 & 0 & 0 \\ 0 & 1 & 0 & 0 \\ 0 & 0 & 1 & 1 \end{pmatrix}$, so that the vanishing polynomial of $\mathbf{y}$ is $p_y(\lambda) = \lambda^3 - \lambda^2 = \lambda^2$

$(\lambda - 1)$, giving $\mathbf{B}$ the eigenvalues 1 and 0 (double). As rank$(\mathbf{B}) = 2$, there are not sufficiently many (only one) corresponding linearly independent eigenvectors for $\lambda = 0$. $\mathbf{B}$ is not diagonalizable.

17. (a) and (b): $\mathbf{x} = \begin{pmatrix} 2 \\ 3 \\ 1 \end{pmatrix}$ is an eigenvector for $\mathbf{A}$.

(c) and (d): $\mathbf{Ay} = \begin{pmatrix} 9 \\ -5 \\ -5 \end{pmatrix} \neq \lambda \begin{pmatrix} 1 \\ -1 \\ 1 \end{pmatrix}$ for any $\lambda \neq 0$, so $\mathbf{y}$ is not an eigenvector of $\mathbf{A}$, and neither is $\mathbf{z} = -\mathbf{y}$.

19. (a) $\mathbf{A}_{\mathcal{U}} = \mathbf{U}^{-1}\mathbf{A}_{\mathcal{E}}\mathbf{U} = \begin{pmatrix} 2 & 0 \\ 0 & 4 \end{pmatrix}$ with $\mathbf{U} = \begin{pmatrix} 1 & 2 \\ -1 & 2 \end{pmatrix}$ and $\mathbf{U}^{-1} = \frac{1}{4}\begin{pmatrix} 2 & -2 \\ 1 & 1 \end{pmatrix}$.

(b) The eigenvalues of $\mathbf{A}$ are 2 and 4, with eigenvectors $\begin{pmatrix} 1 \\ -1 \end{pmatrix}$ and $\begin{pmatrix} 2 \\ 2 \end{pmatrix}$, respectively.

21. For $\mathbf{y} = \mathbf{e}_1$, we have

$\mathbf{y}$	$\mathbf{Ay}$	$\mathbf{A}^2\mathbf{y}$	$\mathbf{A}^3\mathbf{y}$	
$\boxed{1}$	0	$\boxed{1}$	2	$+$ *row*$_3$
0	$\boxed{-1}$	$\boxed{-2}$	-6	-2 *row*$_3$
0	0	$\boxed{-1}$	-6	
$\boxed{1}$	0	0	-4	
0	$\boxed{-1}$	0	6	$\cdot\,(-1)$
0	0	$\boxed{1}$	6	
	I_3		-4	
			-6	
			6	

Hence, $\mathbf{A}^3\mathbf{y} = 6\mathbf{A}^2\mathbf{y} - 6\mathbf{Ay} - 4\mathbf{y}$ and $p_y(\lambda) = \lambda^3 - 6\lambda^2 + 6\lambda + 4$. By trial and error, note that $p_y(2) = 0$, so that $\lambda_1 = 2$ is one eigenvalue of $\mathbf{A}$. Now use long division:

$$
\begin{array}{rrrrl}
(\; \lambda^3 & -6\lambda^2 & +6\lambda & +4 \;) & \div (\lambda - 2) \; = \\
\lambda^3 & \mp 2\lambda^2 & & & \\
\hline
 & -4\lambda^2 & +6\lambda & & \\
 & \mp 4\lambda^2 & \pm 8\lambda & & \\
\hline
 & & -2\lambda & +4 & \\
 & & \mp 2\lambda & \pm 4 & \\
\hline
 & & & 0 &
\end{array}
$$

$= \lambda^2 - 4\lambda - 2$. The factor polynomial $\lambda^2 - 4\lambda - 2 = 0$ has the roots $\lambda_{2,3} = 2 \pm \sqrt{6}$. Hence, the eigenvalues of $\mathbf{A}$ are 2, $2+\sqrt{6}$, $2-\sqrt{6}$, with corresponding eigenvectors $\begin{pmatrix} 1 \\ -2 \\ 1 \end{pmatrix}$, $\begin{pmatrix} -5+2\sqrt{6} \\ -2+\sqrt{6} \\ 1 \end{pmatrix}$, and $\begin{pmatrix} 1 \\ -2+\sqrt{6} \\ -5+2\sqrt{6} \end{pmatrix}$.

$$\mathbf{A}_{\mathcal{U}} = \begin{pmatrix} 2 & 0 & 0 \\ 0 & 2+\sqrt{6} & 0 \\ 0 & 0 & 2-\sqrt{6} \end{pmatrix}.$$

23. If $\mathbf{A}_{nn}$ is singular, then $\ker(\mathbf{A}) = \ker(\mathbf{A} - 0\mathbf{I}_n) = E(0) \neq \{\mathbf{0}\}$, so $\lambda = 0$ is an eigenvalue of $\mathbf{A}$ and conversely.

25. (a) Work with $\mathbf{B} := \frac{\mathbf{A}}{3} = \begin{pmatrix} 1 & 1 & 0 \\ 0 & 3 & -1 \\ 0 & 0 & -2 \end{pmatrix}$ for smaller numbers to compute with. The eigenvalues of $\mathbf{B}$ are one-third the eigenvalues of $\mathbf{A}$; the eigenvectors are identical. Take $\mathbf{y} = \mathbf{e}_3$, for example, and vector iterate:

$\mathbf{y}$	$\mathbf{By}$	$\mathbf{B}^2\mathbf{y}$	$\mathbf{B}^3\mathbf{y}$	
0	0	-1	-2	*swap all rows*
0	-1	-1	-7	
1	-2	4	-8	
$\boxed{1}$	-2	$\boxed{4}$	-8	$+4$ *row*$_3$
0	$\boxed{-1}$	$\boxed{-1}$	-7	$-$ *row*$_3$
0	0	$\boxed{-1}$	-2	
$\boxed{1}$	$\boxed{-2}$	0	-16	-2 *row*$_2$
0	$\boxed{-1}$	0	-5	
0	0	$\boxed{-1}$	-2	$\cdot\,(-1)$
1	0	0	-6	
0	-1	0	-5	$\cdot\,(-1)$
0	0	1	2	
	$\mathbf{I}_3$		-6	
			5	
			2	

Thus, $\mathbf{A}^3\mathbf{y} = 2\mathbf{A}^2\mathbf{y} + 5\mathbf{Ay} - 6\mathbf{y}$, or $p_y(\lambda) = \lambda^3 - 2\lambda^2 - 5\lambda + 6$. Now, $p_\mathbf{y}(1) = 0 = p_y(-2) = p_\mathbf{y}(3)$, by using a little guesswork on the integer factors of the constant term 6. Hence, the eigenvalues of $\mathbf{B}$ are 1, 3, and –2, making those of $\mathbf{A}$ equal to 3, 9, and –6. The corresponding eigenvectors are $\mathbf{e}_1$, $\mathbf{e}_1 + 2\mathbf{e}_2$, and $\begin{pmatrix} -1 \\ 3 \\ 15 \end{pmatrix}$; $\mathbf{A}$ is diagonalizable.

(b) eigenvalues –1 (double) and 2; corresponding eigenvector $\mathbf{e}_1$ (only one) for $\lambda = -1$, and $-\mathbf{e}_1 + 3\mathbf{e}_2$

3. $\mathbf{x}_{\mathcal{E}} = \mathbf{X}_{\mathcal{E}\leftarrow\mathcal{B}}\mathbf{x}_{\mathcal{B}} = \mathbf{B}\mathbf{x}_{\mathcal{B}} = \begin{pmatrix} 6 \\ -13 \\ 1 \end{pmatrix}.$

5. $\mathbf{w}_{\mathcal{U}} = \mathbf{0} \iff \mathbf{w} = 0\mathbf{u}_1 + \ldots + 0\mathbf{u}_n = \mathbf{0} \in \mathbb{R}^n.$

7. (a) $\mathbf{X}_{\mathcal{V}\leftarrow\mathcal{U}} = \mathbf{V}^{-1}\mathbf{U} = \begin{pmatrix} 3/2 & 1/4 & 1/2 \\ -1/2 & -1/4 & 1/2 \\ 1 & 3/2 & 0 \end{pmatrix}.$

(b) $\mathbf{x}_{\mathcal{E}} = \mathbf{X}_{\mathcal{E}\leftarrow\mathcal{U}}\mathbf{x}_{\mathcal{U}} = \mathbf{U}\mathbf{x}_{\mathcal{U}} = \begin{pmatrix} -10 \\ 12 \\ 19 \end{pmatrix}$; $\mathbf{x}_{\mathcal{V}} = \mathbf{X}_{\mathcal{V}\leftarrow\mathcal{U}}\mathbf{x}_{\mathcal{U}}$
$= \frac{1}{4}\begin{pmatrix} -61 \\ 21 \\ -6 \end{pmatrix}.$

9. (a) $\mathbf{w}_{\mathcal{U}} = \mathbf{X}_{\mathcal{U}\leftarrow\mathcal{V}}\mathbf{w}_{\mathcal{V}}$; $\mathbf{w}_{\mathcal{V}} = \mathbf{X}_{\mathcal{V}\leftarrow\mathcal{U}}\mathbf{w}_{\mathcal{U}}.$

(b) $\mathbf{w}_{\mathcal{U}} = \begin{pmatrix} 1 \\ -3 \\ 4 \\ -1 \end{pmatrix}$ solves the linear system

$U\mathbf{w}_{\mathcal{U}} = \mathbf{w}_{\mathcal{E}}.$ (c) $\mathbf{z}_{\mathcal{E}} = \begin{pmatrix} 2 \\ 3 \\ 1 \\ 3 \end{pmatrix}.$

(d) $\mathbf{X}_{\mathcal{E}\leftarrow\mathcal{U}} = \mathbf{U} = \begin{pmatrix} 1 & 1 & 1 & 1 \\ 0 & 1 & 1 & 1 \\ 0 & 0 & 1 & 1 \\ 0 & 0 & 0 & 1 \end{pmatrix}$. $\mathbf{X}_{\mathcal{U}\leftarrow\mathcal{E}} = \mathbf{U}^{-1}$
$= \begin{pmatrix} 1 & -1 & 0 & 0 \\ 0 & 1 & -1 & 0 \\ 0 & 0 & 1 & -1 \\ 0 & 0 & 0 & 1 \end{pmatrix}.$

11. Note that $\begin{pmatrix} | & | & | & | \\ \mathbf{u}_1 & \mathbf{u}_2 & \mathbf{u}_3 & \mathbf{u}_4 \\ | & | & | & | \end{pmatrix} = \begin{pmatrix} | & | & | & | \\ \mathbf{v}_1 & \mathbf{v}_2 & \mathbf{v}_3 & \mathbf{v}_4 \\ | & | & | & | \end{pmatrix}$
$\cdot \begin{pmatrix} 1 & 1 & 0 & 0 \\ 0 & 1 & 0 & 0 \\ 2 & 0 & -1 & 0 \\ -1 & 0 & 0 & 1/2 \end{pmatrix}$, or $\mathbf{U} = \mathbf{V}\mathbf{A}$ for $\mathbf{A}$
$= \begin{pmatrix} 1 & 1 & 0 & 0 \\ 0 & 1 & 0 & 0 \\ 2 & 0 & -1 & 0 \\ -1 & 0 & 0 & 1/2 \end{pmatrix}.$

(a) $\ell = 4.$

(b) $\mathbf{X}_{\mathcal{U}\leftarrow\mathcal{V}} = \mathbf{U}^{-1}\mathbf{V} = \mathbf{A}^{-1} = \begin{pmatrix} 1 & -1 & 0 & 0 \\ 0 & 1 & 0 & 0 \\ 2 & -2 & -1 & 0 \\ 2 & -2 & 0 & 2 \end{pmatrix}.$

(c) $\mathbf{X}_{\mathcal{U}\leftarrow\mathcal{U}} = \mathbf{I}_4.$

(d) $\mathbf{X}_{\mathcal{V}\leftarrow\mathcal{U}} = \mathbf{V}^{-1}\mathbf{U} = \mathbf{X}_{\mathcal{U}\leftarrow\mathcal{V}}^{-1} = \mathbf{A}.$

(e) $\mathbf{X}_{\mathcal{U}\leftarrow\mathcal{U}} = \mathbf{I}_4 = \mathbf{X}_{\mathcal{V}\leftarrow\mathcal{V}}.$

(f) $\mathbf{w}_{\mathcal{U}} = \begin{pmatrix} -1 \\ 2 \\ 3 \\ -1 \end{pmatrix}.$

(g) $\mathbf{w}_{\mathcal{V}} = \mathbf{X}_{\mathcal{V}\leftarrow\mathcal{U}}\begin{pmatrix} -1 \\ 2 \\ 3 \\ -1 \end{pmatrix} = \mathbf{A}\begin{pmatrix} -1 \\ 2 \\ 3 \\ -1 \end{pmatrix} = \begin{pmatrix} 1 \\ 2 \\ -5 \\ 1/2 \end{pmatrix}.$

13. (a) $\mathbf{x}_{\mathcal{V}} = \begin{pmatrix} 7 & -4 & 3 & 1 \end{pmatrix}^T.$

(b), (c) $\mathbf{X}_{\mathcal{U}\leftarrow\mathcal{V}} = \begin{pmatrix} 0 & & 1 \\ & \iddots & \\ 1 & & 0 \end{pmatrix}_{kk} = \mathbf{X}_{\mathcal{U}\leftarrow\mathcal{V}}.$

15. (a) $\mathbf{X} = \mathbf{U}^{-1}\mathbf{C}\mathbf{U}\mathbf{A}^{-1}$;
(b) $\mathbf{X} = (\mathbf{A}^T\mathbf{A} - \mathbf{A}^2)^{-1}.$

17. (a) $\mathbf{X}_{\mathcal{V}\leftarrow\mathcal{U}} = \mathbf{V}^{-1}\mathbf{U} = \begin{pmatrix} 1 & -1 & 1 \\ 0 & 1 & 2 \\ 1 & 1 & 1 \end{pmatrix}$;

(b) $\mathbf{X}_{\mathcal{U}\leftarrow\mathcal{V}} = \mathbf{X}_{\mathcal{V}\leftarrow\mathcal{U}}^{-1} = \frac{1}{4}\begin{pmatrix} 1 & -2 & 3 \\ -2 & 0 & 2 \\ 1 & 2 & -1 \end{pmatrix}$;

(c) $\mathbf{X}_{\mathcal{U}\leftarrow\mathcal{E}} = \begin{pmatrix} | & | & | \\ \mathbf{u}_1 & \mathbf{u}_2 & \mathbf{u}_3 \\ | & | & | \end{pmatrix}^{-1}.$

19. (a) $\mathbf{x} = \begin{pmatrix} 1 \\ 1 \end{pmatrix} \in \ker(\mathbf{A} - 2\mathbf{I})$;

(b) $\mathbf{y} = \begin{pmatrix} 3 \\ 2 \end{pmatrix} \in \ker(\mathbf{A} + \mathbf{I})$;

(c) Impossible, since $\ker(\mathbf{A} - \mathbf{I}) = \{\mathbf{0}\}$.

21. (a) The RREF of $\begin{pmatrix} 1 & 1 & 1 \\ 1 & 1 & 2 \\ 1 & 2 & 3 \end{pmatrix}$ is $\mathbf{I}_3$; hence, the given vectors are linearly independent.

(b) Three linearly independent vectors in $\mathbb{R}^3$ always span all of $\mathbb{R}^3$.

(c) $\mathbf{x}_{\mathcal{B}} = \mathbf{X}_{\mathcal{B}\leftarrow\mathcal{E}}\mathbf{x}_{\mathcal{E}} = \mathbf{B}^{-1}\mathbf{x}_{\mathcal{E}} = \begin{pmatrix} -5 \\ -3 \\ 5 \end{pmatrix}$;

(d) $\mathbf{A}_{\mathcal{B}} = \mathbf{B}^{-1}\mathbf{A}_{\mathcal{E}}\mathbf{B} = \begin{pmatrix} -6 & -8 & -11 \\ 3 & 2 & 6 \\ 1 & 3 & 2 \end{pmatrix}$ for

$\mathbf{A}_{\mathcal{E}} = \begin{pmatrix} -2 & 1 & -1 \\ 0 & -2 & 1 \\ 1 & 0 & 2 \end{pmatrix}.$

23. $\mathbf{A}_{\mathcal{B}} = \mathbf{B}^{-1}\mathbf{A}_{\mathcal{E}}\mathbf{B} = \begin{pmatrix} -6.5 & -7 & -12.5 \\ 1 & 0 & 3 \\ 2.5 & 4 & 4.5 \end{pmatrix}$ for $\mathbf{A}_{\mathcal{E}} = \begin{pmatrix} -2 & 1 & -1 \\ 0 & -2 & 1 \\ 1 & 0 & 2 \end{pmatrix}$ and $\mathbf{B}^{-1} = \begin{pmatrix} 1.5 & -0.5 & -0.5 \\ 0 & 1 & 0 \\ -0.5 & -0.5 & 0.5 \end{pmatrix}$.

25. (a) $\mathbf{A}_{\mathcal{E}} = \begin{pmatrix} 1 & 1 \\ 0 & 2 \end{pmatrix}$.

(b) The RREF of $\begin{pmatrix} -1 & 3 \\ 3 & 1 \end{pmatrix}$ is $\mathbf{I}_2$.

(c) $(T(\mathbf{a}))_{\mathcal{E}} = \mathbf{A}_{\mathcal{E}}\mathbf{a}_{\mathcal{E}} = \begin{pmatrix} -2 \\ -6 \end{pmatrix}$.

(d) $\mathbf{A}_{\mathcal{U}} = \mathbf{U}^{-1}\mathbf{A}_{\mathcal{E}}\mathbf{U} = \begin{pmatrix} 1.6 & 0.2 \\ 1.2 & 1.4 \end{pmatrix}$ for $\mathbf{U} = \begin{pmatrix} -1 & 3 \\ 3 & 1 \end{pmatrix}$ and $\mathbf{U}^{-1} = \frac{1}{10}\begin{pmatrix} -1 & 3 \\ 3 & 1 \end{pmatrix}$.

(e) $\mathbf{a}_{\mathcal{U}} = \mathbf{U}^{-1}\mathbf{a}_{\mathcal{E}} = \begin{pmatrix} -1 \\ 0 \end{pmatrix}$, $(T(\mathbf{a}))_{\mathcal{U}} = \mathbf{A}_{\mathcal{U}}\mathbf{a}_{\mathcal{U}} = \begin{pmatrix} -1.6 \\ -1.2 \end{pmatrix}$.

(f) $(T(\mathbf{a}))_{\mathcal{E}} = -1.6 \cdot \mathbf{u}_1 - 1.2 \cdot \mathbf{u}_2 = \begin{pmatrix} -2 \\ -6 \end{pmatrix}$ from part (e), the same as in part (c).

27. Assume linear dependence—that is, $\begin{pmatrix} | & & | \\ \mathbf{v}_1 & \dots & \mathbf{v}_k \\ | & & | \end{pmatrix}\begin{pmatrix} \alpha_1 \\ \vdots \\ \alpha_k \end{pmatrix} = \begin{pmatrix} 0 \\ \vdots \\ 0 \end{pmatrix} \in \mathbb{R}^n$ for some α_i, not all of which are zero. Since $(\mathbf{v}_i)_{\mathcal{U}} = \mathbf{U}\mathbf{v}_i$, we have $\mathbf{0}_{n,1} = \mathbf{U}\begin{pmatrix} | & & | \\ \mathbf{v}_1 & \dots & \mathbf{v}_k \\ | & & | \end{pmatrix}\begin{pmatrix} \alpha_1 \\ \vdots \\ \alpha_k \end{pmatrix} = \begin{pmatrix} | & & | \\ (\mathbf{v}_1)_{\mathcal{U}} & \dots & (\mathbf{v}_k)_{\mathcal{U}} \\ | & & | \end{pmatrix}\begin{pmatrix} \alpha_1 \\ \vdots \\ \alpha_k \end{pmatrix}$ for the same α_i, not all of which are zero (i.e., the $(\mathbf{v}_i)_{\mathcal{U}}$ are likewise linearly dependent).

29. (a) $\mathbf{X}_{\mathcal{B}} = \mathbf{X}_{\mathcal{B}\leftarrow\mathcal{E}}\mathbf{x}_{\mathcal{E}} = \mathbf{B}^{-1}\mathbf{x}_{\mathcal{E}} = \begin{pmatrix} 1 & 1 \\ -1 & -2 \end{pmatrix}\begin{pmatrix} 1 \\ -3 \end{pmatrix} = \begin{pmatrix} -2 \\ 5 \end{pmatrix}$.

(b) $\mathbf{X}_{\mathcal{C}} = \mathbf{X}_{\mathcal{C}\leftarrow\mathcal{E}}\mathbf{x}_{\mathcal{E}} = \mathbf{C}^{-1}\mathbf{x}_{\mathcal{E}} = \begin{pmatrix} 1 & -1 \\ 1/2 & 0 \end{pmatrix}\begin{pmatrix} 1 \\ -3 \end{pmatrix} = \begin{pmatrix} 4 \\ 1/2 \end{pmatrix}$.

(c) $\mathbf{X}_{\mathcal{D}} = \mathbf{D}^{-1}\mathbf{x}_{\mathcal{E}} = \frac{1}{15}\begin{pmatrix} -1 & 4 \\ 4 & -1 \end{pmatrix}\begin{pmatrix} 1 \\ -3 \end{pmatrix} = \frac{1}{15}\begin{pmatrix} -13 \\ 7 \end{pmatrix}$.

31. $\mathbf{x}_{\mathcal{U}} = (e^{-1}, 1, -\pi)^T$.

T 7. $\mathbf{X}_{\mathcal{V}\leftarrow\mathcal{U}} = \mathbf{V}^{-1}\mathbf{U} = -\begin{pmatrix} 1 & 5 & 10 \\ 2 & 8 & 13 \\ 2 & 7 & 10 \end{pmatrix}$, while $\mathbf{X}_{\mathcal{U}\leftarrow\mathcal{V}} = \mathbf{U}^{-1}\mathbf{V} = \begin{pmatrix} -11 & 20 & -15 \\ 6 & -10 & 7 \\ -2 & 3 & -2 \end{pmatrix}$.

7.2.P p. 217

1. (a) (b) The ranks of **AB** and **BA** are both 3.

(c) rank(**C**) = 1 = rank(**H**).

(d) rank(**C**) = rank$\begin{pmatrix} 1 & 2 \\ 2 & 4 \\ 3 & 6 \end{pmatrix}$; rank(**H**) = rank$\begin{pmatrix} 17 & 101 & -93 \end{pmatrix}$.

3. (a) For $\mathbf{x} = \begin{pmatrix} 2 \\ -2 \\ 0 \end{pmatrix}$, the rank of $\begin{pmatrix} | & | & | & | \\ \mathbf{x} & \mathbf{Ax} & \mathbf{A}^2\mathbf{x} & \mathbf{A}^3\mathbf{x} \\ | & | & | & | \end{pmatrix}$ is 2. For $\mathbf{x} = \begin{pmatrix} 0 \\ 0 \\ -2 \end{pmatrix}$, the rank is 3.

(b) We have the identical results on rank as in part (a).

(c) No: $\mathbf{A}^2\mathbf{x} = \mathbf{A}(\mathbf{Ax})$, etc.

5. (a) From Definition 1, $\mathbf{X}_{2,2} = \mathbf{X}_{\mathcal{U}\leftarrow\mathcal{V}}$ satisfies $\mathbf{V} = \mathbf{UX}$. Thus, $\mathbf{X} = \begin{pmatrix} -1 & 1 \\ 1 & 1 \end{pmatrix}$, since $\mathbf{v}_1 = -\mathbf{u}_1 + \mathbf{u}_2$ and $\mathbf{v}_2 = \mathbf{u}_1 + \mathbf{v}_2$.

(b) $\mathbf{x}_{\mathcal{U}} = \mathbf{X}\mathbf{x}_{\mathcal{V}} = \begin{pmatrix} -5 \\ -1 \end{pmatrix}$.

(c) $\mathbf{x}_{\mathcal{V}} = \mathbf{X}^{-1}\mathbf{x}_{\mathcal{U}} = \frac{1}{2}\begin{pmatrix} 5 \\ 3 \end{pmatrix}$ with $\mathbf{X}^{-1} = \frac{1}{2}\begin{pmatrix} -1 & 1 \\ 1 & 1 \end{pmatrix}$.

7. (a) $\mathbf{x}_{\mathcal{E}} = \mathbf{x} = \mathbf{U}\mathbf{x}_{\mathcal{U}} = \begin{pmatrix} | & | \\ \mathbf{u}_1 & \mathbf{u}_2 \\ | & | \end{pmatrix}\mathbf{x}_{\mathcal{U}}$.

(b) $\mathbf{V}\mathbf{x}_{\mathcal{V}} = \mathbf{x}_{\mathcal{E}} = \mathbf{U}\mathbf{x}_{\mathcal{U}} = \mathbf{UX}\mathbf{x}_{\mathcal{V}}$ implies $\mathbf{V} = \mathbf{UX} = \begin{pmatrix} 5 & -1 \\ -1 & 1 \\ 8 & -2 \end{pmatrix} = \begin{pmatrix} | & | \\ \mathbf{v}_1 & \mathbf{v}_2 \\ | & | \end{pmatrix}$.

9. Form $\mathbf{U}^T = \begin{pmatrix} 1 & -2 & 3 & -1 \\ 0 & 2 & 2 & -2 \end{pmatrix}$ and look at its kernel: $\ker(\mathbf{U}^T) = \text{span}\left\{\begin{pmatrix} -5 \\ -1 \\ 1 \\ 0 \end{pmatrix}, \begin{pmatrix} 3 \\ 1 \\ 0 \\ 1 \end{pmatrix}\right\}$, according to Chapter 3.

11. If $\mathbf{AB} = \mathbf{BA}$ for two symmetric matrices $\mathbf{A}$ and $\mathbf{B}$, then $(\mathbf{AB})^T = \mathbf{B}^T\mathbf{A}^T = \mathbf{BA} = \mathbf{AB}$ (i.e., $\mathbf{AB}$ is symmetric). Conversely, if $\mathbf{AB}$ is symmetric, then $\mathbf{AB} = (\mathbf{AB})^T = \mathbf{B}^T\mathbf{A}^T = \mathbf{BA}$, and $\mathbf{A}$ and $\mathbf{B}$ commute.

7.3.P p. 224

1. (a), (b) Note that

$$\mathbf{A}_{\mathcal{B}\leftarrow\mathcal{C}} = \mathbf{X}_{\mathcal{B}\leftarrow\mathcal{E}_4}\mathbf{A}_{\mathcal{E}_4\leftarrow\mathcal{E}_3}\mathbf{X}_{\mathcal{E}_3\leftarrow\mathcal{C}},$$

where $\mathbf{X}_{\mathcal{B}\leftarrow\mathcal{E}_4} = \mathbf{B}^{-1}$ and $\mathbf{X}_{\mathcal{E}_3\leftarrow\mathcal{C}} = \mathbf{C}$.

$$\mathbf{A}_{\mathcal{B}\leftarrow\mathcal{C}} = \frac{1}{675}\begin{pmatrix} 21 & 4191 & 1814 \\ -1968 & -7428 & -2212 \\ 327 & -1083 & -457 \\ -645 & -2595 & -880 \end{pmatrix}.$$

3. (a) $\alpha \sin x + \beta \cos x = \frac{\beta}{2} g(x) + \left(\frac{3}{4}\beta + \frac{\alpha}{2}\right) h(x)$.

(b) $\mathbf{X}_{\{g,h\}\leftarrow\mathcal{U}} = \begin{pmatrix} 0 & 1/2 \\ 1/2 & 3/4 \end{pmatrix}$. $\mathbf{X}_{\mathcal{U}\leftarrow\{g,h\}} = \begin{pmatrix} -3 & 2 \\ 2 & 0 \end{pmatrix}$.

(c) $\mathbf{k}_{\{g,h\}} = \mathbf{X}_{\{g,h\}\leftarrow\mathcal{U}}\begin{pmatrix} 7 \\ -4 \end{pmatrix} = \begin{pmatrix} -2 \\ 1/2 \end{pmatrix}$.

$\ell_{\mathcal{U}} = \mathbf{X}_{\mathcal{U}\leftarrow\{g,h\}}\ell_{\{g,h\}} = \begin{pmatrix} -29 \\ 14 \end{pmatrix}$.

(d) $\mathbf{D}_{\mathcal{U}} = \begin{pmatrix} 0 & -1 \\ 1 & 0 \end{pmatrix}$.

(e) $g' = -3\cos x - 2\sin x = -\frac{3}{2}g(x) - \frac{13}{4}h(x)$; $h' = 2\cos x = g(x) + \frac{3}{2}h(x)$. Thus,

$\mathbf{D}_{\{g,h\}} = \begin{pmatrix} -3/2 & 1 \\ -13/4 & 3/2 \end{pmatrix}$.

(f) $\mathbf{D}^2_{\mathcal{U}} = \begin{pmatrix} -1 & 0 \\ 0 & -1 \end{pmatrix}$; $\mathbf{D}^2_{\{g,h\}} = \begin{pmatrix} -1 & 0 \\ 0 & -1 \end{pmatrix}$.

(g) $(\mathbf{D}_{\mathcal{U}})^2 = -\mathbf{I}_2 = \mathbf{D}^2_{\mathcal{U}} = \mathbf{D}^2_{\{g,h\}} = (\mathbf{D}_{\{g,h\}})^2$.

5. (a) Each function $\mathbf{v}_i \in \text{span}(U)$ and each $\mathbf{u}_j \in \text{span}(V)$.

(b) From (a), we have $\mathbf{v}_1 = \mathbf{u}_1$, $\mathbf{v}_2 = \mathbf{u}_2$, $\mathbf{v}_3 = 2\mathbf{u}_3 - \mathbf{u}_1$, and $\mathbf{v}_4 = 4\mathbf{u}_4 - 3\mathbf{u}_2$. Thus, $\mathbf{X}_{\mathcal{U}\leftarrow\mathcal{V}}$

$= \begin{pmatrix} 1 & 0 & -1 & 0 \\ 0 & 1 & 0 & -2 \\ 0 & 0 & 2 & 0 \\ 0 & 0 & 0 & 4 \end{pmatrix}$ and

$\mathbf{X}_{\mathcal{V}\leftarrow\mathcal{U}} = \mathbf{X}^{-1}_{\mathcal{U}\leftarrow\mathcal{V}} = \begin{pmatrix} 1 & 0 & 1/2 & 0 \\ 0 & 1 & 0 & 1/2 \\ 0 & 0 & 1/2 & 0 \\ 0 & 0 & 0 & 1/4 \end{pmatrix}$.

(c) $\mathbf{h}_{\mathcal{V}} = \mathbf{X}_{\mathcal{V}\leftarrow\mathcal{U}}\mathbf{h}_{\mathcal{U}} = \begin{pmatrix} -1/2 \\ 13/2 \\ -1/2 \\ 3/4 \end{pmatrix}$ for $\mathbf{h}_{\mathcal{U}} = \begin{pmatrix} 0 \\ 5 \\ -1 \\ 3 \end{pmatrix}$.

Hence, $\int h(x)dx = \int -1/2 + 13/2\cos x - 1/2\cos 2x + 3/4\cos 3x\, dx = -1/2x + 13/2\sin x - 1/4\sin 2x + 1/4\sin 3x + c$, all computed over V, where integration is much simpler than over U.

7. (a) Using Example 6(b), we obtain $\int \ldots\, dt = \mathbf{D}^{-1}_{\mathcal{E}}$ $\begin{pmatrix} -3 \\ 2 \\ 5 \end{pmatrix} = \begin{pmatrix} 1 & 0 & 0 \\ -2 & 1 & 0 \\ 2 & -1 & 1 \end{pmatrix}\begin{pmatrix} -3 \\ 2 \\ 5 \end{pmatrix} = \begin{pmatrix} -3 \\ 8 \\ -3 \end{pmatrix}$ $= -3e^t t^2 + 8te^t - 3e^t (+c)$.

(b) With $e_1 = t^3e^t$, $e_2 = t^2e^t$, $e_3 = te^t$, $e_4 = e^t$,

we have $\mathbf{D}_{\mathcal{E}} = \begin{pmatrix} 1 & 0 & 0 & 0 \\ 3 & 1 & 0 & 0 \\ 0 & 2 & 1 & 0 \\ 0 & 0 & 1 & 1 \end{pmatrix}$ and $\mathbf{D}^{-1}_{\mathcal{E}} =$

$\begin{pmatrix} 1 & 0 & 0 & 0 \\ -3 & 1 & 0 & 0 \\ 6 & -2 & 1 & 0 \\ -6 & 2 & -1 & 1 \end{pmatrix}$. Thus, $\int t^3e^t dt = \mathbf{D}^{-1}_{\mathcal{E}}\begin{pmatrix} 1 \\ 0 \\ 0 \\ 0 \end{pmatrix}$

$= \begin{pmatrix} 1 \\ -3 \\ 6 \\ -6 \end{pmatrix} = t^3e^t - 3t^2e^t + 6te^t - 6e^t (+c)$.

7.R p. 225

3. (a)i) False. ii) True. iii) False. iv) True. v) True. vi) False.

(b) True. (c) False. (d) False. (e) True. (f) False.

5. (b) $\mathbf{X}_{\mathcal{V}\leftarrow\mathcal{U}} = \mathbf{V}^{-1}\mathbf{U} = \frac{1}{2}\begin{pmatrix} 1 & 3 & 1 \\ 3 & -1 & 1 \\ -1 & 1 & 1 \end{pmatrix}$.

(c) $\mathbf{x}_{\mathcal{V}} = (5, -2, 3)^T$.

(d) $\mathbf{X}_{\mathcal{E}\leftarrow\mathcal{V}} = \mathbf{V} = \begin{pmatrix} 1 & 1 & 0 \\ 0 & 1 & 1 \\ 1 & 0 & 1 \end{pmatrix}$.

7. Solving $\mathbf{U}\mathbf{x}_{\mathcal{U}} = \mathbf{x}$ for $\mathbf{x}_{\mathcal{U}}$ is faster.

Chapter 8

8.1.P p. 238

1. Expand along column 1 and use the 2×2 determinant formula:

$\det \mathbf{A} = \det \begin{pmatrix} 4 & 9 \\ 8 & 27 \end{pmatrix} - \det \begin{pmatrix} 2 & 3 \\ 8 & 27 \end{pmatrix} + \det \begin{pmatrix} 2 & 3 \\ 4 & 9 \end{pmatrix} = 108 - 72 - 54 + 24 + 18 - 12 = 12.$

Or use Gaussian elimination (without swaps or scalings) and the 2×2 formula:

$$\begin{array}{ccc|l} \boxed{1} & 2 & 3 & \\ \boxed{1} & 4 & 9 & -\,row_1 \\ \boxed{1} & 8 & 27 & -\,row_1 \\ \hline 1 & 2 & 3 & \\ 0 & 2 & 6 & \\ 0 & 6 & 24 & \end{array}$$

Now expand along column 1: $\det \mathbf{A} = 1 \cdot \det \begin{pmatrix} 2 & 6 \\ 6 & 24 \end{pmatrix} = 48 - 36 = 12.$

3. Expand along row 1: $\det \mathbf{A} = 2 \det \begin{pmatrix} 2 & 1 \\ 1 & 2 \end{pmatrix} - \det \begin{pmatrix} 1 & 1 \\ 1 & 2 \end{pmatrix} = \det \begin{pmatrix} 1 & 2 \\ 1 & 1 \end{pmatrix} = 2(4-1) - 2 + 1 + 1 - 2 = 4.$

5. Via Gaussian elimination and the 2×2 formula:

$$\begin{array}{ccc|l} \boxed{1} & 1 & 1 & \\ \boxed{2} & 4 & 9 & -2\,row_1 \\ \boxed{3} & 27 & 81 & -3\,row_1 \\ \hline 1 & 1 & 1 & \\ 0 & 2 & 7 & \\ 0 & 24 & 78 & \end{array}$$

$\det \mathbf{A} = 2 \cdot 78 - 7 \cdot 24 = -12.$

7. (a) $\mathbf{U}^{-1} = \begin{pmatrix} 1 & -1/2 & -1/2 \\ -1 & 1 & 0 \\ 0 & 1/4 & 1/4 \end{pmatrix}.$

$\mathbf{A}_{\mathcal{U}} = \mathbf{U}^{-1}\mathbf{A}_{\mathcal{E}}\mathbf{U} = \begin{pmatrix} 2 & 1/2 & -2 \\ -2 & -1 & 4 \\ 1/2 & 1/4 & 2 \end{pmatrix}.$

(b) $\{\mathbf{0}\} \subset \mathbb{R}^3$. (c) $\mathcal{E}$.

(d) $\mathbf{A}_{\mathcal{E}} = \begin{pmatrix} 2 & 0 & -1 \\ 1 & 1 & 1 \\ 0 & 1 & 0 \end{pmatrix}$. Expand along row 3:

$\det \mathbf{A}_{\mathcal{E}} = -\det \begin{pmatrix} 2 & -1 \\ 1 & 1 \end{pmatrix} = -(2+1) = -3.$

(e) $\det \mathbf{A}_{\mathcal{U}} = \det \mathbf{A}_{\mathcal{E}}$ does not change from basis to basis, since $\det \mathbf{A}_{\mathcal{U}} = \det(\mathbf{U}^{-1}\mathbf{A}_{\mathcal{E}}\mathbf{U}) = \det \mathbf{U}^{-1} \det \mathbf{A}_{\mathcal{E}} \det \mathbf{U} = \det \mathbf{A}_{\mathcal{E}}$ according to parts 2 and 4 of the Proposition.

9. We have $\det \mathbf{A}_{nn} = 3 \det \mathbf{A}_{n-1,n-1} - \det \mathbf{A}_{n-2,n-2}$ for any size n by expanding along the first row of $\mathbf{A}_{n,n}$, for example. Thus, $\det \mathbf{A}_{6,6} = 3 \det \mathbf{A}_{5,5} - \det \mathbf{A}_{4,4} = 3(3 \det \mathbf{A}_{4,4} - \det \mathbf{A}_{3,3}) - 3 \det \mathbf{A}_{3,3} + \det \mathbf{A}_{2,2} = 9(3 \det \mathbf{A}_{3,3} - \det \mathbf{A}_{2,2}) - 3(3 \det \mathbf{A}_{2,2} - \det \mathbf{A}_{1,1}) - 3(3 \det \mathbf{A}_{2,2} - \det \mathbf{A}_{1,1}) + \det \mathbf{A}_{2,2} = 27(3 \det \mathbf{A}_{2,2} - \det \mathbf{A}_{1,1}) - 9 \cdot 8 - 9 \cdot 8 + 9 - 9 \cdot 8 + 9 + 8 = 27 \cdot 21 - 22 \cdot 9 + 8 = 377$, where $\det \mathbf{A}_{1,1} = \det(3) = 3$ and $\det \mathbf{A}_{2,2} = \det \begin{pmatrix} 3 & 1 \\ 1 & 3 \end{pmatrix} = 9 - 1 = 8.$

11. Expand along row 1 and use the 2×2 formula: $\det \mathbf{A} = -abd$.

13. By row expansion along row 2, $\det \mathbf{A} = 2 \det \begin{pmatrix} x^2 & x \\ 2 & 1 \end{pmatrix} = 2(x^2 - 2x) = 0$ for $x = 0$ or $x = 2$.

15. "Row reduce" crazily and then expand along column 3:

$$\begin{array}{ccc|l} 1 & 2 & -1 & +\,row_3 \\ 3 & 7 & -1 & +\,row_3 \\ -2 & -5 & 1 & \\ \hline -1 & -3 & 0 & \\ 1 & 2 & 0 & \\ -2 & -5 & 1 & \end{array}$$

$\det \mathbf{A} = \det \begin{pmatrix} -1 & -3 \\ 1 & 2 \end{pmatrix} = -2 + 3 = 1 \neq 0$. Thus, $\mathbf{A}$ is invertible due to part 5 of the Proposition.

17. $row_2 + row_3 = (x + y + z) \cdot row_1$ (i.e., the rows are linearly dependent). Thus $\mathbf{A}$ row reduces to a matrix with a zero row. Expand along it and find that $\det \mathbf{A} = 0$.

19. By row expansion along row 1 of $\mathbf{B}$, $\det(\mathbf{B}) = 4a$.

21. (a) i) $\approx n!$ ii) $\approx \frac{n^3}{3}$

(b)

n=2:	2	versus	3
n=5:	120	"	42
n=10:	3,628,800	"	334
n=100:	$\approx 10^{158}$	versus	33,334

23. Expand along column 2 and then along column 3:

$$\det(\mathbf{A}) = -\det\begin{pmatrix} 2 & 0 & 0 \\ -35 & 1 & 0 \\ 734 & 2 & 4 \end{pmatrix} =$$

$$-4\det\begin{pmatrix} 2 & 0 \\ -35 & 1 \end{pmatrix} = -8.$$

25.

$$\begin{array}{ccc|l} 1 & a & -1 & +\,row_3 \\ 2 & 4 & 0 & \\ -1 & b & 1 & \\ \hline 0 & a+b & 0 & \\ 2 & 4 & 0 & \\ -1 & b & 1 & \end{array}$$

Thus, $\det \mathbf{A} = \det\begin{pmatrix} 0 & a+b \\ 2 & 4 \end{pmatrix} = -2(a+b) = 120$ makes $a+b = -60$.

27. (a) $f(x) = x^2 - x - 2 = (x-2)(x+1)$.

(b) all $x \neq 2, -1$. (c) $x = 2$ or -1.

29. $\det(\mathbf{A}) = 55$.

31. (a) $\det(-\mathbf{A}\mathbf{B}^{-1}) = (-1)^5 \det \mathbf{A}(\det \mathbf{B})^{-1} = 8$; (b) 512; (c) $-\frac{1}{2}$; (d) $\frac{1}{32}$; (e) 128; (f) cannot be evaluated from the given information.

33. (a) Both determinants are equal to 6.

(b) Both equal to -4. (c) 42.

35. (a) $\det(\mathbf{A}) = 0$; (b) $\mathbf{D}_{kk}$; (c) $\det(\mathbf{D}) \in \mathbb{R}$. (d) follows from (a).

37. (a) If $(\mathbf{A}+\mathbf{B})^2 = \mathbf{A}^2 + \mathbf{AB} + \mathbf{BA} + \mathbf{B}^2 = \mathbf{A}^2 + \mathbf{B}^2$, then $\mathbf{AB} + \mathbf{BA} = \mathbf{O}_n$ or $\mathbf{AB} = -\mathbf{BA}$.

(b) If $\mathbf{AB} = -\mathbf{BA}$, then $\det \mathbf{A} \cdot \det \mathbf{B} = \det(\mathbf{AB}) = \det(-\mathbf{BA}) = (-1)^n \det \mathbf{B} \cdot \det \mathbf{A}$. Hence, if $\det \mathbf{A}$ and $\det \mathbf{B} \neq 0$, then $(-1)^n = 1$ and n must be even.

(c) Any $\mathbf{A}_{nn}$ and $\mathbf{B} = \mathbf{O}_n$ work.

(d) Take $\mathbf{A} = \begin{pmatrix} -1 & 0 \\ 0 & 1 \end{pmatrix}$ and $\mathbf{B} = \begin{pmatrix} 0 & 1 \\ -1 & 0 \end{pmatrix}$, for example. They are nonsingular and satisfy $\mathbf{AB} = -\mathbf{BA}$.

39. (a) $\det(\mathbf{A}\mathbf{A}^T) = \det \mathbf{A} \cdot \det \mathbf{A}^T = \det \mathbf{A} \cdot \det \mathbf{A} = \det \mathbf{I}_n = 1$ implies $\det \mathbf{A} = \pm 1$, all by using the determinant Proposition.

(b) Take $\mathbf{A}_{nn} := \mathbf{I}_n$, for example.

(c) From (a), we have $(\det \mathbf{A})^2 = \det(3\mathbf{I}_n) = 3^n$. Thus, $\det \mathbf{A} = \pm\sqrt{3^n}$.

41. (a) $\mathbf{I}(\mathbf{x}, 1) = \begin{pmatrix} -1 & 0 \\ 3 & 1 \end{pmatrix}$ and $\mathbf{I}(\mathbf{x}, 2) = \begin{pmatrix} 1 & -1 \\ 0 & 3 \end{pmatrix}$.

(b) $\mathbf{I}(\mathbf{y}, 2) = \begin{pmatrix} 1 & 0 & 0 \\ 0 & 2 & 0 \\ 0 & -1 & 1 \end{pmatrix}$ and

$$\mathbf{I}(\mathbf{y}, 2) = \begin{pmatrix} 1 & 0 & 0 \\ 0 & 1 & 2 \\ 0 & 0 & -1 \end{pmatrix}.$$

(c) $\det(\mathbf{I}(\mathbf{x}, i)) = x_i \det(\mathbf{I}_{n-1}) = x_i$ by expanding along row i.

T 8. (c) Here we have $\mathbf{AB} = \begin{pmatrix} 14 & -21 & 4 \\ 2 & 0 & -4 \\ -1 & 0 & 2 \end{pmatrix}$, which has two linearly dependent rows in positions 2 and 3, and $\mathbf{BA} = \begin{pmatrix} 2 & 7 & 2 \\ 4 & 14 & 10 \\ -2 & -7 & 0 \end{pmatrix}$, with determinant likewise equal to zero.

8.2.P p. 244

1.

σ			ϵ_σ
1	2	3	1
1	3	2	-1
2	1	3	-1
2	3	1	1
3	1	2	1
3	2	1	-1

3. (a) If $(x, y) = \mathbf{a}$ or $(x, y) = \mathbf{b}$, then $f(x, y) = 0$, since the matrix then has two identical rows. Hence, both $\mathbf{a}$ and $\mathbf{b}$ lie on the curve $f(x, y) = 0$.

By expansion, $f(x, y) = x\det\begin{pmatrix} a_2 & 1 \\ b_2 & 1 \end{pmatrix} - y\det\begin{pmatrix} a_1 & 1 \\ b_1 & 1 \end{pmatrix} + \det\begin{pmatrix} a_1 & a_2 \\ b_1 & b_2 \end{pmatrix}$. Thus, $f(x, y) = (a_2 - b_2)x + (b_1 - a_1)y + a_1b_2 - a_2b_1 = 0$. This is the equation of a line in x and y with slope $m = \frac{a_2 - b_2}{a_1 - b_1}$.

(b) This is the same line, since $g(x, y) = -17 \cdot f(x, y) = 0$ by expansion along column 3.

(c) $h(x, y) = 0$ describes the line that is perpendicular to the line in part (a) and that passes through the points $(a_2, -a_1)$ and $(b_2, -b_1) \in \mathbb{R}^2$.

(d) $h(x, y, z) = \det\begin{pmatrix} x & y & z & 1 \\ a_1 & a_2 & a_3 & 1 \\ b_1 & b_2 & b_3 & 1 \\ c_1 & c_2 & c_3 & 1 \end{pmatrix} = 0$ describes the points (x, y, z) on the plane in $\mathbb{R}^3$ that contains the given points $\mathbf{a}$, $\mathbf{b}$, and $\mathbf{c}$.

5. (a) By expansion along row 1, we see that $\det(\mathbf{S}_n) = (-1)^{n+1}2 \cdot \det(\mathbf{S}_{n-1}) = \begin{cases} 2 \cdot \det(\mathbf{S}_{n-1}) & \text{if } n \text{ is odd} \\ -2 \cdot \det(\mathbf{S}_{n-1}) & \text{if } n \text{ is even} \end{cases}$. Thus, $|\det(\mathbf{S}_n)| = 2^n$, and $\det(\mathbf{S}_n)$ switches sign for

every second n. Note that $\det(\mathbf{S}_1) = 2$, $\det(\mathbf{S}_2) = -2^2$, $\det(\mathbf{S}_3) = -2^3$, $\det(\mathbf{S}_4) = 2^4$, $\det(\mathbf{S}_5) = 2^5$, etc. That is, $\det(\mathbf{S}_n) = 2^n$ for $n = 4k$ and $n = 4k+1$, while $\det(\mathbf{S}_n) = -2^n$ for $n = 4k-1$ and $n = 4k-2$.

(b) $\det(\mathbf{S}_n)\det(\mathbf{A}) = \pm 2^n \det(\mathbf{A})$, $\pm 2^{-n}\det(\mathbf{A})$, and $2^{2n}(\det(\mathbf{A}))^2$, where the signs are determined as in part (a).

7. $2 = \det(\mathbf{BC}^{-2}) = \det\mathbf{B}(\det\mathbf{C})^{-2} = \det\mathbf{B}\cdot\frac{1}{16}$ implies that $\det\mathbf{B} = 32$.
$24 = \det(\mathbf{A}^2\mathbf{B}^{-1}) = (\det\mathbf{A})^2(\det\mathbf{B})^{-1} = (\det\mathbf{A})^2\cdot\frac{1}{32}$ implies that $\det\mathbf{A} = \sqrt{32\cdot 24} = 16\sqrt{3}$; thus, $\det\mathbf{A}^3 = 16^3\cdot 3\cdot\sqrt{3} = 12,288\cdot\sqrt{3}$.

9. $\det(\mathbf{A}^{-1}) = \pm 1$.

11.
$$\begin{array}{ccc|l} 2 & 1 & 1 & \\ 1 & k+1 & 0 & \\ -1 & 2 & 1 & -\,row_1 \\ \hline 2 & 1 & 1 & \\ 1 & k+1 & 0 & \\ -3 & 1 & 0 & \end{array}$$

Thus, $\det\mathbf{A} = \det\begin{pmatrix} 1 & k+1 \\ -3 & 1 \end{pmatrix} = 1 + 3(k+1) > 0$ for all $k > -\frac{4}{3}$.

13. Laplace expansion along the modified row cr in $\tilde{\mathbf{A}}$ sums every term of the Laplace expansion of $\mathbf{A}$ along the same row, but multiplied by a factor of c (i.e., $\det(\tilde{\mathbf{A}}) = c\cdot\det\mathbf{A}$).

8.3.P p. 249

1. Area $= \frac{1}{2}\det\begin{pmatrix} 1 & 3 \\ -4 & 17 \end{pmatrix} = \frac{1}{2}(17 + 12) = 14.5$.

3. $\mathbf{A}_{\mathcal{E}} = \frac{1}{3}\begin{pmatrix} -2 & -1 \\ 2 & 7 \end{pmatrix}$, according to Problem 19 of Section 1.1.P. Or, directly from the data,
$\mathbf{A}_{\mathcal{E}}\begin{pmatrix} 4 & 5 \\ 1 & -1 \end{pmatrix} = \begin{pmatrix} -3 & -3 \\ 5 & 1 \end{pmatrix}$ or $\mathbf{A}_{\mathcal{E}} = \begin{pmatrix} -3 & -3 \\ 5 & 1 \end{pmatrix}\begin{pmatrix} 4 & 5 \\ 1 & -1 \end{pmatrix}^{-1} = \frac{1}{3}\begin{pmatrix} -2 & -1 \\ 2 & 7 \end{pmatrix}$.
Also $\det(\mathbf{A}_{\mathcal{E}}) = \left(\frac{1}{3}\right)\cdot(-14 + 2) = -4/3$. Thus,
Area$(T(3\ unit\ square)) = \left|-\frac{4}{3}\right|\cdot 3^2 = 12$.

5. $\mathbf{B}_{\mathcal{E}} = \begin{pmatrix} 3 & 1 \\ -2 & -6 \end{pmatrix}$, according to Chapter 1. $\det\mathbf{B} = -18 + 2 = -16$. Area$(V) = 16\cdot 4 = 64$.

7. $\cos^2 x = \sin^2 x = 1$ (i.e., the three functions are linearly dependent).

9. (a) $W(1 - x^2,\ 1 + x^2) = \det\begin{pmatrix} 1 - x^2 & 1 + x^2 \\ -2x & 2x \end{pmatrix} = 2x(1 - x^2) + 2x(1 + x^2) = 4x \neq 0$ for $x \neq 0$.
Hence, these two functions are linearly independent.

(b) $1 + (4(-x^2)) = 1 - 4x^2$ (i.e. the three functions are linearly dependent).

(c) $W(f, g, f\cdot g) = \det\begin{pmatrix} x+1 & x-1 & x^2-1 \\ 1 & 1 & 2x \\ 0 & 0 & 2 \end{pmatrix}$
$= 2\det\begin{pmatrix} x+1 & x-1 \\ 1 & 1 \end{pmatrix} = 2(x + 1 - x + 1) = 2 \neq 0$. Hence, the three functions are linearly independent.

(d) $W(x^{1/2},\ x^{5/2}) = \det\begin{pmatrix} x^{1/2} & x^{5/2} \\ \frac{1}{2}x^{-1/2} & \frac{5}{2}x^{3/2} \end{pmatrix} = \frac{5}{2}x^2 - \frac{1}{2}x^2 = 2x^2 \neq 0$ for all $x \neq 0$. Thus, these functions are linearly independent.

(e) Linearly dependent precisely for all $\alpha = \pm(2k + 1)\frac{\pi}{2}$ and integers k.

8.R p. 250

1. Add all rows of $\mathbf{A}$ to the first. This does not change the determinant, but creates a first row consisting entirely of zeros. Expand along this row: $\det\mathbf{A} = 0$.

3. (a) Eliminating the (2,1) entry of $\mathbf{A}$ with the first row creates a row consisting entirely of zeros; hence, $\det\mathbf{A} = 0$.

(b) The rows of $\mathbf{A}$ are all identical and hence linearly dependent; thus, $\det\mathbf{A} = 0$.

(c) See part (b).

(d) Expand along the last column 98 times to obtain $\det\mathbf{A} = 2^{98}\cdot\det\begin{pmatrix} 2 & 2 \\ 1 & 2 \end{pmatrix} = 2^{98}(4 - 2) = 2^{99}$.

5. (a) $\det\mathbf{A} = 12 - 10 = 2$. $\det(\mathbf{A} - \lambda\mathbf{I}_2) = \det\begin{pmatrix} 4 - \lambda & -2 \\ -5 & 3 - \lambda \end{pmatrix} = (4 - \lambda)(3 - \lambda) - 10 = 12 - 7\lambda + \lambda^2 - 10 = \lambda^2 - 7\lambda + 2$.

(b) $\mathbf{A} - \lambda\mathbf{I}$ is singular if $\det(\mathbf{A} - \lambda\mathbf{I}) = 0$ (i.e., if $\lambda = 3.5 \pm \sqrt{12.25 - 2} = 3.5 \pm \sqrt{10.25}$).

7. $n!$.

9. $\det(\mathbf{A}) = \det\begin{pmatrix} 1 & 1 \\ 3 & 2 \end{pmatrix} - \det\begin{pmatrix} 5 & 1 \\ 2 & 2 \end{pmatrix} + \det\begin{pmatrix} 5 & 1 \\ 2 & 3 \end{pmatrix} = -1 - 8 + 13 = 4$.

11. $\det\begin{pmatrix} 1 & 2 & a \\ 2 & 2 & b \\ c & 0 & 0 \end{pmatrix} = c \cdot \det\begin{pmatrix} 2 & a \\ 2 & b \end{pmatrix} = 2c(b-a) =$ 0 for $c = 0$ or $a = b$. $\det \mathbf{A} = 0$ precisely when $\mathbf{A}$ is singular (i.e., when $\text{im}(\mathbf{A}) \neq \mathbb{R}^3$). The columns of $\mathbf{A}$ are linearly independent precisely when $\det \mathbf{A} \neq 0$.

13. $\mathbf{A} = \begin{pmatrix} a_{11} & a_{12} \\ a_{21} & a_{22} \\ a_{31} & a_{32} \end{pmatrix}$, $\mathbf{b} = -\begin{pmatrix} a_{10} \\ a_{20} \\ a_{30} \end{pmatrix}$, and $\mathbf{z} = \begin{pmatrix} x \\ y \end{pmatrix}$ $\in \mathbb{R}^2$. If $\mathbf{Az} = \mathbf{b}$ is solvable, then $\det(\mathbf{A} \mid \mathbf{b}) = 0$ by Theorem 3.1. The converse—as attributed erroneously to Leibniz on this Website—is false: simply take $\mathbf{A} = \mathbf{0}_{3\times 2}$ and $\mathbf{b} \neq \mathbf{0} \in \mathbb{R}^2$ arbitrarily.

Chapter 9

9.1.P p. 268

1. Evaluate $\mathbf{Au}_1, \ldots, \mathbf{Au}_4$ and look: $\mathbf{Au}_1 = 6\mathbf{u}_1$. Thus, $\mathbf{u}_1$ is an eigenvector for $\lambda = 6$; and $\mathbf{u}_3$ is for $\lambda = -1$.

3. For $\mathbf{A} = \begin{pmatrix} 1 & -1 \\ 1 & 0 \end{pmatrix}$, for example, and $\mathbf{y} = \begin{pmatrix} 1 \\ 0 \end{pmatrix}$, we have

$\mathbf{y}$	$\mathbf{Ay}$	$\mathbf{A}^2\mathbf{y}$	
$\boxed{1}$	$\boxed{1}$	0	$-$ *row*$_2$
0	$\boxed{1}$	1	
1	0	-1	
0	1	1	

That is, $\mathbf{A}^2\mathbf{y} = \mathbf{Ay} - \mathbf{y}$, or $(\mathbf{A}^2 - \mathbf{A} + \mathbf{I}_2)\mathbf{y} = \mathbf{0}$, and $p_y(\lambda) = \lambda^2 - \lambda + 1 = p_A(\lambda)$, since $\mathbf{A}$ is 2×2. $\mathbf{A}$'s eigenvalues are $\lambda_{1,2} = \frac{1}{2} \pm \sqrt{\frac{1}{4} - 1} \notin \mathbb{R}$.

5. (a) For $\mathbf{y} = \mathbf{e}_1$, we vector iterate and row reduce:

$\mathbf{y}$	$\mathbf{Ay}$	$\mathbf{A}^2\mathbf{y}$	
$\boxed{1}$	1	3	
0	$\boxed{2}$	4	$\div 2$
$\boxed{1}$	$\boxed{1}$	3	$-$ *row*$_2$
0	$\boxed{1}$	2	
1	0	1	
0	1	2	

That is, $\mathbf{A}^2\mathbf{y} = 2\mathbf{Ay} + \mathbf{y}$, or $(\mathbf{A}^2 - 2\mathbf{A} - \mathbf{I}_2)\mathbf{y} = \mathbf{0}$, and $p_y(\lambda) = \lambda^2 - 2\lambda - 1 = p_A(\lambda)$, since $n = 2$. Eigenvalues: $\lambda_{1,2} = 1 \pm \sqrt{2}$.

Now look at $\ker(\mathbf{A} - (1 + \sqrt{2})\mathbf{I})$:

$-\sqrt{2}$	1	0	$\cdot(-\sqrt{2})$
2	$-\sqrt{2}$	0	
$\boxed{2}$	$-\sqrt{2}$	0	
$\boxed{2}$	$-\sqrt{2}$	0	$-$ *row*$_1$
2	$-\sqrt{2}$	0	
0	0	0	

So an eigenvector for the eigenvalue $1 + \sqrt{2}$ of $\mathbf{A}$ is $\begin{pmatrix} 1 \\ \sqrt{2} \end{pmatrix}$; for $1 - \sqrt{2}$, it is $\begin{pmatrix} 1 \\ -\sqrt{2} \end{pmatrix}$.

(d) For $\mathbf{y} = \mathbf{e}_2$, we have

$\mathbf{y}$	$\mathbf{Ay}$	$\mathbf{A}^2\mathbf{y}$	
0	8	0	*swap rows*
1	1	9	
$\boxed{1}$	1	9	
0	$\boxed{8}$	0	$\div 8$
$\boxed{1}$	$\boxed{1}$	9	$-$ *row*$_2$
0	$\boxed{1}$	0	
1	0	9	
0	1	0	

That is, $\mathbf{A}^2\mathbf{y} = 9\mathbf{Ay}$, or $(\mathbf{A}^2 - 9\mathbf{I})\mathbf{y} = \mathbf{0}$, and $p_y(\lambda) = \lambda^2 - 9 = p_A(\lambda)$. Eigenvalues: $\lambda_{1,2} = \pm 3$; eigenvector for $\lambda = 3$: $\ker(\mathbf{A} - 3\mathbf{I}_2)$:

-4	8	0	*swap rows*
1	-2	0	
$\boxed{1}$	-2	0	
$\boxed{-4}$	8	0	$+4$ *row*$_1$
1	-2	0	
0	0	0	

So $\mathbf{x}_1 = \begin{pmatrix} 2 \\ 1 \end{pmatrix}$; for $\lambda = -3$, we find that $\mathbf{x}_2 = \begin{pmatrix} -4 \\ 1 \end{pmatrix}$.

(c) Eigenvalues: $2, -1 \pm \sqrt{2}i$; eigenvectors: $\mathbf{x}_1 = \begin{pmatrix} 1 \\ 1 \\ 1 \end{pmatrix}$, $\mathbf{x}_2 = \begin{pmatrix} 2 - \sqrt{2}i \\ -6 \\ 3 + 3\sqrt{2}i \end{pmatrix}$, $\mathbf{x}_3 = \begin{pmatrix} 2 + \sqrt{2}i \\ -6 \\ 3 - 3\sqrt{2}i \end{pmatrix}$. Note that the vectors

$$\mathbf{z}_2 = \begin{pmatrix} 0.19698 - 0.21944i \\ -0.83626 + 0.71746i \\ -0.46551 - 0.41786i \end{pmatrix} \text{ and}$$

$$\mathbf{z}_3 = \begin{pmatrix} 0.19698 + 0.21944i \\ -0.83626 - 0.71746i \\ -0.46551 + 0.41786i \end{pmatrix}, \text{ obtained by}$$

MATLAB, for example, are also eigenvectors for the two complex eigenvalues $-1 \pm \sqrt{2}i$. Yet these two vectors $\mathbf{z}_1$, $\mathbf{z}_2$ do not appear to be scalar multiples of $\mathbf{x}_2$ and $\mathbf{x}_3$, but instead they are complex multiples.

(d) For $\mathbf{y} = \mathbf{e}_1 + \mathbf{e}_3$, for example, we have

$\mathbf{y}$	$\mathbf{Ay}$	$\mathbf{A}^2\mathbf{y}$	$\mathbf{A}^3\mathbf{y}$	
1	4	10	22	
0	0	0	0	
1	2	4	8	*swap up*
1	4	10	22	
1	2	4	8	$-\ row_1$
0	0	0	0	
1	4	10	*	$+2\ row_2$
0	−2	−6	*	
0	0	0	0	
1	0	−2	−8	
0	−2	−6	−14	$\div(-2)$
0	0	0	0	
1	0	−2	*	
0	1	3	*	

So $\mathbf{A}^2\mathbf{y} = 3\mathbf{Ay} - 2\mathbf{y}$, or $(\mathbf{A}^2 - 3\mathbf{A} + 2\mathbf{I}_3)\mathbf{y} = \mathbf{0}$. Thus, $p_y(\lambda) = \lambda^2 - 3\lambda + 2$ with roots $\lambda_{1,2} = 1.5 \pm \sqrt{2.25 - 2} = 1$ or 2. Now look at $\ker(\mathbf{A} - \mathbf{I}_3) = \text{span}\{\mathbf{e}_1\}$ and $\ker(\mathbf{A} - 2\mathbf{I}_3) = \text{span}\left\{\begin{pmatrix} 3 \\ 0 \\ 1 \end{pmatrix}\right\}$.

Next, we use step 4 of the algorithm: Find a vector $\mathbf{v} \notin \ker(\mathbf{A}^2 - 3\mathbf{A} + 2\mathbf{I})$ if possible. $\mathbf{A}^2 - 3\mathbf{A} + 2\mathbf{I}$ $= \begin{pmatrix} 0 & -4 & 0 \\ 0 & 2 & 0 \\ 0 & 0 & 0 \end{pmatrix}$. So $\mathbf{v} = \mathbf{e}_2$ will do:

$\mathbf{v}$	$\mathbf{Av}$	$\mathbf{A}^2\mathbf{v}$	$\mathbf{A}^3\mathbf{v}$
0	2	2	2
1	0	0	0
0	0	0	0

Thus, $\mathbf{A}^2\mathbf{v} = \mathbf{Av}$, or $(\mathbf{A}^2 - \mathbf{A})\mathbf{v} = \mathbf{0}$, and $p_v(\lambda) = \lambda^2 - \lambda = \lambda(\lambda - 1) = 0$ for $\lambda = 0$ and $\lambda = 1$. The eigenvector $\mathbf{x}_3$ for the newly found eigenvalue 0 of $\mathbf{A}$ lies in the kernel of $\mathbf{A}$ and can be chosen as $\mathbf{x}_3 = \begin{pmatrix} -2 \\ 1 \\ 0 \end{pmatrix}$.

(e) Eigenvalues: 1, 0, 0; only two linearly independent eigenvectors: $\mathbf{e}_1$, $\begin{pmatrix} -2 \\ 1 \\ 0 \end{pmatrix}$; matrix not diagonalizable.

(f) For $\mathbf{y} = \begin{pmatrix} 1 \\ 1 \\ 1 \end{pmatrix}$, we have

$\mathbf{y}$	$\mathbf{Ay}$	$\mathbf{A}^2\mathbf{y}$	$\mathbf{A}^3\mathbf{y}$
1	1	0	0
1	0	0	0
1	1	0	0

Thus, $\mathbf{A}^2\mathbf{y} = 0$, or $p_y(\lambda) = \lambda^2 = 0$ for $\lambda = 0$. A matrix with a quadratic vanishing polynomial is not diagonalizable. $\mathbf{A}$ has only one linearly independent eigenvector for $\lambda = 0$, namely, $\begin{pmatrix} 1 \\ 0 \\ 1 \end{pmatrix}$.

7. (a) If $\mathbf{Ax} = \lambda\mathbf{x}$, then $\mathbf{Bx} = (\mathbf{A} + \alpha\mathbf{I})\mathbf{x} = \mathbf{Ax} + \alpha\mathbf{x} = \lambda\mathbf{x} + \alpha\mathbf{x} = (\lambda + \alpha)\mathbf{x}$.

(b) For $\lambda = 0$, $\ker(\mathbf{A})$ is determined by

4	8	0	$-4\ row_2$
1	2	0	
0	0	0	*swap rows*
1	2	0	
1	2	0	
0	0	0	

Eigenvector: $\mathbf{x}_1 = \begin{pmatrix} -2 \\ 1 \end{pmatrix}$. For $\lambda = 6$, $\ker(\mathbf{A} - 6\mathbf{I}_2)$:

−2	8	0	
1	−4	0	$+1/2\ row_1$
−2	−8	0	
0	0	0	

Eigenvector for $\lambda = 6$: $\mathbf{x}_2 = \begin{pmatrix} 4 \\ 1 \end{pmatrix}$.

$\mathcal{U} = \left\{\begin{pmatrix} -2 \\ 1 \end{pmatrix}, \begin{pmatrix} 4 \\ 1 \end{pmatrix}\right\}$.

(c) Eigenvalues for $\mathbf{B} + 3\mathbf{I}$ are $0 + 3 = 3$ and $6 + 3 = 9$; diagonalizable using same $\mathcal{U}$.

(d) If $\mathbf{Ax} = \lambda\mathbf{x}$, then $\mathbf{A}^2\mathbf{x} = \mathbf{A}(\mathbf{Ax}) = \mathbf{A}(\lambda\mathbf{x}) = \lambda\mathbf{Ax} = \lambda^2\mathbf{x}$, or $\mathbf{A}^2$ has the eigenvalues $0^2 = 0$ and $6^2 = 36$.

9. (a) $\mathbf{x} = \begin{pmatrix} -2 \\ 5 \\ 1 \end{pmatrix} \in \ker(\mathbf{A} - 3\mathbf{I}_3)$.

$$= \begin{pmatrix} 1 & & & \\ & \ddots & & \\ & c_{jk} - c_{jk} & \ddots & \\ & & & 1 \end{pmatrix} = \mathbf{I} \text{ if } j \neq k.$$

$\mathbf{E}_k(c)\mathbf{E}_k(1/c)$

$$= \begin{pmatrix} 1 & & & \\ & \ddots & & \\ & & c & \\ & & & \ddots & \\ & & & & 1 \end{pmatrix} \cdot \begin{pmatrix} 1 & & & \\ & \ddots & & \\ & & 1/c & \\ & & & \ddots & \\ & & & & 1 \end{pmatrix}$$

$= \mathbf{I}$ if $c \neq 0$.

$\mathbf{E}_{jk}\mathbf{E}_{jk}$

$$= \begin{pmatrix} 1 & & & & & \\ & \ddots & & & & \\ & & 0 & \cdots \; \cdots & 1 & \\ & & & 1 & & \\ & & & \ddots & & \\ & & 1 & & 0 & \\ & & & & & 1 \\ & & & & & & \ddots \end{pmatrix} \cdot$$

$$\cdot \begin{pmatrix} 1 & & & & & \\ & \ddots & & & & \\ & & 0 & \cdots \; \cdots & 1 & \\ & & & 1 & & \\ & & & \ddots & & \\ & & 1 & & 0 & \\ & & & & & 1 \\ & & & & & & \ddots \end{pmatrix} = \mathbf{I}.$$

3. (a) $\begin{pmatrix} 1 & & & \\ & 1 & & \\ & 0 & 1 & \\ & 3 & 0 & 1 \end{pmatrix}\begin{pmatrix} 1 & & & \\ & 1 & & \\ & 2 & 1 & \\ & & & 1 \end{pmatrix}$ maps $\begin{pmatrix} 4 \\ -1 \\ 2 \\ 3 \end{pmatrix}$ to $\begin{pmatrix} 4 \\ -1 \\ 0 \\ 0 \end{pmatrix}$.

(b) $\begin{pmatrix} 1 & & & \\ & 1 & & \\ & 0 & 1 & \\ & 3 & 0 & 1 \end{pmatrix}\begin{pmatrix} 1 & & & \\ & 1 & & \\ & 0 & 1 & \\ -1/2 & 0 & 1 & \\ & & & 1 \end{pmatrix}$ also maps $\begin{pmatrix} 4 \\ -1 \\ 2 \\ 3 \end{pmatrix}$ to $\begin{pmatrix} 4 \\ -1 \\ 0 \\ 0 \end{pmatrix}$. (Why are these two matrix products different?)

5. $(\mathbf{A} + \alpha\mathbf{I})(\mathbf{A} - \beta\mathbf{I}) = \mathbf{A}^2 + (\alpha - \beta)\mathbf{A} - \alpha\beta\mathbf{I} = (\mathbf{A} - \beta\mathbf{I})(\mathbf{A} + \alpha\mathbf{I})$.

7. $\mathbf{A}(\mathbf{e}_1) = \begin{pmatrix} 4 \\ 1 \\ 2 \end{pmatrix}$, $\mathbf{A}(\mathbf{e}_2) = \begin{pmatrix} -4 \\ 1 \\ 0 \end{pmatrix}$, and $\mathbf{A}(\mathbf{e}_3)$ $= \begin{pmatrix} 1 \\ 2 \\ 0 \end{pmatrix}$; thus, $\mathbf{A} = \begin{pmatrix} 4 & -4 & 1 \\ 1 & 1 & 2 \\ 2 & 0 & 0 \end{pmatrix}$.

$$\mathbf{A}^{-1} = \begin{pmatrix} 0 & 0 & 1/2 \\ -2/9 & 1/9 & 7/18 \\ 1/9 & 4/9 & -4/9 \end{pmatrix}:$$

4	−4	1	1	0	0	*swap rows 1, 3*
1	1	2	0	1	0	
2	0	0	0	0	1	
[2]	0	0	0	0	1	
[1]	1	2	0	1	0	$-\frac{1}{2}$ *row*$_1$
[4]	−4	1	1	0	0	−2 *row*$_1$
[2]	0	0	0	0	1	÷2
0	[1]	2	0	1	$-\frac{1}{2}$	
0	[−4]	1	1	0	−2	+4 *row*$_2$
[2]	0	0	0	0	1	÷2
0	[1]	2	0	1	$-\frac{1}{2}$	
0	0	[9]	1	4	−4	÷9
[1]	0	0	0	0	$\frac{1}{2}$	
0	[1]	[2]	0	1	$-\frac{1}{2}$	−2 *row*$_3$
0	0	[1]	$\frac{1}{9}$	$\frac{4}{9}$	$-\frac{4}{9}$	
			0	0	$\frac{1}{2}$	
	$\mathbf{I}_3$		$-\frac{2}{9}$	$\frac{1}{9}$	$\frac{7}{18}$	
			$\frac{1}{9}$	$\frac{4}{9}$	$-\frac{4}{9}$	

9. $\begin{pmatrix} d_1 & & \\ & \ddots & \\ & & d_m \end{pmatrix}\begin{pmatrix} - & \mathbf{b}_1^T & - \\ & \vdots & \\ - & \mathbf{b}_m^T & - \end{pmatrix} = \begin{pmatrix} - & d_1\mathbf{b}_1^T & - \\ & \vdots & \\ - & d_m\mathbf{b}_m^T & - \end{pmatrix}$; $\begin{pmatrix} | & & | \\ \mathbf{a}_1 & \ldots & \mathbf{a}_m \\ | & & | \end{pmatrix}$ $\begin{pmatrix} d_1 & & \\ & \ddots & \\ & & d_m \end{pmatrix} = \begin{pmatrix} | & & | \\ d_1\mathbf{a}_1 & \ldots & d_m\mathbf{a}_m \\ | & & | \end{pmatrix}$.

11. $$\begin{pmatrix} \mathbf{A}_{11} & \mathbf{A}_{12} \\ \mathbf{0} & \mathbf{A}_{22} \end{pmatrix}^{-1} = \begin{pmatrix} \mathbf{A}_{11}^{-1} & -\mathbf{A}_{11}^{-1}\mathbf{A}_{12}\mathbf{A}_{22}^{-1} \\ \mathbf{0} & \mathbf{A}_{22}^{-1} \end{pmatrix}.$$

13. (a) Clearly,

$$\mathbf{P} = \begin{pmatrix} 1 & & & & & & & \\ & \ddots & & & & & & \\ & & 0 & \cdots & \cdots & 1 & & \\ & & & 1 & & & & \\ & & & & \ddots & & & \\ & & 1 & & & 0 & & \\ & & & & & & 1 & \\ & & & & & & & \ddots \end{pmatrix} = \mathbf{P}^T \text{ is}$$

symmetric and $\mathbf{P}^2 = \mathbf{P}^T\mathbf{P} = \mathbf{I}_n$.

(b) If $\mathbf{P} = \mathbf{E}_1 \cdots \mathbf{E}_k$ for k swap matrices $\mathbf{E}_\ell$ of the type of Equation (6.5), then $\mathbf{P}^T\mathbf{P} = \mathbf{E}_k^T \cdots \mathbf{E}_1^T\mathbf{E}_1 \cdots \mathbf{E}_k = \mathbf{I}$, according to Theorem 6.3 on the transpose of a matrix product and part (a).

15. $$\mathbf{A}^{-1} = \left(\begin{array}{ccc|cc} 0 & 0 & 1 & 0 & 0 \\ 0 & \frac{1}{2} & 0 & 0 & 0 \\ 1 & 0 & 0 & 0 & 0 \\ \hline 0 & 0 & 0 & 0 & -1 \\ 0 & 0 & 0 & 1 & 0 \end{array}\right).$$

6.R p. 196

1. (a) For an arbitrary vector $\mathbf{Ax} \in \operatorname{im}(\mathbf{A})$, we have $\mathbf{A}(\mathbf{Ax}) = \mathbf{A}^2\mathbf{x} = \mathbf{0}$ [i.e., $\mathbf{Ax} \in \ker(\mathbf{A})$].

(d) $\mathbf{A} = \begin{pmatrix} 1 & 0 \\ 0 & 2 \end{pmatrix}$. For example; $\operatorname{im}(\mathbf{A}) = \mathbb{R}^2$, $\ker(\mathbf{A}) = \{\mathbf{0}\}$.

(f) The two are identical.

3. (c) $\mathbf{A}$ must be square to be invertible.

5. (a) $\mathbf{A}^{-1} = \begin{pmatrix} 1 & -1 & -2 \\ 2 & -3 & -4 \\ -1 & 2 & 1 \end{pmatrix}$:

$$\begin{array}{ccc|ccc|l} 5 & -3 & -2 & 1 & 0 & 0 & \textit{swap rows 3, 1} \\ 2 & -1 & 0 & 0 & 1 & 0 & \\ 1 & -1 & -1 & 0 & 0 & 1 & \\ \hline \boxed{1} & -1 & -1 & 0 & 0 & 1 & \\ \boxed{2} & -1 & 0 & 0 & 1 & 0 & -2\ row_1 \\ \boxed{5} & -3 & -2 & 1 & 0 & 0 & -5\ row_1 \\ \hline \boxed{1} & -1 & -1 & 0 & 0 & 1 & \\ 0 & \boxed{1} & 2 & 0 & 1 & -2 & \\ 0 & \boxed{2} & 3 & 1 & 0 & -5 & -2\ row_2 \\ \hline \boxed{1} & -1 & \boxed{-1} & 0 & 0 & 1 & -\ row_3 \\ 0 & \boxed{1} & \boxed{2} & 0 & 1 & -2 & +2\ row_3 \\ 0 & 0 & \boxed{-1} & 1 & -2 & -1 & \div(-1) \\ \hline \boxed{1} & \boxed{-1} & 0 & -1 & 2 & 2 & +\ row_2 \\ 0 & \boxed{1} & 0 & 2 & -3 & -4 & \\ 0 & 0 & \boxed{1} & -1 & 2 & 1 & \\ \hline & & & 1 & -1 & -2 & \\ & \mathbf{I}_3 & & 2 & -3 & -4 & \\ & & & -1 & 2 & 1 & \end{array}$$

(b) $(\mathbf{A}^T)^{-1} = (\mathbf{A}^{-1})^T = \begin{pmatrix} 1 & 2 & -1 \\ -1 & -3 & 2 \\ -2 & -4 & 1 \end{pmatrix}$.

(c) $\mathbf{x} = (\mathbf{A}^T)^{-1}\mathbf{b} = (-5, 9, 8)^T$.

7. (a) $\mathbf{A}_{\mathcal{E}}\begin{pmatrix} 1 & 1 & 2 \\ -2 & -3 & -5 \\ 1 & 3 & 5 \end{pmatrix} = \begin{pmatrix} 2 & 0 & 1 \\ 0 & 1 & 0 \\ 1 & 4 & -1 \end{pmatrix}$.

(b) $\mathbf{A}_{\mathcal{E}} = \begin{pmatrix} 3 & 0 & -1 \\ -5 & -3 & -1 \\ -23 & -15 & -6 \end{pmatrix} = \begin{pmatrix} 2 & 0 & 1 \\ 0 & 1 & 0 \\ 1 & 4 & -1 \end{pmatrix}$
$\begin{pmatrix} 1 & 1 & 2 \\ -2 & -3 & -5 \\ 1 & 3 & 5 \end{pmatrix}^{-1}$.

(c) $T\begin{pmatrix} x_1 \\ x_2 \\ x_3 \end{pmatrix} = \begin{pmatrix} 3x_1 - x_3 \\ -5x_1 - 3x_2 - x_3 \\ -23x_1 - 15x_2 - 6x_3 \end{pmatrix}$; $T\begin{pmatrix} a \\ 2 \\ b \end{pmatrix}$
$= \begin{pmatrix} 3a - b \\ -5a - 6 - b \\ -23a - 30 - 6b \end{pmatrix}$.

(d) For any real c, the vector $\mathbf{u}$ lies in $\operatorname{im}(T)$.

9. (a) Take $\mathbf{B} = \mathbf{A}^{-1}$, for example, or $\mathbf{A} = \mathbf{B}$.

(b) If $\mathbf{A} = \begin{pmatrix} 0 & 1 \\ -1 & 0 \end{pmatrix}$ and $\mathbf{B} = \begin{pmatrix} 1 & 0 \\ 0 & 2 \end{pmatrix}$, then
$\mathbf{A}^{-1} = \begin{pmatrix} 0 & -1 \\ 1 & 0 \end{pmatrix}$ and $\mathbf{A}^{-1}\mathbf{BA} = \begin{pmatrix} 2 & 0 \\ 0 & 1 \end{pmatrix} \neq \mathbf{B}$.

Chapter 7

7.1.P p. 210

1. $\mathbf{x}_{\mathcal{V}} = \mathbf{X}_{\mathcal{V}\leftarrow\mathcal{U}}\mathbf{x}_{\mathcal{U}} = (\mathbf{X}_{\mathcal{U}\leftarrow\mathcal{V}})^{-1}\,\mathbf{x}_{\mathcal{U}}$ and $\mathbf{x}_{\mathcal{U}} = \begin{pmatrix} 6 \\ 5 \\ -2 \end{pmatrix}$ for $(\mathbf{X}_{\mathcal{U}\leftarrow\mathcal{V}})^{-1} = \begin{pmatrix} -1 & -2 & 2 \\ -1 & -1 & 2 \\ 1 & 1 & -1 \end{pmatrix}$.

for $\lambda = 2$. For the double eigenvalue $\lambda = -1$, we have $\ker(\mathbf{B} + \mathbf{I})$:

$$
\begin{array}{ccc|c|l}
0 & \boxed{1} & 0 & 0 & \\
0 & \boxed{3} & -1 & 0 & -3\ row_1 \\
0 & 0 & 0 & 0 & \\
\hline
0 & \boxed{1} & 0 & 0 & \\
0 & 0 & \boxed{-1} & 0 & \\
0 & 0 & 0 & 0 &
\end{array}
$$

There is only one eigenvector $\mathbf{x}_1 = \mathbf{e}_1$ for $\lambda = -1$; hence, $\mathbf{B}$ is not diagonalizable.

(c) eigenvalues -1 (double) and 2; corresponding eigenvectors $\mathbf{e}_1$, $\mathbf{e}_2$ for -1, and $\begin{pmatrix} -1 \\ 0 \\ 3 \end{pmatrix}$ for 2; $\mathbf{C}$ is diagonalizable.

27. For $\mathbf{y} = \mathbf{e}_1$, we have

$$
\begin{array}{cc|c|l}
\mathbf{y} & \mathbf{A}^2\mathbf{y} & \mathbf{A}^3\mathbf{y} & \\
\hline
1 & 0 & -16 & \\
0 & -8 & 0 & \div(-8) \\
\hline
1 & 0 & -16 & \\
0 & 1 & 0 &
\end{array}
$$

Thus, $\mathbf{A}^2\mathbf{y} = -16\mathbf{A}\mathbf{y}$, or $p_y(\lambda) = \lambda^2 + 16 = 0$ for the eigenvalues $\pm 4i$. Now look at $\ker(\mathbf{A} - 4i\mathbf{I}_2)$:

$$
\begin{array}{cc|c|l}
\boxed{-4i} & 2 & 0 & \\
\boxed{-8} & -4i & 0 & +\ 2i\ row_1 \\
\hline
\boxed{-4i} & 2 & 0 & \\
0 & 0 & 0 &
\end{array}
$$

So $\mathbf{x}_1 = \begin{pmatrix} 1 \\ 2i \end{pmatrix} \in \mathbb{C}^2$ is an eigenvector for $\lambda_1 = 4i$, and $\begin{pmatrix} -1 \\ 2i \end{pmatrix}$ is for $\lambda_2 = -4i$.

29. (a) $\mathbf{A}_{\mathcal{U}} = \mathbf{U}^{-1}\mathbf{A}_{\mathcal{E}}\mathbf{U}$.

(b) $\mathbf{A}_{\mathcal{U}} = \mathbf{U}^{-1}\mathbf{A}\mathbf{U} = \begin{pmatrix} 1 & 0 & 0 \\ 1 & 2 & 0 \\ 2 & -1 & 3 \end{pmatrix}$ for

$\mathbf{U} = \begin{pmatrix} 1 & -1 & 0 \\ 0 & 1 & 1 \\ -2 & 2 & 1 \end{pmatrix}$ and $\mathbf{U}^{-1} = \begin{pmatrix} -1 & 1 & -1 \\ -2 & 1 & -1 \\ 2 & 0 & 1 \end{pmatrix}$.

(c) The eigenvalues of $\mathbf{A}$ appear on the diagonal of the triangular matrix $\mathbf{A}_{\mathcal{U}}$; they are 1, 2, and 3. Since the eigenvalues of $\mathbf{A}$ are distinct, $\mathbf{A}$ is diagonalizable.

31. $\mathbf{A}\mathbf{w} = \begin{pmatrix} 3 \\ 0 \end{pmatrix}$, $\mathbf{A}^2\mathbf{w} = \begin{pmatrix} 9 \\ 6 \end{pmatrix}$, and the RREF of the matrix with columns $\mathbf{w}$, $\mathbf{A}\mathbf{w}$, and $\mathbf{A}^2\mathbf{w}$ is $\begin{pmatrix} 1 & 0 & -6 \\ 0 & 1 & 7 \end{pmatrix}$, just as in Example 2.

33. (a), (b) $\mathbf{u}_2$ and $\mathbf{u}_6$ are eigenvectors for $\lambda = 0$; $\mathbf{u}_5$ is for $\lambda = 8$. $\mathbf{C}$ is diagonalizable.

(c) $\mathbf{C}$ is not invertible. ($\mathbf{u}_2 \neq \mathbf{0} \in \ker(\mathbf{C})$.)

35. We must make sure that $\ker(\mathbf{A} - 2\mathbf{I}_3) \neq \{\mathbf{0}\}$:

$$
\begin{array}{ccc|c|l}
\boxed{-1} & 1 & a & 0 & \\
\boxed{-1} & -2 & 2 & 0 & -\ row_1 \\
\boxed{2} & 1 & 2 & 0 & +2\ row_1 \\
\hline
\boxed{-1} & -1 & a & 0 & \\
0 & \boxed{-3} & 2-a & 0 & \\
0 & \boxed{3} & 2+2a & 0 & +\ row_2 \\
\hline
\boxed{-1} & -1 & a & 0 & \\
0 & \boxed{-3} & 2-a & 0 & \\
0 & 0 & 4+a & 0 &
\end{array}
$$

Thus, $\mathbf{A} - 2\mathbf{I}$ is singular precisely when $a = -4$, in which case 2 is an eigenvalue of $\mathbf{A}$.

37. Observe that $\mathbf{A}^{-1}(\mathbf{A}\mathbf{B})\mathbf{A} = \mathbf{B}\mathbf{A}$. Thus, the two nonsingular matrices $\mathbf{A}\mathbf{B}$ and $\mathbf{B}\mathbf{A}$ have the same eigenvalues.

39. For $\mathbf{y} = \mathbf{e}_1$, we have

$$
\begin{array}{cc|c|l}
\mathbf{y} & \mathbf{A}\mathbf{y} & \mathbf{A}^2\mathbf{y} & \\
\hline
\boxed{1} & \boxed{2} & 5 & -2\ row_2 \\
0 & \boxed{1} & 6 & \\
\hline
1 & 0 & -7 & \\
0 & 1 & 6 &
\end{array}
$$

and

$$
\begin{array}{cc|c|l}
\mathbf{y} & \mathbf{B}\mathbf{y} & \mathbf{B}^2\mathbf{y} & \\
\hline
\boxed{1} & \boxed{6} & 29 & +6\ row_2 \\
0 & \boxed{-1} & -6 & \\
\hline
1 & 0 & -7 & \\
0 & -1 & -6 & \cdot(-1)
\end{array}
$$

so that $\mathbf{A}^2\mathbf{y} = 6\mathbf{A}\mathbf{y} - 7\mathbf{y}$ and $\mathbf{B}^2\mathbf{y} = 6\mathbf{B}\mathbf{y} - 7\mathbf{y}$. That is, both $\mathbf{A}$ and $\mathbf{B}$ have the same vanishing polynomials $p(\lambda) = \lambda^2 - 6\lambda + 7$ with two distinct roots $\lambda_{1,2} = 3 \pm \sqrt{2}$. Hence, they are both similar to the same diagonal matrix. Thus, $\mathbf{A}$ and $\mathbf{B}$ are similar themselves.

41. (a) $(\mathbf{I} - \mathbf{X}^{-1}\mathbf{A}\mathbf{X})^{-1} + (\mathbf{I} - (\mathbf{X}^{-1}\mathbf{A}\mathbf{X})^{-1})^{-1} =$
$(\mathbf{X}^{-1}(\mathbf{I} - \mathbf{A})\mathbf{X})^{-1} + (\mathbf{X}^{-1}(\mathbf{I} - \mathbf{A}^{-1})\mathbf{X})^{-1} =$
$\mathbf{X}^{-1}(\mathbf{I} - \mathbf{A})^{-1}\mathbf{X} + \mathbf{X}^{-1}(\mathbf{I} - \mathbf{A}^{-1})^{-1}\mathbf{X} =$

$\mathbf{X}^{-1}((\mathbf{I}-\mathbf{A})^{-1}+(\mathbf{I}-\mathbf{A}^{-1})^{-1})\mathbf{X} = \mathbf{X}^{-1}\mathbf{I}\mathbf{X} =$ $\mathbf{I}$ if $(\mathbf{I}-\mathbf{A})^{-1}+(\mathbf{I}-\mathbf{A}^{-1})^{-1}=\mathbf{I}$.

(b) Let $\mathbf{X}^{-1}\mathbf{A}\mathbf{X}=\mathbf{D}=\operatorname{diag}(d_i)$. Using part (a) and assuming that all relevant inverses exist, we have $\mathbf{W} := (\mathbf{I}-\mathbf{D})^{-1}+(\mathbf{I}-\mathbf{D}^{-1})^{-1} = \operatorname{diag}\left(\frac{1}{1-d_i}\right) + \operatorname{diag}\left(\left(1-\frac{1}{d_i}\right)^{-1}\right)$. Thus, each diagonal entry of the matrix $\mathbf{W}$ is equal to $\frac{1}{1-d_i}+\frac{1}{1-\frac{1}{d_i}} = \frac{1}{1-d_i}+\frac{d_i}{d_i-1} = \frac{1-d_i}{1-d_i} = 1$. That is, $\mathbf{W}=\mathbf{I}$ as claimed.

(c) Here we have $\mathbf{I}-\mathbf{A} = \begin{pmatrix} -1 & 1 \\ 0 & -1 \end{pmatrix}$; $(\mathbf{I}-\mathbf{A})^{-1} = \begin{pmatrix} -1 & -1 \\ 0 & -1 \end{pmatrix}$; $\mathbf{A}^{-1} = \begin{pmatrix} 1/2 & -1/4 \\ 0 & 1/2 \end{pmatrix}$; $\mathbf{I}-\mathbf{A}^{-1} = \begin{pmatrix} 1/2 & 1/4 \\ 0 & 1/2 \end{pmatrix}$;

$(\mathbf{I}-\mathbf{A}^{-1})^{-1} = \begin{pmatrix} 2 & -1 \\ 0 & 2 \end{pmatrix}$. Thus, $(\mathbf{I}-\mathbf{A})^{-1} + (\mathbf{I}-\mathbf{A}^{-1})^{-1} = \begin{pmatrix} -1 & -1 \\ 0 & -1 \end{pmatrix} + \begin{pmatrix} 2 & -1 \\ 0 & 2 \end{pmatrix}$

$= \begin{pmatrix} 1 & -2 \\ 0 & 1 \end{pmatrix} \neq \mathbf{I}_2$.

T 9. (b) For $\mathbf{B}$ and $\mathbf{y}=\mathbf{e}_2$, for example, the vanishing polynomial is $p_y(x)=(x-1)^2$. Since 1 is a double root of $p_{\mathbf{y}}$, $\mathbf{B}$ is not diagonalizable. Alternatively, we can look at the eigenspace $E(1)=\ker(\mathbf{B}-\mathbf{I})$ for $\mathbf{B}$:

B − I

−4	−1	−6	
−4	−2	−8	
2	1	4	$+\frac{1}{2}$ *row*$_2$
[−4]	−1	−6	
(−4)	−2	−8	− *row*$_1$
0	0	0	
[−4]	−1	−6	
0	[−1]	−2	
0	0	0	

Clearly, $E(1)$ is only one dimensional. (Note the unusual first elimination step.) There are not enough eigenvectors for $\mathbf{B}$; consequently, according to the top of step 4 in our eigenanalysis procedure, $\mathbf{B}$ cannot be diagonalized.

9.1.D.P p. 281

1. Evaluate $\mathbf{A}\mathbf{u}_1, \ldots, \mathbf{A}\mathbf{u}_4$ and look: $\mathbf{A}\mathbf{u}_1 = 6\mathbf{u}_1$. Thus, $\mathbf{u}_1$ is an eigenvector for $\lambda=6$, and $\mathbf{u}_3$ is for $\lambda=-1$.

3. For $\mathbf{A} = \begin{pmatrix} 1 & -1 \\ 1 & 0 \end{pmatrix}$, for example, we have $\det(\lambda\mathbf{I}_2-\mathbf{A}) = \det\begin{pmatrix} \lambda & 1 \\ -1 & \lambda \end{pmatrix} = \lambda^2+1=0$ for $\lambda=\pm i=\pm\sqrt{-1}\notin\mathbb{R}$.

5. (a) $\det(\lambda\mathbf{I}_2-\mathbf{A}) = \det\begin{pmatrix} \lambda-1 & -1 \\ -2 & \lambda-1 \end{pmatrix} = \lambda^2-2\lambda-3=0$ for the eigenvalues $\lambda_{1,2}=1\pm\sqrt{2}$. Now look at $\ker(\mathbf{A}-(1+\sqrt{2})\mathbf{I})$:

$-\sqrt{2}$	1	0	$\cdot(-\sqrt{2})$
2	$-\sqrt{2}$	0	
[2]	$-\sqrt{2}$	0	
(2)	$-\sqrt{2}$	0	− *row*$_1$
[2]	$-\sqrt{2}$	0	
0	0	0	

So an eigenvector for the eigenvalue $1+\sqrt{2}$ of $\mathbf{A}$ is $\begin{pmatrix} 1 \\ \sqrt{2} \end{pmatrix}$; for $1-\sqrt{2}$, it is $\begin{pmatrix} 1 \\ -\sqrt{2} \end{pmatrix}$.

(b) $\det(\lambda\mathbf{I}_2-\mathbf{A}) = \det\begin{pmatrix} \lambda+1 & -8 \\ -1 & \lambda-1 \end{pmatrix} = \lambda^2-9=0$ for the eigenvalues $\lambda_{1,2}=\pm 3$. The eigenvector for 3 is $\ker(\mathbf{A}-3\mathbf{I}_2)$:

−4	8	0	*swap rows*
1	−2	0	
[1]	−2	0	
(−4)	8	0	+4 *row*$_1$
1	−2	0	
0	0	0	

So $\mathbf{x}_1 = \begin{pmatrix} 2 \\ 1 \end{pmatrix}$; for $\lambda=-3$, we find that $\mathbf{x}_2 = \begin{pmatrix} -4 \\ 1 \end{pmatrix}$.

(c) Eigenvalues: 2, $-1\pm\sqrt{2}i$; eigenvectors $\mathbf{x}_1 = \begin{pmatrix} 1 \\ 1 \\ 1 \end{pmatrix}$, $\mathbf{x}_2 = \begin{pmatrix} 2-\sqrt{2}i \\ -6 \\ 3+3\sqrt{2}i \end{pmatrix}$, $\mathbf{x}_3 = \begin{pmatrix} 2+\sqrt{2}i \\ -6 \\ 3-3\sqrt{2}i \end{pmatrix}$. Note that the vectors $\mathbf{z}_2 = \begin{pmatrix} 0.19698-0.21944i \\ -0.83626+0.71746i \\ -0.46551-0.41786i \end{pmatrix}$ and $\mathbf{z}_3 = \begin{pmatrix} 0.19698+0.21944i \\ -0.83626-0.71746i \\ -0.46551+0.41786i \end{pmatrix}$, obtained by MATLAB, for example, are also eigenvectors for the two complex eigenvalues $-1\pm\sqrt{2}i$. Yet these two vectors $\mathbf{z}_1, \mathbf{z}_2$ do not appear to be scalar multiples of $\mathbf{x}_2$ and $\mathbf{x}_3$, but they are complex multiples thereof.

(d) $\det(\lambda\mathbf{I}_3 - \mathbf{A}) = \det\begin{pmatrix} \lambda-1 & -2 & -3 \\ 0 & \lambda & 0 \\ 0 & 0 & \lambda-2 \end{pmatrix}$

$= \lambda(\lambda-1)(\lambda-2)$. So the eigenvalues are 0, 1, and 2, with eigenvectors $\begin{pmatrix} -2 \\ 1 \\ 0 \end{pmatrix}, \begin{pmatrix} 1 \\ 0 \\ 0 \end{pmatrix}$, and $\begin{pmatrix} 3 \\ 0 \\ 1 \end{pmatrix}$.

(e) Eigenvalues: 1, 0, 0; only two linearly independent eigenvectors: $\mathbf{e}_1$ and $\begin{pmatrix} -2 \\ 1 \\ 0 \end{pmatrix}$, matrix not diagonalizable.

(f) $\det(\lambda\mathbf{I}_3 - \mathbf{A}) = \det\begin{pmatrix} \lambda & -1 & 0 \\ -1 & \lambda & 1 \\ 0 & -1 & \lambda \end{pmatrix} = \lambda^3 = 0$

for the triple eigenvalue 0; only one linearly independent eigenvector for $\lambda = 0$, namely, $\begin{pmatrix} 1 \\ 0 \\ 1 \end{pmatrix}$, so $\mathbf{A}$ is not diagonalizable.

7. (a) If $\mathbf{Ax} = \lambda\mathbf{x}$, then $\mathbf{Bx} = (\mathbf{A} + \alpha\mathbf{I})\mathbf{x} = \mathbf{Ax} + \alpha\mathbf{x} = \lambda\mathbf{x} + \alpha\mathbf{x} = (\lambda + \alpha)\mathbf{x}$.

(b) For $\lambda = 0$, $\ker(\mathbf{A})$ is determined by

$$\begin{array}{cc|c l} 4 & 8 & 0 & -4\,row_2 \\ 1 & 2 & 0 & \\ \hline 0 & 0 & 0 & swap\ rows \\ 1 & 2 & 0 & \\ \hline 1 & 2 & 0 & \\ 0 & 0 & 0 & \end{array}$$

Eigenvector: $\mathbf{x}_1 = \begin{pmatrix} -2 \\ 1 \end{pmatrix}$. For $\lambda = 6$, $\ker(\mathbf{A} - 6\mathbf{I}_2)$ is

$$\begin{array}{cc|c l} \boxed{-2} & 8 & 0 & \\ \boxed{1} & -4 & 0 & +1/2\,row_1 \\ \hline \boxed{-2} & -8 & 0 & \\ 0 & 0 & 0 & \end{array}$$

Eigenvector for $\lambda = 6$: $\mathbf{x}_2 = \begin{pmatrix} 4 \\ 1 \end{pmatrix}$.

$\mathcal{U} = \left\{\begin{pmatrix} -2 \\ 1 \end{pmatrix}, \begin{pmatrix} 4 \\ 1 \end{pmatrix}\right\}$.

(c) Eigenvalues for $\mathbf{B} + 3\mathbf{I}$ are $0 + 3 = 3$ and $6 + 3 = 9$; diagonalizable using same $\mathcal{U}$.

(d) If $\mathbf{Ax} = \lambda\mathbf{x}$, then $\mathbf{A}^2\mathbf{x} = \mathbf{A}(\mathbf{Ax}) = \mathbf{A}(\lambda\mathbf{x}) = \lambda\mathbf{Ax} = \lambda^2\mathbf{x}$, or $\mathbf{A}^2$ has the eigenvalues $0^2 = 0$ and $6^2 = 36$.

9. (a) $\mathbf{x} = \begin{pmatrix} -2 \\ 5 \\ 1 \end{pmatrix} \in \ker(\mathbf{A} - 3\mathbf{I}_3)$.

(b) Take $i = 3$, for example. Then $x_3 = 1$, $\frac{1}{x_3}xa_3$

$= \begin{pmatrix} -2 \\ 5 \\ 1 \end{pmatrix}\begin{pmatrix} -11 & -5 & 6 \end{pmatrix}$, and

$\mathbf{B}_0 = \begin{pmatrix} 1 & 0 & -22 \\ -3 & -2 & 4 \\ 0 & 0 & 0 \end{pmatrix}$.

(c) The matrix $\mathbf{B}_0$ has a row of zeros in position i. If $\mathbf{B}_0\mathbf{w} = \mu\mathbf{w}$ then $w_i = 0$. If $\mathbf{Ax} = \lambda\mathbf{x}$, then $\mathbf{B}_0\mathbf{x} = \left(\mathbf{A} - \frac{1}{x_i}\mathbf{x}\mathbf{a}_i\right)\mathbf{x} = \mathbf{Ax} - \frac{1}{x_i}\mathbf{x}\mathbf{a}_i\mathbf{x} = \lambda\mathbf{x} - \lambda\mathbf{x} = 0$, since $\mathbf{a}_i\mathbf{x} = \lambda\mathbf{x}_i$.

(d) $\mathbf{B}_1 = \begin{pmatrix} 1 & 0 \\ -3 & -2 \end{pmatrix}$. $\begin{pmatrix} 1 \\ -1 \end{pmatrix}$ and $\begin{pmatrix} 0 \\ 1 \end{pmatrix}$ are eigenvectors for the eigenvalues 1 and –2 on the diagonal of the triangular matrix of $\mathbf{B}_1$.

(e) If $\mathbf{A}$ is nonsingular and diagonalizable, assume that it has the eigenvalues $\lambda_2, \ldots, \lambda_n \in \mathbb{R}$ for the eigenvectors $\mathbf{y}_2, \ldots, \mathbf{y}_n \in \mathbb{R}^n$. Then $\mathbf{B}_0\left(\mathbf{y}_i - \frac{\mathbf{a}_i\mathbf{y}_i}{\lambda_i x_i}\mathbf{x}\right) = \mathbf{B}_0(\mathbf{y}_i)$, since $\mathbf{B}_0\mathbf{x} = 0$. Moreover, $\mathbf{B}_0\mathbf{y}_i = \left(\mathbf{A} - \frac{1}{x_i}\mathbf{x}\mathbf{a}_i\right)\mathbf{y}_i = \lambda_i\left(\mathbf{y}_i - \frac{\mathbf{a}_i\mathbf{y}_i}{\lambda_i x_i}\mathbf{x}\right) = \mathbf{B}_0\left(\mathbf{y}_i - \frac{\mathbf{a}_i\mathbf{y}_i}{\lambda_i x_i}\mathbf{x}\right)$. So $\mathbf{B}_0$ also has the eigenvalues λ_i of $\mathbf{A}$, provided that the vectors $\mathbf{y}_i - \frac{\mathbf{a}_i\mathbf{y}_i}{\lambda_i x_i}\mathbf{x}$ are nonzero. But this is true, since the $\mathbf{y}_i$ and $\mathbf{x}$ are linearly independent; they form a diagonalizing basis for $\mathbf{A}$.

(f) $\mathbf{A}$ is diagonalizable.

11. (b) and (c) : $\mathbf{A}^2 - 2\mathbf{A} = 3\mathbf{I}$ or $(\mathbf{A} - 2\mathbf{I})\mathbf{A} = 3\mathbf{I}$, thus, $\mathbf{A}^{-1} = \frac{1}{3}(\mathbf{A} - 2\mathbf{I}) = \frac{1}{3}\begin{pmatrix} -1 & 2 \\ 2 & -1 \end{pmatrix}$.

(d) eigenvalues: 3, –1. They are the roots of the characteristic polynomial $\det(\lambda\mathbf{I}_2 - \mathbf{A}) = \det\begin{pmatrix} \lambda-1 & -2 \\ -2 & \lambda-1 \end{pmatrix} = \lambda^2 - 2\lambda - 3 = (\lambda - 3)(\lambda + 1)$ given in part (a) for $\mathbf{A}$.

13. $\mathbf{A}_\mathcal{E} = \begin{pmatrix} 3 & 1 & 0 \\ 0 & 1 & 0 \\ 4 & 2 & 1 \end{pmatrix}$. $\det(\lambda\mathbf{I}_3 - \mathbf{A}_\mathcal{E}) = (\lambda - 1)^2(\lambda - 3)$ with eigenvalues 1, 1, 3. Next, we find the kernel of $\mathbf{A} - \mathbf{I}$:

$$\begin{array}{ccc|c l} \boxed{2} & 1 & 0 & 0 & \\ 0 & 0 & 0 & 0 & \\ \boxed{4} & 2 & 0 & 0 & -2\,row_1 \\ \hline \boxed{2} & 1 & 0 & 0 & \\ 0 & 0 & 0 & 0 & \\ 0 & 0 & 0 & 0 & \end{array}$$

There are two eigenvectors for $\lambda = 1$: $\begin{pmatrix} 0 \\ 0 \\ 1 \end{pmatrix}$ and $\begin{pmatrix} -1 \\ 2 \\ 0 \end{pmatrix}$. For $\lambda = 3$, we find the eigenvector $\begin{pmatrix} 1 \\ 0 \\ 2 \end{pmatrix}$. **A** is diagonalizable, and $\mathbf{A}_{\mathcal{U}} = \mathbf{U}^{-1}\mathbf{A}_{\mathcal{E}}\mathbf{U} = \begin{pmatrix} 1 & 0 & 0 \\ 0 & 1 & 0 \\ 0 & 0 & 3 \end{pmatrix}$ for $\mathbf{U} = \begin{pmatrix} 0 & -1 & 1 \\ 0 & 2 & 0 \\ 1 & 0 & 2 \end{pmatrix}$.

15. For **B**, we compute the characteristic polynomial as $f_B(\lambda) = \lambda^3 - \lambda^2 = \lambda^2(\lambda - 1)$, giving **B** the eigenvalues 1 and 0 (double). Since rank(**B**) = 2, there are not sufficiently many (only one) corresponding linearly independent eigenvectors for $\lambda = 0$. So **B** is not diagonalizable.

17. (a) and (b) : $\mathbf{x} = \begin{pmatrix} 2 \\ 3 \\ 1 \end{pmatrix}$ is an eigenvector for **A**.

(c) and (d) $\mathbf{Ay} = \begin{pmatrix} 9 \\ -5 \\ -5 \end{pmatrix} \neq \lambda \begin{pmatrix} 1 \\ -1 \\ 1 \end{pmatrix}$ for any $\lambda \neq 0$, so **y** is not an eigenvector of **A** and neither is $\mathbf{z} = -\mathbf{y}$.

19. (a) $\mathbf{A}_{\mathcal{U}} = \mathbf{U}^{-1}\mathbf{A}_{\mathcal{E}}\mathbf{U} = \begin{pmatrix} 2 & 0 \\ 0 & 4 \end{pmatrix}$ with $\mathbf{U} = \begin{pmatrix} 1 & 2 \\ -1 & 2 \end{pmatrix}$ and $\mathbf{U}^{-1} = \frac{1}{4}\begin{pmatrix} 2 & -2 \\ 1 & 1 \end{pmatrix}$.

(b) The eigenvalues of **A** are 2 and 4 with eigenvectors $\begin{pmatrix} 1 \\ -1 \end{pmatrix}$ and $\begin{pmatrix} 2 \\ 2 \end{pmatrix}$, respectively.

21. Here $f_A(\lambda) = \lambda^3 - 6\lambda^2 + 6\lambda + 4$. By trial and error, note that $f_A(2) = 0$, so that $\lambda_1 = 2$ is one eigenvalue of **A**. Now use long division:

$$\begin{array}{llllll}
(\lambda^3 & -6\lambda^2 & +6\lambda & +4\ ) & \div (\lambda - 2) = \\
\lambda^3 & \mp 2\lambda^2 & & & \\
\hline
 & -4\lambda^2 & +6\lambda & & \\
 & \mp 4\lambda^2 & \pm 8\lambda & & \\
\hline
 & & -2\lambda & +4 & \\
 & & \mp 2\lambda & \pm 4 & \\
\hline
 & & & 0 &
\end{array}$$

$= \lambda^2 - 4\lambda - 2$. The factor polynomial $\lambda^2 - 4\lambda - 2 = 0$ has the roots $\lambda_{2,3} = 2 \pm \sqrt{6}$. Hence, the eigenvalues of **A** are 2, $2+\sqrt{6}$, $2-\sqrt{6}$, with corresponding eigenvectors $\begin{pmatrix} 1 \\ -2 \\ 1 \end{pmatrix}$, $\begin{pmatrix} -5+2\sqrt{6} \\ -2+\sqrt{6} \\ 1 \end{pmatrix}$, and $\begin{pmatrix} 1 \\ -2+\sqrt{6} \\ -5+2\sqrt{6} \end{pmatrix}$.

$$\mathbf{A}_{\mathcal{U}} = \begin{pmatrix} 2 & 0 & 0 \\ 0 & 2+\sqrt{6} & 0 \\ 0 & 0 & 2-\sqrt{6} \end{pmatrix}.$$

23. If $\mathbf{A}_{nn}$ is singular, then $\ker(\mathbf{A}) = \ker(\mathbf{A} - 0\mathbf{I}_n) = E(0) \neq \{\mathbf{0}\}$, so $\lambda = 0$ is an eigenvalue of **A** and conversely.

25. (a) Work with $\mathbf{B} := \frac{\mathbf{A}}{3} = \begin{pmatrix} 1 & 1 & 0 \\ 0 & 3 & -1 \\ 0 & 0 & -2 \end{pmatrix}$ for smaller numbers to compute with. Its eigenvalues are one-third the eigenvalues of **A**; the eigenvectors are identical. The characteristic polynomial is $f_B(\lambda) = (\lambda - 1)(\lambda + 2)(\lambda - 3)$. Thus, the eigenvalues of **B** are 1, 3, and −2, making those of **A** equal to 3, 9, and –6. The corresponding eigenvectors are $\mathbf{e}_1$, $\mathbf{e}_1 + 2\mathbf{e}_2$, and $\begin{pmatrix} -1 \\ 3 \\ 15 \end{pmatrix}$; **A** is diagonalizable.

(b) eigenvalues are –1 (double) and 2; corresponding eigenvectors are $\mathbf{e}_1$ (only one) for $\lambda = -1$ and $-\mathbf{e}_1 + 3\mathbf{e}_2$ for $\lambda = 2$. For the double eigenvalue $\lambda = -1$, we have $\ker(\mathbf{B} + \mathbf{I})$:

$$\begin{array}{cccc|c|l}
0 & \boxed{1} & 0 & & 0 & \\
0 & \boxed{3} & -1 & & 0 & -3\ row_1 \\
0 & 0 & 0 & & 0 & \\
\hline
0 & \boxed{1} & 0 & & 0 & \\
0 & 0 & \boxed{-1} & & 0 & \\
0 & 0 & 0 & & 0 &
\end{array}$$

There is only one eigenvector $\mathbf{x}_1 = \mathbf{e}_1$ for $\lambda = -1$, hence, **B** is not diagonalizable.

(c) eigenvalues are –1 (double) and 2; corresponding eigenvectors are $\mathbf{e}_1$ and $\mathbf{e}_2$ for −1 and $\begin{pmatrix} -1 \\ 0 \\ 3 \end{pmatrix}$ for 2; so **C** is diagonalizable.

27. We have $\det(\lambda\mathbf{I}_2 - \mathbf{A}) = \det\begin{pmatrix} \lambda & -2 \\ 8 & \lambda \end{pmatrix} = \lambda^2 + 16$. Eigenvalues $\lambda_{1,2} = \pm 4i$. Now look at $\ker(\mathbf{A} - 4i\mathbf{I}_2)$:

$$\begin{array}{cc|c|l}
\boxed{-4i} & 2 & 0 & \\
\boxed{-8} & -4i & 0 & +2i\ row_1 \\
\hline
\boxed{-4i} & 2 & 0 & \\
0 & 0 & 0 &
\end{array}$$

So $\mathbf{x}_1 = \begin{pmatrix} 1 \\ 2i \end{pmatrix} \in \mathbb{C}^2$ is an eigenvector for $\lambda_1 = 4i$, and $\begin{pmatrix} -1 \\ 2i \end{pmatrix}$ is for $\lambda_2 = -4i$.

29. (a) $\mathbf{A}_{\mathcal{U}} = \mathbf{U}^{-1}\mathbf{A}_{\mathcal{E}}\mathbf{U}$.

(b) $\mathbf{A}_{\mathcal{U}} = \mathbf{U}^{-1}\mathbf{A}\mathbf{U} = \begin{pmatrix} 1 & 0 & 0 \\ 1 & 2 & 0 \\ 2 & -1 & 3 \end{pmatrix}$ for $\mathbf{U}$

$= \begin{pmatrix} 1 & -1 & 0 \\ 0 & 1 & 1 \\ -2 & 2 & 1 \end{pmatrix}$ and $\mathbf{U}^{-1} = \begin{pmatrix} -1 & 1 & -1 \\ -2 & 1 & -1 \\ 2 & 0 & 1 \end{pmatrix}$.

(c) The eigenvalues of $\mathbf{A}$ appear on the diagonal of the triangular matrix $\mathbf{A}_{\mathcal{U}}$; they are 1, 2, and 3. Since the eigenvalues of $\mathbf{A}$ are distinct, $\mathbf{A}$ is diagonalizable.

31. $f_A(\lambda) = \lambda^2 - 7\lambda + 6$, with roots 1 and 6.

33. (a), (b) $\mathbf{u}_2$ and $\mathbf{u}_6$ are eigenvectors for $\lambda = 0$; $\mathbf{u}_5$ is for $\lambda = 8$. $\mathbf{C}$ is diagonalizable.

(c) $\mathbf{C}$ is not invertible. ($\mathbf{u}_2 \neq \mathbf{0} \in \ker(\mathbf{C})$.)

35. We must make sure that $\ker(\mathbf{A} - 2\mathbf{I}_3) \neq \{\mathbf{0}\}$:

$$\begin{array}{ccc|c|l}
\boxed{-1} & 1 & a & 0 & \\
\boxed{-1} & -2 & 2 & 0 & -\,row_1 \\
\boxed{2} & 1 & 2 & 0 & +2\,row_1 \\
\hline
\boxed{-1} & -1 & a & 0 & \\
0 & \boxed{-3} & 2-a & 0 & \\
0 & \boxed{3} & 2+2a & 0 & +\,row_2 \\
\hline
\boxed{-1} & -1 & a & 0 & \\
0 & \boxed{-3} & 2-a & 0 & \\
0 & 0 & 4+a & 0 &
\end{array}$$

Thus, $\mathbf{A} - 2\mathbf{I}$ is singular precisely when $a = -4$, in which case 2 is an eigenvalue of $\mathbf{A}$.

37. Observe that $\mathbf{A}^{-1}(\mathbf{AB})\mathbf{A} = \mathbf{BA}$. Thus, the two nonsingular matrices $\mathbf{AB}$ and $\mathbf{BA}$ have the same eigenvalues.

39. Both $\mathbf{A}$ and $\mathbf{B}$ have the same characteristic polynomials $f_A(\lambda) = \lambda^2 - 6\lambda + 7 = f_{\mathbf{B}}(\lambda)$ with two distinct roots $\lambda_{1,2} = 3 \pm \sqrt{2}$. Hence, they are both similar to the same diagonal matrix, and thus, $\mathbf{A}$ and $\mathbf{B}$ are themselves similar.

41. (a) $(\mathbf{I} - \mathbf{X}^{-1}\mathbf{AX})^{-1} + (\mathbf{I} - (\mathbf{X}^{-1}\mathbf{AX})^{-1})^{-1} =$ $(\mathbf{X}^{-1}(\mathbf{I}-\mathbf{A})\mathbf{X})^{-1} + (\mathbf{X}^{-1}(\mathbf{I}-\mathbf{A}^{-1})\mathbf{X})^{-1} =$ $\mathbf{X}^{-1}(\mathbf{I}-\mathbf{A})^{-1}\mathbf{X} + \mathbf{X}^{-1}(\mathbf{I}-\mathbf{A}^{-1})^{-1}\mathbf{X} =$ $\mathbf{X}^{-1}((\mathbf{I}-\mathbf{A})^{-1} + (\mathbf{I}-\mathbf{A}^{-1})^{-1})\mathbf{X} = \mathbf{X}^{-1}\mathbf{I}\mathbf{X} =$ $\mathbf{I}$ if $(\mathbf{I}-\mathbf{A})^{-1} + (\mathbf{I}-\mathbf{A}^{-1})^{-1} =$ $\mathbf{I}$.

(b) Let $\mathbf{X}^{-1}\mathbf{AX} = \mathbf{D} = \text{diag}(d_i)$. Using part (a) and assuming that all relevant inverses exist, we have $\mathbf{W} := (\mathbf{I}-\mathbf{D})^{-1} + (\mathbf{I}-\mathbf{D}^{-1})^{-1} = \text{diag}\left(\frac{1}{1-d_i}\right) + \text{diag}\left(\left(1 - \frac{1}{d_i}\right)^{-1}\right)$. Thus, each diagonal entry of the matrix $\mathbf{W}$ is equal to $\frac{1}{1-d_i} + \frac{1}{1-\frac{1}{d_i}} = \frac{1}{1-d_i} + \frac{d_i}{d_i-1} = \frac{1-d_i}{1-d_i} = 1$. That is, $\mathbf{W} = \mathbf{I}$, as claimed.

(c) Here we have $\mathbf{I} - \mathbf{A} = \begin{pmatrix} -1 & 1 \\ 0 & -1 \end{pmatrix}$; $(\mathbf{I}-\mathbf{A})^{-1} = \begin{pmatrix} -1 & -1 \\ 0 & -1 \end{pmatrix}$; $\mathbf{A}^{-1} = \begin{pmatrix} 1/2 & -1/4 \\ 0 & 1/2 \end{pmatrix}$; $\mathbf{I} - \mathbf{A}^{-1} = \begin{pmatrix} 1/2 & 1/4 \\ 0 & 1/2 \end{pmatrix}$; and $(\mathbf{I}-\mathbf{A}^{-1})^{-1} = \begin{pmatrix} 2 & -1 \\ 0 & 2 \end{pmatrix}$. Thus, $(\mathbf{I}-\mathbf{A})^{-1} + (\mathbf{I}-\mathbf{A}^{-1})^{-1} =$ $\begin{pmatrix} -1 & -1 \\ 0 & -1 \end{pmatrix} + \begin{pmatrix} 2 & -1 \\ 0 & 2 \end{pmatrix} = \begin{pmatrix} 1 & -2 \\ 0 & 1 \end{pmatrix} \neq \mathbf{I}_2$.

T 9. (b) $\mathbf{B}$ has the characteristic polynomial $f_B(x) = (x-1)^2(x+1)$. Look at the eigenspace $E(1) = \ker(\mathbf{B} - \mathbf{I})$ for $\mathbf{B}$:

$$\begin{array}{ccc|l}
\multicolumn{3}{c}{\mathbf{B}-\mathbf{I}} & \\
\hline
-4 & -1 & -6 & \\
-4 & -2 & -8 & \\
2 & 1 & 4 & +\tfrac{1}{2}\,row_2 \\
\hline
\boxed{-4} & -1 & -6 & \\
\boxed{-4} & -2 & -8 & -\,row_1 \\
0 & 0 & 0 & \\
\hline
\boxed{-4} & -1 & -6 & \\
0 & \boxed{-1} & -2 & \\
0 & 0 & 0 &
\end{array}$$

Clearly, $E(1)$ is only one dimensional. (Note the unusual first elimination step.)

9.2.P p. 293

1. (a) The vector iteration matrix is $\begin{pmatrix} | & | & | & | & | \\ \mathbf{y} & \mathbf{Ay} & \mathbf{A}^2\mathbf{y} & \mathbf{A}^3\mathbf{y} & \mathbf{A}^4\mathbf{y} \\ | & | & | & | & | \end{pmatrix}$

$= \begin{pmatrix} 1 & 5 & 24 & 111 & 497 \\ -1 & -4 & -15 & -53 & -171 \\ 1 & 4 & 17 & 70 & 314 \\ 0 & -1 & 0 & -17 & -2 \end{pmatrix}$ with the associated RREF $\begin{pmatrix} 1 & 0 & 0 & 0 & 373 \\ 0 & 1 & 0 & 0 & -151 \\ 0 & 0 & 1 & 0 & -5 \\ 0 & 0 & 0 & 1 & 9 \end{pmatrix}$.

Thus, the vanishing polynomial of $\mathbf{y}$ and $\mathbf{A}$ is $p_y(x) = x^4 - 9x^3 + 5x^2 + 151x - 373$, which is also the minimal and characteristic polynomial of $\mathbf{A}$ because $\mathbf{A}$ is 4×4 and p_y has degree 4.

(b) $f_A(\lambda) = \det(\lambda \mathbf{I} - \mathbf{A}) = \lambda^4 - 9\lambda^3 + 5\lambda^2 + 151\lambda - 373$.

3. If $\mathbf{Bx} = \lambda\mathbf{x}$ for $\mathbf{x} \neq \mathbf{0}$, then $\mathbf{AXx} = \mathbf{XBx} = \lambda\mathbf{Xx}$, and $\mathbf{Xx} \neq \mathbf{0}$ is an eigenvector for the eigenvalue λ of $\mathbf{A}$. If $\mathbf{Xx} = \mathbf{0}$ for $\mathbf{X}_{mn}$ with rows r_i of rank m and $\mathbf{x} \neq \mathbf{0} \in \mathbb{R}^m$, then $\mathbf{x} \perp \text{span}\{\mathbf{r}_1, \dots, \mathbf{r}_n\} = \mathbb{R}^m$ by the assumption that rank $\mathbf{X} = m$. Hence, $\mathbf{x} = \mathbf{0}$, a contradiction.

5. (a) If $\lambda = 0$ is an eigenvalue of $\mathbf{A}_{nn}$, then $\mathbf{A}$ is singular, since $\ker(\mathbf{A} - 0\mathbf{I}_n) = \ker(\mathbf{A}) = E(0) \neq \{0\}$. Hence, all eigenvalues of a nonsingular matrix $\mathbf{A}$ are nonzero.

If $\mathbf{Ax} = \lambda\mathbf{x} \neq \mathbf{0}$ for a nonsingular $\mathbf{A}$, we have $\mathbf{x} = \lambda\mathbf{A}^{-1}\mathbf{x}$, or $\mathbf{A}^{-1}\mathbf{x} = \frac{1}{\lambda}\mathbf{x}$.

(b) If $\mathbf{Bx} = \beta\mathbf{x}$, then $(\mathbf{B} + \alpha\mathbf{I})\mathbf{x} = \mathbf{Bx} + \alpha\mathbf{x} = (\beta + \alpha)\mathbf{x}$. That is, we have the same eigenvector $\mathbf{x}$ and a shifted eigenvalue $\beta + \alpha$.

7. (a) Yes, columns 1 and 3 are proportional and hence linearly dependent.

(b) The eigenvalues are 0, 3, 4, and 5, since $\sum \lambda_i = \text{trace}(\mathbf{A}) = 12$ and we know three eigenvalues—0, 3, and 5—of $\mathbf{A}$.

(c) Yes, $\mathbf{A}$ has four distinct eigenvalues. There are $24 = 4!$ different diagonal matrix representations $\mathbf{D}$; just shuffle the diagonal entries (i.e., the eigenvalues) at will.

(d) $\mathbf{x} = \begin{pmatrix} 77 \\ 0 \\ 11 \\ 0 \end{pmatrix}$, for example, for $\lambda = 0$.

9. (a) $\mathbf{X} = \begin{pmatrix} -4 & -1 \\ 1 & 2 \end{pmatrix}$. The columns of $\mathbf{X}$ are eigenvectors of $\mathbf{A}$ for the diagonal entry eigenvalues 4 and -3. To find those, look at the kernel of $\mathbf{A} - 4\mathbf{I}$ and $\mathbf{A} + 3\mathbf{I}$.

(b) $\mathbf{Y} = \begin{pmatrix} 4 & -3 \\ -1 & 6 \end{pmatrix}$.

(c) $\mathbf{Y} = \mathbf{X} \cdot \mathbf{Z}$ for any nonsingular diagonal matrix $\mathbf{Z}_{2,2}$, since the eigenvector basis of $\mathbf{A}$ (i.e., the eigenvectors of $\mathbf{A}$) are determined only up to nonzero scalar multiples.

11. (a) $\mathbf{A}$ is 6×6.

(b) Eigenvalues: 3, with multiplicity 3; 1, with multiplicity 1; –3, with multiplicity 2.

(c) Not necessarily.

(d) Eigenvalues: 9, with multiplicity 5; 1, with multiplicity 1.

(e) Yes. (f) $f_{A^2}(x) = (x - 9)^5(x - 1)$.

13. (b) Zero is an $n-1$-fold eigenvalue of $\mathbf{B}$; n is a single eigenvalue of $\mathbf{B}$ for the eigenvector $\mathbf{x} = (11 \dots 1)^T$.

(c) $\det \mathbf{C} = \det(\mathbf{B} - \mathbf{I}_n) = \prod \lambda_i = (-1)^{n-1}(n-1)$ for the $n-1$-fold eigenvalue $\lambda_i = -1$, and the single eigenvalue $n-1$ of $\mathbf{C} = \mathbf{B} - \mathbf{I}_n$.

15. (a) and (b) Both $\mathbf{L}$ and $\mathbf{R}$ have the three distinct eigenvalues 12, –17, and –3 that appear on their diagonals. So $\mathbf{L}$ and $\mathbf{R}$ are similar to the same diagonal matrix $\mathbf{D} = \mathbf{U}^{-1}\mathbf{LU} = \mathbf{V}^{-1}\mathbf{RV}$; hence, they are similar to each other.

17. If, for $\mathbf{A}$, $E(\lambda) = \text{span}\{\mathbf{x}_1, \dots, \mathbf{x}_k\}$ with linearly independent eigenvectors $\mathbf{x}_i$, then $(\mathbf{A} + \alpha\mathbf{I})\mathbf{x}_i = \mathbf{Ax}_i + \alpha\mathbf{x}_i = (\lambda + \alpha)\mathbf{x}_i$ for each i. Thus, the geometric multiplicity $\dim E_{\mathbf{A}+\alpha\mathbf{I}}(\lambda + \alpha) = k_\lambda$ as well.

For $\mathbf{A}$, the algebraic multiplicity n_λ of λ is determined by the maximal exponent of $(\lambda - t)$ occurring in the characteristic polynomial $f_A(t) = \det(\mathbf{A} - t\mathbf{I}) = \det((\mathbf{A} + \alpha\mathbf{I}) - (t + \alpha)\mathbf{I}) = f_{A+\alpha I}(t + \alpha)$. Hence, for $\mathbf{A} + \alpha\mathbf{I}$, the algebraic multiplicity of its eigenvalue $t + \alpha$, being determined by the maximal exponent of $(\lambda - (t + \alpha))$ in $f_{A+\alpha I}(t + \alpha)$, is identical to n_λ for λ and $\mathbf{A}$ as well.

9.3.P p. 308

1. $\mathbf{Ax} = \begin{pmatrix} a_{11} & \dots & a_{1n} \\ \vdots & & \vdots \\ a_{n1} & \dots & a_{nn} \end{pmatrix} \begin{pmatrix} x_1 \\ \vdots \\ x_n \end{pmatrix} = \begin{pmatrix} \sum_i a_{1i}x_i \\ \vdots \\ \sum_i a_{ni}x_i \end{pmatrix}$ with component sum $\sum_j \sum_i a_{ji}x_i = \sum_i \left(\sum_j a_{ji}\right) x_i = \sum_i x_i = 1$.

3. The transition matrix is $\mathbf{A} = \begin{pmatrix} 0.8 & 0.4 \\ 0.2 & 0.6 \end{pmatrix}$. $\mathbf{A}^4 \begin{pmatrix} 100 \\ 140 \end{pmatrix} \approx \begin{pmatrix} 158 \\ 82 \end{pmatrix}$, and the "final party attendance" will be 160 guests at the first party with 80 at the second one.

5. Terminal distribution : $\mathbf{x}_\infty = \begin{pmatrix} 3/8 \\ 5/8 \end{pmatrix}$.

7. (a) All eigenvalues of $\mathbf{A}$ and $\mathbf{B}$ are zero, with a one-dimensional eigenspace: Neither $\mathbf{A}$ nor $\mathbf{B}$ is diagonalizable.

(b) $\mathbf{B} = \mathbf{O}_3$. (c) $\mathbf{C} = \mathbf{I}_3$.

9. (a) Eigenvalues of $\mathbf{A}$ are 1 and 6; eigenvectors are $\begin{pmatrix} 1 \\ 2 \end{pmatrix}$ and $\begin{pmatrix} 1 \\ -3 \end{pmatrix}$. $\mathbf{x}(t) = \begin{pmatrix} 1 & 1 \\ 2 & -3 \end{pmatrix} \begin{pmatrix} k_1e^t \\ k_2e^{6t} \end{pmatrix}$.

(b) Eigenvalues are $1 \pm \sqrt{15}i$; eigenvectors are $\begin{pmatrix} 0.433 \\ 0 \end{pmatrix} \pm i \begin{pmatrix} -0.111 \\ 0.894 \end{pmatrix}$. Real solution according to (9.10): $\mathbf{x}(t) = ke^t$

$$\left(\cos(\sqrt{15}t) \begin{pmatrix} 0.433 \\ 0 \end{pmatrix} - \sin(\sqrt{15}t) \begin{pmatrix} -0.111 \\ 0.894 \end{pmatrix}\right).$$

11. Differentiate $\mathbf{x}(t)$ and compare with $\mathbf{A}\mathbf{x}(t)$.

13. (a) "Exploding star":

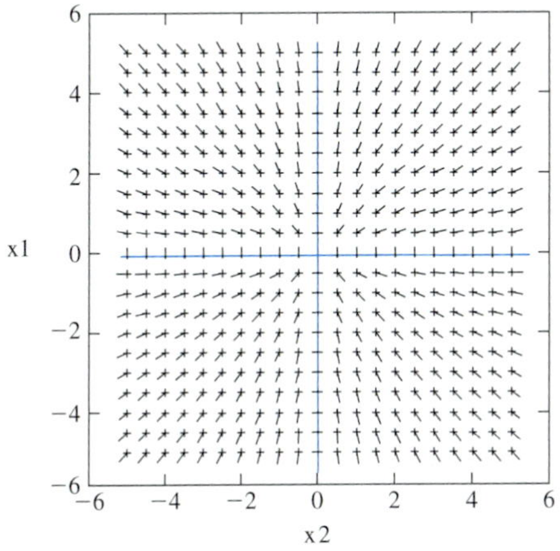

(b) "Imploding star":

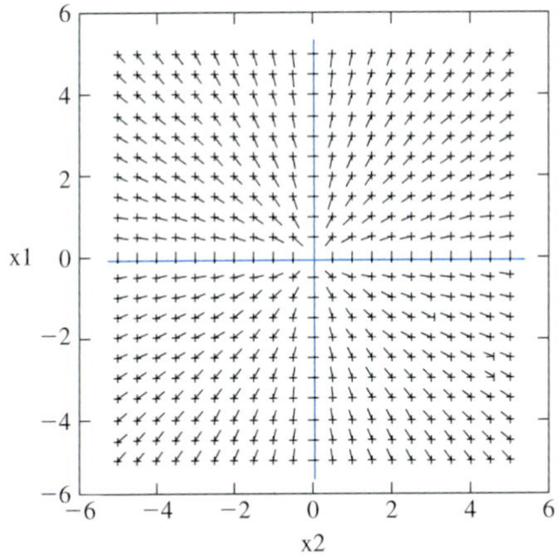

(c)

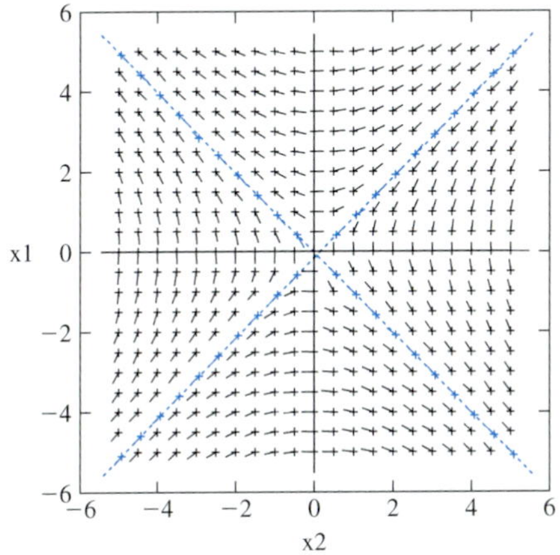

(d) "Just circling":

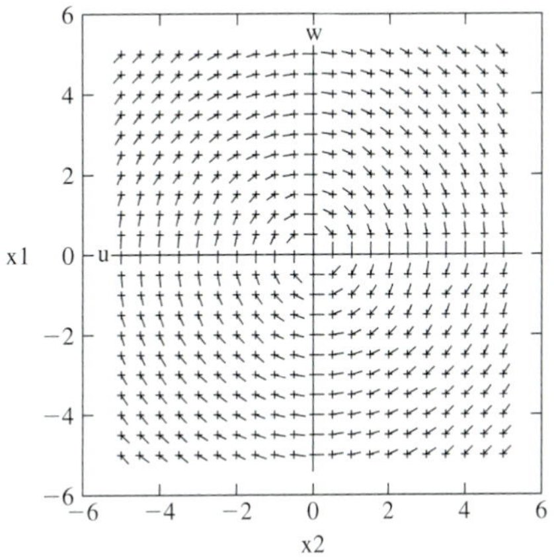

(e) "No movement at all":

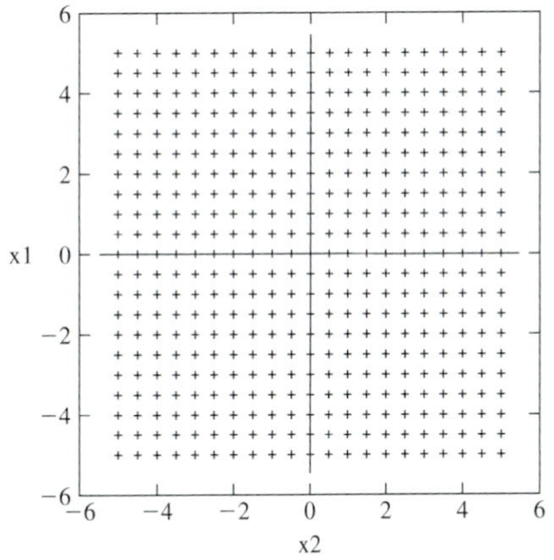

(f) "Fleeing the real axis":

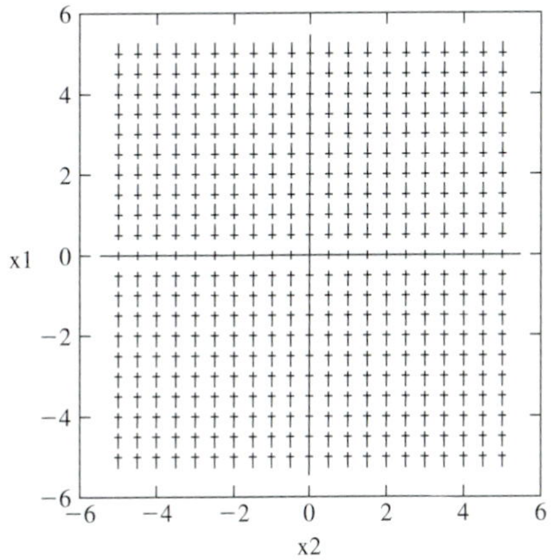

15. (a) Expanding spiral, eigenvalues have positive real parts. Eigenvectors are approximately equal to $\begin{pmatrix} 6 \\ -10 \end{pmatrix} \pm i \begin{pmatrix} 10 \\ 6 \end{pmatrix}$ from the MATLAB plot and the direction of the spiral.

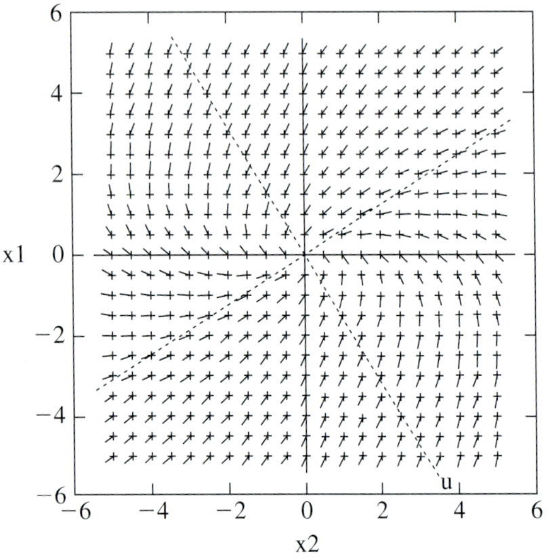

(b) Two real eigenvectors approximately equal to $\begin{pmatrix} 3 \\ 10 \end{pmatrix}; \begin{pmatrix} -10 \\ 3 \end{pmatrix}$ for two negative eigenvalues due to the tightening spiral.

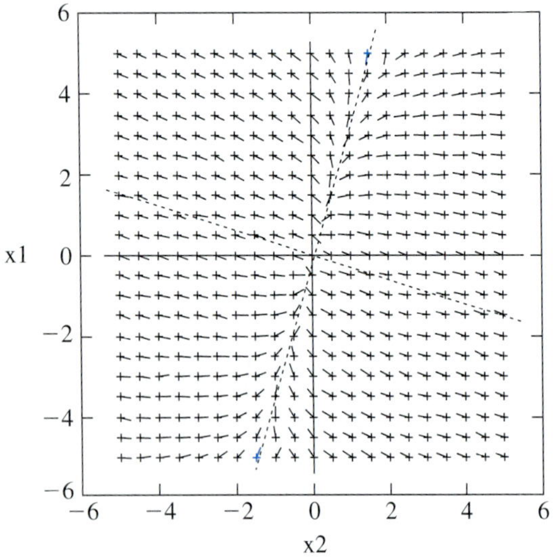

(c) Expanding spiral: two complex eigenvalues with positive real parts. Eigenvectors are approximately $-\mathbf{e}_1 \pm i\mathbf{e}_2$ from the plot.

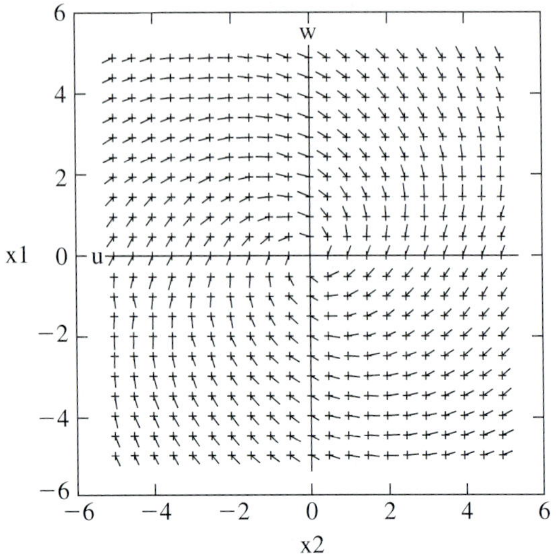

(d) Eigenvectors approximately $\begin{pmatrix} 3 \\ 2 \end{pmatrix}; \begin{pmatrix} -2 \\ 9 \end{pmatrix}$; one set of eigenvalue directions points to the origin, the other points away: We have both a positive and a negative real eigenvalue.

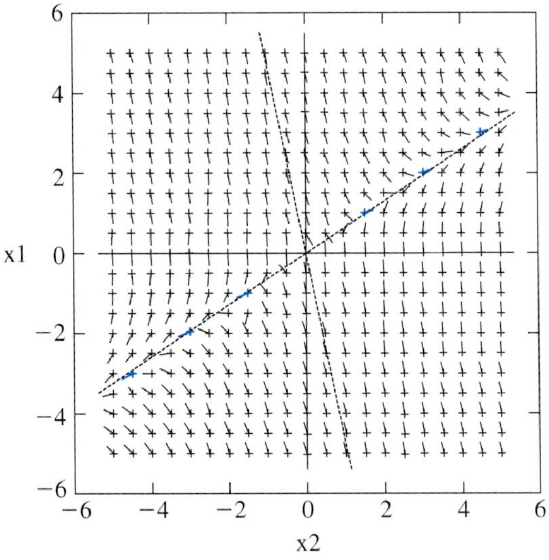

9.R p. 310

3. $\text{trace}(\mathbf{A}_{3,3}) = 0 = \sum_i \lambda_i = 0$ is given. We know that $\mathbf{A}$ has the eigenvalue 0 and a positive one; hence, $\mathbf{A}$ must also have a negative one, so that their sum can equal zero.

5. (a) Take $\mathbf{A} = \begin{pmatrix} 1 & 0 & 0 \\ 0 & 2 & 0 \\ 0 & 0 & 3 \end{pmatrix}$, for example.

(b) Take $\mathbf{A} = \mathbf{I}_3$, for example, with the triple eigenvalue $\lambda = 1$.

(c) $\mathbf{A} = \begin{pmatrix} 1 & 0 & 0 \\ 0 & 1 & 0 \\ 0 & 0 & 2 \end{pmatrix}$ has the eigenvectors $\mathbf{e}_1, \mathbf{e}_2$, and $\mathbf{e}_1 + \mathbf{e}_2$ for the eigenvalue $\lambda = 1$. (Note that these three vectors are not linearly independent.) Likewise, $\mathbf{A}$ has the eigenvectors $\mathbf{e}_1, \mathbf{e}_2$, and $\mathbf{e}_1 - \mathbf{e}_2$ for the eigenvalue $\lambda = 1$.

7. (a) False; $n = 5$. (b) True. (c) False; there are five distinct eigenvalues.

(d) False; $\mathbf{A}$ has the complex eigenvalues $\lambda = \pm 4i$.
(e) True. (f) 1.

9. (a) and (b): The polynomial in both parts is $p(x) = x^n + a_{n-1}x^{n-1} + \cdots + a_1 x + a_0$.

11. Simply verify that $\mathbf{A}\mathbf{u}_i = \lambda_i \mathbf{u}_i$ for $i = 1, 2$.

13. $\mathbf{A}$, $\mathbf{B}$, and $\mathbf{F}$ are diagonalizable; $\mathbf{C}$ is not.

15. $\mathbf{x}'(t) = \mathbf{A}\mathbf{x}(t)$ defines a vector field in $\mathbb{R}^2$. It defines a time dependent continuous trajectory curve for every starting point $\mathbf{x}(0)$.

$\mathbf{x}^{(m)} = \mathbf{A}^m \mathbf{x}^{(0)}$ describes a discrete vector iteration of a starting vector $\mathbf{x}^{(0)} \in \mathbb{R}^2$ under the repeated action by $\mathbf{A}$. Both processes describe motions whose terminal directions are governed by the eigenvectors of $\mathbf{A}$ and depend on the starting point.

17. (a) $\mathbf{A}\begin{pmatrix} 1 \\ 1 \\ -1 \end{pmatrix} = \begin{pmatrix} 0 \\ 0 \\ 0 \end{pmatrix}$ so that $\mathbf{A}$ has the eigenvalue 0.

(b) $\mathbf{B} = \mathbf{A} + 2\mathbf{I}$ then has the eigenvalue $0 + 2 = 2$, while $\mathbf{C} = \mathbf{A} + \mathbf{I}$ has the eigenvalue $0 + 1 = 1$.

(c) The vector $\begin{pmatrix} 1 \\ 1 \\ -1 \end{pmatrix}$ from part (a) is an eigenvector of all three matrices $\mathbf{A}$, $\mathbf{B}$, and $\mathbf{C}$.

(d) $\mathbf{AB} = \mathbf{A}(\mathbf{A} + 2\mathbf{I}) = \mathbf{A}^2 + 2\mathbf{A} = (\mathbf{A} + 2\mathbf{I})\mathbf{A} = \mathbf{BA}$ and likewise $\mathbf{BC} = \mathbf{CB}$.

19. (a) The eigenvalues of both $\mathbf{A}$ and $\mathbf{B}$ are 1, four fold.

(b) Neither $\mathbf{A}$ nor $\mathbf{B}$ is diagonalizable, since the only 4×4 diagonal matrix with the eigenvalues 1, 1, 1, and 1 is $\mathbf{I}_4$.

(c) $\mathbf{A}$ and $\mathbf{B}$ are not similar, since $\dim(E(1)) = 1$ for $\mathbf{A}$, while $\dim(E(1)) = 2$ for $\mathbf{B}$.

21. (a) eigenvalues: 5, 5, 5, – 10.

(b) There is a three-dimensional eigenspace for the repeated eigenvalue $\lambda = 5$: $E(5) = \operatorname{span}\left\{\begin{pmatrix} 1 \\ 0 \\ 0 \\ 2 \end{pmatrix}, \begin{pmatrix} -3 \\ 0 \\ 2 \\ 0 \end{pmatrix}, \begin{pmatrix} -1 \\ 2 \\ 0 \\ 0 \end{pmatrix}\right\}$. Also, $E(10) = \operatorname{span}\left\{\begin{pmatrix} 2 \\ 1 \\ 3 \\ -1 \end{pmatrix}\right\}$.

(c) $\mathbf{A}$ is diagonalizable; it has four linearly independent eigenvectors.

Chapter 10

10.1.P p. 329

1. (a) $(1, -1/2, -1/2)$, for example.

(b) Two, since $\dim \ker(111) = 2$.

3. (a) Not possible. (b) Any basis of $\mathbb{R}^3$ will do.

5. (a) $base = \|\mathbf{v}_1\| = \sqrt{10} \doteq 3.16$.

(b) $height = \|\mathbf{u}_2\|$ for $\mathbf{u}_1 = \mathbf{v}_1$ and $\mathbf{u}_2 = \mathbf{v}_2 - \frac{\mathbf{v}_2 \cdot \mathbf{u}_1}{\mathbf{v}_1 \cdot \mathbf{u}_1}\mathbf{u}_1 = \begin{pmatrix} -8.7 \\ 2.9 \end{pmatrix}$; $height = \|\mathbf{u}_2\| = \sqrt{84.1} \doteq 9.17$.

(c) $Area = \sqrt{10}\sqrt{84.1}/2 = \sqrt{841}/2 = 14.5$.

7. By "integer Gram–Schmidt", $\mathbf{v}_1 = \begin{pmatrix} 1 \\ -1 \\ 2 \end{pmatrix} \perp \mathbf{v}_2 = \mathbf{v}_1 \cdot \mathbf{v}_1 \mathbf{u}_2 - \mathbf{u}_2 \cdot \mathbf{v}_1 \mathbf{v}_1 = 6\begin{pmatrix} 0 \\ 1 \\ 0 \end{pmatrix} + \begin{pmatrix} 1 \\ -1 \\ 2 \end{pmatrix} = \begin{pmatrix} 1 \\ 5 \\ 2 \end{pmatrix}$ spans the column space because $\mathbf{u}_3 = 3\mathbf{u}_1 + 4\mathbf{u}_2$ is linearly dependent on the first two columns $\mathbf{u}_1, \mathbf{u}_2$ in $\mathbf{U}$. Now normalize to $\mathbf{w}_1 := \frac{\mathbf{v}_1}{\sqrt{6}}$ and $\mathbf{w}_2 := \frac{\mathbf{v}_2}{\sqrt{30}}$.

11. First level of Gram–Schmidt:
$\mathbf{v}_1 = \mathbf{u}_1$, $\mathbf{v}_2 = \mathbf{v}_1 \cdot \mathbf{v}_1 \mathbf{u}_2 - \mathbf{u}_2 \cdot \mathbf{v}_1 \mathbf{v}_1 = 50\mathbf{u}_2 \equiv \mathbf{u}_2$, and $\mathbf{v}_3 = \mathbf{v}_1 \cdot \mathbf{v}_1 \mathbf{u}_3 - \mathbf{u}_3 \cdot \mathbf{v}_1 \mathbf{v}_1 = 50\mathbf{u}_3 \equiv \mathbf{u}_3$ by using integer scaling.

Second level:

$$\mathbf{v}_3 = \mathbf{v}_2 \cdot \mathbf{v}_2\mathbf{v}_3 - \mathbf{v}_3 \cdot \mathbf{v}_2\mathbf{v}_2 = 90\mathbf{v}_3 - 32\mathbf{v}_2 = \begin{pmatrix} -166 \\ -354 \\ -302 \\ -142 \end{pmatrix}$$

$\equiv \begin{pmatrix} 83 \\ 177 \\ 151 \\ 71 \end{pmatrix}$. Integer orthogonal basis:

$\left\{ \mathbf{v}_1, \mathbf{v}_2, \begin{pmatrix} 83 \\ 177 \\ 151 \\ 71 \end{pmatrix} \right\}$. Now normalize.

13. $\alpha \mathbf{v}_1 = -\frac{10}{17}\mathbf{v}_1$ for $\alpha = \frac{\mathbf{v}_2 \cdot \mathbf{v}_1}{\mathbf{v}_1 \cdot \mathbf{v}_1} = \frac{-10}{17}$.

15. (a) $\alpha = \frac{13}{17} = \frac{\mathbf{u}_2 \cdot \mathbf{u}_1}{\mathbf{u}_1 \cdot \mathbf{u}_1}$.

(b) $\mathbf{u} = \begin{pmatrix} 4 \\ 1 \end{pmatrix}$, for example.

(c) $d = \|\mathbf{u}_2 - \alpha \mathbf{u}_1\| = \left\| \frac{1}{17} \begin{pmatrix} 4 \\ 1 \end{pmatrix} \right\| = 1.$

(d) Let $\mathbf{u} := \frac{\mathbf{u}_1}{\|\mathbf{u}_1\|}$. Then $\mathbf{P}_\mathcal{U} = \mathbf{u}\mathbf{u}^T = \dfrac{\mathbf{u}_1\mathbf{u}_1^T}{\mathbf{u}_1 \cdot \mathbf{u}_1} =$ $\frac{1}{17}\begin{pmatrix} 1 \\ -4 \end{pmatrix} \begin{pmatrix} 1 & -4 \end{pmatrix} = \frac{1}{17}\begin{pmatrix} 1 & -4 \\ -4 & 16 \end{pmatrix}$.

(e) $\mathbf{P}_\mathcal{U}(\mathbf{u}_2) = \frac{1}{17}\begin{pmatrix} 1 & -4 \\ -4 & 16 \end{pmatrix}\begin{pmatrix} 1 \\ -3 \end{pmatrix} = \frac{13}{17}\mathbf{u}_1$, as in part (a).

17. (a) Any subspace of $\mathbb{R}^n$ has at most dimension n.

(b) $\mathbf{a} = \sum_i \alpha_i \mathbf{u}_i \in U = \text{span}\{\mathbf{u}_i\}$.

(c) For any $\mathbf{u}_j$, we have $(\mathbf{x} - \mathbf{a}) \cdot \mathbf{u}_j = \mathbf{x} \cdot \mathbf{u}_j - \sum_i \left((\mathbf{x}^T\mathbf{u}_i)\mathbf{u}_i \cdot \mathbf{u}_j \right) = \mathbf{x} \cdot \mathbf{u}_j - \mathbf{x} \cdot \mathbf{u}_j = 0$ by orthogonality of the $\mathbf{u}_i$.

(d) From part (b),

$$\mathbf{a} = \begin{pmatrix} 1 & 1 & -2 \\ 1 & 1 & 0 \\ -1 & 2 & 1 \\ -1 & 0 & -3 \end{pmatrix} \begin{pmatrix} \mathbf{x}^T\mathbf{u}_1 \\ \mathbf{x}^T\mathbf{u}_2 \\ \mathbf{x}^T\mathbf{u}_3 \end{pmatrix} = \begin{pmatrix} 1 & 1 & -2 \\ 1 & 1 & 0 \\ -1 & 2 & 1 \\ -1 & 0 & -3 \end{pmatrix} \begin{pmatrix} -7 \\ 0 \\ 1 \end{pmatrix} = \begin{pmatrix} -9 \\ -7 \\ 8 \\ 4 \end{pmatrix},$$

since the given spanning vectors are mutually orthogonal.

(e) $\|\mathbf{x} - \mathbf{a}\| = \left\| \begin{pmatrix} 8 \\ 4 \\ -6 \\ -3 \end{pmatrix} \right\| = \sqrt{125}.$

19. (a) $\mathbf{u}_1$, $\begin{pmatrix} 3 \\ 0 \\ 1 \end{pmatrix}$, $\begin{pmatrix} 2 \\ -1 \\ 0 \end{pmatrix}$, for example.

(b) Take $\mathbf{v}_1$, $1/\sqrt{10}\begin{pmatrix} 3 \\ 0 \\ 1 \end{pmatrix}$, and $1/\sqrt{5}\begin{pmatrix} 2 \\ -1 \\ 0 \end{pmatrix}$, for example.

21. If $\mathbf{U}^T\mathbf{U} = \mathbf{I}_n$ and $\mathbf{V}^T\mathbf{V} = \mathbf{I}_n$, then $(\mathbf{U}\mathbf{V})^T\mathbf{U}\mathbf{V} = \mathbf{V}^T\mathbf{U}^T\mathbf{U}\mathbf{V} = \mathbf{I}_n$.

23. Since $0 \le \|\mathbf{w}\|^2 = \mathbf{w} \cdot \mathbf{w}$ for all vectors $\mathbf{w} \in \mathbb{R}^n$, it follows that for $\mathbf{w} := \mathbf{x} - \frac{\mathbf{x}\cdot\mathbf{y}}{\mathbf{y}\cdot\mathbf{y}}\mathbf{y}$ and $\mathbf{y} \ne \mathbf{0}$,

$$0 \le \left(\mathbf{x} - \frac{\mathbf{x}\cdot\mathbf{y}}{\mathbf{y}\cdot\mathbf{y}}\,\mathbf{y}\right) \cdot \left(\mathbf{x} - \frac{\mathbf{x}\cdot\mathbf{y}}{\mathbf{y}\cdot\mathbf{y}}\,\mathbf{y}\right) = \mathbf{x} \cdot \mathbf{x} - 2\frac{(\mathbf{x}\cdot\mathbf{y})^2}{(\mathbf{y}\cdot\mathbf{y})} +$$

$\frac{(\mathbf{x}\cdot\mathbf{y})^2}{(\mathbf{y}\cdot\mathbf{y})^2}\,\mathbf{y} \cdot \mathbf{y} = \mathbf{x} \cdot \mathbf{x} - \frac{(\mathbf{x}\cdot\mathbf{y})^2}{(\mathbf{y}\cdot\mathbf{y})}$ or $(\mathbf{x} \cdot \mathbf{y})^2 \le \mathbf{x} \cdot \mathbf{x}\ \mathbf{y} \cdot \mathbf{y}$. Thus, $|\mathbf{x} \cdot \mathbf{y}| \le \|\mathbf{x}\|\|\mathbf{y}\|$ if $\mathbf{y} \ne \mathbf{0}$. If $\mathbf{y} = \mathbf{0}$, then the Cauchy–Schwarz inequality simplifies to $0 \le \|\mathbf{x}\|0$, which is always true.

25. (a) $\mathbf{a} = \frac{1}{7}\begin{pmatrix} 2 \\ 4 \\ -2 \\ 2 \\ 2 \end{pmatrix} \in U$ is closest to $\mathbf{b}$, with distance $\|\mathbf{a} - \mathbf{b}\|$ equal to $\frac{1}{7}\sqrt{70}$.

(b) $\mathbf{a} = \frac{2}{5}\begin{pmatrix} 3 \\ 4 \end{pmatrix} \in U$ is closest to $\mathbf{b}$, with the distance to $\mathbf{b}$ equal to 4.

27. (a) If $\mathbf{Y}_{nn}\mathbf{X}_{nn} = \mathbf{I}_n$, then each $\mathbf{y}_j\mathbf{x}_i = 0$ for $i \ne j$, where $\mathbf{y}_j$ denotes the jth row of $\mathbf{Y}$ and $\mathbf{x}_i$ the ith column of $\mathbf{X}$.

(b) Use the identity $\mathbf{XY} = \mathbf{I}_n$ here.

29. (a) $\mathbf{u} \cdot \mathbf{v} = 0$, $\|\mathbf{u}\| = \sqrt{3}$, $\|\mathbf{v}\| = \sqrt{2}$.

(b) ONB $\left\{ \frac{\mathbf{u}}{\sqrt{3}}, \frac{\mathbf{v}}{\sqrt{2}}, \frac{1}{\sqrt{6}}\begin{pmatrix} 1 \\ -2 \\ 1 \end{pmatrix} \right\}$.

31. (a), (b) $\mathbf{v}_2 = 21\mathbf{u}_2 + 7\mathbf{v}_1 = \begin{pmatrix} 70 \\ 7 \\ -14 \end{pmatrix} \perp \mathbf{v}_1$.

(c) $\mathbf{v}_2 \cdot \mathbf{v}_1 = (\mathbf{v}_1 \cdot \mathbf{v}_1\mathbf{u}_2 - \mathbf{u}_2 \cdot \mathbf{v}_1\mathbf{v}_1) \cdot \mathbf{v}_1 = \mathbf{v}_1 \cdot \mathbf{v}_1\mathbf{u}_2 \cdot \mathbf{v}_1 - \mathbf{u}_2 \cdot \mathbf{v}_1\mathbf{v}_1 \cdot \mathbf{v}_1 = 0$ for any vector $\mathbf{u}_2 \in \mathbb{R}^2$.

33. (a) ONB: $\left\{ \mathbf{e}_2, \mathbf{e}_3, \frac{1}{\sqrt{2}}(-\mathbf{e}_1 + \mathbf{e}_4) \right\}$.

(b) $\mathbf{P}_v = \begin{pmatrix} 0 & 0 & -1/\sqrt{2} \\ 1 & 0 & 0 \\ 0 & 1 & 0 \\ 0 & 0 & 1/\sqrt{2} \end{pmatrix}$.

$\cdot \begin{pmatrix} 0 & 1 & 0 & 0 \\ 0 & 0 & 1 & 0 \\ -1/\sqrt{2} & 0 & 0 & 1/\sqrt{2} \end{pmatrix}$

$= \begin{pmatrix} 1/2 & 0 & 0 & -1/2 \\ 0 & 1 & 0 & 0 \\ 0 & 0 & 1 & 0 \\ -1/2 & 0 & 0 & 1/2 \end{pmatrix}$;

(c) $\mathbf{P}_u = \begin{pmatrix} 0 & 0 & -1/\sqrt{2} \\ 1/\sqrt{2} & 1/\sqrt{2} & 0 \\ -1/\sqrt{2} & 1/\sqrt{2} & 0 \\ 0 & 0 & 1/\sqrt{2} \end{pmatrix}$.

$\cdot \begin{pmatrix} 0 & 1/\sqrt{2} & -1/\sqrt{2} & 0 \\ 0 & 1/\sqrt{2} & 1/\sqrt{2} & 0 \\ -1/\sqrt{2} & 0 & 0 & 1/\sqrt{2} \end{pmatrix}$

$= \begin{pmatrix} 1/2 & 0 & 0 & -1/2 \\ 0 & 1 & 0 & 0 \\ 0 & 0 & 1 & 0 \\ -1/2 & 0 & 0 & 1/2 \end{pmatrix} = \mathbf{P}_v$;

(d) Yes, because $\mathbf{P}_v \begin{pmatrix} 2 \\ 1 \\ 3 \\ -2 \end{pmatrix} = \begin{pmatrix} 2 \\ 1 \\ 3 \\ -2 \end{pmatrix}$.

(e) $\|\mathbf{e}_1 - \mathbf{P}_{\mathcal{V}}(\mathbf{e}_1)\| = \frac{\sqrt{2}}{2} = \|\mathbf{e}_4 - \mathbf{P}_{\mathcal{U}}(\mathbf{e}_4)\|$.

37. If $\mathbf{x} \in U$ and $\mathbf{x} \in U^{\perp}$, then $\mathbf{x}^T\mathbf{x} = \sum_i \mathbf{x}_i^2 = 0$ or $\mathbf{x} = \mathbf{0}$.

39. (a) If $\mathbf{x} \in \ker(\mathbf{A})$, then $\mathbf{Ax} = \mathbf{0} \in \mathbb{R}^2$. But $\mathbf{rx} = -2 \neq 0$, so one entry of the vector $\mathbf{Ax}$ cannot be zero, no matter what $\mathbf{A}$ we choose. Therefore, no such $\mathbf{A}$ can exist.

(b) Here, $\mathbf{ry} = 1 + 3a = 0$ makes $a = -\frac{1}{3}$. Hence, we may take $\mathbf{A} = \begin{pmatrix} 1 & -1 & 3 \\ 2 & -2 & -6 \end{pmatrix}$, for example,

and $\mathbf{A} \begin{pmatrix} 2 \\ 1 \\ -\frac{1}{3} \end{pmatrix} = \mathbf{0} \in \mathbb{R}^2$.

T 10.(A) (a) Take $\mathbf{z}_1 = \mathbf{v}_1$.

At the first Gram–Schmidt level, $\mathbf{z}_2 = \mathbf{z}_1 \cdot \mathbf{z}_1\mathbf{v}_2 - \mathbf{v}_2 \cdot \mathbf{z}_1\mathbf{z}_1 = \begin{pmatrix} 4 \\ 4 \\ 4 \\ -12 \end{pmatrix}$ ($\simeq \mathbf{w}_2$ of old).

Also, $\mathbf{z}_3 = \mathbf{z}_1 \cdot \mathbf{z}_1\mathbf{v}_3 - \mathbf{v}_3 \cdot \mathbf{z}_1\mathbf{z}_1 = \begin{pmatrix} 8 \\ 12 \\ 4 \\ -24 \end{pmatrix} \simeq \begin{pmatrix} 2 \\ 3 \\ 1 \\ -6 \end{pmatrix}$.

At level two, $\mathbf{z}_3 = \mathbf{z}_2 \cdot \mathbf{z}_2\mathbf{z}_3 - \mathbf{z}_3 \cdot \mathbf{z}_2\mathbf{z}_2 = \begin{pmatrix} 12 \\ 12 \\ 0 \end{pmatrix}$

($\simeq \mathbf{w}_3$ of old).

(b) First level: $\mathbf{z}_2 \simeq \mathbf{w}_2$ of old, $\mathbf{z}_3 = \begin{pmatrix} -12 \\ -12 \\ -12 \\ 36 \end{pmatrix}$

($\simeq \mathbf{w}_2$ of old).

Second level: $\mathbf{z}_3 = \mathbf{0}$.

10.2.P p. 340

1. (a) $\mathbf{A} = \begin{pmatrix} 1 \\ 0 \\ -2 \end{pmatrix} \begin{pmatrix} 1 & 0 & 2 & 4 \end{pmatrix}$

$= \begin{pmatrix} -2 \\ 0 \\ 4 \end{pmatrix} \begin{pmatrix} -1/2 & 0 & -1 & -2 \end{pmatrix}$, for example.

(b) rank($\mathbf{A}$) = 1; the dyadic representation is not unique: If $\mathbf{A} = \mathbf{uv}^T$, then $\mathbf{A} = \left(\frac{1}{2}\mathbf{u}\right)(2\mathbf{v}^T)$ as well.

3. (a), (b) $\mathbf{u} = \dfrac{1}{\sqrt{42 + 2\sqrt{21}}} \begin{pmatrix} 1 + \sqrt{21} \\ 0 \\ 2 \\ -4 \end{pmatrix} \doteq$

$\begin{pmatrix} 0.7805 \\ 0 \\ 0.2796 \\ -0.5592 \end{pmatrix}$; $\|\mathbf{u}\| = 1$; $\mathbf{H} = \mathbf{I}_3 - 2\mathbf{uu}^T \doteq$

$$\begin{pmatrix} -0.2182 & 0 & -0.4364 & 0.8729 \\ 0 & 1 & 0 & 0 \\ -0.4364 & 0 & 0.8436 & 0.3127 \\ 0.8729 & 0 & 0.3127 & 0.3746 \end{pmatrix}.$$

(c) $* = -\sqrt{21} \doteq -4.5826$.

5. If $\mathbf{Wu} = \mathbf{e}_1$, then $\mathbf{u} = \mathbf{W}^{-1}\mathbf{e}_1$. According to Chapter 1, the first column of $\mathbf{WHW}^{-1}$ is given by $\mathbf{WWH}^{-1}\mathbf{e}_1 = \mathbf{W}(\mathbf{I} - 2\mathbf{uu}^T)\mathbf{u} = \mathbf{W}(\mathbf{u} - 2\mathbf{u}) = -\mathbf{Wu} = -\mathbf{e}_1$. Continue looking at the other columns $\mathbf{WHW}^{-1}\mathbf{e}_j$ of $\mathbf{WHW}^{-1}$: By assumption, we have $\mathbf{W}^{-1}\mathbf{e}_j = \mathbf{W}^T\mathbf{e}_j$. Thus, for $j \neq 1$, we have $(\mathbf{W}^T\mathbf{e}_1)^T\mathbf{W}^T\mathbf{e}_j = \mathbf{e}_1^T\mathbf{WW}^T\mathbf{e}_j = 0$, or $\mathbf{W}^T\mathbf{e}_j \perp \mathbf{u} = \mathbf{W}^T\mathbf{e}_1$ for all $j \neq 1$.

Consequently, $\mathbf{WHW}^{-1}\mathbf{e}_j = \mathbf{W}(\mathbf{I} - 2\mathbf{uu}^T)\mathbf{W}^T\mathbf{e}_j = \mathbf{WW}^T\mathbf{e}_j - 2\mathbf{Wuu}^T\mathbf{W}^T\mathbf{e}_j = \mathbf{e}_j$ for $j \neq 1$. Thus, the columns of $\mathbf{WHW}^{-1}$ are all multiples of the standard unit vectors in order, or $\mathbf{WHW}^{-1}$ is diagonal.

7. $\mathbf{A}^{-1} = \mathbf{A}^T$.

9. If $\mathbf{U}_{nk} = \begin{pmatrix} - & \mathbf{u}_1 & - \\ & \vdots & \\ - & \mathbf{u}_n & - \end{pmatrix}$ with $\mathbf{u}_i^T \mathbf{u}_j = 0$ or 1, according to whether $i \neq j$ or $i = j$, respectively, then $k \geq n$ and $\mathbf{U}_{nk} \left(\mathbf{U}^T\right)_{kn} =$

$$\begin{pmatrix} - & \mathbf{u}_1 & - \\ & \vdots & \\ - & \mathbf{u}_n & - \end{pmatrix} \begin{pmatrix} | & & | \\ \mathbf{u}_1^T & \dots & \mathbf{u}_n^T \\ | & & | \end{pmatrix} = \mathbf{I}_n.$$

11. $\mathbf{V}^T\mathbf{V} = \begin{pmatrix} 1 & \\ & \mathbf{U}^T \end{pmatrix} \begin{pmatrix} 1 & \\ & \mathbf{U} \end{pmatrix} = \begin{pmatrix} 1 & \\ & \mathbf{U}^T\mathbf{U} \end{pmatrix} = \mathbf{I}_{n+1}$.

13. If $\mathbf{u} = \mathbf{0}$ or $\mathbf{v} = \mathbf{0}$, then $\mathbf{uv}^T = \mathbf{O}_n$ has rank zero. Thus, we may assume that $\mathbf{u}^T\mathbf{u} = 1$ and $\mathbf{v} \neq \mathbf{0}$. If $\mathbf{I} - \mathbf{uv}^T$ is orthogonal, then $\mathbf{I} = (\mathbf{I} - \mathbf{uv}^T)^T(\mathbf{I} - \mathbf{uv}^T) = \mathbf{I} - \mathbf{uv}^T - \mathbf{vu}^T + \mathbf{vu}^T\mathbf{uv}^T = \mathbf{I} - \mathbf{uv}^T - \mathbf{vu}^T + \mathbf{vv}^T$. Hence, $(\mathbf{v} - \mathbf{u})\mathbf{v}^T = \mathbf{vu}^T$. By Problem 12, we conclude that $\mathbf{v} - \mathbf{u} = \mathbf{u}$, or $\mathbf{v} = 2\mathbf{u}$, making $\mathbf{I} - \mathbf{uv}^T = \mathbf{I} - 2\mathbf{uu}^T$. Finally, since $\mathbf{u}^T\mathbf{u} = \|\mathbf{u}\|^2 = 1$ was assumed, the statement follows.

For arbitrary nonzero $\mathbf{u}$, the Householder vector $\mathbf{w}$ is $\frac{\mathbf{u}}{\|\mathbf{u}\|}$.

10.3.P p. 345

1. In MATLAB, set up the matrix $\mathbf{A}$ from the given vectors as

```
A =
  2   3   3   1
 -1   1  -4  -1
 -1   4  -7  -1
  2  -1   7   1
  1   2   1   1
  1  -2   5   1
```

The MATLAB command `[Q R]=qr(A)` gives us an ONB for the column space of $\mathbf{A}$ in the first four columns of $\mathbf{Q}$, for which the corresponding diagonal entries in $\mathbf{R}$ are nonzero:

```
Q =
-0.57735  -0.53590   0.53916   0.05984  -0.28851  -0.04433
 0.28868  -0.15513   0.52214  -0.17941   0.57027   0.51242
 0.28868  -0.66283  -0.44602  -0.31472  -0.29237   0.30637
-0.57735   0.14103  -0.24130  -0.70947   0.28851   0.04433
-0.28868  -0.35257  -0.40901   0.53989   0.57726   0.01925
-0.28868   0.32436  -0.11060   0.26524  -0.29935   0.79954
R =
-3.4641e+00    2.8868e-01   -1.0681e+01   -2.3094e+00
          0   -5.9090e+00    5.9090e+00    3.9488e-01
          0             0    2.4526e-15   -2.9786e-01
          0             0             0    6.4963e-01
          0             0             0             0
          0             0             0             0
```

3. For

```
A =
   1   2   0
  -2   0   0
   3   0   1
   4   4  -1
  -1   1   0
```

the MATLAB command `[Q R]=qr(A)` gives us an ONB of U in the first three columns of $\mathbf{Q}$, while the last two columns of $\mathbf{Q}$ contain an ONB for $U^{\perp}$:

```
>> [Q R]=qr(A)
Q =
-1.7961e-01 -4.2479e-01 -4.7393e-01 -7.3102e-01 -1.6822e-01
 3.5921e-01 -3.2095e-01 -2.6676e-01  4.3513e-01 -7.1236e-01
-5.3882e-01  4.8143e-01 -6.2718e-01  2.7848e-01 -8.3641e-02
-7.1842e-01 -5.2863e-01  3.4623e-01  2.7848e-01 -8.3641e-02
 1.7961e-01 -4.5311e-01 -4.3704e-01  3.4810e-01  6.7101e-01
R =
-5.5678e+00  -3.0533e+00   1.7961e-01
          0  -3.4172e+00   1.0101e+00
          0            0  -9.7340e-01
          0            0            0
          0            0            0
```

7. $\mathbf{x}_{LS} = \text{A}\backslash\text{b} = \begin{pmatrix} -0.046794 \\ -0.27210 \\ 0.64818 \end{pmatrix}$ and $\|\mathbf{A}\mathbf{x}_{LS} - \mathbf{b}\| = 4.1795$.

9. For f_5, we compute $a = 0.49687$, $b = 2.9167$, and $c = 1.5$, with the unavoidable error $e_5 = 1.4142$. For f_6, we obtain the best fit for $a = 0.62303$, $b = -0.56446$, and $c = 2.5816$, with $e_6 = 3.2406$. These two model functions do not serve the data well, due to their large residual errors.

11. The respective system matrices are as follows: For f_1: $\mathbf{A} = \begin{pmatrix} 1 & 1 \\ 1 & 5 \\ 1 & 10 \\ 1 & 18 \\ 1 & 20 \end{pmatrix}$, with residual error 1.0022. For f_2 and the data, we have $\mathbf{A} = \begin{pmatrix} 1 & 1 & 1 \\ 1 & 5 & 25 \\ 1 & 10 & 100 \\ 1 & 18 & 324 \\ 1 & 20 & 400 \end{pmatrix}$, with associated error 0.98331. For f_3, we have $\mathbf{A} = \begin{pmatrix} 1 & 1 & 1 & 1 \\ 1 & 5 & 25 & 125 \\ 1 & 10 & 100 & 1000 \\ 1 & 18 & 324 & 5832 \\ 1 & 20 & 400 & 8000 \end{pmatrix}$, with error 0.59687. Finally, for f_4, $\mathbf{A} =$

$$\begin{pmatrix} 1 & 9.6356e-01 & 2.6750e-01 \\ 1 & 2.1512e-01 & 9.7659e-01 \\ 1 & 4.2017e-01 & 9.0745e-01 \\ 1 & -9.8692e-01 & -1.6124e-01 \\ 1 & 7.6256e-01 & 6.4692e-01 \end{pmatrix}$$, with the large unavoidable error 5.9523.

13. $\mathbf{A}_{nk} = \begin{pmatrix} | & & | \\ \mathbf{c}_1 & \dots & \mathbf{c}_k \\ | & & | \end{pmatrix} = \begin{pmatrix} | & & | \\ \frac{\mathbf{c}_1}{\|\mathbf{c}_1\|} & \cdots & \frac{\mathbf{c}_k}{\|\mathbf{c}_k\|} \\ | & & | \end{pmatrix}_{nk} \begin{pmatrix} \|\mathbf{c}_1\| & & \\ & \ddots & \\ & & \|\mathbf{c}_k\| \end{pmatrix}_{kk} = \mathbf{QR}$ with $\mathbf{Q}^T\mathbf{Q} = \mathbf{I}_k$ and diagonal $\mathbf{R}$.

10.R p. 346

1. (d) $\ker(2, -1, 1, 1) = \text{span}\{\mathbf{e}_1 + 2\mathbf{e}_2, \mathbf{e}_1 - 2\mathbf{e}_3, \mathbf{e}_1 - 2\mathbf{e}_4\}$.

(e) $\frac{1}{\sqrt{5}}\begin{pmatrix} 1 \\ 2 \\ 0 \\ 0 \end{pmatrix}, \frac{1}{\sqrt{30}}\begin{pmatrix} 2 \\ -1 \\ -5 \\ 0 \end{pmatrix}, \frac{1}{\sqrt{42}}\begin{pmatrix} -2 \\ 1 \\ -1 \\ 6 \end{pmatrix}$.

(f) If $\mathbf{x} \perp \mathbf{u}$, then $\mathbf{x}^T\mathbf{u} = 0$; that is, $\mathbf{x} \in \ker(\ - \ \mathbf{u} \ - \ )$, which is a subspace according to Chapter 4.

(g) $\ker\begin{pmatrix} - & \mathbf{a} & - \\ - & \mathbf{b} & - \end{pmatrix}$. (h) $\{\mathbf{0}\}$.

3. A matrix $\mathbf{U}$ is orthogonal if $\mathbf{U}^T\mathbf{U} = \mathbf{I}_n$. Thus, for an orthogonal matrix, $\mathbf{U}^T$ is the inverse of $\mathbf{U}$, and $(\mathbf{U}^T)^T\mathbf{U}^T = \mathbf{U}\mathbf{U}^T = \mathbf{I}_n$ as well, since the left and right matrix inverses coincide.

5. ONB $\left\{ \frac{1}{\sqrt{18}}\mathbf{u}_1, \frac{1}{\sqrt{18}}\begin{pmatrix} 2 \\ -1 \\ 3 \\ 2 \end{pmatrix}, \frac{1}{\sqrt{18}}\begin{pmatrix} -2 \\ 1 \\ 3 \\ -2 \end{pmatrix} \right\}$.

7. $U = \{\mathbf{0}\}$.

9. (a) If $\mathbf{A}^T\mathbf{A} = \mathbf{I}_n$, then $\mathbf{A}\mathbf{A}^T = \mathbf{I}_n$ as well, and $\mathbf{A}^T$ is also orthogonal. Note that $\mathbf{A}^T = \mathbf{A}^{-1}$ here.

(b) If $\mathbf{A}^T\mathbf{A} = \mathbf{I}_n$; that is, if $\mathbf{A}$ is orthogonal, then $(\mathbf{A}^2)^T\mathbf{A}^2 = \mathbf{A}^T\mathbf{A}^T\mathbf{A}\mathbf{A} = \mathbf{I}_n$, and $\mathbf{A}^2$ is orthogonal. Since $\mathbf{A}^{-1} = \mathbf{A}^T$, the matrix $\mathbf{A}^{-2} = (\mathbf{A}^T)^2$ is thus also orthogonal by part (a).

Chapter 11

11.1.P p. 357

1. Only $\mathbf{D}$ is symmetric for $a = 1$.

3. $\mathbf{A}$ is singular. (It has repeated rows.) $\ker(\mathbf{A}) = \text{span}\{\mathbf{e}_1 - \mathbf{e}_2, \mathbf{e}_1 - \mathbf{e}_3\}$. Moreover, $\mathbf{A}(\mathbf{e}_1 + \mathbf{e}_2 + \mathbf{e}_3) = 3\mathbf{e}_1 + \mathbf{e}_2 + \mathbf{e}_3$. Thus, $\mathbf{A}$'s eigenvalues are 0 (double) and 3, with corresponding eigenvectors $\begin{pmatrix} 1 \\ -1 \\ 0 \end{pmatrix}, \begin{pmatrix} 1 \\ 0 \\ -1 \end{pmatrix}$, and $\begin{pmatrix} 1 \\ 1 \\ 1 \end{pmatrix}$.

5. Use the standard eigenvalue-finding techniques of Chapter 9: Vector iteration with $\mathbf{y} = \begin{pmatrix} 1 \\ 1 \\ 1 \end{pmatrix}$ yields $p_y(\lambda) = \lambda^3 - 3\lambda^2 - 18\lambda = \lambda(\lambda + 3)(\lambda - 6)$. Or via determinants, $f_A(\lambda) = \lambda^3 - 3\lambda^2 - 18\lambda$ as well. Eigenvalues: 0, –3, 6; corresponding orthonormal eigenvectors: $\frac{1}{\sqrt{6}}\begin{pmatrix} 2 \\ -1 \\ 1 \end{pmatrix}, \frac{1}{\sqrt{3}}\begin{pmatrix} -1 \\ -1 \\ 1 \end{pmatrix}$, and $\frac{1}{\sqrt{2}}\begin{pmatrix} 0 \\ -1 \\ -1 \end{pmatrix}$.

7. If $a = 0$, then any ONB of $\mathbb{R}^2$ is an eigenvector basis for $\mathbf{A}(0) = \mathbf{I}_2$.

For general $a \neq 0$, using Section 9.1.D, we have $f_{A(a)}(x) = x^2 - 2x + 1 - a^2$ with roots $x_{1,2} = 1 \pm a$. Or, by using vector iteration with $\mathbf{y} = \mathbf{e}_1$ for $a \neq 0$, we have

$\mathbf{y}$	$\mathbf{Ay}$	$\mathbf{A}^2\mathbf{y}$	
1	1	$1 + a^2$	
0	a	$2a$	$\div a$
1	1	$1 + a^2$	$-\ row_2$
0	1	2	
1	0	$a^2 - 1$	
0	1	2	

That is, $\mathbf{A}^2\mathbf{y} = 2\mathbf{Ay} + (a^2 - 1)\mathbf{y}$, or $p_y(\lambda) = \lambda^2 - 2\lambda + (1 - a^2)$ with roots $\lambda_{1,2} = 1 \pm a$. A corresponding orthonormal eigenvector basis is

$$\mathcal{U}(a) = \left\{ \frac{1}{\sqrt{2}}\begin{pmatrix} 1 \\ 1 \end{pmatrix}, \frac{1}{\sqrt{2}}\begin{pmatrix} 1 \\ -1 \end{pmatrix} \right\}.$$

Finally, $\mathbf{A}(a)_{\mathcal{U}(a)} = \begin{pmatrix} 1 + a & 0 \\ 0 & 1 - a \end{pmatrix}$.

9. (a) $\mathbf{A} = \mathbf{X}\mathbf{D}\mathbf{X}^T = \begin{pmatrix} | & & | \\ \mathbf{x}_1 & \dots & \mathbf{x}_n \\ | & & | \end{pmatrix} \cdot \begin{pmatrix} \lambda_1 & & \\ & \ddots & \\ & & \lambda_n \end{pmatrix} \begin{pmatrix} - & \mathbf{x}_1^T & - \\ & \vdots & \\ - & \mathbf{x}_n^T & - \end{pmatrix}$

$= \begin{pmatrix} | & & | \\ \lambda_1\mathbf{x}_1 & \dots & \lambda_n\mathbf{x}_n \\ | & & | \end{pmatrix} \begin{pmatrix} - & \mathbf{x}_1^T & - \\ & \vdots & \\ - & \mathbf{x}_n^T & - \end{pmatrix}$

$= \lambda_1 \begin{pmatrix} | \\ \mathbf{x}_1 \\ | \end{pmatrix} \begin{pmatrix} - & \mathbf{x}_1^T & - \end{pmatrix} + \dots + \lambda_n \begin{pmatrix} | \\ \mathbf{x}_n \\ | \end{pmatrix} \begin{pmatrix} - & \mathbf{x}_n^T & - \end{pmatrix}$, since $\begin{pmatrix} | & & | & & | \\ \mathbf{0} & \dots & \lambda_i\mathbf{x}_i & \dots & \mathbf{0} \\ | & & | & & | \end{pmatrix} \begin{pmatrix} 0 & \dots & 0 \\ - & \mathbf{x}_j^T & - \\ 0 & \dots & 0 \\ & \vdots & \\ 0 & \dots & 0 \end{pmatrix} = \mathbf{O}_{nn}$ for $i \neq j$.

(b) The eigenvalues and normalized eigenvectors of $\mathbf{A}$ are 2, –2 and $\frac{1}{\sqrt{2}}\begin{pmatrix} 1 \\ 1 \end{pmatrix}, \frac{1}{\sqrt{2}}\begin{pmatrix} 1 \\ -1 \end{pmatrix}$. Thus, $\mathbf{A} = \begin{pmatrix} 0 & 2 \\ 2 & 0 \end{pmatrix} = \begin{pmatrix} 1 \\ 1 \end{pmatrix} \begin{pmatrix} 1 & 1 \end{pmatrix} - \begin{pmatrix} 1 \\ -1 \end{pmatrix} \begin{pmatrix} 1 & -1 \end{pmatrix}$.

11. $\mathbf{U}^*\mathbf{W}\mathbf{U} = \mathbf{R}$ implies that $\mathbf{U}^*\mathbf{W}^T\mathbf{U} = \mathbf{R}^*$.

(b) From (a), we have $\mathbf{W}^T = \mathbf{U}\mathbf{R}^*\mathbf{U}^*$ and $\mathbf{W} = \mathbf{U}\mathbf{R}\mathbf{U}^*$. Thus, $\mathbf{I} = \mathbf{W}^T\mathbf{W} = \mathbf{U}\mathbf{R}^*\mathbf{U}^*\mathbf{U}\mathbf{R}\mathbf{U}^* = \mathbf{U}\mathbf{R}^*\mathbf{R}\mathbf{U}^*$, or $\mathbf{R}^*\mathbf{R} = \mathbf{I} = \mathbf{R}\mathbf{R}^*$ by multiplying the last equation on the left by $\mathbf{U}^*$ and on the right by $\mathbf{U}$. Hence $\mathbf{R}^*$ is the inverse of $\mathbf{R}$.

(c), (d) If $\mathbf{R}^*\mathbf{R} = \begin{pmatrix} \overline{\lambda_1} & 0 & \dots & 0 \\ \overline{a_{12}} & \overline{\lambda_2} & & \\ & & \ddots & 0 \\ * & & & \overline{\lambda_n} \end{pmatrix} \begin{pmatrix} \lambda_1 & a_{12} & \dots & a_{1n} \\ 0 & \lambda_1 & & \\ & & \ddots & * \\ 0 & & 0 & \lambda_n \end{pmatrix}$

$= \begin{pmatrix} \overline{\lambda_1}\lambda_1 & \overline{\lambda_1}a_{12} & \dots & \overline{\lambda_1}a_{1n} \\ & \ddots & & \\ & & \ddots & \\ & & & \overline{\lambda_n}\lambda_n \end{pmatrix} = \begin{pmatrix} 1 & & 0 \\ & \ddots & \\ 0 & & 1 \end{pmatrix}$

$= \mathbf{I}_n$, then $|\lambda_1| = 1$, $a_{1i} = 0$, and $|\lambda_2| = 1$, $a_{2i} = 0$, etc., for all i. That is, $\mathbf{R}$ is diagonal.

13. $\mathbf{A}$ has only one nonzero eigenvalue $\lambda = 3$ for the eigenvector $\begin{pmatrix} 1 \\ 1 \\ 1 \end{pmatrix}$. Thus, by orthonormal fill-in,

$\mathbf{U} = \begin{pmatrix} 1/\sqrt{3} & 1/\sqrt{2} & 1/\sqrt{6} \\ 1/\sqrt{3} & -1/\sqrt{2} & 1/\sqrt{6} \\ 1/\sqrt{3} & 0 & -2/\sqrt{6} \end{pmatrix}$ is an orthogonal matrix that transforms $\mathbf{A}$ to its Schur normal form $\mathbf{U}^T\mathbf{A}\mathbf{U} = \begin{pmatrix} 3 & 0 & 0 \\ 0 & 0 & 0 \\ 0 & 0 & 0 \end{pmatrix}$. $\mathbf{B}$ has the eigenvalue $\lambda = 3$ for the unit eigenvector $\mathbf{e}_3$. Filling $\mathbf{U}$ in orthonormally makes $\mathbf{U} = \begin{pmatrix} 0 & 1/\sqrt{2} & 1/\sqrt{2} \\ 0 & -1/\sqrt{2} & 1/\sqrt{2} \\ 1 & 0 & 0 \end{pmatrix}$ transform $\mathbf{B}$ to a Schur normal form $\mathbf{U}^T\mathbf{B}\mathbf{U} = \begin{pmatrix} 3 & 0 & \sqrt{2} \\ 0 & 1 & -1 \\ 0 & 0 & 2 \end{pmatrix}$.

15. (a) $\mathbf{C}\begin{pmatrix} 1 \\ 1 \\ 1 \end{pmatrix} = 6\begin{pmatrix} 1 \\ 1 \\ 1 \end{pmatrix}$.

(b) Yes, for $\lambda = 3$, since $\mathbf{C}\mathbf{x} = 3\mathbf{x}$.

(c) Recall the trace equation $\text{trace}(\mathbf{A}) = \sum_i a_{ii} = \sum_i \lambda_i$. So $12 = \lambda_1 + \lambda_2 + \lambda_3 = 6 + 3 + 3$ in our case, making the third eigenvalue of $\mathbf{C}$ also equal to 3 for the eigenvector $\mathbf{y} = \begin{pmatrix} 1 \\ 0 \\ -1 \end{pmatrix}$. $\mathbf{C} = \mathbf{C}^T$ is orthogonally diagonalizable. The three eigenvectors $\mathbf{x}$, $\mathbf{y}$, and $\mathbf{e}_1 + \mathbf{e}_2 + \mathbf{e}_3$ for $\mathbf{C}$ must be orthonormalized.

(d) $\mathbf{U} = \begin{pmatrix} 1/\sqrt{3} & 1/\sqrt{2} & 1/\sqrt{6} \\ 1/\sqrt{3} & -1/\sqrt{2} & 1/\sqrt{6} \\ 1/\sqrt{3} & 0 & -2/\sqrt{6} \end{pmatrix}$.

17. Evaluate $\mathbf{A}\mathbf{A}^T = \mathbf{I}_3$.

19. $\mathbf{A}$ is diagonal and hence symmetric. Its eigenvalues lie on the diagonal: 1, 17, and –31. The standard unit vectors $\mathbf{e}_i$ are the corresponding orthonormal eigenvector basis for $\mathbf{A}$.

21. Look at $\ker(\mathbf{A} - 13\mathbf{I})$: Eigenvector $\mathbf{x} = \begin{pmatrix} -1 \\ 3 \\ 1 \\ 2 \end{pmatrix}$.

23. (a) $\mathbf{B}$ is obviously diagonalizable as a symmetric matrix.

(b) $\mathbf{C}$ is obviously not diagonalizable: $\mathbf{C}$ is upper triangular and has the four-fold eigenvalue zero on its diagonal, but there is only a three-dimensional eigenspace.

(c) **A** will take the most work to decide diagonalizability. (It is diagonalizable with four distinct eigenvalues, two of which are complex.)

25. $\mathbf{A}^T = (\mathbf{UDU}^T)^T = \mathbf{U}^T\mathbf{D}^T\mathbf{U}^T = \mathbf{UDU}^T = \mathbf{A}$.

27. We have $\mathbf{x}^T(\mathbf{A}+\mathbf{B})\mathbf{x} = \mathbf{x}^T\mathbf{A}\mathbf{x} + \mathbf{x}^T\mathbf{B}\mathbf{x}$ for all $\mathbf{x} \in \mathbb{R}^n$. For $\mathbf{x} = \mathbf{U}\mathbf{x}_{\mathcal{U}}$, according to Chapter 7, we have $\mathbf{x}^T\mathbf{B}\mathbf{x} = \mathbf{x}_{\mathcal{U}}^T\mathbf{U}^T\mathbf{B}\mathbf{U}\mathbf{x}_{\mathcal{U}} = \mathbf{x}_{\mathcal{U}}^T\operatorname{diag}(\lambda_1, \dots, \lambda_n)\mathbf{x}_{\mathcal{U}} = \sum_i \lambda_i(\mathbf{x}_{\mathcal{U}})_i^2 \geq 0$. Thus, $\mathbf{x}^T(\mathbf{A}+\mathbf{B})\mathbf{x} \geq \mathbf{x}^T\mathbf{A}\mathbf{x}$ and $\mathbf{x}^T(\mathbf{A}-\mathbf{B})\mathbf{x} \leq \mathbf{x}^T\mathbf{A}\mathbf{x}$ for all $\mathbf{x}$.

29. We have $\mathbf{A}\mathbf{x} = \lambda\mathbf{x}$ and $\mathbf{y}^T\mathbf{A} = \mu\mathbf{y}^T$. Thus, $\lambda\mu\ \mathbf{y}\cdot\mathbf{x} = \mu\mathbf{y}^T\lambda\mathbf{x} = \mathbf{y}^T\mathbf{A}\mathbf{A}\mathbf{x} = (\mathbf{y}^T\mathbf{A}^2)\mathbf{x} = \mu^2\mathbf{y}\cdot\mathbf{x} = \mathbf{y}^T(\mathbf{A}^2\mathbf{x}) = \lambda^2\mathbf{y}\cdot\mathbf{x}$. If $\mathbf{x} \not\perp \mathbf{y}$, then $\lambda\mu = \mu^2 = \lambda^2$ for $\lambda \neq \mu$, leading to contradictions.

31. (a) False: Take $\mathbf{A} = \begin{pmatrix} 1 & 1 \\ 1 & 0 \end{pmatrix}$ and $\mathbf{B} = \begin{pmatrix} 0 & 1 \\ 1 & 1 \end{pmatrix}$.

Then $\mathbf{AB} = \begin{pmatrix} 1 & 2 \\ 0 & 1 \end{pmatrix}$ is not symmetric.

(b), (e) True for $\mathbf{A} = \mathbf{B} = \mathbf{I}$, for example.

(c), (d) True: If $\mathbf{U}^T\mathbf{U} = \mathbf{I}_n$ and $\mathbf{V}^T\mathbf{V} = \mathbf{I}_n$, then $(\mathbf{UV})^T\mathbf{UV} = \mathbf{V}^T\mathbf{U}^T\mathbf{UV} = \mathbf{I}_n$.

33. (a) If $\mathbf{X}^{-1}\mathbf{A}\mathbf{X} = \mathbf{D} = \mathbf{DI}$, then $\mathbf{A} = \mathbf{XDIX}^{-1} = (\mathbf{XDX}^T)(\mathbf{X}^{-T}\mathbf{I}\mathbf{X}^{-1}) = \mathbf{S}_1\mathbf{S}_2$ for $\mathbf{S}_1 = \mathbf{XDX}^T$ and $\mathbf{S}_2 = \mathbf{X}^{-T}\mathbf{X}^{-1}$. The $\mathbf{S}_i$ are both symmetric according to Problem 32(b).

(b) $\mathbf{J} = \begin{pmatrix} 2 & 1 \\ 0 & 1 \end{pmatrix} = \begin{pmatrix} 0 & 1 \\ 1 & 0 \end{pmatrix}\begin{pmatrix} 0 & 2 \\ 2 & 1 \end{pmatrix}$.

35. If, for $\mathbf{A}$, $E(\lambda) = \operatorname{span}\{\mathbf{x}_1, \dots, \mathbf{x}_k\}$ with linearly independent eigenvectors $\mathbf{x}_i$, then $(\mathbf{A}+\alpha\mathbf{I})\mathbf{x}_i = \mathbf{A}\mathbf{x}_i + \alpha\mathbf{x}_i = (\lambda+\alpha)\mathbf{x}_i$ for each i. Thus, the geometric multiplicity $\dim E_{\mathbf{A}+\alpha\mathbf{I}}(\lambda+\alpha) = k_\lambda$ as well.
According to the Schur Normal Form theorem, $\mathbf{A}$ is unitarily similar to an upper triangular matrix $\mathbf{U}^*\mathbf{A}\mathbf{U} = \mathbf{R}$ with $\mathbf{A}$'s eigenvalue λ repeated n_λ times on $\mathbf{R}$'s diagonal. Observe that then $\mathbf{U}^*(\mathbf{A}+\alpha\mathbf{I})\mathbf{U} = \mathbf{U}^*\mathbf{A}\mathbf{U} + \alpha\mathbf{I} = \mathbf{R} + \alpha\mathbf{I}$ also has n_λ copies of $\lambda+\alpha$ on its diagonal. That is, $\mathbf{A}+\alpha\mathbf{I}$ has the eigenvalue $\lambda+\alpha$ with precisely the same algebraic and geometric multiplicity as $\mathbf{A}$ does for λ.

11.2.P p. 367

1. If $\mathbf{A} = \mathbf{A}^T$ is symmetric, then $\mathbf{A}+\mathbf{A}^T = 2\mathbf{A}$ is symmetric, but not skew symmetric, while $\mathbf{A}-\mathbf{A}^T = \mathbf{O}_n$ is both symmetric and skew symmetric.
If $\mathbf{A} = -\mathbf{A}^T$ is skew symmetric, then $\mathbf{A}+\mathbf{A}^T = \mathbf{O}_n$ is both symmetric and skew symmetric, while $\mathbf{A}-\mathbf{A}^T = 2\mathbf{A}$ is skew symmetric, but not symmetric.

3. True. Use the fact that $\mathbf{U}^T$ is the inverse of $\mathbf{U}$.

5. Each $\mathbf{A} = -\mathbf{A}^T$ can be unitarily diagonalized: $\mathbf{U}^*\mathbf{A}\mathbf{U} = \mathbf{D}$. Consequently, $\mathbf{U}^*\mathbf{B}_\alpha\mathbf{U} = \mathbf{U}^*(\mathbf{A}+\alpha\mathbf{I})\mathbf{U} = \mathbf{U}^*\mathbf{A}\mathbf{U} + \alpha\mathbf{U}^*\mathbf{U} = \mathbf{D} + \alpha\mathbf{I}$ is also diagonal for each $\alpha \in \mathbb{C}$. The eigenvectors of $\mathbf{B}_\alpha$ are the eigenvectors of $\mathbf{A}$ with corresponding eigenvalues $\lambda_i + \alpha$ if $\mathbf{A}$ has the eigenvalues $\lambda_1, \dots, \lambda_n$.

7. If $\mathbf{A}\mathbf{x} = \lambda\mathbf{x}$ for $\mathbf{x} \neq \mathbf{0}$, then $(c\mathbf{A})\mathbf{x} = c\mathbf{A}\mathbf{x} = c\lambda\mathbf{x}$.

9. (a) $\mathbf{A} = \dfrac{\mathbf{A}+\mathbf{A}^T}{2} + \dfrac{\mathbf{A}-\mathbf{A}^T}{2} = \mathbf{S} + \mathbf{T}$ with $\mathbf{S} := \dfrac{\mathbf{A}+\mathbf{A}^T}{2} = \mathbf{S}^T$ and $\mathbf{T} := \dfrac{\mathbf{A}-\mathbf{A}^T}{2} = -\mathbf{T}^T$.

(b) $\mathbf{A} = \begin{pmatrix} 4 & 0 & 4 \\ 0 & 4 & 0 \\ 4 & 0 & 6 \end{pmatrix} + \begin{pmatrix} 0 & 2 & 0 \\ -2 & 0 & 0 \\ 0 & 0 & 0 \end{pmatrix} =: \mathbf{S} + \mathbf{T}$.

(c) Eigenvalues of $\mathbf{A}$: $2, 6 \pm \sqrt{8}$; those of $\mathbf{S}$: $4, 5 \pm \sqrt{17}$; those of $\mathbf{T}$: $\pm 2i$ and 0.

11. (a) $b = c$. (b) $a = d = 0$ and $b = -c$.

(c) $b = c$, or $b = -c$ and $a = b$ by looking at the four entries of the equation $\mathbf{A}^T\mathbf{A} = \mathbf{A}\mathbf{A}^T$.

(d) $ad - bc \neq 0$.

(e) $a^2 + c^2 = 1, b^2 + d^2 = 1$, and $ab + cd = 0$.

(f) $ad - bc = 0$.

13. $\mathbf{A}_\varepsilon = \begin{pmatrix} 0 & -1 \\ 1 & 0 \end{pmatrix}$ is skew symmetric and normal.

15. (a) $\mathbf{B}^*\mathbf{B} = \begin{pmatrix} \mathbf{A}^* & \mathbf{A} \\ \mathbf{A} & \mathbf{A}^* \end{pmatrix}\begin{pmatrix} \mathbf{A} & \mathbf{A}^* \\ \mathbf{A}^* & \mathbf{A} \end{pmatrix}$

$= \begin{pmatrix} \mathbf{A}^*\mathbf{A} + \mathbf{A}\mathbf{A}^* & \mathbf{A}^{*2} + \mathbf{A}^2 \\ \mathbf{A}^2 + \mathbf{A}^{*2} & \mathbf{A}^*\mathbf{A} + \mathbf{A}\mathbf{A}^* \end{pmatrix} = \begin{pmatrix} \mathbf{A} & \mathbf{A}^* \\ \mathbf{A}^* & \mathbf{A} \end{pmatrix}$

$\begin{pmatrix} \mathbf{A}^* & \mathbf{A} \\ \mathbf{A} & \mathbf{A}^* \end{pmatrix} = \mathbf{B}\mathbf{B}^*$.

(b) $\mathbf{B} = \begin{pmatrix} \mathbf{A} & \mathbf{A}^* \\ \mathbf{A}^* & \mathbf{A} \end{pmatrix} = \mathbf{B}^* = \begin{pmatrix} \mathbf{A}^* & \mathbf{A} \\ \mathbf{A} & \mathbf{A}^* \end{pmatrix}$ if and only if $\mathbf{A} = \mathbf{A}^*$ is Hermitian.

(c) Take $\mathbf{C} = \begin{pmatrix} \mathbf{O} & -\mathbf{A}^T \\ \mathbf{A}^T & \mathbf{O} \end{pmatrix}$, for example. Then $\mathbf{C}^T = \begin{pmatrix} \mathbf{O} & \mathbf{A} \\ -\mathbf{A} & \mathbf{O} \end{pmatrix} = \begin{pmatrix} \mathbf{O} & -\mathbf{A}^T \\ \mathbf{A}^T & \mathbf{O} \end{pmatrix}$, provided that $\mathbf{A}^T = -\mathbf{A}$.

17. (a) $\mathbf{T} = \begin{pmatrix} 0 & 1 \\ 0 & 0 \end{pmatrix}$ is upper triangular, but not diagonalizable.

(b), (d), (e), (g) $\mathbf{T} = \begin{pmatrix} 0 & 0 \\ 0 & 0 \end{pmatrix}$ is upper triangular, symmetric, skew symmetric, normal, and diagonal.

(c) $\mathbf{T} = \begin{pmatrix} 1 & 1 \\ 0 & -2 \end{pmatrix}$ is upper triangular and diagonalizable, since it has two distinct eigenvalues.

(f) Every upper triangular matrix that is normal is diagonal (i.e., symmetric).

19. $\mathbf{A}^* = (\mathbf{S} + i\mathbf{T})^* = \mathbf{S}^* - i\mathbf{T}^* = \mathbf{S}^T - i\mathbf{T}^T = \mathbf{S} + i\mathbf{T} = \mathbf{A}$ if and only if $\mathbf{S} = \mathbf{S}^T \in \mathbb{R}^{n,n}$ and $\mathbf{T} = -\mathbf{T}^T \in \mathbb{R}^{n,n}$.

11.3.P p. 380

1. (a) $(5) = (1)\cdot(5)$, $(-3) = (-1)\cdot(3)$, $(4i-3) = \left(-\frac{3}{5} + \frac{4}{5}i\right)\cdot(5)$.

(b) Follow the proof of Theorem 11.5: $\mathbf{A}^T\mathbf{A} = \begin{pmatrix} 2 & 1 \\ 1 & 1 \end{pmatrix}$.

Eigenvalues of $\mathbf{A}^T\mathbf{A}$: $\frac{3}{2} \pm \frac{\sqrt{5}}{2}$.

Corresponding eigenvectors: $\mathbf{v}_1 = \begin{pmatrix} 1 \\ -1/2 - \sqrt{5}/2 \end{pmatrix}$ and $\mathbf{v}_2 = \begin{pmatrix} -1 \\ 1/2 - \sqrt{5}/2 \end{pmatrix}$. $\|\mathbf{v}_1\| = \frac{1}{2}\sqrt{10 + 2\sqrt{5}}$, $\|\mathbf{v}_2\| = \frac{1}{2}\sqrt{10 - 2\sqrt{5}}$. Thus, $\mathbf{V} = \begin{pmatrix} \frac{2}{\sqrt{10+2\sqrt{5}}} & \frac{-2}{\sqrt{10-2\sqrt{5}}} \\ \frac{-1-\sqrt{5}}{\sqrt{10+2\sqrt{5}}} & \frac{1-\sqrt{5}}{\sqrt{10-2\sqrt{5}}} \end{pmatrix}$ with $\mathbf{V}^T\mathbf{A}^T\mathbf{A}\mathbf{V} = \mathbf{D}$. Finally, $\mathbf{P} = \mathbf{V}\sqrt{\mathbf{D}}\mathbf{V} = \begin{pmatrix} 1.3416 & 0.44721 \\ 0.44721 & 0.89443 \end{pmatrix}$ is positive definite and $\mathbf{U} = \mathbf{A}\mathbf{P}^{-1} = \begin{pmatrix} 0.89443 & -0.44721 \\ 0.44721 & 0.89443 \end{pmatrix}$ is orthogonal with $\mathbf{A} = \mathbf{U}\mathbf{P}$.

3. (a) $\begin{pmatrix} 2 & -6 \\ -6 & 4 \end{pmatrix}$ or $\begin{pmatrix} 2 & -4 \\ -8 & 4 \end{pmatrix}$.

(b) $\begin{pmatrix} 2 & 0 & 0 \\ 0 & -4 & 0 \\ 0 & 0 & 16 \end{pmatrix}$ or $\begin{pmatrix} 2 & 0 & 7 \\ 0 & -4 & 0 \\ -7 & 0 & 16 \end{pmatrix}$.

(c) $\begin{pmatrix} 2 & -3 & 0 \\ -3 & -4 & 12 \\ 0 & 12 & 16 \end{pmatrix}$ or $\begin{pmatrix} 2 & -3 & 3 \\ -3 & -4 & 12 \\ -3 & 12 & 16 \end{pmatrix}$.

(d) $\begin{pmatrix} 0 & 1 \\ 1 & -7 \end{pmatrix}$ or $\begin{pmatrix} 0 & 20 \\ -18 & -7 \end{pmatrix}$.

5. (a), (b) The discretization matrix equation is

$$\begin{pmatrix} -2-2h^2 & 1 & & \\ 1 & \ddots & \ddots & \\ & \ddots & \ddots & \ddots \end{pmatrix}_{m,m} \begin{pmatrix} f_1 \\ f_2 \\ \vdots \end{pmatrix} = \begin{pmatrix} h^2k(-0.9) - f(-1) \\ h^2k(-0.8) \\ \vdots \end{pmatrix} \text{ for } k(x) := e^x + e^{-x}.$$

(d) Here are two plots with various extensions of the interval of integration to the right beyond the interval of interest, [-1, 1]:

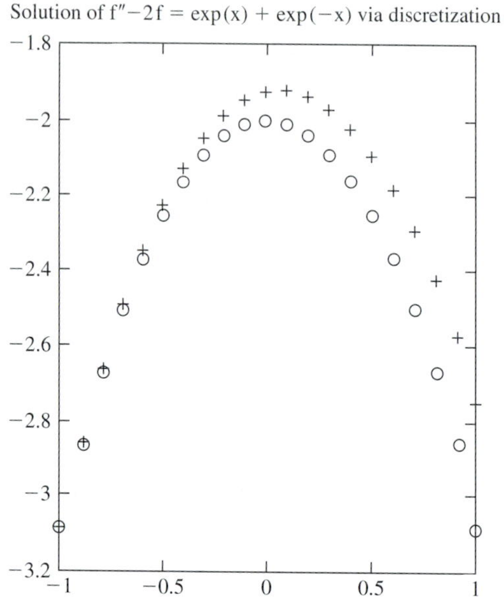

Exact (o) and computed (+) solutions; for h=1/10, integrated to 6

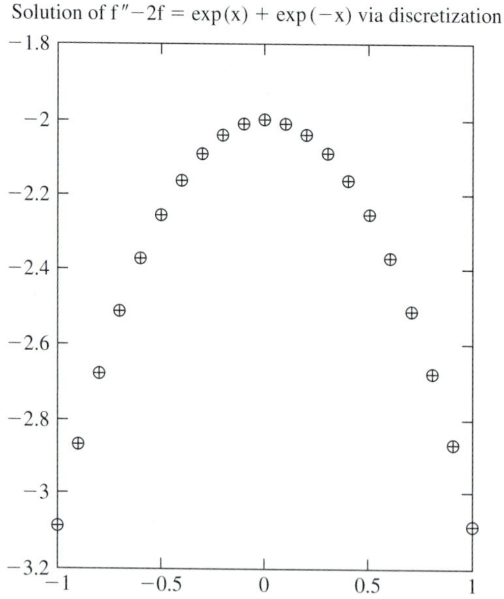

Exact (o) and computed (+) solutions; for h=1/10, integrated to 15

The exact solution is $f(x) = -e^x - e^{-x}$.

7. $\mathbf{S} = \begin{pmatrix} 2 & -2 \\ -2 & -1 \end{pmatrix}$ represents $\mathbf{P}$ symmetrically. $\mathbf{S}$ has the eigenvalues -2 and 4, with corresponding eigenvectors $\mathbf{v}_1 = \begin{pmatrix} 1 \\ 2 \end{pmatrix}$ and $\mathbf{v}_2 = \begin{pmatrix} -2 \\ 1 \end{pmatrix}$. $\mathbf{U}^T\mathbf{S}\mathbf{U} = \begin{pmatrix} -2 & 0 \\ 0 & 3 \end{pmatrix}$ for the ONB in $\mathbf{U} = \frac{1}{\sqrt{3}}\begin{pmatrix} 1 & -2 \\ 2 & 1 \end{pmatrix}$.

Thus, in the $\mathcal{U}$ basis, $-\frac{u_1^2}{\sqrt{3}^2} + \frac{u_2^2}{\sqrt{2}^2} = 1$, or $\mathbf{P}(x) = 6$ represents a hyperbola in $\mathbb{R}^2$ with the two axes of symmetry $\mathbf{u}_i$. It has vertices $\pm\sqrt{2}\mathbf{u}_2$ and opens up to the lower right and upper left of the plane.

9. Assume that $\mathbf{A} = \mathbf{UP}$ with $\mathbf{U}^T\mathbf{U} = \mathbf{I}_n$, $\mathbf{P} = \mathbf{P}^T$ and $\mathbf{UP} = \mathbf{PU}$.

Then $\mathbf{P}^T\mathbf{U}^T = (\mathbf{UP})^T = (\mathbf{PU})^T = \mathbf{U}^T\mathbf{P}^T$, and $\mathbf{U}^T$ and $\mathbf{P}^T$ also commute. Thus, $\mathbf{A}^T\mathbf{A} = (\mathbf{UP})^T\mathbf{UP} = \mathbf{P}^T\mathbf{U}^T\mathbf{UP} = \mathbf{P}^T\mathbf{P} = \mathbf{P}^2$ and $\mathbf{AA}^T = \mathbf{UP}(\mathbf{UP})^T = \mathbf{UPP}^T\mathbf{U}^T = \mathbf{PUU}^T\mathbf{P}^T = \mathbf{PP}^T = \mathbf{P}^2$ as well.

Regarding the converse, observe that without assuming commutativity of $\mathbf{U}$ and $\mathbf{P}$, we still have $\mathbf{A}^T\mathbf{A} = \mathbf{P}^2$ and $\mathbf{AA}^T = \mathbf{UP}^2\mathbf{U}^T$ if $\mathbf{A} = \mathbf{UP}$ is the polar decomposition. Thus, $\mathbf{P}^2\mathbf{U} = \mathbf{UP}^2$ if $\mathbf{A}$ is normal. Now conclude that if $\mathbf{P}^2$ and $\mathbf{U}$ commute, then $\mathbf{P}$ and $\mathbf{U}$ do.

11. $\mathbf{S} = \begin{pmatrix} 6 & -1 \\ -1 & 2 \end{pmatrix}$ represents $\mathbf{Q}$ symmetrically.

The eigenvalues of $\mathbf{S}$ are $4 \pm \sqrt{5} > 0$. Thus, $\mathbf{S}$ is positive definite by Definition 3 and can be written as $\mathbf{S} = \mathbf{VDV}^T$ for a positive diagonal matrix $\mathbf{D}$ and an orthogonal $\mathbf{V}$ by Theorem 11.3. Also, $\mathbf{Q}(\mathbf{x}) = \mathbf{x}^T\mathbf{S}\mathbf{x} = \mathbf{x}^T\mathbf{VDV}^T\mathbf{x} = \|\sqrt{\mathbf{D}}\mathbf{V}^T\mathbf{x}\|^2 > 0$ for all $\mathbf{x} \neq \mathbf{0}$, where $\sqrt{\mathbf{D}}$ is the diagonal matrix made up of the positive square roots $\sqrt{4+\sqrt{5}}$ and $\sqrt{4-\sqrt{5}}$ of the positive diagonal entries $4 \pm \sqrt{5}$ of $\mathbf{D}$.

13. (a) For both $\mathbf{H}$ and $\mathbf{K}$, the eigenvalues are $\pm i$.

(b) For $\mathbf{H}$, $\begin{pmatrix} 1-i \\ i \end{pmatrix}$ and $\begin{pmatrix} 1+i \\ -i \end{pmatrix}$ are the eigenvectors for i and $-i$, respectively. For $\mathbf{K}$, they are $\begin{pmatrix} 1+i \\ i \end{pmatrix}$ and $\begin{pmatrix} 1-i \\ -i \end{pmatrix}$ for i and $-i$.

(c) $\mathbf{X} = \begin{pmatrix} 1 & 2 \\ 0 & 1 \end{pmatrix}$.

15. We have $\begin{pmatrix} \mathbf{AB} & \mathbf{O}_{mn} \\ \mathbf{B} & \mathbf{O}_n \end{pmatrix}\begin{pmatrix} \mathbf{I}_m & \mathbf{A} \\ \mathbf{O}_{nm} & \mathbf{I}_n \end{pmatrix} = \begin{pmatrix} \mathbf{AB} & \mathbf{ABA} \\ \mathbf{B} & \mathbf{BA} \end{pmatrix} = \begin{pmatrix} \mathbf{I}_m & \mathbf{A} \\ \mathbf{O}_{nm} & \mathbf{I}_n \end{pmatrix}\begin{pmatrix} \mathbf{O}_{mm} & \mathbf{O}_{mn} \\ \mathbf{B} & \mathbf{BA} \end{pmatrix}$.

Thus, $\begin{pmatrix} \mathbf{AB} & \mathbf{O}_{mn} \\ \mathbf{B} & \mathbf{O}_n \end{pmatrix}$ and $\begin{pmatrix} \mathbf{O}_{mm} & \mathbf{O}_{mn} \\ \mathbf{B} & \mathbf{BA} \end{pmatrix}$ are similar. Since the eigenvalues of a block triangular matrix are the union of the eigenvalues of the diagonal blocks according to Theorem 9.6, we are done.

17. $\mathbf{A} = \mathbf{A}^T$ is nonsingular with trace$(\mathbf{A}) = 0 = \lambda_1 + \lambda_2$, so that $\mathbf{A}$ has two eigenvalues $\lambda_1 = -\lambda_2 > 0$. If $\mathbf{Ax} = \lambda_2\mathbf{x}$ for $\mathbf{x} \neq \mathbf{0}$, $\lambda_2 < 0$, and $\mathbf{A} = \mathbf{X}^T\mathbf{X}$, then $0 \leq \|\mathbf{Xx}\|^2 = \mathbf{x}^T\mathbf{X}^T\mathbf{Xx} = \mathbf{x}^T\mathbf{Ax} = \lambda_2\mathbf{x}^T\mathbf{x} < 0$, a contradiction.

19. Use Problem 18: $a_{ii} = \mathbf{e}_i^T\mathbf{A}\mathbf{e}_i > 0$.

21. For $\mathbf{x} = \mathbf{e}_j$, a standard unit vector, we have $\mathbf{x}^*\mathbf{Bx} = \mathbf{b}_{jj} = 0$ for all j.

For $\mathbf{x} = \mathbf{e}_j + i\mathbf{e}_\ell$, we have $\mathbf{x}^*\mathbf{Bx} = (\mathbf{x}_j - i\mathbf{e}_\ell)^T\mathbf{B}(\mathbf{e}_j + i\mathbf{e}_\ell) = -i(\mathbf{b}_{\ell j} - \mathbf{b}_{j\ell}) = \mathbf{0}$ for all j and ℓ (i.e., $\mathbf{B} = \mathbf{O}_n$).

23. According to Problem 22, $\mathbf{B}$ is hermitian and has only real eigenvalues. If $\mathbf{Bx} = \lambda\mathbf{x}$ for $\mathbf{x} \neq \mathbf{0}$, then $\mathbf{x}^*\mathbf{Bx} = \lambda\mathbf{x}^*\mathbf{x} = \lambda\|\mathbf{x}\|^2 > 0$ implies that $\lambda > 0$. Therefore, $\mathbf{B}$ is positive definite.

11.R p. 382

1. (a), (c), (d), (e), and (h) are false.

3. If $\mathbf{A}^*\mathbf{A} = \mathbf{AA}^*$, then the Schur Normal Form $\mathbf{R}$ of $\mathbf{A}$ is also normal. The only Schur normal forms that are diagonal are diagonal matrices. Hence, $\mathbf{A}$ is unitarily diagonalizable. The converse also holds.

5. (a) $\mathbf{u}$ and $\mathbf{v}$ are two eigenvectors with two distinct eigenvalues $\lambda = -1$ and $\lambda = 2$ of $\mathbf{A}$; hence, they are linearly independent.

(b) $\mathbf{u} = -\mathbf{Au}$ and $\mathbf{v} = \frac{1}{2}\mathbf{Av}$. So $\mathbf{u} \cdot \mathbf{v} = -\frac{1}{2}\mathbf{u}^T\mathbf{A}^T\mathbf{Av} = -\frac{1}{2}\mathbf{u}^T\mathbf{A}^2\mathbf{v} = -\frac{1}{2}\mathbf{u}^T4\mathbf{v} = -2\mathbf{u}^T\mathbf{v} = -2\mathbf{u} \cdot \mathbf{v}$, forcing $\mathbf{u} \cdot \mathbf{v} = 0$.

(c) $2 = \dim \text{span}\{\mathbf{u}, \mathbf{v}\}$ and $\text{span}\{\mathbf{u}, \mathbf{v}\} \not\subset \ker(\mathbf{A})$, so rank $\mathbf{A} \geq 2$.

7. If $\mathbf{A} = \mathbf{A}^T$, $\mathbf{A} = \mathbf{UDU}^T$, and $\mathbf{D} = \text{diag}(d_i)$ for $d_i > 0$, then set $\sqrt{\mathbf{D}} := \text{diag}(\sqrt{d_i}) \in \mathbb{R}^{n,n}$ and $\mathbf{X} := \sqrt{\mathbf{D}}\mathbf{U}^T$. Then $\mathbf{X}^T\mathbf{X} = \mathbf{U}\sqrt{\mathbf{D}}\sqrt{\mathbf{D}}\mathbf{U}^T = \mathbf{UDU}^T = \mathbf{A}$.

Conversely, if $\mathbf{A} = \mathbf{X}^T\mathbf{X}$ for a nonsingular matrix $\mathbf{X}$ and if $\mathbf{Ax} = \lambda\mathbf{x}$ for a normalized eigenvector $\mathbf{x}$ with $\|\mathbf{x}\| = 1$, then $\lambda = \lambda\mathbf{x}^T\mathbf{x} = \mathbf{x}^T\mathbf{Ax} = \mathbf{x}^T\mathbf{X}^T\mathbf{Xx} = \|\mathbf{Xx}\|^2 > 0$. That is, $\mathbf{A}$ has only positive real eigenvalues.

9. (a) True: $(\mathbf{S}^k)^T = (\mathbf{S}^T)^k = \mathbf{S}^k$ for any nonnegative powers k; and for any k, the matrix $\mathbf{S}^k$ is symmetric if $\mathbf{S}$ is nonsingular.

(b) True: $(\mathbf{U}^k)^T\mathbf{U}^k = (\mathbf{U}^{k-1})^T\mathbf{U}^T\mathbf{U}\mathbf{U}^{k-1} = (\mathbf{U}^{k-1})^T\mathbf{U}^{k-1} = \ \ldots \ = \mathbf{U}^T\mathbf{U} = \mathbf{I}$.

(c) Note that $(\mathbf{S}^2)^T = (\mathbf{S}^T)^2 = (-\mathbf{S})^2 = (-1)^2\mathbf{S}^2 = \mathbf{S}^2$. Similarly, for every even power, the matrix $\mathbf{S}^k$ is symmetric. On the other hand, all odd powers of $\mathbf{S}$ are skew symmetric.

Chapter 12

12.1.P p. 394

1. (a) We have $\mathbf{A}^T\mathbf{A} = \begin{pmatrix} 5 & -5 \\ -5 & 5 \end{pmatrix}$ with eigenvalues 10 and 0. Orthonormal eigenvectors are $\frac{1}{\sqrt{2}}\begin{pmatrix} 1 \\ -1 \end{pmatrix}$ and $\frac{1}{\sqrt{2}}\begin{pmatrix} -1 \\ -1 \end{pmatrix}$. These form the matrix $\mathbf{V}$ columnwise. $\mathbf{U}$ is obtained by evaluating $\frac{\mathbf{A}\mathbf{v}_j}{\sqrt{\lambda_j}}$ as $\frac{1}{\sqrt{5}}\begin{pmatrix} 1 \\ 2 \end{pmatrix}, \frac{1}{\sqrt{5}}\begin{pmatrix} -2 \\ 1 \end{pmatrix}$. Thus, the SVD of $\mathbf{A}$ is $\mathbf{U}\boldsymbol{\Sigma}\mathbf{V}^T = \begin{pmatrix} 1/\sqrt{5} & -2/\sqrt{5} \\ 2/\sqrt{5} & 1/\sqrt{5} \end{pmatrix}\begin{pmatrix} \sqrt{10} & 0 \\ 0 & 0 \end{pmatrix}$
$\begin{pmatrix} 1/\sqrt{2} & -1/\sqrt{2} \\ -1/\sqrt{2} & -1/\sqrt{2} \end{pmatrix}^T$.

(b) $\mathbf{B}^T\mathbf{B} = \begin{pmatrix} 5 & -2 \\ -2 & 8 \end{pmatrix}$; eigenvalues 9 and 4; ONB of eigenvectors $\frac{1}{\sqrt{5}}\begin{pmatrix} 1 \\ -2 \end{pmatrix}, \frac{1}{\sqrt{5}}\begin{pmatrix} 2 \\ 1 \end{pmatrix}$ The SVD of $\mathbf{B}$ is $\mathbf{U}\boldsymbol{\Sigma}\mathbf{V}^T = \begin{pmatrix} -2/\sqrt{5} & -1/\sqrt{5} \\ 1/\sqrt{5} & -2/\sqrt{5} \end{pmatrix}\begin{pmatrix} 3 & 0 \\ 0 & 2 \end{pmatrix}\cdot$
$\cdot\begin{pmatrix} 1/\sqrt{5} & 2/\sqrt{5} \\ -2/\sqrt{5} & 1/\sqrt{5} \end{pmatrix}^T$.

(c) SVD of $\mathbf{C} = \mathbf{U}\boldsymbol{\Sigma}\mathbf{V}^T =$
$$= \begin{pmatrix} 1 & 0 & 0 & 0 \\ 0 & 1/\sqrt{2} & 1/\sqrt{2} & 0 \\ 0 & 1/\sqrt{2} & -1/\sqrt{2} & 0 \\ 0 & 0 & 0 & 1 \end{pmatrix}\cdot$$
$$\cdot\begin{pmatrix} 3 & 0 & 0 \\ 0 & 3 & 0 \\ 0 & 0 & 1 \\ 0 & 0 & 0 \end{pmatrix}\begin{pmatrix} 1 & 0 & 0 \\ 0 & 1/\sqrt{2} & -1/\sqrt{2} \\ 0 & 1/\sqrt{2} & 1/\sqrt{2} \end{pmatrix}^T.$$

(d) SVD of $\mathbf{F} = \mathbf{U}\boldsymbol{\Sigma}\mathbf{V}^T = \begin{pmatrix} 1 & 0 & 0 \\ 0 & 1/\sqrt{2} & -1/\sqrt{2} \\ 0 & 1/\sqrt{2} & 1/\sqrt{2} \end{pmatrix}\cdot$
$$\cdot\begin{pmatrix} 2 & 0 \\ 0 & \sqrt{2} \\ 0 & 0 \end{pmatrix}\mathbf{I}_2\,.$$

(e) $\mathbf{G}$ has the singular values $\sqrt{5.5 \pm \sqrt{5.25}}$.

(f) SVD of $\mathbf{H} = \mathbf{U}\boldsymbol{\Sigma}\mathbf{V}^T =$
$$= \begin{pmatrix} 1/\sqrt{6} & 1/\sqrt{2} & -1/\sqrt{3} \\ 2/\sqrt{6} & 0 & 1/\sqrt{3} \\ -1/\sqrt{6} & 1/\sqrt{2} & 1/\sqrt{3} \end{pmatrix}\begin{pmatrix} \sqrt{3} & 0 \\ 0 & 1 \\ 0 & 0 \end{pmatrix}\cdot$$
$$\cdot\begin{pmatrix} 1/\sqrt{2} & -1/\sqrt{2} \\ 1/\sqrt{2} & 1/\sqrt{2} \end{pmatrix}^T.$$

(g) SVD of $\mathbf{K} = \mathbf{U}\boldsymbol{\Sigma}\mathbf{V}^T = \begin{pmatrix} 1/\sqrt{3} & -2/\sqrt{3} \\ 2/\sqrt{3} & 1/\sqrt{3} \end{pmatrix}\cdot$
$$\cdot\begin{pmatrix} 3 & 0 & 0 \\ 0 & 3 & 0 \end{pmatrix}\begin{pmatrix} 5/\sqrt{45} & 0 & 2/3 \\ 4/\sqrt{45} & -1/\sqrt{3} & -2/3 \\ -2/\sqrt{45} & -2/\sqrt{3} & 1/3 \end{pmatrix}^T.$$

3. If $\mathbf{A}_{nn}$ has the singular values $\sigma_1 \geq \ldots \geq \sigma_n > 0$, set $\boldsymbol{\Sigma}^- := \operatorname{diag}\left(\frac{1}{\sigma_n}, \ldots, \frac{1}{\sigma_1}\right)$ and $\mathbf{E}_{nn} := \begin{pmatrix} 0 & & 1 \\ & \cdot^{\cdot^{\cdot}} & \\ 1 & & 0 \end{pmatrix}$. Then $\mathbf{A}^{-1} = (\mathbf{VE})\boldsymbol{\Sigma}^-(\mathbf{UE})^T$ is the SVD of $\mathbf{A}^{-1}$, since $\mathbf{E}$ just shuffles the left and right singular vectors to order the entries on the diagonal of $\boldsymbol{\Sigma}^-$ appropriately.

5. $\mathbf{U}_{mm}^T\mathbf{O}_{mn}\mathbf{V}_{nn} = \mathbf{O}_{mn}$, for two arbitrary orthogonal matrices $\mathbf{U}$ and $\mathbf{V}$, and $\mathbf{U}_{nn}^T\mathbf{I}_n\mathbf{U}_{nn} = \mathbf{I}_n$, for an arbitrary orthogonal matrix $\mathbf{U}$, are the two singular value decompositions.

7. If $\mathbf{x} \neq \mathbf{0}$ is a column vector, then $\left(-\ \frac{\mathbf{x}}{\|\mathbf{x}\|}\ -\right)\begin{pmatrix} | \\ \mathbf{x} \\ | \end{pmatrix}$
$1 = \|\mathbf{x}\| > 0$ is the SVD for $\mathbf{x}$.

For the column vector $\mathbf{x} = 0$, the SVD is $\mathbf{e}^T\mathbf{x}1 = \mathbf{0}$ for any column vector $\mathbf{e}$ with $\|\mathbf{e}\| = 1$.

If $\mathbf{y} \neq \mathbf{0}$ is a row vector, then $1\left(-\ \mathbf{y}\ -\right)\begin{pmatrix} | \\ \frac{\mathbf{y}}{\|\mathbf{y}\|} \\ | \end{pmatrix}$
$= \|\mathbf{y}\| > 0$ is the SVD for $\mathbf{y}$.

For the row vector $\mathbf{y} = \mathbf{0}$, the SVD is $1\mathbf{y}\mathbf{e}$ for any column vector $\mathbf{e}$ with $\|\mathbf{e}\| = 1$.

9. The right and left singular values and vectors switch position: If $\mathbf{A} = \mathbf{U}\boldsymbol{\Sigma}\mathbf{V}^T$, then $\mathbf{A}^T = \mathbf{V}\boldsymbol{\Sigma}^T\mathbf{U}^T$, according to Chapter 6.

11. Via determinants: $\prod(singular\ values) = \det\boldsymbol{\Sigma} = \det(\mathbf{U}^*\mathbf{AV}) = |\det\mathbf{A}| = |\prod(eigenvalues)|$ since $|\det\mathbf{U}| = 1 = |\det\mathbf{V}|$ for orthogonal matrices.

13. If $\mathbf{x} \in \ker(\mathbf{A})$, then $\mathbf{Ax} = \mathbf{0}$ and $\mathbf{XAx} = \mathbf{X0} = \mathbf{0}$ as well, so that $\mathbf{x} \in \ker(\mathbf{XA})$.

If $\mathbf{x} \in \ker(\mathbf{XA})$, then $\mathbf{XAx} = \mathbf{0}$. Since $\mathbf{X}$ is nonsingular, multiply this equation by $\mathbf{X}^{-1}$ on the left to obtain $\mathbf{Ax} = \mathbf{X}^{-1}\mathbf{XAx} = \mathbf{X}^{-1}\mathbf{0} = \mathbf{0}$. Thus, $\mathbf{x} \in \ker(\mathbf{A})$.

15. Note that the matrices $\frac{1}{\sqrt{5}}\mathbf{V}_0$ and $\frac{1}{\sqrt{6}}\mathbf{U}_0$ are orthogonal. Hence, the SVD of $\mathbf{A}$ is almost known:

(a) Singular values: 3, 2.

(b) Singular vectors of $\mathbf{A}$: the normalized columns of $\mathbf{U}_0$ and $\mathbf{V}_0$.

(c) SVD of $\mathbf{A}$: $\left(\frac{1}{\sqrt{6}}\mathbf{U}_0\right)\boldsymbol{\Sigma}\left(\frac{1}{\sqrt{5}}\mathbf{V}_0^T\right)$.

17. True: (a), (b), (e), (g), and (h). To see that (d) is false, try it for $\mathbf{A} = \begin{pmatrix} 0 & 1 \\ 0 & 0 \end{pmatrix}$ and $\mathbf{A}^2 = \mathbf{O}_2$.

19. (a) Singular values for $\mathbf{A}$: 5, 0; for $\mathbf{A}^2$: 25, 0.

(b) $\mathbf{u} = \frac{1}{\sqrt{5}}\begin{pmatrix} 2 \\ 1 \end{pmatrix}$, the dominant right singular vector.

21. Only (b) and (c) are true.

23. The $|d_{j_i j_i}|$ are the singular values if they are ordered in decreasing order. The unit vectors $\mathbf{e}_{j_i}$, ordered conformally, are the right singular vectors if the left singular vectors are chosen as the unit vectors $\mathbf{e}_i$ in standard order.

25. The SVD of $\mathbf{A}$ is $\mathbf{U}\mathbf{I}_n\mathbf{V}^T$, which is orthogonal as the product of three (or two) orthogonal matrices.

27. All parts are true. (b) follows from the fact that the eigenvalues of $\mathbf{A}^T\mathbf{A}$ are λ_i^2 if $\mathbf{A}$ is symmetric with the eigenvalues λ_i.

29. If $\mathbf{A}_{nn}$ is invertible, then so is $\mathbf{A}^T\mathbf{A}$ by Theorems 6.5 and 7.2, and $\mathbf{A}^T\mathbf{A}$ is positive definite. Hence, by Theorem 12.2, all singular values of $\mathbf{A}$ are nonzero.

If zero is a singular value of $\mathbf{A}$, then in the SVD $\mathbf{A} = \mathbf{U}\boldsymbol{\Sigma}\mathbf{V}^T$ of $\mathbf{A}$, we have $\text{rank}(\mathbf{U}\boldsymbol{\Sigma}\mathbf{V}^T) < n$, since $\text{rank}(\boldsymbol{\Sigma}) < n$. Thus, $\mathbf{A}$ is singular.

31. For $\mathbf{A} = \begin{pmatrix} 1 & 1 \\ 0 & 0 \end{pmatrix}$, we have $\mathbf{A}^2 = \begin{pmatrix} 1 & 1 \\ 0 & 0 \end{pmatrix}\begin{pmatrix} 1 & 1 \\ 0 & 0 \end{pmatrix} = \begin{pmatrix} 1 & 1 \\ 0 & 0 \end{pmatrix} = \mathbf{A}$. Hence, the singular values of $\mathbf{A}$ and $\mathbf{A}^2$ are the same. They are the square roots of the eigenvalues of $\mathbf{A}^T\mathbf{A} = \begin{pmatrix} 1 & 1 \\ 1 & 1 \end{pmatrix}$, or of 0 and 2. But $\sqrt{2}$ is not a singular value of $\mathbf{A}$, while 2 is for $\mathbf{A}^2$, serving as a counterexample.

33. Using Lemma 1 twice, namely, $\|\mathbf{U}\mathbf{x}\| = \|\mathbf{x}\|$ for an orthogonal matrix $\mathbf{U}$, and the SVD $\mathbf{A} = \mathbf{U}\boldsymbol{\Sigma}\mathbf{V}^T$ of $\mathbf{A}$, we obtain

$$\begin{aligned}\min_{\|\mathbf{x}\|=1}\|\mathbf{A}\mathbf{x}\| &= \min_{\|\mathbf{x}\|=1}\|\mathbf{U}\boldsymbol{\Sigma}\mathbf{V}^T\mathbf{x}\| = \min_{\|x\|=1}\|\boldsymbol{\Sigma}\mathbf{V}^T\mathbf{x}\| \\ &= \min_{\|\mathbf{V}^T\mathbf{x}\|=1}\|\boldsymbol{\Sigma}\mathbf{V}^T\mathbf{x}\| = \min_{\|y\|=1}\|\boldsymbol{\Sigma}\mathbf{y}\| \\ &= \min_{\|\mathbf{y}\|=1}\sqrt{\sum_i \sigma_i^2 y_i^2} = \sigma_n,\end{aligned}$$

where the last equation follows in a manner similar to the proof of Lemma 4, but for the minimum.

35. If $\mathbf{A} = \mathbf{U}\boldsymbol{\Sigma}\mathbf{V}^T$, then $\mathbf{A}^T = \mathbf{V}\boldsymbol{\Sigma}^T\mathbf{U}^T$ and $\mathbf{A}^T\mathbf{U} = \mathbf{V}\boldsymbol{\Sigma}^T$. Interpreting the last matrix equation for one column, we thus have $\mathbf{A}^T\mathbf{u} = \mathbf{v}\cdot\sigma = \sigma\mathbf{v}$ if $\mathbf{u}$ and $\mathbf{v}$ are the respective singular vectors associated with $\mathbf{A}$'s singular value σ.

12.2.P p. 404

1. (a) $\|\mathbf{A}\| = \sqrt{10} = \sigma_1$, $\|\mathbf{B}\| = 3$, $\|\mathbf{C}\| = 3$, $\|\mathbf{F}\| = 2$, $\|\mathbf{G}\| = \sqrt{5.5 + \sqrt{5.25}}$, $\|\mathbf{H}\| = \sqrt{3}$, $\|\mathbf{K}\| = 3$.

(b) $\mathbf{A}_1 = \sigma_1\begin{pmatrix} | \\ \mathbf{u}_1 \\ | \end{pmatrix}\begin{pmatrix} - & \mathbf{v}_1^T & - \end{pmatrix} = \mathbf{A}$.

$\mathbf{B}_1 = \sigma_1\begin{pmatrix} | \\ \mathbf{u}_1 \\ | \end{pmatrix}\begin{pmatrix} - & \mathbf{v}_1^T & - \end{pmatrix} = \begin{pmatrix} -1.2 & 2.4 \\ 0.6 & -1.2 \end{pmatrix}$.

$\mathbf{H}_1 = \begin{pmatrix} 0.5 & 0.5 \\ 1 & 1 \\ -0.5 & -0.5 \end{pmatrix}$.

$\mathbf{K}_1 = \begin{pmatrix} 1 & 0.8 & -0.4 \\ 2 & 1.6 & -0.8 \end{pmatrix}$.

(c) $\mathbf{A}_2 = \mathbf{A}$, since $\mathbf{A}$ has rank 1.

$\mathbf{C}_2 = \sigma_1\begin{pmatrix} | \\ \mathbf{u}_1 \\ | \end{pmatrix}\begin{pmatrix} - & \mathbf{v}_1^T & - \end{pmatrix} + \sigma_2\begin{pmatrix} | \\ \mathbf{u}_2 \\ | \end{pmatrix}\cdot\begin{pmatrix} - & \mathbf{v}_2^T & - \end{pmatrix} = \begin{pmatrix} 3 & 0 & 0 \\ 0 & 1.5 & 1.5 \\ 0 & 1.5 & 1.5 \\ 0 & 0 & 0 \end{pmatrix}$.

$\mathbf{F}_2 = \mathbf{F}$, since $\mathbf{F}$ has rank 2.

3. (a) $\mathbf{A}$ is 6 by 6, as are $\mathbf{U}$ and $\mathbf{V}$.

(b) $\|\mathbf{A}\| = 22 = \sigma_1$.

(c) 5 and 0.0002, respectively.

(d) $\mathbf{A}$ is singular. (e) $\text{rank}(\mathbf{A}) = 5$.

5. Singular values of $\mathbf{A}$: $\sqrt{59 \pm \sqrt{3385}}$.

SVD of $\mathbf{A}$: $\mathbf{A} = \mathbf{U}\boldsymbol{\Sigma}\mathbf{V}^T =$

$$\begin{pmatrix} 0.2830 & 0.8679 & 0.4082 \\ 0.5384 & 0.2085 & -0.8165 \\ 0.7938 & -0.4508 & 0.4082 \end{pmatrix}\begin{pmatrix} 10.8250 & 0 \\ 0 & 0.9051 \\ 0 & 0 \end{pmatrix}\cdot\begin{pmatrix} 0.5420 & -0.8404 \\ 0.8404 & 0.5420 \end{pmatrix}^T ; \mathbf{x}_{LS} = \begin{pmatrix} 0.25 \\ 0.25 \end{pmatrix}.$$

7. Since $\mathbf{A} = \mathbf{A}^T$ can be orthogonally diagonalized as $\mathbf{A} = \mathbf{U}\,\text{diag}(\lambda_i)\mathbf{U}^T$ with $\mathbf{U}^T\mathbf{U} = \mathbf{I}_n$, the singular values of $\mathbf{A}$ are $|\lambda_i|$, reordered in decreasing order. The singular vectors are the similarly reordered columns of $\mathbf{U}$, possibly multiplied by a negative sign if $\lambda_i < 0$, so that the dominant eigenvectors come first.

9. $\mathbf{A}$'s singular values are all equal to 1. $\mathbf{A} = \mathbf{A}\mathbf{I}_n\mathbf{I}_n$ or $\mathbf{A} = \mathbf{I}_n\mathbf{I}_n\mathbf{A}$ are two singular value decompositions of $\mathbf{A}$, which is an orthogonal matrix.

$\mathbf{B}$ has the single nonzero singular value n^2 for the right singular vector $\mathbf{u}_1 = \frac{1}{\sqrt{n}}\begin{pmatrix}1\\ \vdots\\ 1\end{pmatrix}$.

$\mathbf{C}$ has the single singular value 1 and an $(n-1)$-fold singular value $n-1$ with right singular vectors $\sum_i \frac{1}{\sqrt{n}}\mathbf{e}_i$ and $\frac{1}{\sqrt{2}}(\mathbf{e}_1 - \mathbf{e}_j)$ for $j = 2, \dots, n$, respectively. All other singular values are zero, and the corresponding singular vectors can be chosen as any orthonormal fill–in.

11. (a) $\mathbf{J}^T\mathbf{J} = \begin{pmatrix} 0 & & & 0\\ & 0 & 1 & \\ & & & \ddots & \\ & & & & 1\end{pmatrix}$. Thus, $\mathbf{J}$ has the singular value 1, $(n-1)$-fold, and zero once.

(b) $\|\mathbf{J}\| = 1 = \sigma_1$.

(c) $\mathbf{J}^k$ is the $n \times n$ matrix of all zeros, except for a superdiagonal of all ones in the kth superdiagonal for each $1 \le k \le n-1$. Thus, $\mathbf{J}^{n-1} \neq \mathbf{O}_n = \mathbf{J}^n$.

(d) 1 $(n-k)$-fold and 0 k–fold. Simply compute $(\mathbf{J}^k)^T\mathbf{J}^k$. It is a diagonal matrix with k zeros and $n-k$ ones on its diagonal.

(e) $\|\mathbf{J}^k\| = \begin{cases} 1 & \text{if } k < n\\ 0 & \text{if } k \ge n\end{cases}$.

13. $\mathbf{A} = \sigma_1\mathbf{u}_1\mathbf{v}_1^T + \sigma_2\mathbf{u}_2\mathbf{v}_2^T$ from the SVD

$$\mathbf{U\Sigma V}^T = \begin{pmatrix} 0.95857 & -0.09857 & 0.26726\\ -0.1644 & 0.57454 & 0.80178\\ 0.23259 & 0.81252 & -0.5345\end{pmatrix} \cdot \begin{pmatrix} 2.3268 & 0 & 0\\ 0 & 1.6080 & 0\\ 0 & 0 & 0\end{pmatrix}\begin{pmatrix} 0 & 0 & 1\\ 0.92388 & 0.38268 & 0\\ -0.38268 & 0.92388 & 0\end{pmatrix}^T$$

of $\mathbf{A}$; or $\mathbf{A} = \begin{pmatrix}-1\\ 1\\ 1\end{pmatrix}\begin{pmatrix}0 & 0 & 1\end{pmatrix} + \begin{pmatrix}2\\ 0\\ 1\end{pmatrix}\begin{pmatrix}0 & 1 & 0\end{pmatrix}$.

15. $\mathbf{A}_1 = \begin{pmatrix}1\\ 0\\ 0\end{pmatrix}\begin{pmatrix}0 & 1 & 0\end{pmatrix}$, $\mathbf{B}_1 = \begin{pmatrix}0\\ 1\\ 0\end{pmatrix}\begin{pmatrix}0 & 0 & 1\end{pmatrix}$ and $\mathbf{C}_1 = \begin{pmatrix}0\\ -2\\ 0\end{pmatrix}\begin{pmatrix}0 & 0 & -1/2\end{pmatrix}$ approximate $\mathbf{A}$ with $\|\mathbf{A} - \mathbf{A}_1\| = \|\mathbf{A} - \mathbf{B}_1\| = \|\mathbf{A} - \mathbf{C}_1\| = 1$.

17. $\mathbf{A}_{nn} = \sum_{i=1}^{n} \lambda_i \begin{pmatrix} | \\ \mathbf{x}_i \\ | \end{pmatrix}\begin{pmatrix} - & \mathbf{y}_i^T & - \end{pmatrix}$ for the eigenvalues λ_i of $\mathbf{A}$, the eigenvectors $\mathbf{x}_i$, and the row vectors $\mathbf{y}_i^T$ of the inverse $\mathbf{X}^{-1}$ of the eigenvector matrix $\mathbf{X} = \begin{pmatrix} | & & | \\ \mathbf{x}_1 & \dots & \mathbf{x}_n \\ | & & | \end{pmatrix}$.

12.3.P p. 412

1. (a) The MATLAB commands `A=[1 -2 7 13;-2 -4 0 2;3 12 -6 -9;0 0 5 0];` `[U S V]=svd(A)` produce the left and right singular vector matrices $\mathbf{U}$ and $\mathbf{V}$ and $\mathbf{\Sigma} = \mathbf{U}^T\mathbf{AV}$ as follows:

```
U =
-6.3908e-01  -7.2683e-01   2.3000e-01   -1.0194e-01
-1.7481e-01   3.4681e-01   2.1284e-01   -8.9659e-01
 7.4149e-01  -5.6862e-01   1.1380e-01   -3.3750e-01
-1.0587e-01  -1.6766e-01  -9.4279e-01   -2.6802e-01
S =
2.1128e+01            0            0             0
         0   8.6594e+00            0             0
         0            0   4.4814e+00             0
         0            0            0    7.3180e-01
V =
 9.1585e-02  -3.6103e-01   3.2516e-02    9.2748e-01
 5.1473e-01  -7.8031e-01   1.2106e-02   -3.5500e-01
-4.4736e-01  -2.9036e-01  -8.4500e-01   -3.9228e-02
-7.2563e-01  -4.2008e-01   5.3365e-01   -1.1058e-01
```

(b) $\text{rank}(\mathbf{A}) = 4$.

(c) The command

`A2 = U(:,1:2)*S(1:2,1:2)*V(:,1:2)'` produces the best rank 2 approximation $\mathbf{A}_2$ of $\mathbf{A}$ as

```
A2 =
 1.0357e+00  -2.0390e+00   7.8680e+00   1.2442e+01
-1.4225e+00  -4.2445e+00   7.8025e-01   1.4184e+00
 3.2125e+00   1.1906e+01  -5.5787e+00  -9.2995e+00
 3.1929e-01  -1.8482e-02   1.4222e+00   2.2330e+00
```

Its distance to $\mathbf{A}$ is equal to the third-largest singular value 4.4814e+00 of $\mathbf{A}$.

(d), (e) The column space of $\mathbf{A}$ is all of $\mathbb{R}^4$; any ONB will span it. The null space of $\mathbf{A}$ is $\{\mathbf{0}\}$, which does not have a basis.

For the image of $\mathbf{A}_2$ an ONB is given by the first two columns of $\mathbf{U}$, while the last two columns of $\mathbf{V}$ form an ONB for $\mathbf{A}_2$'s kernel.

3. Here is a useful MATLAB m.file † for this problem:

†Download from `http://www.auburn.edu/~uhligfd/TLA/download`

```
function mess = SVDword(word)
[u,s,v]= svd(word); [n,m] = size(word);
mess = zeros(n,m); d = diag(s);
format +, disp(word), format, pause
for k = 1:min(n,m)
  format short, fprintf('k = %g \n \n',k),
  mess = mess + d(k) * u(:,k) * v(:,k)';
  messk = mess;
   for i = 1:n
     for j = 1:m
      if abs(messk(i,j)) < 0.5
        messk(i,j) = 0; end; end; end
  format +, disp(messk), format, pause,
end
```

5. A\b yields $\mathbf{x}_{LS} = \begin{pmatrix} 4 \\ 1 \end{pmatrix}$ with error $\|\mathbf{A}\mathbf{x}_{LS} - \mathbf{b}\| = 7.0711$.

7. (b) The MATLAB commands `A=[5 3 -1;0 2 1;1 1 4];  [ua sa va]=svd(A); As=ua(:,1)*sa(1,1)*va(:,1)'` compute **As** as

```
As =
4.7181e+00   3.2989e+00   2.7070e-01
9.7560e-01   6.8214e-01   5.5976e-02
1.2926e+00   9.0374e-01   7.4161e-02
```

Similarly for **B**: `B=[5 1;8 -2], [ub sb vb]=svd(B), Bs=ub(:,1)*sb(1,1)*vb(:,1)'` gives

```
Bs =
4.7918e+00   -6.1708e-01
8.1228e+00   -1.0461e+00
```

The commands
`A=[5 3 -1;0 2 1;1 1 4];`
`[va ea]=eig(A);`
`iva=inv(va);`
`Ae=va(:,1)*ea(1,1)*iva(:,1)'` produce

```
Ae =
3.3333e+00   3.2733e+00   -6.2994e-01
1.6667e+00   1.6366e+00   -3.1497e-01
5.0000e+00   4.9099e+00   -9.4491e-01
```

The commands `B=[5 1;8 -2],`
`[vb eb]=eig(B),`
`ivb=inv(vb); k=1;`
`Be=vb(:,k)*eb(k,k)*ivb(:,k)'` produce the dominant eigenvalue–eigenvector dyad **Be** for **B** as

```
Be =
5.3333e+00   -3.8006e+00
5.3333e+00   -3.8006e+00
```

For example, $\|\mathbf{B} - \mathbf{Be}\| \doteq 5.178$ and $\|\mathbf{B} - \mathbf{Bs}\| = \sigma_2 \doteq 1.893$, while $\|\mathbf{A} - \mathbf{Ae}\| \doteq 7.709$ and $\|\mathbf{A} - \mathbf{As}\| = \sigma_2 \doteq 4.3066$, indicating that the SVD dyadic approximation of a matrix is much better than the eigenvalue–eigenvector based one.

12.R p. 414

1. Use $\mathbf{\Sigma}$ from the SVD for $\mathbf{A}_\varepsilon$, for example.

3. (a) $\text{rank}(\mathbf{A}) = 3$.

(b) $\mathbf{A} = \mathbf{U}\mathbf{\Sigma}\mathbf{V}^T =$

$$\begin{pmatrix} -0.0506 & -0.2444 & 0.0097 & 0.7121 & 0.6562 \\ -0.1422 & -0.0100 & -0.9898 & 0 & 0 \\ 0.8383 & 0.4198 & -0.1247 & -0.0837 & 0.3137 \\ 0.2167 & -0.7675 & -0.0234 & -0.5234 & 0.2992 \\ 0.4769 & -0.4182 & -0.0643 & 0.4605 & -0.6177 \end{pmatrix} \cdot$$

$$\begin{pmatrix} 76.7048 & 0 & 0 & 0 & 0 \\ 0 & 26.1799 & 0 & 0 & 0 \\ 0 & 0 & 1.7288 & 0 & 0 \\ 0 & 0 & 0 & 0 & 0 \\ 0 & 0 & 0 & 0 & 0 \end{pmatrix} \cdot$$

$$\begin{pmatrix} 0.4192 & -0.2522 & 0.8721 & 0 & 0 \\ -0.1285 & 0.3911 & 0.1748 & 0.8944 & -0.0034 \\ 0 & 0 & 0 & -0.0038 & -1 \\ 0.8612 & 0.4144 & -0.2942 & 0 & 0 \\ -0.2569 & 0.7821 & 0.3496 & -0.4472 & 0.0017 \end{pmatrix}^T .$$

(c) The first three columns of **U** span the column space of **A**, as do the three columns in the defining dyad representation of **A**.

(d) $\mathbf{A}^T = \mathbf{V}\mathbf{\Sigma}^T\mathbf{U}^T$.

(e) The first three columns of **V** span the column space of $\mathbf{A}^T$ (i.e., the row space of **A** is spanned by the transposed first three columns of **V**). Likewise, the three rows that define **A** dyadically span its row space.

(f) The last two columns of **V** span the kernel of **A**.

(g) $\|\mathbf{A}\| = 76.7048$.

(h) $\mathbf{A}_2 =$

$$\begin{pmatrix} -0.0506 & -0.2444 \\ -0.1422 & -0.0100 \\ 0.8383 & 0.4198 \\ 0.2167 & -0.7675 \\ 0.4769 & -0.4182 \end{pmatrix} \begin{pmatrix} 76.7048 & 0 \\ 0 & 26.1799 \end{pmatrix} \cdot$$

$$\cdot \begin{pmatrix} 0.4192 & -0.1285 & 0 & 0.8612 & -0.2569 \\ -0.2522 & 0.3911 & 0 & 0.4144 & 0.7821 \end{pmatrix} .$$

(i) $\mathbf{x}_{LS} = \begin{pmatrix} -0.0032 \\ 0 \\ 0 \\ -0.0057 \\ -0.0355 \end{pmatrix}$; $\mathbf{y}_{LS} = \begin{pmatrix} 0.8919 \\ 0 \\ 0 \\ -0.3032 \\ 0.3113 \end{pmatrix}$.

5. (a) The singular values of $\mathbf{B} = \mathbf{UAV}$ are identical to the singular values of $\mathbf{A}$.

(b) They are not related at all.

7. According to Chapter 11, the positive definite symmetric matrix $\mathbf{A}$ can be orthogonally diagonalized as $\mathbf{U}^T\mathbf{AU} = \mathbf{D}$, with $\mathbf{D}$ having positive diagonal entries. They are the eigenvalues of $\mathbf{A}$. Thus, $\mathbf{A} = \mathbf{UDU}^T$ is also an SVD of $\mathbf{A}$. And $\mathbf{A}$'s eigenvalues and singular values coincide.

Chapter 13

13.1.P p. 420

1. The QR iterates of $\mathbf{A}$ do not converge to upper triangular form. The (1,1) entry converges to the eigenvalue 2.3532 of $\mathbf{A}$, but the lower right diagonal 2×2 submatrix oscillates. It represents the two complex conjugate eigenvalues of $\mathbf{A}$ and obviously $\mathbf{A}$ cannot be triangularized over the reals.

For $\mathbf{B}$, the unshifted basic QR algorithm converges to triangular form with $\mathbf{B}$'s three real eigenvalues on the diagonal.

Hence, the QR algorithm converges to upper triangular, or triangular Schur form, only if all eigenvalues are real. If they are not, it converges to a block upper triangular matrix with square diagonal blocks of sizes 1 or 2, the latter appearing for complex conjugate pairs of eigenvalues.

3. (d) Since only the subdiagonal entries in positions (3,2) and (5,4) (together with the lower triangular block for the QR iterates of $\mathbf{A}$) seem to converge to zero in the basic QR iterations, $\mathbf{A}$ likely has three complex conjugate pairs of eigenvalues.

(e) The singular values of $\mathbf{A}$ compute as 3.7426, 1.7997, 1.5701, 1.5130, 1.4142, and 1.4142.

Appendix A

A.A.P p. 430

1. $|(a+bi)(c+di)| = |ac - bd + (ad+bc)i|$
$= \sqrt{(ac-bd)^2 + (ad+bc)^2}$
$= \sqrt{a^2c^2 + b^2d^2 + a^2d^2 + b^2c^2}$ and $|a+bi||c+di|$
$= \sqrt{a^2+b^2}\sqrt{c^2+d^2} = \sqrt{a^2c^2 + a^2d^2 + b^2c^2 + b^2d^2}$
coincide.

3. $zw = 8 + i$, $\dfrac{z}{w} = \dfrac{1}{13}(4 - 7i)$, $\dfrac{w}{z} = \dfrac{1}{5}(4 + 7i)$, and $z - 2w = -4 - 5i$.

5. (a) $\|\mathbf{w}\| = \sqrt{4} = 2$; (b) $\|\mathbf{z}\| = \sqrt{10}$; (c) $\sqrt{22}$;

(d) $\mathbf{u} = \begin{pmatrix} -3+2i \\ 7+8i \end{pmatrix}$, $\|\mathbf{u}\| = \sqrt{126}$;

(e) $\|\mathbf{u} - \mathbf{w}\| = \sqrt{134}$; (f) $\|\mathbf{x}\| = \sqrt{14}$; (g) $\|\mathbf{y}\| = \sqrt{62}$.

7. $\mathbf{A}(\alpha)\mathbf{A}(\beta) =$
$\begin{pmatrix} \cos\alpha & -\sin\alpha \\ \sin\alpha & \cos\alpha \end{pmatrix} \begin{pmatrix} \cos\beta & -\sin\beta \\ \sin\beta & \cos\beta \end{pmatrix} =$
$\begin{pmatrix} \cos\alpha\cos\beta - \sin\alpha\sin\beta & -\sin\alpha\cos\beta - \cos\alpha\sin\beta \\ \sin\alpha\cos\beta + \cos\alpha\sin\beta & \cos\alpha\cos\beta - \sin\alpha\sin\beta \end{pmatrix}$
$= \begin{pmatrix} \cos(\alpha+\beta) & -\sin(\alpha+\beta) \\ \sin(\alpha+\beta) & \cos(\alpha+\beta) \end{pmatrix} = \mathbf{A}(\alpha+\beta) = \mathbf{A}(\beta + \alpha) = \mathbf{A}(\beta)\mathbf{A}(\alpha)$ by symmetry.

9. (a) Use the facts that $z^{-1} = \frac{\bar{z}}{|z|^2}$ with its sustaining angle equal to the negative one for z and that $\cos(-\alpha) = \cos(\alpha)$ and $\sin(-\alpha) = -\sin(\alpha)$ for all α.

(b) $\mathbf{ZZ}^{-1}$
$= |z| \begin{pmatrix} \cos\alpha & -\sin\alpha \\ \sin\alpha & \cos\alpha \end{pmatrix} \frac{1}{|\bar{z}|} \begin{pmatrix} \cos\alpha & \sin\alpha \\ -\sin\alpha & \cos\alpha \end{pmatrix}$
$= \begin{pmatrix} \cos^2(\alpha) + \sin^2(\alpha) & 0 \\ 0 & \cos^2(\alpha) + \sin^2(\alpha) \end{pmatrix} = \mathbf{I}_2,$
since $|z| = \sqrt{a^2+b^2} = |\bar{z}|$ for any $z = a + bi$ and $\bar{z} = a - bi \in \mathbb{C}$.

(c) Clearly, $|z| = |\bar{z}|$, and for the two matrix representations we have $\mathbf{Z}^T = |z| \begin{pmatrix} \cos\alpha & \sin\alpha \\ -\sin\alpha & \cos\alpha \end{pmatrix} =$
$|\bar{z}| \begin{pmatrix} \cos(-\alpha) & -\sin(-\alpha) \\ \sin(-\alpha) & \cos(-\alpha) \end{pmatrix} = \bar{\mathbf{Z}}$ with $\bar{z} = a - bi$ for $z = a + bi$.

11. (a) $\overline{zw} = (a - bi)(c - di) = ac - bd - (ad + bc)i$, while $\overline{zw} = \overline{(a+bi)(c+di)} = \overline{ac - bd + (ad+bc)i}$ $= ac - bd - (ad+bc)i$.

(b) $\overline{z + w} = \overline{a + bi + c + di} = \overline{a + c + (b+d)i} = a + c - (b+d)i$ and $\bar{z} + \bar{w} = \overline{a + bi} + \overline{c + di} = a + c - (b+d)i$.

(c) Use (a): $\overline{z^m} = \overline{z \cdot z^{m-1}} = \bar{z}\overline{z^{m-1}} = \cdots = \bar{z}^m$.

13. (a) $|z| = \sqrt{10}$, $\alpha = \arctan(3)$.

(b) $|w| = \sqrt{53}$, $\alpha = -\arctan(3.5)$.

(c) $|u| = 2\sqrt{2}$, $\alpha = 45^o$.

(d) $|v| = \sqrt{148}$, $\alpha = -\arctan(1/3)$.

(e) $|r| = 23.73$, $\alpha = 0^o$.

(f) $|s| = 14.3$, $\alpha = 90^o$.

15. (a) $x_1 = 1, x_2 = 1.5, x_3 = 1.4, x_4 = 1.4167, x_5 = 1.4138, x_6 = 1.4143, x_7 = 1.4142$. ($\sqrt{2} = 1.4142$)

(b) Upon convergence to x^*, we have $x^* = 1 + \frac{1}{x^*+1}$, or $x^*(x^*+1) = x^* + 1 + 1$, and $x^{*^2} = 2$. Thus, $x^* = \sqrt{2}$.

Appendix B

A.B.P p. 435

1. (a) By guessing, we obtain $p(2) = 8 + 24 - 2 - 30 = 0$. Using long division, we get $(x^3 + 6x^2 - x - 30) \div (x-2) = x^2 + 8x + 15$. And $x^2 + 8x + 15 = (x+3)(x+3)$. So the roots of p are 2, –3, –5.

(b) Evaluate $q(x)$ for $x = \pm 1, \pm 2, \pm 3, \pm 5, \pm 10, \pm 15, \pm 30$. None of these integer factors x of the constant term 30 of q makes $q(x)$ equal to zero.

(c) q is a real polynomial of odd degree; hence, it has at least one real root, since its complex roots must occur in complex conjugate pairs.

3. By guessing for the integer factors of p's constant term -7, we get $p(1) = 1 + 7 - 1 - 7 = 0$, $p(-1) = -1 + 7 + 1 - 7 = 0$ and $p(-7) = -7^5 + 7^5 + 7 - 7 = 0$. Now use long division of p by $(x+7)(x-1)(x+1) = x^3 + 7x^2 - x - 7$ to obtain $(x^2 + 7x^4 - x - 7) \div (x^3 + 7x^2 - x - 7) = x^2 + 1$. Hence, the roots of p are 1, –1, i, $-i$, and –7.

5. We have $x^4 - 1 = (x^2+1)(x^2-1)$, so that the roots are ± 1 and $\pm i$.

7. (a), (b) $p(x) = x^n + a_{n-1}x^{n-1} + \ldots + a_1 x + a_0 = (x - x_1)\cdots(x - x_n) = x^n - \sum_i x_i x^{n-1} + \ldots + (-1)^n \prod_i x_i$. Thus, $a_{n-1} = -\sum_i x_i$ and $a_0 = (-1)^n \prod_i x_i$.

(c) We have $x_1 + x_2 = -\frac{a}{2} + \sqrt{\ldots} - \frac{a}{2} - \sqrt{\ldots} = -a$ and $x_1 \cdot x_2 = \left(-\frac{a}{2} + \sqrt{\frac{a^2}{4} - b}\right)\left(-\frac{a}{2} - \sqrt{\frac{a^2}{4} - b}\right) = \frac{a^2}{4} - (\frac{a^2}{4} - b) = b$.

9.

A		y	Ay	A^2y	
10	7	1	10	149	
7	5	0	7	105	$\div 7$
		1	10	149	$-10\ r_2$
		0	1	15	
		1	0	−1	
		0	1	15	

Thus, $p_y(x) = x^2 - 15x + 1$ with eigenvalues $x_{1,2} = 7.5 \pm \sqrt{55.25} = 7.5 \pm 7.433$.

Next, $f_B(\lambda) = \det(\mathbf{B} - \lambda \mathbf{I}_2) = \det\begin{pmatrix} 7-\lambda & 10 \\ 5 & 7-\lambda \end{pmatrix} = \lambda^2 - 14\lambda - 1$ with roots $\lambda_{1,2} = 7 \pm \sqrt{50} = 7 \pm 7.071$.

11. (a) Note that $p(0) = 0 + \ldots + 0 + a_0 \neq 0$.

(b) If $p(x_1) = a_n x_1^n + a_{n-1}x_1^{n-1} + \ldots + a_1 x_1 + a_0 = 0$ for $x_1 \neq 0$, then we can divide by $x_1^n \neq 0$ to obtain $\frac{p(x_1)}{x_1^n} = a_n + a_{n-1}\frac{1}{x_1} + \ldots + a_1 \frac{1}{x_1^{n-1}} + a_0 \frac{1}{x_1^n} = 0$. Now $\frac{p(x_1)}{x_1^n} = q\left(\frac{1}{x_1}\right)$. Therefore, $p(x) = 0$ for $x \neq 0$ if and only if $q\left(\frac{1}{x}\right) = 0$.

Appendix C

A.C.P p. 440

1. (a) This is not a vector space, since it is not closed under scalar multiplication; (b), (c), and (d) are vector spaces.

3. (b), (d) are subspaces; (a) and (c) are not, since in both cases $f \equiv 1$ belongs to the respective sets, but $2f$ does not.

5. Check the conditions for a subspace: From calculus, if $r(x)$ and $s(x)$ are power series with the same radius of convergence, so is $(\alpha r + \beta s)(x)$ for all α and $\beta \in \mathbb{R}$.

7. (a), (b), (d), and (f) are subspaces;

(c) and (e) are not.

9. (a) Sums and scalar multiples of diagonal matrices are also diagonal.

(b) This is not a vector space. When two such matrices are added, their (1,1) entry becomes equal to $6 \neq 3$.

(c) is a vector space.

John Todd and Olga Taussky–Todd
Two pioneers of modern Matrix Theory

List of Photographs

Frank Uhlig
Department of Mathematics
Auburn University
Auburn, AL 36849–5310
tla@auburn.edu
http://www.auburn.edu/~uhligfd

Leaving

Index

The gate is open